Modelling climate change impacts on agricultural systems

It is widely recognised that agriculture is a significant contributor to global warming and climate change. Agriculture needs to reduce its environmental impact and adapt to current climate change whilst still feeding a growing population, i.e. become more 'climate-smart'. Burleigh Dodds Science Publishing is playing its part in achieving this by bringing together key research on making the production of the world's most important crops and livestock products more sustainable.

Based on extensive research, our publications specifically target the challenge of climate-smart agriculture. In this way we are using 'smart publishing' to help achieve climate-smart agriculture.

Burleigh Dodds Science Publishing is an independent and innovative publisher delivering high quality customer-focused agricultural science content in both print and online formats for the academic and research communities. Our aim is to build a foundation of knowledge on which researchers can build to meet the challenge of climate-smart agriculture.

For more information about Burleigh Dodds Science Publishing simply call us on +44 (0) 1223 839365, email info@bdspublishing.com or alternatively please visit our website at www.bdspublishing.com.

Related titles:

Assessing the environmental impact of agriculture
Print (ISBN 978-1-78676-228-3); Online (ISBN 978-1-78676-230-6, 978-1-78676-231-3)

Advances in crop modelling for a sustainable agriculture
Print (ISBN 978-1-78676-240-5); Online (ISBN 978-1-78676-243-6, 978-1-78676-242-9)

Climate change and agriculture
Print (ISBN 978-1-78676-320-4); Online (ISBN 978-1-78676-322-8, 978-1-78676-323-5)

Improving data management and decision support systems in agriculture
Print (ISBN 978-1-78676-340-2); Online (ISBN 978-1-78676-343-3, 978-1-78676-342-6)

Chapters are available individually from our online bookshop:
https://shop.bdspublishing.com

climate•SMART•publishing
IN AGRICULTURAL SCIENCE

BURLEIGH DODDS SERIES IN AGRICULTURAL SCIENCE
NUMBER 131

Modelling climate change impacts on agricultural systems

Edited by Professor Claas Nendel, Leibniz Centre for Agricultural Landscape Research (ZALF), Germany

burleigh dodds
SCIENCE PUBLISHING

Published by Burleigh Dodds Science Publishing Limited
82 High Street, Sawston, Cambridge CB22 3HJ, UK
www.bdspublishing.com

Burleigh Dodds Science Publishing, 1518 Walnut Street, Suite 900, Philadelphia, PA 19102-3406, USA

First published 2023 by Burleigh Dodds Science Publishing Limited

Library of Congress Control Number: 2022949277

British Library Cataloguing in Publication Data
A catalogue record for this book is available from the British Library

ISBN 978-1-80146-174-0 (print)
ISBN 978-1-80146-177-1 (PDF)
ISBN 978-1-80146-176-4 (ePub)
ISSN 2059-6936 (print)
ISSN 2059-6944 (online)

DOI: 10.19103/AS.2022.0115

Typeset by Deanta Global Publishing Services, Dublin, Ireland

Contents

Series list xiv
Introduction xxiv
Acknowledgements xxx

Part 1 Advances in modelling

1 Advances in integrating different models assessing the impact of climate change on agriculture 3
Jacques-Eric Bergez, Julie Constantin, Philippe Debaeke, Hélène Raynal and Sophie Plassin, INRAE, France; Magali Willaume, Toulouse INP-ENSAT, France; and Raphael Martin, INRAE, France

1 Introduction 3
2 Understanding models and their application to problems such as climate change 6
3 The 'ideal model' to assess the impact of climate change on agriculture 8
4 Implementing the 'ideal model': current approaches 11
5 From ideal to pragmatic modelling of climate change: coupling sub-models 13
6 Integrating new knowledge and facilitating interoperability 15
7 Integrating adaptation management practices into models 17
8 Incorporating inter- and intra-species diversity into models 19
9 From white to black box models: hybrid approaches and gridded simulations 20
10 Ensemble modelling 21
11 Conclusion and future trends 22
12 References 26

2 Improving data flow and integration in models assessing the impact of climate change on agriculture 39
Claas Nendel, Roland Baatz, Michael Berg-Mohnicke and Gohar Ghazaryan, Leibniz Centre for Agricultural Landscape

Research (ZALF), Germany; Sander Janssen, Wageningen Environmental Research, The Netherlands; Pierre Martre, INRAE, France; and Cheryl Porter, University of Florida, USA

1 Introduction 39
2 Model-data integration 40
3 Informing spatiotemporal simulations 44
4 Assimilation of data in spatiotemporal simulations 48
5 Workflows for massive parallel computing 51
6 Model-model integration 54
7 Granularity and modular design for model improvement, reuse, exchange and interoperability 56
8 Concepts for distributed modelling 58
9 Future trends and conclusion 61
10 Where to look for further information 61
11 References 62

3 Incorporating genetics into crop models to identify new phenotypes adapted to climate change 75
Carlos D. Messina, Institute of Food and Agricultural Sciences (IFAS), University of Florida, USA and The ARC Centre of Excellence for Plant Success in Nature and Agriculture, Australia; and Mark Cooper, The Queensland Alliance for Agriculture and Food Innovation (QAAFI) and The ARC Centre of Excellence for Plant Success in Nature and Agriculture, Australia

1 Introduction 75
2 Systems analyses and models of crop growth 77
3 Linking genetics and physiology for phenotypic prediction 80
4 Models that predict emergent phenotypes 81
5 Integrating environmental and genomic predictions using crop models 82
6 Assessing impacts of traits and their underpinning genetics on crop performance 84
7 Managing genotype × environment interactions to optimize rather than hamper genetic gain in current and future climates 86
8 Predictive breeding enables iterative impact assessment and adaptation to climate change 89
9 Crop growth models can enable a rethinking of breeding objectives to adapt to and combat climate change 93
10 Conclusion and future trends 95
11 References 96

4 Developing more integrated approaches in models assessing the impact of climate change on agriculture 105
Ian P. Holman, Cranfield University, UK and Global Change Research Centre, Czech Republic; and Paula A. Harrison, UK Centre for Ecology & Hydrology, UK and Global Change Research Centre, Czech Republic

1 Introduction 105
2 Addressing key challenges in integrated modelling 108
3 Case study 117
4 Conclusion 120
5 Future trends in research 120
6 Acknowledgements 122
7 Where to look for further information 122
8 References 122

5 Accounting for uncertainties in modeling the impact of climate change on agriculture 131
Fulu Tao, Chinese Academy of Sciences, China and Natural Resources Institute Finland (Luke), Finland; Taru Palosuo, Natural Resources Institute Finland (Luke), Finland; and Reimund Paul Rötter, Tropical Plant Production and Agricultural Systems Modelling (TROPAGS) and University of Göttingen, Germany

1 Introduction 131
2 Managing uncertainties from greenhouse gas emission and climate change scenarios 134
3 Managing uncertainties from crop model input data 136
4 Managing uncertainties from model parameters and human error 136
5 Managing uncertainties from model structures 137
6 Managing uncertainties from adaptations in agricultural practices 140
7 Reducing model uncertainties 141
8 Conclusion and future trends 142
9 Acknowledgements 142
10 Where to look for further information 142
11 References 143

Part 2 Modelling climate change impacts on particular aspects of agricultural systems

6 Modeling climate change impacts on crop growth and yield formation 155
Davide Cammarano and Johannes Wilhelmus Maria Pullens, Aarhus University, Denmark; and Pierre Martre, Université de Montpellier, France

1 Introduction 155
2 Concept of modeling crop yield 157
3 Crop modeling response: the role of data and model structure 161
4 Assessing climate impacts on yield with model ensembles 169
5 Conclusion and future trends in research 177
6 Where to look for further information 178
7 References 178

7 Modelling climate change impacts on livestock production 187
Ryan G. McGuire, Shannon M. McLaughlin, Tanisha L. Waring, Kayley D. Barnes, Sharon A. Huws and Nigel. D. Scollan, Queen's University Belfast, UK

1 Introduction 187
2 Enteric methane mitigation 190
3 Grassland agriculture: a net zero livestock system 203
4 Conclusion and future trends 219
5 Where to look for further information 220
6 References 221

8 Modeling climate change impact on low-input smallholder farming systems 235
Myriam Adam, Centre de coopération internationale en recherche agronomique pour le développement (CIRAD), UMR AGAP Institute, Univ Montpellier, CIRAD, INRAE, Institut Agro, Montpellier, France and National University of Battambang, Cambodia; Gatien Falconnier, CIRAD, UPR AIDA, F-34398 Montpellier and AIDA, Univ Montpellier, CIRAD, Montpellier, France and International Maize and Wheat Improvement Centre (CIMMYT), Zimbabwe; David Berre, CIRAD, UPR AIDA, F-34398 Montpellier and AIDA, Univ Montpellier, CIRAD, Montpellier, France; Katrien Descheemaeker, Wageningen University & Research, The Netherlands; Juliette Lairez, CIRAD, UPR AIDA, F-34398 Montpellier and AIDA, Univ Montpellier, CIRAD, Montpellier, France and Institut de l'Environnement et de Recherche Agricole (INERA), Burkina Faso; and Louise Leroux, CIRAD, UPR AIDA, F-34398 Montpellier and AIDA, Univ Montpellier, CIRAD, Montpellier, France, CIRAD, UPR AIDA, Nairobi and IITA, Kenya

1 Introduction 235
2 Modeling climate change effect on crops, livestock, grasslands and trees 237
3 Toward an integrative perspective 248
4 Conclusion 255
5 Where to look for further information 256
6 References 256

9 Modeling climate change impact on agro-ecosystem services 265
Charlotte Weil, École Polytechnique Fédérale de Lausanne, Switzerland; Justin A. Johnson, University of Minnesota, USA; and Rebecca Chaplin-Kramer, University of Minnesota and SPRING, USA

1 Introduction 265
2 Climate change impacts on ecosystem services in agricultural systems 267
3 Climate change impacts on food-provisioning ecosystem services 275
4 From models to decision-relevant metrics: integrating demand 283
5 Conclusion 291
6 Future trends in research 293
7 Where to look for further information 294
8 References 295

10 Modelling climate change impacts on agricultural commodity markets 311
Ignacio Pérez Domínguez, Jordan Hristov, Christian Elleby and Thomas Fellmann, European Commission, Joint Research Centre, Spain; Andrea Toreti, European Commission, Joint Research Centre, Italy; Thomas Chatzopoulos and Ana Luisa Barbosa, European Commission, Joint Research Centre, Spain; and Frank Dentener, European Commission, Joint Research Centre, Italy

1 Introduction 311
2 Measuring direct impacts of climate change on agriculture 313
3 Economic adaptation potential 315
4 The role of technologies for mitigation and adaptation 317
5 Conclusion 320
6 Future trends in research 321
7 Where to look for further information 324
8 References 324

11 Modelling transition of agricultural systems in response to climate change 327
Arnout van Soesbergen and Fiona Jones, King's College London, UK

1 Introduction 327
2 The transition of agricultural systems in response to climate change 328
3 The role of models in informing adaptation responses 330
4 Capturing the transition in modelling approaches and climate change impact studies 331
5 Overarching limitations of current modelling approaches 341
6 Future trends in research 344
7 Where to look for further information 349
8 References 350

Part 3 Modelling climate change impacts on regional agricultural systems

12 Modelling the impact of climate change on agriculture in Europe 361
Mirek Trnka and Rudolf Brázdil, Global Change Research Institute CAS, Czech Republic; Lorenzo Brilli, Institute for BioEconomy (IBE) – National Research Council (CNR), Italy; Hannah C. von Czettritz und Neuhaus, Leibniz Centre for Agricultural Landscape Research (ZALF), Germany; Sergi Costafreda-Aumedes, Institute for BioEconomy (IBE) – National Research Council (CNR), Italy; Petr Dobrovolný, Petr Holub and Karel Klem, Global Change Research Institute CAS, Czech Republic; Matthias Kuhnert, University of Aberdeen, UK; Luisa Leolini, University of Florence, Italy; Marco Moriondo, Institute for BioEconomy (IBE) – National Research Council (CNR), Italy; Nina Muntean, Czech University of Life Sciences Prague, Czech Republic; Jørgen Eivind Olesen, Aarhus University, Denmark; Markéta Poděbradská, Vera Potopová, Petr Štěpánek and Otmar Urban, Global Change Research Institute CAS, Czech Republic; and Peter Zander and Claas Nendel, Leibniz Centre for Agricultural Landscape Research (ZALF), Germany

1 Introduction 361
2 Past climate and future projections for Europe 366
3 Agricultural systems across Europe 371
4 Agricultural system modelling: models and platforms 374
5 Modelling and experimental work on impacts of climate change: case studies 376
6 Future trends and conclusion 393
7 Where to look for further information 395
8 Acknowledgements 395
9 References 396

13 Modeling the impact of climate change on agriculture in the United States 409
Bruno Basso, Neville Millar and Lydia Price, Michigan State University, USA

1 Introduction 409
2 Climate change impact on crop-based agriculture 410
3 Can new genetics mitigate climate change impacts on yields? 424
4 Overview of incentives and payment mechanisms for carbon markets 428
5 References 432

14 Modeling the impact of climate change on agriculture in Latin America 435
Fábio R. Marin and Ivo Zution Gonçalves, University of São Paulo, Brazil; Francisco J Meza, Forestal. Pontificia Universidad Católica

de Chile, Chile; Néstor M. Riaño H., Ministry of Agriculture and Rural Development, Colombia; Andrés J. Peña Q, Colombian Corporation for Agricultural Research (AGROSAVIA), Colombia; Alexandre B. Heinneman, EMBRAPA, Brazil; Nereu A. Streck and Alencar Jr Zanon, Universidade Federal de Santa Maria, Brazil; Murilo S. Vianna, University of Bonn, Germany; and Evandro H. F. Silva and Nilson A. Vieira Jr, University of São Paulo, Brazil

1 Introduction 435
2 Developments in crop modeling and climate change in Chile 437
3 Developments in crop modeling and climate change in Colombia 441
4 Developments in crop modeling in Brazil: upland rice production and climate change 442
5 Developments in crop modeling in Brazil: irrigated rice production and climate change 446
6 Developments in crop modeling in Brazil: common bean production and climate change 448
7 Developments in crop modeling in Brazil: soybean production and climate change 451
8 Developments in crop modeling in Brazil: sugarcane production and climate change 454
9 Developments in crop modeling in Brazil: grass forage production for livestock and climate change 458
10 Conclusion 460
11 Where to look for further information 461
12 References 461

15 Modelling the impact of climate change on agriculture in Australia and Oceania 481
Enli Wang, CSIRO Agriculture and Food, Australia; Edmar Teixeira, The New Zealand Institute for Plant & Food Research Limited, New Zealand; Bangyou Zheng, CSIRO Agriculture and Food, Australia; Neal Hughes, Australian Bureau of Agricultural and Resource Economics and Sciences (ABARES), Australia; Karine Chenu, The University of Queensland, Queensland Alliance for Agriculture and Food Innovation (QAAFI), Australia; James Hunt, University of Melbourne, Australia; Afshin Ghahramani, University of Southern Queensland, Australia; Andries B. Potgieter, The University of Queensland, Queensland Alliance for Agriculture and Food Innovation (QAAFI), Australia; Junqi Zhu and Rogerio Cichota, The New Zealand Institute for Plant & Food Research Limited, New Zealand; and Neil Huth, CSIRO Agriculture and Food, Australia

1 Introduction 481
2 Past climate change and future projections 487

3 Challenges to agriculture 492
4 Agricultural systems modelling 495
5 Modelling the impact of climate change: case studies 499
6 Future trends in research 522
7 Conclusion 524
8 Where to look for further information 525
9 References 526

16 Modelling the impact of climate change on agriculture in South Asia 541
S. Naresh Kumar, ICAR-Indian Agricultural Research Institute, India

1 Introduction 541
2 Climate-change scenarios for agriculture: the case of India 544
3 How climactic stresses affect field crops: key mechanisms 546
4 Methods for quantifying the impact of climate change on crop production: overview 548
5 Methods for quantifying the impact of climate change on crop production: the role of simulation modelling 555
6 Projected impacts of climate change on key crops: India 558
7 Projected impacts of climate change on key crops: Pakistan, Bangladesh, Nepal, Sri Lanka, Bhutan and the Maldives 566
8 Projected impacts of climate change on other horticultural crops 570
9 Assessing the impact of climate change on pests and diseases 572
10 Conclusion and future trends 575
11 Where to look for further information 576
12 Acknowledgements 577
13 References 578

17 Modelling the impact of climate change on agriculture in East Asia 589
Bing Liu, Zi Ye, Yuan Cao, Bo Liu and Yan Zhu, Nanjing Agricultural University, China

1 Introduction 589
2 Modelling the impact of climate change on rice 590
3 Modelling the impact of climate change on wheat 598
4 Modelling the impact of climate change on maize 605
5 Conclusion and future trends 609
6 Where to look for further information 615
7 References 615

18 Modelling the impact of climate change on agriculture in North Africa and Southwest Asia 623
Ehsan Eyshi Rezaei , Leibniz Centre for Agricultural Landscape Research (ZALF), Germany; and Sevim Seda Yamaç, Konya Food and Agriculture University, Turkey

1 Introduction 623
2 North Africa: geography, economy, climate and agriculture 624
3 Southwest Asia: geography, economy, climate and agriculture 629
4 Future climate projections and their effects on regional crop production 633
5 Modelling tools to quantify crop response to climate change 638
6 Current climate change impact assessments in North Africa and Southwest Asia 640
7 Accounting for the potential of adaptation strategies 642
8 Limitations of climate change impact assessments 646
9 Conclusion and future trends 648
10 Where to look for further information 650
11 References 650

19 Modelling the impact of climate change on agriculture in West Africa 667

D. S. MacCarthy, University of Ghana, Ghana; P. B. I. Akponikpe, Université de Parakou (UP), Benin; F. M. Akinseye, International Crop Research Institute for the Semi-Arid Tropics (ICRISAT), Nigeria and Centre d'Etude Régional pour l'Amélioration de l'Adaptation à la Sécheresse (CERAAS), Sénégal; M. Ly, Cheikh, Anta Diop University, Senegal; E. C. Timpong-Jones, University of Ghana, Ghana; I. Hathie, Initiative Prospective Agricole et Rurale (IPAR), Senegal; and S. G. K. Adiku, University of Ghana, Ghana

1 Introduction 667
2 The West African region 668
3 The West African climate: measurement, characteristics and trends 669
4 Modelling climate change in West Africa 672
5 Crop modelling in West African farming systems 678
6 Modelling cereal and legume production in West Africa 681
7 Accounting for the evolution of agricultural systems in climate change impact assessments 684
8 Modelling climate change adaptation strategies 686
9 Key issues in modelling climate change impacts on livestock production 689
10 Modelling livestock systems 693
11 Integrated climate change impact assessments 699
12 Conclusion and future trends 701
13 Where to look for further information 703
14 References 704

Index 713

Series list

Title	Series number
Achieving sustainable cultivation of maize - Vol 1 From improved varieties to local applications Edited by: Dr Dave Watson, CGIAR Maize Research Program Manager, CIMMYT, Mexico	001
Achieving sustainable cultivation of maize - Vol 2 Cultivation techniques, pest and disease control Edited by: Dr Dave Watson, CGIAR Maize Research Program Manager, CIMMYT, Mexico	002
Achieving sustainable cultivation of rice - Vol 1 Breeding for higher yield and quality Edited by: Prof. Takuji Sasaki, Tokyo University of Agriculture, Japan	003
Achieving sustainable cultivation of rice - Vol 2 Cultivation, pest and disease management Edited by: Prof. Takuji Sasaki, Tokyo University of Agriculture, Japan	004
Achieving sustainable cultivation of wheat - Vol 1 Breeding, quality traits, pests and diseases Edited by: Prof. Peter Langridge, The University of Adelaide, Australia	005
Achieving sustainable cultivation of wheat - Vol 2 Cultivation techniques Edited by: Prof. Peter Langridge, The University of Adelaide, Australia	006
Achieving sustainable cultivation of tomatoes Edited by: Dr Autar Mattoo, USDA-ARS, USA and Prof. Avtar Handa, Purdue University, USA	007
Achieving sustainable production of milk - Vol 1 Milk composition, genetics and breeding Edited by: Dr Nico van Belzen, International Dairy Federation (IDF), Belgium	008
Achieving sustainable production of milk - Vol 2 Safety, quality and sustainability Edited by: Dr Nico van Belzen, International Dairy Federation (IDF), Belgium	009
Achieving sustainable production of milk - Vol 3 Dairy herd management and welfare Edited by: Prof. John Webster, University of Bristol, UK	010
Ensuring safety and quality in the production of beef - Vol 1 Safety Edited by: Prof. Gary Acuff, Texas A&M University, USA and Prof. James Dickson, Iowa State University, USA	011
Ensuring safety and quality in the production of beef - Vol 2 Quality Edited by: Prof. Michael Dikeman, Kansas State University, USA	012
Achieving sustainable production of poultry meat - Vol 1 Safety, quality and sustainability Edited by: Prof. Steven C. Ricke, University of Arkansas, USA	013
Achieving sustainable production of poultry meat - Vol 2 Breeding and nutrition Edited by: Prof. Todd Applegate, University of Georgia, USA	014

Achieving sustainable production of poultry meat - Vol 3 015
Health and welfare
Edited by: Prof. Todd Applegate, University of Georgia, USA

Achieving sustainable production of eggs - Vol 1 016
Safety and quality
Edited by: Prof. Julie Roberts, University of New England, Australia

Achieving sustainable production of eggs - Vol 2 017
Animal welfare and sustainability
Edited by: Prof. Julie Roberts, University of New England, Australia

Achieving sustainable cultivation of apples 018
Edited by: Dr Kate Evans, Washington State University, USA

Integrated disease management of wheat and barley 019
Edited by: Prof. Richard Oliver, Curtin University, Australia

Achieving sustainable cultivation of cassava - Vol 1 020
Cultivation techniques
Edited by: Dr Clair Hershey, formerly International Center for Tropical Agriculture (CIAT), Colombia

Achieving sustainable cultivation of cassava - Vol 2 021
Genetics, breeding, pests and diseases
Edited by: Dr Clair Hershey, formerly International Center for Tropical Agriculture (CIAT), Colombia

Achieving sustainable production of sheep 022
Edited by: Prof. Johan Greyling, University of the Free State, South Africa

Achieving sustainable production of pig meat - Vol 1 023
Safety, quality and sustainability
Edited by: Prof. Alan Mathew, Purdue University, USA

Achieving sustainable production of pig meat - Vol 2 024
Animal breeding and nutrition
Edited by: Prof. Julian Wiseman, University of Nottingham, UK

Achieving sustainable production of pig meat - Vol 3 025
Animal health and welfare
Edited by: Prof. Julian Wiseman, University of Nottingham, UK

Achieving sustainable cultivation of potatoes - Vol 1 026
Breeding improved varieties
Edited by: Prof. Gefu Wang-Pruski, Dalhousie University, Canada

Achieving sustainable cultivation of oil palm - Vol 1 027
Introduction, breeding and cultivation techniques
Edited by: Prof. Alain Rival, Center for International Cooperation in Agricultural Research for Development (CIRAD), France

Achieving sustainable cultivation of oil palm - Vol 2 028
Diseases, pests, quality and sustainability
Edited by: Prof. Alain Rival, Center for International Cooperation in Agricultural Research for Development (CIRAD), France

Achieving sustainable cultivation of soybeans - Vol 1 029
Breeding and cultivation techniques
Edited by: Prof. Henry T. Nguyen, University of Missouri, USA

Achieving sustainable cultivation of soybeans - Vol 2 030
Diseases, pests, food and non-food uses
Edited by: Prof. Henry T. Nguyen, University of Missouri, USA

Achieving sustainable cultivation of sorghum - Vol 1 031
Genetics, breeding and production techniques
Edited by: Prof. William Rooney, Texas A&M University, USA

Achieving sustainable cultivation of sorghum - Vol 2 032
Sorghum utilization around the world
Edited by: Prof. William Rooney, Texas A&M University, USA

Achieving sustainable cultivation of potatoes - Vol 2 033
Production, storage and crop protection
Edited by: Dr Stuart Wale, Potato Dynamics Ltd, UK

Achieving sustainable cultivation of mangoes 034
Edited by: Prof. Víctor Galán Saúco, Instituto Canario de Investigaciones Agrarias (ICIA), Spain and Dr Ping Lu, Charles Darwin University, Australia

Achieving sustainable cultivation of grain legumes - Vol 1 035
Advances in breeding and cultivation techniques
Edited by: Dr Shoba Sivasankar et al., formerly International Crops Research Institute for the Semi-Arid Tropics (ICRISAT), India

Achieving sustainable cultivation of grain legumes - Vol 2 036
Improving cultivation of particular grain legumes
Edited by: Dr Shoba Sivasankar et al., formerly International Crops Research Institute for the Semi-Arid Tropics (ICRISAT), India

Achieving sustainable cultivation of sugarcane - Vol 1 037
Cultivation techniques, quality and sustainability
Edited by: Prof. Philippe Rott, University of Florida, USA

Achieving sustainable cultivation of sugarcane - Vol 2 038
Breeding, pests and diseases
Edited by: Prof. Philippe Rott, University of Florida, USA

Achieving sustainable cultivation of coffee 039
Edited by: Dr Philippe Lashermes, Institut de Recherche pour le Développement (IRD), France

Achieving sustainable cultivation of bananas - Vol 1 040
Cultivation techniques
Edited by: Prof. Gert H. J. Kema, Wageningen University and Research, The Netherlands and Prof. André Drenth, University of Queensland, Australia

Global Tea Science 041
Current status and future needs
Edited by: Dr V. S. Sharma, formerly UPASI Tea Research Institute, India and Dr M. T. Kumudini Gunasekare, Coordinating Secretariat for Science Technology and Innovation (COSTI), Sri Lanka

Integrated weed management 042
Edited by: Prof. Emeritus Robert L. Zimdahl, Colorado State University, USA

Achieving sustainable cultivation of cocoa 043
Edited by: Prof. Pathmanathan Umaharan, Cocoa Research Centre – The University of the West Indies, Trinidad and Tobago

Robotics and automation for improving agriculture 044
Edited by: Prof. John Billingsley, University of Southern Queensland, Australia

Water management for sustainable agriculture 045
Edited by: Prof. Theib Oweis, ICARDA, Jordan

Improving organic animal farming 046
Edited by: Dr Mette Vaarst, Aarhus University, Denmark and Dr Stephen Roderick, Duchy College, UK

Improving organic crop cultivation 047
Edited by: Prof. Ulrich Köpke, University of Bonn, Germany

Managing soil health for sustainable agriculture - Vol 1 048
Fundamentals
Edited by: Dr Don Reicosky, Soil Scientist Emeritus USDA-ARS and University of Minnesota, USA

Managing soil health for sustainable agriculture - Vol 2 049
Monitoring and management
Edited by: Dr Don Reicosky, Soil Scientist Emeritus USDA-ARS and University of Minnesota, USA

Rice insect pests and their management 050
E. A. Heinrichs, Francis E. Nwilene, Michael J. Stout, Buyung A. R. Hadi and Thais Freitas

Improving grassland and pasture management in temperate agriculture 051
Edited by: Prof. Athole Marshall and Dr Rosemary Collins, IBERS, Aberystwyth University, UK

Precision agriculture for sustainability 052
Edited by: Dr John Stafford, Silsoe Solutions, UK

Achieving sustainable cultivation of temperate zone tree fruit and berries - Vol 1 053
Physiology, genetics and cultivation
Edited by: Prof. Gregory A. Lang, Michigan State University, USA

Achieving sustainable cultivation of temperate zone tree fruit and berries - Vol 2 054
Case studies
Edited by: Prof. Gregory A. Lang, Michigan State University, USA

Agroforestry for sustainable agriculture 055
Edited by: Prof. María Rosa Mosquera-Losada, Universidade de Santiago de Compostela, Spain and Dr Ravi Prabhu, World Agroforestry Centre (ICRAF), Kenya

Achieving sustainable cultivation of tree nuts 056
Edited by: Prof. Ümit Serdar, Ondokuz Mayis University, Turkey and Emeritus Prof. Dennis Fulbright, Michigan State University, USA

Assessing the environmental impact of agriculture 057
Edited by: Prof. Bo P. Weidema, Aalborg University, Denmark

Critical issues in plant health: 50 years of research in African agriculture 058
Edited by: Dr Peter Neuenschwander and Dr Manuele Tamò, IITA, Benin

Achieving sustainable cultivation of vegetables 059
Edited by: Emeritus Prof. George Hochmuth, University of Florida, USA

Advances in breeding techniques for cereal crops 060
Edited by: Prof. Frank Ordon, Julius Kuhn Institute (JKI), Germany and Prof. Wolfgang Friedt, Justus-Liebig University of Giessen, Germany

Advances in Conservation Agriculture - Vol 1 061
Systems and Science
Edited by: Prof. Amir Kassam, University of Reading, UK and Moderator, Global Conservation Agriculture Community of Practice (CA-CoP), FAO, Rome, Italy

Advances in Conservation Agriculture - Vol 2 062
Practice and Benefits
Edited by: Prof. Amir Kassam, University of Reading, UK and Moderator, Global Conservation Agriculture Community of Practice (CA-CoP), FAO, Rome, Italy

Achieving sustainable greenhouse cultivation 063
Edited by: Prof. Leo Marcelis and Dr Ep Heuvelink, Wageningen University, The Netherlands

Achieving carbon-negative bioenergy systems from plant materials 064
Edited by: Dr Chris Saffron, Michigan State University, USA

Achieving sustainable cultivation of tropical fruits 065
Edited by: Prof. Elhadi M. Yahia, Universidad Autónoma de Querétaro, Mexico

Advances in postharvest management of horticultural produce 066
Edited by: Prof. Chris Watkins, Cornell University, USA

Pesticides and agriculture 067
Profit, politics and policy
Dave Watson

Integrated management of diseases and insect pests of tree fruit 068
Edited by: Prof. Xiangming Xu and Dr Michelle Fountain, NIAB-EMR, UK

Integrated management of insect pests 069
Current and future developments
Edited by: Emeritus Prof. Marcos Kogan, Oregon State University, USA and Emeritus Prof. E. A. Heinrichs, University of Nebraska-Lincoln, USA

Preventing food losses and waste to achieve food security and sustainability 070
Edited by: Prof. Elhadi M. Yahia, Universidad Autónoma de Querétaro, Mexico

Achieving sustainable management of boreal and temperate forests 071
Edited by: Dr John Stanturf, Estonian University of Life Sciences , Estonia

Advances in breeding of dairy cattle 072
Edited by: Prof. Julius van der Werf, University of New England, Australia and Prof. Jennie Pryce, Agriculture Victoria and La Trobe University, Australia

Improving gut health in poultry 073
Edited by: Prof. Steven C. Ricke, University of Arkansas, USA

Achieving sustainable cultivation of barley 074
Edited by: Prof. Glen Fox, University of California-Davis, USA and The University of Queensland, Australia and Prof. Chengdao Li, Murdoch University, Australia

Advances in crop modelling for a sustainable agriculture 075
Edited by: Emeritus Prof. Kenneth Boote, University of Florida, USA

Achieving sustainable crop nutrition 076
Edited by: Prof. Zed Rengel, University of Western Australia, Australia

Achieving sustainable urban agriculture 077
Edited by: Prof. Johannes S. C. Wiskerke, Wageningen University, The Netherlands

Climate change and agriculture 078
Edited by Dr Delphine Deryng, NewClimate Institute/Integrative Research Institute on Transformations of Human-Environment Systems (IRI THESys), Humboldt-Universität zu Berlin, Germany

Advances in poultry genetics and genomics 079
Edited by: Prof. Samuel E. Aggrey, University of Georgia, USA, Prof. Huaijun Zhou, University of California-Davis, USA, Dr Michèle Tixier-Boichard, INRAE, France and Prof. Douglas D. Rhoads, University of Arkansas, USA

Achieving sustainable management of tropical forests 080
Edited by: Prof. Jürgen Blaser, Bern University of Life Sciences, Switzerland and Patrick D. Hardcastle, Forestry Development Specialist, UK

Improving the nutritional and nutraceutical properties of wheat and other cereals 081
Edited by: Prof. Trust Beta, University of Manitoba, Canada

Achieving sustainable cultivation of ornamental plants 082
Edited by: Emeritus Prof. Michael Reid, University of California-Davis, USA

Improving rumen function 083
Edited by: Dr C. S. McSweeney, CSIRO, Australia and Prof. R. I. Mackie, University of Illinois, USA

Biostimulants for sustainable crop production 084
Edited by: Youssef Rouphael, Patrick du Jardin, Patrick Brown, Stefania De Pascale and Giuseppe Colla

Improving data management and decision support systems in agriculture 085
Edited by: Dr Leisa Armstrong, Edith Cowan University, Australia

Achieving sustainable cultivation of bananas - Volume 2 086
Germplasm and genetic improvement
Edited by: Prof. Gert H. J. Kema, Wageningen University, The Netherlands and Prof. Andrè Drenth, The University of Queensland, Australia

Reconciling agricultural production with biodiversity conservation 087
Edited by: Prof. Paolo Bàrberi and Dr Anna-Camilla Moonen, Institute of Life Sciences - Scuola Superiore Sant'Anna, Pisa, Italy

Advances in postharvest management of cereals and grains 088
Edited by: Prof. Dirk E. Maier, Iowa State University, USA

Biopesticides for sustainable agriculture 089
Edited by: Prof. Nick Birch, formerly The James Hutton Institute, UK and Prof. Travis Glare, Lincoln University, New Zealand

Understanding and improving crop root function 090
Edited by: Emeritus Prof. Peter J. Gregory, University of Reading, UK

Understanding the behaviour and improving the welfare of chickens 091
Edited by: Prof. Christine Nicol, Royal Veterinary College - University of London, UK

Advances in measuring soil health 092
Edited by: Prof. Wilfred Otten, Cranfield University, UK

The sustainable intensification of smallholder farming systems 093
Edited by: Dr Dominik Klauser and Dr Michael Robinson, Syngenta Foundation for Sustainable Agriculture, Switzerland

Advances in horticultural soilless culture 094
Edited by: Prof. Nazim S. Gruda, University of Bonn, Germany

Reducing greenhouse gas emissions from livestock production 095
Edited by: Dr Richard Baines, Royal Agricultural University, UK

Understanding the behaviour and improving the welfare of pigs 096
Edited by: Emerita Prof. Sandra Edwards, Newcastle University, UK

Genome editing for precision crop breeding 097
Edited by: Dr Matthew R. Willmann, Cornell University, USA

Understanding the behaviour and improving the welfare of dairy cattle 098
Edited by: Dr Marcia Endres, University of Minnesota, USA

Developing sustainable food systems 099
Dave Watson

Plant genetic resources 100
A review of current research and future needs
Edited by: Dr M. Ehsan Dulloo, Bioversity International, Italy

Developing animal feed products 101
Edited by: Dr Navaratnam Partheeban, formerly Royal Agricultural University, UK

Improving dairy herd health 102
Edited by: Prof. Émile Bouchard, University of Montreal, Canada

Understanding gut microbiomes as targets for improving pig gut health 103
Edited by: Prof. Mick Bailey and Emeritus Prof. Chris Stokes, University of Bristol, UK

Advances in Conservation Agriculture - Vol 3 104
Adoption and Spread
Edited by: Professor Amir Kassam, University of Reading, UK and Moderator, Global Conservation Agriculture Community of Practice (CA-CoP), FAO, Rome, Italy

Advances in precision livestock farming 105
Edited by: Prof. Daniel Berckmans, Katholieke University of Leuven, Belgium

Achieving durable disease resistance in cereals 106
Edited by: Prof. Richard Oliver, formerly Curtin University, Australia

Seaweed and microalgae as alternative sources of protein 107
Edited by: Prof. Xin Gen Lei, Cornell University, USA

Microbial bioprotectants for plant disease management 108
Edited by: Dr Jürgen Köhl, Wageningen University & Research, The Netherlands and Dr Willem Ravensberg, Koppert Biological Systems, The Netherlands

Improving soil health 109
Edited by: Prof. William R. Horwath, University of California-Davis, USA

Improving integrated pest management in horticulture 110
Edited by: Prof. Rosemary Collier, Warwick University, UK

Climate-smart production of coffee 111
Improving social and environmental sustainability
Edited by: Prof. Reinhold Muschler, CATIE, Costa Rica

Developing smart agri-food supply chains 112
Using technology to improve safety and quality
Edited by: Prof. Louise Manning, Royal Agricultural University, UK

Advances in integrated weed management 113
Edited by: Prof. Per Kudsk, Aarhus University, Denmark

Understanding and improving the functional and nutritional properties of milk 114
Edited by: Prof. Thom Huppertz, Wageningen University & Research, The Netherlands and Prof. Todor Vasiljevic, Victoria University, Australia

Energy-smart farming 115
Efficiency, renewable energy and sustainability
Edited by: Emeritus Prof. Ralph Sims, Massey University, New Zealand

Understanding and optimising the nutraceutical properties of fruit and vegetables 116
Edited by: Prof. Victor R. Preedy, King's College London, UK and Dr Vinood B. Patel, University of Westminster, UK

Advances in plant phenotyping for more sustainable crop production 117
Edited by: Prof. Achim Walter, ETH Zurich, Switzerland

Optimising pig herd health and production 118
Edited by: Prof. Dominiek Maes, Ghent University, Belgium and Prof. Joaquim Segalés, Universitat Autònoma de Barcelona and IRTA-CReSA, Spain

Optimising poultry flock health 119
Edited by: Prof. Sjaak de Wit, Royal GD and University of Utrecht, The Netherlands

Advances in seed science and technology for more sustainable crop production 120
Edited by: Dr Julia Buitink, INRAE, France and Prof. Olivier Leprince, L'Institut Agro Rennes Angers, France

Understanding and fostering soil carbon sequestration 121
Edited by: Dr Cornelia Rumpel, CNRS, Sorbonne University, Institute of Ecology and Environmental Sciences Paris, France

Advances in sensor technology for sustainable crop production 122
Edited by: Dr Craig Lobsey, University of Southern Queensland, Australia and Prof. Asim Biswas, University of Guelph, Canada

Achieving sustainable cultivation of bananas - Vol 3 123
Diseases and pests
Edited by: Prof. André Drenth, The University of Queensland, Australia and Prof. Gert H. J. Kema, Wageningen University and Research, The Netherlands

Developing drought-resistant cereals 124
Edited by: Prof. Roberto Tuberosa, University of Bologna, Italy

Achieving sustainable turfgrass management 125
Edited by: Prof. Michael Fidanza, Pennsylvania State University, USA

Promoting pollination and pollinators in farming 126
Edited by: Emeritus Prof. Peter Kevan and Dr D. Susan Willis Chan, University of Guelph, Canada

Improving poultry meat quality 127
Edited by: Prof. Massimiliano Petracci, Alma Mater Studiorum - Università di Bologna, Italy and Dr Mario Estévez, Universidad de Extremadura, Spain

Advances in monitoring of native and invasive insect pests of crops 128
Edited by: Dr Michelle Fountain, NIAB-EMR, UK and Dr Tom Pope, Harper Adams University, UK

Advances in understanding insect pests affecting wheat and other cereals 129
Edited by: Prof. Sanford Eigenbrode and Dr Arash Rashed, University of Idaho, USA

Understanding and improving crop photosynthesis 130
Edited by: Dr Robert Sharwood, Western Sydney University, Australia

Modelling climate change impacts on agricultural systems 131
Edited by: Prof. Claas Nendel, Leibniz Centre for Agricultural Landscape Research (ZALF), Germany

Understanding and minimising fungicide resistance 132
Edited by: Dr Fran Lopez-Ruiz, Curtin University, Australia

Advances in sustainable dairy cattle nutrition 133
Edited by: Prof. Alexander N. Hristov, The Pennsylvania State University, USA

Embryo development and hatchery practice in poultry production 134
Edited by: Dr Nick French

Developing circular agricultural production systems 135
Edited by: Prof. Barbara Amon, Leibniz Institute for Agricultural Engineering and Bioeconomy (ATB), Germany

Advances in ensuring the microbiological safety of fresh produce 136
Edited by: Prof. Karl Matthews, Rutgers University, USA

Frontiers in agri-food supply chains 137
Frameworks and case studies
Edited by: Prof. Sander de Leeuw, Dr Renzo Akkerman and Dr Rodrigo Romero Silva, Wageningen University, The Netherlands

Improving water management in agriculture 138
Edited by: Prof. Jerry Knox, Cranfield University, UK

Advances in agri-food robotics 139
Edited by: Prof. Eldert van Henten, Wageningen University, The Netherlands and Prof. Yael Edan, Ben-Gurion University of the Negev, Israel

Key issues in agricultural ethics 140
Edited by: Prof. Emeritus Robert L. Zimdahl, Colorado State University, USA

Advances in plant factories 141
New technologies in indoor vertical farming
Edited by: Emeritus Prof. Toyoki Kozai, Chiba University and Japan Plant Factory Association, Japan and Dr Eri Hayashi, Vice President – Japan Plant Factory Association, Japan

Improving the quality of apples 142
Edited by: Prof. Fabrizio Costa, University of Trento, Italy

Protecting natural capital and biodiversity in the agri-food sector 143
Edited by: Prof. Jill Atkins, Cardiff University, UK

Consumers and food 144
Understanding and shaping consumer behaviour
Edited by: Professor Marian Garcia Martinez, The University of Kent, UK

Advances in cultured meat technology 145
Edited by: Prof. Mark Post, Maastricht University, The Netherlands, Prof. Che Connon, Newcastle University, UK and Dr Chris Bryant, University of Bath, UK

Understanding and preventing soil erosion 146
Edited by: Dr Karl Manuel Seeger, University of Trier, Germany

Smart farms 147
Improving data-driven decision making in agriculture
Edited by: Prof. Claus Sørensen, Aarhus University, Denmark

Improving standards and certification in agri-food supply chains 148
Ensuring safety, sustainability and social responsibility
Edited by: Prof. Louise Manning, University of Lincoln, UK

Managing biodiversity in agricultural landscapes 149
Edited by: Prof. Nick Reid, University of New England, Australia, Dr Rhiannon Smith, University of New England, Australia and Adjunct Associate Prof. David C. Paton, University of Adelaide, Australia

Improving nitrogen use efficiency in crop production 150
Edited by: Prof. Jagdish Kumar Ladha, University of California-Davis, USA

Understanding and utilising soil microbiomes for a more sustainable agriculture 151
Edited by: Prof. Kari Dunfield, University of Guelph, Canada

Advances in pig breeding and reproduction 152
Edited by: Prof. Jason Ross, Iowa State University, USA

Advances in organic dairy cattle farming 153
Edited by: Dr Mette Vaarst, Aarhus University, Denmark, Dr Stephen Roderick, Duchy College, UK and Dr Lindsay Whistance, Organic Research Centre, UK

Insects as alternative sources of protein for food and feed 154
Edited by: Ms Adriana Casillas, Tebrio, Spain

Advances in pig nutrition 155
Edited by: Prof. Julian Wiseman, University of Nottingham, UK

Advances in temperate agroforestry 156
Edited by: Prof. Maria Rosa Mosquera-Losada, Universidade de Santiago de Compostela, Spain, Dr Ladislau Martin, Embrapa, Brazil, Prof. Anastasia Pantera, Agricultural University of Athens, Greece and Dr Allison Chatrchyan, Cornell University, USA

Sustainable production and postharvest handling of avocado 157
Edited by: Emeritus Prof. Elhadi M. Yahia, Autonomous University of Querétaro, Mexico

Advances in bioprotection against plant diseases 158
Edited by: Prof. Shashi Sharma, Murdoch University, Australia and Dr Minshad Ansari, Bionema UK

Advances in poultry nutrition 159
Edited by: Prof. Todd Applegate, University of Georgia, USA

Introduction

Climate change is a key threat to agriculture. Its increasing impact requires people working in the agricultural sector to come up with rapid solutions both to adapt to its effects whilst also reducing the sector's contribution to the greenhouse gas (GHG) emissions which cause climate change. Unfortunately, solutions to one problem may compromise solutions to another. Finding the optimum solution requires testing of many alternatives but current projections suggest time is running out to find the best way forward.

Modelling is fundamental to assessing the potential impacts of climate change in a quick and efficient way, at both the field level and on a global scale. Models are available for testing potential mitigation and adaptation measures in a range of scenarios, allowing the agricultural sector to work towards a system transformation that is more climate-adapted and sustainable. This collection summarises the current state of research on improving climate impact models and their use in assessing impacts on different aspects of the agricultural system, on different regions and on specific climate-related problems in agriculture that demand urgent solutions.

The chapters are split into three parts: Part 1 chapters focus on various advances in modelling, such as model integration, improving data flow and integration, the use of crop models to identify new phenotypes adapted to climate change, the development of more integrated approaches in models assessing the impact of climate change and accounting for uncertainties in modelling. Chapters in Part 2 examine modelling climate change impacts on particular aspects of agricultural systems, focusing specifically on crop growth and yield formation, livestock production, low-input smallholder farming systems and agro-ecosystem services. Chapters also discuss modelling climate change impacts on agricultural commodity markets and modelling transition of agricultural systems in response to climate change. Part 3 chapters provide specific examples of modelling climate change impacts on regional agricultural systems, drawing attention to Europe, the United States, Latin America, Australia and Oceania, South and East Asia, North Africa and Southwest Asia and finally, West Africa.

Part 1 Advances in modelling

The first chapter of the book reviews advances in integrating different models that assess the impact of climate change on agriculture. Chapter 1 begins by first reviewing the application of models to problems such as climate change. Based on an 'ideal' modelling archetype, the chapter then goes on to discuss different solutions such as a complete and balanced integration of models, coupling sub-models, integrating new knowledge and facilitating interoperability, integrating behaviours and decision-making to adapt to climate change, incorporating inter- and intra-species diversity into models, from white to black box models and ensemble modelling. The chapter includes examples of all these solutions. The last section explains why these modelling solutions may be insufficient and discusses future trends in modelling.

The subject of Chapter 2 is improving data flow and integration in models assessing the impact of climate change on agriculture. It starts by first describing model-data integration, focusing on multi-criteria calibration of mechanistic agro-ecosystem models. The chapter moves on to review informing spatio-temporal simulations through methods such as remote sensing, proximal sensing and distributed data. A section on the assimilation of data in spatio-temporal simulations is also provided, followed by an analysis of workflows for massive parallel computing. The chapter reviews model-model integration as well as granularity and modular design for model improvement, reuse, exchange and interoperability. A section on the concepts for distributed modelling is also provided.

The next chapter focuses on incorporating genetics into crop models to identify new phenotypes adapted to climate change. Chapter 3 begins by providing an overview of systems analyses and models of crow growth, then goes on to review linking genetics and physiology for phenotypic prediction. The chapter also examines models that predict emergent phenotypes and how crop models can be used to integrate environmental and genomic predictions. A section on the impact of traits and their underpinning genetics on crop performance is also included, followed by an overview of managing genotype x environment interactions to optimise genetic gain in current and future climates. The chapter also discusses how predictive breeding enables iterative impact assessment and adaptation to climate change as well as how crop growth models can enable rethinking of breeding objectives to adapt to and combat climate change.

Chapter 4 draws attention to developing more integrated approaches in models assessing the impact of climate change on agriculture. The chapter identifies four key challenges in regional integrated modelling of agricultural systems: 1) better characterisation of model structural uncertainty and its propagation through the modelled system; 2) improved representation of the importance of non-climate drivers on both the supply and demand sides of agriculture; 3) enabling adaptive capacity to be modified by evolving future socio-economic conditions; and 4) incorporating objective targets or goals for adaptation. The chapter also discusses the IMPRESSIONS Integrated Assessment Platform (IAP) through the form of a case study and identifies four potential areas for future research in improving integrated modelling.

The final chapter of Part 1 reviews accounting for uncertainties in modelling the impact of climate change on agriculture. Chapter 5 provides in-depth discussions on five different forms of uncertainties and how they can be managed, starting with greenhouse gas emission and climate change scenarios. This is then followed by an overview of uncertainties from crop model input data. The chapter also examines managing uncertainties from model structures, then finally discusses managing uncertainties from adaptations in agricultural practices. A section on reducing model uncertainties is also provided.

Part 2 Modelling climate change impacts on particular aspects of agricultural systems

Part 2 opens with a chapter that focuses on modelling climate change impacts on crop growth and yield formation. Chapter 6 begins by introducing the concept of modelling crop yield in agriculture. It then moves on to review crop modelling responses, focusing on structures of crop growth models, data and soil-crop-climate interactions. The chapter also examines assessing climate change impacts on yield with model ensembles, drawing specific attention to early model comparisons and Agricultural Model Intercomparison and Improvement Project (AgMIP) intercomparisons. The chapter includes examples of AgMIP studies such as wheat, maize, barley, rice, potato, soybean and other comparisons such as sugarcane, livestock and grassland as a more general discussion followed by a review of a pest and disease model.

The next chapter of Part 2 examines modelling climate change impacts on livestock production. Chapter 7 first outlines the crucial role of livestock production systems within global food security, economic growth, employment and society. The chapter then goes on to argue the need for rapid implementation of climate mitigations for livestock production systems, namely ruminant production systems. It also places focus on two of the most crucial mitigation measures available for livestock production systems, enteric methane mitigation (such as breeding and dietary interventions) and carbon sequestration and storage to deliver net zero livestock systems.

Chapter 8 draws attention to modelling climate change impact on low-input smallholder farming systems. The chapter presents the effect of climate change on the different components of the farming systems (crop, livestock, grassland and tree), in the context of low-input systems. After this account, the chapter argues the need for a more integrative approach to assess the effect of climate change on farming systems as a whole. The chapter exposes different approaches: coupling of models and the use of an optimisation model. The chapter also highlights the importance of involving farmers in the process of using models to co-design relevant adaptations to climate change in the context of low input systems.

The subject of Chapter 9 is modelling climate change impact on agro-ecosystem services. The chapter reviews how multiple ecosystem services may be impacted by climate change and provides an example for modelling benefit-relevant metrics that can assess the potential impacts of future global change on human wellbeing. The chapter also reviews the state of the art of climate impacts on supporting services for agricultural production and co-benefits of agroecosystems, then moves on to emphasise the need for a systemic global approach for modelling food provision as an ecosystem service. The chapter also examines and integrating climate change and societal changes cohesively, with a focus on one approach to modelling nutrition under cohesive climate and societal changes getting to benefit-relevant indicators. The chapter also goes on to discuss benefit-relevant indicators, putting nutrition further in its context by integrating projections of demand.

Chapter 10 focuses on modelling climate change impacts on agricultural commodity markets. The chapter provides some insights on how climate change can affect agricultural commodity markets in the medium to long term. It includes some methodological aspects on how different quantitative tools, ranging from earth system models to farm-level production decisions, can be combined in the analysis of climate change impacts on agricultural markets. The analysis presented in the chapter is supported by several agro-economic studies carried out by the Joint Research Centre of the European Commission.

The final chapter of Part 2 discusses modelling the transition of agricultural systems in response to climate change. Chapter 11 first examines natural responses, then moves on to highlight two forms of man-made responses: incremental short-term and transformative long-term. The chapter also looks at the role of models in informing adaptation responses, followed by ways to capture the transition in modelling approaches and climate change impact studies, such as empirical crop models and mechanistic models. It highlights models of elements of the agricultural system, specifically drawing attention to crop rotation models, ruminant digestion, soil organic matter turnover and nitrate leaching. Sections on agro-economic models, integrated assessment models and suitability models are also provided, followed by an overview of modelling future pathways of agriculture. Overarching limitations of current modelling approaches are discussed, such as uncertainty and uncertainty propagation, interactions and feedbacks and data limitations.

Part 3 Modelling climate change impacts on regional agricultural systems

The first chapter of Part 3 reviews modelling the impact of climate change on agriculture in Europe. Chapter 12 starts by discussing past climate and future projects for Europe, then goes on to highlight agricultural systems currently in use across the region. A section on agricultural system modelling is also provided, focusing specifically on models and platforms, which is followed by an overview of modelling and experimental work on impacts of climate change through various case studies. These case studies draw attention to crop yields, changes in soil organic carbon, greenhouse gas emissions, farm economics, animal production systems, grapevine and olives and vegetables. The chapter also emphasises how the development of European agroecosystem models should be seen as being one of the research priorities within the region's agricultural research agenda.

Chapter 13 focuses on modelling the impact of climate change on agriculture in the United States. The chapter begins by providing an overview of the recent past and predicted near future changes in climate at the national and regional scale in the US, then moves on to examine the impacts of this on crop-based agriculture, in particular the direct effect on soil, water and major cash crops. The chapter also provides a case study that uses a crop and soil biogeochemical process-based model to investigate potential responses of corn yield to continued trends in temperature and precipitation for the near future using adapted cultivar. A summary of some of the incentives and payment mechanisms related to carbon markets, especially those that focus on agricultural product supply chains, is also provided.

The subject of Chapter 14 is modelling the impact of climate change on agriculture in Latin America. It focuses first on developments in crop modelling and climate change in Chile and Colombia, assessing model characteristics, strengths and weaknesses and areas for future development. The chapter then includes developments in modelling the impact of climate change on key crops in Brazil: upland and irrigated rice, common beans, soybeans, sugarcane and grass forage production for livestock.

Chapter 15 looks at modelling the impact of climate change on agriculture in Australia and Oceania. The chapter summarises climate change in the recent past and future projections for the Oceanian region. It also reviews how agriculture has responded to past climate changes and how systems modelling has helped to quantify the impact of future climate change, evaluate adaptation strategies and influence policy. Case studies in Australia, New Zealand and Papua New Guinea are presented and future research needs for modelling are discussed.

The next chapter reviews the vulnerability to climate change of the key agricultural sector in South Asian countries. After looking at India as an example of how to assess the specific impact of climate change on agriculture, Chapter 16 summarises the ways climate change affects basic plant processes. It then looks at the range of methods used to analyse climate change impacts and their relative strengths and weaknesses, including mechanistic models derived from historic data, the use of field trials, the role of remote sensing data and controlled environment experiments. The chapter also reviews the combined use of different models (e.g. crop, pest, soil and hydrological models) in impact assessments. It reviews the range of current projections for climate change impacts for a wide range of crops across a number of South Asian countries.

Chapter 17 examines modelling the impact of climate change on Agriculture in East Asia. The chapter describes recent research involving climate change impact assessments for the production of primary food crops in the region, focusing specifically on rice, wheat and maize. An overview of the historical impact of climate change on each crop is included, as well as a review of the projected impact of future climate change. The chapter also highlights how modelling extreme climate impact on crop production is becoming more crucial to ensure regional and global food security, as well as how modelling climate change impact on crop disease, weeds and pests should also be developed in order to allow crops to adapt. Sections on modelling climate change impact on crop quality and for diverse and intensified cropping systems are also provided.

The next chapter focuses on modelling the impact of climate change on agriculture in North Africa and Southwest Asia. Chapter 18 begins by examining the geography, economy, climate and agriculture of both regions. It then goes on to describe future climate projections and their effects on regional crop production, followed by an overview of current modelling tools to quantity crop responses to climate change. A section on current climate change impact assessments in North Africa and Southwest Asia is provided. The chapter also examines accounting for the potential of adaptation strategies as well as the limitations of climate change impact assessments.

The final chapter of the book focuses on modelling the impact of climate change on agriculture in West Africa. Chapter 21 begins with an overview of West Africa as a region, followed by an overview of West African climatic conditions. The chapter then provides a meta-analysis of studies investigating the effects of climate change on West Africa. It also highlights challenges in modeling the varied and distinctive characteristics of West African farming systems as well as potential adaptation options. The next section reviews the impact of climate change on livestock production followed by a brief section on integrated climate change impact assessment. The chapter ends with future developments and links to further reading.

Acknowledgements

Cover image acknowledgement:

- Cover image: Drought-struck rye crop. 2008. © Claas Nendel.

We wish to acknowledge the following for their help in reviewing a particular chapter:

- Chapter 2: Dr Ioannis Athanasiadis, Wageningen University, The Netherlands

Part 1

Advances in modelling

Chapter 1

Advances in integrating different models assessing the impact of climate change on agriculture

Jacques-Eric Bergez[1], Julie Constantin[1], Philippe Debaeke[1], Hélène Raynal[1] and Sophie Plassin[1], INRAE, France; Magali Willaume[2], Toulouse INP-ENSAT, France; and Raphael Martin[3], INRAE, France

1 Introduction
2 Understanding models and their application to problems such as climate change
3 The 'ideal model' to assess the impact of climate change on agriculture
4 Implementing the 'ideal model': current approaches
5 From ideal to pragmatic modelling of climate change: coupling sub-models
6 Integrating new knowledge and facilitating interoperability
7 Integrating adaptation management practices into models
8 Incorporating inter- and intra-species diversity into models
9 From white to black box models: hybrid approaches and gridded simulations
10 Ensemble modelling
11 Conclusion and future trends
12 References

1 Introduction

Climate change has become a major concern, especially for agriculture (Wiebe et al., 2019). The IPCC Sixth Assessment Report (AR6) projects that global surface temperature will continue to increase until 2100 by +1.0°C to +1.8°C under the very low greenhouse gases (GHG) emissions scenario (SSP1-1.9) and up to +3.3°C to +5.7°C under the very high GHG emissions scenario (SSP5-8.5), compared with 1850–1900 (Porter et al., 2014; IPCC, 2021). As global

[1] AGIR, University of Toulouse, INRAE, France.
[2] AGIR, University of Toulouse, INRAE, Toulouse INP-ENSAT, France.
[3] Ecosystème Prairial, University of Clermont Auvergne, Vetagro Sup, INRAE, France.

http://dx.doi.org/10.19103/AS.2022.0115.01

temperature rises, the rainfall regime will be significantly altered. Precipitation is likely to become more variable across regions, with an increase in high latitudes and a decrease in parts of the subtropics. Changes in atmospheric circulation and wind speed are also expected (Zeng et al., 2019; Abell et al., 2021). Beyond the changes in mean variables, the intensity, frequency and severity of extreme weather events such as heavy precipitation, flooding, drought and heatwaves are likely to become amplified (IPCC, 2021), but these events are generally difficult to predict (Sillmann et al., 2017).

Although there is uncertainty with regard to the nature, timing and extent of climate change impact, there is a general consensus that the combined effects of increasing temperatures, changing precipitation patterns and elevated carbon dioxide (CO_2) concentrations will affect agricultural and livestock production (Escarcha et al., 2018; Makowski et al., 2020), and that the effects of extreme climate events are likely to exceed those estimated from changes in mean variables (Tubiello et al., 2007). On a biophysical level, increased temperatures accelerate plant growth and development, leading to shifts in plant phenology and growing periods (Gong et al., 2021). Rising temperatures also increase soil evaporation (Abtew and Melesse, 2013), diminish soil moisture (Grillakis, 2019) and increase the water requirement of plants. Changing climatic conditions may also affect the geographical distribution of crops and crop pests (i.e. insects, pathogens and weeds) and outbreak severity (Lamichhane et al., 2015), and facilitate the spread of animal diseases and pests from low to mid-latitudes (Tubiello et al., 2007). The occurrence of climate shocks, such as heat or water stress during critical stages of plant development, could have deleterious effects on yields (Zampieri et al., 2020). A meta-analysis estimated that without implementing adaptation measures, future climate trends and variability are likely to result in higher crop yield variability and aggregated yield losses for the most commonly grown cereals (rice, wheat and maize) (Challinor et al., 2009). It may also impair crop quality, e.g. reduced grain protein and mineral nutrient concentrations and altered lipid composition (DaMatta et al., 2010). Climate change may also be a threat for livestock production because of its impacts on feed crop and forage quantity and quality, animal growth and milk production, animal reproduction, health and mortality (Thornton et al., 2009; Nardone et al., 2010; Rojas-Downing et al., 2017a). Beyond the impact on plant and animal physiology and food quality, climate change may also alter land suitability for crop production (Ramirez-Cabral et al., 2017; Paola, 2018) and changes in hydrological cycles may drive an increased risk of soil erosion (Borrelli et al., 2020).

Climate change is also likely to have socio-economic impacts at farm, local and global levels. Changes in crop productivity could affect farmers' income and threaten food security (Schmidhuber and Tubiello, 2007). In

Published by Burleigh Dodds Science Publishing Limited, 2023.

addition, adverse weather shocks can result in an increase in food prices and price volatility (Hertel, 2010; Nelson et al., 2014), which could affect farmers' income and their cropping decisions, and disrupt food supply and the global agricultural market (Porfirio et al., 2018). Studies have also shown that extreme climate events such as drought can alter farmers' risk preferences regarding climate change (Bozzola and Finger, 2021), which may lead to investments in low-risk, low-return activities (Dercon and Christiaensen, 2011).

Climate-smart agriculture (CSA) is promoted as an approach to transform and reorient agricultural development to address the impacts of climate change (Lipper et al., 2014). CSA is defined by the FAO as a set of actions that (FAO, 2010):

1 sustainably increases agricultural productivity;
2 builds resilience to climate change from the farm to national levels (adaptation); and
3 reduces or removes GHG emissions (mitigation) while enhancing the achievement of national food security and development goals.

Mitigation encompasses crop and livestock management strategies, increases soil and biomass carbon storage (e.g. by introducing cover crops and developing pasture-based systems) and reduces energy and fertiliser nitrogen (N) use (e.g. reduced tillage or using legumes; better housing and manure management in livestock systems). Adaptation strategies range from small adjustments to the redesign of entire systems (Willaume et al., 2014) that aim to reduce the vulnerability of agricultural systems to climate change, i.e. 'the degree to which a system is susceptible to, and unable to cope with, adverse effects of climate change', which is a function of exposure, sensitivity and adaptive capacity (IPCC, 2022).

For livestock systems, adaptation measures include modifying periods and frequencies of grazing, stocking rates, timing of reproduction, using more heat-tolerant livestock breeds (Rojas-Downing et al., 2017b) and increasing animal and grassland diversity (Martin and Magne, 2015). Adaptation strategies in cropping systems encompass the selection of species and varieties that are better adapted to extreme climate events and resistant to pests and diseases, the use of climate forecasting, adjustment of fertiliser rates, amount and timing of irrigation, the location and timing of cropping activities, diversification of cropping systems (e.g. more diverse rotations) and the use of new technologies and practices that improve water use efficiency, maintain soil moisture and prevent water erosion (e.g. through Conservation Agriculture) (Tubiello et al., 2007; Debaeke et al., 2017).

2 Understanding models and their application to problems such as climate change

Models are partisan and simplified representations of reality (Bouleau, 1999). By partisan, we mean that the model is the result of choices made by the modeller for a given purpose and hypothesis. These are *a priori* choices about the representation of the processes to be described, the state variables chosen and the form of the equations describing the behaviour of these variables. By simplified, we mean that a model is a representation or an abstraction of a piece of reality which does not represent the overall complexity of the systems studied and the many ways elements in the system function and interact.

Since the infancy of modelling (Whisler et al., 1986; Sinclair and Seligman, 1996), when models were mainly a way to gather knowledge and test assumptions, the use of models has become more widespread. Models make it possible (Van Ittersum and Donatelli, 2003):

1 to represent and order new knowledge not yet acquired (conceptual operating models);
2 to synthesise knowledge (acquired through experimentation or bibliographical research);
3 to predict the evolution of systems under untested conditions (virtual experiments, forecasting);
4 to challenge different knowledge (points of view) and identify new ideas for reflection and research; and
5 to provide a tool for understanding and negotiating trade-offs as well as for capacity-building.

Use of the possibilities offered by the models thus depends on the purpose of the model: knowledge (scientific world) or innovation (transfer in the socio-economic world). Other aspects related to modelling work that have also changed significantly include:

1 new sources of data to build models (e.g. remote sensing, field sensors, high-throughput phenotyping in platforms, etc.) and (in some cases) a huge flow of data obtained at high speed or on a large scale (crowdsourcing);
2 evolution of software engineering, including the development of coupling solutions, interoperability and modelling framework;
3 new ways of using models (from on-lab modelling to participatory modelling, ensemble modelling, surrogate models, etc.); and
4 improvement of computer performance, development of cloud computing and the evolution of programming languages.

When discussing modelling in this chapter, we do not include only classical differential or difference equation types of model. All types of modelling (statistical models, process-based models, data-driven models, Bayesian model, finite-state and Petri network models, agent-based models and so on) share similar basic issues such as a necessary level of simplification and abstraction.

The impact of climate change on agricultural systems, as for other processes, requires specific understanding, hypotheses and types of knowledge. Models are well designed to reach such goals. The use of simulation models to explore the impact of climate change and assess the effects of mitigation and adaptation strategies has grown over time, and a review for two crop system models (APSIM and DSSAT) showed that climate change has become one of the largest fields of application (Keating and Thorburn, 2018). However, when dealing with climate change, certain modelling aspects need to be considered:

1 New processes and an extended range of environmental factors have to be integrated as current knowledge is not sufficient to deal with the expected climatic impacts on living organisms and agricultural production.
2 A transition phase is an important element as the adaptation to climate change is a long, dynamic process compared with the annual cycle of farming systems and related incremental decisions. Simulated indicators, showing the evolution of the states of the farming systems, have to be developed.
3 Adaptation of farming systems requires multi-criteria assessment and a focus on some processes that require new models.
4 Mitigation of farming practices to reduce climate change is an important feature that needs modelling, as well as the synergies and trade-offs involved in adaptation strategies.
5 The climate in coming years is still uncertain regarding e.g. the spatial and temporal distribution of rainfall or extreme events. Therefore, stochastic modelling needs to be taken into account.
6 In the case of adaptation and mitigation, the needs of decision-makers are crucial. Specific modelling of decision-making processes and adaptive behaviour modelling are therefore necessary.
7 New types of simulations for different scenarios (to reflect the needs of different stakeholders, institutions and policymakers) require innovative approaches to simulation.

This chapter has the following structure. In the first section, we describe an 'ideal' conceptual model to consider every aspect of the effects of climate change on agriculture. We will highlight the operational challenges of such an approach considering the general climate change issues we described earlier. In the following sections, we describe advances in modelling relating to climate change and agricultural production. These sections are based on real modelling exercises and show current advances in integrating different models assessing the impact of climate change on agriculture. In the final section, we highlight missing elements and provide some guidance to moving towards a more complete toolbox for modelling impacts, adaptation and mitigation of agricultural systems facing climate change.

In the following sections, we will use the terms 'system' and 'sub-system' to describe the conceptual entities of the modelled entity (systemic approach), and module and sub-module for the implemented model. The resulting model is, therefore, composed of modules and sub-modules. Simulation is used when a model is run.

3 The 'ideal model' to assess the impact of climate change on agriculture

The agricultural system is composed of a central sub-system, the farm, which interacts with its local environment but is also dependent on more global factors. In theory, to comprehensively model the impact of climate change on agriculture, all these components and their interactions should be represented in a conceptual framework. On each scale (farm, local and global), biophysical and socio-economical sub-systems should be identified (Fig. 1).

At the farm level, biophysical systems are numerous and complex. Plants, animals and soil behaviour can be modelled at different scales (plant to field, animal to herd). They can be modelled within the cropping system or livestock system separately or combined (fodder/manure system). Management is also an important part of this farm system and is driven by the socio-economical sub-system of the farm, i.e. the farmer and resources (financial, workforce, land, equipment etc.). This can be represented by fixed values (i.e. amount of fertiliser, date of sowing, grazing periods, feed ration) or decision rules (i.e. irrigation strategy, crop and variety choice, breeding strategy for animals). Farmer's decisions occur at several levels in time (i.e. planning and investment, strategy and tactical responses) and in space (i.e. field level, crop planning on the farm, animal stocking rate). Each component of these systems can be described in detail or in outline.

The farm is an open system interacting with biophysical and socio-economical sub-systems that can also be affected by climate change. Farms are nested in a landscape and within a hydrological system. These three sub-systems affect, and are impacted by, surrounding biodiversity, either wanted (e.g. pollinators) or unwanted (pests and diseases). The surrounding socio-economical environment of the farms should also be considered in an

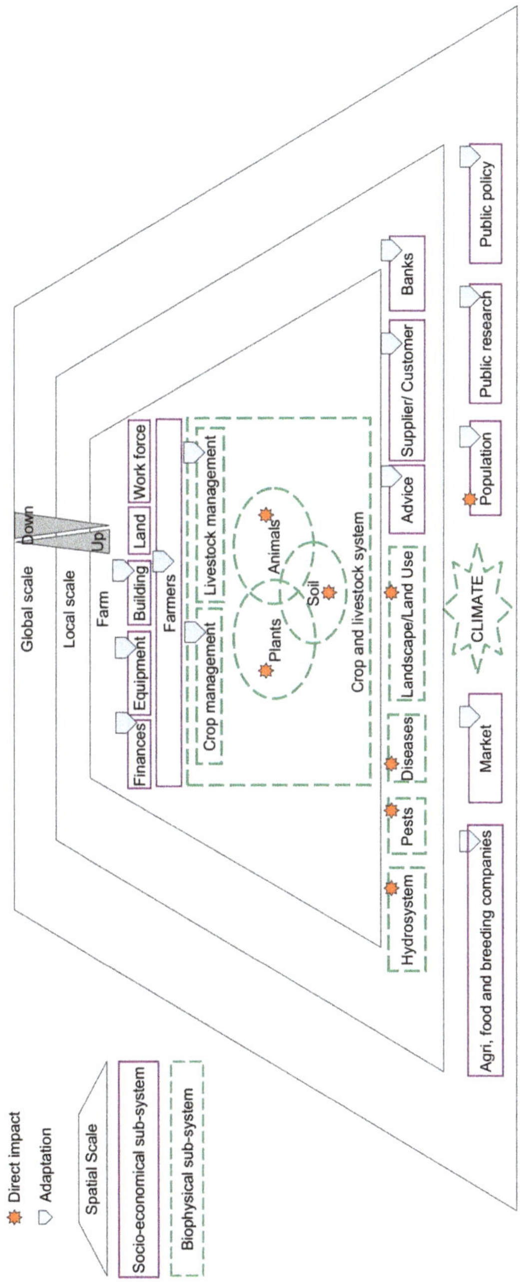

Figure 1 'Ideal' conceptual model to deal with climate changes. All sub-systems and their interactions should be represented.

'ideal model', including interaction with advisors, support from banks and relationships with suppliers/consumers, taking into account their organisation and dynamics. The farm is also influenced by more global phenomena such as market dynamics and regulation, population growth and food security, public policies and subsidies and advances in research. Breeding companies influence farms by disseminating cultivars better adapted to new environments, while agro-food chain companies also provide inputs to the system. Climate change also operates at a global scale. This complex environment represents a system in nested scales with different levels of organisation to model, raising the issue of upscaling and downscaling. Modelling on such a large spatial scale also implies representing a large temporal scale since both are linked (Ewert et al., 2011).

The impact of climate change on agriculture will mainly modify the biophysical sub-systems at various scales (field level, landscape level, etc.). These effects should be modelled whatever the climatic event (temperature rising, drought, flooding, storm, increased carbon dioxide, etc.). The 'ideal model' should also be able to combine the effects of adaptation and mitigation measures designed and applied by all socio-economical actors at the various scales. Time is also a crucial component, since some effects may be cumulative over time or transitory.

Ideally, this model should be parsimonious, i.e. it should be as simple as possible in design and computing power required, simple to understand and easy to handle and use. It should be easy to calibrate ; be robust in a wide range of soil, climate and agronomic conditions ; and require little data to initialise new simulations. The calculation time should also be minimised to allow multi-simulation over large ranges of situations or spatial areas.

To build such a model is quite challenging, as it is like a 'labyrinthine system'. However, this representation could be used as a toolbox in which one can select the necessary sub-systems to consider in order to address a specific question. Depending on the objective, it is not necessary to model all sub-systems with the same level of detail.

We have described the different components and organisation levels of the agricultural system with a focus on the farm. Numerous interactions exist between the system components and the environment, which have not been accounted for in this approach. Besides climate change, many biophysical and socio-economic dynamics influence farms leading to emerging properties and unexpected behaviour of the whole system. The complexity of this system has generally to be simplified for modelling purposes.

The impact of climate change is often quantified by simulating variables of interest (e.g. crop production, GHG fluxes, etc.) under climate projections at regional or national levels considering near- or far-future climatic scenarios by way of statistical or process-based models applied to croplands or grasslands

(Challinor et al., 2009; Lobell and Burke, 2010; Rosenzweig et al., 2014). In finer resolution approaches, management (sowing date, crop species, variety, grazing period, stocking rate, irrigation, etc.) is used as input data for site-specific or gridded approaches (White et al., 2011; Lehmann et al., 2013). Less frequently, models also include the decision process, the cropping plan or the farm structure, which allow evaluation of incremental or systemic adaptations to climate change in terms of production, management and land use (Holman et al., 2019). All these approaches simulate the effect of climate change on crop production but at different levels, with differing resolutions and based on various sources of information.

It is often not necessary to represent every part of the system to assess the impact of climate change on a specific part of the agricultural system (such as crop yield, nitrate leaching or calf production, for instance) or to evaluate the impact of adaptation measures. Only some sub-systems are needed to make this assessment depending on the specific objectives of the study or the scenarios to be tested. The level of detail in each sub-system varies and depends on the objectives and accuracy requirements in the simulation, from a very simplified to a more refined representation. The end use of the model is an important element to consider when designing such models to keep them useful (Prost et al., 2012).

Building a model on a large spatial and temporal scale by integrating all the different levels of organisation is extremely challenging. It raises issues regarding lack of knowledge of all the different processes (and quite often interactions between entities), of having data of sufficient accuracy and on a sufficient scale to calibrate, evaluate and run the model, all of which could compromise the accuracy of the results. Nevertheless, some attempts have been made to integrate several to almost every dimension(s) of the agricultural system. The various modelling choices have different advantages and limits that will be detailed in the following sections.

4 Implementing the 'ideal model': current approaches

Implementing the 'ideal model' is a dual challenge given:

1. the integration of a large number of scientific disciplines; and
2. technical implementation requirements.

Nevertheless, there have been several examples developed over the last few decades. The majority have been initiated because stakeholders need an estimation of the impact of climate change on agriculture using multi-criteria assessment. They also need information to build new public policies (e.g. the EU Green Deal for a more climate-smart European agriculture) and devise adaptation and mitigation solutions.

The SEAMLESS-IF modelling framework (Van Ittersum et al., 2008) is a good example of the implementation of such an 'ideal model'. It is the result of exchanges between the scientific community of agronomic and economist modellers and the European Commission. The SEAMLESS-IF framework is based on the concept of Integrated Assessment and Modelling (IAM) (Parson, 1995; Harris, 2002; Parker et al., 2002). Operational IAM is part of the systems analysis paradigm which is seen as a way to model the biophysical, economic, social and institutional aspects of a system in an integrated and balanced way.

Other models include AROPAj (Barberis et al., 2021) and GlobAgri-WRR (Le Mouël et al., 2016) used in different European studies. GlobAgri-WRR is a global accounting and biophysical model that is designed to quantify GHG emissions and land-use demands related to agricultural production for specified levels of diets, population, non-food uses, food loss and waste and production systems (techniques, crop yields) in every country of the world. It incorporates several biophysical models (e.g. a global livestock industry model, a European Commission Joint Research Centre [JRC] land-use model, a nitrogen emissions model, etc.) but does not consider economic feedback effects. GlobAgri-WRR has primarily been used to estimate how changes in demand or production (from climate change impacts or new production methods) might impact land-use demands and GHG emissions (Searchinger et al., 2019). Such frameworks also make the links between micro-level (field-farm-small region) and macro-level (market or sector).

More recently, 'user-friendly' tools have been developed, such as the Integrated Assessment Platform (IMPRESSION) (https://climate-adapt.eea.europa.eu/metadata/tools/climsave-integrated-assessment-ia-platform). This tool offers a web interface, directly usable by European stakeholders, to explore the complex multi-sectoral issues surrounding impact, vulnerability and adaptation to climate and socio-economic change across Europe.

At the regional or watershed scale, some models have been developed following the philosophy of the 'ideal model', such as the MAELIA platform (Catarino et al., 2021) or MOSAICA model (Chopin et al., 2015). The MAELIA platform enables the assessment of the environmental, economic and social impacts of combined changes in agricultural activities (e.g. recycling of biomass), natural resource management strategies (e.g. water) and global drivers (e.g. dynamics of land cover and climate change). It aims to design sustainable strategies for water resource management within a watershed in the uncertain context of global change. The MOSAICA model aims to understand the effects of policy changes on cropping systems at the regional scale and their contribution to the sustainable development of regions. It explicitly incorporates information at field, farm, sub-regional and regional scales to provide cropping system mosaics by way of regional optimisation of the sum of the individual farmer's options under field, farm and territory biophysical and socio-economic constraints.

It is important that integration is achieved without favouring one aspect of the system over another, since biases are often observed in integrated assessments towards economic, biophysical or environmental issues (Van Ittersum et al., 2008). Britz et al. (2014) distinguish several approaches in a context of design of public policies: bio-economic farm models (BEFMs), multi-agent models, life cycle analysis and agri-environmental impact simulation. For example, BEFMs have been widely used and are a specific form of model that aims to optimise resource management decisions by the farmer, in relation to inputs and outputs. Used for public policy issues, they integrate different scales and are linked to a partial equilibrium model, e.g. CAPRI-FT (Britz and Witzke, 2014), IFM-CAP (Louhichi et al., 2017), FARMDYN (Britz et al., 2014) and FSSIM (Kanellopoulos et al., 2014). Even if these models are operational and have been used in different projects, they are still based on basic assumptions and simplifications (e.g. an inflexible representation template of farm processes, a limited number of farms not necessarily representative of all the farms in the territory, etc.), which may be a limitation to their use in a climate change context.

5 From ideal to pragmatic modelling of climate change: coupling sub-models

Modellers have to adopt a systemic approach but limit the complexity of the system representation by selecting the relevant spatial and temporal granularity or the appropriate interactions, depending on the issue they want to address. Over the last decade, we have observed a trend in promoting this approach to develop pragmatic modelling approaches to climate change analysis (Gouttenoire et al., 2011; van Oosterzee et al., 2014; Ewert et al., 2015; Ghahramani and Bowran, 2018; Naulleau et al., 2022). This pragmatism often leads to selecting and linking specific existing models. The subsequent model is therefore the result of coupling:

1 Sub-modules coming from other models (recycling existing knowledge);
2 New sub-modules developed as required by a project (new knowledge); and
3 specific interfaces for the engineering aspects of coupling and management of inputs and outputs.

Coupling refers to communication and interchange of information between sub-modules. As noted by Siad et al. (2019), the way the sub-modules are coupled varies greatly, depending on the expected outcomes but also on the possible computational techniques.

We have distinguished a gradient in the levels of coupling (Fig. 2). The first case concerns where models are chained: the principle of modelling

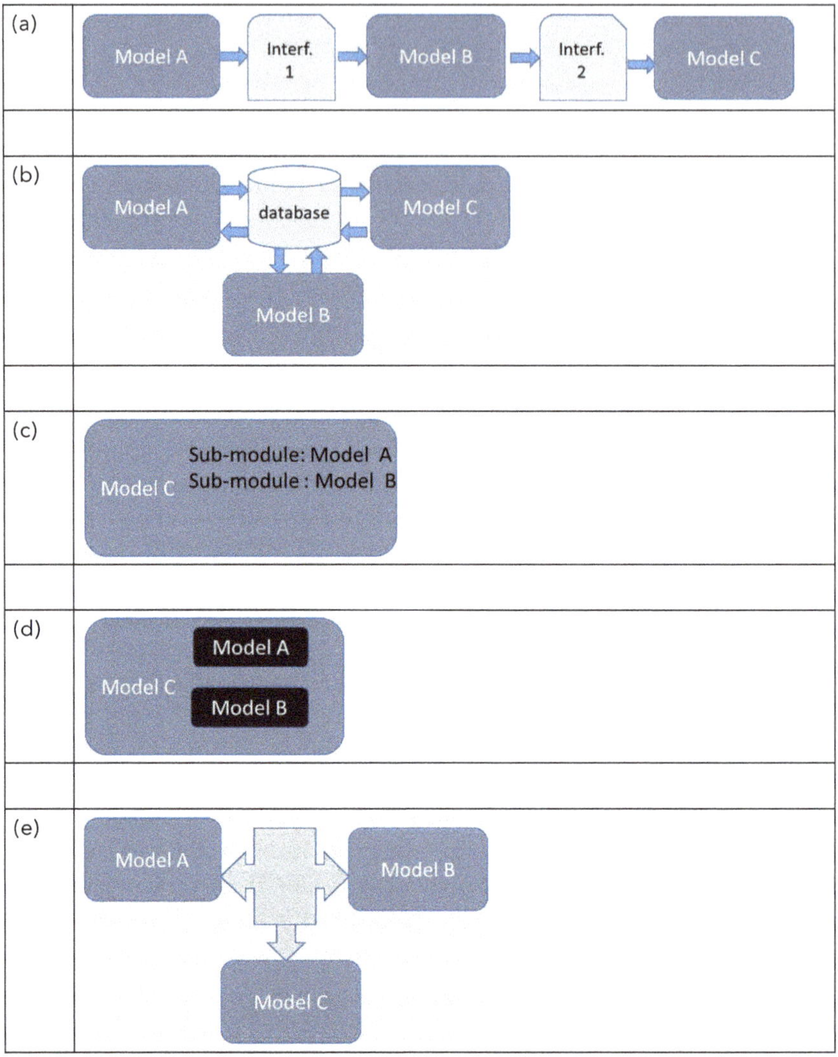

Figure 2 The main five ways to couple models and modules: (a) modelling chain, (b) weak coupling, (c) sub-model integrated as a sub-module, (d) code integration, and (e) model coupling served by the modelling framework engine.

chains or workflow. In this case, there is no real coupling as each model is fully independent. The second case concerns so-called weak coupling where the models will exchange information through files or a shared database. The third case consists of completely integrating models so that the included model becomes a simple module or sub-module. The fourth case consists of integrating different modules within the same model, at the level of the computer code. This case also includes models developed from scratch and fully rewriting codes. The result is a coherent whole from a computer science point of view and generally allows easier representation of interactions, in particular, feedback type between the various components of the model. Finally, the fifth case consists of developing the model in a framework, with coupling ensured by the coupler engine of the platform.

The sub-models to combine can be selected using different approaches. For instance, the decision can be driven by scientific considerations (e.g. relevant processes included in the sub-model), operational constraints (e.g. software simulation performance) or a participatory modelling approach (Voinov and Gaddis, 2008). In the latter case, scientists and key actors in designing adaptation and mitigation strategies work together to build the model. Together they select the sub-models by introducing biodiversity into biogeochemistry models, for instance as Van Oijen et al. (2020a), and scenarios to test in order to address climate change issues (Rodriguez et al., 2014; Cradock-Henry et al., 2020; Van Oijen et al., 2020b; Naulleau et al., 2022). Today, a large variety of coupled models arise from completeness of the modelled processes, method of coupling used or IT solutions. As the 'ideal model' emphasises the importance of representing the farm level, we include some examples where the farm model is designed by coupling sub-models: Robert et al. (2016); Lieffering et al. (2016); Fitzgerald et al. (2009); and Kalaugher et al. (2017). Later we discuss techniques commonly used to represent farmer behaviour.

6 Integrating new knowledge and facilitating interoperability

Coupling technically facilitates the development of new models by combining existing ones, but it is often not enough. New issues related to climate change assessment, adaptation and mitigation necessitate the improvement of models through the integration of advanced knowledge and processes. For example, over the last few decades, the cropland and grassland modelling community has been organised into two international knowledge hubs:

1 The Agricultural Model Intercomparison and Improvement Project (AgMIP; www.agmip.org); and
2 The Modelling Agriculture with Climate Change for Food Security project (MACSUR; www.macsur.eu).

These networks have been important arenas of the exchange of knowledge for the agricultural modelling community and have contributed significantly to the improvement of farming systems models, both in their design and in their computational implementation (Thorburn et al., 2018). To account for the impact of climate change on plant development, crop modellers have been obliged to revisit current crop models. This exercise was chiefly initiated by crop model intercomparison exercises for major crops, which showed important differences between models (Fig. 3). This was mainly explained by the fact that climate change (e.g. elevation of the CO_2 rate, temperature increase), by exploring unexperimented conditions, requires significant modifications in the representation of some biophysical processes currently implemented in most crop growth models (e.g. effect of priming and microbial diversity in ecosystem functioning as presented in Perveen et al. (2014)). It soon became clear that this work was easier when the original model had been built in a modular way, with a module associated with each process. If modelling frameworks such as APSIM (Holzworth et al., 2018), BioMA (Donatelli et al., 2010), DSSAT (Hoogenboom et al., 2019), OpenAlea (Pradal et al., 2015), RECORD (Bergez et al., 2013) and Simplace (Gaiser et al., 2013) helped to produce models with a good level of

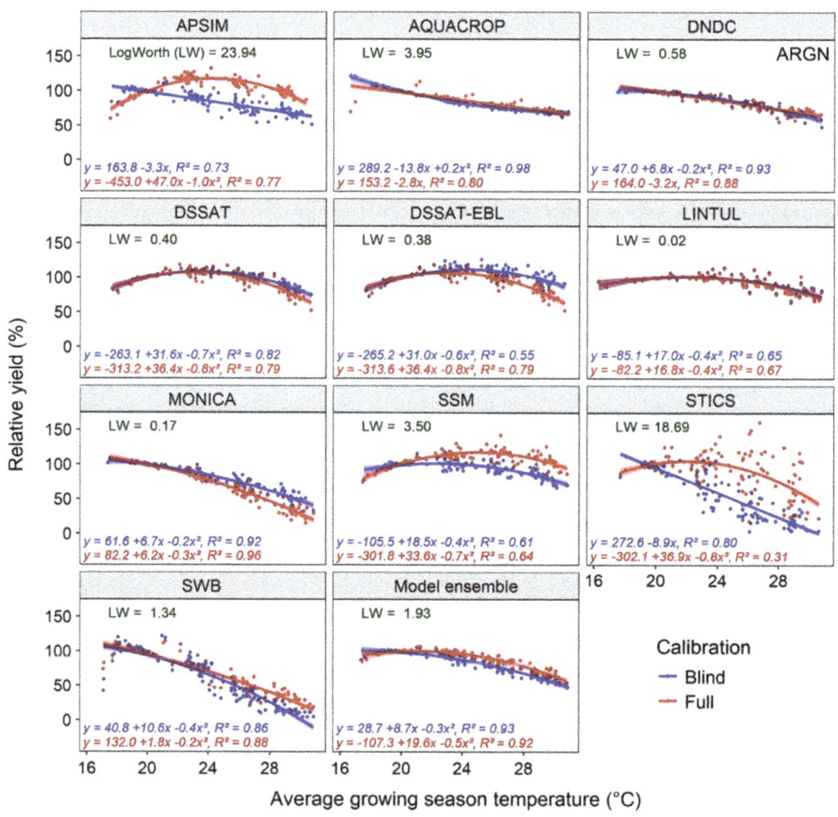

Figure 3 The use of different crop models to test the impact of climate change on soya relative yield. Instead of developing a new model, the AgMIP approach provides new knowledge on comparing different existing well-known models. From Kothari et al. (2022).

modularity, this was rarely the case for models with legacy code. Modularity has long been a recommended practice in the crop modelling community (Reynolds and Acock, 1997; Jones et al., 2001).

Most models today have developed a modular approach (Brown et al., 2018), which facilitates their readability, their evolution by adding new modules and their sharing with the modeller community.

In addition, to facilitate model sharing, these communities have worked for the implementation of semantic interoperability. The interoperability is also essential because simulation results are, and will remain, uncertain despite all the progress on the models themselves. This is reinforced by the fact that, in order to assess the impact of climate change or to explore possible adaptation pathways, it is necessary to consider uncertainties in future climate predictions. One way to reduce uncertainty is to cross-reference information sources and this relies, in particular, on semantic interoperability. In connection with AgMIP, researchers have worked on the definition of standards to facilitate the interoperability of modules of models and data. Interoperability corresponds to one of the FAIR principles (Wilkinson et al., 2016), now extensively promoted in the agrifood community (Top et al., 2022), and uses the International Consortium for Agricultural Systems Applications (ICASA) vocabulary (White et al., 2013). Recently, a module specification standard and a metalanguage Crop2ML (Midingoyi et al., 2021) were proposed as a standard to express module equations.

Beyond the AgMIP community, we also observe the development of ontologies and their use in modelling work. They help in sharing data used as model inputs for model calibration and documentation. The agronomic community has developed many controlled vocabularies and ontologies such as AGROV OC (Subirats-Coll et al., 2022) from FAO or AGROPORTAL (Jonquet et al., 2018). Some ontologies are very specific, such as OntoHydroAgro, which is dedicated to the impact of climatic changes and agricultural activities on water resources (Bonacin et al., 2016). To facilitate the access to cross-referenced sources, it is also necessary to have open and collaborative software systems enabling the sharing of large-scale, multifaceted data and models (e.g. knowledge hub). Interoperability also relies on progress in software engineering with the development of virtualisation which allows model simulation whatever the computer environment. This involves, for example, distributing virtual images, using software engineering tools (e.g. docker) (Anderson, 2015).

7 Integrating adaptation management practices into models

These improvements, although necessary, do not consider changes in management practices that will result from global change. In particular, during a working session of the MACSUR knowledge hub (Kipling et al., 2019), modellers emphasised the importance of integrating biophysical/economy/ management interactions into models when working on adaptation to, or mitigation of, climate change. This triple interaction involves a wide range of actors such as farmers and government decision-makers, each of them with a different perspective on what they want from models. This included different spatial scales, from the agricultural plot, the farm up to country or region. Models must consider operational, tactical and strategic decision-making processes.

However, these models require other approaches than those classically used to build biophysical models, with the latter based on mathematical formulae such as differential equations. An alternative approach e.g. has been to use decision rules to represent management aspects of cropland/ grassland. For each crop practice, a set of rules is defined which, during the simulation, will allow triggering of a decision (e.g. sowing, harvesting, irrigation, fertilisation, grazing) according to the model state variable (e.g. available biomass, leaf area index, soil moisture, air temperature) or even by calculating, for example, the rate of nitrogen fertiliser or the optimal animal stocking rate (Aubry et al., 1998; Bergez et al., 2006; Vuichard et al., 2007; Martin-Clouaire and Rellier, 2011; Dury et al., 2012; Chabrier et al., 2015). Using this approach in modelling work related to climate change is particularly important since climate change makes the use of fixed calendar dates for particular agronomic practices based on the current context obsolete. For example, in one model intercomparison study (Constantin et al., 2019), sowing dates were adapted using decision rules to accommodate climatic variability and trends (Fig. 4).

For tactical decisions at the plot or farm level, there are fewer examples than for operational decisions. Nevertheless, we can refer to a number of works related to crop rotation or crop planning (Robert et al., 2018; Hajimirzajan, 2021; Pahmeyer, 2021). In these studies, the crop model is coupled to an economic model built to optimise an objective function (such as profit) under different constraints (e.g. water availability). Aghajanzadeh-Darzi et al. (2017) coupled PaSim to AROPAj in order to estimate the economic effect of climate change on grassland. Others have developed applications for optimisation of fertilisation, such as in Ramos-Castillo et al. (2021). In the case of approaches at a broader (e.g. territorial) scale accounting for interactions between different actors, multi-agent formalism is often used (Burli et al., 2021). Multi-agent-based models are used when assessing the impact of policymakers' decisions on climate change mitigation policies (Huber et al., 2018).

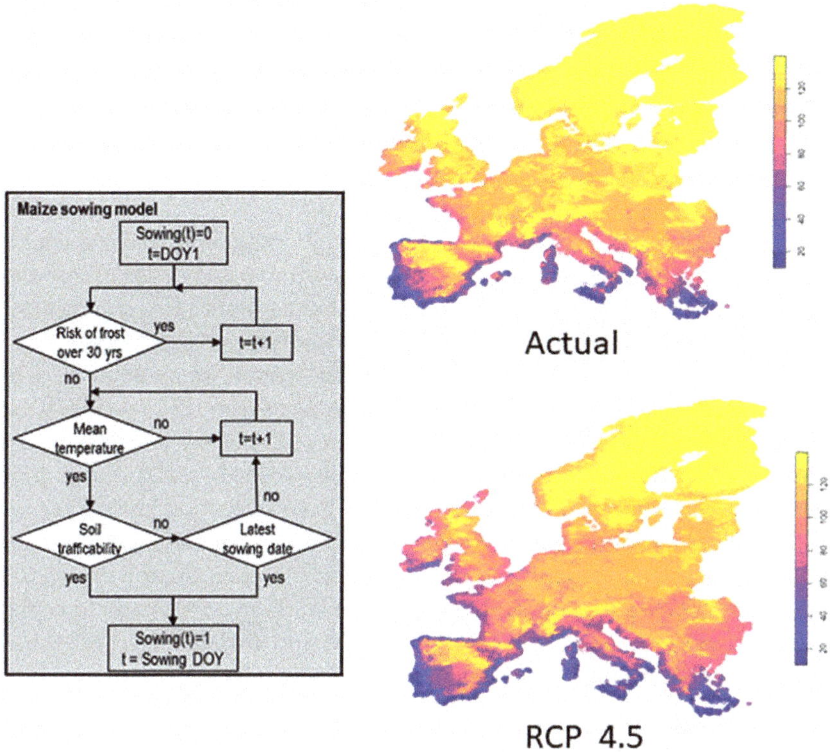

Figure 4 Modelling decision-making is important when considering climate change. In Constantin et al. (2019), a simple maize sowing decision model was created to analyse the impact of the sowing dates.

8 Incorporating inter- and intra-species diversity into models

The consideration of more global processes can overlook more local processes which, if not taken into account, could distort simulations. For example, there is now evidence that climate change impacts species diversity and that this diversity is also a major way to mitigate the effects of climate change (Van Oijen et al., 2018). More and more modelling work is taking into account this inter-specific diversity (Li et al., 2011; Movedi et al., 2019). It is important to take account of genetics in models because designing new varieties able to deal with abiotic and biotic stresses is a way to adapt to climatic extremes and variability (Martre et al., 2015). White and Hoogenboom (2003) have described five ways of considering genetics in models:

1 No reference to species, making the model generic;
2 The model is species-specific without references to genotypes;

3 Genetic differences are represented by cultivar-specific parameters of the model;
4 Genetic differences are represented by specific alleles and gene action is represented through linear effects on model parameters; and
5 Genetic differences are represented by genotypes, with gene action explicitly simulated.

Until recently, most crop models have only integrated cultivar-specific parameters and based calibration on field experiments, in order to compare several varieties and determine those that are best adapted to a given climate change scenario (e.g. Zhang et al. (2022) for rice cultivars in China; Tao et al. (2017) for barley; Gérardeaux et al. (2018) for cotton, etc.). Extending the inclusion of genetics in models is a promising approach for breeders facing climate change. It implies integrating more ecophysiology into the crop or animal model, and making the link to quantitative trait loci (QTL) and genes (Hammer et al., 2002). More recently, models have been customised in order to simulate response to climate change depending on genes/QTL. Martre et al. (2017) have provided an overview of progress in this area. Guitton et al. (2018) worked on sorghum to develop a photosensitive ideotype adapted to climate change using a crop and a combination of QTL, and Wallach et al. (2016) have predicted the flowering time of beans.

9 From white to black box models: hybrid approaches and gridded simulations

The multiplicity of processes involved at heterogeneous scales has tended to make models more complex in their implementation and use more resources (e.g. more input required, more computing power and time needed). In recent years, statistical models have emerged from the expansion of machine learning techniques (Liakos et al., 2018). They are constructed by relating historical variables (e.g. yield) which are then simulated with other variables such as climate, soil or management and then applied in a context of future climate. While the use of statistical models beyond observed conditions raises the issue of uncertainty associated with the results (Lischeid et al., 2022), they are useful in high-resolution simulations over a large area. They require less data than mechanistic models and decrease the computational load.

A hybrid approach using mechanistic models and a metamodel emulating the mechanistic model is one method to address the issue of climate change impact, adaptation and mitigation on a large scale. Successful examples of this hybrid approach include choosing a cultivar to adapt to climate change (Zhang et al., 2022; Tao et al., 2017; de los Campos et al., 2020) simulating soil organic carbon (SOC) (Luo et al., 2019), crop yield prediction and mapping (Corrales et al., 2022; Guilpart et al., 2022) and climate impact assessment (Blanc, 2017). This hybrid approach can also be used successfully in optimisation processes used to mimic farmer tactical decision-making (Nguyen et al., 2019). Modellers have often used a linear regression method to build the surrogate model. Advanced statistical methods can be used to improve performance (e.g. learning methods such as random decision forests).

However, the use of statistical models is not always appropriate because they are generally built on historical data, limiting their ability to predict the future. Statistical models built from simulated data also require adding and updating the knowledge base whenever the model is modified. One approach is the use of gridded simulations. This involves a map divided into cells. In Launay et al. (2021) e.g. cells are the intersection of climate and soil data which are heterogeneous in terms of surface. It is then necessary, for each of these cells, to run one or more instances of the model with its own inputs based on soil, climate, management, etc. Gridded simulations require increasingly large computing resources. This can range from a high-performance workstation to a cluster (e.g. where information needs to be provided in real time, such as for water flows in a watershed) or a computing grid (independent simulations between cells, with their own resources on each node). Moreover, these gridded simulations require specific databases at the territory scale. There are numerous sources for these databases and there is ongoing work to develop them, e.g. soil data at 10 km^2 resolution (Han et al., 2019), irrigation mapping (Zajac et al., 2022), testing (e.g. light radiation data) (Araghi et al., 2022) and climate forcing data (Ruane et al., 2021).

Specific platforms have been developed to supervise these gridded simulations. These platforms are responsible for formatting the data for the model, distributing model operations over the available resources and then aggregating the outputs in an easily manipulated format (i.e. NetCDF). These platforms can be model-specific: Jang et al. (2019) for EPIC, Eza et al. (2015) for PaSim or multi-model, Shelia et al. (2019) for DSSAT, APSIM and SARRA-H and Kim et al. (2020) for ORYZA2000.

10 Ensemble modelling

Given the multiplicity of models and easy access to computational resources needed to run simulations, many studies have compared model outputs with each other and shown that even similar models differ in predictions relating to climate change (Sándor et al., 2016) and in sensitivity to different types of climatic hazards. In some fields of research, it is well known that models, due to their structure, can produce highly variable outputs. Pearson et al. (2006) e.g. showed that the prediction of species range distribution can vary, for a specific species, from -92% to +322% depending on the model used.

The multi-model ensemble (MME) technique is used to reduce prediction errors and has been described in detail by Wallach et al. (2016). It consists of aggregating the independent predictions of the MME to define a global prediction and an associated degree of uncertainty. Several models can be included in this MME. In addition, several sets of parameters and input data sets can be used for the same model.

Numerous studies have shown the efficiency of such a process, such as Martre et al. (2015) who concluded that the result of such an MME (by using mean or median) is better than that of its best model. Similarly, recent work (Ehrhardt et al., 2018; Sándor et al., 2020) has shown the reduction of uncertainties by using the MME median for the simulation of yield and carbon-nitrogen (C–N) cycles in field crops and grasslands. This reduction in uncertainty is also found for the calculation of soil C by Farina et al. (2021). Sándor et al. (2018) show that this approach can be used to test the impact of grassland management (reduced animal stocking and N input) on GHG emissions. These types of projections can be greatly improved in terms of accuracy and robustness, as shown by Araujo and New (2007). This can also be seen in Ruane et al. (2021) who evaluated 14 models and 11 climate data sets.

11 Conclusion and future trends

As we have seen, there are many differing features that models can (should) integrate, as well as different uses of models within agricultural systems in relation to climate change (mitigation, adaptation and impact). Figure 5 provides an overview of the required properties of the models using a structured list based on the 5W1H approach:

1. What is the target for the stakeholder?
2. Why do we want to model?
3. What about the climatic change dimension?
4. What about the sustainability dimensions?
5. What biophysical processes are involved?
6. What decision-making processes are needed?
7. What geographical scale needs to be considered?
8. What timescale should be used?
9. How to model?
10. How to use the model?
11. Any view on the simulation output analysis?

All these questions have to be answered before modelling can begin. In Fig. 5, we have highlighted some of the different examples discussed in this chapter. However, not all the cases mentioned in the figure have been discussed. We

Target/ stakeholdersMM	Researcher M A N G	Farmer	Advisors N	Branch organisation G	decision-makers M A N G		
Why ?	Understand A	Produce Knowledge A G	Warn /alert/ educate M N	Investigate/forecast M N G	Anticipate/ react	Conceive M N	
What about CC ?	Impact A G	Mitigation M	Adaptation N	Opportunities M	Risks M A N	Resilience	
What about sustainability ?	Agronomical M A N	Environment M A N G	Biodiversity M	Economical M A?N G	Social M G		
What biophysical processes ?	Physiology A N	Genotype Environment Interaction A	Hydrology M N	Soil M A N	Abiotic M A N	Biotic M A(?)	Farming practices M N G
What decision making processes ?	Operational (ex: Agricultural practices) M N	Tactical Ex: Crop rotation, location M N	Strategical ex: Ressources sharing investiment M N	Stakeholders interaction M			
Spatial scale	Field/animal M A N	Crop and/or livestock system M N	Gridded A	Farm M N	Local N	Regional M A	Global G
Time scale	Short (2030) M N	Middle (2050 - 2060) A G	Long term (2100)	Historic M A N			
How to model ?	Reuse/ Benchmark M A	Add CC specific features in current models A	Model chaining / weak coupling	Models integration/ strong coupling M N	Development from scratch	Participatory modeling	
How to use ?	Ensemble modeling simulation A	Participatory simulation M	Scenario test M A N G	Optimization	Support decision tools M	Surrogate simulation	

Figure 5 Specific properties of agricultural models dedicated to climate change. Letters are related to examples given in this chapter: M, MAELIA; A, AgMIP; N, Namaste; G, GlobAgri.

have not e.g. discussed the development of a direct farmer decision-system support tool to deal with climate change. Mitigation is also a large source of uncertainty regarding modelling. The 'what if?' and 'how to?' queries require a specific modelling approach that is currently lacking in the literature. The feedback loops between climate change and different sub-systems have not really been considered. To date, mitigation actions have not impacted the climate and therefore modified the properties of the modelled systems. A larger list of such feedback loops could be created, but this is not the topic of this chapter.

Figure 5 highlights a range of new challenges in relation to climate change when discussing modelling. Some of these are listed below:

1 Adaptation and learning processes in a long-term series simulation. If one wants to simulate a long-term series, and especially when dealing with climate change, which is quite a lengthy process, models should take into account adaptation and learning processes. Farmers often change their practices depending on conditions (climate, economic, pest pressure, etc.). Ecological processes also change depending on environmental pressure. Progress has been poor with regard to integrating such changes into a model.
2 Feedback effects due to mitigation. A large body of research is available on mitigation options (reducing GHG, increasing C storage, etc.) at different scales (plot, farm, territory, nation and even globally). Several EU projects are concerned with this (see for example ClieNFarms, https://www.inrae.fr/sites/default/files/pdf/ClieNFarms_PR.pdf). One of the big challenges when mitigation is modelled is to integrate feedback effects. Depending on differing levels of some variables, different processes can be triggered or not, but this dynamic modelling is far from simple.
3 Systemic and holistic approach. As explained throughout this chapter, a larger systemic and holistic approach to integrate climate change and effects on farming systems is required, which needs a more interdisciplinary approach. 'Interdisciplinary' is a buzz word in current research but is complicated to achieve in practice.
4 New end users involved. As explained earlier, modelling is the process of creating a model in order to simulate a specific scenario. When dealing with climate change, some new end users and types of simulation may be required. As explained in the discussion of the ideal model, policymakers e.g. are a key group who need models to simulate climate change processes. Integrating the modelling process and education of users on the use of models may also be a challenge.

Other chapters of this book will give more information on the different models to tackle some specific questions (Fig. 6), and will cover a large range of

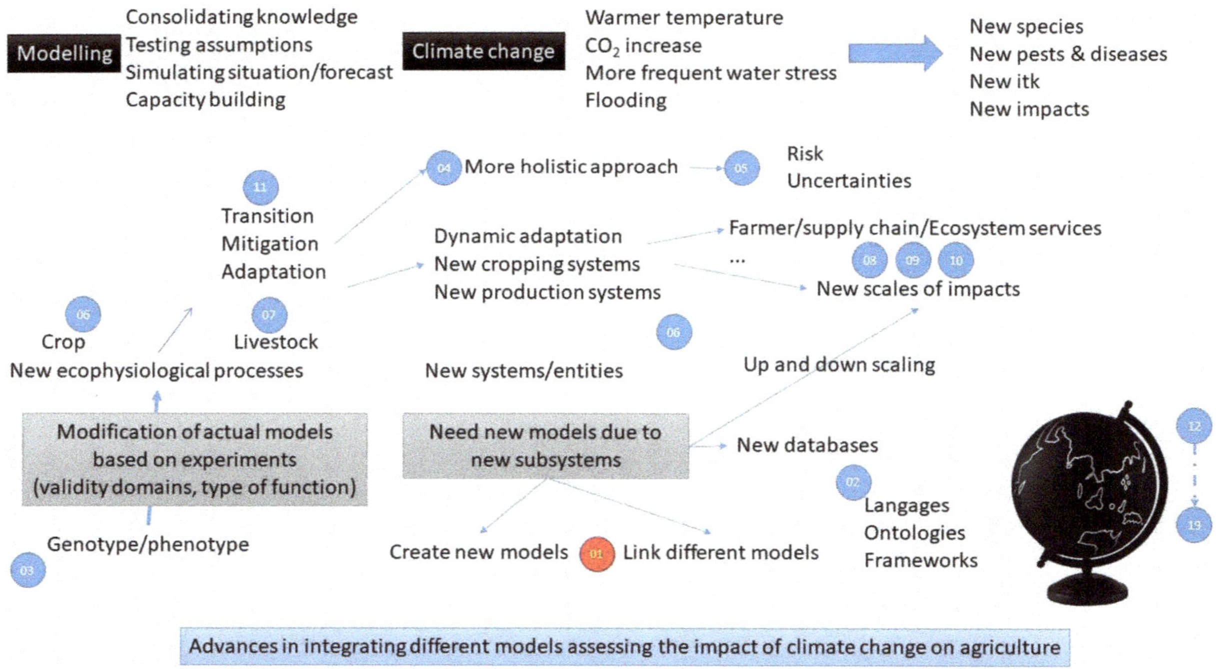

Figure 6 The general positioning of this chapter in the general layout of the book. Circled numbers refer to the chapter number.

applications from gene to region, from biophysical processes to decision-making aspects, from north to south. This introductory chapter sets the scene and provides a broad overview of the impressive advances in modelling over the last few decades to tackle climate change issues impacting agricultural systems.

12 References

Abell, J. T., Winckler, G., Anderson, R. F. and Herbert, T. D. (2021). Poleward and weakened westerlies during Pliocene warmth, *Nature* 589(7840), 70–75. doi: 10.1038/s41586-020-03062-1.

Abtew, W. and Melesse, A. (2013). Climate change and evapotranspiration. In: Abtew, W. and Melesse, A. (Eds.), *Evaporation and Evapotranspiration: Measurements and Estimations*. Dordrecht: Springer Netherlands, pp. 197–202. doi: 10.1007/978-94-007-4737-1_13.

Aghajanzadeh-Darzi, P., Martin, R., Laperche, S. and Jayet, P. (2017). Climate change impacts on European agriculture revisited: Adding the economic dimension of grasslands, *Regional Environmental Change* 17(1), 261–272. doi: 10.1007/s10113-016-1018-z.

Anderson, C. (2015). Docker [Software engineering], *IEEE Software* 32(3), 102–103. doi: 10.1109/MS.2015.62.

Araghi, A., Martinez, C. J. and Olesen, J. E. (2022). Evaluation of multiple gridded solar radiation data for crop modeling, *European Journal of Agronomy* 133, 126419. doi: 10.1016/j.eja.2021.126419.

Araujo, M. B. and New, M. (2007). Ensemble forecasting of species distributions, *Trends in Ecology and Evolution* 22(1), 42–47. doi: 10.1016/j.tree.2006.09.010.

Aubry, C., Papy, F. and Capillon, A. (1998). Modelling decision-making processes for annual crop management, *Agricultural Systems* 56(1), 45–65. doi: 10.1016/S0308-521X(97)00034-6.

Barberis, D., Chiadmi, I., Humblot, P., Jayet, P., Lungarska, A. and Ollier, M. (2021). Climate change and irrigation water: Should the North/south hierarchy of impacts on agricultural systems be reconsidered?, *Environmental Modeling and Assessment* 26(1), 13–36. doi: 10.1007/s10666-020-09724-8.

Bergez, J., Garcia, F. and Wallach, D. (2006). *Representing and Optimizing Management Decisions*.

Bergez, J.-E., Chabrier, P., Gary, C., Jeuffroy, M. H., Makowski, D., Quesnel, G., Ramat, E., Raynal, H., Rousse, N., Wallach, D., Debaeke, P., Durand, P., Duru, M., Dury, J., Faverdin, P., Gascuel-Odoux, C. and Garcia, F. (2013). An open platform to build, evaluate and simulate integrated models of farming and agro-ecosystems, *Environmental Modelling and Software* 39, 39–49. doi: 10.1016/J.ENVSOFT.2012.03.011.

Blanc, É. 2017. Statistical emulators of maize, rice, soybean and wheat yields from global gridded crop models. *Agric. For. Meteorol.* 236, 145–161. https://doi.org/10.1016/J.AGRFORMET.2016.12.022.

Bonacin, R., Nabuco, O. F. and Pierozzi Junior, I. (2016). Ontology models of the impacts of agriculture and climate changes on water resources: Scenarios on interoperability and information recovery, *Future Generation Computer Systems* 54, 423–434. doi: 10.1016/j.future.2015.04.010.

Borrelli, P., Robinson, D. A., Panagos, P., Lugato, E., Yang, J. E., Alewell, C., Wuepper, D., Montanarella, L. and Ballabio, C. (2020). Land use and climate change impacts on global soil erosion by water (2015-2070), *Proceedings of the National Academy of Sciences of the United States of America* 117(36), 21994–22001. doi: 10.1073/pnas.2001403117.

Bouleau, N. (1999). Philosophies des mathématiques et de la modélisation, *Harmattan*. Available at: https://www.editions-harmattan.fr/index.asp?navig=catalogue&obj=livre&no=9577 (Accessed: 18 December 2018).

Bozzola, M. and Finger, R. (2021). Stability of risk attitude, agricultural policies and production shocks: Evidence from Italy, *European Review of Agricultural Economics* 48(3), 477–501. doi: 10.1093/erae/jbaa021.

Britz, W., Lengers, B., Kuhn, T. and Schäfer, D. (2014). *A Highly Detailed Template Model for Dynamic Optimization of Farms-FARMDYN*. Institute for Food and Resource Economics, University of Bonn.

Britz, W. and Witzke, P. (2014). *Capri Model Documentation 2014*. University of Bonn, Institute for Food and Resource Economics.

Brown, H., Huth, N. and Holzworth, D. 2018. Crop model improvement in APSIM: Using wheat as a case study. *Eur. J. Agron.* 100, 141–150. https://doi.org/10.1016/j.eja.2018.02.002.

Burli, P. H., Nguyen, R. T., Hartley, D. S., Griffel, L. M., Vazhnik, V. and Lin, Y. 2021. Farmer characteristics and decision-making: A model for bioenergy crop adoption. *Energy* 234, 121235. https://doi.org/10.1016/J.ENERGY.2021.121235.

Catarino, R., Therond, O., Berthomier, J., Miara, M., Mérot, E., Misslin, R., Vanhove, P., Villerd, J. and Angevin, F. (2021). Fostering local crop-livestock integration via legume exchanges using an innovative integrated assessment and modelling approach based on the MAELIA platform, *Agricultural Systems* 189, 103066. doi: 10.1016/j.agsy.2021.103066.

Chabrier, P., Martin, O., Raynal, H. and Bergez, J. (2015). A new graphical modeling plugin based on Forrester diagrams for the RECORD platform, *Computers and Electronics in Agriculture* 111, 103–106. doi: 10.1016/j.compag.2014.12.013.

Challinor, A. J., Ewert, F., Arnold, S., Simelton, E. and Fraser, E. (2009). Crops and climate change: Progress, trends, and challenges in simulating impacts and informing adaptation, *Journal of Experimental Botany* 60(10), 2775–2789. doi: 10.1093/jxb/erp062.

Chopin, P., Doré, T., Guindé, L. and Blazy, J. (2015). MOSAICA: A multi-scale bioeconomic model for the design and ex ante assessment of cropping system mosaics, *Agricultural Systems* 140, 26–39. doi: 10.1016/j.agsy.2015.08.006.

Constantin, J., Raynal, H., Casellas, E., Hoffmann, H., Bindi, M., Doro, L., Eckersten, H., Gaiser, T., Grosz, B., Haas, E., Kersebaum, K., Klatt, S., Kuhnert, M., Lewan, E., Maharjan, G. R., Moriondo, M., Nendel, C., Roggero, P. P., Specka, X., Trombi, G., Villa, A., Wang, E., Weihermüller, L., Yeluripati, J., Zhao, Z., Ewert, F. and Bergez, J. (2019). Management and spatial resolution effects on yield and water balance at regional scale in crop models, *Agricultural and Forest Meteorology* 275, 184–195. doi: 10.1016/j.agrformet.2019.05.013.

Corrales, D. C., Schoving, C., Raynal, H., Debaeke, P., Journet, E. and Constantin, J. (2022). A surrogate model based on feature selection techniques and regression learners to improve soybean yield prediction in southern France, *Computers and Electronics in Agriculture* 192, 106578. doi: 10.1016/J.COMPAG.2021.106578.

Cradock-Henry, N. A., Blackett, P., Hall, M., Johnstone, P., Teixeira, E. and Wreford, A. (2020). Climate adaptation pathways for agriculture: Insights from a participatory process, *Environmental Science and Policy* 107, 66–79. doi: 10.1016/J.ENVSCI.2020.02.020.

DaMatta, F. M., Grandis, A., Arenque, B. C. and Buckeridge, M. S. (2010). Impacts of climate changes on crop physiology and food quality, *Food Research International* 43(7), 1814–1823. doi: 10.1016/j.foodres.2009.11.001.

de los Campos, G., Pérez-Rodríguez, P., Bogard, M., Gouache, D. and Crossa, J. (2020). A data-driven simulation platform to predict cultivars' performances under uncertain weather conditions, *Nature Communications* 11(1), 4876. doi: 10.1038/s41467-020-18480-y.

Debaeke, P., Pellerin, S. and Scopel, E. (2017). Climate-smart cropping systems for temperate and tropical agriculture: Mitigation, adaptation and trade-offs, *Cahiers Agricultures* 26(3). doi: 10.1051/cagri/2017028.

Dercon, S. and Christiaensen, L. (2011). Consumption risk, technology adoption and poverty traps: Evidence from Ethiopia, *Journal of Development Economics* 96(2), 159–173. doi: 10.1016/j.jdeveco.2010.08.003.

Di Paola, A. (2018). The expansion of wheat thermal suitability of Russia in response to climate change, *Land Use Policy* 2018, 8. Available at: http://files/22090/PaolaRuss.pdf.

Donatelli, M., Russell, G., Rizzoli, A. E., Acutis, M., Adam, M., Athanasiadis, I. N., Balderacchi, M., Bechini, L., Belhouchette, H., Bellocchi, G. and Bergez, J. E. (2010). *A Component-Based Framework for Simulating Agricultural Production and Externalities, Environmental and Agricultural Modelling: Integrated Approaches for Policy Impact Assessment*. doi: 10.1007/978-90-481-3619-3_4.

Dury, J., Schaller, N., Garcia, F., Reynaud, A. and Bergez, J. E. (2012). Models to support cropping plan and crop rotation decisions: A review, *Agronomy for Sustainable Development* 32(2), 567–580. doi: 10.1007/s13593-011-0037-x.

Ehrhardt, F., Soussana, J. F., Bellocchi, G., Grace, P., McAuliffe, R., Recous, S., Sándor, R., Smith, P., Snow, V., de Antoni Migliorati, M., Basso, B., Bhatia, A., Brilli, L., Doltra, J., Dorich, C. D., Doro, L., Fitton, N., Giacomini, S. J., Grant, B., Harrison, M. T., Jones, S. K., Kirschbaum, M. U. F., Klumpp, K., Laville, P., Léonard, J., Liebig, M., Lieffering, M., Martin, R., Massad, R. S., Meier, E., Merbold, L., Moore, A. D., Myrgiotis, V., Newton, P., Pattey, E., Rolinski, S., Sharp, J., Smith, W. N., Wu, L. and Zhang, Q. (2018). Assessing uncertainties in crop and pasture ensemble model simulations of productivity and N_2O emissions, *Global Change Biology* 24(2), e603–e616. doi: 10.1111/gcb.13965.

Escarcha, J. F., Lassa, J. A. and Zander, K. K. (2018). Livestock under climate change: A systematic review of impacts and adaptation, *Climate* 6(3), 54. doi: 10.3390/cli6030054.

Ewert, F., Rötter, R. P., Bindi, M., Webber, H., Trnka, M., Kersebaum, K. C., Olesen, J. E., van Ittersum, M. K., Janssen, S., Rivington, M., Semenov, M. A., Wallach, D., Porter, J. R., Stewart, D., Verhagen, J., Gaiser, T., Palosuo, T., Tao, F., Nendel, C., Roggero, P. P., Bartošová, L. and Asseng, S. (2015). Crop modelling for integrated assessment of risk to food production from climate change, *Environmental Modelling and Software* 72, 287–303. doi: 10.1016/J.ENVSOFT.2014.12.003.

Ewert, F., van Ittersum, M. K., Heckelei, T., Therond, O., Bezlepkina, I. and Andersen, E. (2011). Scale changes and model linking methods for integrated assessment of agri-environmental systems, *Agriculture, Ecosystems and Environment* 142(1–2), 6–17. doi: 10.1016/j.agee.2011.05.016.

Eza, U., Shtiliyanova, A., Borras, D., Bellocchi, G., Carrère, P. and Martin, R. (2015). An open platform to assess vulnerabilities to climate change: An application to agricultural systems, *Ecological Informatics* 30, 389–396. doi: 10.1016/j.ecoinf.2015.10.009.

FAO. (2010). *Climate-Smart, Agriculture: Policies, Practices and Financing for Food Security, Adaptation and Mitigation*. Rome: FAO. Available at: https://www.fao.org/3/i1881e/i1881e00.htm.

Farina, R., Sándor, R., Abdalla, M., Álvaro-Fuentes, J., Bechini, L., Bolinder, M. A., Brilli, L., Chenu, C., Clivot, H., De Antoni Migliorati, M., Di Bene, C., Dorich, C. D., Ehrhardt, F., Ferchaud, F., Fitton, N., Francaviglia, R., Franko, U., Giltrap, D. L., Grant, B. B., Guenet, B., Harrison, M. T., Kirschbaum, M. U. F., Kuka, K., Kulmala, L., Liski, J., McGrath, M. J., Meier, E., Menichetti, L., Moyano, F., Nendel, C., Recous, S., Reibold, N., Shepherd, A., Smith, W. N., Smith, P., Soussana, J. F., Stella, T., Taghizadeh-Toosi, A., Tsutskikh, E. and Bellocchi, G. (2021). Ensemble modelling, uncertainty and robust predictions of organic carbon in long-term bare-fallow soils, *Global Change Biology* 27(4), 904–928. doi: 10.1111/gcb.15441.

Fitzgerald, J. B., Brereton, A. J. and Holden, N. M. 2009. Assessment of the adaptation potential of grass-based dairy systems to climate change in Ireland–The maximised production scenario. *Agric. For. Meteorol.* 149(2), 244–255. https://doi.org/10.1016/J.AGRFORMET.2008.08.006.

Gaiser, T., Perkons, U., Küpper, P. M., Kautz, T., Uteau-Puschmann, D., Ewert, F., Enders, A. and Krauss, G. 2013. Modeling biopore effects on root growth and biomass production on soils with pronounced sub-soil clay accumulation. *Ecol. Modell.* 256, 6–15. https://doi.org/10.1016/J.ECOLMODEL.2013.02.016.

Gérardeaux, E., Loison, R., Palaï, O. and Sultan, B. (2018). Adaptation strategies to climate change using cotton (*Gossypium hirsutum* L.) ideotypes in rainfed tropical cropping systems in sub-Saharan Africa: A modeling approach, *Field Crops Research* 226, 38–47. doi: 10.1016/J.FCR.2018.07.007.

Ghahramani, A. and Bowran, D. (2018). Transformative and systemic climate change adaptations in mixed crop-livestock farming systems, *Agricultural Systems* 164, 236–251. doi: 10.1016/J.AGSY.2018.04.011.

Gong, L., Tian, B., Li, Y. and Wu, S. (2021). Phenological changes of soybean in response to climate conditions in frigid region in China over the past decades, *International Journal of Plant Production* 15(3), 363–375. doi: 10.1007/s42106-021-00145-5.

Gouttenoire, L., Cournut, S. and Ingrand, S. (2011). Modelling as a tool to redesign livestock farming systems: A literature review, *Animal: An International Journal of Animal Bioscience* 5(12), 1957–1971. doi: 10.1017/S175173111100111X.

Grillakis, M. G. (2019). Increase in severe and extreme soil moisture droughts for Europe under climate change, *Science of the Total Environment* 660, 1245–1255. doi: 10.1016/j.scitotenv.2019.01.001.

Guilpart, N., Iizumi, T. and Makowski, D. (2022). Data-driven projections suggest large opportunities to improve Europe's soybean self-sufficiency under climate change, *Nature Food* 3(4), 255–265. doi: 10.1038/s43016-022-00481-3.

Guitton, B., Théra, K., Tékété, M. L., Pot, D., Kouressy, M., Témé, N., Rami, J. F. and Vaksmann, M. 2018. Integrating genetic analysis and crop modeling: A major QTL can finely adjust photoperiod-sensitive sorghum flowering. F. Crop. *Res.* 221, 7–18. https://doi.org/10.1016/J.FCR.2018.02.007.

Hajimirzajan, A., Vahdat, M., Sadegheih, A., Shadkam, E. and Bilali, H. El 2021. An integrated strategic framework for large-scale crop planning: sustainable climate-smart crop planning and agri-food supply chain management. *Sustain. Prod. Consum.* 26, 709-732. https://doi.org/10.1016/J.SPC.2020.12.016.

Hammer, G. L., Kropff, M. J., Sinclair, T. R. and Porter, J. R. (2002). Future contributions of crop modelling–From heuristics and supporting decision making to understanding genetic regulation and aiding crop improvement, *European Journal of Agronomy* 18(1-2), 15-31. doi: 10.1016/S1161-0301(02)00093-X.

Han, E., Ines, A. V. M. and Koo, J. (2019). Development of a 10-km resolution global soil profile dataset for crop modeling applications, *Environmental Modelling and Software: With Environment Data News* 119, 70-83. doi: 10.1016/j.envsoft.2019.05.012.

Harris, G. (2002). Integrated assessment and modelling: An essential way of doing science1A summary of a workshop on Integrated Assessment and Modelling, held at EcoSummit 2000: Integrating the Sciences, Halifax, June 18-22, 2000. See Costanza and Jorgensen (2001) for a further report on Ecosummit, *Environmental Modelling and Software* 17(3), 201-207. doi: 10.1016/S1364-8152(01)00058-5.

Hertel, T. W. (2010). The poverty implications of climate-induced crop yield changes by 2030, *Global Environmental Change*, 9. doi: 10.1016/j.gloenvcha.2010.07.001.

Holman, I. P., Brown, C., Carter, T. R., Harrison, P. A. and Rounsevell, M. (2019). Improving the representation of adaptation in climate change impact models, *Regional Environmental Change* 19(3), 711-721. doi: 10.1007/s10113-018-1328-4.

Holzworth, D., Huth, N. I., Fainges, J., Brown, H., Zurcher, E., Cichota, R., Verrall, S., Herrmann, N. I., Zheng, B. and Snow, V. (2018). APSIM next generation: Overcoming challenges in modernising a farming systems model, *Environmental Modelling and Software* 103, 43-51. doi: 10.1016/J.ENVSOFT.2018.02.002.

Hoogenboom, G., Porter, C. H., Boote, K. J., Shelia, V., Wilkens, P. W., Singh, U., White, J. W., Asseng, S., Lizaso, J. I., Moreno, L. P. and Pavan, W. (2019). The DSSAT crop modeling ecosystem. In Boote, K. (Ed.). (2019). Advances in Crop Modelling for a Sustainable Agriculture. Cambridge: Burleigh Dodds Science Publishing, pp. 173-216. doi: https://doi.org/10.1201/9780429266591.

Huber, R., Bakker, M., Balmann, A., Berger, T., Bithell, M., Brown, C., Grêt-Regamey, A., Xiong, H., Le, Q. B., Mack, G., Meyfroidt, P., Millington, J., Müller, B., Polhill, J. G., Sun, Z., Seidl, R., Troost, C. and Finger, R. (2018). Representation of decision-making in European agricultural agent-based models, *Agricultural Systems* 167, 143-160. doi: 10.1016/j.agsy.2018.09.007.

IPCC. (2021). The physical science basis: Contribution of working group I to the sixth assessment report of the Intergovernmental Panel on Climate Change. In Masson-Delmotte, S. L. and Zhai, P. (Eds.), *Climate Change*. Cambridge: Cambridge University Press.

IPCC. (2022). *Impacts, Adaptation, and Vulnerability. Contribution of Working Group II to the Sixth Assessment Report of the Intergovernmental Panel on Climate Change*. Cambridge: Cambridge University Press.

Jang, W. S., Lee, Y., Neff, J. C., Im, Y., Ha, S. and Doro, L. (2019). Development of an EPIC parallel computing framework to facilitate regional/global gridded crop modeling with multiple scenarios: A case study of the United States, *Computers and Electronics in Agriculture* 158, 189-200. doi: 10.1016/j.compag.2019.02.004.

Jones, J. W., Keating, B. A. and Porter, C. H. 2001. Approaches to modular model development. *Agric. Syst.* 70(2-3), 421-443. https://doi.org/10.1016/S0308-521X(01)00054-3.

Jonquet, C., Toulet, A., Arnaud, E., Aubin, S., Dzalé Yeumo, E., Emonet, V., Graybeal, J., Laporte, M., Musen, M. A., Pesce, V. and Larmande, P. (2018). AgroPortal: A vocabulary and ontology repository for agronomy, *Computers and Electronics in Agriculture* 144, 126–143. doi: 10.1016/j.compag.2017.10.012.

Kalaugher, E., Beukes, P., Bornman, J. F., Clark, A. and Campbell, D. I. 2017. Modelling farm-level adaptation of temperate, pasture-based dairy farms to climate change. *Agric. Syst.* 153, 53–68. https://doi.org/10.1016/J.AGSY.2017.01.008.

Kanellopoulos, A., Reidsma, P., Wolf, J. and van Ittersum, M. K. (2014). Assessing climate change and associated socio-economic scenarios for arable farming in the Netherlands: An application of benchmarking and bio-economic farm modelling, *European Journal of Agronomy* 52, 69–80. doi: 10.1016/j.eja.2013.10.003.

Keating, B. A. and Thorburn, P. J. (2018). Modelling crops and cropping systems—Evolving purpose, practice and prospects, *European Journal of Agronomy* 100, 163–176. doi: 10.1016/j.eja.2018.04.007.

Kim, J., Park, J., Hyun, S., Fleisher, D. H. and Kim, K. S. (2020). Development of an automated gridded crop growth simulation support system for distributed computing with virtual machines, *Computers and Electronics in Agriculture* 169, 105196. doi: 10.1016/j.compag.2019.105196.

Kipling, R. P., Topp, C. F. E., Bannink, A., Bartley, D. J., Blanco-Penedo, I., Cortignani, R., Del Prado, A., Dono, G., Faverdin, P., Graux, A. I., Hutchings, N. J., Lauwers, L., Özkan Gülzari, Ş, Reidsma, P., Rolinski, S., Ruiz-Ramos, M., Sandars, D. L., Sándor, R., Schönhart, M., Seddaiu, G., van Middelkoop, J., Shrestha, S., Weindl, I. and Eory, V. (2019). To what extent is climate change adaptation a novel challenge for agricultural modellers?, *Environmental Modelling and Software: With Environment Data News* 120, 104492. doi: 10.1016/J.ENVSOFT.2019.104492.

Kothari, K., Battisti, R., Boote, K. J., Archontoulis, S. V., Confalone, A., Constantin, J., Cuadra, S. V., Debaeke, P., Faye, B., Grant, B., Hoogenboom, G., Jing, Q., van der Laan, M., Macena da Silva, F. A., Marin, F. R., Nehbandani, A., Nendel, C., Purcell, L. C., Qian, B., Ruane, A. C., Schoving, C., Silva, E. H. F. M., Smith, W., Soltani, A., Srivastava, A., Vieira, N. A., Slone, S. and Salmerón, M. (2022). Are soybean models ready for climate change food impact assessments?, *European Journal of Agronomy* 135, 126482. doi: 10.1016/J.EJA.2022.126482.

Lamichhane, J. R., Barzman, M., Booij, K., Boonekamp, P., Desneux, N., Huber, L., Kudsk, P., Langrell, S. R. H., Ratnadass, A., Ricci, P., Sarah, J. and Messéan, A. (2015). Robust cropping systems to tackle pests under climate change: A review, *Agronomy for Sustainable Development* 35(2), 443–459. doi: 10.1007/s13593-014-0275-9.

Launay, C., Constantin, J., Chlebowski, F., Houot, S., Graux, A. I., Klumpp, K., Martin, R., Mary, B., Pellerin, S. and Therond, O. (2021). Estimating the carbon storage potential and greenhouse gas emissions of French arable cropland using high-resolution modeling, *Global Change Biology* 27(8), 1645–1661. doi: 10.1111/gcb.15512.

Le Mouël, C., Mora, O. and De Lattre-Gasquet, M. (2016). *Agrimonde-Terra, a Foresight Exercise on Land Use and Food Security in 2050: Scenario Building Method and Conceptual Framework*. doi: 10.15454/txjf-2e77.

Lehmann, N., Finger, R., Klein, T., Calanca, P. and Walter, A. (2013). Adapting crop management practices to climate change: Modeling optimal solutions at the field scale, *Agricultural Systems* 117, 55–65. doi: 10.1016/j.agsy.2012.12.011.

Li, F., Snow, V. and Holzworth, D. (2011). Modelling the seasonal and geographical pattern of pasture production in New Zealand, *New Zealand Journal of Agricultural Research* 54(4), 331–352. doi: 10.1080/00288233.2011.613403.

Liakos, K. G., Busato, P., Moshou, D., Pearson, S. and Bochtis, D. (2018). Machine learning in agriculture: A review, *Sensors* 18(8), 2674. doi: 10.3390/s18082674.

Lieffering, M., Newton, P. C. D., Vibart, R. and Li, F. Y. 2016. Exploring climate change impacts and adaptations of extensive pastoral agriculture systems by combining biophysical simulation and farm system models. *Agric. Syst.* 144, 77–86. https://doi.org/10.1016/J.AGSY.2016.01.005.

Lipper, L., Thornton, P., Campbell, B. M., Baedeker, T., Braimoh, A., Bwalya, M., Caron, P., Cattaneo, A., Garrity, D., Henry, K., Hottle, R., Jackson, L., Jarvis, A., Kossam, F., Mann, W., McCarthy, N., Meybeck, A., Neufeldt, H., Remington, T., Sen, P. T., Sessa, R., Shula, R., Tibu, A. and Torquebiau, E. F. (2014). Climate-smart agriculture for food security, *Nature Climate Change* 4(12), 1068–1072. doi: 10.1038/nclimate2437.

Lischeid, G., Webber, H., Sommer, M., Nendel, C. and Ewert, F. (2022). Machine learning in crop yield modelling: A powerful tool, but no surrogate for science, *Agricultural and Forest Meteorology* 312, 108698. doi: 10.1016/J.AGRFORMET.2021.108698.

Lobell, D. B. and Burke, M. B. (2010). On the use of statistical models to predict crop yield responses to climate change, *Agricultural and Forest Meteorology* 150(11), 1443–1452. doi: 10.1016/j.agrformet.2010.07.008.

Louhichi, K., Ciaian, P., Espinosa, M., Colen, L., Perni, A. and Paloma, S. Gy (2017). Does the crop diversification measure impact EU farmers' decisions? An assessment using an Individual Farm Model for CAP Analysis (IFM-CAP), *Land Use Policy* 66, 250–264. doi: 10.1016/J.LANDUSEPOL.2017.04.010.

Luo, Z., Eady, S., Sharma, B., Grant, T., Liu, D. L., Cowie, A., Farquharson, R., Simmons, A., Crawford, D., Searle, R. and Moore, A. 2019. Mapping future soil carbon change and its uncertainty in croplands using simple surrogates of a complex farming system model. *Geoderma* 337, 311–321. https://doi.org/10.1016/J.GEODERMA.2018.09.041.

Makowski, D., Marajo-Petitzon, E., Durand, J. and Ben-Ari, T. (2020). Quantitative synthesis of temperature, CO_2, rainfall, and adaptation effects on global crop yields, *European Journal of Agronomy* 115, 126041. doi: 10.1016/j.eja.2020.126041.

Martin, G. and Magne, M. A. (2015). Agricultural diversity to increase adaptive capacity and reduce vulnerability of livestock systems against weather variability – A farm-scale simulation study, *Agriculture, Ecosystems and Environment* 199, 301–311. doi: 10.1016/j.agee.2014.10.006.

Martin-Clouaire, R. and Rellier, J.-P. (2011). A generic framework for simulating agricultural production systems, *Modelling Nutrient Digestion and Utilisation in Farm Animals*, 13–21. doi: 10.3920/978-90-8686-712-7_1.

Martre, P., Quilot-Turion, B., Luquet, D., Memmah, M. M. O. S., Chenu, K. and Debaeke, P. (2015). Model-assisted phenotyping and ideotype design. In *Crop Physiology*. Elsevier, pp. 349–373. doi://10.1016/B978-0-12-417104-6.00014-5.

Martre, P., Wallach, D., Asseng, S., Ewert, F., Jones, J. W., Rötter, R. P., Boote, K. J., Ruane, A. C., Thorburn, P. J., Cammarano, D., Hatfield, J. L., Rosenzweig, C., Aggarwal, P. K., Angulo, C., Basso, B., Bertuzzi, P., Biernath, C., Brisson, N., Challinor, A. J., Doltra, J., Gayler, S., Goldberg, R., Grant, R. F., Heng, L., Hooker, J., Hunt, L. A., Ingwersen, J., Izaurralde, R. C., Kersebaum, K. C., Müller, C., Kumar, S. N., Nendel, C., O'Leary, G., Olesen, J. E., Osborne, T. M., Palosuo, T., Priesack, E., Ripoche, D., Semenov, M. A.,

Shcherbak, I., Steduto, P., Stöckle, C. O., Stratonovitch, P., Streck, T., Supit, I., Tao, F., Travasso, M., Waha, K., White, J. W. and Wolf, J. (2015). Multimodel ensembles of wheat growth: Many models are better than one, *Global Change Biology* 21(2), 911–925. doi: 10.1111/gcb.12768.

Martre, P., Yin, X. and Ewert, F. (2017). Modeling crops from genotype to phenotype in a changing climate, *Field Crops Research* 202, 1–4. doi: 10.1016/J.FCR.2017.01.002.

Midingoyi, C. A., Pradal, C., Enders, A., Fumagalli, D., Raynal, H., Donatelli, M., Athanasiadis, I. N., Porter, C., Hoogenboom, G., Holzworth, D., Garcia, F., Thorburn, P. and Martre, P. (2021). Crop2ML: An open-source multi-language modeling framework for the exchange and reuse of crop model components, *Environmental Modelling and Software* 142, 105055. doi: 10.1016/j.envsoft.2021.105055.

Movedi, E., Bellocchi, G., Argenti, G., Paleari, L., Vesely, F., Staglianò, N., Dibari, C. and Confalonieri, R. (2019). Development of generic crop models for simulation of multi-species plant communities in mown grasslands, *Ecological Modelling* 401, 111–128. doi: 10.1016/j.ecolmodel.2019.03.001.

Nardone, A., Ronchi, B., Lacetera, N., Ranieri, M. S. and Bernabucci, U. (2010). Effects of climate changes on animal production and sustainability of livestock systems, *Livestock Science* 130(1–3), 57–69. doi: 10.1016/j.livsci.2010.02.011.

Naulleau, A., Gary, C., Prévot, L., Berteloot, V., Fabre, J., Crevoisier, D., Gaudin, R. and Hossard, L. (2022). Participatory modeling to assess the impacts of climate change in a Mediterranean vineyard watershed, *Environmental Modelling and Software* 150, 105342. doi: 10.1016/j.envsoft.2022.105342.

Nelson, G. C., Valin, H., Sands, R. D., Havlík, P., Ahammad, H., Deryng, D., Elliott, J., Fujimori, S., Hasegawa, T., Heyhoe, E., Kyle, P., Von Lampe, M., Lotze-Campen, H., Mason d'Croz, D., van Meijl, H., van der Mensbrugghe, D., Müller, C., Popp, A., Robertson, R., Robinson, S., Schmid, E., Schmitz, C., Tabeau, A. and Willenbockel, D. (2014). Climate change effects on agriculture: Economic responses to biophysical shocks, *Proceedings of the National Academy of Sciences of the United States of America* 111(9), 3274–3279. doi: 10.1073/pnas.1222465110.

Nguyen, T. H., Nong, D. and Paustian, K. 2019. Surrogate-based multi-objective optimization of management options for agricultural landscapes using artificial neural networks. *Ecol. Modell.* 400, 1–13. https://doi.org/10.1016/J.ECOLMODEL.2019.02.018.

Parker, P., Letcher, R., Jakeman, A., Beck, M. B., Harris, G., Argent, R. M., Hare, M., Pahl-Wostl, C., Voinov, A., Janssen, M., Sullivan, P., Scoccimarro, M., Friend, A., Sonnenshein, M., Barker, D., Matejicek, L., Odulaja, D., Deadman, P., Lim, K., Larocque, G., Tarikhi, P., Fletcher, C., Put, A., Maxwell, T., Charles, A., Breeze, H., Nakatani, N., Mudgal, S., Naito, W., Osidele, O., Eriksson, I., Kautsky, U., Kautsky, E., Naeslund, B., Kumblad, L., Park, R., Maltagliati, S., Girardin, P., Rizzoli, A., Mauriello, D., Hoch, R., Pelletier, D., Reilly, J., Olafsdottir, R. and Bin, S. (2002). Progress in integrated assessment and modelling1A Summary of a workshop on Integrated Assessment and Modelling, held at EcoSummit 2000: Integrating the Sciences, Halifax, June 18–22, 2000. See Costanza and Jorgensen (2001) for a further report on Ecosummit, *Environmental Modelling and Software* 17(3), 209–217. doi: 10.1016/S1364-8152(01)00059-7.

Pahmeyer, C., Kuhn, T. and Britz, W. 2021. 'Fruchtfolge': A crop rotation decision support system for optimizing cropping choices with big data and spatially explicit modeling. *Comput. Electron. Agric.* 181, 105948. https://doi.org/10.1016/J.COMPAG.2020.105948.

Parson, E. A. (1995). Integrated assessment and environmental policy making, *Energy Policy* 23(4–5), 463–475. doi: 10.1016/0301-4215(95)90170-C.

Pearson, R. G., Thuiller, W., Araújo, M. B., Martinez-Meyer, E., Brotons, L., McClean, C., Miles, L., Segurado, P., Dawson, T. P. and Lees, D. C. (2006). Model-based uncertainty in species range prediction, *Journal of Biogeography* 33(10), 1704–1711. doi: 10.1111/j.1365-2699.2006.01460.x.

Perveen, N., Barot, S., Alvarez, G., Klumpp, K., Martin, R., Rapaport, A., Herfurth, D., Louault, F. and Fontaine, S. (2014). Priming effect and microbial diversity in ecosystem functioning and response to global change: A modeling approach using the SYMPHONY model, *Global Change Biology* 20(4), 1174–1190. doi: 10.1111/gcb.12493.

Porfirio, L. L., Newth, D., Finnigan, J. J. and Cai, Y. (2018). Economic shifts in agricultural production and trade due to climate change, *Palgrave Communications* 4(1), 111. doi: 10.1057/s41599-018-0164-y.

Porter, J. R., Xie, L., Challinor, A. J., Cochrane, K., Howden, S. M., Iqbal, M. M., Lobell, D. B. and Travasso, M. I. (2014). *Food Security and Food Production Systems*. Cambridge: Cambridge University Press.

Pradal, C., Fournier, C., Valduriez, P. and Cohen-Boulakia, S. (2015). OpenAlea: Scientific workflows combining data analysis and simulation. In *Proceedings of the 27th International Conference on Scientific and Statistical Database Management 2015 Jun 29*. New York, NY: ACM, pp. 1–6. doi: 10.1145/2791347.2791365.

Prost, L., Cerf, M. and Jeuffroy, M.-H. (2012). Lack of consideration for end-users during the design of agronomic models. A review, *Agronomy for Sustainable Development* 32(2), 581–594. doi: 10.1007/s13593-011-0059-4.

Ramirez-Cabral, N. Y. Z., Kumar, L. and Shabani, F. (2017). Global alterations in areas of suitability for maize production from climate change and using a mechanistic species distribution model (CLIMEX), *Scientific Reports* 7(1), 5910. doi: 10.1038/s41598-017-05804-0.

Ramos-Castillo, M., Orvain, M., Naves-Maschietto, G., de Faria, A. B. B., Chenu, D. and Albuquerque, M. (2021). Optimal agricultural spreading scheduling through surrogate-based optimization and MINLP models, *Information Processing in Agriculture* 8(1), 159–172. doi: 10.1016/J.INPA.2020.02.005.

Reynolds, J. F. and Acock, B. (1997). Modularity and genericness in plant and ecosystem models, *Ecological Modelling* 94(1), 7–16. doi: 10.1016/S0304-3800(96)01924-2.

Robert, M., Thomas, A., Sekhar, M., Badiger, S., Ruiz, L., Raynal, H. and Bergez, J. 2016. Adaptive and dynamic decision-making processes: A conceptual model of production systems on Indian farms. AGSY, 1–13. https://doi.org/10.1016/j.agsy.2016.08.001.

Robert, M., Thomas, A., Sekhar, M., Raynal, H., Casellas, É, Casel, P., Chabrier, P., Joannon, A. and Bergez, J. (2018). A dynamic model for water management at the farm level integrating strategic, tactical and operational decisions, *Environmental Modelling and Software* 100, 123–135. doi: 10.1016/J.ENVSOFT.2017.11.013.

Rodriguez, D., Cox, H., deVoil, P. and Power, B. (2014). A participatory whole farm modelling approach to understand impacts and increase preparedness to climate change in Australia, *Agricultural Systems* 126, 50–61. doi: 10.1016/J.AGSY.2013.04.003.

Rojas-Downing, M. M., Nejadhashemi, A. P., Harrigan, T. and Woznicki, S. A. (2017). Climate change and livestock: Impacts, adaptation, and mitigation, *Climate Risk Management* 16, 145–163. doi: 10.1016/j.crm.2017.02.001.

Rosenzweig, C., Elliott, J., Deryng, D., Ruane, A. C., Müller, C., Arneth, A., Boote, K. J., Folberth, C., Glotter, M., Khabarov, N., Neumann, K., Piontek, F., Pugh, T. A., Schmid, E., Stehfest, E., Yang, H. and Jones, J. W. (2014). Assessing agricultural risks of climate change in the 21st century in a global gridded crop model intercomparison, *Proceedings of the National Academy of Sciences of the United States of America* 111(9), 3268–3273. doi: 10.1073/pnas.1222463110.

Ruane, A. C., Phillips, M., Müller, C., Elliott, J., Jägermeyr, J., Arneth, A., Balkovic, J., Deryng, D., Folberth, C., Iizumi, T., Izaurralde, R. C., Khabarov, N., Lawrence, P., Liu, W., Olin, S., Pugh, T. A. M., Rosenzweig, C., Sakurai, G., Schmid, E., Sultan, B., Wang, X., de Wit, A. and Yang, H. (2021). Strong regional influence of climatic forcing datasets on global crop model ensembles, *Agricultural and Forest Meteorology* 300, 108313. doi: 10.1016/j.agrformet.2020.108313.

Sándor, R., Barcza, Z., Hidy, D., Lellei-Kovács, E., Ma, S. and Bellocchi, G. (2016). Modelling of grassland fluxes in Europe: Evaluation of two biogeochemical models, *Agriculture, Ecosystems and Environment* 215, 1–19. doi: 10.1016/j.agee.2015.09.001.

Sándor, R., Ehrhardt, F., Brilli, L., Carozzi, M., Recous, S., Smith, P., Snow, V., Soussana, J. F., Dorich, C. D., Fuchs, K., Fitton, N., Gongadze, K., Klumpp, K., Liebig, M., Martin, R., Merbold, L., Newton, P. C. D., Rees, R. M., Rolinski, S. and Bellocchi, G. (2018). The use of biogeochemical models to evaluate mitigation of greenhouse gas emissions from managed grasslands, *Science of the Total Environment* 642, 292–306. doi: 10.1016/j.scitotenv.2018.06.020.

Sándor, R., Ehrhardt, F., Grace, P., Recous, S., Smith, P., Snow, V., Soussana, J., Basso, B., Bhatia, A., Brilli, L., Doltra, J., Dorich, C. D., Doro, L., Fitton, N., Grant, B., Harrison, M. T., Kirschbaum, M. U. F., Klumpp, K., Laville, P., Léonard, J., Martin, R., Massad, R., Moore, A., Myrgiotis, V., Pattey, E., Rolinski, S., Sharp, J., Skiba, U., Smith, W., Wu, L., Zhang, Q. and Bellocchi, G. (2020). Ensemble modelling of carbon fluxes in grasslands and croplands, *Field Crops Research* 252, 107791. doi: 10.1016/j.fcr.2020.107791.

Schmidhuber, J. and Tubiello, F. N. (2007). Global food security under climate change, *Proceedings of the National Academy of Sciences of the United States of America* 104(50), 19703–19708. doi: 10.1073/pnas.0701976104.

Searchinger, T., Waite, R., Hanson, C., Ranganathan, J., Dumas, P., Matthews, E. and Klirs, C. (2019). *Creating a Sustainable Food Future: A Menu of Solutions to Feed Nearly 10 Billion People by 2050*. Final report. Washington Research Institute.

Shelia, V., Hansen, J., Sharda, V., Porter, C., Aggarwal, P., Wilkerson, C. J. and Hoogenboom, G. (2019). A multi-scale and multi-model gridded framework for forecasting crop production, risk analysis, and climate change impact studies, *Environmental Modelling and Software* 115, 144–154. doi: 10.1016/j.envsoft.2019.02.006.

Siad, S. M., Iacobellis, V., Zdruli, P., Gioia, A., Stavi, I. and Hoogenboom, G. (2019). A review of coupled hydrologic and crop growth models, *Agricultural Water Management* 224, 105746. doi: 10.1016/j.agwat.2019.105746.

Sillmann, J., Thorarinsdottir, T., Keenlyside, N., Schaller, N., Alexander, L. V., Hegerl, G., Seneviratne, S. I., Vautard, R., Zhang, X. and Zwiers, F. W. (2017). Understanding, modeling and predicting weather and climate extremes: Challenges and opportunities, *Weather and Climate Extremes* 18, 65–74. doi: 10.1016/j.wace.2017.10.003.

Sinclair, T. R. and Seligman, N. G. (1996). Crop modeling: From infancy to maturity, *Agronomy Journal* 88(5), 698–704. doi: 10.2134/agronj1996.00021962008800050004x.

Subirats-Coll, I., Kolshus, K., Turbati, A., Stellato, A., Mietzsch, E., Martini, D. and Zeng, M. (2022). AGROVOC: The linked data concept hub for food and agriculture, *Computers and Electronics in Agriculture* 196, 105965. doi: 10.1016/j.compag.2020.105965.

Tao, F., Rötter, R. P., Palosuo, T., Díaz-Ambrona, C. G. H., Mínguez, M. I., Semenov, M. A., Kersebaum, K. C., Nendel, C., Cammarano, D., Hoffmann, H., Ewert, F., Dambreville, A., Martre, P., Rodríguez, L., Ruiz-Ramos, M., Gaiser, T., Höhn, J. G., Salo, T., Ferrise, R., Bindi, M. and Schulman, A. H. (2017). Designing future barley ideotypes using a crop model ensemble, *European Journal of Agronomy* 82, 144–162. doi: 10.1016/J.EJA.2016.10.012.

Thorburn, P. J., Boote, K. J., Nendel, C., Rötter, R. P. and Ewert, F. (2018). Recent advances in crop modelling to support sustainable agricultural production and food security under global change, *European Journal of Agronomy* 100, 1–3. doi: 10.1016/J.EJA.2018.10.009.

Thornton, P. K., van de Steeg, J., Notenbaert, A. and Herrero, M. (2009). The impacts of climate change on livestock and livestock systems in developing countries: A review of what we know and what we need to know, *Agricultural Systems* 101(3), 113–127. doi: 10.1016/j.agsy.2009.05.002.

Top, J., Janssen, S., Boogaard, H., Knapen, R. and Şimşek-Şenel, G. (2022). Cultivating FAIR principles for agri-food data, *Computers and Electronics in Agriculture* 196, 106909. doi: 10.1016/J.COMPAG.2022.106909.

Tubiello, F. N., Soussana, J. F. and Howden, S. M. (2007). Crop and pasture response to climate change, *Proceedings of the National Academy of Sciences of the United States of America* 104(50), 19686–19690. doi: 10.1073/pnas.0701728104.

Van Ittersum, M. K. and Donatelli, M. (2003). Modelling cropping systems - Highlights of the symposium and preface to the special issues, *European Journal of Agronomy* 18(3–4), 187–197. doi: 10.1016/S1161-0301(02)00095-3.

van Ittersum, M. K., Ewert, F., Heckelei, T., Wery, J., Alkan Olsson, J., Andersen, E., Bezlepkina, I., Brouwer, F., Donatelli, M., Flichman, G., Olsson, L., Rizzoli, A. E., van der Wal, T., Wien, J. E. and Wolf, J. (2008). Integrated assessment of agricultural systems - A component-based framework for the European Union (SEAMLESS), *Agricultural Systems* 96(1–3), 150–165. doi: 10.1016/j.agsy.2007.07.009.

Van Oijen, M., Barcza, Z., Confalonieri, R., Korhonen, P., Kröel-Dulay, G., Lellei-Kovács, E., Louarn, G., Louault, F., Martin, R., Moulin, T., Movedi, E., Picon-Cochard, C., Rolinski, S., Viovy, N., Wirth, S. B. and Bellocchi, G. (2020). Incorporating biodiversity into biogeochemistry models to improve prediction of ecosystem services in temperate grasslands: Review and roadmap, *Agronomy* 10(2), 259. doi: 10.3390/agronomy10020259.

van Oijen, M., Bellocchi, G. and Höglind, M. (2018). Effects of climate change on grassland biodiversity and productivity: The need for a diversity of models, *Agronomy* 8(2), 14. doi: 10.3390/agronomy8020014.

van Oosterzee, P., Dale, A. and Preece, N. D. (2014). Integrating agriculture and climate change mitigation at landscape scale: Implications from an Australian case study, *Global Environmental Change* 29, 306–317. doi: 10.1016/j.gloenvcha.2013.10.003.

Voinov, A. and Gaddis, E. J. B. (2008). Lessons for successful participatory watershed modeling: A perspective from modeling practitioners, *Ecological Modelling* 216(2), 197–207. doi: 10.1016/J.ECOLMODEL.2008.03.010.

Vuichard, N., Soussana, J., Ciais, P., Viovy, N., Ammann, C., Calanca, P., Clifton-Brown, J., Fuhrer, J., Jones, M. and Martin, C. (2007). Estimating the greenhouse gas fluxes of European grasslands with a process-based model: 1. Model evaluation from in situ measurements, *Global Biogeochemical Cycles* 21(1). doi: 10.1029/2005GB002611.

Wallach, D., Mearns, L. O., Ruane, A. C., Rötter, R. P. and Asseng, S. (2016). Lessons from climate modeling on the design and use of ensembles for crop modeling, *Climatic Change* 139(3), 551–564. doi: 10.1007/s10584-016-1803-1.

Whisler, F. D., Acock, B., Baker, D. N., Fye, R. E., Hodges, H. F., Lambert, J. R., Lemmon, H. E., McKinion, J. M. and Reddy, V. R. (1986). Crop simulation models in agronomic systems, *Advances in Agronomy* 40, 141–208. doi: 10.1016/S0065-2113(08)60282-5.

White, J. W. and Hoogenboom, G. (2003). Gene-based approaches to crop simulation, *Agronomy Journal* 95(1), 52–64. doi: 10.2134/agronj2003.5200.

White, J. W., Hoogenboom, G., Kimball, B. A. and Wall, G. W. (2011). Methodologies for simulating impacts of climate change on crop production, *Field Crops Research* 124(3), 357–368. doi: 10.1016/j.fcr.2011.07.001.

White, J. W., Hunt, L. A., Boote, K. J., Jones, J. W., Koo, J., Kim, S., Porter, C. H., Wilkens, P. W. and Hoogenboom, G. (2013). Integrated description of agricultural field experiments and production: The ICASA version 2.0 data standards, *Computers and Electronics in Agriculture* 96, 1–12. doi: 10.1016/j.compag.2013.04.003.

Wiebe, K., Robinson, S. and Cattaneo, A. (2019). Climate change, agriculture and food security. In *Sustainable Food and Agriculture*. Elsevier, pp. 55–74. doi: 10.1016/B978-0-12-812134-4.00004-2.

Wilkinson, M. D., Dumontier, M., Aalbersberg, I. J., Appleton, G., Axton, M., Baak, A., Blomberg, N., Boiten, J. W., da Silva Santos, L. B., Bourne, P. E., Bouwman, J., Brookes, A. J., Clark, T., Crosas, M., Dillo, I., Dumon, O., Edmunds, S., Evelo, C. T., Finkers, R., Gonzalez-Beltran, A., Gray, A. J., Groth, P., Goble, C., Grethe, J. S., Heringa, J., 't Hoen, P. A., Hooft, R., Kuhn, T., Kok, R., Kok, J., Lusher, S. J., Martone, M. E., Mons, A., Packer, A. L., Persson, B., Rocca-Serra, P., Roos, M., van Schaik, R., Sansone, S. A., Schultes, E., Sengstag, T., Slater, T., Strawn, G., Swertz, M. A., Thompson, M., van der Lei, J., van Mulligen, E., Velterop, J., Waagmeester, A., Wittenburg, P., Wolstencroft, K., Zhao, J. and Mons, B. (2016). The FAIR guiding principles for scientific data management and stewardship, *Scientific Data* 3(1), 160018. doi: 10.1038/sdata.2016.18.

Willaume, M., Rollin, A. and Casagrande, M. (2014). Farmers in southwestern France think that their arable cropping systems are already adapted to face climate change, *Regional Environmental Change* 14(1), 333–345. doi: 10.1007/s10113-013-0496-5.

Zajac, Z., Gomez, O., Gelati, E., van der Velde, M., Bassu, S., Ceglar, A., Chukaliev, O., Panarello, L., Koeble, R., van den Berg, M., Niemeyer, S. and Fumagalli, D. (2022). Estimation of spatial distribution of irrigated crop areas in Europe for large-scale modelling applications, *Agricultural Water Management* 266, 107527. doi: 10.1016/j.agwat.2022.107527.

Zampieri, M., Weissteiner, C. J., Grizzetti, B., Toreti, A., van den Berg, M. and Dentener, F. (2020). Estimating resilience of crop production systems: From theory to practice, *Science of the Total Environment* 735, 139378. doi: 10.1016/j.scitotenv.2020.139378.

Zeng, Z., Ziegler, A. D., Searchinger, T., Yang, L., Chen, A., Ju, K., Piao, S., Li, L. Z. X., Ciais, P., Chen, D., Liu, J., Azorin-Molina, C., Chappell, A., Medvigy, D. and Wood, E. F. (2019).

A reversal in global terrestrial stilling and its implications for wind energy production, *Nature Climate Change* 9(12), 979–985. doi: 10.1038/s41558-019-0622-6.
Zhang, L., Zhang, Z., Tao, F., Luo, Y., Zhang, J. and Cao, J. (2022). Adapting to climate change precisely through cultivars renewal for rice production across China: When, where, and what cultivars will be required? *Agric. For. Meteorol.* 316, 108856. https://doi.org/10.1016/J.AGRFORMET.2022.108856.

Published by Burleigh Dodds Science Publishing Limited, 2023.

Chapter 2

Improving data flow and integration in models assessing the impact of climate change on agriculture

Claas Nendel, Roland Baatz, Michael Berg-Mohnicke and Gohar Ghazaryan, Leibniz Centre for Agricultural Landscape Research (ZALF), Germany; Sander Janssen, Wageningen Environmental Research, The Netherlands; Pierre Martre, INRAE, France; and Cheryl Porter, University of Florida, USA

1 Introduction
2 Model-data integration
3 Informing spatiotemporal simulations
4 Assimilation of data in spatiotemporal simulations
5 Workflows for massive parallel computing
6 Model-model integration
7 Granularity and modular design for model improvement, reuse, exchange and interoperability
8 Concepts for distributed modelling
9 Future trends and conclusion
10 Where to look for further information
11 References

1 Introduction

A growing world population depends for more and more of its daily needs on the food and agriculture system, while planetary resources are infinite and bounded (O'Neill et al., 2018; Steffen et al., 2015). Today's food and agriculture systems are facing multiple challenges (e.g. climate change, biodiversity loss, and soil degradation), so with the same or even fewer resources in terms of fossil fuel and land, more demands have to be met. Some relationships between the abovementioned challenges and agriculture systems have been more extensively studied than other relationships. For example, there is a fairly good and quantitative understanding of the interconnection between climate

http://dx.doi.org/10.19103/AS.2022.0115.03

change and agricultural production (Asseng et al., 2015), i.e. temperature increase and productivity effects, which also implies that these have been extensively modelled (e.g. Ewert et al., 2015; Rosenzweig et al., 2013). This is different for biodiversity loss, where there is still a limited understanding of how species loss and species diversity would impact agricultural systems and their productivity. Even the challenges are themselves interconnected, so a holistic approach towards the agricultural system is required to capture relevant development pathways and not be blinded by a deep understanding of one domain (i.e. agronomy, economy, supply chains), while missing important interactions across domains.

With these challenges occurring at the same time, and the food and agriculture system as multi-scale, multi-domain and multi-sectoral, integrated modelling approaches for agriculture and food systems are crucial (Harris, 2002; Parker et al., 2002). The integrated modelling approaches and the consequent modelling results can be used as a method for deepening the understanding of the different system analysis, an ex-ante tool to evaluate (policy) options for future development, and for sustainability analysis of existing agricultural systems. In the past these integrated modelling approaches have been most actively developed by linking models presenting domains together into an integrated framework, especially in the 2000s; see, for example, van Ittersum et al. (2008) and Janssen et al. (2011). In the early 2010s the focus shifted from model linking to ensemble analysis of the system components models, like economic and crop models, through initiatives such as the Agricultural Model Intercomparison and Improvement Project (AgMIP; Rosenzweig et al., 2013) or the Inter-Sectoral Impact Model Intercomparison Project (ISIMIP; Warszawski et al., 2014). At the same time there has been a lot of focus on the potential of digitalisation and data-driven technologies to release new data sources, and new analysis techniques (see e.g. Holzworth et al., 2015; Janssen et al., 2017; Lokers et al., 2016), which could be beneficial to new approaches to (integrated) modelling.

It is therefore timely to review the issues involved in integrated modelling, especially with respect to an already well-studied relationship between climate change and agricultural systems. Even so, many elements require further elaboration and discovery in this relationship between climate change and agricultural systems.

2 Model-data integration

Every model is created on data. Beyond the model-building process, the classical way of informing any kind of model with data is to provide the data as input for the model before the run. While statistical models would need input in exactly the form with which the models were created, mechanistic models often provide a bit more flexibility with respect to the amount of data that they could use for their run. For this chapter, we concentrate on mechanistic, i.e. process-based simulation models, as at present they still represent the vast majority of models in use for climate change impact assessments in agriculture. For these models, we will first look at data that are needed during model development, starting with the calibration process, and then move on to the initialisation of models. In this section, we will concentrate on data provisioning for large-area applications. Finally, we will address assimilation of data as a method to inform models while running the simulations, a technique that allows integrating more and better data as soon as they become available and improve the simulation underway.

2.1 Multi-criteria calibration of mechanistic agro-ecosystem models

Mechanistic simulation models for agro-ecosystems (AEMs) are commonly composed of a multitude of algorithms that make up subcomponents (modules), which then are merged to build the system that the model shall represent. Each of these algorithms has been derived from individual experiments that were set up to find relationships between two or sometimes more variables that interact. These experiments are mostly designed in a way that other influencing factors are excluded, creating more or less artificial conditions, which as such do not appear in the real system. At each level of aggregation, from algorithms to modules and from modules to model, distinct target variables are computed. To ensure that algorithm, module and model do what they are supposed to do, the performance of the component needs to be tested, against experimentally observed target variables, or target variables monitored under real-world conditions. Looking back into the historical literature on model calibration, often studies report on such performance checks against one variable only, while others consider multiple variables, but no systematic procedure to arrive at the best calibration. Furthermore, there are studies that calibrate for optimal performance and then test against previously unused (independent) data obtained from the same system under similar conditions. Very often, the model testing yields very satisfying results. However, agro-ecosystem models compute several potentially interesting target variables, and all of them are output of the same simulated system and therefore dependent on each other. Calibrating only to the optimal simulation of one of them often comes at the expense of low performances in simulating others, often without recognising that there is a problem ('right for the wrong reason'). It is therefore necessary to test the performance of the model in simulating several target variables simultaneously to make sure that the whole system's behaviour is well captured by the model (Archontoulis et al., 2014).

2.1.1 Methods

Simultaneous calibration for multiple target variables requires appropriate methods, which differ from the concepts used for calibration towards one single target variable. Multi-objective approaches account for compensating effects in the calibration procedure. Considering multiple variables in global optimisation decreases the risk of the search algorithm getting trapped in a local minimum, which may result in a good fit for one variable but in an otherwise inappropriate parameter set for others. In many studies with integrative aspects, different observed variables are optimised in a single-objective calibration process. The multi-objective approach (Groh et al., 2018; Wöhling et al., 2015) has so far received less attention in the application of agro-ecosystem models, although widely accepted in the hydrology community (Hernandez-Suarez et al., 2021; Houska et al., 2017; Mostafaie et al., 2018).

Multi-objective optimisation approaches use trade-offs to determine a set of non-dominated parameters that cannot be improved for one objective without compromising the other objective. For this purpose, they spread the search within the parameter space to identify feasible parameter sets (solutions) with acceptable trade-offs along the Pareto front (Kamali et al., 2022). Since different parameters belong to different processes in the model, the solutions along the Pareto front are all realistic and robust parameter estimates yielding similar model performances.

The most common algorithms for multi-objective calibration include particle swarm optimisation (Kennedy and Eberhart, 1995), genetic algorithms (Fonseca and Fleming, 1993) and complex evolution (Yapo et al., 1998). While most multi-objective methods offer superior performance compared to single-objective calibration (Kamali et al., 2013), they also suffer from their inability to provide information on the uncertainty of model predictions. Bayesian approaches can account for parameter uncertainty in optimisation, and their superior performance has recently been demonstrated for eco-hydrological models (Tang et al., 2018; Wöhling et al., 2013) where mostly two variables (leaf area index and soil moisture) have been targeted.

2.1.2 Harmonised data sets

Preparation of data for use in process-based crop models requires that the data are aggregated to the correct spatial and temporal scales; transformed into the correct dimensional units; and checked for completeness, semantic and syntactic alignment, and to ensure quality (Janssen et al., 2017). With the increased availability of data from remote and proximal sensing, the need for interoperability of these data sources increases. The primary data requirements for process-based crop models are weather, typically daily records of rainfall, temperature and solar radiation; soil physical and chemical properties for surface and subsoil; management event details such as land preparation, sowing, application of inputs, and organic matter management; and genetic parameters (Hoogenboom et al., 2012).

Data preparation for large-scale, coordinated modelling activities, including ensemble and gridded modelling, requires some level of automation of these data preparation processes, but there are currently no interoperability standards for setting temporal and spatial scales, variable notation and units, file formats or data quality in large-scale crop modelling activities (van Evert, 2019). The choice of spatial scale is generally determined by the resolution of available data and the intent of the modelling activity. Temporal scale is limited by the time step of the model, generally daily, but in most large-scale applications, seasonal or annual values are reported due to data volume and analysis computational constraints.

Harmonisation of the semantics and syntax of agricultural datasets was addressed by the AgMIP (agmip.org) (Porter et al., 2014) to streamline ensemble modelling activities for Regional Integrated Assessments (Antle et al., 2015). The AgMIP approach describes a standardised data exchange mechanism using the ICASA vocabulary (White et al., 2013) implemented in a flexible data schema and allowing gap-filling of missing data. However, the AgMIP data interoperability protocols were developed for use with site-based field crop experimental data or farm survey data and were thus associated with individual plots or farms rather than spatially distributed data and these standards have not been adapted for gridded modelling activities.

Numerous vocabularies and ontologies have been developed for agricultural research, but no standard has emerged for use in spatial modelling activities. The robust agronomic vocabulary developed with the ICASA standards has limited provision for spatial applications. Several ontologies (Jonquet et al., 2015) have been developed for agricultural use, but none contain the full range of terms required for crop modelling activities yet.

The Global Gridded Crop modelling Intercomparisons (GGCMI phase 1) (Müller et al., 2017); GGCMI phase 2 (Franke et al., 2020); GGCMI phase 3 (Jagermeyr et al., 2021) may be the best examples of large-scale, ensemble modelling activities using harmonised data. These model intercomparisons were performed at a spatial resolution of 0.5 arc-degrees. Much of the data preparation was not automated and required extensive effort on the part of the study organisers. However, the resulting datasets have been shared under open-source licences.

For these types of gridded modelling efforts, the model input data are stored in geo-referenced files, typically one variable per file. NetCDF (Network Common Data Form) has emerged as the *de facto* standard for efficiently storing multidimensional arrays for crop modelling inputs which vary spatially but often also with time (weather data) or depth (soil data). In the case of GGCMI simulations, model input data were provided to teams in NetCDF format but each team was responsible for converting data to their specific model formats for simulations. Model outputs from each team were converted to NetCDF format for analysis and intercomparison.

3 Informing spatiotemporal simulations

Once a model has been successfully calibrated and declared fit for purpose, it can be applied in very different manners and at different scales. This implies that the model will be applied at sites to which it has not been explicitly calibrated. The quality of the simulations then often depends entirely on the quality of the input information, if no further calibration as applied at larger scales with aggregated observed data. The most fundamental drivers for mechanistic agro-ecosystem models are weather and soil data. For any spatial application of models, soil data needs to be available in sufficient detail, so that for each grid cell or point that shall be simulated, a realistic representation of soil information can be provided. This includes not only a soil type but also soil texture information and soil organic matter contents (and sometimes more, depending on the need of the model) for each soil horizon down the profile. The challenge here is that despite an often very sophisticated soil survey and sampling has been applied, the information product that is then produced from this data is often less detailed. In Germany, as one example, the most detailed soil survey was conducted during the 1930s for land use planning and taxation. Only in individual federal states, contemporary soil surveys produce soil maps at a similar degree of detail. For many other areas, detailed soil surveys have not been conducted at all, and available soil maps build on much coarser information, e.g. the Brazilian soil survey (RADAMBRASIL) of the 1970s (De Negreiros et al., 2009). And then, these maps are often not available in a digital format, ready to be used with models. Recently, first attempts have been presented to build finer resolved digital soil maps using methods of artificial intelligence to fill gaps (e.g. https://soilgrids.org; Hengl et al., 2017).

For weather data, a similar problem exists. Fine-resolved weather data products are available for areas in which meteorological observation networks are dense. For areas that are less densely covered, the quality of the weather data is lower *per se* and relies a lot on the skills of the meteorologists to interpolate reasonably the available data. This applies not only to spatial gap filling but also to temporal, as met stations sometimes fail to record. In addition to this, many met stations are not fully equipped and record temperatures and precipitation amounts, but not air humidity and solar radiation, two variables often required by crop and agro-ecosystem models. This aggravates the gap-filling problem. Especially for model applications to inform farmers on what happens on their fields or in their region (early-warning systems, irrigation and fertilisation recommendation, etc.), fine-scaled weather information is in high demand, such as the AgERA5 10 × 10 km^2 for global applications (Boogaard et al., 2020) or the 1 × 1 km^2 weather data grid of the German Weather Service for Germany. While observed weather and short-term forecasts may be available in high resolution, long-term forecasts and climate projections are often not. Here, science and engineering are still underway towards better and more detailed information products. Until then, modellers have to live with lower-resolution weather products, which comes at the expense of a higher uncertainty. How this influences the quality of the model predictions is discussed elsewhere in this book.

Beyond soil and weather data, agro-ecosystem models require more information, e.g. on crop types, crop and soil management and other items. For these highly dynamic variables, no static maps can help, but frequent observations. For this purpose, different data sources are available, of which we illustrate some in the following subsections.

3.1 Remote sensing

Remote sensing (RS) is one of the main sources of information across different spatial and temporal scales that can be used to monitor agricultural systems and the impacts of climate change. Several parameters describing vegetation growth and surface characteristics can be derived and used to inform the mechanistic models (Novelli et al., 2019). The increasing availability of RS data with improved spatial, spectral and temporal properties allows mapping and extraction of RS-based metrics over large areas. Such metrics can be used to derive proxies of vegetation status and condition, which can be the basis for deriving parameters describing crop growth stage, condition and phenology (Meroni et al., 2021). Examples of such parameters are the start (greening), peak, amplitude and the end of the growing season (senescence; Younes et al., 2021). The parameters in turn can be further used to derive information on growing conditions (Guo and Gu, 2022) and management practices (e.g. sowing and harvest date; Rezaei et al., 2021) and crop type (Blickensdörfer et al., 2022; Griffiths et al., 2019).

Crop type information is one of the essential datasets that can be generated from RS. Different data, such as Landsat, Sentinel and MODIS, has been used to derive spatially explicit crop maps. Nevertheless, availability of the crop maps is not at the same level globally and datasets covering large areas often specify only agricultural practices such as arable land or rangeland (Hoefsloot et al., 2012). Based on the freely available RS time series and use of different machine learning models, large-scale crop maps are available for USA (Boryan et al., 2011), Germany (Blickensdörfer et al., 2022), as well as large-area mapping initiatives such as Sentinel-based crop mapping (http://www.esa-sen2agri.org/).

Besides agricultural land cover characterisation, further information on crop and land management can be derived from RS, such as irrigation events and nutrient application, e.g. nitrogen (N) concentration in the plant tissue. Irrigation is one of the main tools for increasing crop productivity and the information can improve models performance and assist the estimation of water resource demands. Several studies have shown the applicability of RS data for mapping irrigation extent and amount (Ren et al., 2021; Zappa et al., 2021). Although some advances have been made, the majority of these studies focus on the local scale, with the use of a large amount of ground data, and are concentrated on arid and semiarid regions. Many studies have applied RS data for N estimation, based on the correlation between N and some leaf traits. Fewer studies assessed P and K, mostly focusing on vegetation index-based assessment based on hyperspectral and multispectral data.

RS has also been used for soil tillage assessment, mainly by estimation of parameters related to residue cover and surface roughness using optical multi- or hyperspectral and Synthetic Aperture Radar (SAR) sensors (Begue et al., 2018). There is also great interest in the detection and early warning of crop pests and diseases (Schirrmann et al., 2022). However, such a task requires high-resolution images and high-frequency fly-overs (Ibrahim et al., 2023; Oerke et al., 2014), and only a few models are yet ready to include this information (Bregaglio et al., 2021). Lastly, RS has been used to monitor highly dynamic grassland systems and to identify mowing and grazing events based on intra-seasonal dynamics of optical and SAR-based data, such as backscatter and interferometric coherence (Andreatta et al., 2022; Schwieder et al., 2022).

3.2 Proximal sensing

Proximal sensing is another source of information about site-specific conditions of soil and crop, which can deliver model input parameters. Compared to RS, where sensors are deployed on aerial or satellite platforms, proximal sensors are often placed within 2 m from the target (Adamchuk et al., 2018; Deery et al., 2017). Visible, near and mid-infrared spectrometry is one of the most commonly used methods to measure soil and crop properties. For soil, important variables include organic matter, soil nitrogen, particle size distribution, pH and moisture (Adamchuk et al., 2018; Dhawale et al., 2015). Ion-selective potentiometric sensors have also been used to measure the chemical properties of the soil, such as pH and nutrient content (Lobsey et al., 2010). Among the soil variables, moisture is probably the most important variable, as it critically influences crop production (Kashyap and Kumar, 2021). Proximal sensing methods such as ground penetrating radar (Minet et al., 2013), electromagnetic induction (Hedley et al., 2010; Tabbagh et al., 2022), ground-based radiometer and radar systems as well as gamma-ray spectroscopy (Baldoncini et al., 2019) have been

Published by Burleigh Dodds Science Publishing Limited, 2023.

developed for field-scale soil moisture determination (Babaeian et al., 2019; Kashyap and Kumar, 2021).

Optical sensors that record radiation reflected or absorbed by vegetation have been used to collect data non-destructively to assess plant properties such as nutrient content or leaf area index (LAI). While there is a large number of studies that estimate variables related to nitrogen content, such as crop N uptake and concentration (Corti et al., 2018; Morari et al., 2021; Padilla et al., 2018), other macro- and micronutrients remain comparably less investigated. The reflectance measurements can often be complemented with fluorescence data that can give additional information regarding plant characteristics, such as plant stress (Maxwell and Johnson, 2000). The advantage of proximal fluorescence sensing is that it is not sensitive to soil backgrounds, environmental light or biomass conditions. Aside from nitrogen, different fluorescence indices can be useful for estimating nutrient statuses such as potassium, magnesium and calcium (Holland et al., 2019).

Besides optical sensors, light detection and ranging (LiDAR) data have been used to assess the plant growth and canopy properties (El-Naggar et al., 2021). LiDAR has been shown to be useful for deriving information such as LAI (Hosoi et al., 2011), crop density and volume (Saeys et al., 2009) and water use (El-Naggar et al., 2021). In addition, infrared thermography, fluorescence and spectral sensors have been used for the assessment of the plant stress and disease and their spatiotemporal variation (Baker and Rosenqvist, 2004; Oerke et al., 2014). Lastly, there are several research activities towards an integrated use of proximal data with satellite based-observations for upscaling (Munnaf et al., 2021; Wolters et al., 2021).

3.3 Distributed data

Some information that is considered useful to drive AEM is difficult to sense and is not available in maps and other static information products. It often comprises information on crop impacts with limited spatial extent, e.g. scattered flooding, infections of pests and diseases, the appearance of rodents or larger animals that destroy fields while browsing, or farmer's preferences for certain management that is not immediately visible, as e.g. the choice of crop varieties, or the agents being used to battle pests and diseases. However, such information may be available through campaigns, surveys or observations that are distributed unevenly in space and time. Citizen science projects, for example, provide such information, but the nature of this data collection mode entails a sampling design that is far from ideal for data analysis. Experience with distributed data collection through laypersons revealed that people observe where they sojourn and not where their targets occur (Hampf et al., 2021; Kamp et al., 2016), or prefer easy-to-observe targets to others (Callaghan et al., 2021)

Published by Burleigh Dodds Science Publishing Limited, 2023.

and thus produce biased data. Analysing such biased data requires knowledge on the quality issues and appropriate methods that deal with them (Hochachka et al., 2012).

4 Assimilation of data in spatiotemporal simulations

The great variety of input data that AEM consume for their simulations leads to a very large state-space of possible results. Uncertainties afflicted with individual data sources add up and propagate through the course of the simulation, and any method that contributes to reducing these errors is highly welcomed. An idea that emerged in early model applications was the frequent testing against observed data, and instantaneous correction of the simulations in case it went off the observed course of the respective variable (Dorigo et al., 2007; Wallach et al., 2018; de Wit and van Diepen, 2007). The process of regularly updating a running simulation is referred to as data assimilation. The key motivation for data assimilation is the optimal estimation of the modelled state-space at any point in time an observation is available.

Data assimilation into spatiotemporal simulations enables the prediction to consider a wider range of uncertainties than traditional forward, optimisation and calibration simulations. The employed AEM has been constructed on a large empirical basis and they contain complex multivariate relationships and numerical solutions for high-dimensional problems. Data assimilation considers the often neglected uncertainties in atmospheric forcing conditions to the agro-ecosystem models, uncertainties in initial and boundary conditions, model structure (process representation, process dynamics) and model parameterisation (vegetation, soil) and most importantly observation uncertainty. Data assimilation as understood in this chapter is the optimal merging of model predictions and observations under consideration of the uncertainties in both simulation and observation. The consideration of these uncertainties in the model prediction is the key difference of data assimilation to traditional optimisation, inverse modelling, calibration and machine learning approaches. The following section outlines the two main data assimilation methods used in agro-ecosystem modelling: Ensemble Kalman Filters (EnKFs) and Particle Filters.

The original Kalman Filter estimates the optimal position of a system state variable based on observations and observation uncertainty (Kalman, 1960). Usually a numerical model provides further constraints on the estimated system state. The EnKF is the most commonly used and most generalised variant of Kalman Filters in spatiotemporal simulations (Evensen, 1994). The EnKF estimates the system state from the spatiotemporal model forecast, the observation at forecast time, and the uncertainty of both. The update process is schematically represented in Fig. 1. Model uncertainty is calculated

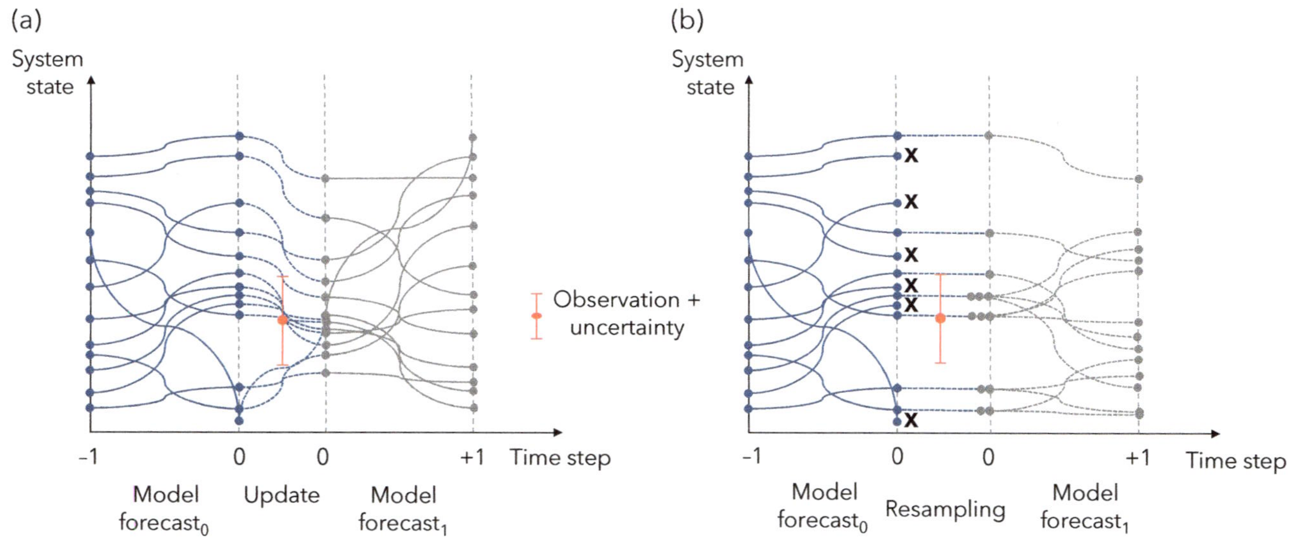

Figure 1 Updating scheme of the Ensemble Kalman Filter (a) at the time $t = 0$, when the observation (red) becomes available with known uncertainty (red bar). The resampling of the Particle Filter in (b) shows the removal of model realisations marked with X and recreation of new realisations based on weights. The update (a) actively alters the model state while the resampling (b) reproduces existing model states hence maintaining mass and energy balance.

from an ensemble of model realisations, which were created from uncertain and often randomly perturbed initial conditions, parameters and forcing. The observational uncertainty and model forecast uncertainty are optimally combined in the cost function to calculate the Kalman gain and hence the updated model state, from which the model is re-initialised until the next observation becomes available. Parameter updates can also be achieved by extending the state vector with the parameter vector (e.g. Baatz et al., 2017). Parameter updates usually require further assumptions and numerically robust control of parameter updates through e.g. damping factors. Further variants of the Kalman Filter introduce spatial constraints through localisation (e.g. Local Ensemble Transform). The Kalman Filter by Hunt et al. (2007) requires inflation i.e. deterioration of the forecast accuracy or simplifying the calculation of the Kalman gain via the Extended Kalman Filter (e.g. Anderson et al., 2009). This can be useful, as often the main assumption of Gaussian distribution in model and observation uncertainty is not accurate, and localisation of updates avoids non-causal geospatial correlations and inflation prevents overfitting. The EnKF is particularly well suited for large numerically complex models with small ensemble sizes due to the high cost of running numerically expensive models and calculating the Kalman gain. It comes at the expense of potentially altering model assumptions (e.g. energy and mass balance) through model state updates to continuous numerically calculated solutions.

Particle Filters optimise the model forecast with the observation under consideration of both uncertainties as is done by many Kalman Filter variants. Particle Filters consider an ensemble of model realisations as a swarm of particles where each particle is represented by one model realisation (Doucet et al., 2000). Particle Filters propagate the model forward until an observation is available. At the time of observation, the particles i.e. the ensemble of model states are resampled based on the probability density function of the model forecasts and the observation uncertainty. The resampled particle swarm is re-initialised with updated states until the next time step with an available observation. Resampling maintains, removes and reproduces model realisations based on the cost function and the weights assigned to individual realisations. The resampling process is schematically represented in Fig. 1b. Although the main principle remains the one described, a large number of Particle Filter variations were developed mainly by addressing the resampling scheme to account for various problem formulations (e.g. Penny and Miyoshi, 2016; van Leeuwen et al., 2019). Resampling was made subject to localisation, inflation was introduced to avoid overfitting, and the resampling algorithm was extended to include further time steps beyond the one at the time of observation. Particle Filters are independent of the assumption that uncertainties follow Gaussian distribution. Particle Filters typically require a larger ensemble size than EnKF approaches to quantify model uncertainty. This makes Particle Filters suitable for numerically

Published by Burleigh Dodds Science Publishing Limited, 2023.

less complex models. Particle Filters do not alter the conservation laws in the model ensemble due to the resampling of realisations.

Many examples exist where data assimilation was applied to merge forecasts of agro-ecosystem model states and remotely sensed observations in an optimal way (Ling et al., 2019). de Wit and van Diepen (2007) used the EnKF to assimilate the soil water index obtained from the ERS scatterometer at 0.25° spatial resolution into the WOFOST agro-ecosystem model, with the aim to improve maize and winter wheat yield predictions across the European continent. They found limited benefits of the data assimilation, as there were too large errors originating from sources not considered in this study that lowered the predictive performance of the model. Ines et al. (2013) investigated the combined assimilation of LAI and soil water content into DSSAT crop model forecasts with the EnKF. Interestingly, they found the LAI-only assimilation yielded favourable results under high-biomass and wet conditions, and favourable results of the combined assimilation under low-biomass and dry conditions. De Bernardis et al. (2016) applied the Particle Filter with MODIS NDVI observations to update plant phenology of a simple crop growth model, while Zare et al. (2022) applied it with Landsat 7 and 8 and Sentinel-2A and B to update LAI. While studies on vegetation indices other than LAI remain rare, consideration of the range of spectral indices is a promising avenue (e.g. Silvestro et al., 2021). Further advances can also be anticipated when considering problem-focused alternative Particle Filter developments and even merged EnKF variants with the Particle Filter (van Leeuwen et al., 2019). Given the increasing availability of high-performance computing infrastructures, preconditions for numerically complex model-data-fusion frameworks are rapidly improving.

5 Workflows for massive parallel computing

The need for executing large-scale simulations or simulations that also include data assimilation efforts in a reasonable amount of time necessitates the application of large computing resources. These are often situated in HPC facilities or can be accessed via cloud services. To utilise these resources efficiently, the user needs to understand the general structure in which these computing powers are available. In most cases the user has to deal with a potentially very large amount of single machines (tens to thousands) with each machine offering usually a certain amount of computing cores (tens - right now - to hundreds - in the near future) within one machine. In order to use all available cores within a single machine, either many programs can be executed in parallel or one single program can run many execution threads on top of the operating system's threading mechanism. As long as the need for computational resources can be satisfied with a single machine, the only restriction comes from limited access to storage or limited parallelisation potential of the program. Communication of parallel processes can be fast via the multi-threading approach where code runs in the same address space or somewhat slower if multiple programs communicate via inter-process communication. In both cases, latency due to communication overhead is low and the potential for broken communication channels is reduced. This all changes if the need for computational resources rises and multiple machines have to be involved in the simulations. The first question to be answered is whether a simulation consists of many independent pieces. If not, users have to try to move parts of the simulation set-up closer together to speed up communication, if possible onto the same machine or even process. Independent computations can be distributed more freely. Currently, this is true for most simulations in the agricultural domain. They often use heavyweight point-based (one-dimensional) models running on a grid, with each grid cell being represented by an independent simulation (Elliott et al., 2015; Zhao et al., 2016). In this case, after some initial set-up, the model spends most of its time computing, and less for communication. The smaller the computational part becomes, the larger the ratio of communication (set-up, loading and storing of data) to computation (simulation) gets. These concepts have to be kept in mind in order to understand and evaluate the trade-offs of the general parallelisation methods described below.

Parallelisation concepts can be ordered along multiple dimensions. The first dimension distinguishes data from task parallelism. Data parallelism refers to large amounts of data being processed using a single program. Here, the user wants to apply parallel algorithms within the program to process the data efficiently. From the user's point of view, this may be seen as a single processing pipeline, but the functions being applied to the data work internally in parallel. In contrast, task parallelism is a concept applied to larger tasks (e.g. the abovementioned point model application), which run as separate programs. For a large simulation, the focus is more on how to create the necessary set-up for each task and how to run it. In the agricultural domain, this often means running complex models on millions of set-ups to reflect the changes at different locations (e.g. soil and climate data).

There is a constant gradient between data and task parallelism. Somewhere in the middle, one can find frameworks and environments for (massively) parallel data flows, e.g. Apache Spark (https://en.wikipedia.org/wiki/Apache_Spark) or Storm (https://en.wikipedia.org/wiki/Apache_Storm). Recalling the image of data parallelism as a single pipeline, each execution step may exhibit the potential for parallelisation. If datasets get larger and at some point do not fit into memory anymore, data flow pipelines can be constructed in which each component within the pipeline is specialised for particular processing steps. These components do not all live necessarily within a single machine, as they can also be distributed over many different machines. In that way, a particular data flow can be computationally scaled to the demands of the data being processed. An example of a particular short data flow is the Map-Reduce pattern (https://en.wikipedia.org/wiki/MapReduce), in which a large amount of independent data is being 'mapped' and then reduced in one or more further steps. Mapping in this case refers to the application of a function (or an equivalent to it) to every single data piece. As the data are independent, this map function can be executed on arbitrary amounts of computational units (cores, processors, machines). The results of the applied function will usually need some aggregation steps, which sequentially follow the mapping step.

The distribution of tasks to the available computational resources can be described using established mechanisms. The most popular is the manual distribution of the tasks and data to the computational units (mostly machines or nodes in a HPC or cloud). On these machines, a program or script will be executed to run the actual program on the set of data assigned to the machine. At the machine level, this resembles batch processing. To distribute the work automatically, cluster management and job scheduling tools like SLURM (https://slurm.schedmd.com/documentation.html) have been designed, which will acquire the necessary resources and run shell scripts on the involved compute nodes. The same tools can also be used to distribute programs, which use a message-passing interface (MPI, https://en.wikipedia.org/wiki/Message_Passing_Interface). Instead of manually taking care of distributing data and code and organising workflows at the operating system level, the user writes a program, which uses standardised interfaces to do this task. Using MPI, the same program will be started on all computational nodes, but one of these nodes is assigned to coordinate the initial distribution of work and the final aggregation of results. The author of the program does not have to be concerned with the peculiarities of the infrastructure anymore and rather writes a single program that uses the underlying MPI implementation of the infrastructure to distribute data and do the necessary communication. Depending on the kind of task, programs using MPI can scale very easily to large amounts of computing units, but with the drawback of being more difficult to debug due to its indirect execution.

Somewhere between manual batch execution and MPI-style program distribution, one can think in terms of clouds of programs. If the time to transfer the input and output data is sufficiently low compared to running the code itself, it can be advantageous to start many instances of a program as a service on remote machines, provided by either a cloud provider or an HPC cluster. Subsequently, a short pipeline is created, consisting of at least one program producing jobs (Producer) for one of the models and another program receiving the results of the model run for further processing (Consumer). This is similar to the Map-Reduce pattern but offers more flexibility in what the Producer can do, as the act of creating a job can be arbitrarily complex. This approach introduces bottlenecks in the sense that in the naive case, one task is producing jobs for many processes and their results have to be funnelled back into a single process for aggregation. However, the advantages emerge from the fact that the Producer and Consumer processes can be developed incrementally and can be run and even debugged locally as long as remote access to the cloud of models is available. Optimising the job creation side will lead towards the MPI-approach or more complex parallel data flows. However, optimising this concept has a few trade-offs to consider, where computation speed stands against process responsibilities within the pipeline.

6 Model-model integration

Agricultural systems include a range of very different subsystems, including biophysical, socioeconomic and even mixed systems. Simulation models have been developed for almost all of these subsystems, but each based on different methods and modelling philosophies. To analyse high-level feedback loops and system behaviour, it is sometimes desired to couple some of these subsystems, represented by individual models. A simple example is the handshake between a mechanistic agro-ecosystem model providing crop yields and an agent-based economic model that uses these yields to simulate market dynamics or farm-level decision-making. Here, a so-called 'hard coupling' of the two models physically combined in computer code is often not required, as the passing-on of the yield values already suffices for the purpose ('soft coupling'). However, there are more applications imaginable, and some of them become complex. In the following subchapters, we will address some of these challenges and provide some solutions.

6.1 Obstacles for model-model integration

6.1.1 Resolution in time and space

One common issue with coupling models is the individual resolution of the processes represented in the model. As for the temporal resolution, the processes describing the dynamics in agro-ecosystems are typically resolved in days, as a compromise between the need to reflect short-term, event-based management and weather shocks (e.g. heat or frost events) and the availability of weather data to drive the models and computation time. Some of the process descriptions may benefit from a finer-temporal resolution, which is why some models have internal modules that run in an hourly or even minutely resolution. Such models are able to describe day and night processes, or the response of crop and soil to heavy rain and wind gusts. Coupling a daily resolved agro-ecosystem model with an erosion model, in order to assess the impact of a rainfall event on erosion and crop growth simultaneously, requires that both models match in their temporal resolution.

If not, the two remaining options are (i) simplifying the finer resolved model so that it matches the coarser or (ii) running the two models at different paces. As an example, EROSION 3D (Schmidt, 1991) is a mechanistic model that simulates erosion events on agricultural fields in very high temporal resolution, also on marginal slopes. For this purpose, it resolves rainfall data in minutes and requires this data as input. While coupling to an agro-ecosystem model with daily resolution would be feasible at two different paces at the field scale, this concept would hit computational limitations when scaling to larger areas, as for each AEM simulation time step, 1440 simulations of EROSION 3D need to be executed. On the other hand, if reducing the resolution of EROSION 3D to daily time steps, the model would miss important information on rainfall intensity to drive the simulation of erosion events. What remains is a more general erosion approach, such as the Universal Soil Loss Equation (Wischmeier and Smith, 1965), which, however, often underestimates event-based erosion.

At the other end, feedback between soil and plant processes to explain the genesis of soils in agricultural landscapes, elaborate sinks and sources for carbon and subsequently assess carbon sequestration potentials requires the consideration of hundreds if not thousands of years. Models designed for this purpose often use only yearly time steps and thus look at soil and plant management only through frosted glass rather than a binocular. Using an AEM with a soil formation model therefore requires AEM runs for a long time, with the need to make assumptions on crop and soil management (Lugato et al., 2014).

Spatial resolution has played a more prominent role in considerations (Ewert et al., 2011) and a considerable body of literature has grown on the effect of using weather (Hoffmann et al., 2015; Nendel et al., 2013; Zhao et al., 2015), soil (Grosz et al., 2017; Hoffmann et al., 2016; Zhao et al., 2016) or management data (Constantin et al., 2019) at different scales to drive AEM. Even though the general impression from these studies is that scaling between 1 km^2 to 100 km^2 does not result in drastic effects on the simulation of crop yields or other variables, the scales below 1 km^2 have not yet been intensively studied. Suitable data products, especially for soils, are still rare and may not even contain the information that would be relevant, e.g. for investigating macro pore fluxes of infiltrating water after rainfall events, and the respective fast transport of water deep into the soil along cracks, earthworm passages or root channels (Christiansen et al., 2004; Jarvis et al., 1991), which has consequences for the water budget in the topsoil. In general, AEMs are sensitive to soil heterogeneity also at the subfield scale (Wallor et al., 2018). However, their application on grid cells larger than 100 km^2 comes always with massive simplifications in the assumptions that have to be made for soil distribution and crop management and that – despite a low scaling effect – contribute to the limitations in the conclusions that can be drawn from such model applications (e.g. Heinicke et al., 2022).

6.1.2 Model philosophies

Different models often clash because of different philosophies being used as the foundation of their development. Using the erosion and crop model coupling as an example again, the one-dimensional nature of many crop or agro-ecosystem models makes it difficult to employ them for science cases in which proximity relationships play a role in the processes considered. While on crop yields, the condition of the neighbouring plants and soils have often little influence, an erosion event means explicit matter exchange with the neighbouring spatial unit. Coupling an AEM with an erosion model therefore requires a mass balance across space, and the ability to add or subtract soil mass to or from the soil profile, with consequences for some soil properties. Burying organic matter under accumulating substrate, or gradually exposing horizons low in organic matter to frequent organic matter inputs through erosion can then be simulated with such a coupled model.

Another example of clashing philosophies is the coupling of an AEM with a farm economy model that requires annual production from the farmer's fields as input. As such, yields can be simulated using an AEM, this coupling seems an easy undertaking to begin with, but the devil lies in the detail. Farm economy models, e.g. MPMAS (Berger, 2001), often do not care about the actual locality of the fields but only about the statistical coherence. Any decision the virtual farmer takes in such a model yields a new statistical distribution of crops being produced on that farm in the new season. However, whether crop rotation rules on individual fields are followed is often beyond focus. In contrast, AEMs do need to make sure that one crop follows another explicitly and plausibly on one field to track carry-over effects of water and nutrients. This spatial coherence is key to any simulation of irrigation water requirements, N fertiliser applications and other cost-afflicted management, which would be of interest to the farm economy model besides yields.

7 Granularity and modular design for model improvement, reuse, exchange and interoperability

The first computational models of crop and soil processes were developed in the 1960s (Jones et al., 2017). Their long history has produced a large number of crop models, which have evolved from a few landmark models such as CERES and EPIC in the USA or SUCROS and ARCWHEAT in Europe (Muller and Martre, 2019). Crop models differ according to their choice of relationships and hypotheses regarding process functioning and feedback loops and their combinations of mechanistic components. In the last decade, the AgMIP

(Rosenzweig et al., 2013) and MACSUR (Rötter et al., 2013) networks of crop modellers have carried out several benchmarking crop simulation model intercomparison studies. These studies highlighted both the capacities and the limits of current models (e.g. Asseng et al., 2013; Bassu et al., 2014; Kollas et al., 2015; Li et al., 2015; Martre et al., 2015) and have led to important model improvements (e.g. Maiorano et al., 2017; Wang et al., 2017). This process highlighted also the importance of equally considering soil processes, which is done in AEM, but often not sufficiently in crop models. However, it became also obvious that model improvement and extension (e.g. to consider crop disease and extreme weather event impacts) is complicated by the different software architectures and programming languages in which they are implemented.

When developing a model or choosing a model for a specific analysis, a key decision to be made is on the level of granularity that is required. Individual choices have to be made regarding how to represent real-world entities in the model, and the boundaries and scope of the system must be defined. All these choices may have to be balanced, including a number of non-functional requirements such as availability and performance. All these contribute to define the granularity of a model and will have important implications on its reusability, transparency, and reproducibility, and on the interpretability of the simulation results (Neveu et al., 2020).

The internal granularity of a component must be distinguished from the granularity of the whole model. The internal granularity of a component is easier to define in most cases and should be at the scale of unitary processes (often defined by a single controlling equation). The choice of the granularity of a modelling solution, which defines the number of components, should be primarily guided by reusability aspects. A model component that simulates the hourly or daily energy balance of canopies is more likely to be reused than a component that simulates the canopies' latent heat flux.

Most crop modelling platforms have a modular architecture (e.g. Brown et al., 2014; Donatelli et al., 2014; Enders et al., 2010; Jones et al., 2001; Wang et al., 2002), which facilitates the development, reuse and extension of models within each of these frameworks. However, their specificities and the different programming languages they use make the reuse of components outside the platform in which they were developed, or the extension of an existing model with components from a different platform, a difficult task. Recently, the Agricultural Model Exchange Initiative has coordinated the development of a centralised framework (Crop2ML, Crop Modelling Meta Language) to support the development of modular model and facilitate the exchange and reuse of soil, plant, and crop process components between modelling platforms (Midingoyi et al., 2020; Midingoyi et al., 2021). Crop2ML reduces the emphasis on the software part of the model, bridging the gap between computer programmers and model developers, who can better capture the biophysical processes represented in the model. Models in the open-source format Crop2ML can be stored, retrieved, and shared on an easily accessible online open model repository (CropMRespository; http://crop2ml.org). Crop2ML does not aim at replacing existing modelling platforms or at simulating components within large modelling solutions; rather Crop2ML provides a solution to create transparent and well-documented components, which can be easily extended, reused, and shared between crop modelling platforms.

Several other initiatives in other modelling communities are also developing solutions to facilitate the exchange and the coupling of models. For example, in the earth's surface modelling, the Community Surface Dynamics Modelling System (CSDMS; Tucker et al., 2022) provides cyber-infrastructure to promote the quantitative modelling of earth surface processes and distributes software tools and models. The CSDMS developed the Basic Model Interface (BMI), a set of standard query and control functions that make models both easier to learn and easier to couple with other software (Hutton et al., 2020; Peckham et al., 2013). As each community of modellers is developing its own system for coupling and exchanging modelling solutions or model components within its own community, a challenge for more integrated modelling is to ease the coupling of models developed by different communities. The Open Modelling Foundation (https://openmodelingfoundation.github.io) is working in this direction by coordinating and administrating a common, community-developed body of standards and best practices among diverse communities of modelling scientists to support the exchange, reuse, and interoperability of human and natural system models across communities.

8 Concepts for distributed modelling

In the last decades, the world of computing has moved from single machines, which execute program code sequentially, to a world of interconnected devices of all kinds and sizes, in which the mode of operation is increasingly concurrent, even within the devices themselves. There are already thousands of operational systems available, in which continuous data streams (e.g. from sensors) are fed into cloud-computing infrastructures which run large-scale computations in a highly parallel manner. Those systems can be accessed using mobile devices, which as well are able to do multiple things in parallel, for instance accessing different data sources or interacting with remote environments and services.

In today's networked world all components of such a workflow can potentially be distributed, i.e. made living on separate processes, machines or networks. In the extreme case, this distribution is even applied at the level of model components, even though there is a limit to a meaningful granularity of the distributed components. A distributed system needs a way to address and access its remote components. To do this uniformly, a defined protocol has to

be used. In the internet of today, such a widely used protocol is the HyperText Transfer Protocol (HTTP), which allows remote resources to be addressed via Uniform Resource Locators (URL) or more generally as Uniform Resource Identifiers (URI). Taking the HTTP protocol to its logical conclusion leads to the well-established REST-based (REpresentational State Transfer) web-services. This allows the creation of very flexible services built upon the widely supported HTTP mechanism. A disadvantage though lies in its text-based nature, which is less space-efficient and requires many latency-increasing network round trips for even simple service interactions. Diverging from the clean and pure REST mechanism, optimising for some aspects, e.g. transfer efficiency, leads to the RPC (Remote Procedure Call) paradigm. Here, a remote component offers some kind of interface that defines a set of procedures, equivalent to a function in a programming language, to be called. RPC implementations do not have to comply with HTTP rules or text-based data transfer, which is why they often use binary data encoding. Depending on the design goals, these data representations can be more compact for reducing the required transfer bandwidth or optimised for efficient handling, e.g. for minimising copy operations. In many cases, schema languages are used to describe the remote interfaces (set of procedures) as well as the data structures.

The divergence from REST allows for more specialised and optimised interfaces, reducing latency. Nevertheless, complex remote Application Programming Interfaces (API) still suffer from many necessary network round trips. However, the CapTP protocol (http://www.erights.org/elib/distrib/captp/index.html) and its implementations offer a way out, regaining some of the efficiency local procedure calls offer and even getting back some of the flexibility and cleanliness the REST-based approach exhibited, but without the drawbacks. One such implementation is Cap'n Proto, a fast data interchange format and capability-based RPC system (https://capnproto.org). A Cap'n Proto schema file will describe the messages that a remote object will understand and the structure data in these messages will have. To make a model or component remotely accessible it has to implement the according interface. The possibility to have interfaces implemented by different models and components allows them to interoperate. Contrary to most RPC systems, Cap'n Proto allows interfaces themselves to be sent or received in messages. In this context, the interfaces are called capabilities and are the powerful foundation for the Capability Security paradigm (Miller et al., 2003).

One challenge neglected so far is how to manage access and authority to remote objects (models, components and services). In a traditional RPC or REST-based setting access to an URL/URI has to be authenticated. This can happen via usernames and passwords, systems like OAuth (https://en.wikipedia.org/wiki/OAuth) or access tokens (a kind of capability). Except for the last one, these methods make use of ambient authority, i.e. the authority and the means to exercise this authority are separated. A consequence of this is that a system hosting remote objects and employing ambient authority has to rely on a central authority (an administrative user) to manage access rights. This makes it difficult and sometimes impossible to apply the Principle of Least Authority (https://en.wikipedia.org/wiki/Principle_of_least_privilege), which states that to execute a certain task a user should be given the required authority, but nothing more than that. In our context, that task could refer to running a remote computation on some input data. The actual authority needed is just the ability to send a message to the remote object, which represents this computation. This logic is well supported by Object Capability systems (e.g. Cap'n Proto), in which a capability is the reference to a remote object. Possessing this reference means being authorised to communicate with the remote object. Figure 2 abstractly illustrates this concept: the object *Alice* holds the right to communicate with remote object *Carol*. To allow also remote object *Bob* to talk to remote object *Carol*, *Alice* just has to send *Bob* a message referencing *Carol* (Berg-Mohnicke and Nendel, 2022).

With these foundations in place, ubiquitous connected systems, fast data interchange and capability-secured remote procedure calls it is conceivable to allow for distributed modelling, model application, and more generally distributed workflows. Models or data flows can be composed of remote components without the need to know where they are executed as long as they conform to a common interface and can understand each other. All that is needed is a capability to a remote component in order to use it. Of course, this comes at the cost of efficiency and can be prohibitive in some contexts, for example in applications due to the demand for large amounts of data (potentially without

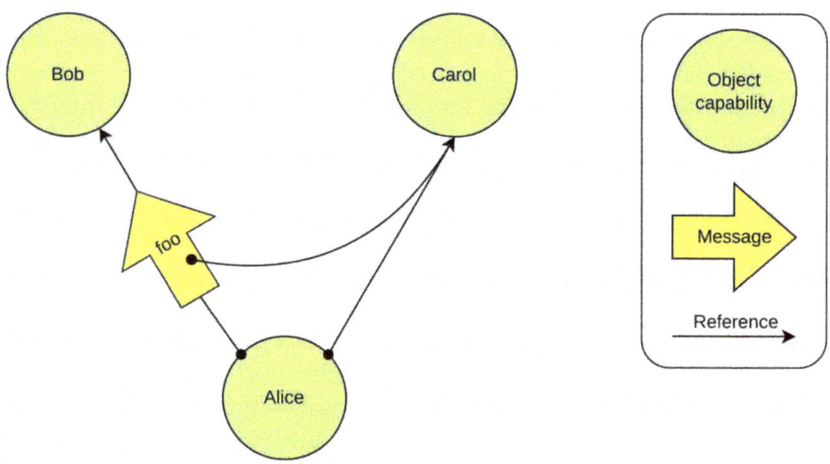

Figure 2 Three owners of an object capability (*Alice*, *Bob* and *Carol*) communicating via the message *foo*. Source: Miller et al. (2003).

data locality), tight coupling with a lot of interactions per time unit (latency being the limiting factor) or reliance on remote availability and less control over the execution environment (e.g. slow hardware, no parallelisation or optimisation possible). Some issues can be mitigated for instance by moving code and data close together. However, while speed is often an issue with distributed modelling, distributing models and components offer new possibilities, such as enabling new ways of collaboration, local testing and debugging, access to large remote datasets or the participation in remote workflows.

9 Future trends and conclusion

Simulation models for different components of the agricultural system have been available already for a while, and often model developers hesitate about touching again a model that has once been satisfactorily coded, parameterised and calibrated. However, increasingly available high-performing computing resources and big data gradually offer more and more application opportunities for models, and this increases the need to adapt structure, code and framework of the models to benefit from this development. Here, standards for data interoperability and cross-platform communication at the model-data and model-model interface are needed and currently under development. Big data especially opens opportunities for large-area applications of models in sufficiently high resolution to allow not only simulations that are closer to the real world that the models represent but also better investigations of the model's representativeness in time and space and the error that the models and their drivers still carry along. Continuous data streams in principle allow data assimilation to improve simulations under limited system knowledge, but the multivariate and highly intertwined nature of the models that we deal with in the agricultural domain often poses challenges with regard to multi-objective optimisation problems. Some driving variables may remain limited in the future, but in the bid to reproduce spatiotemporal patterns of target variables, the use of artificial intelligence may soon help to replace some of the unknown input data. Machine learning has already now demonstrated its power in this realm (Guilpart et al., 2022; Shahhosseini et al., 2021; van Klompenburg et al., 2020; Webber et al., 2020) but remains difficult to interpret (Lischeid et al., 2022).

10 Where to look for further information

Current developments in NetCDF:

- https://www.unidata.ucar.edu/software/netcdf/.

Cap'n Proto overview

- https://capnproto.org.

11 References

Adamchuk, V. I., Ji, W., Viscarra Rossel, R., Gebbers, R. and Trembley, N. (2018). Proximal soil and plant sensing. In: Sudduth, K. and Clay, D. (Eds.), *Precision Agriculture Basics* (pp. 119–140). American Society of Agronomy, Madison, WI.

Anderson, J., Hoar, T., Raeder, K., Liu, H., Collins, N., Torn, R. and Avellano, A. (2009). The data assimilation research testbed - A community facility. *Bulletin of the American Meteorological Society* 90(9), 1283–1296. https://doi.org/10.1175/2009BAMS2618.1.

Andreatta, D., Gianelle, D., Scotton, M., Vescovo, L. and Dalponte, M. (2022). Detection of grassland mowing frequency using time series of vegetation indices from Sentinel-2 imagery. *GIScience and Remote Sensing* 59(1), 481–500. https://doi.org/10.1080/15481603.2022.2036055.

Antle, J. M., Valdivia, R. O., Boote, K. J., Janssen, S., Jones, J. W., Porter, C. H., Rosenzweig, C., Ruane, A. C. and Thorburn, P. J. (2015). AgMIP's transdisciplinary agricultural systems approach to regional integrated assessment of climate impacts, vulnerability, and adaptation. In: Rosenzweig, C. and Hillel, D. (Eds.), *Handbook of Climate Change and Agroecosystems* (pp. 27–44). Imperial College Press & American Society of Agronomy. The Agricultural Model Intercomparison and Improvement Project. https://doi.org/10.1142/p970.

Archontoulis, S. V., Miguez, F. E. and Moore, K. J. (2014). A methodology and an optimization tool to calibrate phenology of short-day species included in the APSIM PLANT model: Application to soybean. *Environmental Modelling and Software* 62, 465–477. https://doi.org/10.1016/j.envsoft.2014.04.009.

Asseng, S., Ewert, F., Martre, P., Rötter, R. P., Lobell, D. B., Cammarano, D., Kimball, B. A., Ottman, M. J., Wall, G. W., White, J. W., Reynolds, M. P., Alderman, P. D., Prasad, P. V. V., Aggarwal, P. K., Anothai, J., Basso, B., Biernath, C., Challinor, A. J., De Sanctis, G., Doltra, J., Fereres, E., Garcia-Vila, M., Gayler, S., Hoogenboom, G., Hunt, L. A., Izaurralde, R. C., Jabloun, M., Jones, C. D., Kersebaum, K. C., Koehler, A., Müller, C., Naresh Kumar, S., Nendel, C., O'Leary, G., Olesen, J. E., Palosuo, T., Priesack, E., Eyshi Rezaei, E., Ruane, A. C., Semenov, M. A., Shcherbak, I., Stöckle, C., Stratonovitch, P., Streck, T., Supit, I., Tao, F., Thorburn, P. J., Waha, K., Wang, E., Wallach, D., Wolf, J., Zhao, Z. and Zhu, Y. (2015). Rising temperatures reduce global wheat production. *Nature Climate Change* 5(2), 143–147. https://doi.org/10.1038/nclimate2470.

Asseng, S., Ewert, F., Rosenzweig, C., Jones, J. W., Hatfield, J. L., Ruane, A. C., Boote, K. J., Thorburn, P. J., Rotter, R. P., Cammarano, D., Brisson, N., Basso, B., Martre, P., Aggarwal, P. K., Angulo, C., Bertuzzi, P., Biernath, C., Challinor, A. J., Doltra, J., Gayler, S., Goldberg, R., Grant, R., Heng, L., Hooker, J., Hunt, L. A., Ingwersen, J., Izaurralde, R. C., Kersebaum, K. C., Muller, C., Kumar, S. N., Nendel, C., O'Leary, G., Olesen, J. E., Osborne, T. M., Palosuo, T., Priesack, E., Ripoche, D., Semenov, M. A., Shcherbak, I., Steduto, P., Stockle, C., Stratonovitch, P., Streck, T., Supit, I., Tao, F., Travasso, M., Waha, K., Wallach, D., White, J. W., Williams, J. R. and Wolf, J. (2013). Uncertainty in simulating wheat yields under climate change. *Nature Climate Change* 3(9), 827–832. https://doi.org/10.1038/nclimate1916.

Baatz, R., Franssen, H. J. H., Han, X. J., Hoar, T., Bogena, H. R. and Vereecken, H. (2017). Evaluation of a cosmic-ray neutron sensor network for improved land surface model prediction. *Hydrology and Earth System Sciences* 21(5), 2509–2530. https://doi.org/10.5194/hess-21-2509-2017.

Published by Burleigh Dodds Science Publishing Limited, 2023.

Babaeian, E., Sadeghi, M., Jones, S. B., Montzka, C., Vereecken, H. and Tuller, M. (2019). Ground, proximal, and satellite remote sensing of soil moisture. *Reviews of Geophysics* 57(2), 530–616. https://doi.org/10.1029/2018RG000618.

Baker, N. R. and Rosenqvist, E. (2004). Applications of chlorophyll fluorescence can improve crop production strategies: An examination of future possibilities. *Journal of Experimental Botany* 55(403), 1607–1621. https://doi.org/10.1093/jxb/erh196.

Baldoncini, M., Alberi, M., Bottardi, C., Chiarelli, E., Raptis, K. G. C., Strati, V. and Mantovani, F. (2019). Biomass water content effect on soil moisture assessment via proximal gamma-ray spectroscopy. *Geoderma* 335, 69–77. https://doi.org/10.1016/j.geoderma.2018.08.012.

Bassu, S., Brisson, N., Durand, J. L., Boote, K., Lizaso, J., Jones, J. W., Rosenzweig, C., Ruane, A. C., Adam, M., Baron, C., Basso, B., Biernath, C., Boogaard, H., Conijn, S., Corbeels, M., Deryng, D., De Sanctis, G., Gayler, S., Grassini, P., Hatfield, J., Hoek, S., Izaurralde, C., Jongschaap, R., Kemanian, A. R., Kersebaum, K. C., Kim, S. H., Kumar, N. S., Makowski, D., Muller, C., Nendel, C., Priesack, E., Pravia, M. V., Sau, F., Shcherbak, I., Tao, F., Teixeira, E., Timlin, D. and Waha, K. (2014). How do various maize crop models vary in their responses to climate change factors? *Global Change Biology* 20(7), 2301–2320. https://doi.org/10.1111/gcb.12520.

Begue, A., Arvor, D., Bellon, B., Betbeder, J., de Abelleyra, D., Ferraz, R. P. D., Lebourgeois, V., Lelong, C., Simoes, M. and Veron, S. R. (2018). Remote sensing and cropping practices: A review. *Remote Sensing* 10(2), Article 99. https://doi.org/10.3390/rs10010099.

Berg-Mohnicke, M. and Nendel, C. (2022). A case for object capabilities as the foundation of an environmental model and simulation infrastructure. *Environmental Modelling and Software* 156, 105471. https://doi.org/10.1016/j.envsoft.

Berger, T. (2001). Agent-based spatial models applied to agriculture: A simulation tool for technology diffusion, resource use changes and policy analysis. *Agricultural Economics* 25(2–3), 245–260. https://doi.org/10.1111/j.1574-0862.2001.tb00205.x.

Blickensdörfer, L., Schwieder, M., Pflugmacher, D., Nendel, C., Erasmi, S. and Hostert, P. (2022). Mapping of crop types and crop sequences with combined time series of Sentinel-1, Sentinel-2 and Landsat 8 data for Germany. *Remote Sensing of Environment* 269, Article 112831. https://doi.org/10.1016/j.rse.2021.112831.

Boogaard, H., Schubert, J., De Wit, A., Lazebnik, J., Hutjes, R. and Van der Grijn, G. (2020). *Agrometeorological Indicators From 1979 to Present Derived From Reanalysis Version 1.0*. https://doi.org/10.24381/cds.6c68c9bb.

Boryan, C., Yang, Z. W., Mueller, R. and Craig, M. (2011). Monitoring US agriculture: The US Department of Agriculture, National Agricultural Statistics Service, Cropland Data Layer Program. *Geocarto International* 26(5), 341–358. https://doi.org/10.1080/10106049.2011.562309.

Bregaglio, S., Willocquet, L., Kersebaum, K. C., Ferrise, R., Stella, T., Ferreira, T. B., Pavan, W., Asseng, S. and Savary, S. (2021). Comparing process-based wheat growth models in their simulation of yield losses caused by plant diseases. *Field Crops Research* 265, Article 108108. https://doi.org/10.1016/j.fcr.2021.108108.

Brown, H. E., Huth, N. I., Holzworth, D. P., Teixeira, E. I., Zyskowski, R. F., Hargreaves, J. N. G. and Moot, D. J. (2014). Plant modelling framework: Software for building and running crop models on the APSIM platform. *Environmental Modelling and Software* 62, 385–398. https://doi.org/10.1016/j.envsoft.2014.09.005.

Callaghan, C. T., Poore, A. G. B., Hofmann, M., Roberts, C. J. and Pereira, H. M. (2021). Large-bodied birds are over-represented in unstructured citizen science data. *Scientific Reports* 11(1), Article 19073. https://doi.org/10.1038/s41598-021-98584-7.

Constantin, J., Raynal, H., Casellas, E., Hoffman, H., Bindi, M., Doro, L., Eckersten, H., Gaiser, T., Grosz, B., Haas, E., Kersebaum, K. C., Klatt, S., Kuhnert, M., Lewan, E., Maharjan, G. R., Moriondo, M., Nendel, C., Roggero, P. P., Specka, X., Trombi, G., Villa, A., Wang, E. L., Weihermuller, L., Yeluripati, J., Zhao, Z. G., Ewert, F. and Bergez, J. E. (2019). Management and spatial resolution effects on yield and water balance at regional scale in crop models. *Agricultural and Forest Meteorology* 275, 184–195. https://doi.org/10.1016/j.agrformet.2019.05.013.

Corti, M., Cavalli, D., Cabassi, G., Gallina, P. M. and Bechini, L. (2018). Does remote and proximal optical sensing successfully estimate maize variables? A review. *European Journal of Agronomy* 99, 37–50. https://doi.org/10.1016/j.eja.2018.06.008.

De Bernardis, C., Vicente-Guijalba, F., Martinez-Marin, T. and Lopez-Sanchez, J. M. (2016). Particle filter approach for real-time estimation of crop phenological states using time series of NDVI images. *Remote Sensing* 8(7), Article 610. https://doi.org/10.3390/rs8070610.

De Negreiros, G. H., Schlesinger, P., Nepstad, D. C. and Lefebvre, P. A. (2009). Pre-LBA RADAMBRASIL project data. https://doi.org/10.3334/ORNLDAAC/941.

de Wit, A. J. W. and van Diepen, C. A. (2007). Crop model data assimilation with the Ensemble Kalmann filter for improving regional crop yield forecasts. *Agricultural and Forest Meteorology* 146(1–2), 38–56. https://doi.org/10.1016/j.agrformet.2007.05.004.

Deery, D., Jimenez-Berni, J., Jones, H., Sirault, X. and Furbank, R. (2017). Proximal remote sensing buggies and potential applications for field-based phenotyping. *Agronomy* 4(3), 349–379. https://doi.org/10.3390/agronomy4030349.

Dhawale, N. M., Adamchuk, V. I., Prasher, S. O., Rossel, R. A. V., Ismail, A. A. and Kaur, J. (2015). Proximal soil sensing of soil texture and organic matter with a prototype portable mid-infrared spectrometer. *European Journal of Soil Science* 66(4), 661–669. https://doi.org/10.1111/ejss.12265.

Donatelli, M., Bregaglio, S., Confalonieri, R., De Mascellis, R. and Acutis, M. (2014). A generic framework for evaluating hybrid models by reuse and composition - A case study on soil temperature simulation. *Environmental Modelling and Software* 62, 478–486. https://doi.org/10.1016/j.envsoft.2014.04.011.

Dorigo, W. A., Zurita-Milla, R., de Wit, A. J. W., Brazile, J., Singh, R. and Schaepman, M. E. (2007). A review on reflective remote sensing and data assimilation techniques for enhanced agroecosystem modeling. *International Journal of Applied Earth Observation and Geoinformation* 9(2), 165–193. https://doi.org/10.1016/j.jag.2006.05.003.

Doucet, A., Godsill, S. and Andrieu, C. (2000). On sequential Monte Carlo sampling methods for Bayesian filtering. *Statistics and Computing* 10(3), 197–208. https://doi.org/10.1023/A:1008935410038.

El-Naggar, A. G., Jolly, B., Hedley, C. B., Horne, D., Roudier, P. and Clothier, B. E. (2021). The use of terrestrial LiDAR to monitor crop growth and account for within-field variability of crop coefficients and water use. *Computers and Electronics in Agriculture* 190, 106416.

Elliott, J., Muller, C., Deryng, D., Chryssanthacopoulos, J., Boote, K. J., Buchner, M., Foster, I., Glotter, M., Heinke, J., Iizumi, T., Izaurralde, R. C., Mueller, N. D., Ray, D. K.,

Rosenzweig, C., Ruane, A. C. and Sheffield, J. (2015). The global gridded crop model intercomparison: Data and modeling protocols for phase 1 (v1.0). *Geoscientific Model Development* 8(2), 261–277. https://doi.org/10.5194/gmd-8-261-2015.

Enders, A., Diekkrüger, B., Laudien, R., Gaiser, T. and Bareth, G. (2010). The IMPETUS spatial decision support systems. In: Speth, P., Christoph, M. and Diekkrüger, B. (Eds.), *Impacts of Global Change on the Hydrological Cycle in West and Northwest Africa* (pp. 360–393). Springer.

Evensen, G. (1994). Sequential data assimilation with a nonlinear quasi-geostrophic model using Monte-Carlo methods to forecast error statistics. *Journal of Geophysical Research* 99(C5), 10143–10162. https://doi.org/10.1029/94JC00572.

Ewert, F., Rotter, R. P., Bindi, M., Webber, H., Trnka, M., Kersebaum, K. C., Olesen, J. E., van Ittersum, M. K., Janssen, S., Rivington, M., Semenov, M. A., Wallach, D., Porter, J. R., Stewart, D., Verhagen, J., Gaiser, T., Palosuo, T., Tao, F., Nendel, C., Roggero, P. P., Bartosova, L. and Asseng, S. (2015). Crop modelling for integrated assessment of risk to food production from climate change. *Environmental Modelling and Software* 72, 287–303. https://doi.org/10.1016/j.envsoft.2014.12.003.

Ewert, F., van Ittersum, M. K., Heckelei, T., Therond, O., Bezlepkina, I. and Andersen, E. (2011). Scale changes and model linking methods for integrated assessment of agri-environmental systems. *Agriculture, Ecosystems and Environment* 142(1–2), 6–17. https://doi.org/10.1016/j.agee.2011.05.016.

Fonseca, C. M. and Fleming, P. J. (1993). Genetic algorithms for multiobjective optimization: Formulation, discussion, and generalization. In: *Proceedings of the 5th International Conference on Genetic Algorithms*.

Franke, J. A., Muller, C., Elliott, J., Ruane, A. C., Jagermeyr, J., Balkovic, J., Ciais, P., Dury, M., Falloon, P. D., Folberth, C., Francois, L., Hank, T., Hoffmann, M., Izaurralde, R. C., Jacquemin, I., Jones, C., Khabarov, N., Koch, M., Li, M., Liu, W. F., Olin, S., Phillips, M., Pugh, T. A. M., Reddy, A., Wang, X. H., Williams, K., Zabel, F. and Moyer, E. J. (2020). The GGCMI Phase 2 experiment: Global gridded crop model simulations under uniform changes in CO2, temperature, water, and nitrogen levels (protocol version 1.0). *Geoscientific Model Development* 13(5), 2315–2336. https://doi.org/10.5194/gmd-13-2315-2020.

Griffiths, P., Nendel, C. and Hostert, P. (2019). Intra-annual reflectance composites from Sentinel-2 and Landsat for national-scale crop and land cover mapping. *Remote Sensing of Environment* 220, 135–151. https://doi.org/10.1016/j.rse.2018.10.031.

Groh, J., Stumpp, C., Lucke, A., Putz, T., Vanderborght, J. and Vereecken, H. (2018). Inverse estimation of soil hydraulic and transport parameters of layered soils from water stable isotope and Lysimeter data. *Vadose Zone Journal* 17(1), Article 170168. https://doi.org/10.2136/vzj2017.09.0168.

Grosz, B., Dechow, R., Gebbert, S., Hoffmann, H., Zhao, G., Constantin, J., Raynal, H., Wallach, D., Coucheney, E., Lewan, E., Eckersten, H., Specka, X., Kersebaum, K. C., Nendel, C., Kuhnert, M., Yeluripati, J., Haas, E., Teixeira, E., Bindi, M., Trombi, G., Moriondo, M., Doro, L., Roggero, P. P., Zhao, Z. G., Wang, E. L., Tao, F. L., Rotter, R., Kassie, B., Cammarano, D., Asseng, S., Weihermuller, L., Siebert, S., Gaiser, T. and Ewert, F. (2017). The implication of input data aggregation on up-scaling soil organic carbon changes. *Environmental Modelling and Software* 96, 361–377. https://doi.org/10.1016/j.envsoft.2017.06.046.

Guilpart, N., Iizumi, T. and Makowski, D. (2022). Data-driven projections suggest large opportunities to improve Europe's soybean self-sufficiency under climate change. *Nature Food* 3(4), 255–265. https://doi.org/10.1038/s43016-022-00481-3.

Guo, W. X. and Gu, H. B. (2022). Advances in remote/aerial sensing of crop water status. In: Lobsey, C. and Biswas, A. (Eds.), *Advances in Sensor Technology for Sustainable Crop Production*. Burleigh Dodds Science Publishing, Cambridge, UK.

Hampf, A. C., Nendel, C., Strey, S. and Strey, R. (2021). Biotic yield losses in the Southern Amazon, Brazil: Making use of smartphone-assisted plant disease diagnosis data. *Frontiers in Plant Science* 12, Article 621168. https://doi.org/10.3389/fpls.2021.621168.

Harris, G. (2002). Integrated assessment and modelling: An essential way of doing science. *Environmental Modelling and Software* 8152(1), 201–207. https://doi.org/10.1016/s1364-8152(01)00058-5.

Hedley, C. B., Yule, I. J., Zuohy, M. P. and Kusomo, B. H. (2010). *Proximal Sensing Methods for Mapping Soil Water Status in an Irrigated Maize Field*. Springer. https://doi.org/10.1007/978-90-481-8859-8_32.

Heinicke, S., Frieler, K. A., Jagermeyr, J. and Mengel, M. (2022). Global gridded crop models underestimate yield responses to droughts and heatwaves. *Environmental Research Letters* 17(4), Article 044026. https://doi.org/10.1088/1748-9326/ac592e.

Hengl, T., Mendes de Jesus, J., Heuvelink, G. B., Ruiperez Gonzalez, M., Kilibarda, M., Blagotić, A., Shangguan, W., Wright, M. N., Geng, X., Bauer-Marschallinger, B., Guevara, M. A., Vargas, R., MacMillan, R. A., Batjes, N. H., Leenaars, J. G., Ribeiro, E., Wheeler, I., Mantel, S. and Kempen, B. (2017). SoilGrids250m: Global gridded soil information based on machine learning. *PLoS ONE* 12(2), Article e0169748. https://doi.org/10.1371/journal.pone.0169748.

Hernandez-Suarez, J. S., Nejadhashemi, A. P. and Deb, K. (2021). A novel multi-objective model calibration method for ecohydrological applications. *Environmental Modelling and Software* 144, Article 105161. https://doi.org/10.1016/j.envsoft.2021.105161.

Hochachka, W. M., Fink, D., Hutchinson, R. A., Sheldon, D., Wong, W. K. and Kelling, S. (2012). Data-intensive science applied to broad-scale citizen science. *Trends in Ecology and Evolution* 27(2), 130–137. https://doi.org/10.1016/j.tree.2011.11.006.

Hoefsloot, P., Ines, A. V. M., van Dam, J., Duveiller, G., Kayitakire, F. and Hansen, J. W. (2012). *Combining Crop Models and Remote Sensing for Yield Prediction: Concepts, Applications and Challenges for Heterogeneous Smallholder Environments*.

Hoffmann, H., Zhao, G., Asseng, S., Bindi, M., Biernath, C., Constantin, J., Coucheney, E., Dechow, R., Doro, L., Eckersten, H., Gaiser, T., Grosz, B., Heinlein, F., Kassie, B. T., Kersebaum, K. C., Klein, C., Kuhnert, M., Lewan, E., Moriondo, M., Nendel, C., Priesack, E., Raynal, H., Roggero, P. P., Rotter, R. P., Siebert, S., Specka, X., Tao, F. L., Teixeira, E., Trombi, G., Wallach, D., Weihermuller, L., Yeluripati, J. and Ewert, F. (2016). Impact of spatial soil and climate input data aggregation on regional yield simulations. *PLoS ONE* 11(4), Article e0151782. https://doi.org/10.1371/journal.pone.0151782.

Hoffmann, H., Zhao, G., van Bussel, L. G. J., Enders, A., Specka, X., Sosa, C., Yeluripati, J., Tao, F., Constantin, J., Raynal, H., Teixeira, E., Grosz, B., Doro, L., Zhao, Z., Wang, E., Nendel, C., Kersebaum, K. C., Haas, E., Kiese, R., Klatt, S., Eckersten, H., Vanuytrecht, E., Kuhnert, M., Lewan, E., Rotter, R., Roggero, P. P., Wallach, D., Cammarano, D., Asseng, S., Krauss, G., Siebert, S., Gaiser, T. and Ewert, F. (2015). Variability of effects of spatial climate data aggregation on regional yield simulation by crop models. *Climate Research* 65, 53–69. https://doi.org/10.3354/cr01326.

Holland, J. E., Cammarano, D., Fitzgerald, G. J., Perry, E. M., Poile, G. and Conyers, M. K. (2019). Proximal fluorescence sensing of potassium responsive crops to develop improved predictions of biomass, yield and grain quality of wheat and barley. *Precision Agriculture* 20(2), 379–397. https://doi.org/10.1007/s11119-018-09629-3.

Holzworth, D. P., Snow, V., Janssen, S., Athanasiadis, I. N., Donatelli, M., Hoogenboom, G., White, J. W. and Thorburn, P. (2015). Agricultural production systems modelling and software: Current status and future prospects. *Environmental Modelling and Software* 72, 276–286. https://doi.org/10.1016/j.envsoft.2014.12.013.

Hoogenboom, G., Jones, J. W., Traore, P. C. S. and Boote, K. J. (2012). Experiments and data for model evaluation and application. In: Kihara, J., Fatondji, D., Jones, J. W., Hoogenboom, G., Tabo, R. and Bationo, A. (Eds.), *Improving Soil Fertility Recommendations in Africa Using the Decision Support System for Agrotechnology Transfer (DSSAT)*. Springer. https://doi.org/10.1007/978-94-007-2960-5_2.

Hosoi, F., Nakabayashi, K. and Omasa, K. (2011). 3-D modeling of tomato canopies using a high-resolution portable scanning Lidar for extracting structural information. *Sensors* 11(2), 2166–2174.

Houska, T., Kraft, P., Liebermann, R., Klatt, S., Kraus, D., Haas, E., Santabarbara, I., Kiese, R., Butterbach-Bahl, K., Müller, C. and Breuer, L. (2017). Rejecting hydro-biogeochemical model structures by multi-criteria evaluation. *Environmental Modelling and Software* 93, 1–12.

Hunt, B. R., Kostelich, E. J. and Szunyogh, I. (2007). Efficient data assimilation for spatiotemporal chaos: A local ensemble transform Kalman filter. *Physica D: Nonlinear Phenomena* 230(1–2), 112–126. https://doi.org/10.1016/j.physd.2006.11.008.

Hutton, E. W. H., Piper, M. D. and Tucker, G. E. (2020). The basic model interface 2.0: A standard interface for coupling numerical models in the geosciences. *Journal of Open Source Software* 5(51), 2317.

Ibrahim, E. S., Nendel, C., Kamali, B., Gajere, E. N. and Hostert, P. (2023). Predicting potato diseases for smallholder agriculture in Nigeria using remote sensing and climate data. *PhytoFrontiers*.

Ines, A. V. M., Das, N. N., Hansen, J. W. and Njoku, E. G. (2013). Assimilation of remotely sensed soil moisture and vegetation with a crop simulation model for maize yield prediction. *Remote Sensing of Environment* 138, 149–164. https://doi.org/10.1016/j.rse.2013.07.018.

Jagermeyr, J., Muller, C., Ruane, A. C., Elliott, J., Balkovic, J., Castillo, O., Faye, B., Foster, I., Folberth, C., Franke, J. A., Fuchs, K., Guarin, J. R., Heinke, J., Hoogenboom, G., Iizumi, T., Jain, A. K., Kelly, D., Khabarov, N., Lange, S., Lin, T. S., Liu, W. F., Mialyk, O., Minoli, S., Moyer, E. J., Okada, M., Phillips, M., Porter, C., Rabin, S. S., Scheer, C., Schneider, J. M., Schyns, J. F., Skalsky, R., Smerald, A., Stella, T., Stephens, H., Webber, H., Zabel, F. and Rosenzweig, C. (2021). Climate impacts on global agriculture emerge earlier in new generation of climate and crop models. *Nature Food* 2(11), 875–. https://doi.org/10.1038/s43016-021-00400-y.

Janssen, S., Athanasiadis, I. N., Bezlepkina, I., Knapen, R., Li, H. T., Dominguez, I. P., Rizzoli, A. E. and van Ittersum, M. K. (2011). Linking models for assessing agricultural land use change. *Computers and Electronics in Agriculture* 76(2), 148–160. https://doi.org/10.1016/j.compag.2010.10.011.

Janssen, S. J. C., Porter, C. H., Moore, A. D., Athanasiadis, I. N., Foster, I., Jones, J. W. and Antle, J. M. (2017). Towards a new generation of agricultural system data, models and knowledge products: Information and communication technology. *Agricultural Systems* 155, 200–212. https://doi.org/10.1016/j.agsy.2016.09.017.

Jarvis, N. J., Jansson, P. E., Dik, P. E. and Messing, I. (1991). Modeling water and solute transport in macroporous soil. 1. Model description and sensitivity analysis. *Journal of Soil Science* 42(1), 59–70. https://doi.org/10.1111/j.1365-2389.1991.tb00091.x.

Jones, J. W., Antle, J. M., Basso, B., Boote, K. J., Conant, R. T., Foster, I., Godfray, H. C. J., Herrero, M., Howitt, R. E., Janssen, S., Keating, B. A., Munoz-Carpena, R., Porter, C. H., Rosenzweig, C. and Wheeler, T. R. (2017). Brief history of agricultural systems modeling. *Agricultural Systems* 155, 240–254. https://doi.org/10.1016/j.agsy.2016.05.014.

Jones, J. W., Keating, B. A. and Porter, C. H. (2001). Approaches to modular model development. *Agricultural Systems* 70(2–3), 421–443. https://doi.org/10.1016/S0308-521X(01)00054-3.

Jonquet, C., Dzalé-Yeumo, E., Arnaud, E. and Larmande, P. (2015). *AgroPortal: A proposition for ontology-based services in the agronomic domain. IN-OVIVE'15: atelier INtégration de sources/masses de données hétérogènes et ontologies.* Montpellier: Dans: le Domaine des Sciences du VIVant et de l'Environnement.

Kalman, R. E. (1960). A new approach to linear filtering and prediction problems. *Journal of Basic Engineering* 82(1), 35–45. https://doi.org/10.1115/1.3662552.

Kamali, B., Mousavi, S. J. and Abbaspour, K. C. (2013). Automatic calibration of HEC-HMS using single-objective and multi-objective PSO algorithms. *Hydrological Processes* 27(26), 4028–4042. https://doi.org/10.1002/hyp.9510.

Kamali, B., Stella, T., Berg-Mohnicke, M., Pickert, J., Groh, J. and Nendel, C. (2022). Improving simulations of permanent grasslands across Germany by multi-objective uncertainty-based calibration of plant-water dynamics. *European Journal of Agronomy* 134, 126464. https://doi.org/10.1016/j.eja.2022.126464.

Kamp, J., Oppel, S., Heldbjerg, H., Nyegaard, T. and Donald, P. F. (2016). Unstructured citizen science data fail to detect long-term population declines of common birds in Denmark. *Diversity and Distributions* 22(10), 1024–1035. https://doi.org/10.1111/ddi.12463.

Kashyap, B. and Kumar, R. (2021). Sensing methodologies in agriculture for soil moisture and nutrient monitoring. *IEEE Access* 9, 14095–14121. https://doi.org/10.1109/ACCESS.2021.3052478.

Kennedy, J. and Eberhart, R. (1995). Particle swarm optimization. In: *Proceedings of the ICNN'95 - International Conference on Neural Networks* (Vol. 4, pp. 1942–1948).

Kollas, C., Kersebaum, K. C., Nendel, C., Manevski, K., Muller, C., Palosuo, T., Armas-Herrera, C. M., Beaudoin, N., Bindi, M., Charfeddine, M., Conradt, T., Constantin, J., Eitzinger, J., Ewert, F., Ferrise, R., Gaiser, T., de Cortazar-Atauri, I. G., Giglio, L., Hlavinka, P., Hoffmann, H., Hoffmann, M. P., Launay, M., Manderscheid, R., Mary, B., Mirschel, W., Moriondo, M. and Olesen, J. E. (2015). Crop rotation modelling: A European model intercomparison. *European Journal of Agronomy* 70, 98–111. https://doi.org/10.1016/j.eja.2015.06.007.

Li, T., Hasegawa, T., Yin, X. Y., Zhu, Y., Boote, K., Adam, M., Bregaglio, S., Buis, S., Confalonieri, R., Fumoto, T., Gaydon, D., Marcaida, M., Nakagawa, H., Oriol, P., Ruane, A. C., Ruget, F., Singh, B., Singh, U., Tang, L., Tao, F. L., Wilkens, P., Yoshida, H., Zhang, Z. and Bouman, B. (2015). Uncertainties in predicting rice yield by current crop models under a wide range of climatic conditions. *Global Change Biology* 21(3), 1328–1341. https://doi.org/10.1111/gcb.12758.

Ling, X. L., Fu, C. B., Yang, Z. L. and Guo, W. D. (2019). Comparison of different sequential assimilation algorithms for satellite-derived leaf area index using the data assimilation testbed. *Geological Science Development* 12, 3119–3133.

Lischeid, G., Webber, H., Sommer, M., Nendel, C. and Ewert, F. (2022). Machine learning in crop yield modelling: A powerful tool, but no surrogate for science. *Agricultural and Forest Meteorology* 312, Article 108698. https://doi.org/10.1016/j.agrformet.2021.108698.

Lobsey, C. R., Viscara Rossel, R. A. and McBratney, A. B. (2010). Proximal soil nutrient sensing using electrochemical sensors. In: Viscarra Rossel, R. A., McBratney, A. B. and Minasny, B. (Eds.), *Proximal Soil Sensing: Progress in Soil Science* (pp. 77–88). Springer. https://doi.org/10.1007/978-90-481-8859-8_6.

Lokers, R., Knapen, R., Janssen, S., van Randen, Y. and Jansen, J. (2016). Analysis of Big Data technologies for use in agro-environmental science. *Environmental Modelling and Software* 84, 494–504. https://doi.org/10.1016/j.envsoft.2016.07.017.

Lugato, E., Panagos, P., Bampa, F., Jones, A. and Montanarella, L. (2014). A new baseline of organic carbon stock in European agricultural soils using a modelling approach. *Global Change Biology* 20(1), 313–326. https://doi.org/10.1111/gcb.12292.

Maiorano, A., Martre, P., Asseng, S., Ewert, F., Muller, C., Rotter, R. P., Ruane, A. C., Semenov, M. A., Wallach, D., Wang, E. L., Alderman, P. D., Kassie, B. T., Biernath, C., Basso, B., Cammaran, D., Challinor, A. J., Doltra, J., Dumont, B., Rezaei, E. E., Gayler, S., Kersebaum, K. C., Kimball, B. A., Koehler, A. K., Liu, B., O'Leary, G. J., Olesen, J. E., Ottman, M. J., Priesack, E., Reynolds, M., Stratonovitch, P., Streck, T., Thorburn, P. J., Waha, K., Wall, G. W., White, J. W., Zhao, Z. G. and Zhu, Y. (2017). Crop model improvement reduces the uncertainty of the response to temperature of multi-model ensembles. *Field Crops Research* 202, 5–20. https://doi.org/10.1016/j.fcr.2016.05.001.

Martre, P., Wallach, D., Asseng, S., Ewert, F., Jones, J. W., Rotter, R. P., Boote, K. J., Ruane, A. C., Thorburn, P. J., Cammarano, D., Hatfield, J. L., Rosenzweig, C., Aggarwal, P. K., Angulo, C., Basso, B., Bertuzzi, P., Biernath, C., Brisson, N., Challinor, A. J., Doltra, J., Gayler, S., Goldberg, R., Grant, R. F., Heng, L., Hooker, J., Hunt, L. A., Ingwersen, J., Izaurralde, R. C., Kersebaum, K. C., Muller, C., Kumar, S. N., Nendel, C., O'Leary, G., Olesen, J. E., Osborne, T. M., Palosuo, T., Priesack, E., Ripoche, D., Semenov, M. A., Shcherbak, I., Steduto, P., Stockle, C. O., Stratonovitch, P., Streck, T., Supit, I., Tao, F. L., Travasso, M., Waha, K., White, J. W. and Wolf, J. (2015). Multimodel ensembles of wheat growth: Many models are better than one. *Global Change Biology* 21(2), 911–925. https://doi.org/10.1111/gcb.12768.

Maxwell, K. and Johnson, G. N. (2000). Chlorophyll fluorescence - A practical guide. *Journal of Experimental Botany* 51(345), 659–668. https://doi.org/10.1093/jxb/51.345.659.

Meroni, M., d'Andrimont, R., Vrieling, A., Fasbender, D., Lemoine, G., Rembold, F., Seguini, L. and Verhegghen, A. (2021). Comparing land surface phenology of major European crops as derived from SAR and multispectral data of Sentinel-1 and-2. *Remote Sensing of Environment* 253, Article 112232. https://doi.org/10.1016/j.rse.2020.112232.

Midingoyi, C. A., Pradal, C., Athanasiadis, I. N., Donatelli, M., Enders, A., Fumagalli, D., Garcia, F., Holzworth, D., Hoogenboom, G., Porter, C., Raynal, H., Thorburn, P. J. and

Martre, P. (2020). Reuse of process-based models: Automatic transformation into many programming languages and simulation platforms. *In Silico Plants* 2(1).

Midingoyi, C. A., Pradal, C., Enders, A., Fumagalli, D., Raynal, H., Donatelli, M., Athanasiadis, I. N., Porter, C., Hoogenboom, G., Holzworth, D., Garcia, F., Thorburn, P. and Martre, P. (2021). Crop2ML: An open-source multi-language modeling framework for the exchange and reuse of crop model components. *Environmental Modelling and Software* 142, Article 105055. https://doi.org/10.1016/j.envsoft.2021.105055.

Miller, M. S., Yee, K. P. and Shapiro, J. (2003). *Capability Myths Demolished*.

Minet, J., Verhoest, N. E. C., Lambot, S. and Vanclooster, M. (2013). Temporal stability of soil moisture patterns measured by proximal ground-penetrating radar. *Hydrology and Earth System Sciences Discussions* 10(4), 4063–4097.

Morari, F., Zanella, V., Gobbo, S., Bindi, M., Sartori, L., Pasqui, M., Mosca, G. and Ferrise, R. (2021). Coupling proximal sensing, seasonal forecasts and crop modelling to optimize nitrogen variable rate application in durum wheat. *Precision Agriculture* 22(1), 75–98. https://doi.org/10.1007/s11119-020-09730-6.

Mostafaie, A., Forootan, E., Safari, A. and Schumacher, M. (2018). Comparing multi-objective optimization techniques to calibrate a conceptual hydrological model using in situ runoff and daily GRACE data. *Computational Geosciences* 22(3), 789-814. https://doi.org/10.1007/s10596-018-9726-8.

Müller, C., Elliott, J., Chryssanthacopoulos, J., Arneth, A., Balkovic, J., Ciais, P., Deryng, D., Folberth, C., Glotter, M., Hoek, S., Iizumi, T., Izaurralde, R. C., Jones, C., Khabarov, N., Lawrence, P., Liu, W. F., Olin, S., Pugh, T. A. M., Ray, D. K., Reddy, A., Rosenzweig, C., Ruane, A. C., Sakurai, G., Schmid, E., Skalsky, R., Song, C. X., Wang, X. H., de Wit, A. and Yang, H. (2017). Global gridded crop model evaluation: Benchmarking, skills, deficiencies and implications. *Geoscientific Model Development* 10(4), 1403–1422. https://doi.org/10.5194/gmd-10-1403-2017.

Muller, B. and Martre, P. (2019). Plant and crop simulation models: Powerful tools to link physiology, genetics, and phenomics. *Journal of Experimental Botany* 70(9), 2339–2344. https://doi.org/10.1093/jxb/erz175.

Munnaf, M. A., Haesaert, G. and Mouazen, A. M. (2021). Map-based site-specific seeding of seed potato production by fusion of proximal and remote sensing data. *Soil and Tillage Research* 206, Article 104801. https://doi.org/10.1016/j.still.2020.104801.

Nendel, C., Wieland, R., Mirschel, W., Specka, X., Guddat, C. and Kersebaum, K. C. (2013). Simulating regional winter wheat yields using input data of different spatial resolution. *Field Crops Research* 145, 67–77. https://doi.org/10.1016/j.fcr.2013.02.014.

Neveu, P., David, R. and Jonquet, C. (2020). Improving data identification and tagging for more effective decision making in agriculture. In: Armstrong, L. (Ed.), *Improving Data Management and Decision Support Systems in Agriculture*. Burleigh Dodds Science Publishing, Cambridge, UK.

Novelli, F., Spiegel, H., Sandén, T. and Vuolo, F. (2019). Assimilation of Sentinel-2 leaf area index data into a physically-based crop growth model for yield estimation. *Agronomy* 9(5), Article 255. https://doi.org/10.3390/agronomy9050255.

O'Neill, D. W., Fanning, A. L., Lamb, W. F. and Steinberger, J. K. (2018). A good life for all within planetary boundaries. *Nature Sustainability* 1(2), 88–95. https://doi.org/10.1038/s41893-018-0021-4.

Published by Burleigh Dodds Science Publishing Limited, 2023.

Oerke, E.-C., Mahlein, A.-K. and Steiner, U. (2014). Proximal sensing of plant diseases. In: Gullino, M. L. and Bonants, P. J. M. (Eds.), *Detection and Diagnostics of Plant Pathogens* (pp. 55–68). Springer Netherlands.

Padilla, F. M., Gallardo, M., Pena-Fleitas, M. T., de Souza, R. and Thompson, R. B. (2018). Proximal optical sensors for nitrogen management of vegetable crops: A review. *Sensors* 18(7), Article 2083. https://doi.org/10.3390/s18072083.

Parker, P., Letcher, R., Jakeman, A., Beck, M. B., Harris, G., Argent, R. M., Hare, M., Pahl-Wostl, C., Voinov, A., Janssen, M., Sullivan, P., Scoccimarro, M., Friend, A., Sonnenshein, M., Baker, D., Matejicek, L., Odulaja, D., Deadman, P., Lim, K., Larocque, G., Tarikhi, P., Fletcher, C., Put, A., Maxwell, T., Charles, A., Breeze, H., Nakatani, N., Mudgal, S., Naito, W., Osidele, O., Eriksson, I., Kautsky, U., Kautsky, E., Naeslund, B., Kumblad, L., Park, R., Maltagliati, S., Girardin, P., Rizzoli, A., Mauriello, D., Hoch, R., Pelletier, D., Reilly, J., Olafsdottir, R. and Bin, S. (2002). Progress in integrated assessment and modelling. *Environmental Modelling and Software* 8152(1), 209–217. https://doi.org/10.1016/s1364-8152(01)00059-7.

Peckham, S. D., Hutton, E. W. H. and Norris, B. (2013). A component-based approach to integrated modeling in the geosciences: The design of CSDMS. *Computers and Geosciences* 53, 3–12. https://doi.org/10.1016/j.cageo.2012.04.002.

Penny, S. G. and Miyoshi, T. (2016). A local particle filter for high-dimensional geophysical systems. *Nonlinear Processes in Geophysics* 23(6), 391–405. https://doi.org/10.5194/npg-23-391-2016.

Porter, C. H., Villalobos, C., Holzworth, D., Nelson, R., White, J. W., Athanasiadis, I. N., Janssen, S., Ripoche, D., Cufi, J., Raes, D., Zhang, M., Knapen, R., Sahajpal, R., Boote, K. and Jones, J. W. (2014). Harmonization and translation of crop modeling data to ensure interoperability. *Environmental Modelling and Software* 62, 495–508. https://doi.org/10.1016/j.envsoft.2014.09.004.

Ren, J., Shao, Y., Wan, H., Xie, Y. H. and Campos, A. (2021). A two-step mapping of irrigated corn with multi-temporal MODIS and Landsat analysis ready data. *ISPRS Journal of Photogrammetry and Remote Sensing* 176, 69–82. https://doi.org/10.1016/j.isprsjprs.2021.04.007.

Rezaei, E. E., Ghazaryan, G., Gonzalez, J., Cornish, N., Dubovyk, O. and Siebert, S. (2021). The use of remote sensing to derive maize sowing dates for large-scale crop yield simulations. *International Journal of Biometeorology* 65(4), 565–576. https://doi.org/10.1007/s00484-020-02050-4.

Rosenzweig, C., Jones, J. W., Hatfield, J. L., Ruane, A. C., Boote, K. J., Thorburn, P., Antle, J. M., Nelson, G. C., Porter, C., Janssen, S., Asseng, S., Basso, B., Ewert, F., Wallach, D., Baigorria, G. and Winter, J. M. (2013). The Agricultural Model Intercomparison and Improvement Project (AgMIP): Protocols and pilot studies. *Agricultural and Forest Meteorology* 170, 166–182. https://doi.org/10.1016/j.agrformet.2012.09.011.

Rötter, R. P., Ewert, F., Palosuo, T., Bindi, M., Kersebaum, K. C., Olesen, J. E., Trnka, M., van Ittersum, M. K., Janssen, S., Rivington, M., Semenov, M., Wallach, D., Porter, J. R., Stewart, D., Verhagen, J., Angulo, C., Gaiser, T., Nendel, C., Martre, P. and de Wit, A. (2013, 27–30 May). Challenges for agro-ecosystem modelling in climate change risk assessment for major European crops and farming systems. In: *Impacts World Conference Potsdam*, Germany.

Saeys, W., Lenaerts, B., Craessaerts, G. and De Baerdemaeker, J. (2009). Estimation of the crop density of small grains using LiDAR sensors. *Biosystems Engineering* 102(1), 22–30.

Schirrmann, M. (2023). Advances in remote sensing technologies for assessing crop health. In: Lobsey, C. and Biswas, A. (Eds.), *Advances in Sensor Technology for Sustainable Crop Production* (pp. 43–64). Burleigh Dodds Science Publishing, Cambridge, UK.

Schmidt, J. (1991). A mathematical model to simulate rainfall erosion. In: Bork, H.-R., De Ploey, J. and Schick, A. P. (Eds.), *Erosion, Transport and Deposition Processes–Theories and Models CATENA* 19.

Schwieder, M., Wesemeyer, M., Frantz, D., Pfoch, K., Erasmi, S., Pickert, J., Nendel, C. and Hostert, P. (2022). Mapping grassland mowing events across Germany based on combined Sentinel-2 and Landsat 8 time series. *Remote Sensing of Environment* 269, Article 112795. https://doi.org/10.1016/j.rse.2021.112795.

Shahhosseini, M., Hu, G. P., Huber, I. and Archontoulis, S. V. (2021). Coupling machine learning and crop modeling improves crop yield prediction in the US Corn Belt. *Scientific Reports* 11(1), Article 1606. https://doi.org/10.1038/s41598-020-80820-1.

Silvestro, P. C., Casa, R., Hanus, J., Koetz, B., Rascher, U., Schuettemeyer, D., Siegmann, B., Skokovic, D., Sobrino, J. and Tudoroiu, M. (2021). Synergistic use of multispectral data and crop growth modelling for spatial and temporal evapotranspiration estimations. *Remote Sensing* 13(11), Article 2138. https://doi.org/10.3390/rs13112138.

Skovdal Christiansen, J., Thorsen, M., Clausen, T., Hansen, S. and Christian Refsgaard, J. (2004). Modelling of macropore flow and transport processes at catchment scale. *Journal of Hydrology* 299(1–2), 136–158. https://doi.org/10.1016/j.jhydrol.2004.04.029.

Steffen, W., Richardson, K., Rockstrom, J., Cornell, S. E., Fetzer, I., Bennett, E. M., Biggs, R., Carpenter, S. R., de Vries, W., de Wit, C. A., Folke, C., Gerten, D., Heinke, J., Mace, G. M., Persson, L. M., Ramanathan, V., Reyers, B. and Sorlin, S. (2015). Sustainability. Planetary boundaries: Guiding human development on a changing planet. *Science* 347(6223), Article 1259855. https://doi.org/10.1126/science.1259855.

Tabbagh, A., Seger, M. and Cousin, I. (2022). Using proximal electromagnetic/electrical resistivity/electrical sensors to assess soil health. In: Lobsey, C. and Biswas, A. (Eds.), *Advances in Sensor Technology for Sustainable Crop Production*. Burleigh Dodds Science Publishing, Cambridge, UK.

Tang, Y. T., Marshall, L., Sharma, A. and Ajami, H. (2018). A Bayesian alternative for multi-objective ecohydrological model specification. *Journal of Hydrology* 556, 25–38. https://doi.org/10.1016/j.jhydrol.2017.07.040.

Tucker, G. E., Hutton, E. W. H., Piper, M. D., Campforts, B., Gan, T., Barnhart, K. R., Kettner, A. J., Overeem, I., Peckham, S. D., McCready, L. and Syvitski, J. (2022). CSDMS: A community platform for numerical modeling of Earth surface processes. *Geoscientific Model Development* 15(4), 1413–1439. https://doi.org/10.5194/gmd-15-1413-2022.

Van Evert, F. K. (2019). Data for developing, testing and applying crop and farm models. In: *Boote, Advances in Crop Modelling for a Sustainable Agriculture* (pp. 385–418). Burleigh Dodds Science Publishing, Cambridge, UK.

van Ittersum, M. K., Ewert, F., Heckelei, T., Wery, J., Olsson, J. A., Andersen, E., Bezlepkina, I., Brouwer, F., Donatelli, M., Flichman, G., Olsson, L., Rizzoli, A. E., van der Wal, T., Wien, J. E. and Wolf, J. (2008). Integrated assessment of agricultural systems - A component-based framework for the European Union (SEAMLESS). *Agricultural Systems* 96(1–3), 150–165. https://doi.org/10.1016/j.agsy.2007.07.009.

van Klompenburg, T., Kassahun, A. and Catal, C. (2020). Crop yield prediction using machine learning: A systematic literature review. *Computers and Electronics in Agriculture* 177, Article 105709. https://doi.org/10.1016/j.compag.2020.105709.

van Leeuwen, P. J., Kunsch, H. R., Nerger, L., Potthast, R. and Reich, S. (2019). Particle filters for high-dimensional geoscience applications: A review. *Quarterly Journal of the Royal Meteorological Society. Royal Meteorological Society* 145(723), 2335–2365. https://doi.org/10.1002/qj.3551.

Wallach, D., Makowski, D., Jones, J. W. and Brun, F. (2018). Data assimilation for dynamic models. In: Wallach, D., Makowski, D., Jones, J. W. and Brun, F. (Eds.), *Working With Dynamic Crop Models* (3rd edn., pp. 487–518). Academic Press.

Wallor, E., Kersebaum, K. C., Ventrella, D., Bindi, M., Cammarano, D., Coucheney, E., Gaiser, T., Garofalo, P., Giglio, L., Giola, P., Hoffmann, M. P., Iocola, I., Lana, M., Lewan, E., Maharjan, G. R., Moriondo, M., Mula, L., Nendel, C., Pohankova, E., Roggero, P. P., Trnka, M. and Trombi, G. (2018). The response of process-based agro-ecosystem models to within-field variability in site conditions. *Field Crops Research* 228, 1–19. https://doi.org/10.1016/j.fcr.2018.08.021.

Wang, E., Martre, P., Zhao, Z., Ewert, F., Maiorano, A., Rötter, R. P., Kimball, B. A., Ottman, M. J., Wall, G. W., White, J. W., Reynolds, M. P., Alderman, P. D., Aggarwal, P. K., Anothai, J., Basso, B., Biernath, C., Cammarano, D., Challinor, A. J., De Sanctis, G., Doltra, J., Dumont, B., Fereres, E., Garcia-Vila, M., Gayler, S., Hoogenboom, G., Hunt, L. A., Izaurralde, R. C., Jabloun, M., Jones, C. D., Kersebaum, K. C., Koehler, A., Liu, L., Müller, C., Naresh Kumar, S., Nendel, C., O'Leary, G., Olesen, J. E., Palosuo, T., Priesack, E., Eyshi Rezaei, E., Ripoche, D., Ruane, A. C., Semenov, M. A., Shcherbak, I., Stöckle, C., Stratonovitch, P., Streck, T., Supit, I., Tao, F., Thorburn, P., Waha, K., Wallach, D., Wang, Z., Wolf, J., Zhu, Y. and Asseng, S. (2017). The uncertainty of crop yield projections is reduced by improved temperature response functions. *Nature Plants* 3(8), Article 17102. https://doi.org/10.1038/nplants.2017.102.

Wang, E., Robertson, M. J., Hammer, G. L., Carberry, P. S., Holzworth, D., Meinke, H., Chapman, S. C., Hargreaves, J. N. G., Huth, N. I. and McLean, G. (2002). Development of a generic crop model template in the cropping system model APSIM. *European Journal of Agronomy* 18(1–2), 121–140. https://doi.org/10.1016/S1161-0301(02)00100-4.

Warszawski, L., Frieler, K., Huber, V., Piontek, F., Serdeczny, O. and Schewe, J. (2014). The inter-sectoral impact model intercomparison project (ISI-MIP): Project framework. *Proceedings of the National Academy of Sciences of the United States of America* 111(9), 3228–3232. https://doi.org/10.1073/pnas.1312330110.

Webber, H., Lischeid, G., Sommer, M., Finger, R., Nendel, C., Gaiser, T. and Ewert, F. (2020). No perfect storm for crop yield failure in Germany. *Environmental Research Letters* 15(10), Article 104012. https://doi.org/10.1088/1748-9326/aba2a4.

White, J. W., Hunt, L. A., Boote, K. J., Jones, J. W., Koo, J., Kim, S., Porter, C. H., Wilkens, P. W. and Hoogenboom, G. (2013). *Integrated Description of Agricultural Field Experiments and Production: A ICASA Version 2.0 Data Standards. Computers and Electronics in Agriculture* (Vol. 96, pp. 1–12). https://doi.org/10.1016/j.compag.2013.04.003.

Wischmeier, W. C. and Smith, D. D. (1965). *Predicting Rainfall-Erosion Losses From Cropland East of the Rocky Mountains* (Vol. 282). United States Department of Agriculture.

Wöhling, T., Gayler, S., Priesack, E., Ingwersen, J., Wizemann, H. D., Hogy, P., Cuntz, M., Attinger, S., Wulfmeyer, V. and Streck, T. (2013). Multiresponse, multiobjective calibration as a diagnostic tool to compare accuracy and structural limitations of five coupled soil-plant models and CLM3.5. *Water Resources Research* 49(12), 8200-8221. https://doi.org/10.1002/2013WR014536.

Wöhling, T., Schöniger, A., Gayler, S. and Nowak, W. (2015). Bayesian model averaging to explore the worth of data for soil-plant model selection and prediction. *Water Resources Research* 51(4), 2825–2846. https://doi.org/10.1002/2014WR016292.

Wolters, S., Soderstrom, M., Piikki, K., Reese, H. and Stenberg, M. (2021). Upscaling proximal sensor N-uptake predictions in winter wheat (*Triticum aestivum* L.) with Sentinel-2 satellite data for use in a decision support system. *Precision Agriculture* 22(4), 1263–1283. https://doi.org/10.1007/s11119-020-09783-7.

Yapo, P. O., Gupta, H. V. and Sorooshian, S. (1998). Multi-objective global optimization for hydrologic models. *Journal of Hydrology* 204(1–4), 83–97. https://doi.org/10.1016/S0022-1694(97)00107-8.

Younes, N., Joyce, K. E. and Maier, S. W. (2021). All models of satellite-derived phenology are wrong, but some are useful: A case study from northern Australia. *International Journal of Applied Earth Observation and Geoinformation* 97, Article 102285. https://doi.org/10.1016/j.jag.2020.102285.

Zappa, L., Schlaffer, S., Bauer-Marschallinger, B., Nendel, C., Zimmerman, B. and Dorigo, W. (2021). Detection and quantification of irrigation water amounts at 500 m using Sentinel-1 surface soil moisture. *Remote Sensing* 13(9), Article 1727. https://doi.org/10.3390/rs13091727.

Zare, H., Weber, T. K. D., Ingwersen, J., Nowak, W., Gayler, S. and Streck, T. (2022). Combining crop modelling with remote sensing data using a particle filtering technique to produce real-time forecasts of winter wheat yields under uncertain boundary conditions. *Remote Sensing* 14(6), 1360.

Zhao, G., Hoffmann, H., van Bussel, L. G. J., Enders, A., Specka, X., Sosa, C., Yeluripati, J., Tao, F. L., Constantin, J., Raynal, H., Teixeira, E., Grosz, B., Doro, L., Zhao, Z. G., Nendel, C., Kiese, R., Eckersten, H., Haas, E., Vanuytrecht, E., Wang, E., Kuhnert, M., Trombi, G., Moriondo, M., Bindi, M., Lewan, E., Bach, M., Kersebaum, K. C., Rotter, R., Roggero, P. P., Wallach, D., Cammarano, D., Asseng, S., Krauss, G., Siebert, S., Gaiser, T. and Ewert, F. (2015). Effect of weather data aggregation on regional crop simulation for different crops, production conditions, and response variables. *Climate Research* 65, 141–157. https://doi.org/10.3354/cr01301.

Zhao, G., Hoffmann, H., Yeluripati, J., Xenia, S., Nendel, C., Coucheney, E., Kuhnert, M., Tao, F. L., Constantin, J., Raynal, H., Teixeira, E., Grosz, B., Doro, L., Kiese, R., Eckersten, H., Haas, E., Cammarano, D., Kassie, B., Moriondo, M., Trombi, G., Bindi, M., Biernath, C., Heinlein, F., Klein, C., Priesack, E., Lewan, E., Kersebaum, K. C., Rotter, R., Roggero, P. P., Wallach, D., Asseng, S., Siebert, S., Gaiser, T. and Ewert, F. (2016). Evaluating the precision of eight spatial sampling schemes in estimating regional means of simulated yield for two crops. *Environmental Modelling and Software* 80, 100–112. https://doi.org/10.1016/j.envsoft.2016.02.022.

Chapter 3

Incorporating genetics into crop models to identify new phenotypes adapted to climate change

Carlos D. Messina, Institute of Food and Agricultural Sciences (IFAS), University of Florida, USA and The ARC Centre of Excellence for Plant Success in Nature and Agriculture, Australia; and Mark Cooper, The Queensland Alliance for Agriculture and Food Innovation (QAAFI) and The ARC Centre of Excellence for Plant Success in Nature and Agriculture, Australia

1 Introduction
2 Systems analyses and models of crop growth
3 Linking genetics and physiology for phenotypic prediction
4 Models that predict emergent phenotypes
5 Integrating environmental and genomic predictions using crop models
6 Assessing impacts of traits and their underpinning genetics on crop performance
7 Managing genotype × environment interactions to optimize rather than hamper genetic gain in current and future climates
8 Predictive breeding enables iterative impact assessment and adaptation to climate change
9 Crop growth models can enable a rethinking of breeding objectives to adapt to and combat climate change
10 Conclusion and future trends
11 References

1 Introduction

There is consensus that climate change will have a net negative and inequitable impact on agriculture (Lobell et al., 2011; Challinor et al., 2014; Webber et al., 2018; IPCC, 2021). The increase in temperature, vapor pressure deficit (VPD), and shifting water balances worldwide will likely change the geographical patterns of farming (Rippke et al., 2016; Ficklin and Novick, 2017). Disease pressure will increase in high latitudes, challenging further crop production

http://dx.doi.org/10.19103/AS.2022.0115.04

and the design of stable agricultural systems (Chaloner et al., 2021). There is agreement that crop improvement will be key to coping with climate change effects on food security (Cairns et al., 2012; Atliln et al., 2017; Hernandez-Ochoa et al., 2019; IPCC, 2021; Kalihuoto et al., 2020; Snowdon et al., 2021; Kholová et al., 2021) and approaches have been proposed (Ceccarelli et al., 2010; Ramirez-Villegas et al., 2015, 2018, 2020; Ceccarelli and Grando, 2020).

Modeling approaches that integrate physiology, genetics, and environment can enable society to define G × M technologies and practices to make agriculture a key part of the solution to climate change (Brummer et al., 2011; Messina et al., 2022c). Surprisingly, very few efforts were dedicated to answering the question of how to breed crops for climate change while regenerating natural resources and reducing greenhouse gas emissions (Brummer et al., 2011, Messina et al., 2022c; Cooper and Messina, 2023). Crop growth models (CGM, Jones et al., 2003; Holzworth et al., 2014) that link genetics, physiology, and breeding are necessary to conduct impact assessments of climate change on agriculture while considering adaptation strategies.

Genotype (G) by environment (E) by management (M) interactions (G × E × M) are ubiquitous in agriculture. These are expressed as changes in rankings among genotypes when exposed to different environments and agronomic management practices. Therefore, selecting genotypes for an inadequately defined target population of environments (TPE) and/or management inevitably leads to lower than the attainable rates of genetic gain and, thus, adaptation to the target environment (Cooper and Hammer, 1996; Kholová et al., 2021). There is current evidence that genetic gain in wheat has been hampered by climate change (Morgounov et al., 2013; Xiong et al., 2021). This is particularly critical because, whenever the future TPE differs from the current TPE due to climate change, there is a risk that implementing breeding programs for future climates may decrease the current rate of genetic gain for current environments and thus compromise current food security.

There is evidence that wheat breeding may be falling into this trap as rankings of cultivars are changing with climate change (Morgounov et al., 2013; Xiong et al., 2021). Because the traits and trait network interactions underpinning adaptation (at least to drought) change with levels of evapotranspiration (Messina et al., 2011; Borrell et al., 2014; Gleason et al., 2022; Cooper and Messina, 2023), the environmental distance between the current and future TPE could become significant drivers of G × E × M interactions. In wheat, increased spring maximum temperatures led to both increased and decreased yields (Morgounov et al., 2013). In maize, the conductance response to vapor pressure deficit is a trait underpinning drought adaptation (Shekoofa et al., 2015; Messina et al., 2015). The use of CGM enables us to estimate how the current and future TPEs may match (Harrison et al., 2014), and which plant traits, genes, and genomic regions can explain the adaptation or lack thereof to a mixture of expected environments (Harrison et al., 2014; Messina et al., 2015; Cooper and Messina, 2023).

A progressive intensification of drought under climate change would magnify the selection pressure applied to the germplasm for elevated levels of limited conductance. The normal operation of the breeding program is expected to be sufficient to increase the frequency of preferred alleles for the required trait levels, and thus the breeding program would smoothly match the expression levels of the trait to the changing TPE. In contrast, such as in well-watered dryland systems, a sudden intensification of drought due to the occurrence of extreme conditions will be conducive to G × E × M interactions, and dedicated breeding and agronomy efforts will be required to breed new genotypes for the changing TPE (Messina et al., 2011; Messina et al., 2015; Cooper et al., 2022; Cooper and Messina, 2023). The opposite could be said about improvements in radiation use efficiency as evidenced in maize breeding (Messina et al., 2022b). It is anticipated that the requirements for the creation of adapted germplasm and cultivars will be inevitably different for different geographies and cropping systems (Kholová et al., 2021). A key question is then,

> how to harmonize breeding efforts for agricultural systems that regenerate the environment while providing nutrition security to society and improved adaptation to climate change; and what technologies are required to conduct impact assessments and design adaptation strategies?

In this chapter, we expand on the framework proposed by Cooper and Messina (2023). We first review crop growth models and how these can link physiology and genetics with the environment. Then, we use the breeder's equation framework and theoretical principles of G × E interactions to demonstrate the critical need to properly time the pace of crop improvement to the pace of climate change and the opportunity to use advanced prediction methodologies based on the use of crop growth models to accomplish this goal. Then we articulate the need to rethink breeding objectives to enable agriculture to become part of the solution to climate change (Brummer et al., 2011; Kholová et al., 2021; Messina et al., 2022c). We finally discuss the need to enable breeders with dynamic gene-to-phenotype platforms to implement the needed changes in plant breeding and agronomy so society can meet current nutritional and ecosystem regeneration demands without compromising current and/or future societal needs.

2 Systems analyses and models of crop growth

The formalization of system concepts in engineering spurred the development of dynamic models of crop growth and development and their implementation in various software packages to enable various systems analyses ranging from how the environment affects the physiology of the plant to the whole agricultural system response to climate change. The state variable approach, organized in modular code, enables modeling systems with clear system boundaries, inputs, outputs, process rates, stocks, parameters, and so on. Figure 1a illustrates a simple model with one state variable, one process affected by the state variable and regulated by the inputs (*I*) from a source, and a set of known parameters $\mathbf{x}_k$. Figure 2b shows the translation of the graph into mathematical equations. Here it is made explicit that the parameter set could be any combination of parameters that are measured, and thus we call it phenotypic (subscript *P*), or that could be estimated using genetic predictors using genomic information, and thus we call it genotypic (subscript *G*).

Messina et al. (2011) introduced formalisms to represent any combination of these sets including known, unknown, and interactions between genes. Figure 1c illustrates how these representations to model soil and crop processes could be integrated in a modular way (Jones et al., 2003; Holzworth et al., 2014) to facilitate the modeling of system components that are of interest to the modeler. Integration of modules leads to the creation of crop growth models (CGM), which represent and aim at predicting certain aspects of the real

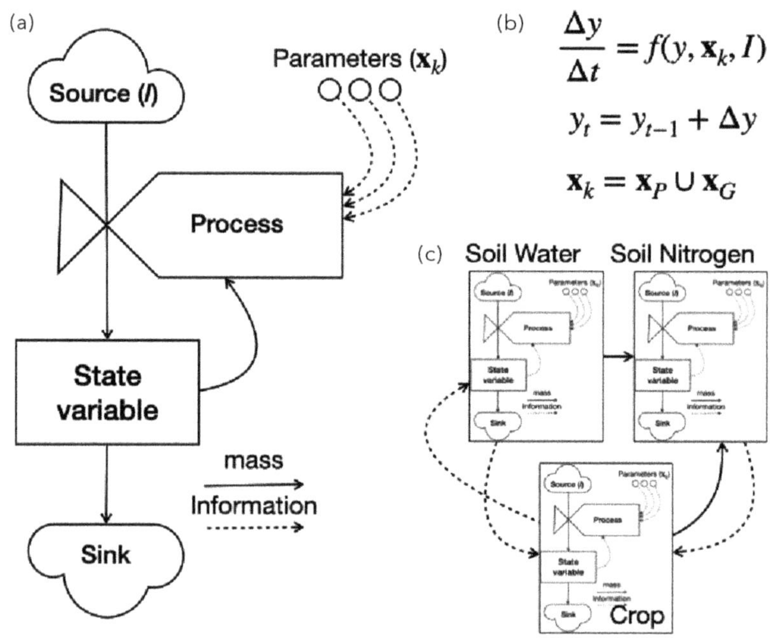

Figure 1 Systems modeling and simulation. (a) Forrester diagram for a simple system. (b) mathematical representation of the system. (c) Simple crop model schematic with flows of information and mass.

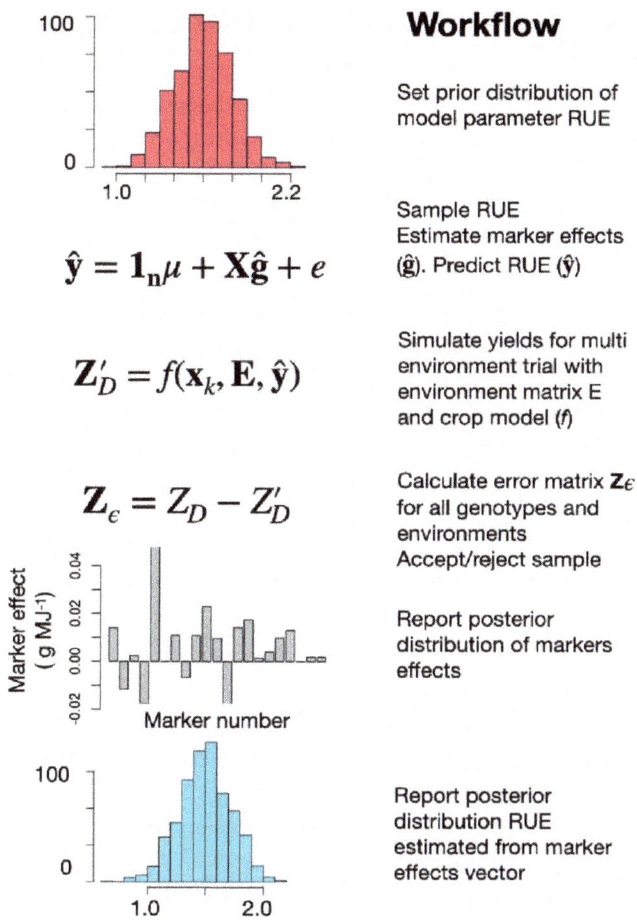

Figure 2 Pseudo algorithm to describe the workflow of a crop growth model - genomic selection approach for a single crop and a single trait (radiation use efficiency, RUE) that uses genomic best linear predictors at the trait level **y** from makers (**X**), marker effects **g**, environment inputs **E**, and set of fixed crop model parameters $\mathbf{x}_k$.

system under study. Here we illustrate flows of mass, for example, the transfer or mass residue from the crop module to the soil nitrogen module, the transfer of information in the form of nitrogen and water stress indices from the soil and nitrogen modules to the crop module, and the transfer of information about crop water demand from crop module to soil module. Many other flows (here we illustrate just a few examples) are required to create dynamic CGMs that can reproduce nonlinear relations between environment and management inputs, harness genetic information, and simulate with time steps from hourly to daily, the soil carbon, nitrogen, and water balance, and the plant carbon and nitrogen balance (Jones et al., 2003; Holzworth et al., 2014).

3 Linking genetics and physiology for phenotypic prediction

Genetic predictors of performance (e.g. yield) are often based on regression or association within different statistical frameworks (Meuwissen et al., 2001; Yu et al., 2006; Heffner et al., 2009; Zhang et al., 2010; Lorenz et al., 2011),

$$\hat{y}_i = \mu + \sum_j v_{ij}\kappa_j \tag{1}$$

where the trait phenotype y for individual i is predicted based on the population mean μ and the sum over the genotype markers $j(v_{ij})$ times the marker effects K_j. Other models extend this model to include environmental covariates (Boer et al., 2007; Heslot et al., 2014; Jarquin et al., 2014; Millet et al., 2019), consider nonlinear associations (Collantonio et al., 2022) or are integrated with dynamical models (Technow et al., 2015; Messina et al. 2018; Diepenbrock et al., 2022). In this latter case, dynamic CGMs are considered link functions (Cooper et al., 2021) between genotype and phenotype through the interaction of physiological and environmental processes to approximate the phenotypic function,

$$F(\cdot)_T = Q(\cdot)_D + H(\cdot)_U \tag{2}$$

where the function F represents perfect knowledge of the observable phenotype as determined by nature, the function Q represents the phenotypes that are predictable based on our current scientific knowledge and phenotyping systems (Fig. 1), and H is the function that represents what we don't know, it is not knowable/predictable (Day, 1976), or that we do not want to include in the model. If we define function Q as a CGM, the CGM becomes a cognitive construct that we use to represent processes, states, and the topology of relations among biological processes (Cooper et al., 2009). CGMs can span several levels of the organization; however, there is consensus that empiricisms are needed whenever we seek to link or use the information for more than two levels of the organization away from the target level of the system to be modeled (e.g. cell to crop). This is an important consideration in how to link genetics and physiology, as we want to decrease the gene-to-phenotype distance but not make a mechanistic connection between the transcription and translation of genes on each cell and the performance of a cultivar at the regional level.

CGMs have a set of parameters that are known, unknown, and estimable (u) and a set of inputs, $CGM(x_k, x_u, I)$. The quantitative genetics model of the form described in equation 1 has proved useful to estimate x_k and x_u from field observations and state variables (e.g. Hoogenboom, 2004; Messina et al., 2006; Diepenbrock et al., 2022).

4 Models that predict emergent phenotypes

For a CGM to deliver on the promise to add information by improving impact assessments, designing adaptation strategies, and predicting genotypes with higher skill beyond the current sample of environments, they must be able to predict what we define as emergent phenotypes. These are observable phenotypes that are different from what we can predict by understanding the parts in isolation or as independent components (Anderson, 1972; Roeder et al., 2022). It has been shown that simple equations can have complex behavior. For example, the logistic equations that model populations of plants can manifest complex emergent behaviors such as bifurcations and chaos (May 1976; Roeder et al., 2022) that conform well with the observation of bimodal distributions of maize barren plants (Edmeades and Daynard, 1979).

Routines recently implemented in the crop model APSIM (Holzworth et al., 2014) can simulate emergent patterns such as the relationship between anthesis-sinking interval and yield in maize or the relationship between plant growth and kernel numbers (Messina et al., 2019). While the routines in the current APSIM model do not yet have the capacity to simulate the observed bifurcation behavior of barren and non-barren plants in a crop, the capacity to predict emergent phenotypes suggests that dynamical models are tools that are evolving and are capable of predicting phenotypes different from what is expected from the analyses of the parts in isolation. For example, understanding a system to reduce photorespiration in leaves in isolation leads to an overestimate of the plausible impacts on productivity at the crop scale and for the mixture of environment types that comprise the TPE (Hammer et al., 2019). This prediction using APSIM conforms well with the limited success of large numbers of single gene transformation technologies for yield improvement (Simmons et al., 2021).

The ability of current CGMs to simulate some emergent behaviors can enable breeders to use this kind of tool to inform selections to improve germplasm to meet the needs of the adjacent environment space of the future TPE while considering the expected change in frequencies of environments in the selection and on-farm production situations (Snowdon et al., 2021; Messina et al., 2022a, c). Because the CGM integrates scientific knowledge, for many situations, these tools can be better equipped than purely empirical models to predict genotype performance in environments that were never included in the training sets and, more importantly, environments that the crops have never experienced before (Battisti and Naylor, 2009).

5 Integrating environmental and genomic predictions using crop models

Equation 1 corresponds to a simple linear model for a genomic predictor, and this could yield across environments or more than one environment type. While we often think about y_i as yield for genotype *i*, breeders often predict other traits such as flowering time, time to maturity, or grain moisture at maturity. Objective and/or subjective indices are used to integrate these predictions for multiple adapted traits. Millet et al. (2019) used various approaches to predict yield and yield components and were able to build a static crop model and use environmental covariates to predict some important G × E interactions for yield.

These static crop models are often linear and unable to generate unexpected or emergent behavior that may surface when predicting new genotypes in new environments, that is, into a G × E space that was not included in the training set or is beyond the environments that the germplasm was generally grown (Battisti and Naylor, 2009; Hammer et al., 2019). Recall that the relationship between growth and kernel number in maize is nonlinear (Andrade et al., 1999), so linear approximations, albeit useful for some applications, are limited from their conception. However, the same linear static model applied to the prediction of the rates of change in physiological processes in response to environmental variation can lead to emergent patterns of G × E upon numerical integration over the growing season (Messina et al., 2019; Hammer et al., 2019) and over cycles of selection (Chapman et al., 2003; Messina et al., 2011). This approach was demonstrated in various crops including soybean (Messina et al., 2006), dry bean (Hoogenboom et al., 2004), Maize (Messina et al., 2011), sorghum (Chapman et al., 2003), and Barley (Yin et al., 2003) among others. The experimental demands often limit the applicability of these models when many loci underpin the control of the traits of interest.

Bayesian approaches were proposed to overcome this limitation (Technow et al., 2015; Messina et al., 2018), estimate genetic and physiological models simultaneously, and deal explicitly with the levels of uncertainty,

$$P(\boldsymbol{g} \mid D, \boldsymbol{x}_k) \propto P(D \mid \boldsymbol{x}_k, \boldsymbol{g}) \times P\left(\boldsymbol{g}\right) \tag{3}$$

where $P(\boldsymbol{g} \mid D, \boldsymbol{x}_k)$ is the posterior distribution, conditional to the vector of traits (CGM parameters) that were kept constant ($\mathbf{x}_k$), $P(D \mid \boldsymbol{x}_k, \boldsymbol{g})$ is the likelihood function with data (*D*) generated by the crop models with $\mathbf{x}_k$ estimable parameters (marker effects, $\boldsymbol{g}$), and $P(\mathbf{g})$ is the prior distribution of the traits and the marker effects for the traits. Nonlinear responses of state variables such as yield to the mean substitution effect at each locus emerge from the interplay of the equations used to model physiological processes and their

relationships with the environment. Figure 2 shows a diagram of a workflow for the estimation of a single trait, RUE, which is adapted from Technow et al. (2015) and implemented as an Approximate Bayesian Computation approach. The procedure consists in sampling the posterior distribution of RUE to assign values to the various genotypes in the training set. A genomic predictor for RUE based on markers (Eq. 2) is estimated; Genomic Best Linear Predictors (GBLUPs) are generated for each genotype. Phenotypes for which the breeder collected data (yield, flowering time, kernel numbers, etc.) are predicted using the CGM with environmental inputs for the trial and model parameter/s estimated using genomic predictors such as GBLUP for the physiological trait, here RUE. The distance between the observed and the predicted phenotypes is calculated. Using a metric such as root mean square error, others are also valid for individual traits or trait combinations, and a decision is made based on a rejection algorithm to keep the sample and build the posterior distribution.

In the example presented in Fig. 2, the mean (μ) and vector of marker effects for RUE were modeled as,

$$\begin{pmatrix} \mu \\ \boldsymbol{g} \end{pmatrix} = \begin{pmatrix} 1_n' 1_n & 1_n \boldsymbol{X} \\ \boldsymbol{X}' 1_n & \boldsymbol{X}'\boldsymbol{X} + \boldsymbol{I}\lambda \end{pmatrix}^{-1} \begin{pmatrix} 1_n' y \\ \boldsymbol{X}' y \end{pmatrix} \tag{4}$$

where $\mathbf{X}_{nxm}$ is a matrix of n genotypes by m molecular markers, $\mathbf{y}$ is a vector of RUE values, and λ is a parameter to describe the signal-to-noise ratio in the data. The RUE for each genotype is estimated from the molecular markers as,

$$\boldsymbol{y} = 1_n \mu + \boldsymbol{X} \boldsymbol{g} + e \tag{5}$$

The posterior distribution is built from marker vectors for which the root mean square error for the yield prediction was less or equal to a constant. In the example shown in Fig. 2, we choose 20% of the mean. Pseudo-code for this algorithm is presented in Technow et al. (2015).

Figure 3a shows the simulation experiment whereby a sample of genotypes for various levels of RUE was generated. The map between RUE and simulated yields for two environments contrasting for water availability is shown (Fig. 3a). The bifurcation is generated by how the model simulated the pattern of water use and thus, how much water was allocated to vegetative growth vs. reproductive growth. In the water-sufficient environment E1 (Fig. 3a), yield increased with increasing RUE. The opposite occurs in E2, where high RUE leads to high water use preflowering and induces drought during grain-fill (Fig. 3a). This result agrees with experimental observations (Cooper et al., 2014), large-scale simulation for the limited transpiration trait in maize in the United States (Messina et al., 2015) and yields data (Adee et al., 2016). When RUE is predicted using markers, the predicted range of RUE is lower than

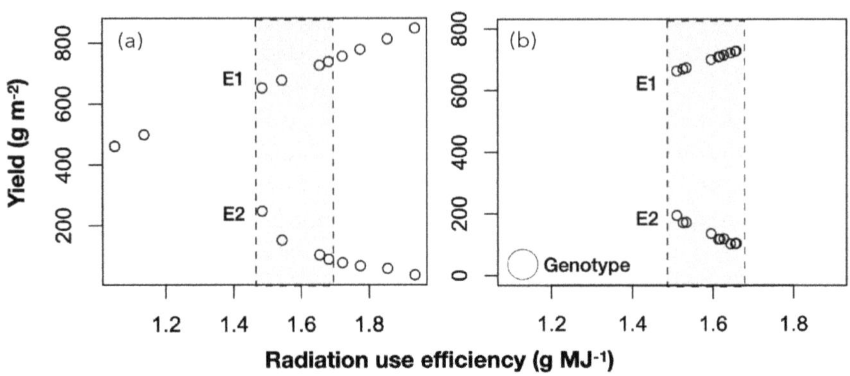

Figure 3 Simulated (a) and predicted using crop model-genomic selection (b) yield response to radiation use efficiency for ten genotypes (circles) in two environments (E1, E2)

the observation, as expected from the use of the GBLUP method that shrinks predictions toward the mean in proportion to the signal-to-noise ratio (eq. 4). However, the CGM linked to a whole genome prediction (WGP) method such as GBLUP (CGM-WGP) can regenerate the emergent relation between RUE and yield (Fig. 3b).

The results shown presented in Fig. 3b are for a very simple G × E system but these results are comparable with those obtained for much-complex cases. Using models in the form of eq. 2 to link genetics and physiology to predict thousands of genotypes created within a maize breeding program and grown in several environments, Messina et al. (2018, 2022c) and Diepenbrock et al. (2022) showed a) overall advantage in prediction accuracy of CGM-WGP over WGP (Bayes A) for yield, b) use of multiple traits further increases the predictive skill of CGM-WGP, and c) CGM-WGP can borrow information by using the trait relations embedded in the CGM to make predictions on phenotypes that were not measured or used in the training of the model. The ability of CGM-WGP to predict time to silking and kernel numbers in the absence of silking and kernel number data (Messina et al., 2022c) when the model was trained on yield alone is a promising result to encourage further development of approaches to predict other state variables that are more difficult to measure such as effective water use, soil nitrogen losses, soil water recharge, and soil carbon accumulation.

6 Assessing impacts of traits and their underpinning genetics on crop performance

CGMs are also useful tools to explore the relevance of traits and trait networks (Messina et al., 2020; Ramirez-Villegas et al., 2020; Gleason et al., 2022) and genetic networks (Messina et al., 2011; Powell et al., 2022) on yield performance under current and future climates. Figure 4a shows the contribution of individual traits or trait networks (T × T) to the fraction of phenotypic variance. At the environmental extremes of high and low evapotranspiration (ET), trait interactions are less relevant with either growth traits or reproductive resiliency traits, explaining most of the variation in this maize example. At intermediate levels of ET, the interactions of traits as trait networks underpin most of the phenotypic variation. In the presence of these trait interactions and interactions with the environment, we have observed that a CGM can perform better than empirical models (Fig. 4b, Technow et al., 2015; Messina et al., 2018; Diepenbrock et al. 2022; Messina et al., 2022c). Cooper et al. (2009), using simulation approaches, argued that the benefit of a molecular breeding strategy over phenotypic selection increases with increasing complexity of the genotype and environment system. Diepenbrock et al. (2022) showed that improvements in predictive skill by incorporating biological knowledge into the prediction algorithm increase with decreasing ET (Fig. 4b) in agreement with theoretical predictions (Cooper et al., 2005; Cooper et al., 2009; Messina et al., 2011).

The emergence of G × E interactions also depends on the distance between environments (E1 to E6, Fig. 4a). We can illustrate various types of G × E interactions depending on the tradeoffs determined by biophysics embedded in the CGM. If climate change determines a rapid shift in ET from E2 to E1, we should not expect major changes in G × E, and thus the normal operation

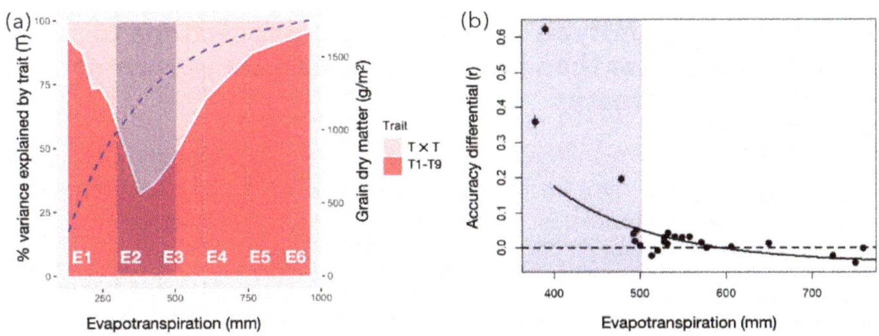

Figure 4 Percent of variance in maize yield explained by the sum of traits (T) or trait interactions (a) and accuracy difference (correlation coefficient, r) between a symbolic-sub symbolic fusion approach (crop model + genomic selection) minus sub-symbolic approach (genomic selection) along an evapotranspiration gradient (b). Purple areas in A and B highlight the increase in prediction accuracy difference between methods correlates with the importance of trait × trait interactions on the determination of yields. Environment E1-6 posted along the evapotranspiration gradient is shown as an example to illustrate that the distance between these is key to assessing the impact of climate change in the context of the relevance of T × T interactions to explain phenotypic variation in the target population of environments.

of a breeding program will suffice to adapt the germplasm to the changing TPE; the physiological basis of adaptation moves from complex (high fraction of the variance explained by trait × trait interactions) to simple (few interactions underpin adaptation). In contrast, if climate change determines a shift from E5 to E3, a drastic increase in G × E interactions can emerge underpinned by new trait × trait interactions, and deliberate breeding augmented by CGM and CGM-WGP can be expected to enhance the potential to harmonize improved germplasm adaptation with the rate of environmental change. Yields in E5 are largely determined by growth (e.g. RUE, Messina et al., 2022b) that requires large quantities of water. Genotypes with high RUE that will rank high in E5 will rapidly consume soil water and yield poorly in E3 under severe water deficits. Similar examples could be drawn for soybeans (Sinclair, 2011), chickpeas (Sivasakthi et al., 2017), and other crops. Traits such as root angle can also affect the dynamics of water use, and thus the expression of emergent phenotypes and trait interactions; genomic regions associated with root angle were implicated in the maintenance of green canopies and thus biomass assimilation through photosynthesis postflowering (Manschadi et al., 2008; Mace et al., 2012). In E1 environments, reproductive resilience and not water capture or pattern of water use is the main determinant of yield under stress (Messina et al., 2021). The use of genomic and environmental predictors integrated via CGM not only can assess the impact of climate change in performance for any genotype but also can assess the impact of climate change in the presence and absence of agronomic and genetic adaptation to the changing climate.

7 Managing genotype × environment interactions to optimize rather than hamper genetic gain in current and future climates

Decomposing the heritability into variance components (genetic (G) σ_g^2; G × E σ_{ge}^2; error σ_ϵ^2; n denotes the number of environments e and replicates r) illuminates the potential for G × E interactions to hamper genetic gain (Comstock and Moll, 1963; Cooper and Hammer, 1996),

$$h^2 = \frac{\sigma_g^2}{\sigma_g^2 + \frac{\sigma_{ge}^2}{n_e} + \frac{\sigma_\epsilon^2}{n_e n_r}} \tag{6}$$

However, the various forms of G × E can have differential impacts on genetic gain (Cooper and Hammer, 1996) depending on the importance of G × E due to heterogeneity of variances $V\left(\sigma_{e(env)}\right)$ or lack of correlation (Fig. 5a). The latter could be decomposed further into r_g, the pooled genetic correlation among environments, and $\overline{\sigma_{g(j)}\sigma_{g(j')}}$, the arithmetic average over all pairwise geometric means among all the genotypic variance for environment *j*'s. Figure 5a shows how a lack of correlation can cause changes in the ranking of genotypes while heterogeneity of variances does not. Therefore, to maximize genetic gain, it is important to know which environments generate cross-over interactions and the frequencies of these environment types within the TPE. In the context of breeding for climate change, it is of critical importance to determine whether future climates will contribute to cross-over interactions or not. The encapsulation of crop, soil, and environmental science within crop models enables the assessment of traits to understand traits undermining adaptation to current and future climates (Hammer et al., 2014, 2020; Ramirez-Villegas et al., 2020; Cooper and Messina, 2023). The results presented in Figs 4 and 5 suggest that the strategies for improving crop adaptation will be regional in nature, and the climate change-induced distances among environment types and their frequency of occurrence are expected to vary across geographies.

The research conducted over two decades in maize in the United States (Messina et al., 2020; Cooper and Messina, 2023) is informative to consider what the scientific community and breeders working in other crops and geographies can expect. Our research studies how predictive skill differentials between pure statistical approaches compared with those augmented by dynamic models

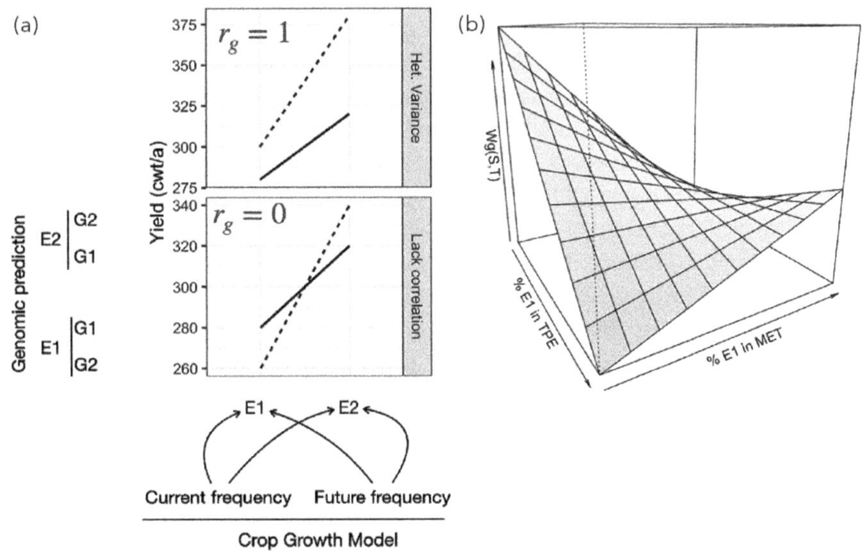

Figure 5 Types of genotype × environment interactions (G × E) as determined by the heterogeneity of variance or lack of correlation (a) and their implication given an adequate (inadequate) sampling of environments (%E1) in the target population of environments (TPE) and the multienvironment trial (MET) on the covariance of performance of the selected genotypes (Wg(S,T), Sample, Target) (b). Opportunities for the use of crop growth modeling to estimate current and future frequencies of environment types (e.g. E1), and the use of genomic prediction for genotypes G1 and G2, are shown.

(Diepenbrock et al., 2022) and how trait networks and trait interactions underpin the variance explained by these components (Messina et al., 2020, Gleason et al., 2022; Cooper and Messina, 2023) indicates that CGM-WGP approaches can be expected to have an increasingly important role in breeding for the impacts of climate change. The exacerbation of climate extremes within the TPE because of a changing climate (IPCC, 2021), will be conducive to increasing the distance between environment types encountered within the TPE. In Fig. 4, we show how different traits underpin adaptation to these extreme environment types and how these traits can lead to the emergence of a lack of correlation G × E interactions (Fig. 5), and thus an increasing challenge to maintain or increase rates of genetic gain within the TPE.

The development and deployment of platforms capable of predicting G × E × M interactions (APSIM, Holzworth et al., 2014; DSSAT, Jones et al., 2003) by harnessing environmental and genomic predictors within a physiological framework (e.g. Peng et al., 2020; Diepenbrock et al., 2022) for many different crops and geographies is a research imperative in the years to come. Evaluating and improving CGM-WGP will require engaging breeders, agronomists, and physiologists/modelers, who can expand the CGM-WGP capability beyond maize and make it accessible to the broader community. Albeit limited, the evidence available to date suggests that yield-trait performance landscapes for agricultural systems are complex (Chapman et al., 2003; Messina et al., 2011; Hammer et al., 2006, 2014, 2020) and there will be a need for resources and information technologies to enable simulation at very large scale to enable the world community of breeders to explore and leverage the knowledge hidden within a massive gene-to-phenotype space (Cooper et al., 2009; Ramstein et al., 2019; Cooper et al., 2020; Cooper et al., 2022).

CGM-WGP technology on its own will be limited in the type of solutions it can bring to address the climate change challenge. Gap analyses, the study of actual crop yields in farmer's fields in the context of resource use and attainable yields for that level of resource use (Lobell et al., 2009; Van Ittersum et al., 2013; Cooper et al., 2020), provide a framework within which the breeder and the agronomist can search the G × E × M state space and develop G × M technologies for the patterns of change they expect in the TPE as a consequence and in line with the rate of change due to climate change; G × M technologies that can close the yield gap taking into consideration the expected frequencies of E environment types in the current and future climate-affected TPE. This plant breeding-agronomy gap analyses framework could be a foundational method to translate CGM-WGP technology into practical G × M technology applications by answering the questions posed by breeders (what are the best genotypes for the TPE), the agronomists (which genotype subset fits best the cropping systems in my geography, which sets of genotypes would enable me to innovate at the cropping system level?), and the farmer question

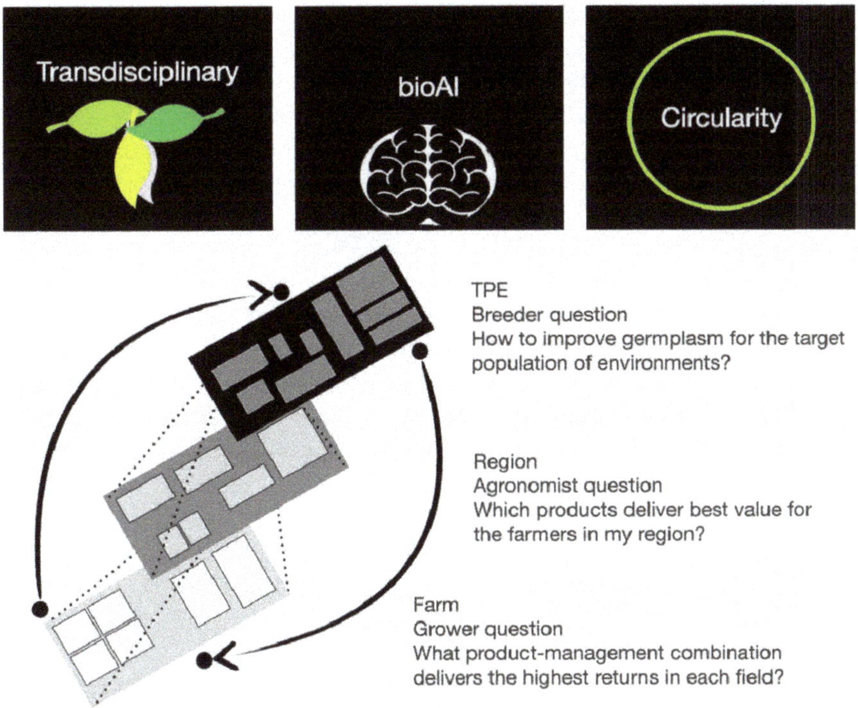

Figure 6 Transdisciplinary teams will be required to keep a dialogue going to answer questions from breeder to farmer, which answers will enable society to create systems adapted to climate change. Given the complexity of the methodologies that will be required to identify and design genotype, management, and agricultural systems, we foresee the need to develop a biologically grounded symbolic and sub-symbolic integration of artificial intelligence methods (bioAI). Principles of circular economies would enable the design of agricultural systems that could be part of the solution to climate change.

(which genotypes perform in my operation with the resources available to me for production?, Fig. 6). Answering these questions will require harnessing AI, involve transdisciplinary thinking, and embrace circularity in agricultural production.

8 Predictive breeding enables iterative impact assessment and adaptation to climate change

The goal of a plant breeding program is to create germplasm that solves problems in agriculture and thus creates value for producers, consumers, and society (Kholová et al., 2021). Hence, plant breeders define objectives that can seek to improve productivity and quality and reduce negative environmental footprints, among others. Important breeding objectives in crop improvement programs for row-crop agriculture are yield potential, disease tolerance, and yield stability. In fruits and vegetables, breeding objectives also include postharvest traits, flavor, and appearance, among others (Tieman et al., 2017; Colantonio et al., 2022). To achieve these objectives, breeders create and evaluate a sample of the germplasm in a sample of the TPE over various stages of evaluation and selection (Fig. 7a). The number of individuals evaluated in fields decreases from 10^6 to 10^2 as the germplasm cohorts advance through the stages of product development, selection, and testing, while the number of environments and agronomic management practices sampled increases from 10^2 to 10^6 at the same time (Fig. 7a). By sampling the TPE and the target population of genotypes (Fig. 7b), breeders create prediction models based on statistical and/or dynamic models. The use of managed stress environments helps to expose the germplasm to a set of environments that are of key interest to the breeder (Fig. 7b). These models help them use a relatively small sample

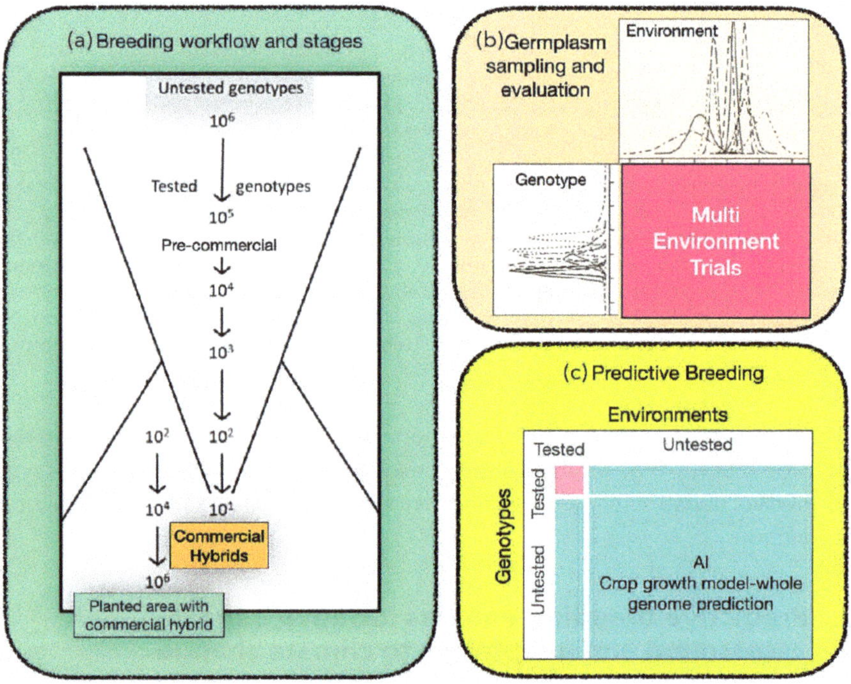

Figure 7 Representation of a breeding program from an operational perspective (a), quantitative genetics view of the sampling of environments within the target population of environments, and sampling genotypes from the target population of genotypes (b), and prediction opportunities that harness the data generated in (b) to augment the size of the breeding program (c) by predicting the possible performance of untested genotypes (a) in untested current and/or future environments (b).

of germplasm and environments to explore the complex dimensions of a much larger G × E state space (Fig. 7c; Ramstein et al., 2019). Various genotype-to-phenotype prediction methodologies were developed, some of which are described in this review (Lorenz et al., 2011; Jarquín et al., 2014; Messina et al., 2018; Ramirez-Villegas et al., 2020). Plant breeders can use the framework of the breeder's equation (Lynch and Walsh, 1998) to evaluate genetic progress and optimize breeding systems,

$$\Delta S = i \times h^2 \times \sigma_P \tag{7}$$

Where ΔS is a response to selection for one cycle of directional selection, i is the standardized selection differential applied to a trait, h_2 is heritability or fraction of the total phenotypic variation that could be attributed to genetic variation, and σ_p is the expected standard deviation for the observed on-farm values of the selection units for the same trait. Prediction methodologies such as WGP seek to use genomic information to maximize the correlation $r_{A(Mu,TPE)}$ between the observed values or breeding values of traits measured on the selection units in the multi-environment trial (*Mu*) and the true trait values or breeding values of the selection units in the TPE (Cooper and Messina, 2023),

$$\Delta S = i_{t,Mu} \times h^2_{t,Mu} \times h^2_{t,TPE} \times r_{A(Mu,TPE)} \times \sigma_{P(t,TPE)} \tag{8}$$

where $i_{t,Mu}$ is the standardized selection differential applied to the trait *t* based on the data and/or predictions from models constructed using the sample of environments obtained in trials *Mu,* the heritabilities are for the trait estimated for *Mu* ($h^2_{t,Mu}$) and the TPE ($h^2_{t,TPE}$). The goal of prediction methods is to enable breeders to expand their breeding program by adding a virtual component or to maintain the size of the breeding program while using less resources. The efficacy of the application of prediction within large commercial breeding programs for maize has been demonstrated (Cooper et al., 2014; Cooper et al., 2016). The experimental design seeks to reduce experimental errors and thus seeks to increase h^2, and increase σ_p by exposing the germplasm to environments and management conditions conducive to expressing variation in adaptive traits in both selection (S) and target (T) environments (Cooper and Hammer, 1996; Cooper and Messina, 2023).

Figure 5b shows a theoretical representation of the plausible consequences of inadequate decisions regarding selection for climate change in the presence of both types of G × E interaction and the mismatch between expected frequencies of selection and target environment (Fig. 5a). Taking E1 as current climates, the percent frequency of E1 in the TPE indicates the transition from current to future climates. When the frequency of environment type E1 matches the frequency in the TPE, say evaluation in the current climates for selection in the current climates, the genetic covariance *Wg*(S, T) between S and T environments is highest (Cooper and Hammer, 1996; Lynch and Walsh, 1998). The difference at the extremes is due to the heterogeneity in the variance in E1 and E2 (Fig. 5a). Because of the lack of correlation between environment types, selecting for E1 when the target is E2, that is selecting in current environments for performance in future environments, can lead to negative genetic covariances between environments. Most of the arguments to start breeding for climate change are based on this concept (Cairns et al., 2012; Atliln et al., 2017). Lack of action can lead to global food insecurities and geographical famines with consequent starvation of future human populations; there is merit in this argument because of the ubiquitous G × E in major cropping systems subject to drought stress (Cooper and Messina, 1996; Messina et al., 2015; Xiong et al., 2021). However, the opposite is also possible, and premature selection for future climates (E2) can hamper genetic progress for current climates contributing to food insecurity for current populations.

Selection strategies that accommodate G × E interactions were developed (Cooper et al., 1995), and weighted selection strategies have been proposed (Podlich et al., 1999) to improve the use of the phenotypic data created within the testing system within the breeding program. In this case, weights are estimated from the relative frequency of sampled environments over the expected frequencies of the environment types that comprise the mixture of environments of the TPE. The frequency of environments could be defined based on climatology or more sophisticated methods (Chapman et al., 2000; Löffler et al., 2005; Kholová et al., 2013; Ramirez-Villegas et al., 2020; Cooper and Messina, 2021; Carcedo et al., 2022). Breeding for climate change can embed this thinking by creating weights for current and future climates and applying these weights to phenotypic and simulated yield data to inform selection decisions.

This analysis leads to the following propositions:

- it is critical to time correctly breeding strategies with expected changes in climate,
- computational methods in the form of genomic prediction could be used in combination with weighted selection and crop models to create cultivars and germplasm for a dynamic TPE,
- predictive breeding connected with climate predictions via genetic-enabled CGMs can time the required adjustments within breeding programs correctly to produce cultivars ready for use by farmers in alignment with changes in the TPE (Challinor et al., 2016).

Because breeding is a dynamic process, careful decisions need to be made, and the evolution of the germplasm monitored. Simulation studies empowered by CGMs have demonstrated that breeding programs are sensitive to the founding germplasm used to start the breeding program (initial conditions) and can manifest temporal patterns that could diverge with increasing cycles of selection (Messina et al., 2011) in similar ways as the weather and climate systems do (IPCC, 2021). The prediction of the plausible trajectories of breeding programs seeking to create the germplasm required for adapting agricultural systems to climate change adds another layer of complexity and uncertainty. This requires careful consideration of breeding objectives for climate change.

9 Crop growth models can enable a rethinking of breeding objectives to adapt to and combat climate change

In the face of climate change and the need to regenerate natural resources, we advocate for a new framework in which breeding objectives are defined by answering the question: how to use genetic and agronomic levers together to maximize the societal benefit of a unit of resource use (Cooper et al., 2020; Hunt et al., 2019, 2021; Zhao et al., 2022), and how to minimize environmental degradation (Brummer et al., 2011; Rodell et al., 2018; Messina et al., 2022a,c)? Reimagining breeding to enable cropping systems that can improve the efficiency of water use (Blum, 2009; Cooper et al., 2022) is paramount in the context of the temporal changes in patterns of seasonal water availability for food production worldwide (Rosegrant et al., 2009; Richey et al., 2015; Rodell et al., 2018; Caparas et al., 2021). Similarly, given the importance of nitrogen in food production, nitrogen oxides emissions from agricultural fields on global warming, and nitrates on water pollution (Robertson and Vitousek, 2009; Bowles et al., 2018; Chang et al., 2021), it is imperative to articulate breeding objectives to create germplasm and cultivars that enable the creation of systems with high efficiency of nitrogen use, and low nitrogen oxides emissions.

CGM which uses genetic information to predict phenotypes and states of the environment, would enable breeders to incorporate effective water use, nitrogen losses and emissions, soil carbon, and other metrics as components of the breeding objectives. Unlike in the past, breeders would have the information needed to create cultivars to minimize externalities, maximize effective water and nitrogen use, soil carbon accumulation, or a combination that contributes to producing food while combating climate change (Messina et al., 2022c; Cooper and Messina, 2023). While improving confidence in the models to simulate the metrics of interest is an important and likely costly undertaking, these investments pale relative to the massive socioeconomic consequences of no action. Traits that contribute to the improvement of both productivity and reduced externalities need not be directly related to the biological efficiency of resource use, as shown for nitrogen (Mueller et al., 2019) and water in maize (Cooper et al., 2014; Messina et al., 2022a), but at the cropping system level. It has been proposed there is a need to increase the circularity of nutrient cycling in maize by improving maize tolerance to low temperatures and remobilization to roots (Buckler, pers. comm., Fig. 8). Figure 8 illustrates a concept where tolerance to cold stress in maize could increase yields, increase light interception, and thus reduce externalities while increasing productivity by enabling farmers to plant early in the season. Such system-level thinking enabled by CGMs could be conducive to a shift in mindsets on how to define breeding objectives that move us away from plant/crop-centric idiotypes with a focus on productivity (Perego et al., 2014; Rötter et al., 2015; Paleari et al.,

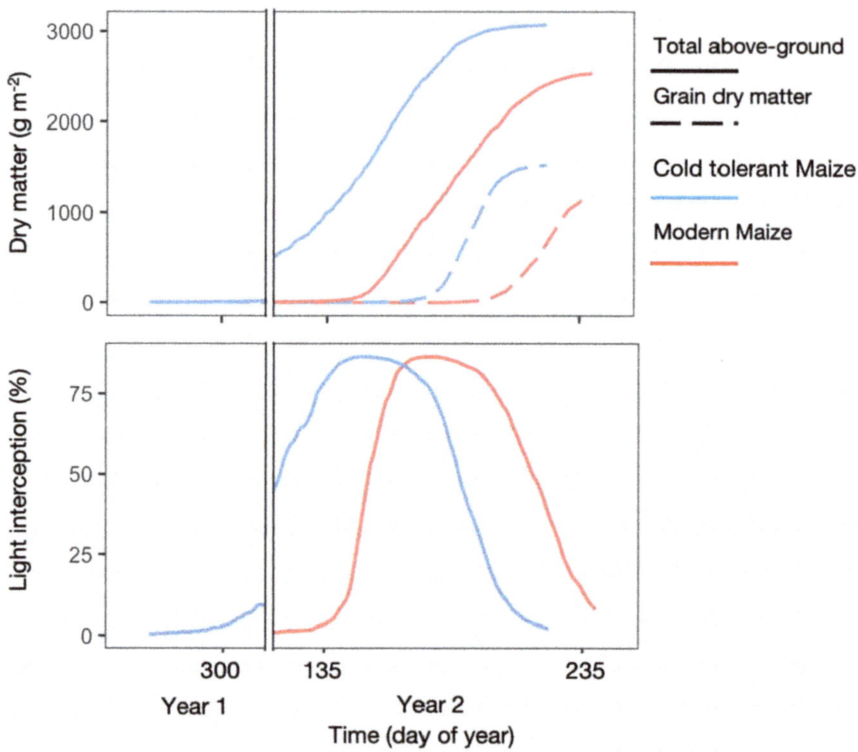

Figure 8 Total above-ground and grain dry matter and light interception percent for modern maize planted at normal planting date and for a conceptual novel maize genotype tolerance to cold planted at an ultra-early planting date. Cold-tolerant maize could enable higher yields due to increased light interception. Early growth shifts the pattern of nitrogen demand and thus may decrease the pool of nitrogen susceptible to denitrification and leaching.

2017; Hammer et al., 2020; Ramirez-Villegas et al., 2020) toward system-centric thinking with a focus on balance between providing nutrient security while minimizing environmental externalities and their negative contributions that hasten climate change; these are two important dimensions contributing to current and future human health and well-being.

10 Conclusion and future trends

Genetics for crop improvement ranks in the top set of technologies that can contribute to human adaptation to climate change. Plant breeding has contributed to yield improvement since the domestication of crop species, and it was most evident in the last century. In this chapter, we have shown not only the need to develop new G × M technologies to adapt to climate change but also the need to transition from current to future G × M technologies in a way that future food security does not come at the expense of current food security. To spur scientific and technological development for climate change adaptation, we have demonstrated an approach to integrate genomic, agronomic, and environmental predictors to accomplish three critical goals:

- predict emergent phenotypes that stem from the dynamic interplay between G, E, and M, and thus enable the breeder to consider the behavior of genetic and trait combinations in environments that plants have not been exposed to or tested within before, and/or in more variable and extreme environments,
- identify G × M technologies that could increase food and nutritional security while regenerating natural and production resources; outputs from crop model-based approaches that consider genetic and genomic predictors for physiological processes could be used to create multi-dimensional selection indices,
- enable breeders and agronomists with tools to harness multiple sources of information to create G × M technologies to address the inequitable impacts of climate change, diverse cultural backgrounds, and geographically granular societal needs.

Breeding enables not only incremental adaptation by improving upon existing crops but also transformational adaptation by enabling new farming systems and new crops. The same toolset, approaches, and needs introduced in this chapter for single crops, could be used by a team focused on crop improvement for new agricultural systems. Soybean-maize production systems like Safrinha in Brazil took 40 years of research in breeding and agronomy. Society can no longer afford long time-lags to develop sustainable technologies to adapt cropping systems to climate change. The implementation of dynamic gene-to-phenotype platforms, such as CGM-WGP, within farming systems simulators such as DSSAT and APSIM seems a natural next development to shorten the development time for discovery, refinement, and optimization of novel agricultural systems. To effectively search the extremely large search spaces, now G is expanded to G within crop within a crop sequence, and the investment will be required in both artificial intelligence and information technologies to search these spaces effectively, manage very large data sets, and run the simulations at an unprecedentedly large scale.

As society continues to make environmentally and health-conscious decisions and increase the demand and consumption of plant-based foods in diets, breeding objectives will continue to evolve to deliver more nutritional value over quantities and reduce environmental footprints. The expansion will be needed to hasten breeding in fruits, vegetables, and pulses, which can benefit from applying breeding technologies previously primarily developed for row crops. The framework introduced in this chapter is based on linking genetics with physiology and environment (e.g. CGM-WGP), or, more generally, AI could have the largest impacts on breeding programs with small footprints by maximizing information extraction from data and enabling prediction for future environments.

11 References

Adee, E., Roozeboom, K., Balboa, G. R., Schlegel, A. and Ciampitti, I. A. (2016). Drought-tolerant corn hybrids yield more in drought-stressed environments with no penalty in non-stressed environments. *Front. Plant Sci.* 7: 1534.

Anderson, P. W. (1972). More is different. *Science* 177(4047): 393–396.

Andrade, F. H., Vega, C., Uhart, S., Cirilo, A., Cantarero, M. and Valentinuz, O. (1999). Kernel number determination in maize. *Crop Sci.* 39(2): 453–459.

Atliln, G. N., Cairns, J. E. and Das, B. (2017). Rapid breeding and varietal replacement are critical to adaptation of cropping systems in the developing world to climate change. *Glob. Food Sec.* 12: 31–37.

Battisti, D. S. and Naylor, R. L. (2009). Historical warnings of future food insecurity with unprecedented seasonal heat. *Science* 323(5911): 240–244.

Binns, C. W., Lee, M. K., Maycock, B., Torheim, L. E., Nanishi, K. and Duong, D. T. T. (2021). Climate change, food supply and dietary guidelines. *Annu. Rev. Public Health* 42: 233–255.

Blum, A. (2009). Effective use of water (EUW) and not water-use efficiency (WUE) is the target of crop yield improvement under drought stress. *Field Crops Res.* 112(2–3): 119–123.

Boer, M. P., Wright, D., Feng, L., Podlich, D. W., Luo, L., Cooper, M. and van Eeuwijk, F. A. (2007). A mixed-model quantitative trait loci (QTL) analysis for multiple-environment trial data using environmental covariables for QTL-by-environment interactions, with an example in maize. *Genetics* 177(3): 1801–1813.

Borrell, A. K., Mullet, J. E., George-Jaeggli, B., van Oosterom, E. J., Hammer, G. L., Klein, P. E. and Jordan, D. R. (2014). Drought adaptation of stay-green sorghum is associated with canopy development, leaf anatomy, root growth, and water uptake. *J. Exp. Bot.* 65(21): 6251–6263.

Bowles, T. M., Atallah, S. S., Campbell, E. E., Gaudin, A. C. M., Wieder, W. R. and Grandy, A. S. (2018). Addressing agricultural nitrogen losses in a changing climate. *Nat. Sustain.* 1(8): 399–408. https://doi.org/10.1038/s41893-018-0106-0.

Brummer, E. C., Barber, W. T., Collier, S. M., Cox, T. S., Johnson, R., Murray, S. C., Olsen, R. T., Pratt, R. C. and Thro, A. M. (2011). Plant breeding for harmony between agriculture and the environment. *Front. Ecol. Environ.* 9(10): 561–568.

Cairns, J. E., Sonder, K., Zaidi, P. H., Verhulst, N., Mahuku, G., Babu, R., Nair, S. K., Das, B., Govaerts, B., Vinayan, M. T., Rashid, Z., Noor, J. J. and Devi, P. (2012). Maize production in a changing climate: Impacts, adaptation, and mitigation strategies. *Adv. Agron.* 114: 1–58.

Caparas, M., Zobel, Z., Castanho, A. D. A. and Schwalm, C. R. (2021). Increasing risks of crop failure and water scarcity in global breadbaskets by 2030. *Environ. Res. Lett.* 16(10): 104013.

Carcedo, A. J. P., Mayor, L., Demarco, P., Morris, G. P., Lingenfelser, J., Messina, C. D. and Ciampitti, I. A. (2022). Environment characterization in Sorghum (*Sorghum bicolor* L.) by modeling water-deficit and heat patterns in the great plains region, United States. *Front. Plant Sci.* 13: 768610.

Carter, E. K., Rita, S. J., Melkonian, J. and Steinschneider, S. (2018). Yield response to climate, management, and genotype: A large-scale observational analysis to identify climate-adaptive crop management practices in high-input maize systems. *Environ. Res. Lett.* 13(11): 114006.

Ceccarelli, S. and Grando, S. (2020). Evolutionary plant breeding as a response to the complexity of climate change. *iScience* 23(12): 101815.

Ceccarelli, S., Grando, S., Maatougui, M., Michael, M., Slash, M., Haghparast, R., Rahmanian, M., Taheiri, A., Al-yassin, A., Benbelkacems, A., Labdi, M., Mimoun, H. and Nachit, M. (2010). Plant breeding and climate changes. *J. Agric. Sci.* 148(6): 627–637.

Challinor, A. J., Koehler, A.-K., Ramirez-Villegas, J., Whitfield, S. and Das, B. (2016). Current warming will reduce yields unless maize breeding and seed systems adapt immediately. *Nat. Clim. Change* 6(10): 954–958.

Challinor, A. J., Watson, J., Lobell, D. B., Howden, S. M., Smith, D. R. and Chhetri, N. (2014). A meta-analysis of crop yield under climate change and adaptation. *Nat. Clim. Change* 4(4): 287–291.

Chaloner, T. M., Gurr, S. J. and Bebber, D. P. (2021). Plant pathogen infection risk tracks global crop yields under climate change. *Nat. Clim. Chang.* 11(8): 710–715. https://doi.org/10.1038/s41558-021-01104-8.

Chang, J., Havlík, P., Leclère, D., de Vries, W., Valin, H., Deppermann, A., Hasegawa, T. and Obersteiner, M. (2021). Reconciling regional nitrogen boundaries with global food security. *Nat. Food* 2(9): 700–711. https://doi.org/10.1038/s43016-021-00366-x.

Chapman, S., Cooper, M., Podlich, D. and Hammer, G. (2003). Evaluating plant breeding strategies by simulating gene action and dryland environment effects. *Agron. J.* 95(1): 99–113.

Chapman, S. C., Hammer, G. L., Butler, D. G. and Cooper, M. (2000). Genotype by environment interactions affecting grain sorghum. III. Temporal sequences and spatial patterns in the target population of environments. *Aust. J. Agric. Res.* 51(2): 223-233.

Christopher, J., Christopher, M., Jennings, R., Jones, S., Fletcher, S., Borrell, A., Manschadi, A. M., Jordan, D., Mace, E. and Hammer, G. (2013). QTL for root angle and number in a population developed from bread wheats (*Triticum aestivum*) with contrasting adaptation to water-limited environments. *Theor. Appl. Genet.* 126(6): 1563–1574.

Colantonio, V., Ferrão, L. F. V., Tieman, D. M., Bliznyuk, N., Sims, C., Klee, H. J., Munoz, P. and Resende, M. F. R. (2022). Metabolomic selection for enhanced fruit flavor. *Proc. Natl Acad. Sci. U. S. A.* 119(7): e2115865119.

Comstock, R. E. and Moll, R. H. (1963). Genotype-environment interactions. In: Hanson, W. D. and Robinson, H. F. (Eds.), *Statistical Genetics and Plant Breeding*. Publication 982. National Academy of Sciences – National Research Council, Washington, DC, pp. 164–196.

Cooper, M., Technow, F., Messina, C., Gho, C. and Totir, R. (2016). Use of crop growth models with whole genome prediction: application to a maize multi-environment trial. *Crop Science.* 56(5): 2141–2156.

Cooper, M., Gho, C., Leafgren, R., Tang, T. and Messina, C. (2014). Breeding drought-tolerant maize hybrids for the US corn-belt: Discovery to product. *J. Exp. Bot.* 65(21): 6191–6204.

Cooper, M. and Hammer, G. L. (1996). *Plant Adaptation and Crop Improvement*. CAB International, Wallingford, UK.

Cooper, M. and Messina, C. D. (2023). Breeding crops for drought affected environments and improved climate resilience. *Plant Cell* 35(1): 162–186.

Cooper, M., Messina, C. D., Tang, T., Gho, C., Powell, O. M., Podlich, D. W., Technow, F. and Hammer, G. L. (2022). Predicting genotype x environment x management (GExM) interactions for design of crop improvement strategies: Integrating breeder, agronomist, and farmer perspectives. *Plant Breed. Rev.* 46: 467–585.

Cooper, M., Podlich, D. W. and Smith, O. S. (2005). Gene-to-phenotype models and complex trait genetics. *Aust. J. Agric. Res.* 56(9): 895–918.

Cooper, M., Powell, O., Voss-Fels, K. P., Messina, C. D., Gho, C., Podlich, D. W., Technow, F., Chapman, S. C., Beveridge, C. A., Ortiz-Barrientos, D. and Hammer, G. L. (2021). Modelling selection response in plant breeding programs using crop models as mechanistic gene-to-phenotype (CGM-G2P) multi-trait link functions. *In Silico Plants* 3: diaa016.

Cooper, M., Tang, T., Gho, C., Hart, T., Hammer, G. and Messina, C. (2020). Integrating genetic gain and gap analysis to predict improvements in crop productivity. *Crop Sci.* 60(2): 582–604.

Cooper, M., van Eeuwijk, F. A., Hammer, G. L., Podlich, D. W. and Messina, C. (2009). Modeling QTL for complex traits: Detection and context for plant breeding. *Curr. Opin. Plant Biol.* 12(2): 231–240.

Cooper, M., Woodruff, D. R., Eisemann, R. L., Brennan, P. S. and DeLacy, I. H. (1995). A selection strategy to accommodate genotype-by-environment interaction for grain yield of wheat: Managed-environments for selection among genotypes. *Theor. Appl. Genet.* 90(3–4): 492–502.

de Felipe, M., Gerde, J. A. and Rotundo, J. L. (2016). Soybean genetic gain in maturity groups III to V in Argentina from 1980 to 2015. *Crop Sci.* 56(6): 3066–3077. https://doi.org/10.2135/cropsci2016.04.0214.

Diepenbrock, C. H., Tang, T., Jines, M., Technow, F., Lira, S., Podlich, D., Cooper, M. and Messina, C. (2022). Can we harness digital technologies and physiology to hasten genetic gain in U.S. maize breeding? *Plant Physiol.* 188(2): 1141–1157.

Edmeades, G. O. and Daynard, T. B. (1979). The development of plant-to-plant variability in maize at different planting densities. *Can. J. Plant Sci.* 59(3): 561–576.

Ficklin, D. L. and Novick, K. A. (2017). Historic and projected changes in vapor pressure deficit suggest a continental-scale drying of the United States atmosphere. *J. Geophys. Res. Atmos.* 122(4): 2061–2079. https://doi.org/10.1002/2016JD025855.

Fischer, T., Byerlee, D. and Edmeades, G. (2014). *Crop Yields and Global Food Security: Will Yield Increase Continue to Feed the World?* ACIAR Monograph No. 158. Australian Centre for International Agricultural Research, Canberra.

Fukai, S. and Cooper, M. (1995). Development of drought-resistant cultivars using physio-morphological traits in rice. *Field Crops Res.* 40(2): 67–86.

Gaffney, J., Schussler, J., Löffler, C., Cai, W., Paszkiewicz, S., Messina, C., Groeteke, J., Keaschall, J. and Cooper, M. (2015). Industry-scale evaluation of maize hybrids selected for increased yield in drought-stress conditions of the US corn belt. *Crop Sci.* 55(4): 1608–1618.

Gleason, S. M., Barnard, D. M., Green, T. R., Mackay, S., Wang, D. R., Ainsworth, E. A., Altenhofen, J., Brodribb, T. J., Cochard, H., Comas, L. H., Cooper, M., Creek, D., DeJonge, K. C., Delzon, S., Fritschi, F. B., Hammer, G., Hunter, C., Lombardozzi, D., Messina, C. D., Ocheltree, T., Stevens, B. M., Stewart, J. J., Vadez, V., Wenz, J., Wright, I. J., Yemoto, K. and Zhang, H. (2022). Physiological trait networks enhance understanding of crop growth and water use in contrasting environments. *Plant Cell Environ.* 45(9): 2554–2572.

Hammer, G., Cooper, M., Tardieu, F., Welch, S., Walsh, B., van Eeuwijk, F., Chapman, S. and Podlich, D. (2006). Models for navigating biological complexity in breeding improved crop plants. *Trends Plant Sci.* 11(12): 587–593.

Hammer, G. L., Dong, Z., McLean, G., Doherty, A., Messina, C., Schussler, J., Zinselmeier, C., Paszkiewicz, S. and Cooper, M. (2009). Can changes in canopy and/or root system architecture explain historical maize yield trends in the US Corn Belt? *Crop Sci.* 49(1): 299–312.

Hammer, G. L., McLean, G., Chapman, S., Zheng, B., Doherty, A., Harrison, M. T., van Oosterom, E. and Jordan, D. (2014). Crop design for specific adaptation in variable dryland production environments. *Crop Pasture Sci.* 65(7): 614–626.

Hammer, G. L., McLean, G., van Oosterom, E., Chapman, S., Zheng, B., Wu, A., Doherty, A. and Jordan, D. (2020). Designing crops for adaptation to the drought and high-temperature risks anticipated in future climates. *Crop Sci.* 60(2): 605–621.

Hammer, G. L., Messina, C., Wu, A. and Cooper, M. (2019). Biological reality and parsimony in crop models – Why we need both in crop improvement. *In Silico Plants* 2019: diz010.

Harrison, M. T., Tardieu, F., Dong, Z., Messina, C. D. and Hammer, G. L. (2014). Characterizing drought stress and trait influence on maize yield under current and future conditions. *Glob. Change Biol.* 20(3): 867–878.

Heffner, E. L., Sorrells, M. E. and Jannink, J.-L. (2009). Genomic selection for crop improvement. *Crop Sci.* 49(1): 1–12.

Hernandez-Ochoa, I. M., Luz Pequeno, D. N., Reynolds, M., Babar, M. A., Sonder, K., Milan, A. M., Hoogenboom, G., Robertson, R., Gerber, S., Rowland, D. L., Fraisse, C. W. and Asseng, S. (2019). Adapting irrigated and rainfed wheat to climate change in semi-arid environments: Management, breeding options and land use change. *Eur. J. Agron.* 109: 125915.

Heslot, N., Akdemir, D., Sorrells, M. E. and Jannink, J. L. (2014). Integrating environmental covariates and crop modeling into the genomic selection framework to predict genotype by environment interactions. *Theor. Appl. Genet.* 127(2): 463–480.

Holzworth, D. P., Huth, N. I., deVoil, P. G., Zurcher, E. J., Herrmann, N. I., McLean, G., Chenu, K., van Oosterom, E. J., Snow, V., Murphy, C., Moore, A. D., Brown, H., Whish, J. P. M., Verrall, S., Fainges, J., Bell, L. W., Peake, A. S., Poulton, P. L., Hochman, Z., Thorburn, P. J., Gaydon, D. S., Dalgliesh, N. P., Rodriguez, D., Cox, H., Chapman, S., Doherty, A., Teixeira, E., Sharp, J., Cichota, R., Vogeler, I., Li, F. Y., Wang, E., Hammer, G. L., Robertson, M. J., Dimes, J. P., Whitbread, A. M., Hunt, J., van Rees, H., McClelland, T., Carberry, P. S., Hargreaves, J. N. G., MacLeod, N., McDonald, C., Harsdorf, J., Wedgwood, S. and Keating, B. A. (2014). APSIM – Evolution towards a new generation of agricultural systems simulation. *Environ. Modell. Softw.* 62: 327–350.

Hoogenboom, G., White, J. W. and Messina, C. D. (2004). From genome to crop: integration through simulation modelling. *Field Crops Research* 90: 145–163.

Howden, S. M., Soussana, J. F., Tubiello, F. N., Chhetri, N., Dunlop, M. and Meinke, H. (2007). Adapting agriculture to climate change. *Proc. Natl Acad. Sci. U. S. A.* 104(50): 19691–19696.

Hunt, J. R., Kirkegaard, J. A., Harris, F. A., Porker, K. D., Rattey, A. R., Collins, M. J., Celestina, C., Cann, D. J., Hochman, Z., Lilley, J. M. and Flohr, B. M. (2021). Exploiting genotype x management interactions to increase rainfed crop production: A case study from south-eastern Australia. *J. Exp. Bot.* 72(14): 5189–5207.

Hunt, J. R., Lilley, J. M., Trevaskis, B., Flohr, B. M., Peake, A., Fletcher, A., Zward, A. B., Gobbett, D. and Kirkegaard, J. A. (2019). Early sowing systems can boost Australian wheat yields despite recent climate change. *Nat. Clim. Change* 9(3): 244–247.

IPCC. (2021). 2021: The physical science basis. Contribution of Working Group I to the sixth assessment report of the Intergovernmental Panel on Climate Change. In: *Climate Change*. Cambridge University Press, Cambridge, UK.

Jarquín, D., Crossa, J., Lacaze, X., Du Cheyron, P., Daucourt, J., Lorgeou, J., Piraux, F., Guerreiro, L., Perez, P., Calus, M., Burgueño, J. and de los Campos, G. (2014). A reaction norm model for genomic selection using high-dimensional genomic and environmental data. *Theor. Appl. Genet.* 127(3): 595–607.

Jones, J. W., Hoogenboom, G., Porter, C. H., Boote, K. J., Batchelor, W. D., Hunt, L. A., Wilkens, P. W., Singh, U., Gijsman, A. J. and Ritchie, J. T. (2003). The DSSAT cropping system model. *Eur. J. Agron.* 18(3–4): 235–265.

Jordan, D. R., Hunt, C. H., Cruickshank, A. W., Borrell, A. K. and Henzell, R. G. (2012). The relationship between the stay-green trait and grain yield in elite sorghum hybrids grown in a range of environments. *Crop Sci.* 52(3): 1153–1161.

Kholová, J., McLean, G., Vadez, V., Craufurd, P. and Hammer, G. L. (2013). Drought stress characterization of post-rainy season (rabi) sorghum in India. *Field Crops Res.* 141: 38–46.

Kholová, J., Urban, M. O., Cock, J., Arcos, J., Arnaud, E., Aytekin, D., Acevedo, V., Barnes, A. P., Ceccarelli, S., Chavarriaga, P., Cobb, J. N., Connor, D., Cooper, M., Craufurd, P., Debouck, D., Fungo, R., Grando, S., Hammer, G. L., Jara, C. E., Messina, C., Mosquera, G., Nchanji, E., Ng, E. H., Prager, S., Sankaran, S., Selvaraj, M., Tardieu, F., Thornton, P., Valdes-Gutierrez, S. P., van Etten, J., Wenzl, P. and Xu, Y. (2021). In pursuit of a better world: Crop improvement and the CGIAR. *J. Exp. Bot.* 72(14): 5158–5179.

Lobell, D. B., Cassman, K. G. and Field, C. B. (2009). Crop yield gaps: Their importance, magnitudes and causes. *Annu. Rev. Environ. Resour.* 34(1): 179-204.

Lobell, D. B., Schlenker, W. and Costa-Roberts, J. (2011). Climate trends and global crop production since 1980. *Science* 333(6042): 616-620.

Löffler, C. M., Wei, J., Fast, T., Gogerty, J., Langton, S., Bergman, M., Merrill, B. and Cooper, M. (2005). Classification of maize environments using crop simulation and geographic information systems. *Crop Sci.* 45(5): 1708-1716.

Lorenz, A. J., Chao, S., Asoro, F. G., Heffner, E. L., Hayashi, T., Iwata, H., Smith, K. P., Sorrells, M. E. and Jannink, J.-L. (2011). Genomic selection in plant breeding. In: Donald, L. S. (Ed.), *Advances in Agronomy*. Academic Press, The Netherlands, pp. 77-123.

Lynch, M. and Walsh, B. (1998). *Genetics and Analysis of Quantitative Traits*. Sinauer Associates, Inc., Sunderland, MA.

Mace, E. S., Singh, V., Van Oosterom, E. J., Hammer, G. L., Hunt, C. H. and Jordan, D. R. (2012). QTL for nodal root angle in sorghum (Sorghum bicolor L. Moench) co-locate with QTL for traits associated with drought adaptation. *Theor. Appl. Genet.* 124(1): 97-109.

Manschadi, A. M., Hammer, G. L., Christopher, J. T. and Devoil, P. (2008). Genotypic variation in seedling root architectural traits and implications for drought adaptation in wheat (*Triticum aestivum* L.). *Plant Soil* 303(1-2): 115-129.

May, R. M. (1976). Simple mathematical models with very complicated dynamics. *Nature* 261(5560): 459-467.

Messina, C., Cooper, M., Hammer, G., Berning, D., Ciampitti, I., Clark, R., Diepenbrock, C. H., Gho, C., Jines, M., Lee, T., McCormick, R. F., Mihura, E., Podlich, D., Rotundo, J., Smalley, M., Tang, T., Truong, S. and van Eeuwijk, F. (2020). Two decades of creating drought tolerant maize and underpinning prediction technologies in the US corn-belt: Review and perspectives on the future of crop design. *bioRxiv* 2020/1/1.

Messina, C., McDonald, D., Poffenbarger, H., Clark, R., Salinas, A., Fang, Y., Gho, C., Tang, T., Graham, G., Hammer, G. L. and Cooper, M. (2021). Reproductive resilience but not root architecture underpins yield improvement under drought in maize. *J. Exp. Bot.* 72(14): 5235-5245.

Messina, C. D., Ciampitti, I. A., Berning, D., Bubeck, D., Hammer, G. L. and Cooper, M. (2022a). Sustained improvement in tolerance to water deficit accompanies maize yield increase in temperate environments. *Crop Sci.* 62(6): 2138-2150. https://doi.org/10.1002/csc2.20781.

Messina, C. D., Rotundo, J., Hammer, G. L., Gho, C., Reyes, A., Fang, Y., van Oosterom, E., Borras, L. and Cooper, M. (2022b). Radiation use efficiency increased over a century of maize (*Zea mays* L.) breeding in the US corn belt. *J. Exp. Bot.* 73(16): 5503-5513. https://doi.org/10.1093/jxb/erac212.

Messina, C. D., van Eeuwijk, F., Tang, T., Truong, S. K., McCormick, R. F., Technow, F., Powell, O., Mayor, L., Gutterson, N., Jones, J. W., Hammer, G. L. and Cooper, M. (2022c). Crop improvement for circular bioeconomy systems. *J. ASABE* 65(3): 491-504.

Messina, C. D., Hammer, G. L., McLean, G., Cooper, M., van Oosterom, E. J., Tardieu, F., Chapman, S. C., Doherty, A. and Gho, C. (2019). On the dynamic determinants of reproductive failure under drought in maize. *In Silico Plants* 1(1): diz003.

Messina, C. D., Podlich, D., Dong, Z., Samples, M. and Cooper, M. (2011). Yield-trait performance landscapes: From theory to application in breeding maize for drought tolerance. *J. Exp. Bot.* 62(3): 855-868.

Messina, C. D., Sinclair, T. R., Hammer, G. L., Curan, D., Thompson, J., Oler, Z., Gho, C. and Cooper, M. (2015). Limited-transpiration trait may increase maize drought tolerance in the US corn belt. *Agron. J.* 107(6): 1978–1986.

Messina, C. D., Technow, F., Tang, T., Totir, R., Gho, C. and Cooper, M. (2018). Leveraging biological insight and environmental variation to improve phenotypic prediction: Integrating crop growth models (CGM) with whole genome prediction (WGP). *Eur. J. Agr.* 100: 151–162.

Meuwissen, T. H., Hayes, B. J. and Goddard, M. E. (2001). Prediction of total genetic value using genome-wide dense marker maps. *Genetics* 157(4): 1819–1829.

Millet, E. J., Kruijer, W., Couple-Ledru, A., Prado, S. A., Cabrera-Bosquet, L., Lacube, S., Charcosset, A., Welcker, C., van Eeuwijk, F. and Tardieu, F. (2019). Genomic prediction of maize yield across European environmental conditions. *Nat. Genet.* 51(6): 952–956.

Morgounov, A., Haun, S., Lang, L., Martynov, S. and Sonder, K. (2013). Climate change at winter wheat breeding sites in central Asia eastern Europe, and USA, and implications for breeding. *Euphytica* 194(2): 277–292.

Mueller, S. M., Messina, C. D. and Vyn, T. J. (2019). Simultaneous gains in grain yield and nitrogen efficiency over 70 years of maize genetic improvement. *Sci. Rep.* 9(1): 9095.

Nurmberg, P. L., Brito, A., Zimmer, C. M., Schwantes, D. O., Mantovani, E. E., Bosetti, F., Rotundo, J., Truong, S. K., Borras, L., de la Vega, A., Messina, C. and Mccormick, R. (2022). A commercial breeding perspective of maize improvement for drought stress tolerance. *Rev. Bras Milho Sorgo* 21: e1277.

Paleari, L., Movedi, E., Cappelli, G., Wilson, L. T. and Confalonieri, R. (2017). Surfing parameter hyperspaces under climate change scenarios to design future rice ideotypes. *Glob. Change Biol.* 23(11): 4651–4662.

Peng, B., Guan, K., Tang, J., Ainsworth, E. A., Asseng, S., Bernacchi, C. J., Cooper, M., Delucia, E. H., Elliott, J. W., Ewert, F., Grant, R. F., Gustafson, D. I., Hammer, G. L., Jin, Z., Jones, J. W., Kimm, H., Lawrence, D. M., Li, Y., Lombardozzi, D. L., Marshall-Colon, A., Messina, C. D., Ort, D. R., Schnable, J. C., Vallejos, C. E., Wu, A., Yin, X. and Zhou, W. (2020). Towards a multiscale crop modelling framework for climate change adaptation assessment. *Nat. Plants* 6(4): 338–348.

Perego, A., Sanna, M., Giussani, A., Chiodini, M. E., Fumagalli, M., Pilu, S. R., Bindi, M., Moriondo, M. and Acutis, M. (2014). Designing a high-yielding maize ideotype for a changing climate in Lombardy plain (northern Italy). *Sci. Total Environ.* 499: 497–509.

Podlich, D. W., Cooper, M., Basford, K. E. and Geiger, H. H. (1999). Computer simulation of a selection strategy to accommodate genotype-environment interactions in a wheat recurrent selection programme. *Plant Breed.* 118(1): 17–28.

Powell, O. M., Barbier, F., Voss-Fels, K. P., Beveridge, C. and Cooper, M. (2022). Investigations into the emergent properties of gene-to-phenotype networks across cycles of selection: A case study of shoot branching in plants. *In Silico Plants* 4: diac006.

Prasanna, B. M., Cairns, J. E., Zaidi, P. H., Beyene, Y., Makumbi, D., Gowda, M., Magorokosho, C., Zaman-Allah, M., Olsen, M., Das, A., Worku, M., Gethi, J., Vivek, B. S., Nair, S. K., Rashid, Z., Vinayan, M. T., Issa, A. B., San Vicente, F., Dhliwayo, T. and Zhang, X. (2021). Beat the stress: Breeding for climate resilience in maize for the tropical rainfed environments. *Theor. Appl. Genet.* 134(6): 1729–1752.

Ramirez-Villegas, J., Heinemann, A. B., Pereira de Castro, A., Breseghello, F., Navarro-Racines, C., Li, T., Rebolledo, M. C. and Challinor, A. J. (2018). Breeding implications of drought stress under future climate for upland rice in Brazil. *Glob. Change Biol.* 24(5): 2035–2050.

Ramirez-Villegas, J., Molero Milan, A., Alexandrov, N., Asseng, S., Challinor, A. J., Crossa, J., Eeuwijk, F., Ghanem, M. E., Grenier, C., Heinemann, A. B., Wang, J., Juliana, P., Kehel, Z., Kholova, J., Koo, J., Pequeno, D., Quiroz, R., Rebolledo, M. C., Sukumaran, S., Vadez, V., White, J. W. and Reynolds, M. (2020). CGIAR modeling approaches for resource-constrained scenarios: I. Accelerating crop breeding for a changing climate. *Crop Sci.* 60(2): 547–567.

Ramirez-Villegas, J., Watson, J. and Challinor, A. J. (2015). Identifying traits for genotypic adaptation using crop models. *J. Exp. Bot.* 66(12): 3451–3462.

Ramstein, G. P., Jensen, S. E. and Buckler, E. S. (2019). Breaking the curse of dimensionality to identify causal variants in Breeding 4. *Theor. Appl. Genet.* 132(3): 559–567.

Rattis, L., Brando, P. M., Macedo, M. N., Spera, S. A., Castanho, A. D. A., Marques, E. Q., Costa, N. Q., Silverio, D. V. and Coe, M. T. (2021). Climatic limit for agriculture in Brazil. *Nat. Clim. Change* 11(12): 1098–1104.

Ray, D. K., Mueller, N. D., West, P. C. and Foley, J. A. (2013). Yield trends are insufficient to double global crop production by 2050. *PLoS ONE* 8(6): e66428.

Richey, A. S., Thomas, B. F., Lo, M. H., Reager, J. T., Famiglietti, J. S., Voss, K., Swenson, S. and Rodell, M. (2015). Quantifying renewable groundwater stress with GRACE. *Water Resour. Res.* 51(7): 5217–5238. https://doi.org/10.1002/2015WR017349.

Rippke, U., Ramirez-Villegas, J., Jarvis, A., Vermeulen, S. J., Parker, L., Mer, F., Diekkrüger, B., Challinor, A. J. and Howden, M. (2016). Timescales of transformational climate change adaptation in sub-Saharan African agriculture. *Nat. Clim. Change* 6(6): 605–609. https://doi.org/10.1038/nclimate2947.

Robertson, G. P. and Vitousek, P. M. (2009). Nitrogen in agriculture: Balancing the cost of an essential resource. *Annu. Rev. Environ. Resour.* 34(1): 97–125.

Rodell, M., Famiglietti, J. S., Wiese, D. N., Reager, J. T., Beaudoing, H. K., Landerer, F. W. and Lo, M. H. (2018). Emerging trends in global freshwater availability. *Nature* 557(7707): 651–659.

Roeder, A. H. K., Otegui, M. S., Dixit, R., Anderson, C. T., Faulkner, C., Zhang, Y., Harrison, M. J., Kirchhelle, C., Goshima, G., Coate, J. E., Doyle, J. J., Hamant, O., Sugimoto, K., Dolan, L., Meyer, H., Ehrhardt, D. W., Boudaoud, A. and Messina, C. (2022). Fifteen compelling open questions in plant cell biology. *Plant Cell* 34(1): 72–102.

Rosegrant, M. W., Ringler, C. and Zhu, T. (2009). Water for agriculture: Maintaining food security under growing scarcity. *Annu. Rev. Environ. Resour.* 34(1): 205–222.

Rötter, R. P., Tao, F., Höhn, J. G. and Palosuo, T. (2015). Use of crop simulation modelling to aid ideotype design of future cereal cultivars. *J. Exp. Bot.* 66(12): 3463–3476.

Shekoofa, A., Sinclair, T. R., Messina, C. D. and Cooper, M. (2015). Variation among maize hybrids in response to high vapor pressure deficit at high temperatures. *Crop Sci.* 55: 392–396.

Simmons, C. R., Lafitte, H. R., Reimann, K. S., Brugiere, N., Rösler, K., Albertsen, M. C., Greene, T. W. and Habben, J. E. (2021). Successes and insights of an industry biotech program to enhance maize agronomic traits. *Plant Sci.* 307: 110899.

Sinclair, T. R. (2011). Challenges in breeding for yield increase for drought. *Trends Plant Sci.* 16(6): 289–293.

Sivasakthi, K., Tharanya, M., Kholová, J., Wangari Muriuki, R., Thirunalasundari, T. and Vadez, V. (2017). Chickpea genotypes contrasting for vigor and canopy conductance also differ in their dependence on different water transport pathways. *Front. Plant Sci.* 8: 1663. https://doi.org/10.3389/fpls.2017.01663.

Snowdon, R. J., Wittkop, B., Chen, T. W. and Stahl, A. (2021). Crop adaptation to climate change as a consequence of long-term breeding. *Theor. Appl. Genet.* 134(6): 1613–1623.

Technow, F., Messina, C. D., Totir, L. R. and Cooper, M. (2015). Integrating crop growth models with whole genome prediction through approximate Bayesian computation. *PLoS ONE* 10(6): e0130855.

The Rockefeller Foundation. (2021). True cost of food: Measuring what matters to transform the US food system. Available at: https://www.rockefellerfoundation.org/report/true-cost-of-food-measuring-what-matters-to-transform-the-u-s-food-system/.

Tieman, D., Zhu, G., Resende Jr., M. F., Lin, T., Nguyen, C., Bies, D., Rambla, J. L., Beltran, K. S. O., Taylor, M., Zhang, B., Ikeda, H., Liu, Z., Fisher, J., Zemach, I., Monforte, A., Zamir, D., Granell, A., Kirst, M., Huang, S. and Klee, H. (2017). A chemical genetic roadmap to improved tomato flavor. *Science* 355(6323): 391–394.

Van Ittersum, M. K., Cassman, K. G., Grassini, P., Wolf, J., Tittonell, P. and Hochman, Z. (2013). Yield gap analysis with local to global relevance – A review. *Field Crops Res.* 143: 4–17.

Webber, H., Ewert, F., Olesen, J. E., Müller, C., Fronzek, S., Ruane, A. C., Bourgault, M., Martre, P., Ababaei, B., Bindi, M., Ferrise, R., Finger, R., Fodor, N., Gabaldón-Leal, C., Gaiser, T., Jabloun, M., Kersebaum, K. C., Lizaso, J. I., Lorite, I. J., Manceau, L., Moriondo, M., Nendel, C., Rodríguez, A., Ruiz-Ramos, M., Semenov, M. A., Siebert, S., Stella, T., Stratonovitch, P., Trombi, G. and Wallach, D. (2018). Diverging importance of drought stress for maize and winter wheat in Europe. *Nat. Commun.* 9(1): 4249.

Xiong, W., Reynolds, M. P., Crossa, J., Schulthess, U., Sonder, K., Montes, C., Addimando, N., Singh, R. P., Ammar, K., Gerard, B. and Payne, T. (2021). Increased ranking change in wheat breeding under climate change. *Nat. Plants* 7(9): 1207–1212.

Yin, X., Stam, P., Kropff, M. J. and Schapendonk, A. H. C. M. (2003). Crop modeling, QTL mapping, and their complementary role in plant breeding. *Agron. J.* 95(1): 90–98.

Yu, J., Pressoir, G., Briggs, W. H., Bi, I. V., Yamasaki, M., Doebley, J. F., McMullen, M. D., Gaut, B. S., Nielsen, D. M., Holland, J. B., Kresovich, S. and Buckler, E. S. (2006). A unified mixed-model method for association mapping that accounts for multiple levels of relatedness. *Nat. Genet.* 38(2): 203–208.

Zhang, Z., Ersoz, E., Lai, C. Q., Todhunter, R. J., Tiwari, H. K., Gore, M. A., Bradbury, P. J., Yu, J., Arnett, D. K., Ordovas, J. M. and Buckler, E. S. (2010). Mixed linear model approach adapted for genome-wide association studies. *Nat. Genet.* 42(4): 355–360.

Zhao, Z., Wang, E., Kirkegaard, J. A. and Rebetzke, G. J. (2022). Novel wheat varieties facilitate deep sowing to beat the heat of changing climates. *Nat. Clim. Change* 12(3): 291–296.

Chapter 4

Developing more integrated approaches in models assessing the impact of climate change on agriculture

Ian P. Holman, Cranfield University, UK and Global Change Research Centre, Czech Republic; and Paula A. Harrison, UK Centre for Ecology & Hydrology, UK and Global Change Research Centre, Czech Republic*

1 Introduction
2 Addressing key challenges in integrated modelling
3 Case study
4 Conclusion
5 Future trends in research
6 Acknowledgements
7 Where to look for further information
8 References

1 Introduction

Agriculture is widely recognised as being highly sensitive to climate change (Sultan and Gaetani, 2016; Ciscar et al., 2018; Anderson et al., 2020), with potential impacts extending from primary production to the food supply chain (Duchenne-Moutien and Neetoo, 2021; Niles et al., 2018). The likelihood of 'high-end' climate change that is associated with global temperature rises of greater than +2°C is increasingly plausible, despite some strengthening of agreements at the Glasgow COP26, unless deep reductions in carbon dioxide and other greenhouse gas emissions occur in the coming decades (IPCC, 2021).

Over multi-decadal to century timescales, the likelihood of significant and unforeseen socio-economic changes, in addition to climate change, becomes increasingly likely, which may enable or hinder climate change adaptation and mitigation in the agricultural sector and across food supply systems. Consequently, to understand the vulnerability of agricultural systems to future

* Corresponding author: i.holman@cranfield.ac.uk.

http://dx.doi.org/10.19103/AS.2022.0115.05

change, assessments are needed that take account of uncertainties due to climate change combined with plausible future societal change, including how collectively these may affect potential land-ocean-atmosphere tipping points (Cai et al., 2016). These interlinked drivers of change are captured within the scenario framework of Representative Concentration Pathways (RCPs) and Shared Socioeconomic Pathways (SSPs) developed by the scientific community for use in the Intergovernmental Panel on Climate Change assessments.

Modelling tools, such as process models, statistical models and integrated assessment models, are commonly used to explore agricultural impacts of, and adaptations to, climate change under scenarios such as the RCPs and SSPs (Rötter et al., 2018; Ciscar et al., 2018). At the global scale, Integrated Assessment Models (IAMs) (e.g. van Vuuren et al., 2011), which combine representations of the economic, social and natural systems, are used to project impacts across a range of sectors, including agriculture, but are limited in the context of agricultural simulation by spatial differentiation and process representation (Rounsevell et al., 2014). At finer scales, a larger range of modelling methodologies are applied that span 'top-down' (such as statistical, optimisation, equilibrium or other equation-based models; e.g. Alexander et al., 2017a) and 'bottom-up' (such as agent-based, multi-agent or cellular-automata; e.g. Brown et al., 2016; Brown et al., 2021) approaches.

Previous modelling studies have demonstrated that non-climatic pressures such as future socio-economic change can be as, if not more, important as climate change (Harrison et al., 2016). Climate change impacts on agriculture will interact with those arising from socio-economic changes at scales from (sub)national (e.g. Padle et al., 2021) to regional (e.g. Mitter et al., 2020; Kok et al., 2019) to global (O'Neill et al., 2020) in complex, non-additive ways (Harrison et al., 2016; Adger et al., 2018). In a similar manner to climate change, socio-economic change can act on agriculture directly through changing agricultural policy (Mitter et al., 2020), food demands (e.g. population growth: Terama et al., 2019; dietary change: Hess et al., 2015; Weindl et al., 2017), food supply networks and agricultural innovation. They can also act indirectly through changing land availability (e.g. through urbanisation: Li et al., 2017a; Terama et al., 2019) and cross-sectoral policy impacts (e.g. flood risk management: Arnell and Lloyd-Hughes, 2014; afforestation: Lee et al., 2019; water resource management: Sabbaghi et al., 2020; biodiversity policy: Ekroos et al., 2016).

Interactions between water resources, water allocation and agricultural water demand are particularly important within assessments of irrigated croplands (Winter et al., 2017) and ignoring such cross-sectoral interdependencies will lead to an incomplete and potentially misleading understanding of climate change impacts. Wimmer et al. (2015) integrated metamodels of water resource availability, non-agricultural water demand (domestic, manufacturing and electricity generation) and agricultural irrigation demand to evaluate the relative performance of a range of water allocation schemes in meeting water demands, whilst reducing cross-sectoral impacts on water use sectors and aquatic ecosystems across climate and socio-economic scenarios. Their findings illustrate the benefits of combined solutions of water demand management and cross-sectoral water allocation and the need for local-level water allocation policies to consider displacement effects.

Lee et al. (2019) explored the challenges of increasing European forest extent as part of land-based mitigation whilst meeting minimum levels of European food provision (from domestic production and imports) under RCP2.6. Their results showed significant land-use changes and regional trade-offs in the 2050s in Europe and highlighted that substantial changes in both the supply side (e.g. crop yield and irrigation efficiency improvement) and demand side (dietary change to diets with less or no meat) of the European food system are needed if increased food imports and associated displacement of food production and potential deforestation are to be avoided.

Understanding the consequences of global environmental changes on agriculture therefore requires understanding intrinsically regional phenomena (e.g. climate change, climate impacts, land-use decision-making, policy responses, etc.) within an integrated (holistic) framework (Hibbard and Janetos, 2013) that takes account of interactions between sectors as they compete for land and water, as well as interactions and feedbacks between subsystems and the agricultural sector (e.g. flood risk, afforestation, food demands and diet, etc.). Taking a sectoral or 'silo-ed' agricultural modelling perspective ignores such interactions and risks the over- or underestimation of future impacts, which may in turn influence decisions regarding the need for adaptation (Harrison et al., 2016).

In developing more integrated/holistic modelling approaches, we identify four key challenges for improving the application of regional integrated modelling for projecting future impacts of, and adaptation to, climate change in agricultural systems:

- Better characterisation of the importance of model structural uncertainty, and its propagation through the modelled system, in integrated modelling;
- Combining socio-economic drivers of impacts with climate change drivers to better reflect the importance of non-climate drivers on both the supply and demand sides of agriculture;

- Enabling socio-economic scenarios to condition adaptation responses to better recognise that adaptive capacity is not constant but constrained by evolving socio-economic conditions;
- Representing adaptation as an objective purposive response to impacts with specific targets or goals, rather than an arbitrary response to change.

2 Addressing key challenges in integrated modelling

2.1 Taking account of model structural uncertainty in integrated modelling

There are many sources of uncertainty in integrated modelling (Rounsevell et al., 2021), but we focus on model structural uncertainty (and its propagation) because of its significance for the necessary modelling of many subsystems within integrated approaches (Kirchner et al., 2020) - from crop and timber yields to farm management, from hydrology to water resources management, from urban expansion to biodiversity impacts. The diversity of model paradigms and models means that uncertainty due to model structure is ubiquitous and significant, whether simulating crop yield (e.g. Tao et al., 2018; Fronzek et al., 2018), livestock emissions (e.g. Hutchings et al., 2018) or land use-land cover change (e.g. Alexander et al., 2017a; Brown et al., 2021). Whilst there are a considerable range of broad-scale agricultural systems and farm models (Reidsma et al., 2018; Huber et al., 2018), the underpinning biophysical crop models are dominated by the bulk commodity crops (wheat: Rodriguez et al., 2019; barley: Tao et al., 2020; rice: Zhang et al., 2017; potatoes: Fleisher et al., 2017; oilseeds, etc.) and intensive grasslands (e.g. Korhonen et al., 2018; Hjelkrem et al., 2017), with fewer model options for horticultural crops (Marcelis et al., 1998), rotational break crops and heterogeneous extensive swards (Kipling et al., 2016), which are important for being able to represent farm-level rotations and multifunctional, heterogeneous landscapes.

As an example of exploring the consequences of model structural uncertainty, Tao et al. (2018) used an ensemble of seven process-based crop models to simulated barley yields under RCP8.5 using eight contrasting Global Climate Models (GCMs) from the CMIP5 (Climate Model Intercomparison Project 5) ensemble for field sites in Finland (Boreal climatic zone) and Spain (Mediterranean climatic zone). Across the eight climate projections, the median simulated yield change in the 2050s ranged from 34% to +8% in Finland and 39% to +138% in Spain, showing that the models introduced uncertainty in the magnitude of change and inconsistency in the direction of change. Using a single set of calibrated crop model parameters, crop model structure contributed 60% and 65% of the total variance of the ensemble output for the Finnish and Spanish sites, respectively.

However, such uncertainty is not confined to field-scale simulation. Hutchings et al. (2018) compared the behaviour of four farm-scale models under eight dairy farming scenarios. Despite specifying key production characteristics within the scenario (such as milk yield per cow; nitrogen fertilisation rates), model differences in process representation, feedback loops driving production and emissions, and the extent to which farm management is internalised led to significant differences between models in key outputs, such as annual milk production per hectare, the amounts of N fertiliser and concentrate feed imported, simulated dairy cow numbers and fodder maize area.

At a higher scale, Brown et al. (2021) compared European land-use simulations using models with contrasting paradigms but common inputs (including crop yields and boundary condition food trade flows) under a range of climatic and socio-economic scenarios. The models were a 'top-down' constrained optimising economic-equilibrium model and a 'bottom-up' stochastic agent-based model. The models produced very different results in some scenarios with regard to simulated food production and the extent of intensive agriculture. The agent-based model simulated food shortages, whilst the optimisation model maintained the required food supply through expansion/contraction of agricultural production in the most profitable areas, sometimes leading to agricultural land abandonment or major land-use change in extensive areas of Europe. An analysis of global and European land cover projections from 18 models by Alexander et al. (2017a) showed a similar wide range in land cover area projections, with the greatest variability in future cropland areas, which was influenced by model typology. Fitton et al. (2019) further analysed the global model simulations from Alexander et al. (2017a) to assess the regional uncertainty in the projected changes in agricultural land area for selected future scenarios, showing that model uncertainty differed between scenarios and regions (Fig. 1).

These studies demonstrate the need to recognise the importance of model structural uncertainty at multiple scales within integrated modelling, particularly as uncertainties propagate through the modelled system (Dunford et al., 2015; Brown et al., 2015). At the level of individual models, determining which model or model type is 'best' for a specific purpose is challenging given the (1) limited historic time series and (2) that the ability of a given model to reproduce historic data does not ensure that future conditions will be adequately simulated (Alexander et al., 2017a). Where possible, therefore, a diverse set of models and approaches should be used to more fully quantify model uncertainty and to ensure that outcomes from particular models or approaches do not dominate (Alexander et al., 2017a). Previous research has suggested that the median value of a multimodel ensemble is more accurate

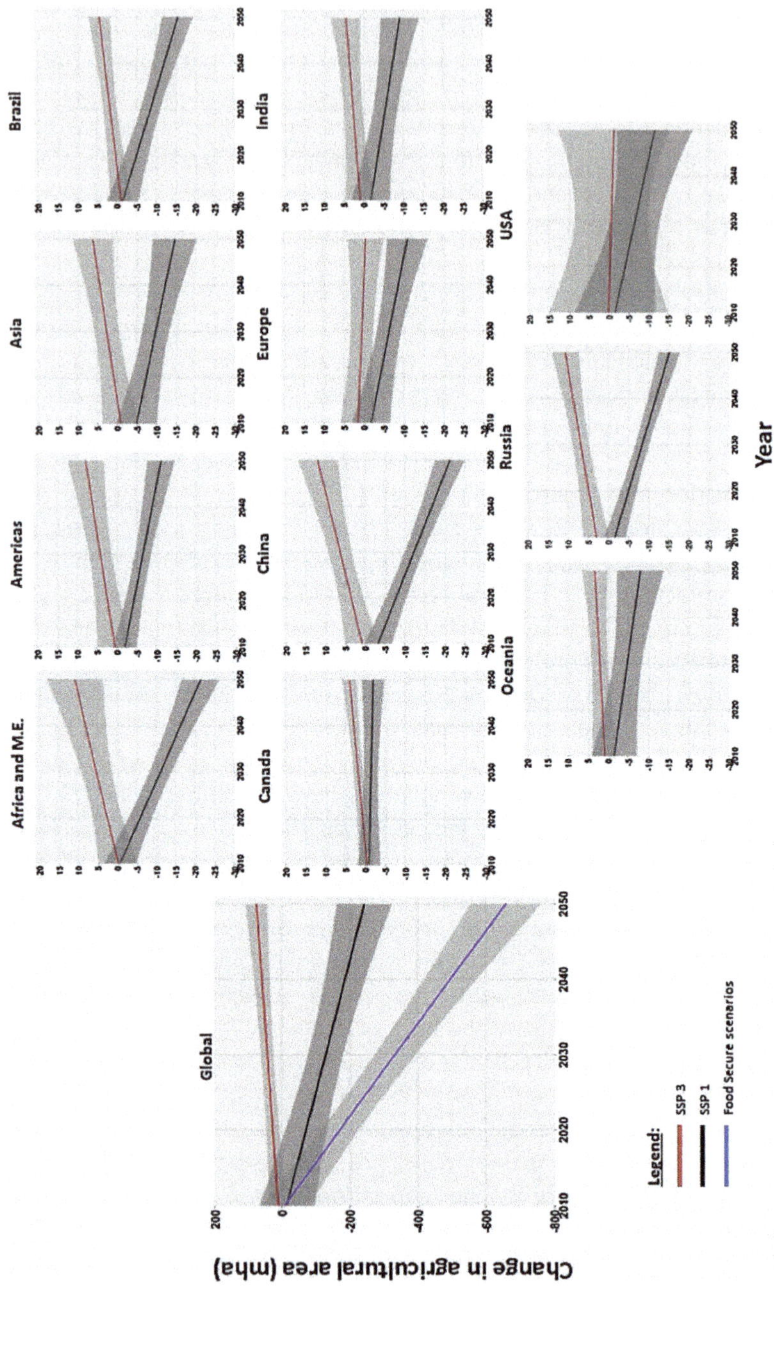

Figure 1 Change in the total agricultural area (cropland and pasture) between 2010 and 2050 for SSP1 and SSP3, and Food Secure scenarios relative to the SSP2 (baseline) scenario. All values are in millions of hectares (Mha). Straight lines were fitted between 2010 and 2050 using the mean of all land use change models, with grey areas representing the 95% C.I. across models. Source: Adapted from Fitton et al. (2019).

in simulating the crop temperature response and yields than any single model (Asseng et al., 2013; Martre et al., 2015), but care must be taken to avoid hiding important uncertainty such as when there is divergence in the directions of change of individual models (Tao et al., 2018).

Within integrated modelling, the nature of the coupling between individual biophysical and/or socio-economic models strongly influences the ability to assess the consequences of model structural uncertainty and how such uncertainties (or errors) propagate through the model chain. Soft coupled models, where data is passed between individual but separate models, enable model substitution (at least in theory) such that the consequences of individual model structure can be assessed. In contrast, hard coupling where models are embedded within a fixed software architecture provides more limited opportunities – this is the case with most global Integrated Assessment Models where structural uncertainty assessment is done at the level of the whole model (rather than individual model components) within Model Intercomparison Projects (MIPs). However, Dunford et al. (2015) combined fuzzy set methods (applied to modeller assessment of the 'holistic uncertainty' of each model variable) and network analysis methods applied to a variable-to-variable network of the regional Integrated Assessment Platform (IAP: Harrison et al., 2016) to assess compound uncertainty through the model. Also using the IAP, Brown et al. (2015) systematically sampled across the (uncertain) model input parameter ranges for four European socio-economic scenarios to show that data, model and scenario uncertainties did not entirely erode differences between scenarios.

2.2 Incorporating socio-economic drivers of impacts

Agricultural production, whether considered at the scale of a farm, a nation or a continent, is subject to multiple environmental and human system interactions that occur across multiple scales (Fig. 2). These include (1) competition for land between agriculture, urban and leisure development and bioenergy (Hendry et al., 2018), but also between space for food, forests and biodiversity; (2) competition for water in irrigated farming with public water supply, industry, navigation and the environment (Wimmer et al., 2015) and also between upstream and downstream river catchments (Henriques et al., 2008; Winter et al., 2017); (3) policy spillover effects, for example, flood and coastal defence policy affecting floodplain land suitability for agriculture (Mokrech et al., 2008; Toimil et al., 2017) or water quality regulation limiting agricultural nutrient use; (4) changing endogenous agricultural demand as a consequence of dietary preferences (Hess, 2021; Henry et al., 2019; Alexander et al., 2017b) and global food supply chains (Hess and Sutcliffe,2018).

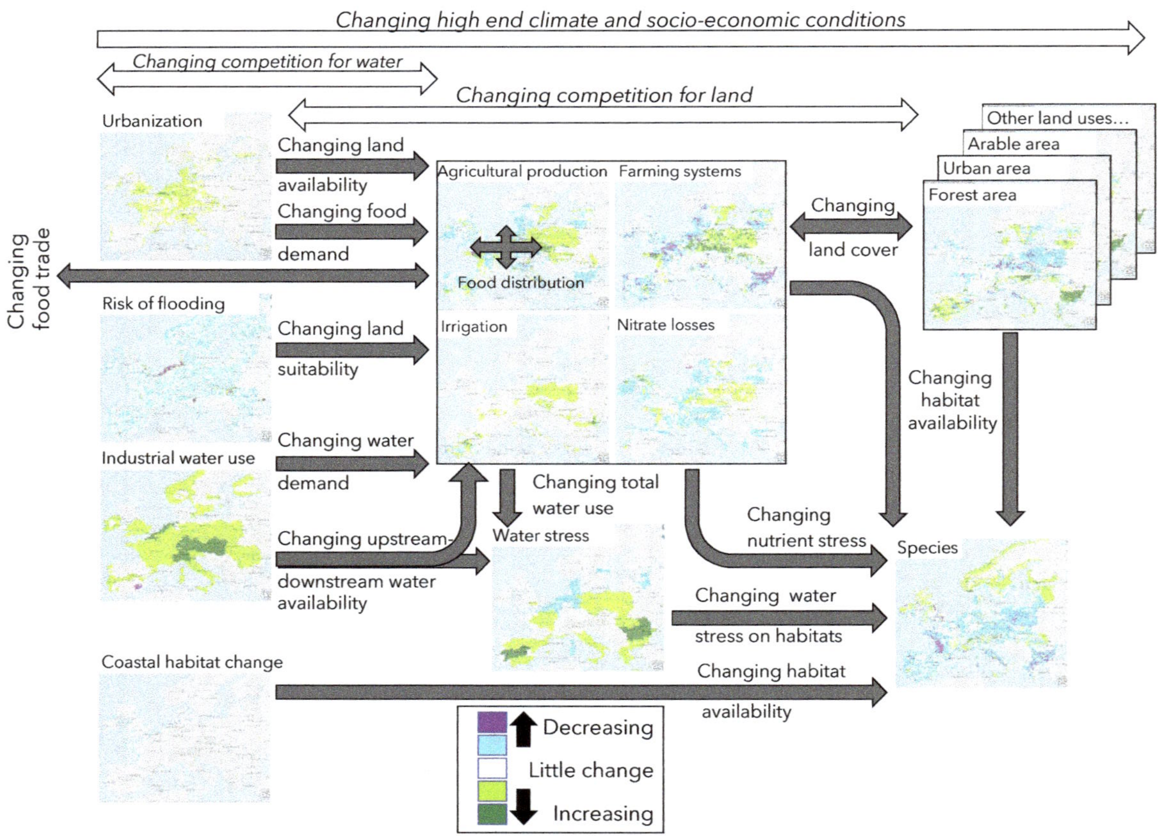

Figure 2 Illustrative cross-sectoral interactions and competition within the European Integrated Assessment Platform (IAP) that lead to complex trade-offs and synergies between water, land, biodiversity and people. Source: Adapted from www.highendsolutions.eu.

Many of these interactions are influenced by non-climate drivers as well as climate change. The Shared Socioeconomic Pathways, which replace the earlier Special Report on Emissions Scenario (SRES), describe five plausible scenarios of societal development out to 2100 based on different socio-economic challenges to climate change adaptation and mitigation. Each scenario describes internally consistent assumptions about future trajectories of a wide range of socio-economic indicators, such as population growth, access to education, urbanisation, economic growth, resource availability, technological development and drivers of demand, such as lifestyle. They include worlds of sustainability-focused growth and equality (SSP1); 'middle of the road' continuation of past trends (SSP2); fragmentation and nationalism (SSP3); increasing inequality (SSP4); and rapid and unconstrained growth (SSP5). The contrasting nature of the different SSP futures means that they will differ in all factors that influence the demand for food (e.g. population, demographics, diet) and agricultural production (e.g. food trade, legislative constraints, agricultural subsidies, agricultural technology and R&D).

The global SSPs provide qualitative narratives and sub-global quantification of selected variables (e.g. population, GDP, etc.) to inform integrated assessments at large scales, but their spatial resolution and sectoral richness are insufficient to directly be used in regional agriculture and land-use studies (Li et al., 2017b; Audsley et al., 2015). This has led to many exercises in downscaling and enriching the SSPs for specific regions (e.g. Padle et al., 2020; Kok et al., 2019) and sectors (e.g. Mitter et al., 2020; Lehtonen et al., 2021). For example, the Shared Socioeconomic Pathways for European agriculture and food systems (or Eur-Agri-SSPs) of Mitter et al. (2020) provide rich narrative storylines for each of the five Eur-Agri-SSPs and directions of change for key storyline elements (Fig. 3). However, such outputs are descriptive and qualitative, requiring complementary methods such as fuzzy sets (e.g. Pedde et al., 2019; Kok et al., 2014) and Bayesian reasoning (Kemp-Benedict et al., 2013) approaches to address the 'conversion challenge' of quantifying the narratives that is required

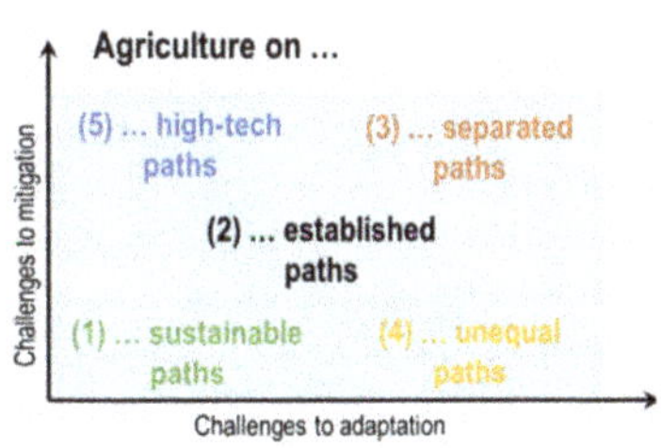

Storyline elements [and no.]

- Population, urbanisation and farming demographics [11]
- Economic model [3]
- International trade and global markets [5]
- Domestic food demand patterns [5]
- Costs, employment and prices [7]
- Policies and institutions [9]
- Agricultural and rural development support [4]
- Technology [3]
- Environment and natural resources [5]

Figure 3 The five Eur-Agri-SSPs and the qualitative storyline elements. Source: Adapted from Mitter et al. (2020).

to objectively represent the scenarios within modelling. Nevertheless, assuming that only the climate will change and that future socio-economics will either explicitly or implicitly follow a Business-As-Usual projection is indefensible.

2.3 Enabling socio-economic scenarios to condition adaptation

Given the extreme contrasts in socio-economic futures represented by the five SSPs, it is inevitable that they will pose differing challenges for adaptation within agriculture. Each SSP future will impose different economic, technological, knowledge, social, ethical, cultural and governance constraints on the uptake and effectiveness of adaptation options (Moser and Ekstrom, 2010). Adaptation will not occur instantaneously, but will involve multiple time lags within policy development and on-the-ground uptake (Brown et al., 2019). Uptake of adaptation actions is likely to be a progressive, spatially structured process that depends upon knowledge, sociocultural context, personal experience and adaptation champions (Hamilton-Webb et al., 2017). These time lags in uptake mean that adaptation actions can fail to achieve their expected adoption in some contexts (Brown et al., 2019).

Consequently, there needs to be a more rigorous incorporation of adaptation effectiveness within modelling. Holman et al. (2019) found few land and water modelling studies allowed scenario constraints (Fig. 4) to condition adaptation over time (e.g. Papadimitriou et al., 2019a; Brown et al.,

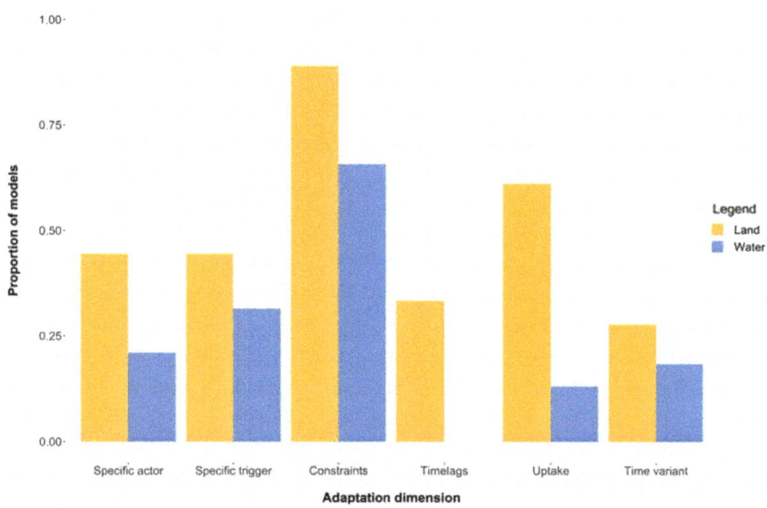

Figure 4 The proportion of reviewed land and water sector models incorporating different adaptation dimensions. Source: Adapted from Holman et al. (2018).

2016; Steinbuks and Hertel, 2016; Murray-Rust et al., 2014). Furthermore, Adger et al. (2018) and Patt et al. (2010) considered the current generation of integrated assessment models to be deficient in elucidating the role of adaptation in reducing risks. Therefore, the limited consideration of the adaptation challenges posed by the socio-economic scenario context in most studies suggests that both the uptake of adaptation measures and the benefits obtained will be overestimated (Harrison et al., 2016; Patt et al., 2010).

2.4 Representing adaptation as an objective purposive response

The conventional approach in climate change impact studies is to combine climate change projections (through RCPs) with socio-economic projections (through SSPs), so as to recognise the interdependency of climatic, economic, social and environmental drivers of change. However, using such exploratory scenarios does not provide the normative preferences or desires that inform adaptation planning, thereby providing no explicit goal for modelled adaptation. Holman et al. (2019) found that adaptation responses were commonly modelled as a consequence of either (1) triggers relating to underlying model assumptions (such as the enforced equilibrium between food supply and demand) or (2) the modellers' subjective interpretation of reductions/increases in model output variables, thereby equating a change in state as an unacceptable impact.

Transformative Climate Science (Tàbara et al., 2019) aims to move the narrative of 'climate change impact, adaptation and vulnerability' (CCIAV) away from focusing on responses to the negative impacts of a given scenario towards driving action to achieve a positive future, articulated in a scenario-independent vision (Fig. 5). In an agricultural context, such a vision may express desires regarding adoption of healthy diets, agricultural GHG emissions, domestic food production, etc. By making values, aspirations and choices for a faraway future explicit (Wiek and Iwaniec, 2014; Constanza, 2000) visions can be important tools and reference points for guiding, assessing and inspiring strategic choices and action in the short- and midterm to facilitate radical and systemic change in the long-term (Pereira et al., 2018). By articulating a desired scenario-independent outcome, visions support moving away from scenario optimisation approaches that implicitly minimise future scenario uncertainty towards identifying robust no/low regret adaptation options that are beneficial across the scenario uncertainty space.

Deriving vision targets that can guide action requires adequate stakeholder representation to engender legitimacy within the vision and associated target. This can be gained from participation of stakeholders that provide

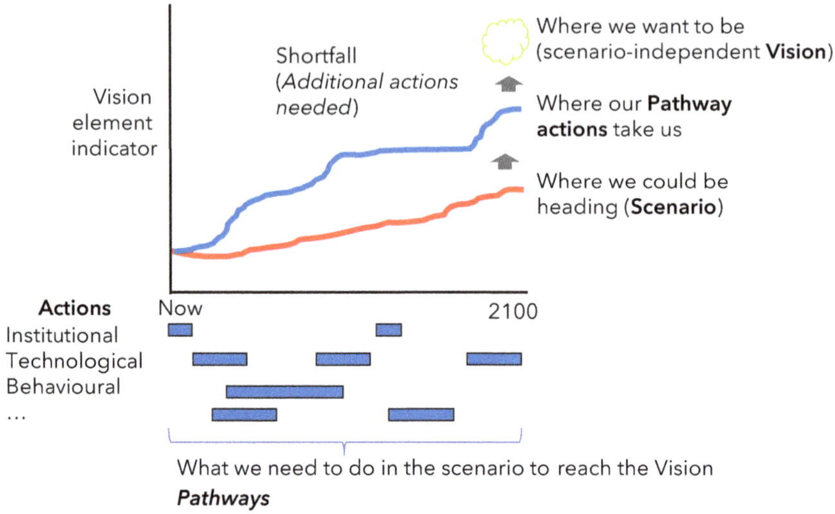

Figure 5 Illustration of how visions and pathways can enable more deliberative identification of adaptation needs.

gender-balance across different societal groups (decision-making, business and industry, civil society organisations), age groups, sectors and geographical regions (Gramberger et al., 2015). Vision targets may be explicit in the vision or may need a translation and quantification approach (e.g. Pedde et al., 2019; Schönenberg et al., 2017; Kok et al., 2014).

Aligned to this, the 'pathways' framing from the transformations literature has been increasingly adopted within climate change adaptation research as a means of enabling a decision-focused approach (Wise et al., 2014; Haasnoot et al., 2013; Rosenbloom, 2017; Werners et al., 2021). In a modelling context, pathways represent a shift away from simulating independent incremental changes in model inputs that represent a series of adaptation actions (such as increasing irrigation efficiency, potential yields, etc.) towards developing a purposive set of scheduled actions (which include addressing the technological, social, governance, etc. enablers of such actions) that move states towards the desired outcome (the Vision). They can also help to design a shift from incremental adaptation that may be inadequate to deal with rapid shifts and tipping points for food production under climate change to transformational adaptation in agriculture (Vermeulen et al., 2018; Panda, 2018. Frantzeskaki et al. (2019) describe the co-development of a series of pathways encapsulating mitigation, adaptation and transformational actions to move Europe (including its agricultural sector) closer to a sustainability-oriented vision.

3 Case study

The IMPRESSIONS Integrated Assessment Platform (IAP) is a freely available online integrated modelling platform for Europe that contains downscaled European SSPs (Kok et al., 2019) and RCP-based climate projections (Madsen et al., 2016). The IAP combines a series of linked models that simulate the response of urban development, flooding, agriculture, forestry, water resources, water demands and biodiversity to climate and socio-economic change. It recognises that agricultural production in Europe is influenced by changing endogenous and exogenous food demand; that there are competing demands for land to meet human development needs, timber production and biodiversity protection; that the quality of agricultural land is affected by the changing climate and flood risk; and that irrigated agricultural production is in competition for water with drinking water supply, industry and environmental needs. The IAP has been extensively reported (e.g. Harrison et al., 2016; Holman et al., 2016; Papadimitriou et al., 2019 a & b) and tested, including sensitivity analyses (Kebede et al., 2015; Fronzek et al., 2019) and uncertainty assessments (Dunford et al., 2014; Brown et al., 2015). The IAP takes a novel approach (Fig. 6) to conditioning adaptation with scenario-specific limits to the model inputs that take account of compatibility of the modelled adaptation response with the underlying scenario logic and the availability of the associated limiting capital (financial, social, human or manufactured). The degree of effort invested in a given adaptation determines proximity to these scenario-specific limits, which the modeller can base on expert-judgement or on the effort implicit in adaptation pathways (e.g. Frantzeskaki et al., 2019).

Papadimitriou et al. (2019b) used the IAP to assess the combined impacts of future climate and socio-economic change within the Czech Republic to help inform national adaptation planning. By

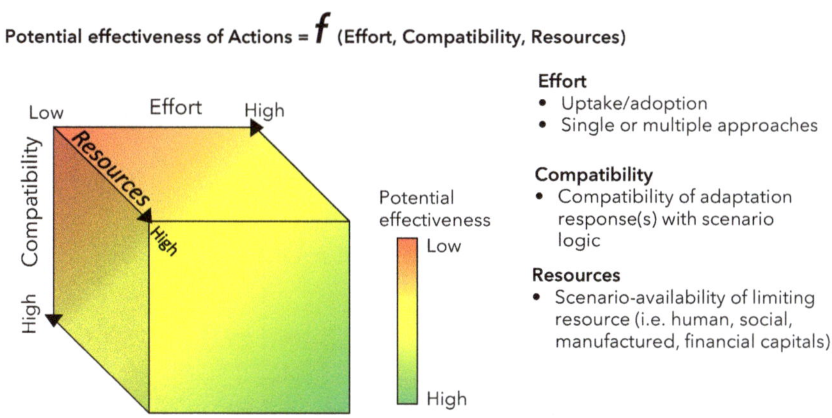

Figure 6 Schematic illustration of how the IAP reflects the potential effectiveness of adaptation actions in model input changes.

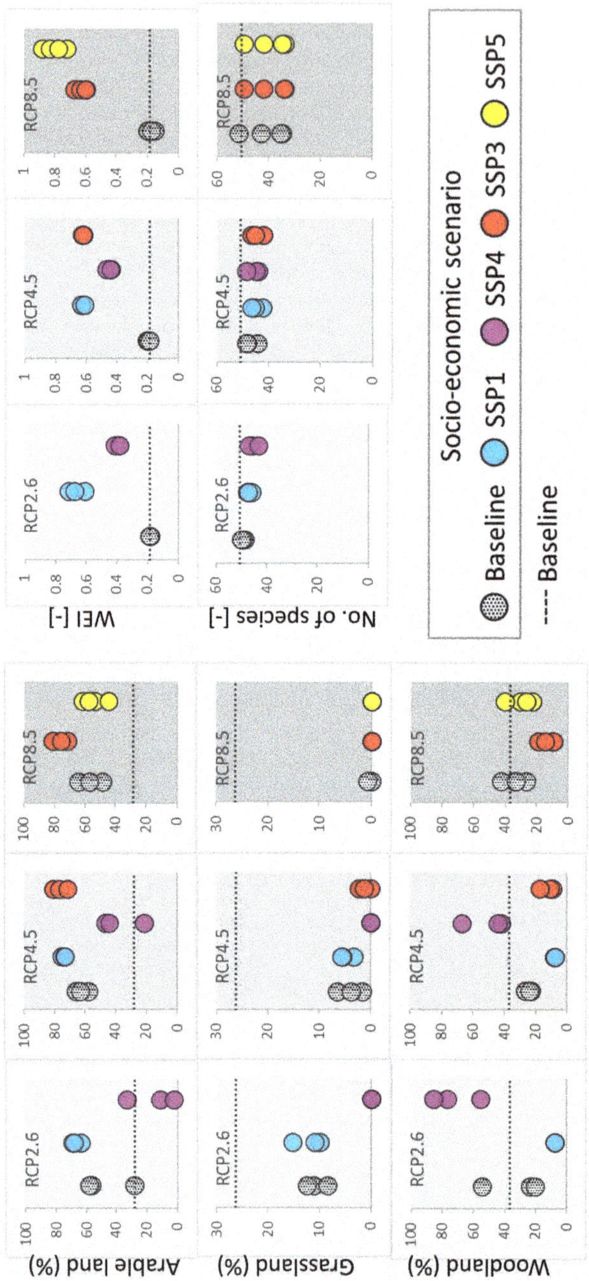

Figure 7 Land use, water and biodiversity simulations from the Integrated Assessment Platform 2 (IAP2), for the baseline (dashed line) and 2080s future time period (coloured dots), aggregated for the Czech Republic. Results are shown per Representative Concentration Pathway (RCP) and Shared Socioeconomic Pathway. Dots of the same colour represent different global climate model (GCM)-regional climate model (RCM) combinations of the ensemble used in each RCP simulation. Source: Adapted from Papadimitriou et al. (2019b).

evaluating future national agricultural production within a European-scale modelling framework, the IAP illustrated the changing role of the country's agricultural sector within its regional and continental domain. The Czech Republic was projected to have increasing importance as a crop-growing region in Europe across most futures, with arable land expanding at the expense of livestock farming and forestry. This reflected the country gaining a future competitive advantage in European agriculture within the assumed frictionless food market, as southern agricultural regions become limited by aridity, temperature and water resource availability constraints and northern regions gained improved but still less productive agroclimates. However, the simulated agricultural expansion was associated with increased water exploitation and reduced biodiversity highlighting the need for policymakers to develop national multisectoral adaptation actions to reduce adverse impacts and realise opportunities (Fig. 7).

Papadimitriou et al. (2019a) used the IAP to perform an integrated multi-objective assessment of the ability of adaptation strategies to improve a spectrum of indicators, relating to sustainable development, under the adaptation constraints imposed by the SSPs. The adaptation strategies included sustainable water management, forest expansion, land sharing, agricultural

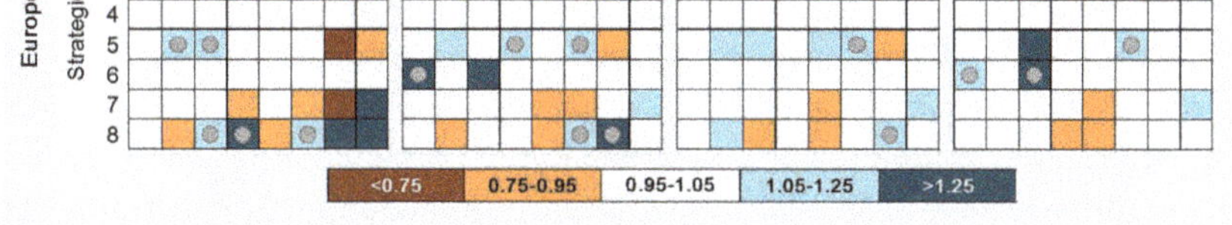

Figure 8 Effects of adaptation strategies on Sustainable Development Indicators (SDI) for different socio-economic scenarios (SSPs) combined with RCP2.6 for Europe expressed as Strategy X/No adaptation. Adaptation strategies correspond to the following: 1. Sustainable water management; 2. Maximising forest area; 3. Land sharing; 4. Flood protection; 5. Sustainable behaviours; 6. Human and social capital; 7. Bioenergy; 8. Agricultural intensification for land sparing. Blue colour hues represent improvements in the SDIs (greater than 5%) and orange hues deteriorations (greater than 5%). The grey dots indicate that the improved SDI is at the same or better state as at the baseline period. Source: Adapted from Papadimitriou et al. (2019a).

intensification for land sparing and sustainable behaviours. Whilst the implementation of some adaptation strategies enabled selected Sustainable Development Indicators (SDIs) (those marked with grey dots in Fig. 8) to return to their baseline state, most had residual impacts. Almost all of the adaptation strategies had unintended consequences on selected SDIs suggesting the existence of unavoidable trade-offs.

Both applications of the IAP demonstrate the socio-economic scenario dependency of impact and adaptation outcomes, emphasising the need for socio-economic futures to be included within CCIAV studies. They also show the need to develop holistic adaptation strategies that take account of the synergies and trade-offs between sectoral adaptation strategies, sectors and regions. Both are essential if researchers are to avoid providing over-certain and over-optimistic assessments.

4 Conclusion

Agriculture and food production systems are complex and multiscale with multiple interacting relationships with other environmental and human systems and policy arenas. Agriculture is also directly and indirectly impacted by both climate and non-climate drivers of change. Understanding how agriculture will be affected by, and respond to, future climate and socio-economic changes requires integrated models that recognise these complex dependencies rather than siloed sectoral modelling approaches that provide an incomplete representation.

Integrated modelling approaches take a more holistic or systemic approach to evaluating the autonomous and planned responses to future change, which recognises that agricultural adaptation does not act in isolation from, and can be affected by, behavioural, technological and policy responses in other sectors.

Integrated modelling studies that recognise both the uncertainty in climate and socio-economic change and the sensitivity of the uptake and effectiveness of agricultural adaptation options to socio-economic change will provide a more comprehensive understanding of the future challenges and opportunities facing agriculture and an improved basis for planning responses.

5 Future trends in research

We identify four future trends in research:

- Improved integration of climate and socio-economic scenarios within regional integrated modelling: This will be associated with a move away from the simple combination of independent climate and non-climate (socio-economic) scenarios with no interactions, towards integrated scenarios which incorporate co-evolution of drivers (Adger et al., 2018).

- Multiscale model coupling to better represent boundary conditions: The globalisation of food supply and the international trade of commodities as well as environmental interactions gives rise to tele-coupling effects, where changes occurring in other regions of the globe may influence land use in a particular region or country of interest. Whilst many downscaled socio-economic scenarios provide a descriptive representation of cross-boundary flows, these provide an incomplete and uncertain representation of cross-scale flows of risks and materials (Challinor et al., 2018). This requires the increased use of global modelling outputs as inputs or boundary conditions for more detailed regional/national models. However, current global models that rely on the inherent assumption of equilibrium, such as CGE (computable general equilibrium) and PE (partial equilibrium) models, raise concerns regarding their applicability in divergent climate and/or socio-economic scenarios in which tipping points and regime shifts are likely (Müller et al., 2020).
- Increased recognition of the importance of shocks and extreme events in integrated modelling The previously discussed interactions between the climate, biophysical systems and socio-economic drivers can lead to cascading indirect effects and systemic risks, especially during and following extreme events (Adger et al., 2018). Farming, agricultural systems and food supply chains are particularly susceptible to shocks or extreme events (Hamilton et al., 2020), which can be the trigger event for change (Sutherland et al., 2012; Pedde et al., 2019). This requires improved characterisation of extreme events (climate) and shocks (socio-economic) in scenarios; modelling approaches that are able to simulate shocks, system transformations and regime shifts in socio-ecological systems (Brown and Rounsevell, 2021); and improved representation of autonomous responses to changes in variability rather than in the mean from individuals (e.g. farmers) to supply chains.
- Transition from PC-based to cloud-based modelling infrastructure to facilitate greater science-stakeholder-policy engagement: Integrated modelling platforms or frameworks that incorporate multisectoral interactions to represent system behaviour are typically inflexible and inaccessible; have runtimes that hinder rapid simulation and interactive engagement; and focus on a one-way outward flow of information (Holman and Harman, 2008). A transition to cloud-based modelling infrastructure that facilitates the use of large data volumes and complex analytical processes within high-performance computational platforms holds the potential to build trust and user relevance (Salter et al., 2010; Saltelli et al., 2020). Collaborative cloud-based environments, such as virtual labs (Hollaway et al., 2020), offer a

range of building blocks to support the development of integrated models by supporting the sharing of vocabularies and ontologies that are needed to achieve interoperability between different programming languages - potentially contributing to the ambition of digital twins of the natural environment (Blair, 2021). Such environments can also provide a modular approach to the application and integration of models that aims to eliminate reliance on a single (and inevitably uncertain) model by using 'plug-and-play' technology. This allows the interchange and intercomparison of different models, allowing the consequences of unavoidable structural uncertainty to be reflected in decision-making.

6 Acknowledgements

We acknowledge the SustES project (Adaptation strategies for sustainable ecosystem services and food security under adverse environmental conditions, CZ.02.1.01/0.0/0.0/16_019/0000797) for part-funding this work.

7 Where to look for further information

The following articles provide a good overview of key issues:

- Harrison, P. A., Dunford, R., Holman, I. P. and Rounsevell, M. D. A. (2016). Climate change impact modelling needs to include cross-sectoral interactions. Nature Climate Change 6(9): 885+.
- Brown, C., Holman, I. and Rounsevell, M. (2021). How modelling paradigms affect simulated future land-use change. Earth System Dynamics 12: 211–231.
- Kirchner, M., Mitter, H., Schneider, U. A., Sommer, M., Falkner, K. and Schmid, E. (2020). Uncertainty concepts for integrated modeling - Review and application for identifying uncertainties and uncertainty propagation pathways. Environmental Modelling and Software 135: 104905.
- Müller et al. (2020). Modelling food security: Bridging the gap between the micro and the macro scale. Global Environmental Change 63: 102085.
- Holman, I. P., Brown, C., Carter, T. R., Harrison, P. A. and Rounsevell, M. D. A. (2019). Improving the representation of adaptation in climate change impact models. Regional Environmental Change 19(3): 711–721.

8 References

Adger, W. N., Brown, I. and Surminski, S. (2018). Advances in risk assessment for climate change adaptation policy. *Philosophical Transactions: Series A, Mathematical, Physical, and Engineering Sciences* 376(2121): 20180106. DOI: 10.1098/rsta.2018.0106.

Alexander, P., Prestele, R., Verberg, P. H., Arneth, A., Baranzelli, C., Batista E. Silva, F., Brown, C., Butler, A., Calvin, K., Dendoncker, N., Doelman, J. C., Dunford, R., Engström, K., Eitelberg, D., Fujimori, S., Harrison, P. A., Hasegawa, T., Havlik, P., Holzhauer, S., Humpenöder, F., Jacobs-Crisioni, C., Jain, A. K., Krisztin, T., Kyle, P., Lavalle, C., Lenton, T., Liu, J., Meiyappan, P., Popp, A., Powell, T., Sands, R. D., Schaldach, R., Stehfest, E., Steinbuks, J., Tabeau, A., van Meijl, H., Wise, M. A. and Rounsevell, M. D. (2017a). Assessing uncertainties in land cover projections. *Global Change Biology* 23(2): 767–781. DOI: 10.1111/gcb.13447.

Alexander, P., Brown, C., Arneth, A., Dias, C., Finnigan, J., Moran, D. and Rounsevell, M. D. A. (2017b). Could consumption of insects, cultured meat or imitation meat reduce global agricultural land use? *Global Food Security* 15: 22–32. DOI: 10.1016/j.gfs.2017.04.001.

Anderson, R., Bayer, P. E., PE and Edwards, D. (2020). Climate change and the need for agricultural adaptation. *Current Opinion in Plant Biology* 56: 197–202. DOI: 10.1016/j.pbi.2019.12.006.

Arnell, N. W. and Lloyd-Hughes, B. (2014). The global-scale impacts of climate change on water resources and flooding under new climate and socio-economic scenarios. *Climatic Change* 122(1–2): 127–140. DOI: 10.1007/s10584-013-0948-4.

Asseng, S., Ewert, F., Rosenzweig, C., Jones, J. W., Hatfield, J. L., Ruane, A. C., Boote, K. J., Thorburn, P. J., Rötter, R. P., Cammarano, D., Brisson, N., Basso, B., Martre, P., Aggarwal, P. K., Angulo, C., Bertuzzi, P., Biernath, C., Challinor, A. J., Doltra, J., Gayler, S., Goldberg, R., Grant, R., Heng, L., Hooker, J., Hunt, L. A., Ingwersen, J., Izaurralde, R. C., Kersebaum, K. C., Müller, C., Naresh Kumar, S., Nendel, C., O'Leary, G., Olesen, J. E., Osborne, T. M., Palosuo, T., Priesack, E., Ripoche, D., Semenov, M. A., Shcherbak, I., Steduto, P., Stöckle, C., Stratonovitch, P., Streck, T., Supit, I., Tao, F., Travasso, M., Waha, K., Wallach, D., White, J. W., Williams, J. R. and Wolf, J. (2013). Uncertainty in simulating wheat yields under climate change. *Nature Climate Change* 3(9): 827–832. DOI: 10.1038/nclimate1916.

Audsley, E., Trnka, M., Sabaté, S., Maspons, J., Sanchez, A., Sandars, D., Balek, J. and Pearn, K. (2015). Interactive modelling of land profitability to estimate European agricultural and forest land use under future scenarios of climate, socio-economics and adaptation. *Climatic Change* 128(3–4): 215–227. DOI: 10.1007/s10584-014-1164-6.

Blair, G. S. (2021). Digital twins of the natural environment. *Patterns* 2(10): 100359. DOI: 10.1016/j.patter.2021.100359.

Brown, C., Alexander, P., Almuth, A., Holman, I. and Rounsevell, M. (2019). Achievement of Paris climate goals unlikely due to time lags in the land system. *Nature Climate Change* 9(3): 203–208.

Brown, C., Brown, E., Murray-Rust, D., Cojocaru, G., Savin, C. and Rounsevell, M. (2015). Analysing uncertainties in climate change impact assessment across sectors and scenarios. *Climatic Change* 128(3–4): 293–306. DOI: 10.1007/s10584-014-1133-0.

Brown, C., Brown, K. and Rounsevell, M. (2016). A philosophical case for process-based modelling of land use change. *Modeling Earth Systems and Environment* 2(2): 50. DOI: 10.1007/s40808-016-0102-1.

Brown, C., Holman, I. and Rounsevell, M. (2021). How modelling paradigms affect simulated future land-use change. *Earth System Dynamics* 12(1): 211-231.

Brown, C. and Rounsevell, M. (2021). How can social–ecological system models simulate the emergence of social–ecological crises? *People and Nature* 3(1): 88–103. DOI: 10.1002/pan3.10167.

Cai, Y., Lenton, T. M. and Lontzek, T. S. (2016). Risk of multiple interacting tipping points should encourage rapid CO_2 emission reduction. *Nature Climate Change* 6(5): 520-525. DOI: 10.1038/nclimate2964.

Challinor, A. J., Adger, W. N., Benton, T. G., Conway, D., Joshi, M. and Frame, D. (2018). Transmission of climate risks across sectors and borders. *Philosophical Transactions. Series A, Mathematical, Physical, and Engineering Sciences* 376(2121): 20170301. DOI: 10.1098/rsta.2017.0301.

Ciscar, J. C., Fisher-Vanden, K. and Lobell, D. B. (2018). Synthesis and review: An inter-method comparison of climate change impacts on agriculture. *Environmental Research Letters* 13(7): 070401. DOI: 10.1088/1748-9326/aac7cb.

Constanza, R. (2000). Visions of alternative (unpredictable) futures and their use in policy analysis. *Conservation Ecology* 4(1): 5–22.

Duchenne-Moutien, R. A. and Neetoo, H. (2021). Climate change and emerging food safety issues: A review. *Journal of Food Protection* 84(11): 1884–1897. DOI: 10.4315/JFP-21-141.

Dunford, R., Harrison, P. A. and Rounsevell, M. D. A. (2015). Exploring scenario and model uncertainty in cross-sectoral integrated assessment approaches to climate change impacts. *Climatic Change* 132(3): 417–432. DOI: 10.1007/s10584-014-1211-3.

Ekroos, J., Odman, A. M., Andersson, G. K. S., Birkhofer, K., Herbertsson, L., Klatt, B. K., Olsson, O., Olsson, P. A., Persson, A. S., Prentice, H. C., Rundlöf, M. and Smith, H. G. (2016). Sparing land for biodiversity at multiple spatial scales. *Frontiers in Ecology and Evolution* 3: 145. DOI: 10.3389/fevo.2015.00145.

Fitton, N., Alexander, P., Arnell, N., Bajzelj, B., Calvin, K., Doelman, J., Gerber, J. S., Havlik, P., Hasegawa, T., Herrero, M., Krisztin, T., van Meijl, H., Powell, T., Sands, R., Stehfest, E., West, P. C. and Smith, P. (2019). The vulnerabilities of agricultural land and food production to future water scarcity. *Global Environmental Change* 58: 101944. DOI: 10.1016/j.gloenvcha.2019.101944.

Fleisher, D. H., Condori, B., Quiroz, R., Alva, A., Asseng, S., Barreda, C., Bindi, M., Boote, K. J., Ferrise, R., Franke, A. C., Govindakrishnan, P. M., Harahagazwe, D., Hoogenboom, G., Naresh Kumar, S., Merante, P., Nendel, C., Olesen, J. E., Parker, P. S., Raes, D., Raymundo, R., Ruane, A. C., Stockle, C., Supit, I., Vanuytrecht, E., Wolf, J. and Woli, P. (2017). A potato model intercomparison across varying climates and productivity levels. *Global Change Biology* 23(3): 1258–1281.

Frantzeskaki, N., Holscher, K., Holman, I. P., Pedde, S., Jaeger, J., Kok, K. and Harrison, P. A. (2019). Transition pathways to sustainability in greater than 2°C climate futures of Europe. *Regional Environmental Change* 19(3): 777–789.

Fronzek, S., Carter, T. R., Pirttioja, N., Alkemade, R., Audsley, E., Bugmann, H., Flörke, M., Holman, I., Honda, Y., Ito, A., Janes-Bassett, V., Lafond, V., Leemans, R., Mokrech, M., Nunez, S., Sandars, D., Snell, R., Takahashi, K., Tanaka, A., Wimmer, F. and Yoshikawa, M. (2019). Determining sectoral and regional sensitivity to climate and socio-economic change in Europe using impact response surfaces. *Regional Environmental Change* 19(3): 679–693.

Fronzek, S., Pirttioja, N., Carter, T. R., Bindi, M., Hoffmann, H., Palosuo, T., Ruiz-Ramos, M., Tao, F., Trnka, M., Acutis, M., Asseng, S., Baranowski, P., Basso, B., Bodin, P., Buis, S., Cammarano, D., Deligios, P., Destain, M., Dumont, B., Ewert, F., Ferrise, R., François, L., Gaiser, T., Hlavinka, P., Jacquemin, I., Kersebaum, K. C., Kollas, C., Krzyszczak, J., Lorite, I. J., Minet, J., Minguez, M. I., Montesino, M., Moriondo, M., Müller, C., Nendel, C., Öztürk, I., Perego, A., Rodríguez, A., Ruane, A. C., Ruget, F., Sanna, M., Semenov,

M. A., Slawinski, C., Stratonovitch, P., Supit, I., Waha, K., Wang, E., Wu, L., Zhao, Z. and Rötter, R. P. (2018). Classifying multi-model wheat yield impact response surfaces showing sensitivity to temperature and precipitation change. *Agricultural Systems* 159: 209–224.

Gramberger, M., Zellmer, K., Kok, K. and Metzger, M. J. (2015). Stakeholder integrated research (STIR): A new approach tested in climate change adaptation research. *Climatic Change* 128(3–4): 201–214. DOI: 10.1007/s10584-014-1225-x.

Haasnoot, M., Kwakkel, J. H., Walker, W. E. and ter Maat, J. (2013). Dynamic adaptive policy pathways: A method for crafting robust decisions for a deeply uncertain world. *Global Environmental Change* 23(2): 485–498. DOI: 10.1016/j.gloenvcha.2012.12.006.

Hamilton, H., Henry, R., Rounsevell, M., Moran, D., Cossar, F., Allen, K., Boden, L. and Alexander, P. (2020). Exploring global food system shocks, scenarios and outcomes. *Futures* 123: 102601. DOI: 10.1016/j.futures.2020.102601.

Hamilton-Webb, A., Manning, L., Naylor, R. and Conway, J. (2017). The relationship between risk experience and risk response: A study of farmers and climate change. *Journal of Risk Research* 20(11): 1379–1393.

Harrison, P. A., Dunford, R. W., Holman, I. P. and Rounsevell, M. D. A. (2016). Climate change impact modelling needs to include cross-sectoral interactions. *Nature Climate Change* 6(9): 885–890.

Hendry, R. C., Engström, K., Olin, S., Alexander, P., Arneth, A. and Rounsevell, M. D. A. (2018). Food supply and bioenergy production within the global cropland planetary boundary. *PLoS ONE* 13(3): e0194695. DOI: 10.1371/journal.pone.0194695.

Henry, R. C., Alexander, P., Rabin, S., Anthoni, P., Rounsevell, M. D. A. and Arneth, A. (2019). The role of global dietary transitions for safeguarding biodiversity. *Global Environmental Change* 58: 101956. DOI: 10.1016/j.gloenvcha.2019.101956.

Hess, T. (2021). How can we avoid eating ourselves out of water? *Nature Food* 2(4): 225–226.

Hess, T. and Sutcliffe, C. (2018). The exposure of a fresh fruit and vegetable supply chain to global water-related risks. *Water International* 43(6): 746–761.

Hess, T. M., Andersson, U., Mena, C. and Williams, A. (2015). The impact of healthier dietary scenarios on the global blue water scarcity footprint of food consumption in the UK. *Food Policy* 50: 1–10.

Hibbard, K. A. and Janetos, A. C. (2013). The regional nature of global challenges: A need and strategy for integrated regional modelling. *Climatic Change* 118(3–4): 565–577.

Holman, I. P., Brown, C., Carter, T. R., Harrison, P. A. and Rounsevell, M. (2019). Improving the representation of adaptation in climate change impact models. *Regional Environmental Change* 19(3): 711–721.

Hollaway, M. J., Dean, G., Blair, G. S., Brown, M., Henrys, P. A. and Watkins, J. (2020). Tackling the challenges of 21st-century open science and beyond: A data science lab approach. *Patterns* 1(7): 100103. DOI: 10.1016/j.patter.2020.100103.

Holman, I. P., Brown, C., Carter, T. R., Harrison, P. A. and Rounsevell, M. (2019). Improving the representation of adaptation in climate change impact models. *Regional Environmental Change* 19(3): 711–721. DOI: 10.1007/s10113-018-1328-4.

Holman, I. P. and Harman, J. (2008). Preliminary evaluation of the benefits of a participatory regional integrated assessment software. *Climatic Change* 90(1–2): 169–187.

Holman, I. P., Harrison, P. A. and Metzger, M. J. (2016). Cross-sectoral impacts of climate and socio-economic change in Scotland – implications for adaptation policy. *Regional Environmental Change* 16(1): 97–109. DOI: 10.1007/s10113-014-0679-8.

Huber, R., Bakker, M., Balmann, A., Berger, T., Bithell, M., Brown, C., Grêt-Regamey, A., Xiong, H., Le, Q. B., Mack, G., Meyfroidt, P., Millington, J., Müller, B., Polhill, J. G., Sun, Z., Seidl, R., Troost, C. and Finger, R. (2018). Representation of decision-making in European agricultural agent-based models. *Agricultural Systems* 167: 143–160. DOI: 10.1016/j.agsy.2018.09.007.

Hutchings, N. J., Özkan Gülzari, Ş., de Haan, M. and Sandars, D. (2018). How do farm models compare when estimating greenhouse gas emissions from dairy cattle production? *Animal: An International Journal of Animal Bioscience* 12(10): 2171-2180. DOI: 10.1017/S175173111700338X.

IPCC. (2021). Summary for Policymakers. In: Masson-Delmotte, V., Zhai, P., Pirani, A., Connors, S. L., Péan, C., Berger, S., Caud, N., Chen, Y., Goldfarb, L., Gomis, M. I., Huang, M., Leitzell, K., Lonnoy, E., Matthews, J. B. R., Maycock, T. K., Waterfield, T., Yelekçi, O., Yu, R. and Zhou, B. (Eds.), *Climate Change 2021: The Physical Science Basis. Contribution of Working Group I to the Sixth Assessment Report of the Intergovernmental Panel on Climate Change*. Cambridge University Press, Cambridge, UK and New York, NY, 3–32.

Kebede, A. S., Dunford, R., Mokrech, M., Audsley, E., Harrison, P. A., Holman, I. P., Nicholls, R. J., Rickebusch, S., Rounsevell, M. D. A., Sabaté, S., Sallaba, F., Sanchez, A., Savin, C., Trnka, M. and Wimmer, F. (2015). Direct and indirect impacts of climate and socio-economic change in Europe: A sensitivity analysis for key land- and water-based sectors. *Climatic Change* 128(3-4): 261–277. DOI: 10.1007/s10584-014-1313-y.

Kemp-Benedict, E. (2013). Going from narrative to number: Indicatordriven scenario quantification. In: *Recent Developments in Foresight Methodologies*. New York: Springer, pp. 123–131. DOI: 10.1007/978-1-4614-5215-7_8.

Kipling, R. P., Virkajärvi, P., Breitsameter, L., Curnel, Y., De Swaef, T., Gustavsson, A. M., Hennart, S., Höglind, M., Järvenranta, K., Minet, J., Nendel, C., Persson, T., Picon-Cochard, C., Rolinski, S., Sandars, D. L., Scollan, N. D., Sebek, L., Seddaiu, G., Topp, C. F. E., Twardy, S., Van Middelkoop, J., Wu, L. and Bellocchi, G. (2016). Key challenges and priorities for modelling European grasslands under climate change. *Science of the Total Environment* 566–567: 851–864.

Kirchner, M., Mitter, H., Schneider, U. A., Sommer, M., Falkner, K. and Schmid, E. (2021). Uncertainty concepts for integrated modeling – Review and application for identifying uncertainties and uncertainty propagation pathways <Modelling and Software 135>. *Environmental Modelling and Software* 135: 104905. DOI: 10.1016/j.envsoft.2020.104905.

Kok, K., Bärlund, I., Flörke, M., Holman, I., Gramberger, M., Sendzimir, J., Stuch, B. and Zellmer, K. (2014). European participatory scenario development: Strengthening the link between stories and models. *Climatic Change* 128(1-14): 187–200. DOI: 10.1007/s10584-014-1143-y.

Kok, K., Pedde, S., Gramberger, M., Harrison, P. A. and Holman, I. P. (2019). New European socio-economic scenarios for climate change research: Operationalising concepts to extend the shared socio-economic pathways. *Regional Environmental Change* 19(3): 643–654.

Korhonen, P., Palosuo, T., Persson, T., Höglind, M., Jégo, G., Van Oijen, M., Gustavsson, A., Bélanger, G. and Virkajärvi, P. (2018). Modelling grass yields in northern climates – A comparison of three growth models for timothy. *Field Crops Research* 224: 37–47.

Lee, H., Brown, C., Seo, B., Holman, I., Audsley, E., Cojocaru, G. and Rounsevell, M. (2019). Implementing land-based mitigation to achieve the Paris Agreement in Europe requires food system transformation. *Environmental Research Letters* 14(10): 104009.

Lehtonen, H. S., Aakkula, J., Fronzek, S., Helin, J., Hilden, M., Huttunen, S., Kaljonen, M., Niemi, J., Palosuo, T., Pirttioja, N., Rikkonen, P., Varho, V. and Carter, T. R. (2021). Shared socioeconomic pathways for climate change research in Finland: co-developing extended SSP narratives for agriculture. *Regional Environmental Change* 21(1): 7.

Li, S., Juhász-Horváth, L., Pedde, S., Pintér, L., Rounsevell, M. D. A. and Harrison, P. A. (2017a). Integrated modelling of urban spatial development under uncertain climate futures: A case study in Hungary. *Environmental Modelling and Software* 96: 251–264. DOI: 10.1016/j.envsoft.2017.07.005.

Li, S., Juhász-Horváth, L. and Harrison, P. A. (2017b). Modelling regional cropping patterns under scenarios of climate and socio-economic change in Hungary. *Science of the Total Environment* 622–623: 1611–1620. DOI: 10.1016/j.scitotenv.2017.10.038.

Madsen, M. S., Maule, C. F., Christensen, J. H., Fronzek, S. and Carter, T. R. (2016). IMPRESSIONS climate scenarios. EU FP7 IMPRESSIONS project deliverable D2.3. Available at: http://www.impressions-project.eu/documents/1/.

Martre, P., Wallach, D., Asseng, S., Ewert, F., Jones, J. W., Rötter, R. P., Boote, K. J., Ruane, A. C., Thorburn, P. J., Cammarano, D., Hatfield, J. L., Rosenzweig, C., Aggarwal, P. K., Angulo, C., Basso, B., Bertuzzi, P., Biernath, C., Brisson, N., Challinor, A. J., Doltra, J., Gayler, S., Goldberg, R., Grant, R. F., Heng, L., Hooker, J., Hunt, L. A., Ingwersen, J., Izaurralde, R. C., Kersebaum, K. C., Müller, C., Kumar, S. N., Nendel, C., O'leary, G., Olesen, J. E., Osborne, T. M., Palosuo, T., Priesack, E., Ripoche, D., Semenov, M. A., Shcherbak, I., Steduto, P., Stöckle, C. O., Stratonovitch, P., Streck, T., Supit, I., Tao, F., Travasso, M., Waha, K., White, J. W. and Wolf, J. (2015). Multimodel ensembles of wheat growth: many models are better than one. *Global Change Biology* 21(2): 911–925. DOI: 10.1111/gcb.12768.

Marcelis, L. F. M., Heuvelink, E. and Goudriaan, J. (1998). Modelling biomass production and yield of horticultural crops: A review. *Scientia Horticulturae* 74(1–2): 83–111.

Mitter, H., Techen, A. K., Sinabell, F., Helming, K., Schmid, E., Bodirsky, B. L., Holman, I., Kok, K., Lehtonen, H., Leip, A., Le Mouël, C., Mathijs, E., Mehdi, B., Mittenzwei, K., Mora, O., Øistad, K., Øygarden, L., Priess, J. A., Reidsma, P., Schaldach, R. and Schönhart, M. (2020). Shared Socio-economic Pathways for European agriculture and food systems: The Eur-Agri-SSPs. *Global Environmental Change: Human and Policy Dimensions* 65: 102159.

Mokrech, M., Nicholls, R. J., Richards, J. A., Henriques, C., Holman, I. P. and Shackley, S. (2008). Regional impact assessment of flooding under future climate and socio-economic scenarios for East Anglia and North West England. *Climatic Change* 90(1–2): 31–55.

Moser, S. C. and Ekstrom, J. A. (2010). A framework to diagnose barriers to climate change adaptation. *Proceedings of the National Academy of Sciences of the United States of America* 107(51): 22026–22031. DOI: 10.1073/pnas.1007887107.

Müller, B., Hoffmann, F., Heckelei, T., Müller, C., Hertel, T. W., Polhill, J. G., van Wijk, M., Achterbosch, T., Alexander, P., Brown, C., Kreuer, D., Ewert, F., Ge, J., Millington, J. D. A., Seppelt, R., Verburg, P. H. and Webber, H. (2020). Modelling food security: Bridging the gap between the micro and the macro scale. *Global Environmental Change* 63: 102085. DOI: 10.1016/j.gloenvcha.2020.102085.

Murray-Rust, D., Brown, C., van Vliet, J., Alam, S. J., Robinson, D. T., Verburg, P. H. and Rounsevell, M. (2014). Combining agent functional types, capitals and services to model land use dynamics. *Environmental Modelling and Software* 59: 187–201. DOI: 10.1016/j.envsoft.2014.05.019.

Niles, M. T., Ahuja, R., Barker, T., Esquivel, J., Gutterman, S., Heller, M. C., Mango, N., Portner, D., Raimond, R., Tirado, C. and Vermeulen, S. (2018). Climate change mitigation beyond agriculture: A review of food system opportunities and implications. *Renewable Agriculture and Food Systems* 33(3): 297–308. DOI: 10.1017/S1742170518000029.

O'Neill, B. C., Carter, T. R., Ebi, K., Harrison, P. A., Kemp-Benedict, E., Kok, K., Kriegler, E., Preston, B. L., Riahi, K., Sillmann, J., van Ruijven, B. J., van Vuuren, D., Carlisle, D., Conde, C., Fuglestvedt, J., Green, C., Hasegawa, T., Leininger, J., Monteith, S. and Pichs-Madruga, R. (2020). Achievements and needs for the climate change scenario framework. *Nature Climate Change* 10(12): 1074–1084. DOI: 10.1038/s41558-020-00952-0.

Padel, S., Levidow, L. and Pearce, B. (2020). UK farmers' transition pathways towards agroecological farm redesign: Evaluating explanatory models. *Agroecology and Sustainable Food Systems* 44(2): 139–163. DOI: 10.1080/21683565.2019.1631936.

Panda, A. (2018). Transformational adaptation of agricultural systems to climate change. *WIREs Climate Change* 9(4): e520. DOI: 10.1002/wcc.520.

Papadimitriou, L., Holman, I. P., Dunford, R. and Harrison, P. A. (2019a). Trade-offs are unavoidable in multi-objective adaptation even in a post-Paris Agreement world. *Science of the Total Environment* 696: 134027. DOI: 10.1016/j.scitotenv.2019.134027.

Papadimitriou, L., Trnka, M., Harrison, P. and Holman, I. (2019b). Cross-sectoral and trans-national interactions in national-scale climate change impacts assessment—The case of the Czech Republic. *Regional Environmental Change* 19(8): 2453–2464.

Patt, A. G., van Vuuren, D. P., Berkhout, F., Aaheim, A., Hof, A. F., Isaac, M. and Mechler, R. (2010). Adaptation in integrated assessment modeling: Where do we stand? *Climatic Change* 99(3–4): 383–402. DOI: 10.1007/s10584-009-9687-y.

Pedde, S., Harrison, P. A., Holman, I. P., Powney, G. D., Lofts, S., Schmucki, R., Gramberger, M. and Bullock, J. M. (2021). Enriching the shared socioeconomic pathways to co-create consistent multi-sector scenarios for the UK. *Science of the Total Environment* 756: 143172.

Pedde, S., Kok, K., Onigheit, J., Brown, C., Holman, I. P. and Harrison, P. A. (2019). Bridging uncertainty concepts across narratives and simulations in environmental scenarios. *Regional Environmental Change* 19(3): 655–666. DOI: 10.1007/s10113-018-1338-2.

Pereira, L. M., Hichert, T., Hamann, M., Preiser, R. and Biggs, R. (2018). Using futures methods to create transformative spaces: visions of a good Anthropocene in southern Africa. *Ecology and Society* 23(1): 19. DOI: 10.5751/ES-09907-230119.

Reidsma, P., Janssen, S., Jansen, J. and van Ittersum, M. K. (2018). On the development and use of farm models for policy impact assessment in the European Union – A review. *Agricultural Systems* 159: 111–125. DOI: 10.1016/j.agsy.2017.10.012.

Rodriguez, A., Ruiz-Ramos, M., Palosuo, T., Carter, T. R., Fronzek, S., Lorite, I. J., Ferrise, R., Pirttioja, N., Bindi, M., Baranowski, P., Buis, S., Cammarano, D., Chen, Y., Dumont, B., Ewert, F., Gaiser, T., Hlavinka, P., Hoffmann, H., Höhn, J. G., Jurecka, F., Kersebaum, K. C., Krzyszczak, J., Lana, M., Mechiche-Alami, A., Minet, J., Montesino, M., Nendel, C., Porter, J. R., Ruget, F., Semenov, M. A., Steinmetz, Z., Stratonovitch, P., Supit, I., Tao, F., Trnka, M., de Wit, A. and Rötter, R. P. (2019). Implications of crop model ensemble size and composition for estimates of adaptation effects and agreement of recommendations. *Agricultural and Forest Meteorology* 264: 351–362.

Rosenbloom, D. (2017). Pathways: an emerging concept for the theory and governance of low-carbon transitions. *Global Environmental Change* 43: 37–50. DOI: 10.1016/j.gloenvcha.2016.12.011.

Rötter, R. P., Hoffmann, M. P., Koch, M. and Müller, C. (2018). Progress in modelling agricultural impacts of and adaptations to climate change. *Current Opinion in Plant Biology* 45(B): 255–261. DOI: 10.1016/j.pbi.2018.05.009.

Rounsevell, M. D. A., Arneth, A., Alexander, P., Brown, D. G., de Noblet-Ducoudré, N., Ellis, E., Finnigan, J., Galvin, K., Grigg, N., Harman, I., Lennox, J., Magliocca, N., Parker, D., O'Neill, B. C., Verburg, P. H. and Young, O. (2014). Towards decision-based global land use models for improved understanding of the Earth system. *Earth System Dynamics* 5(1): 117–137. DOI: 10.5194/esd-5-117-2014.

Rounsevell, M. D. A., Arneth, A., Brown, C., Cheung, W. W. L., Gimenez, O., Holman, I., Leadley, P., Luján, C., Mahevas, S., Maréchaux, I., Pélissier, R., Verburg, P. H., Vieilledent, G., Wintle, B. A. and Shin, Y. (2021). Identifying uncertainties in scenarios and models of socio-ecological systems in support of decision-making. *One Earth* 4(7): 967–985.

Sabbaghi, M. A., Nazari, M., Araghinejad, S. and Soufizadeh, S. (2020). Economic impacts of climate change on water resources and agriculture in Zayandehroud river basin in Iran. *Agricultural Water Management* 241: 106323. DOI: 10.1016/j.agwat.2020.106323.

Saltelli, A., Bammer, G., Bruno, I., Charters, E., Di Fiore, M., Didier, E., Nelson Espeland, W., Kay, J., Lo Piano, S., Mayo, D., Pielke, R., Portaluri, T., Porter, T. M., Puy, A., Rafols, I., Ravetz, J. R., Reinert, E., Sarewitz, D., Stark, P. B., Stirling, A., van der Sluijs, J. and Vineis, P. (2020). Five ways to ensure that models serve society: A manifesto. *Nature* 582(7813): 482–484. DOI: 10.1038/d41586-020-01812-9.

Salter, J., Robinson, J. and Wiek, A. (2010). Participatory methods of integrated assessment – A review. *WIREs Climate Change* 1(5): 697–717.

Schönenberg, R., Schaldach, R., Lakes, T., Göpel, J. and Gollnow, F. (2017). Inter- and transdisciplinary scenario construction to explore future land-use options in southern Amazonia. *Ecology and Society* 22(3): 13. DOI: 10.5751/ES-09032-220313.

Steinbuks, J. and Hertel, T. W. (2016). Confronting the food-energy-environment trilemma: Global land use in the long run. *Environmental and Resource Economics* 63(3): 545–570. DOI: 10.1007/s10640-014-9848-y.

Sultan, B. and Gaetani, M. (2016). Agriculture in West Africa in the twenty-first century: Climate change and impacts scenarios, and potential for adaptation. *Frontiers in Plant Science* 7: 1262. DOI: 10.3389/fpls.2016.01262.

Sutherland, L. A., Burton, R. J. F., Ingram, J., Blackstock, K., Slee, B. and Gotts, N. (2012). Triggering change: Towards a conceptualisation of major change processes in farm decision-making. *Journal of Environmental Management* 104: 142–151. DOI: 10.1016/j.jenvman.2012.03.013.

Tàbara, J. D., Jäger, J., Mangalagiu, D. and Grasso, M. (2019). Defining transformative climate science to address high-end climate change. *Regional Environmental Change* 19(3): 807–818. DOI: 10.1007/s10113-018-1288-8.

Tao, F., Palosuo, T., Roetter, R. P., Díaz-Ambrona, C. G. H., Inés Mínguez, M., Semenov, M. A., Kersebaum, K. C., Cammarano, D., Specka, X., Nendel, C., Srivastava, A. K., Ewert, F., Padovan, G., Ferrise, R., Martre, P., Rodríguez, L., Ruiz-Ramos, M., Gaiser,

T., Höhn, J. G., Salo, T., Dibari, C. and Schulman, A. H. (2020). Why do crop models diverge substantially in climate impact projections? A comprehensive analysis based on eight barley crop models. *Agricultural and Forest Meteorology* 281: 107851.

Tao, F., Rötter, R. P., Palosuo, T., Gregorio Hernández Díaz-Ambrona, C., Mínguez, M. I., Semenov, M. A., Kersebaum, K. C., Nendel, C., Specka, X., Hoffmann, H., Ewert, F., Dambreville, A., Martre, P., Rodríguez, L., Ruiz-Ramos, M., Gaiser, T., Höhn, J. G., Salo, T., Ferrise, R., Bindi, M., Cammarano, D. and Schulman, A. H. (2018). Contribution of crop model structure, parameters and climate projections to uncertainty in climate change impact assessments. *Global Change Biology* 24(3): 1291-1307. DOI: 10.1111/gcb.14019.

Terama, E., Clarke, E., Rounsevell, M. D. A., Fronzek, S. and Carter, T. R. (2019). Modelling population structure in the context of urban land use change in Europe. *Regional Environmental Change* 19(3): 667–677. DOI: 10.1007/s10113-017-1194-5.

Toimil, A., Losada, I. J., Diaz-Simal, P., Izaguirre, C. and Camus, P. (2017). Multi-sectoral, high-resolution assessment of climate change consequences of coastal flooding. *Climatic Change* 145(3–4): 431–444. DOI: 10.1007/s10584-017-2104-z.

van Vuuren, D. P., Lowe, J., Stehfast, E., Gohar, L., Hof, A. F., Hope, C., Warren, R., Meinshausen, M. and Plattner, G. K. (2011). How well do integrated assessment models simulate climate change? *Climatic Change* 104(2): 255–285. DOI: 10.1007/s10584-009-9764-2.

Vermeulen, S. J., Dinesh, D., Howden, S. M., Cramer, L. and Thornton, P. K. (2018). Transformation in practice: A review of empirical cases of transformational adaptation in agriculture under climate change. *Frontiers in Sustainable Food Systems* 2: 65. DOI: 10.3389/fsufs.2018.00065.

Weindl, I., Popp, A., Bodirsky, B. L., Rolinski, S., Lotze-Campen, H., Biewald, A., Humpenöder, F., Dietrich, J. P. and Stevanović, M. (2017). Livestock and human use of land: Productivity trends and dietary choices as drivers of future land and carbon dynamics. *Global and Planetary Change* 159: 1–10.

Werners, S. E., Wise, R. M., Butler, J. R. A., Totin, E. and Vincent, K. (2021). Adaptation pathways: A review of approaches and a learning framework. *Environmental Science and Policy* 116: 266–275. DOI: 10.1016/j.envsci.2020.11.003.

Wiek, A. and Iwaniec, D. M. (2014). Quality criteria for visions and visioning in sustainability science. *Sustainability Science* 9(4): 497–512. DOI: 10.1007/s11625-013-0208-6.

Wimmer, F., Audsley, E., Malsy, M., Savin, C., Dunford, R., Harrison, P. A., Schaldach, R. and Flörke, M. (2015). Modelling the effects of cross-sectoral water allocation schemes in Europe. *Climatic Change* 128(3–4): 229–244. DOI: 10.1007/s10584-014-1161-9.

Winter, J. M., Lopez, J. R., Ruane, A. C., Young, C. A., Scanlon, B. R. and Rosenzweig, C. (2017). Representing water scarcity in future agricultural assessments. *Anthropocene* 18: 15–26. DOI: 10.1016/j.ancene.2017.05.002.

Wise, R. M., Fazey, I., Stafford Smith, M., Park, S. E., Eakin, H. C., Archer Van Garderen, E. R. M. and Campbell, B. (2014). Reconceptualising adaptation to climate change as part of pathways of change and response. *Global Environmental Change* 28: 325–336. DOI: 10.1016/j.gloenvcha.2013.12.002.

Zhang, S., Tao, F. and Zhang, Z. (2017). Uncertainty from model structure is larger than that from model parameters in simulating rice phenology in China. *European Journal of Agronomy* 87: 30–39.

Chapter 5

Accounting for uncertainties in modeling the impact of climate change on agriculture

Fulu Tao, Chinese Academy of Sciences, China and Natural Resources Institute Finland (Luke), Finland; Taru Palosuo, Natural Resources Institute Finland (Luke), Finland; and Reimund Paul Rötter, Tropical Plant Production and Agricultural Systems Modelling (TROPAGS) and University of Göttingen, Germany

1 Introduction
2 Managing uncertainties from greenhouse gas emission and climate change scenarios
3 Managing uncertainties from crop model input data
4 Managing uncertainties from model parameters and human error
5 Managing uncertainties from model structures
6 Managing uncertainties from adaptations in agricultural practices
7 Reducing model uncertainties
8 Conclusion and future trends
9 Acknowledgements
10 Where to look for further information
11 References

1 Introduction

Impacts of climate change on future agricultural production and food security are a key concern because both climate risk and global demand for agricultural production are both expected to increase (Godfray et al., 2010). There have been many studies since the 1990s assessing the future impact of climate change on agricultural crop production using statistical and process-based crop modeling methods (e.g. Rötter and Van de Geijn, 1999; Lobell and Burke, 2008; White et al., 2011; Porter et al., 2014; Asseng et al., 2015). Statistical approaches have intrinsic limitations in predicting future climate change impacts since they are constrained by the historical data from which they were developed. Process-based crop models built on understanding processes and

http://dx.doi.org/10.19103/AS.2022.0115.07

interlinkages in agroecosystems have been considered more promising and are routinely applied to simulate the potential impacts of climate change on crop growth and productivity from local to global scales (e.g. Asseng et al., 2015; Challinor, et al., 2009; Porter et al., 2014; Rosenzweig et al., 2014; Rötter et al., 2011; Tao et al., 2008, 2009; White et al., 2011). Simulations have been usually driven by climate projections from global climate models (GCMs) downscaled using statistical methods or regional climate models (RCMs) (White et al., 2011).

All climate change impact assessments include a degree of uncertainty due to the number and complexity of the physical, biological and socioeconomic processes involved (Asseng et al., 2013, 2015; Challinor et al., 2009; Lobell and Burke, 2008; Rötter et al., 2011, 2012; Tao et al., 2009a, 2009b, 2018; Wallach et al., 2016, 2017). Among other things, uncertainties can originate from (Fig. 1):

- variability in future development of greenhouse gas (GHG) emissions;
- variability in climate responses to global emissions;
- gaps or errors in input data;
- limitations in model structures and parameters; and
- human error.

Model uncertainty can be defined as the distribution of simulated values arising from these variables (Wallach, 2020). It is critical to understand and account for these uncertainties since they may substantially affect the accuracy of climate change impact assessments and resulting adaptation strategies.

There has been significant progress in dealing with uncertainties in models. One example is a study of future climate impacts on rice productivity

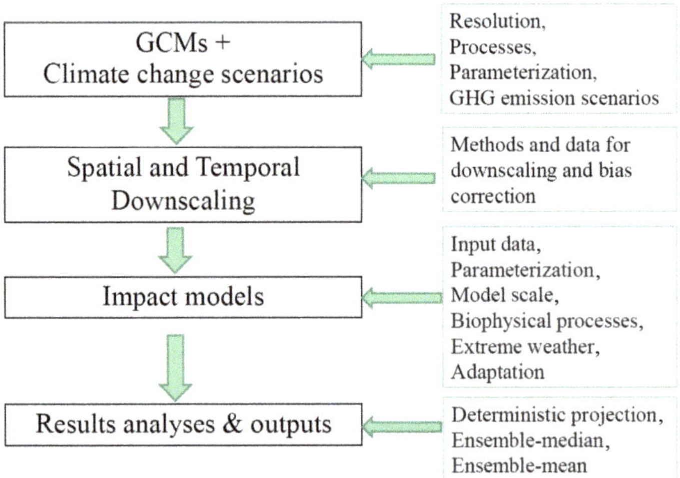

Figure 1 Sources of uncertainty in climate change impact assessment procedures.

and water use in China which used 20 climate change scenarios and a Monte Carlo simulation to test each scenario as a way of accounting for uncertainty in individual projections (Tao et al., 2008). Another approach has been to try to account for uncertainties associated with biophysical parameters used in a crop model, in this case simulating groundnut yield in India (Challinor et al., 2009). Probability distributions of biophysical parameter values in a crop model were inferred using Bayesian probability inversion and a Markov Chain Monte Carlo (MCMC) technique together with long-term data on crop phenology and yield (Iizumi et al., 2009; Tao et al., 2009b). Another study used double-ensemble probabilistic assessment to evaluate climate change impacts on maize productivity and water use in China (Tao et al., 2009b). Using a single crop model, 60 sets of crop model parameters and 10 climate scenarios, the study suggested that uncertainties in climate projection generally contributed more uncertainty to climate impact assessments than crop model parameters.

A recent focus has been uncertainty deriving from the structures and parameters of different models (Asseng et al., 2013, 2015; Bassu et al., 2014; Li et al., 2015; Martre et al., 2015; Palosuo et al., 2011; Pirttioja et al., 2015; Rötter et al., 2011, 2012; Wang et al., 2017; Tao et al., 2018, 2020). As an example, Palosuo et al. (2011) estimated uncertainty from model structure by comparing eight widely used crop models for winter wheat under current climatic conditions across Europe. A similar study has been carried out for barley using nine crop models (Rötter et al., 2012). International initiatives have included the Modelling European Agriculture with Climate Change for Food Security (MACSUR) project (Ewert et al., 2015) and the Agricultural Model Intercomparison and Improvement Project (AgMIP) (Rosenzweig et al., 2014). Both MACSUR and AgMIP have made great progresses in quantifying and reducing uncertainty from model structure in simulating the response of crop yields to climate change, through ensemble modeling approaches and suggested improvements to individual models (e.g. Asseng et al., 2013, 2015; Bassu et al., 2014; Li et al., 2015; Martre et al., 2015; Pirttioja et al., 2015; Maiorano et al., 2017; Wallach et al., 2016; Wang et al., 2017; Tao et al., 2018, 2020). The results showed that simulated climate change impacts on crop yield varied significantly across models due to differences in crop model structure and parameters. One study of 26 wheat simulation models found that degrees of uncertainty in these models were larger than those from downscaled GCMs (Asseng et al., 2013).

Previous attempts to account for uncertainty have tended to focus on one or two key sources of variability such as:

- climate projections (e.g. Tao et al., 2008);
- crop model parameters (e.g. Challinor et al., 2009; Iizumi et al., 2009; Tao et al., 2009a);

- crop model parameters and climate projections combined (Tao et al., 2009b);
- crop model structure (Palosuo et al., 2011; Rötter et al., 2012);
- crop model structure and climate projections combined (Asseng et al., 2013, 2015; Bassu et al., 2014; Li et al., 2015).

Building on this, Tao et al. (2018) developed a triple-ensemble probabilistic assessment to account for uncertainties in seven crop models from three sources: crop model structure, crop model parameters and climate change projections. The results showed that the contribution of crop model structure to the total variance of ensemble output was larger than that from downscaled climate projections and model parameters. The relative contribution of crop model parameters and downscaled climate projections to the total variance of ensemble output varied greatly between the seven crop models. The contribution of downscaled climate projections was on average larger than that of crop model parameters.

This information on degrees of uncertainty from different sources is essential for model users to decide where to put the most effort when developing or improving models for impact analyses. The study concluded that triple-ensemble probabilistic approach incorporating uncertainties from multiple sources provides more comprehensive information for quantifying uncertainties than studies that only account for uncertainties from one or two sources. The following sections review each of the major sources of uncertainty in model-based climate impact studies for crop production and how levels of uncertainty can be reduced.

2 Managing uncertainties from greenhouse gas emission and climate change scenarios

Future climate impacts will depend on the future development of GHG emissions and the climate response to rising GHG concentrations. Socioeconomic and emission scenarios have been used to provide plausible descriptions of how the future may evolve. These scenarios involve assumptions about changes in socioeconomic conditions, technologies, energy requirements, land use as well as related emissions. Various scenarios have been developed, starting with the IS92 scenarios published in the 1992 Supplementary Report to the IPCC Assessment (Leggett et al., 1992), the Special Report on Emissions Scenarios (SRES) scenarios (Nakicenovic et al., 2000), the Representative Concentration Pathways (RCPs) (Van Vuuren et al., 2011) and, more recently, a framework (incorporating Shared Socioeconomic Pathways (SSPs) as discussed in O'Neill et al., 2014) to facilitate production of integrated scenarios based on combinations of climate model projections, socioeconomic conditions and assumptions about climate policies (Van Vuuren et al., 2011; O'Neill et al., 2020). Since completion of the IPCC's Fifth Assessment Report (AR5) in 2014, concentration pathways have been considered together with SSPs (Riahi et al., 2017). RCPs are greenhouse gas concentration (rather than emission) trajectories adopted by the IPCC. Four pathways were created to describe different climate scenarios, depending on the future volume of GHGs emitted. The RCPs - originally RCP2.6, RCP4.5, RCP6 and RCP8.5 - set total radiative forcing targets (the difference between incoming and radiation in the upper atmosphere, measured in watts per square meter) to the year 2100 (2.6, 4.5, 6 and 8.5 W/m^2, respectively).

GCMs contain large uncertainties in projecting future climate due to limited understanding of various earth processes, temporal and spatial resolution, and various biophysical parameters (Knutti et al., 2013). The Coupled Model Intercomparison Project, which began in 1995, is now in its sixth phase (CMIP6). CMIP6 coordinates independent GCM intercomparison activities and experiments and has established a common infrastructure for collecting, organizing and distributing output from climate models based on common sets of experiments (https://esgf-node.llnl.gov/projects/cmip6/). Multi-model results provide insights into errors and uncertainty in individual climate model simulations (Collins et al., 2011). GCMs have been continuously improved to reflect the latest developments in climate science. An example is refinement of estimates of equilibrium climate sensitivity, an important measure of long-term temperature rise due to a doubling of atmospheric CO_2. The current AR6 estimate is 3°C with a likely range of 2.5–4°C, compared to 1.5–4.5°C assessed in AR5. According to IPCC AR6 (2021), compared to 1850–1900, global surface temperature averaged over 2081-2100 is projected to range from 1.0°C to 1.8°C under the very low GHG emissions scenario (SSP1-1.9) to 3.3-5.7°C under the very high GHG emissions scenario (SSP5-8.5).

Since spatial resolution of GCMs is generally too coarse, spatial downscaling and bias correction are necessary before using the data for impact assessments (Laux et al., 2021). Different approaches for spatial downscaling and bias correction have been developed to facilitate regional climate projections (Semenov and Stratonovitch, 2015). It should be noted that climate change not only affects mean climate variables but also overall climate variability (IPCC, 2012). Robust climate change impacts studies need to consider climate variability as well. Unfortunately, uncertainties in projections of extreme events are largely unknown (IPCC, 2021).

One way to quantify uncertainties in predicting climate change is to use multiple GCMs with different emissions scenarios (e.g. Tao et al., 2008, Liu et al., 2019, Palosuo et al., 2021, Webber et al., 2018). This reduces uncertainty though some unquantifiable uncertainty will always remain (Rötter, 2014). An alternative approach is to use probabilistic impact estimation using impact response surfaces. This technique combines impact modeling with probabilistic climate projections that can be applied to any climate scenario (Pirttioja et al., 2015, 2019).

3 Managing uncertainties from crop model input data

The quality of crop model input data, e.g. related to weather, soil and agricultural management, including issues such as how data is obtained and data resolution, is critical for robust model. At a site scale, high-quality field experiment data on soil properties, environmental conditions, agricultural managements, as well as crop development, growth, biomass and yield, are essential for model calibration and validation (Kersebaum et al., 2015). At a regional scale, crop simulations frequently use low-resolution input data. Climate and soil data are often generated via averaging or sampling data. This may bias simulated yields at large scales (Angulo et al., 2013).

Research quantifying the aggregation effect found that the relative mean absolute error of most models using aggregated soil data was similar to or larger than inter-annual or inter-model variability in yields. This error increased further when both climate and soil data were aggregated (Hoffmann et al., 2015). Similarly, weather data and management data resolution also affect results. Quality of soil and management data can cause significant uncertainty where a model developed from conditions in one area is applied to a new area. Soil type-related yield variability generally outweighs inter-annual variability in yield due to weather variables (Folberth, et al., 2016). Due to nonlinear responses within models, input data should be as accurate as possible to reduce uncertainties in climate change impact assessment. Although good agricultural management data is critical for modeling climate change impacts, it is notoriously difficult to obtain. An alternative approach is to derive agricultural management information from high-resolution remote sensing data, e.g. relating to leaf area index (LAI) or soil moisture status and incorporate it into crop models to improve simulation performance (Chen and Tao, et al., 2022).

4 Managing uncertainties from model parameters and human error

Model parameters affect model simulation results and related uncertainties substantially. Model parameters need to be calibrated, i.e. adapted to the target cultivar and environment using empirical data, so that model performance is consistent with what is known about the behavior and management of a particular crop (Iizumi et al., 2009; Seidel et al., 2018;

Tao et al., 2009a, b; Gao et al., 2020, 2021; Wallach et al., 2021a). Currently, there is no standard way of undertaking calibration and calibration practices among the crop models vary (Seidel et al., 2018). Practices vary from ad hoc, i.e. typically trial and error approaches, to least squares and Bayesian parameter estimation. For example, biophysical parameter values in a crop model were optimized using Bayesian probability inversion and MCMC by Iizumi et al. (2009) and Tao et al. (2009a). Gao et al. (2020) compared three calibration methods including Ordinary Least Square (OLS), MCMC and Generalized Likelihood Uncertainty Estimation (GLUE) and found that MCMC for model calibration, coupled with estimation of model error variance, is a promising method for quantifying prediction uncertainty. Where goodness-of-fit was the main criterion, OLS was the fastest and most effective method. More generally, the process of model calibration underlines the importance of the (skills of) the modeler or model user. Model comparison studies have shown variability in results from the same model when applied by different users (e.g. Confalonieri et al., 2016; Salo et al., 2016) and that much of the variability comes from the calibration process (Wallach et al., 2021a, 2021b). Based on an analysis of approaches to calibration by multiple modeling groups, Wallach et al. (2021b) have proposed a standardized approach to calibration of the phenology component of crop models to reduce model uncertainty due to calibration.

Another source of uncertainty is human error which can affect all aspects of model design, collection and input of data and model use. Quantifying uncertainties due to human error is challenging. It is possible to reduce likelihood of error by, e.g., training, creating modeling protocols, routines for checking key stages, visualization and plausibility tests.

5 Managing uncertainties from model structures

Recent studies have consistently shown that uncertainty from issues relating to crop model structure is greater than that from variability in climate projections and crop model parameters (Asseng et al., 2013, 2015; Bassu et al., 2014; Li et al., 2015; Tao et al., 2018). A key area in reducing uncertainty is improving crop model structure (Wang et al., 2017; Challinor et al., 2018; Tao et al., 2018; Rötter et al., 2018). A recent analysis found that differing assumptions about temperature and CO_2 relationships in models caused large discrepancies in climate impact projections between models (Tao et al., 2020). In particular, the impacts of increases in temperature and CO_2 on leaf area development were identified as the major causes of uncertainty in simulating changes in evapotranspiration (ET), above-ground biomass and grain yield.

5.1 Model improvements to better simulate impacts of extreme temperature

Most of the main crop simulation models for predicting wheat grain yield are highly uncertain in simulating crop responses to extreme temperatures. Asseng et al. (2015) tested 30 different wheat crop models against field experiments in which growing season mean temperatures ranged from 15°C to 32°C. Many models simulated yields well at conventional temperatures but were less accurate at higher temperatures. Schewe et al. (2019), using the 2003 European heat wave and drought as a historical analogue for comparable events in the future, found that a majority of models underestimate the scale of impacts. Crop models can be improved, e.g. through re-parameterization and/or incorporating or modifying heat stress effects on phenology, leaf growth and senescence, biomass growth, and grain number and size in the light of field experimental data (Maiorano et al., 2017).

Wang et al. (2017) showed that the variations in the mathematical functions currently used to simulate temperature responses of physiological processes account for >50% of uncertainty in simulated grain yields for mean growing season temperatures ranging from 14°C to 33°C. Improving temperature response functions can improve modeling of crops under increasing temperatures, leading to improved crop yield projections.

5.2 Model improvements to improve simulation of crop responses to elevated CO_2

Elevated CO_2 concentrations have positive effects on the productivity of C3 crops, such as wheat and rice, by increasing photosynthesis, biomass and grain yield and by reducing crop transpiration and increasing crop water use efficiency (Leakey et al., 2009). They can partially offset the negative impacts of increasing temperature (Rosenzweig et al., 2014).

Current crop models differ in simulating major crop processes including LAI development, photosynthesis, ET and biomass production, as well as effects of elevated CO_2 (Asseng et al., 2013; Tao et al., 2020). Commonly used methods for modeling these processes include (Farquhar et al., 1980; Hasegawa et al., 2017):

1. the coarse-grained concept of radiation use efficiency (RUE);
2. the light response curve (LRC) of instantaneous leaf photosynthesis scaled to hourly or daily canopy photosynthesis and crop respiration; and
3. the Farquhar, von Caemmerer & Berry (FvCB) biochemical model of C3 photosynthesis.

The diversity in model algorithms, combined with inconsistency in model parameterization procedures, can produce significant uncertainties in model projections (Asseng et al., 2013, 2015; Bassu et al., 2014; Li et al., 2015; Tao et al., 2020).

The performance of 21 maize models in assessing the impact of elevated CO_2 levels on maize yield and water use has been assessed using a 2-year free-air carbon dioxide enrichment (FACE) experiment (Durand et al., 2018). The models matched experimental results in predicting yield responses to elevated CO_2 under well-watered conditions, as well as to the impact of water deficit at ambient CO_2. However, the models captured only 30% of the impact of exceptionally high CO_2 levels on observed yields under water deficit conditions. Models were unable to simulate low soil water content at anthesis and the increase of soil water and grain number brought about by the elevated CO_2 under dry conditions. Models incorporating stomatal control of transpiration tended to perform better, highlighting the need for model improvement in simulating transpiration water use and its impact on water status during the kernel-set phase (Durand et al., 2018).

The performance of 16 rice models in predicting yield under elevated CO_2 was investigated using FACE and chamber experiments by Hasegawa et al. (2017). The model ensemble reproduced experimental results well. However, yield prediction in response to elevated CO_2 varied significantly among models. The variation was not associated with model structure or magnitude of photosynthetic response to elevated CO_2 but was significantly associated with simulated morphological development, primarily leaf area, under elevated CO_2. Cammarano et al. (2016) compared 16 crop models for wheat water use and found that uncertainties in simulated water use and transpiration were greater with increased temperatures and elevated atmospheric CO_2, which were associated with different approaches to simulating transpiration.

Leaf area development is fundamental to biomass accumulation and yield formation, directly affecting canopy intercepted radiation, photosynthesis rate, ET and the balance of evaporation and transpiration. The impacts of temperature and CO_2 on LAI development are major causes of uncertainty in simulating above-ground biomass and grain yield of crops (Tao et al., 2020).

Many crop models use a biomass-dependent approach to simulate LAI development (Bannyayan et al., 2005; Ratjen et al., 2018). While this biomass-dependent approach has been shown to simulate leaf area development fairly well in normal conditions, it may greatly overestimate leaf area development under elevated CO_2 (Bannyayan et al., 2005). The approach depends on simulating biomass, leaf-stem partitioning processes and specific leaf area (SLA). Most importantly, it does not account for the uncoupling of leaf expansion and photosynthesis under water deficit (Muller et al., 2011) and elevated CO_2 (Ainsworth and Long, 2005) and the very different temperature sensitivities of these two processes (Asseng et al., 2013, 2015; Wang et al., 2017). Assumptions about developmental stages may not capture the allometric relationship between LAI and biomass (Ratjen et al., 2018). Leaf-stem partitioning and SLA are not independent given the allometric relationship between LAI and biomass, and LAI may explain canopy SLA better than developmental traits (Ratjen et al., 2018). However, these mechanisms are rarely represented in current crop models.

As a possible solution, leaf-stem partitioning and SLA could be simulated based on LAI dynamics instead of developmental stage or thermal time (Ratjen et al., 2018). Alternatively, leaf area development could be simulated based on the availability of nitrogen rather than biomass (Jamieson and Semenov, 2000; Sinclair et al., 2003; Martre and Dambreville, 2018), which could improve the simulations of CO_2 responses and capture their secondary processes (Vanuytrecht and Thorburn, 2017). Temperature impact functions and cardinal temperatures in simulating crop phenological development, LAI development, photosynthesis rate, ET and yield formation need to be further validated based on data from free-air temperature-increase experiments (Wang et al., 2017). Advances in understanding and simulating the basic processes and thresholds by which increases in temperature and elevated CO_2 will affect leaf area development, crop ET, photosynthesis and grain formation in contrasting environments should be incorporated to improve model performance. An alternative suggested approach is the use of fundamental biology and mathematical principles to develop simpler equations that can better model crop growth (Yin et al., 2021).

6 Managing uncertainties from adaptations in agricultural practices

Projected climate change impacts on crop growth and yields need to take account of adaptation strategies in modeling (IPCC, 2022). The adaptation settings in models to account for variations in sowing dates, cultivars, fertilization and irrigation regimes will affect projected climate impacts on crop growth and yields. A fixed set of model parameters fails, e.g., to take into account factors such as crop genotypic variability (Tao et al., 2009b). Since it can be challenging to obtain information on genotypes when models are applied to new situations, ensemble simulations using multiple sets of model parameters provide a sound solution (Tao et al., 2017). This is true for other adaptation options such as sowing dates, fertilization and irrigation.

A key challenge for models is that farming practices continue to evolve and improve, a process that will only accelerate in response to climate change (Hampf et al., 2020). However, there have been relatively few studies comparing the performance of different cropping systems to account for developments in agricultural practices (Xin and Tao, 2020). And an approach for forecasting technology-driven yield increases based on biophysical yield maxima has been developed to account for improvements in breeding and crop management (Hampf et al., 2020). The design and use of Representative Agricultural Pathways (RAPs) in regional integrated assessments of climate impacts provides one way to account for broader socioeconomic developments (Antle et al., 2017). There is also a need for integrated biophysical and economic modeling of agricultural production systems (Antle and Stöckle, 2017).

7 Reducing model uncertainties

Probabilistic approaches combining multiple climate change scenarios (Tao et al., 2008), multiple sets of model parameters (Iizumi et al., 2009; Tao et al., 2009a, b) and ensembles of crop models (Palosuo et al., 2011; Rötter et al., 2011; Asseng et al., 2013, 2015; Bassu et al., 2014; Li et al., 2015; Pirttioja et al., 2015; Tao et al., 2017) have become well-established approaches in model-based climate change impact assessments. These have been combined into super-ensemble-based probabilistic assessments (Tao et al., 2009b; Tao et al., 2020). These approaches help to account for uncertainty in model predictions (Tao et al., 2018) and produce more robust predictions, e.g. using an ensemble mean or median (Martre et al., 2015; Wallach et al., 2018). Some studies, e.g., suggest that the median value of a multi-model ensemble is more accurate in simulating the crop temperature response and yields than any single model (Asseng et al., 2013; Martre et al., 2015).

In working with crop multi-model ensembles, several key questions should be considered. These include defining criteria for acceptance of models in a crop multi-model ensembles, the influence of ensemble size and composition on model outcomes, and possible differential weighting of models in an ensemble (Wallach et al., 2016, 2018; Rodríguez et al., 2019). The predictive values of an ensemble mean (e-mean) and median (e-median) have also been investigated (Wallach et al., 2018). They found mean squared error of the e-mean decreases monotonically with the size of the ensemble if models are added at random but has a minimum with 2-6 models if best-fit models are added first (Wallach et al., 2018). An Ensemble Outcome Agreement (EOA) index developed by Rodríguez et al. (2019) has analyzed the effect of changes in composition and size of a multi-model ensemble (MME). Typically, each individual model in an ensemble is assigned an equal weight by simply averaging the predictions over all the models. Use of Bayesian model averaging (BMA), which assigns different weights to individual models according to their performance in reproducing historical data, has been shown to achieve better results (Gao et al., 2021).

8 Conclusion and future trends

Model-based climate change impact assessments involve uncertainties given the many physical, biological and socioeconomic processes involved. Among other things, uncertainties arise from unknown future development of GHG scenarios and uncertainties in climate change scenarios, input data, model structures and parameters and human error. Recent developments with crop models have helped to quantify or reduce these uncertainties.

Since uncertainty due to crop model structure is a particular problem, improvements to the model structure are key to reducing uncertainty. Advances in understanding the basic processes through which increasing temperatures and CO_2 may affect leaf area development, crop ET, photosynthesis and grain formation in contrasting environments will improve model accuracy. The impacts of different levels of temperature and CO_2 changes on many physiological and developmental phenomena need to be studied in contrasting environments. The three-way or four-way interactions of multiple climate change factors on crop development, growth, water use and yield formation need to be investigated experimentally. Advanced biological insights and mathematical methods should be combined to derive simple but theoretically sound equations that apply to different processes of crop growth. Another area of improvement is improving calibration techniques, e.g. to reduce human error. RAPs in regional integrated assessments can help to take account of socioeconomic factors.

9 Acknowledgements

Fulu Tao was supported by the National Natural Science Foundation of China (Project No. 42061144003). Fulu Tao and Taru Palosuo were supported by the Diversifying cropping system for Climate Smart Agriculture (DivCSA) project funded by the Academy of Finland (decision no. 316215). Reimund Paul Rötter received financial report from the Advanced tools for breeding BARley for Intensive and SusTainable Agriculture under climate change scenarios (BARISTA) project (Grant number: 031B0811A).

10 Where to look for further information

Further information is available from:

- https://www.macsur.eu/;
- https://agmip.org/;
- https://www.ipcc.ch/assessment-report/ar6/; and
- https://www.researchgate.net/profile/Fulu-Tao.

11 References

Ainsworth, E. A. and Long, S. P. 2005. What have we learned from 15 years of free-air CO_2 enrichment (FACE)? A meta-analytic review of the responses of photosynthesis, canopy properties and plant production to rising CO_2. *New Phytologist* 165(2), 351–371.

Angulo, C., Rötter, R., Trnka, M., Pirttioja, N., Gaiser, T., Hlavinka, P. and Ewert, F. 2013. Characteristic 'fingerprints' of crop model responses to weather input data at different spatial resolutions. *European Journal of Agronomy* 49, 104–114. https://doi.org/10.1016/j.eja.2013.04.003.

Antle, J. M., Mu, J. E., Zhang, H., Capalbo, S. M., Diebel, P. L., Eigenbrode, S. D., Kruger, C. E., Stöckle, C. O., Wulfhorst, J. D. and Abatzoglou, J. T. 2017. Design and use of representative agricultural pathways for integrated assessment of climate change in U. S. Pacific northwest cereal-based systems. *Frontiers in Ecology and Evolution* 5, 99. https://doi.org/10.3389/fevo.2017.00099.

Antle, J. M. and Stöckle, C. O. 2017. Climate impacts on agriculture: Insights from agronomic-economic analysis. *Review of Environmental Economics and Policy* 11(2), 299–318.

Asseng, S., Ewert, F., Martre, P., Rötter, R. P., Lobell, D. B., Cammarano, D., Kimball, B. A., Ottman, M. J., Wall, G. W., White, J. W., Reynolds, M. P., Alderman, P. D., Prasad, P. V. V., Aggarwal, P. K., Anothai, J., Basso, B., Biernath, C., Challinor, A. J., De Sanctis, G., Doltra, J., Fereres, E., Garcia-Vila, M., Gayler, S., Hoogenboom, G., Hunt, L. A., Izaurralde, R. C., Jabloun, M., Jones, C. D., Kersebaum, K. C., Koehler, A., Müller, C., Naresh Kumar, S., Nendel, C., O'Leary, G., Olesen, J. E., Palosuo, T., Priesack, E., Eyshi Rezaei, E., Ruane, A. C., Semenov, M. A., Shcherbak, I., Stöckle, C., Stratonovitch, P., Streck, T., Supit, I., Tao, F., Thorburn, P. J., Waha, K., Wang, E., Wallach, D., Wolf, J., Zhao, Z. and Zhu, Y. 2015. Rising temperatures reduce global wheat production. *Nature Climate Change* 5(2), 143–147.

Asseng, S., Ewert, F., Rosenzweig, C., Jones, J. W., Hatfield, J. L., Ruane, A. C., Boote, K. J., Thorburn, P. J., Rötter, R. P., Cammarano, D., Brisson, N., Basso, B., Martre, P., Aggarwal, P. K., Angulo, C., Bertuzzi, P., Biernath, C., Challinor, A. J., Doltra, J., Gayler, S., Goldberg, R., Grant, R., Heng, L., Hooker, J., Hunt, L. A., Ingwersen, J., Izaurralde, R. C., Kersebaum, K. C., Müller, C., Naresh Kumar, S., Nendel, C., O'Leary, G., Olesen, J. E., Osborne, T. M., Palosuo, T., Priesack, E., Ripoche, D., Semenov, M. A., Shcherbak, I., Steduto, P., Stöckle, C., Stratonovitch, P., Streck, T., Supit, I., Tao, F., Travasso, M., Waha, K., Wallach, D., White, J. W., Williams, J. R. and Wolf, J. 2013. Uncertainty in simulating wheat yields under climate change. *Nature Climate Change* 3(9), 827–832.

Bannyayan, M., Kobayashi, K., Kim, H., Lieffering, M., Okada, M. and Mirza, S. 2005. Modelling the interactive effects of atmospheric CO_2 and N on rice growth and yield. *Field Crops Research* 93(2–3), 237–251.

Bassu, S., Brisson, N., Durand, J. L., Boote, K., Lizaso, J., Jones, J. W., Rosenzweig, C., Ruane, A. C., Adam, M., Baron, C., Basso, B., Biernath, C., Boogaard, H., Conijn, S., Corbeels, M., Deryng, D., De Sanctis, G., Gayler, S., Grassini, P., Hatfield, J., Hoek, S., Izaurralde, C., Jongschaap, R., Kemanian, A. R., Kersebaum, K. C., Kim, S. H., Kumar, N. S., Makowski, D., Müller, C., Nendel, C., Priesack, E., Pravia, M. V., Sau, F., Shcherbak, I., Tao, F., Teixeira, E., Timlin, D. and Waha, K. 2014. How do various maize crop models vary in their responses to climate change factors? *Global Change Biology* 20(7), 2301–2320.

Cammarano, D., Rötter, R. P., Asseng, S., Ewert, F., Wallach, D., Martre, P., Hatfield, J. L., Jones, J. W., Rosenzweig, C., Ruane, A. C., Boote, K. J., Thorburn, P. J., Kersebaum, K. C., Aggarwal, P. K., Angulo, C., Basso, B., Bertuzzi, P., Biernath, C., Brisson, N., Challinor, A. J., Doltra, J., Gayler, S., Goldberg, R., Heng, L., Hooker, J., Hunt, L. A., Ingwersen, J., Izaurralde, R. C., Müller, C., Kumar, S. N., Nendel, C., O'Leary, G. J., Olesen, J. E., Osborne, T. M., Palosuo, T., Priesack, E., Ripoche, D., Semenov, M. A., Shcherbak, I., Steduto, P., Stöckle, C. O., Stratonovitch, P., Streck, T., Supit, I., Tao, F., Travasso, M., Waha, K., White, J. W. and Wolf, J. 2016. Uncertainty of wheat water use: Simulated patterns and sensitivity to temperature and CO_2. *Field Crops Research* 198, 80–92. https://doi.org/10.1016/j.fcr.2016.08.015.

Challinor, A. J., Müller, C., Asseng, S., Deva, C., Nicklin, K. J., Wallach, D., Vanuytrecht, E., Whitfield, S., Ramirez-Villegas, J. and Koehler, A. K. 2018. Improving the use of crop models for risk assessment and climate change adaptation. *Agricultural Systems* 159, 296–306.

Challinor, A. J., Wheeler, T., Hemming, D. and Upadhyaya, H. D. 2009. Ensemble yield simulations: Crop and climate uncertainties, sensitivity to temperature and genotypic adaptation to climate change. *Climate Research* 38, 117–127.

Chen, Y. and Tao, F. 2022. Potential of remote sensing data-crop model assimilation and seasonal weather forecasts for early-season crop yield forecasting over a large area. *Field Crops Research* 276, 108398.

Collins, M., Booth, B. B. B., Bhaskaran, B., Harris, G. R., Murphy, J. M., Sexton, D. M. H. and Webb, M. J. 2011. Climate model errors, feedbacks and forcings: A comparison of perturbed physics and multi-model ensembles. *Climate Dynamics* 36(9–10), 1737–1766. https://doi.org/10.1007/s00382-010-0808-0.

Confalonieri, R., Bregaglio, S., Adam, M., Ruget, F., Li, T., Hasegawa, T., Yin, X., Zhu, Y., Boote, K., Buis, S., Fumoto, T., Gaydon, D., Lafarge, T., Marcaida, M., Nakagawa, H., Ruane, A. C., Singh, B., Singh, U., Tang, L., Tao, F., Fugice, J., Yoshida, H., Zhang, Z., Wilson, L. T., Baker, J., Yang, Y., Masutomi, Y., Wallach, D., Acutis, M. and Bouman, B. 2016. A taxonomy-based approach to shed light on the babel of mathematical models for rice simulation. *Environmental Modelling and Software* 85, 332–341.

Confalonieri, R., Orlando, F., Paleari, L., Stella, T., Gilardelli, C., Movedi, E., Pagani, V., Cappelli, G., Vertemara, A., Alberti, L., Alberti, P., Atanassiu, S., Bonaiti, M., Cappelletti, G., Ceruti, M., Confalonieri, A., Corgatelli, G., Corti, P., Dell'Oro, M., Ghidoni, A., Lamarta, A., Maghini, A., Mambretti, M., Manchia, A., Massoni, G., Mutti, P., Pariani, S., Pasini, D., Pesenti, A., Pizzamiglio, G., Ravasio, A., Rea, A., Santorsola, D., Serafini, G., Slavazza, M. and Acutis, M. 2016. Uncertainty in crop model predictions: What is the role of users? *Environmental Modelling and Software* 81, 165–173. https://doi.org/10.1016/j.envsoft.2016.04.009.

Durand, J. L., Delusca, K., Boote, K., Lizaso, J., Manderschei, R., Weigel, H. J., Ruane, A. C., Rosenzweig, C., Jones, J., Ahuja, L., Anapalli, S., Basso, B., Baron, C., Bertuzzi, P., Biernath, C., Deryng, D., Ewert, F., Gaiser, T., Gayler, S., Heinlein, F., Kersebaum, K. C., Kim, S., Müller, C., Nendel, C., Olioso, A., Priesack, E., Villegas, J. R., Ripoche, D., Rötter, R. P., Seidel, S. I., Srivastava, A., Tao, F., Timlin, D., Twine, T., Wang, E., Webber, H. and Zhao, Z. 2018. How accurately do maize crop models simulate the interactions of atmospheric CO_2 concentration levels with limited water supply on water use and yield? *European Journal of Agronomy* 100, 67–75.

Ewert, F., Rötter, R. P., Bindi, M., Webber, H., Trnka, M., Kersebaum, K. C., Olesen, J. E., van Ittersum, M. K., Janssen, S., Rivington, M., Semenov, M. A., Wallach, D., Porter,

J. R., Stewart, D., Verhagen, J., Gaiser, T., Palosuo, T., Tao, F., Nendel, C., Roggero, P. P., Bartošová, L. and Asseng, S. 2015. Crop modelling for integrated assessment of risk to food production from climate change. *Environmental Modelling & Software* 72, 287-303.

Farquhar, G. D., von Caemmerer, S. and Berry, J. A. 1980. A biochemical model of photosynthesis CO_2 assimilation in leaves of C_3 species. *Planta* 149(1), 78-90.

Folberth, C., Skalský, R., Moltchanova, E., Balkovič, J., Azevedo, L. B., Obersteiner, M. and van der Velde, M. 2016. Uncertainty in soil data can outweigh climate impact signals in global crop yield simulations. *Nature Communications* 7, 11872.

Gao, Y., Wallach, D., Hasegawa, T., Tang, L., Zhang, R., Asseng, S., Kahveci, T., Liu, L., He, J. and Hoogenboom, G. 2021. Evaluation of crop model prediction and uncertainty using Bayesian parameter estimation and Bayesian model averaging. *Agricultural and Forest Meteorology* 311, 108686.

Gao, Y., Wallach, D., Liu, B., Dingkuhn, M., Boote, K. J., Singh, U., Asseng, S., Kahveci, T., He, J., Zhang, R., Confalonieri, R. and Hoogenboom, G. 2020. Comparison of three calibration methods for modeling rice phenology. *Agricultural and Forest Meteorology* 280, 107785.

Godfray, H. C. J., Beddington, J. R., Crute, I. R., Haddad, L., Lawrence, D., Muir, J. F., Pretty, J., Robinson, S., Thomas, S. M. and Toulmin, C. 2010. Food security: The challenge of feeding 9 billion people. *Science* 327(5967), 812-818.

Hampf, A. C., Stella, T., Berg-Mohnicke, M., Kawohl,T., Kilian,M. and Nendel, C. 2020. Future yields of double-cropping systems in the Southern Amazon, Brazil, under climate change and technological development. *Agricultural Systems* 177, 102707.

Hasegawa, T., Li, T., Yin, X., Zhu, Y., Boote, K., Baker, J., Bregaglio, S., Buis, S., Confalonieri, R., Fugice, J., Fumoto, T., Gaydon, D., Kumar, S. N., Lafarge, T., Marcaida, M., Masutomi, Y., Nakagawa, H., Oriol, P., Ruget, F., Singh, U., Tang, L., Tao, F., Wakatsuki, H., Wallach, D., Wang, Y., Wilson, L. T., Yang, L., Yang, Y., Yoshida, H., Zhang, Z. and Fumoto, T. 2017. Causes of variation among rice models in yield response to CO_2 examined with free-air CO_2 enrichment and growth chamber experiments. *Scientific Reports* 7(1), 14858. https://doi.org/10.1038/s41598-017-13582-y.

Hoffmann, H., Zhao, G., Bussel, L. G. J. van, Enders, A., Specka, X., Sosa, C., Yeluripati, J., Tao, F., Constantin, J., Raynal, H., Teixeira, E., Grosz, B., Doro, L., Zhao, Z., Wang, E., Nendel, C., Kersebaum, K. C., Haas, E., Kiese, R., Klatt, S., Eckersten, H., Vanuytrecht, E., Kuhnert, M., Lewan, E., Rötter, R., Roggero, P. P., Wallach, D., Cammarano, D., Asseng, S., Krauss, G., Siebert, S., Gaiser, T. and Ewert, F. 2015. Variability of effects of spatial climate data aggregation on regional yield simulation by crop models. *Climate Research* 65, 53-69. https://doi.org/10.3354/cr01326.

Iizumi, T., Yokozawa, M. and Nishimori, M. 2009. Parameter estimation and uncertainty analysis of a large-scale crop model for paddy rice: Application of a Bayesian approach. *Agricultural and Forest Meteorology* 149(2), 333-348.

IPCC. 2012. *Managing the Risks of Extreme Events and Disasters to Advance Climate Change Adaptation. A Special Report of Working Groups I and II of the Intergovernmental Panel on Climate Change*. New York: Cambridge University Press.

IPCC. 2021. 2021: The physical science basis. Contribution of working group I to the sixth assessment report of the Intergovernmental Panel on Climate Change. In: Masson-Delmotte, V., Zhai, P., Pirani, A., Connors, S. L., Péan, C., Berger, S., Caud, N., Chen,

Y., Goldfarb, L., Gomis, M. I., et al. (Eds.), *Climate Change*. Cambridge: Cambridge University Press.

IPCC. 2022. 2022: Impacts, adaptation, and vulnerability. Contribution of working Group II to the sixth assessment report of the Intergovernmental Panel on Climate Change. In: Pörtner, H.-O., Roberts, D. C., Tignor, M., Poloczanska, E. S., Mintenbeck, K., Alegría, A., Craig, M., Langsdorf, S., Löschke, S., Möller, V., et al. (Eds.), *Climate Change*. Cambridge: Cambridge University Press.

Jamieson, P. D. and Semenov, M. A. 2000. Modelling nitrogen uptake and redistribution in wheat. *Field Crops Research* 68(1), 21–29.

Kersebaum, K. C., Boote, K. J., Jorgenson, J. S., Nendel, C., Bindi, M., Frühauf, C., Gaiser, T., Hoogenboom, G., Kollas, C., Olesen, J. E., Rötter, R. P., Ruget, F., Thorburn, P. J., Trnka, M. and Wegehenkel, M. 2015. Analysis and classification of data sets for calibration and validation of agro-ecosystem models. *Environmental Modelling and Software* 72, 402–417. https://doi.org/10.1016/j.envsoft.2015.05.009.

Knutti, R. and Sedláček, J. 2013. Robustness and uncertainties in the new CMIP5 climate model projections. *Nature Climate Change* 3(4), 369–373. https://doi.org/10.1038/nclimate1716.

Laux, P., Rötter, R. P., Webber, H., Dieng, D., Rahimi, J., Wie, J., Faye, B., Srivastava, A. K., Bliefernicht, J., Adeyeri, O., Arnault, J. and Kunstmann, H. 2021. To bias correct or not to bias correct? An agricultural impact modelers' perspective on regional climate model data. *Agricultural and Forest Meteorology* 304–305, 108406. https://doi.org/10.1016/j.agrformet.2021.108406.

Leakey, A. D. B., Ainsworth, E. A., Bernacchi, C. J., Rogers, A., Long, S. P. and Ort, D. R. 2009. Elevated CO_2 effects on plant carbon, nitrogen, and water relations: Six important lessons from FACE. *Journal of Experimental Botany* 60(10), 2859–2876.

Leggett, J., Pepper, W. and Swart, R. J. 1992. The supplementary report to the IPCC scientific assessment: Emissions scenarios for the IPCC: An update. In: Houghton, J. T., Callander, B. A. and Varney, S. K. (Eds.), *Climate Change*. Cambridge: Cambridge University Press, 71–95.

Li, T., Hasegawa, T., Yin, X., Zhu, Y., Boote, K., Adam, M., Bregaglio, S., Buis, S., Confalonieri, R., Fumoto, T., Gaydon, D., Marcaida, M., Nakagawa, H., Oriol, P., Ruane, A. C., Ruget, F., Singh, B., Singh, U., Tang, L., Tao, F., Wilkens, P., Yoshida, H., Zhang, Z. and Bouman, B. 2015. Uncertainties in predicting rice yield by current crop models under a wide range of climatic conditions. *Global Change Biology* 21(3), 1328–1341.

Liu, B., Martre, P., Ewert, F., Porter, J. R., Challinor, A. J., Müller, C., Ruane, A. C., Waha, K., Thorburn, P. J., Aggarwal, P. K., Ahmed, M., Balkovič, J., Basso, B., Biernath, C., Bindi, M., Cammarano, D., De Sanctis, G., Dumont, B., Espadafor, M., Eyshi Rezaei, E., Ferrise, R., Garcia-Vila, M., Gayler, S., Gao, Y., Horan, H., Hoogenboom, G., Izaurralde, R. C., Jones, C. D., Kassie, B. T., Kersebaum, K. C., Klein, C., Koehler, A. K., Maiorano, A., Minoli, S., Montesino San Martin, M., Naresh Kumar, S., Nendel, C., O'Leary, G. J., Palosuo, T., Priesack, E., Ripoche, D., Rötter, R. P., Semenov, M. A., Stöckle, C., Streck, T., Supit, I., Tao, F., Van der Velde, M., Wallach, D., Wang, E., Webber, H., Wolf, J., Xiao, L., Zhang, Z., Zhao, Z., Zhu, Y. and Asseng, S. 2019. Global wheat production with 1.5 and 2.0°C above pre-industrial warming. *Global Change Biology* 25(4), 1428–1444. https://doi.org/10.1111/gcb.14542.

Lobell, D. B. and Burke, M. B. 2008. Why are agricultural impacts of climate change so uncertain? The importance of temperature relative to precipitation. *Environmental Research Letters* 3(3), 034007.

Maiorano, A., Martre, P., Asseng, S., Ewert, F., Müller, C., Rötter, R. P., Ruane, A. C., Semenov, M. A., Wallach, D., Wang, E., Alderman, P. D., Kassie, B. T., Biernath, C., Basso, B., Cammarano, D., Challinor, A. J., Doltra, J., Dumont, B., Rezaei, E. E., Gayler, S., Kersebaum, K. C., Kimball, B. A., Koehler, A., Liu, B., O'Leary, G. J., Olesen, J. E., Ottman, M. J., Priesack, E., Reynolds, M., Stratonovitch, P., Streck, T., Thorburn, P. J., Waha, K., Wall, G. W., White, J. W., Zhao, Z. and Zhu, Y. 2017. Crop model improvement reduces the uncertainty of the response to temperature of multi-model ensembles. *Field Crops Research* 202, 5-20.

Martre, P. and Dambreville, A. 2018. A model of leaf coordination to scale-up leaf expansion from the organ to the canopy. *Plant Physiology* 176(1), 704-716.

Martre, P., Wallach, D., Asseng, S., Ewert, F., Jones, J. W., Rötter, R. P., Boote, K. J., Ruane, A. C., Thorburn, P. J., Cammarano, D., Hatfield, J. L., Rosenzweig, C., Aggarwal, P. K., Angulo, C., Basso, B., Bertuzzi, P., Biernath, C., Brisson, N., Challinor, A. J., Doltra, J., Gayler, S., Goldberg, R., Grant, R. F., Heng, L., Hooker, J., Hunt, L. A., Ingwersen, J., Izaurralde, R. C., Kersebaum, K. C., Müller, C., Kumar, S. N., Nendel, C., O'leary, G., Olesen, J. E., Osborne, T. M., Palosuo, T., Priesack, E., Ripoche, D., Semenov, M. A., Shcherbak, I., Steduto, P., Stöckle, C. O., Stratonovitch, P., Streck, T., Supit, I., Tao, F., Travasso, M., Waha, K., White, J. W. and Wolf, J. 2015. Multimodel ensembles of wheat growth: Many models are better than one. *Global Change Biology* 21(2), 911-925.

Muller, B., Pantin, F., Génard, M., Turc, O., Freixes, S., Piques, M. and Gibon, Y. 2011. Water deficits uncouple growth from photosynthesis, increase C content, and modify the relationships between C and growth in sink organs. *Journal of Experimental Botany* 62(6), 1715-1729.

Nakicenovic. 2000. *Special Report on Emissions Scenarios (SRES)*. Cambridge: Cambridge University Press.

O'Neill, B. C., Carter, T. R., Ebi, K., Harrison, P. A., Kemp-Benedict, E., Kok, K., Kriegler, E., Preston, B. L., Riahi, K., Sillmann, J., van Ruijven, B. J., van Vuuren, D., Carlisle, D., Conde, C., Fuglestvedt, J., Green, C., Hasegawa, T., Leininger, J., Monteith, S. and van Ruijven, B. J. 2020. Achievements and needs for the climate change scenario framework. *Nature Climate Change*, 1-11.

Palosuo, T., Hoffmann, M. P., Rötter, R. P. and Lehtonen, H. S. 2021. Sustainable intensification of crop production under alternative future changes in climate and technology: The case of the North Savo region. *Agricultural Systems* 190, 103135. https://doi.org/10.1016/j.agsy.2021.103135.

Palosuo, T., Kersebaum, K. C., Angulo, C., Hlavinka, P., Moriondo, M., Olesen, J. E., Patil, R. H., Ruget, F., Rumbaur, C., Takáč, J., Trnka, M., Bindi, M., Çaldağ, B., Ewert, F., Ferrise, R., Mirschel, W., Şaylan, L., Šiška, B. and Rötter, R. 2011. Simulation of winter wheat yield and its variability in different climates of Europe: A comparison of eight crop growth models. *European Journal of Agronomy* 35(3), 103-114.

Pirttioja, N., Carter, T. R., Fronzek, S., Bindi, M., Hoffmann, H., Palosuo, T., Ruiz-Ramos, M., Tao, F., Trnka, M., Acutis, M., Asseng, S., Baranowski, P., Basso, B., Bodin, P., Buis, S., Cammarano, D., Deligios, P., Destain, M., Dumont, B., Ewert, F., Ferrise, R., François, L., Gaiser, T., Hlavinka, P., Jacquemin, I., Kersebaum, K., Kollas, C., Krzyszczak, J., Lorite, I., Minet, J., Minguez, M., Montesino, M., Moriondo, M., Müller, C., Nendel, C., Öztürk, I., Perego, A., Rodríguez, A., Ruane, A., Ruget, F., Sanna, M., Semenov, M., Slawinski, C., Stratonovitch, P., Supit, I., Waha, K., Wang, E., Wu, L., Zhao, Z. and Rötter, R. P. 2015. A crop model ensemble analysis of temperature and precipitation effects on wheat yield across a European transect using impact response surfaces. *Climate Research* 65, 87–105.

Pirttioja, N., Palosuo, T., Fronzek, S., Räisänen, J., Rötter, R. P. and Carter, T. R. 2019. Using impact response surfaces to analyse the likelihood of impacts on crop yield under probabilistic climate change. *Agricultural and Forest Meteorology* 264, 213–224. https://doi.org/10.1016/j.agrformet.2018.10.006.

Porter, J. R., Xie, L., Challinor, A., Cochrane, K., Howden, S. M., Iqbal, M. M., Tompkins, E. L. and Garrett, K. 2014. Impacts, adaptation and vulnerability. Contribution of working group II to the fifth assessment report of the intergovernmental panel on climate change (pp. 485–533). Food security and food production systems. In: Field, C. B., Barros, V. R., Dokken, D. J., Mach, K. J., Mastrandrea, M. D., Bilir, T. E., Chatterjee, M., Ebi, K. L., Estrada, Y. O., Genova, R. C., et al. (Eds.), *Climate Change*. Cambridge, UK: Cambridge University Press.

Ratjen, A. M., Lemaire, G., Kage, H., Plénet, D. and Justes, E. 2018. Key variables for simulating leaf area and N status: Biomass based relations versus phenology driven approaches. *European Journal of Agronomy* 100, 110–117. https://doi.org/10.1016/j.eja.2018.04.008.

Riahi, K., van Vuuren, D. P., Kriegler, E., Edmonds, J., O'Neill, B. C., Fujimori, S., Bauer, N., Calvin, K., Dellink, R., Fricko, O., Lutz, W., Popp, A., Cuaresma, J. C., Kc, S., Leimbach, M., Jiang, L., Kram, T., Rao, S., Emmerling, J., Ebi, K., Hasegawa, T., Havlik, P., Humpenöder, F., Da Silva, L. A., Smith, S., Stehfest, E., Bosetti, V., Eom, J., Gernaat, D., Masui, T., Rogelj, J., Strefler, J., Drouet, L., Krey, V., Luderer, G., Harmsen, M., Takahashi, K., Baumstark, L., Doelman, J. C., Kainuma, M., Klimont, Z., Marangoni, G., Lotze-Campen, H., Obersteiner, M., Tabeau, A. and Tavoni, M. 2017. The shared socioeconomic pathways and their energy, land use, and greenhouse gas emissions implications: An overview. *Global Environmental Change* 42, 153–168.

Rodríguez, A., Ruiz-Ramos, M., Palosuo, T., Carter, T. R., Fronzek, S., Lorite, I. J., Ferrise, R., Pirttioja, N., Bindi, M., Baranowski, P., Buis, S., Cammarano, D., Chen, Y., Dumont, B., Ewert, F., Gaiser, T., Hlavinka, P., Hoffmann, H., Höhn, J. G., Jurecka, F., Kersebaum, K. C., Krzyszczak, J., Lana, M., Mechiche-Alami, A., Minet, J., Montesino, M., Nendel, C., Porter, J. R., Ruget, F., Semenov, M. A., Steinmetz, Z., Stratonovitch, P., Supit, I., Tao, F., Trnka, M., de Wit, A. and Rötter, R. P. 2019. Implications of crop model ensemble size and composition for estimates of adaptation effects and agreement of recommendations. *Agricultural and Forest Meteorology* 264, 351–362.

Rosenzweig, C., Elliott, J., Deryng, D., Ruane, A. C., Müller, C., Arneth, A., Boote, K. J., Folberth, C., Glotter, M., Khabarov, N., Neumann, K., Piontek, F., Pugh, T. A., Schmid, E., Stehfest, E., Yang, H. and Jones, J. W. 2014. Assessing agricultural risks of climate change in the 21st century in a global gridded crop model intercomparison. *Proceedings of the National Academy of Sciences of the United States of America* 111(9), 3268–3273.

Rötter, R. P. 2014. Robust uncertainty. *Nature Climate Change* 4(4), 251–252.

Rötter, R. P., Appiah, M., Fichtler, E., Kersebaum, K. C., Trnka, M. and Hoffmann, M. P. 2018. Linking modelling and experimentation to better capture crop impacts of agro-climatic extremes - A review. *Field Crops Research* 221, 142–156.

Rötter, R. P., Carter, T. R., Olesen, J. E. and Porter, J. R. 2011. Crop-climate models need an overhaul. *Nature Climate Change* 1(4), 175–177.

Rötter, R. P., Palosuo, T., Kersebaum, K. C., Angulo, C., Bindi, M., Ewert, F., Ferrise, R., Hlavinka, P., Moriondo, M., Nendel, C., Olesen, J. E., Patil, R. H., Ruget, F., Takáč, J. and Trnka, M. 2012. Simulation of spring barley yield in different climatic zones of

Northern and Central Europe: A comparison of nine crop growth models. *Field Crops Research* 133, 23–36.

Rötter, R. P. and Van de Geijn, S. C. 1999. Climate change effects on plant growth, crop yield and livestock. *Climatic Change* 43(4), 651–681.

Salo, T. J., Palosuo, T., Kersebaum, K. C., Nendel, C., Angulo, C., Ewert, F., Bindi, M., Calanca, P., Klein, T., Moriondo, M., Ferrise, R., Olesen, J. E., Patil, R. H., Ruget, F., Takac, J., Hlavinka, P., Trnka, M. and Röttter, R. P. 2016. Comparing the performance of 11 crop simulation models in predicting yield response to nitrogen fertilization. *Journal of Agricultural Science* 154(7), 1218–1240. https://doi.org/10.1017/S0021859615001124.

Schewe, J., Gosling, S. N., Reyer, C., Zhao, F., Ciais, P., Elliott, J., Francois, L., Huber, V., Lotze, H. K., Seneviratne, S. I., van Vliet, M. T. H., Vautard, R., Wada, Y., Breuer, L., Büchner, M., Carozza, D. A., Chang, J., Coll, M., Deryng, D., de Wit, A., Eddy, T. D., Folberth, C., Frieler, K., Friend, A. D., Gerten, D., Gudmundsson, L., Hanasaki, N., Ito, A., Khabarov, N., Kim, H., Lawrence, P., Morfopoulos, C., Müller, C., Müller Schmied, H., Orth, R., Ostberg, S., Pokhrel, Y., Pugh, T. A. M., Sakurai, G., Satoh, Y., Schmid, E., Stacke, T., Steenbeek, J., Steinkamp, J., Tang, Q., Tian, H., Tittensor, D. P., Volkholz, J., Wang, X. and Warszawski, L. 2019. State-of-the-art global models underestimate impacts from climate extremes. *Nature Communications* 10(1), 1005. https://doi.org/10.1038/s41467-019-08745-6.

Seidel, S. J., Palosuo, T., Thorburn, P. and Wallach, D. 2018. Towards improved calibration of crop models – Where are we now and where should we go? *European Journal of Agronomy* 94, 25–35. https://doi.org/10.1016/j.eja.2018.01.006.

Semenov, M. A. and Stratonovitch, P. 2015. Adapting wheat ideotypes for climate change: Accounting for uncertainties in CMIP5 climate projections. *Climate Research* 65, 123–139.

Sinclair, T. R., Farias, J. R., Neumaier, N. and Nepomuceno, A. L. 2003. Modeling nitrogen accumulation and use by soybean. *Field Crops Research* 81(2–3), 149–158.

Tao, F., Hayashi, Y., Zhang, Z., Sakamoto, T. and Yokozawa, M. 2008. Global warming, rice production, and water use in China: Developing a probabilistic assessment. *Agricultural and Forest Meteorology* 148(1), 94–110.

Tao, F., Palosuo,T., Rötter, R. P., Díaz-Ambrona, C. G. H., Inés Mínguez, M., Semenov, M. A., Kersebaum, K. C., Cammarano, D., Specka, X., Nendel, C., Srivastava, A. K., Ewert, F., Padovan, G., Ferrise, R., Martre, P., Rodríguez, L., Ruiz-Ramos, M., Gaiser, T., Höhn, J. G., Salo, T., Dibari, C. and Schulman, A. H. 2020. Why do crop models diverge substantially in climate impact projections? A comprehensive analysis based on eight barley crop models. *Agricultural and Forest Meteorology* 281, 107851.

Tao, F., Rötter, R. P., Palosuo, T., Díaz-Ambrona, C. G. H., Mínguez, M. I., Semenov, M. A., Kersebaum, K. C., Nendel, C., Cammarano, D., Hoffmann, H., Ewert, F., Dambreville, A., Martre, P., Rodríguez, L., Ruiz-Ramos, M., Gaiser, T., Höhn, J. G., Salo, T., Ferrise, R., Bindi, M. and Schulman, A. H. 2017. Designing future barley ideotypes using a crop model ensemble. *European Journal of Agronomy* 82, 144–162.

Tao, F., Rötter, R. P., Palosuo, T., Gregorio Hernández Díaz-Ambrona, C., Mínguez, M. I., Semenov, M. A., Kersebaum, K. C., Nendel, C., Specka, X., Hoffmann, H., Ewert, F., Dambreville, A., Martre, P., Rodríguez, L., Ruiz-Ramos, M., Gaiser, T., Höhn, J. G., Salo, T., Ferrise, R., Bindi, M., Cammarano, D. and Schulman, A. H. 2018. Contribution of crop model structure, parameters and climate projections to uncertainty in climate change impact assessments. *Global Change Biology* 24(3), 1291–1307.

Tao, F., Yokozawa, M. and Zhang, Z. 2009a. Modeling the impacts of weather and climate variability on crop productivity over a large area: A new process-based model development, optimization, and uncertainties analysis. *Agricultural and Forest Meteorology* 149(5), 831–850.

Tao, F., Zhang, Z., Liu, J. and Yokozawa, M. 2009b. Modeling the impacts of weather and climate variability on crop productivity over a large area: A new super-ensemble based probability projection. *Agricultural and Forest Meteorology* 149(8), 1266–1278.

Van Vuuren, D. P., Edmonds, J., Kainuma, M., Riahi, K., Thomson, A., Hibbard, K., Hurtt, G. C., Kram, T., Krey, V., Lamarque, J.-F., Masui, T., Meinshausen, M., Nakicenovic, N., Smith, S. J. and Rose, S. K. 2011. The representative concentration pathways: An overview. *Climatic Change* 109(1–2), 5–31. https://doi.org/10.1007/s10584-011-0148-z.

Vanuytrecht, E. and Thorburn, P. J. 2017. Responses to atmospheric CO_2 concentrations in crop simulation models: A review of current simple and semicomplex representations and options for model development. *Global Change Biology* 23(5), 1806–1820.

Wallach, D. 2019. Dealing with uncertainty in crop models. In: Boote, K. (Ed.), *Advances in Crop Modelling for Sustainable Agriculture*. Cambridge, UK: Burleigh Dodds Science Publishing, 419–448.

Wallach, D., Martre, P., Liu, B., Asseng, S., Ewert, F., Thorburn, P. J., van Ittersum, M., Aggarwal, P. K., Ahmed, M., Basso, B., Biernath, C., Cammarano, D., Challinor, A. J., De Sanctis, G., Dumont, B., Eyshi Rezaei, E., Fereres, E., Fitzgerald, G. J., Gao, Y., Garcia-Vila, M., Gayler, S., Girousse, C., Hoogenboom, G., Horan, H., Izaurralde, R. C., Jones, C. D., Kassie, B. T., Kersebaum, K. C., Klein, C., Koehler, A. K., Maiorano, A., Minoli, S., Müller, C., Naresh Kumar, S., Nendel, C., O'Leary, G. J., Palosuo, T., Priesack, E., Ripoche, D., Rötter, R. P., Semenov, M. A., Stöckle, C., Stratonovitch, P., Streck, T., Supit, I., Tao, F., Wolf, J. and Zhang, Z. 2018. Multimodel ensembles improve predictions of crop–environment–management interactions. *Global Change Biology* 24(11), 5072–5083. https://doi.org/10.1111/gcb.14411.

Wallach, D., Mearns, L. O., Ruane, A. C., Rötter, R. P. and Asseng, S. 2016. Lessons from climate modeling on the design and use of ensembles for crop modeling. *Climatic Change* 139(3), 551–564.

Wallach, D., Palosuo, T., Thorburn, P., Gourdain, E., Asseng, S., Basso, B., Buis, S., Crout, N., Dibari, C., Dumont, B., Ferrise, R., Gaiser, T., Garcia, C., Gayler, S., Ghahramani, A., Hochman, Z., Hoek, S., Hoogenboom, G., Horan, H., Huang, M., Jabloun, M., Jing, Q., Justes, E., Kersebaum, K. C., Klosterhalfen, A., Launay, M., Luo, Q., Maestrini, B., Mielenz, H., Moriondo, M., Nariman Zadeh, H., Olesen, J. E., Poyda, A., Priesack, E., Pullens, J. W. M., Qian, B., Schütze, N., Shelia, V., Souissi, A., Specka, X., Srivastava, A. K., Stella, T., Streck, T., Trombi, G., Wallor, E., Wang, J., Weber, T. K. D., Weihermüller, L., de Wit, A., Wöhling, T., Xiao, L., Zhao, C., Zhu, Y. and Seidel, S. J. 2021. How well do crop modeling groups predict wheat phenology, given calibration data from the target population? *European Journal of Agronomy* 124, 126195. https://doi.org/10.1016/j.eja.2020.126195.

Wallach, D., Palosuo, T., Thorburn, P., Hochman, Z., Andrianasolo, F., Asseng, S., Basso, B., Buis, S., Crout, N., Dumont, B., Ferrise, R., Gaiser, T., Gayler, S., Hiremath, S., Hoek, S., Horan, H., Hoogenboom, G., Huang, M., Jabloun, M., Jansson, P.-E., Jing, Q., Justes, E., Kersebaum, K. C., Launay, M., Lewan, E., Luo, Q., Maestrini, B., Moriondo, M., Olesen, J. E., Padovan, G., Poyda, A., Priesack, E., Pullens, J. W. M., Qian, B., Schütze,

N., Shelia, V., Souissi, A., Specka, X., Kumar Srivastava, A., Stella, T., Streck, T., Trombi, G., Wallor, E., Wang, J., Weber, T. K. D., Weihermüller, L., de Wit, A., Wöhling, T., Xiao, L., Zhao, C., Zhu, Y. and Seidel, S. J. 2021. Multi-model evaluation of phenology prediction for wheat in Australia. *Agricultural and Forest Meteorology* 298–299, 108289. https://doi.org/10.1016/j.agrformet.2020.108289.

Wallach, D., Palosuo,T., Thorburn, P., Hochman, Z., Gourdain, E., Andrianasolo, F., Asseng, S., Basso, B., Buis, S., Crout, N., Dibari, C., Dumont, B., Ferrise, R., Gaiser, T., Garcia, C., Gayler, S., Ghahramani, A., Hiremath, S., Hoek, S., Horan, H., Hoogenboom, G., Huang, M., Jabloun, M., Jansson, P., Jing, Q., Justes, E., Kersebaum, K. C., Klosterhalfen, A., Launay, M., Lewan, E., Luo, Q., Maestrini, B., Mielenz, H., Moriondo, M., Nariman Zadeh, H., Padovan, G., Olesen, J. E., Poyda, A., Priesack, E., Pullens, J. W. M., Qian, B., Schütze, N., Shelia, V., Souissi, A., Specka, X., Srivastava, A. K., Stella, T., Streck, T., Trombi, G., Wallor, E., Wang, J., Weber, T. K. D., Weihermüller, L., de Wit, A., Wöhling, T., Xiao, L., Zhao, C., Zhu, Y. and Seidel, S. J. 2021. The chaos in calibrating crop models: Lessons learned from a multi-model calibration exercise. *Environmental Modelling and Software* 145, 105206.

Wang, E., Martre, P., Zhao, Z., Ewert, F., Maiorano, A., Rötter, R. P., Kimball, B. A., Ottman, M. J., Wall, G. W., White, J. W., Reynolds, M. P., Alderman, P. D., Aggarwal, P. K., Anothai, J., Basso, B., Biernath, C., Cammarano, D., Challinor, A. J., De Sanctis, G., Doltra, J., Dumont, B., Fereres, E., Garcia-Vila, M., Gayler, S., Hoogenboom, G., Hunt, L. A., Izaurralde, R. C., Jabloun, M., Jones, C. D., Kersebaum, K. C., Koehler, A. K., Liu, L., Müller, C., Kumar, S. N., Nendel, C., O'Leary, G., Olesen, J. E., Palosuo, T., Priesack, E., Rezaei, E. E., Ripoche, D., Ruane, A. C., Semenov, M. A., Shcherbak, I., Stöckle, C., Stratonovitch, P., Streck, T., Supit, I., Tao, F., Thorburn, P., Waha, K., Wallach, D., Wang, Z., Wolf, J., Zhu, Y. and Asseng, S. 2017. Author correction: The uncertainty of crop yield projections is reduced by improved temperature response functions. *Nature Plants* 3(10), 833, Article Number: 17102.

Webber, H., Ewert, F., Olesen, J. E., Müller, C., Fronzek, S., Ruane, A. C., Bourgault, M., Martre, P., Ababaei, B., Bindi, M., Ferrise, R., Finger, R., Fodor, N., Gabaldón-Leal, C., Gaiser, T., Jabloun, M., Kersebaum, K. C., Lizaso, J. I., Lorite, I. J., Manceau, L., Moriondo, M., Nendel, C., Rodríguez, A., Ruiz-Ramos, M., Semenov, M. A., Siebert, S., Stella, T., Stratonovitch, P., Trombi, G. and Wallach, D. 2018. Diverging importance of drought stress for maize and winter wheat in Europe. *Nature Communications* 9(1), 4249. https://doi.org/10.1038/s41467-018-06525-2.

White, J. W., Hoogenboom, G., Kimball, B. A. and Wall, G. W. 2011. Methodologies for simulating impacts of climate change on crop production. *Field Crops Research* 124(3), 357–368.

Xin, Y. and Tao, F. 2020. Developing climate-smart agricultural systems in the North China Plain. *Agriculture, Ecosystems and Environment* 291, 106791.

Yin, X., Struik, P. C. and Goudriaan, J. 2021. On the needs for combining physiological principles and mathematics to improve crop models. *Field Crops Research* 271, 108254.

Part 2

Modelling climate change impacts on particular aspects of agricultural systems

Chapter 6

Modeling climate change impacts on crop growth and yield formation

Davide Cammarano and Johannes Wilhelmus Maria Pullens, Aarhus University, Denmark; and Pierre Martre, Université de Montpellier, France

1 Introduction
2 Concept of modeling crop yield
3 Crop modeling response: the role of data and model structure
4 Assessing climate impacts on yield with model ensembles
5 Conclusion and future trends in research
6 Where to look for further information
7 References

1 Introduction

Current and projected trends in human population and food consumption indicate that the global food supply must increase by around 45% in 2050 to meet demand (Godfray et al., 2010). This target must be achieved by (a) producing on the same (or less) amount of cropped land while maintaining soil health, (b) reducing negative environmental issues such as pollution of the aquatic ecosystem and greenhouse gas emissions, and (c) keeping farming economically viable (Davis et al., 2021; Tian et al., 2021). Unprecedented changes to the climate system and projected trends (IPCC, 2021) will make such an accomplishment harder to achieve.

The impacts of climate variability and change on crop growth, development, and yield have been quantified by various modeling approaches that can be clustered into two main types - statistical- and process-based (Lobell and Asseng, 2017). The statistical-based model takes into account the functional relationship between observed weather data and observed crop yield (Lobell and Burke, 2010). The latter simulates daily growth, development, and yield as influenced by daily weather patterns, soil chemical and physical properties, crop features, and agronomic management (Boote et al., 2013). Detailed discussion regarding the main differences between these two approaches can be found elsewhere (Liu et al., 2016; Lobell and Asseng, 2017) and is not the focus of this chapter.

http://dx.doi.org/10.19103/AS.2022.0115.08

Crop growth models (CMs) are used for a variety of applications on varying spatial and temporal scales. They are used to quantify climate change impacts at local, national, and global scales (Asseng et al., 2015; Rosenzweig et al., 2014; Rötter et al., 2018) to evaluate adaptation strategies (Ruiz-Ramos et al., 2018), to design future ideotypes with several crop traits (Tao et al., 2017), linking them with genetic information (Bustos-Korts et al., 2019; Zheng et al., 2012), and to evaluate the spatial variability of agronomic practices (Cammarano et al., 2021).

White et al. (2011) observed how the use of CM for climate change studies should be carried out with care as these models were not originally designed for such purposes and could miss key processes related to climate conditions. Comparing CMs or detailed sub-routines is not a new practice (Jamieson et al., 1998; Rötter et al., 2012; Verburg, 1995) and rarely leads to understanding the cause of uncertainty and divergences between models and observations and to significantly improving modeling sub-routines. In the last ten years, the Agricultural Modelling Intercomparison and Improvement Project (AgMIP) (Rosenzweig et al., 2013) has organized many different process-based modelers for different crop-specific model intercomparison studies and researchers, breeders, and crop physiologists have worked with modelers on aspects of modeling improvement (Alderman et al., 2013). Due to a common protocol in terms of comparing and planning for model improvements, a regular outcome is that the median of the multi-models' prediction is often better than a single model (Asseng et al., 2013; Martre et al., 2015;; Wallach et al., 2016). Such effects are not uncommon among models, as has been observed in the climate community for quite some time (Doblas-Reyes et al., 2005; Pierce et al., 2009).

The ultimate goal across the different CM intercomparison activities is to improve models' sub-routines in order to both minimize variability among the models and reduce bias among observations. To do so, several process-specific comparisons have been made for various crops, which is expected to improve understanding of specific processes and ultimately improve crop modeling (Cammarano et al., 2016; Kimball et al., 2019; Webber et al., 2017). However, several challenges make such goals difficult to achieve, many of which are operational since these activities are carried out by researchers, whose available time is often stretched. Other challenges include a lack of detailed experimental data for a given simulated process, the difficulties of managing a large group of modelers with models of several degrees of mechanisms, and finding meaningful collaborations with experts from different disciplines.

The aim of this chapter is to briefly illustrate current practices in modeling crop yield as a response to climate change and what has been learned so far, especially in terms of multi-model assessment and comparisons. For additional information, readers are directed to Section 6 for links and references.

2 Concept of modeling crop yield

Photosynthesis is among the most important biological processes as it enables crop production (and energy) for humans. Agricultural yields are linked to the capacity of plants to photosynthesize. Plants convert energy from sunlight into products directly used for human consumption (e.g. rice, wheat, corn, canola, and cassava) or indirectly through their use as livestock feed to produce milk, eggs, and meat (e.g. silage corn, forages, and coarse grains).

The first step is photosynthesis, in which CO_2 and water in the chloroplasts (components of green plant cells) combine with the energy from sunlight to produce carbohydrates (or glucose sugar), releasing oxygen and water. Carbohydrates (glucose) are used for growth processes and to form harvesting products. The chemical reaction of the process is:

$$6CO_2 + 6H_2O \underset{\rightarrow}{\text{light}} C_6H_{12}O_6 + 6O_2 \tag{8.1}$$

which can be simplified as:

$$CO_2 + H_2O \rightarrow CH_2O + O_2 \tag{8.2}$$

One other principle, which goes with photosynthetic processes and can impact yield, is respiration. In a simplified way, the rate of crop biomass growth can be represented as follows:

$$\frac{dW}{dt} = E_f P_g - R_m W \tag{8.3}$$

where $\frac{dW}{dt}$ is the rate of dry weight growth of any given part of the plant, e.g. leaves, stem, roots, or total plant (g of dry weight m^{-2} h^{-1}), W is the total plant dry weight (g m^{-2}), E_f is the conversion efficiency of CH_2O into a plant tissue (g tissue g^{-1} CH_2O), P_g is the canopy gross photosynthesis rate (g CH_2O m^{-2} h^{-1}), and R_m is the maintenance respiration rate (g CH_2O g^{-1} tissue h^{-1}). Respiration therefore translates into no gains in plant biomass, and in the case of maintenance respiration, the CO_2 released or the O_2 consumed during the generation of usable energy for plant maintenance also includes dark respiration at night for plant maintenance. More details of photosynthesis processes can be found elsewhere (Connor et al., 2011).

In crop growth models, photosynthesis and yield processes are simulated by a variety of approaches (Table 1). For photosynthesis processes, there are four main approaches of varying complexity, namely Farquhar, canopy photosynthesis (Acock), radiation use efficiency (RUE), and transpiration efficiency (TE).

Table 1 Crop growth modeling approaches of the 27 wheat models used in the AgMIP wheat study

Model	Leaf area/ light interc.[a]	Light utilization[b]	Yield formation[c]	Model relative[d]	Model type[e]	APSIM-Nwheat
S	RUE	Prt	C	P	APSIM-wheat	S
RUE	Prt/Gn/B	C	P	AquaCrop	S	TE
HI/B	None	P	CropSyst	S	TE/RUE	HI/B
None	P	DSSAT-CERES	S	RUE	B/Gn	C
P	DSSAT-CROPSIM	S	RUE	Prt	None	p
Ecosys	D	P-R	Gn-Prt	None	P	EPIC wheat
S	RUE	HI	E	P/G	Expert-N-CERES	S
RUE	B/Gn	C	P	Expert-N-GECROS	D	P-R/TE
Gn/Prt	S	P	Expert-N-SPASS	D	P-R	Gn/Prt
C/S	P	Expert-N-SUCROS	D	P-R	Prt	S
P	FASSET	D	RUE	HI/B	None	P
GLAM-wheat	S	RUE/TE	B/HI	None	G	HERMES
D	P-R	Prt	S/C	P	InfoCrop	D
RUE	Prt/Gn	S	P	LINTUL-4	D	RUE
Prt/B	L	P	LINTUL-FAST	D	RUE	Prt
L	P	LPJmL	S	P-R	HI_mws/B	None
G	MCWLA-wheat	S	P-R	HI/B	None	G
MONICA	S	P-R	Prt	H	P	O'Leary-model
S	TE	Gn/Prt	None	P	SALUS	S
RUE	Prt/HI	C	P	Sirius	D	RUE
B/Prt	None	P	SiriusQuality	D	RUE	B/Prt
I	P	STICS	D	RUE	Gn/B	C
P	WOFOST	D	P-R	Prt/B	S	G

[a]S, simple approach (e.g. LAI); D, detailed approach (e.g. canopy layers).

[b]RUE, radiation use efficiency approach; P-R, gross photosynthesis-respiration; TE, transpiration efficiency biomass growth.

[c]HI, fixed harvest index; B, total (aboveground) biomass; Gn, number of grains; Prt, partitioning during reproductive stages; HI_mw, harvest index modified by water stress.

[d]C, CERES; L, LINTUL; E, EPIC; S, SUCROS; I, Sirius; H, HERMES.

[e]P, point model; G, global or regional model (regarding the main purpose of the model).

Source: Adapted from Asseng et al. (2013) to represent the processes of photosynthesis, yield, and relative models.

The Farquhar model (Farquhar et al., 1980) computes photosynthesis as a function of the demand and supply of CO_2. In this model, photosynthesis is regulated by radiation, transpiration, air humidity, leaf temperature, CO_2 availability, and leaf nitrogen content. In addition, the plant is impacted by radiation saturation at high levels of radiation.

However, the model requires driving variables as input to the simulation in hourly steps. This model is only applicable to individual leaves in many crop models. Farquhar's equations have been applied directly to plant canopies by assuming a plant canopy to function as a big leaf. The big-leaf approach can estimate seasonal trends of canopy photosynthesis but is less accurate for simulations of daily variations.

On the other end of the spectrum, there is the RUE approach in which the main assumption is that the photosynthetic rate of crops is a function of the quantity of radiation they intercept and the efficiency of utilization. Therefore, the RUE is defined as the g of biomass produced per MJ of light intercepted, relative to incident/intercepted radiation.

$$\frac{dW}{dt} = \text{RUE} * I_0 \left[1 - exp\left(-k * \text{LAI}\right)\right] \tag{8.4}$$

where I_0 is the incident irradiance, and as Connor et al. (2011) reported, the expression $I_0\left[1 - exp\left(-k * \text{LAI}\right)\right]$ is the interception; k is the extinction coefficient, which relates leaf angle to the ability of unit leaf area to intercept radiation; and LAI is the leaf area index.

Grain yield can be simulated in a variety of ways (Table 1). Some approaches use grain number determined as a function of the difference between aboveground biomass at the end of the anthesis and that at the early developmental stages, which is usually taken at germination. In this way, the dry matter factor is equal to the total biomass at the end of anthesis, and there is a slope factor that accounts for genotypic differences. The potential grain dry matter accumulation is a function of grain number and potential growth per individual grain, which is genotypic-specific. The individual grain depends on the maximum grain weight of the cultivar and the duration of the grain filling, as well as the simulation of grain number for each plant.

Another approach is when grain number per plant is calculated at anthesis based on the following factors - a cultivar-specific (and subjected to calibration) coefficient, canopy weight, average rate of carbohydrate accumulation during anthesis, temperature, water, and nitrogen stresses. Using this modeling approach, the potential kernel number is also a user-defined coefficient reflecting cultivar differences.

For the Harvest Index (HI) approach, the yield (Y) if function on total aboveground biomass at maturity (W_m) and the HI is as follows:

$$Y = W_m HI \tag{8.5}$$

HI is calculated based on a 'potential' HI, which is then modified as a function of the water and nitrogen stresses during the reproductive stages.

The main conceptual difference among these approaches depends on the distribution of carbohydrates in the plant. The HI approach, the complex relationships, and assumptions regarding grain number and size are not taken into account. On the other hand, approaches based on the partitioning of carbohydrates to different organs need to consider the dynamic changes in carbohydrate distribution among organs at different growth stages. In these models, correctly simulating crop phenology (that can impact carbohydrate distribution among organs) and having accurate partitioning coefficients are fundamental to accurate yield simulation.

In addition, different modeling approaches consider grain mass supply/demand in various ways, such as source-limited, sink-limited, and a combination of the two. The former considers grain mass assimilation to be limited by the ability of other organs (e.g. leaves) to supply sufficient assimilates to satisfy grain demand. Sink-limited, on the other hand, considers grain growth to be limited by the ability of the grains to absorb the available assimilates. In this case, grain number and potential single grain size are important parameters for determining the final yield. The last approach simulates, separately, supply and demand, and grain growth is considered the minimum of the two. Along with carbon, N is also simulated using two approaches - the HI approach in which an N harvest index has a linear increase as a function of thermal time. The other approach is the sink-source model.

While each approach has its pros and cons, the main advantage of a given approach should be as a function of the objectives of the simulations. While the HI approach is robust in simulating yield, the grain number/grain size considers a more sophisticated response. For example, most crop simulation models simulate grain N and therefore protein concentration, but grain protein fractions simulations require an additional mechanism. The grain N accumulation is source regulated and so the rate of N transferal to the grain impacts the main protein fractions (Martre et al., 2006; Martre et al., 2003).

The simulation of quantity vs. quality is an important aspect of many crops. Most simulation studies focus on the simulation of yield quantity as impacted, for example, by climate change. However, for some crops - such as wheat, barley, processing tomato, and sugar beet - the quality components can be more important than the dry yield. For example, in wheat, the protein concentration varies according to usage, from 8% for pastries to >14% for pasta and bread (Asseng et al., 2019), while for barley, for malting purposes the grain N% is often required to be between 1.45% and 1.85% (Cammarano et al., 2021). In a study comparing multiple wheat models, Asseng et al. (2019) demonstrated the value of adaptation to climate change in terms of grain yield increase, which does not translate into keeping protein content at desired levels. In contrast, Xie et al. (2018) reported the negative impacts of climate change on beer supply but looked only at changes in dry yield with no insight into N%, which is required for malting. In tomato, Giuliani et al. (2019) modified a crop model to include the simulation of lycopene content, an important feature for tomato quality, and how this is impacted by climate change.

For wheat protein, concentration and composition are the most important components that need to be simulated and that are currently missing in most crop models. For barley, a better representation of grain N accumulation and dynamics is also required as there are insufficient studies showing the impacts of projected changes in temperature, CO_2 concentration, and extreme events on barley grain quality, as well as how current crop models capture those climate patterns and how they impact on grain quality. Such limitations are more marked for crops subject to fewer simulation studies, such as sugar content in fruits, lycopene in processing tomato, and quality traits of grapes (important for wine making).

3 Crop modeling response: the role of data and model structure

It has been observed how changes in air temperature, the amount and frequency of precipitation, atmospheric carbon dioxide (CO_2), and ozone (O_3) concentrations impact crop growth and productivity (Ainsworth and Long, 2020; Emberson et al., 2018; Hatfield et al., 2011; Hatfield and Prueger, 2015; Tai et al., 2021). Single crop models have been used to assess the impacts of climate change on crop growth and yield and to evaluate adaptation and mitigation strategies (Cammarano et al., 2019a; Kim et al., 2014; Nendel et al., 2014). The use of a single crop growth model should not be dismissed as it can still provide useful information on climate impacts and adaptation strategies. In addition, it is easier to carry out by one research group rather than learning and running new models. However, it is worth pointing out that the AgMIP activity is also working on a platform that allows the standardization of input data in order to run multiple crop models, as carried out by Cammarano et al. (2020).

However, the number of crop growth models to use for climate impact studies depends on the objective of the study per se. If the aim is to truly understand how climate change impacts the simulation of growth and development processes and which process requires improvement, a multi-model approach serves the purpose. Wang et al. (2017) provide a good example of how multi-model intercomparison leads to a better understanding of specific processes. For example, they showed that variations in the mathematical functions used by a number of wheat models to simulate temperature responses of physiological processes account for more than 50% of uncertainty in simulated grain yields. Wang et al. (2017) constructed new temperature response functions, which replaced the original ones in four wheat models, and decreased the bias in grain yield simulations by 42% on average across global locations.

The multi-model ensemble of crop growth models draws from the experience of other scientific fields and has been evaluated in terms of mean and median with respect to single models with encouraging results (Rötter et al., 2012). Martre et al. (2015) performed a quantitative analysis of the quality of simulators based on crop model ensembles with respect to single models. The study focused on several key growth variables rather than yield alone, concluding that a multi-crop growth model ensemble can be used to generate new estimators with better accuracy and consistency for quantifying crop growth dynamics.

Ultimately the driver for multi-model ensemble is better prediction, which is fine for climate impact studies, but does not necessarily lead to a better understanding of the underlying processes unless the ensemble is evaluated in terms of uncertainty (Asseng et al., 2013) and actions are taken (Alderman et al., 2013; Wang et al., 2017).

3.1 Structures of crop growth models

CMs vary in their degree of complexity and the processes they simulate, which affects their response to climate impacts. The crop growth model simulates interactions (daily or hourly) between soil-plant-atmosphere-agronomic management, but the mechanisms and the sub-routines implemented can vary. Asseng et al. (2013) and Wang et al. (2017) provide a good example of the differences between processes simulated across crop growth models. For example, across 27 wheat models, light utilization was calculated via three different sub-routines: (a) the radiation use efficiency (RUE) approach, (b) gross photosynthesis–respiration, and (c) transpiration efficiency. Regardless of the approach, the noise in the daily simulated aboveground dry biomass patterns is evident (Fig. 1). In addition to the various methods for calculating aboveground biomass, there are different ways to account for stresses (e.g. water and nitrogen stresses). Each method has its pros and cons and originally was implemented in crop growth models to serve a specific purpose.

Webber et al. (2017) compared nine process-based crop growth models, which employed three approaches to simulate canopy temperature (empirical, energy balance with neutral atmospheric stability, and energy balance with correction for atmospheric stability conditions). An interesting finding was

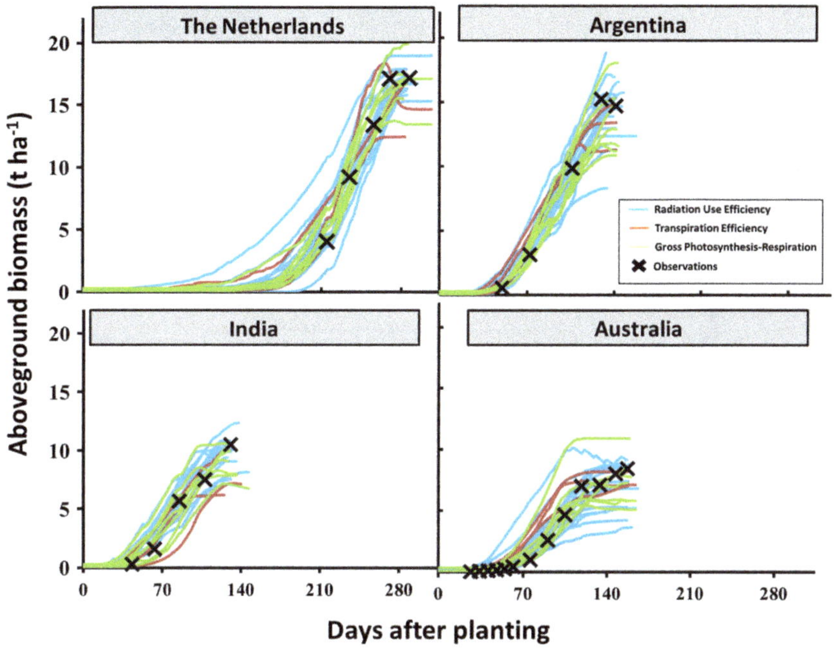

Figure 1 Simulated (lines) and observed (cross) daily patterns of aboveground biomass from the AgMIP Wheat dataset (Asseng et al., 2013) for the four locations of the study and plotted for the different light utilization approaches: radiation use efficiency (blue lines); transpiration efficiency (red lines); and gross photosynthesis/respiration (green lines).

that additional mechanisms did not improve the simulation of the final yield. Comparing simpler and complex approaches was also undertaken for potential evapotranspiration and soil water balance with similar findings (Cammarano et al., 2016; Verburg, 1995).

This does not mean that simpler sub-routines are better, as it will depend on the data that modelers use to implement and test their sub-routines. To date, it has not been possible to fully identify which processes are responsible for the high uncertainty in simulated outputs that cause some models to be outliers. One of the best-known reasons is that models contain many feedbacks, making it difficult to trace the exact cause of deviation (Goudriaan, 1996).

For example, the potential evapotranspiration (ET_p) can be calculated by a variety of approaches (Fig. 2) while other models do not calculate ET_p but use an hourly energy balance to simulate the equilibrium of canopy and soil temperature and leaf conductance (Kimball et al., 2019). If ET_p is calculated, the partitioning into actual crop transpiration (T_a) and soil evaporation (E_a) is carried out by employing several different approaches, as described elsewhere (Kimball et al., 2019). From ET_p, the potential crop transpiration (T_p) and soil evaporation (E_p) are computed first (again a variety of approaches exist) and

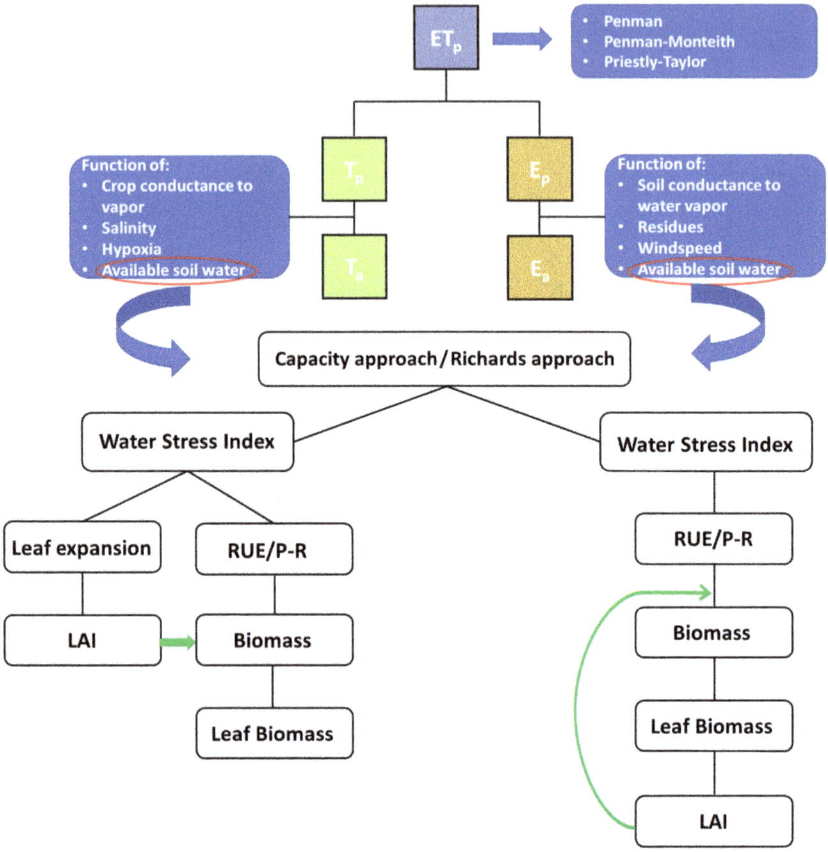

Figure 2 Schematic summary of a process that involves simulating several components of the water balance. The top part is just a representation of a class of models since other approaches are also employed. The simulation of potential evapotranspiration (ET_p), which can be carried out by a number of available approaches (just a few are reported in the box next to ET_p), is then partitioned into potential crop transpiration (T_p) and potential soil evaporation (E_p). The actual crop transpiration (T_a) and soil evaporation (E_a) are often lower than the potential counterparts as a function of several factors. Considering one of those factors (e.g. soil water content), this can be simulated by a number of approaches that are used to calculate the Water Stress Index. Depending on the crop model approach, the water stress impact on biomass can be calculated by either of the two approaches where RUE is radiation use efficiency, P-R is photosynthesis-respiration, and LAI is leaf area index.

then Ta and Ea are calculated. The latter passage is a function of several factors. The available soil water content is among the factors considered for both transpiration and evaporation and, again, different crop models use different methods to calculate water in the soil (Capacitance or Richards' approaches). In many crop modeling approaches, the soil water content is used, along with the atmospheric evaporative demand to calculate a Water Stress Index that is used as a multiplier factor affecting biomass accumulation and expansive growth processes. This Water Stress Index can be calculated in a variety of ways, depending on the crop models, and the way this index can impact biomass depends also on the way expansive growth processes are simulated (Fig. 2). Figure 2 shows an overview of the process described above. Assuming that from the T_p to the calculation of the Water Stress Index a single modeling approach is used, it is easy to see that differences in biomass simulations can still occur due to the way in which expansive growth processes are simulated.

When researchers, crop physiologists, and modelers work together, progress can be achieved as shown by the successful temperature sub-routine workshop on wheat (Alderman et al., 2013) and the subsequent application of the findings (Asseng et al., 2015; Wang et al., 2017). The success of that work also lay in the fact that good data was available, which is not often the case as modelers deal with incomplete datasets and need to make assumptions about critical model inputs.

Porter et al. (1993) highlighted the importance of comparing models in order to identify which sections of models produce systematic errors and require improvement. Fixing these errors can be undertaken through steps that aim at improving the performance of the simulations, based on a better analysis of the processes involved. Ultimately, there is value in having multiple models as the diversity in approach for different processes is still useful.

3.2 Data

CMs require a variable number of data used as inputs to the model or to quantify the agreement and consistency of CM outputs against observed values. Over the years, concepts of data quality and the amount needed for specific CM applications have been refined as discussed extensively by Kersebaum et al. (2015) and Boote et al. (2016) and in other sections of this book.

Calibration of grain yield is performed after calibrating for phenology and growth (e.g. biomass). The most common observations used to calibrate the final yield depend on the process of yield simulation. However, the most used measurement is dry yield. For some models, grain number and grain size can also be used.

In theory, if all inputs are measured and added in a CM, the only items to calibrate are crop parameters like phenology, growth, and yield. However, in practice, datasets are often incomplete as some types of data are not available or not measured. There are cases in which to calibrate for yield a modeler will have only anthesis dates, maturity date, and final yield, with an incomplete dataset regarding initial conditions and soil information. This lack of information can impact the way a crop model simulates yield, and it is up to the modeler to assume and/or derive inputs.

Depending on the CM, the calibration of a crop yield is usually carried out under optimal or near-potential conditions. This is because some CMs simulate potential growth as non-impacted by abiotic and biotic factors and then apply reduction functions to account for water and nutrient stresses. However, this adds an additional layer of uncertainty as not all experiments are carried out under non-stressed conditions such as, for example, nitrogen agronomic trials run in rainfed environments.

In addition, the calibration and evaluation of grain yield for a new cultivar (e.g. a new maize hybrid not previously calibrated) cannot be properly performed if boundary conditions are not accurate. In fact, it is of little use to parameterize the yield of a CM if the boundary conditions are not correct because the simulated growth patterns might be impacted by incorrect boundary conditions and not by a given crop parameter. Cammarano et al. (2021) evaluated a CM in the spatial context in which all the initial conditions, soil, management, and weather information were measured on site. The cultivar was calibrated in an independent study under optimal conditions (Cammarano et al., 2019b) and those crop parameters were used in the abovementioned study.

Therefore, users' assumptions and/or calculation of data that are missing can play an important part in how the CM accurately reproduces observations. Experimental data are not free from errors and, if they have not been carefully screened with the help of researchers, the calibration of a CM can also be affected.

When modeling improvement of grain yield, data availability can be a significant limitation as it is not easy to find datasets that contain all the input and initial conditions plus the in-season detailed data that need to be used for improving the models. For example, when improving the modeling of grain protein composition, Martre et al. (2003) utilized a series of detailed experiments allowing the development of a more mechanistic yield quality simulation.

Another limitation of data driving CM concerns simulating yield for multiple cultivars/agronomic management situations in studies predicting yields across larger areas. In fact, such studies, along with maps of agricultural land, are important for ensuring policy planning regarding future land use. While sensors (e.g. satellite data) have been used to define where crops (and which crop type) actually grow (Blickensdörfer et al., 2022), the identification of different cultivars within a crop type is challenging. At the same time, it is difficult to collect information on the nitrogen and eventual irrigation supplied, which might only be available from farmers' survey data. Those surveys might also suffer from a lack of accuracy and are often unavailable for a large area. Even if CMs are able to simulate yield differences between varieties at the plot level, that information is scarce for scaling (see Chapter 2 of this book).

3.3 Soil-crop-climate interactions

The need for a given amount of detailed input data, cited as one of the biggest limitations in CM applications, is because CMs simulate the interaction of a system – the soil-plant-atmosphere management. This interconnection means that a given process is not simulated in isolation but in connection with others. Improvements to a given model sub-routine can be challenging as it is difficult to completely isolate one sub-routine from another.

Crop models have been used to evaluate the impact of projected climate and extreme events on yield, and to test adaptation and mitigation strategies (Basso et al., 2018; Nendel et al., 2019). The impact of soil fertility and water content on both yield and adaptation/mitigation strategies is not often considered, mostly because models tend to be re-initialized every year because the main focus is to quantify climate impacts on yield. Soil organic matter (SOM) dynamics, its N mineralization rates, and soil water content are all important elements that can impact crop growth processes and final yield. Rötter et al. (2012) concluded that yield variability as impacted by soil variability is significant.

Cammarano et al. (2016), as part of a larger study, compared three different potential evapotranspiration equations (Priestly-Taylor, Penman-Monteith, and the Penmann approach) across 26 wheat models and found there was a statistically significant difference among the three methods, as well as a slight divergence in results (Fig. 3). This last result was surprising because if a set of given CMs use the Penman-Monteith approach, it should produce the same results. However, the reason for such discrepancy lay in the way the equation was translated into the model's code and what assumptions were made for various parameter estimations and approximations. In that work, the comparison was easier because the sub-routine was less linked to plant processes. However, if the same method is applied to simulate actual evapotranspiration, then it is challenging because crop models have different and complex ways of linking the different components of water balance and plant growth. Improving potential evapotranspiration equations should be much easier, but the main problem is data availability. The same conclusion was reached in an evapotranspiration comparison for maize models, as illustrated in detail by Kimball et al. (2019).

In other cases, it is hard to improve a sub-routine without affecting other model processes. Figure 4 shows the relationship between simulated yield and cumulative evapotranspiration from sowing to harvest of 27 wheat models across four contrasting environments. Depending on location, there is a large noise in either simulated evapotranspiration or in simulated yield. However, the simulation of evapotranspiration requires a specific platinum dataset and an understanding of which processes are involved in the simulation of the water balance, as illustrated by Kimball et al. (2019).

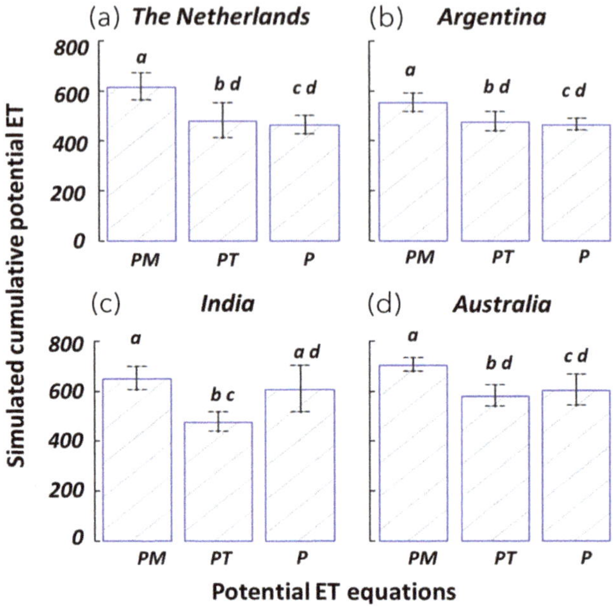

Figure 3 Cumulative potential evapotranspiration as calculated from the Penman-Monteith (PM), Priestly-Taylor (PT), and the Penmann (P) among 26 wheat models for four different agro-environments with different climatic patterns: (a) the Netherlands; (b) Argentina; (c) India; and (d) Australia. The error bars represent the standard deviations of the values and the letters the statistical significance at lsd = 0.05.

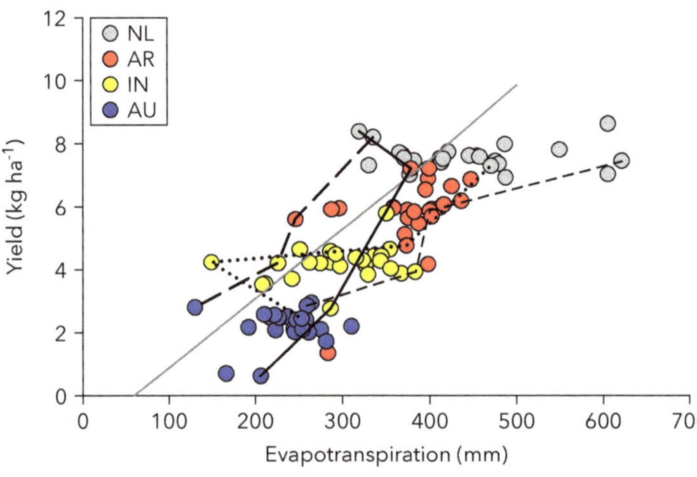

Figure 4 Simulated values of cumulative growing season evapotranspiration and grain yield from the AgMIP Wheat dataset (Asseng et al., 2013) for 27 wheat models in the Netherlands (gray dots), Argentina (red dots), India (yellow dots), and Australia (blue dots). The connecting lines indicate the same model outputs for the four different locations. The gray line is the French and Schultz (1984) fitted line for potential yield calculation.

Simulated grain yield is the result of many interactive growth and environmental factors and, to date, the scientific community has not been able to identify clearly which processes, or which data sources, are responsible for models not delivering satisfactory results.

Nendel et al. (2019) pointed out how the N coming from mineralization of SOM can be a significant source of plant nutrition but with environmental implications, if not used by crops. Currently, measuring N mineralization *in situ* is challenging and so the way in which models simulate SOM dynamics and mineralization processes has not been subjected to comparison and improvements (Nendel et al., 2019). Accurate soil process simulations are an important link for the interconnection between yield and mitigation strategies. Basso et al. (2018), using an ensemble of seven wheat and five maize models, simulated the soil organic carbon dynamics and soil N under different temperatures and atmospheric CO_2 concentrations. The increase in atmospheric CO_2 tends to compensate for the increase in air temperature in terms of impacts on yield, but higher air temperatures caused less biomass and fewer residues returned to the soil, meaning lower soil organic carbon levels and lower N availability. These findings agree with another study in which a global estimation of soil N mineralization was made and where air temperature and soil moisture impacted the SOM processes (Nendel et al., 2019).

4 Assessing climate impacts on yield with model ensembles

The impact of climate on growth, development, and yield as simulated using models has been addressed in many scientific works at different spatial scales and is not the focus of this chapter. Here, the focus is to highlight the climate impacts as reported in various intercomparison studies on the main crops in the major published studies. The aim is to understand how model ensembles respond to climate sensitivity and if studies to follow up on specific aspects of uncertainty should be made. The intention is to understand if the models of the main crops agree with one another and if there is room for improvement.

4.1 Early comparisons

The idea of linking researchers with modelers, and also modelers with other modelers, is not new and early efforts have highlighted the importance of doing so in order to improve models and their assumptions (Goudriaan, 1996; Porter et al., 1993).

In wheat, an early model intercomparison was performed by Porter et al. (1993), which showed the divergency of the models in simulating phenology, growth, and yield in potential production conditions. They pointed out that the simulation of yield, especially grain and final ear number, required particular improvement. In addition, the authors concluded that the different assumptions regarding the effects of temperature and day length on crop phenology and leaf area development were the main cause of models' variability with respect to the observations (Porter et al., 1993). Goudriaan (1996) showed how comparing ten models in different agro-environmental conditions, run with the same setup, led to a large discrepancy among the models. In fact, it was reported that aboveground biomass at harvest ranged between 6 and 20.6 t ha^{-1} and a yield between 2.5 and 8 t DM ha^{-1} for the Northern American site (Goudriaan, 1996). Such a large deviation was not significantly reduced when the simulation of LAI was the same across models. An interesting outcome was that these models were used beyond the conditions for which they were developed and were therefore unable to replicate observations. In addition, model complexity was not correlated with better results, i.e. complex models did not work better than simpler ones, an outcome found also in recent intercomparison work, as highlighted earlier. There was good agreement among models for CO_2 response in relative terms, but the degree of response to CO_2 varied greatly among models. However, the response to temperature was where models disagreed the most. Jamieson et al. (1998) used five wheat models simulating drought conditions. Comparison with observed data showed that variation among models was due to several factors, including to what degree the water stress index (as a ratio between water supply and demand) should begin to impact leaf area and photosynthesis. In addition, assumptions regarding root distribution and soil dryness on the capability of extracting water caused a divergence in simulations. The setup of the study by Palosuo et al. (2011) could be considered the precursor of the current AgMIP intercomparison studies (which are detailed below). They compared eight wheat models in different agro-ecological zones in Europe and provided modelers with limited calibration data. This meant that the simulated yield between years and sites had high degrees of uncertainty while the mean prediction from the ensemble of eight models agreed with observations.

Wolf (2002) compared two potato models across six sites in Europe, finding that changes in climate variability do not affect the irrigated and water-limited yield and that simulated yields were not affected by soil type but more by the crop model used.

In barley, Rötter et al. (2012) compared nine crop models to simulate spring-sown barley in seven northern and central Europe sites. As with the study by Palosuo et al. (2011), the models were supplied with low calibration information and the accuracy of the ensemble mean vs. individual models for simulating observed yield. While the mean of the ensemble was a better predictor of observed yield, the study also highlighted how the models that were best at simulating phenology were not necessarily those that could accurately simulate yield (Rötter et al., 2012). In a follow-up study, Tao et al.

(2020) observed how simulated yield is impacted by multiple biophysical processes and proposed a method to identify why crop models diverge. Comparing eight barley models, they found that temperature and CO_2 are the major factors behind the discrepancies among models, concluding how basic processes and the threshold by which those two factors affect leaf development, evapotranspiration, photosynthesis, and yield formation are in need of further improvement. In one of those intercomparisons on barley, it was also found that variations in soil N mineralization and LAI development, as impacted by weather variability, are not well captured by models, causing significant divergence (Salo et al., 2016a).

A European consortium - the Modeling Agriculture with Climate Change for Food Security (MACSUR) - has produced several intercomparison works, as well as some interesting research into intercropping and spatial field simulations. Some of the MACSUR intercomparison work presented below was carried out in conjunction with AgMIP activity and is presented together.

4.2 AgMIP intercomparisons

In the AgMIP crop intercomparison and improvement activities, several crops and crop models were subjected to comparison activities (and in some cases improvements). Overall, stage 1 of those activities shared a common protocol that was used for maize, potato, rice, and wheat simulations. Sensitivities to temperature and atmospheric CO_2 helped identify areas of weakness among crop models which were subject to further improvement (see Section 4.3 and references herein). Overall, across the different locations, there was a negative impact from temperature and a positive one due to increased atmospheric CO_2. The uncertainty of the single models could be quite large, as shown in Fig. 5.

4.3 AgMIP wheat

Building on these early efforts, Asseng et al. (2013) compared 26 wheat models to quantify the uncertainty of wheat yield simulations. The main findings were that response to temperature caused large disagreement with respect to CO_2, as found in Goudriaan (1996), and that most of the variability for climate change impact came from crop models rather than climate models. Another significant finding was that the model median was a better predictor of the different observations than the single models (Martre et al., 2015). The use of the model's ensemble median was then applied to quantify the global wheat yield decline to temperature (Asseng et al., 2015). Additional steps included experimental data with heat shock, high temperature, and elevated CO_2 concentration effects, which are important datasets for testing the models' ability to simulate climate change impacts. Another source of model uncertainty was adaptation

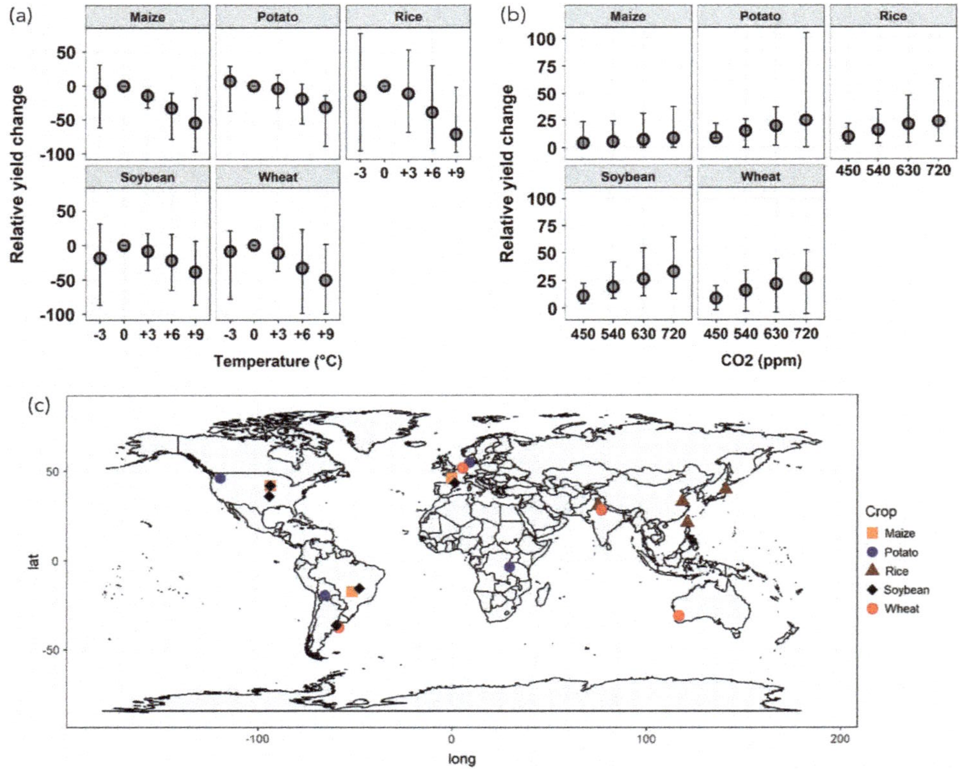

Figure 5 Mean relative yield changes for maize, potato, rice, and wheat, averaged across the four locations) for (a) sensitivity to air temperature; (b) sensitivity to atmospheric carbon dioxide concentration (CO_2); and (c) locations of the AgMIP sites. The data points are from the AgMIP crop modeling intercomparison and the error bars represent the 10th and 90th percentiles in terms of lowest and highest values among the changes. The raw data for this information is reported in the boxplots and/or the associated published raw data by Asseng et al. (2013), Bassu et al. (2014), Fleisher et al. (2017), Li et al. (2015), and Kothari et al. (2022).

to climate change for each model, especially in simulating grain quality at low- to mid-latitudes (Asseng et al., 2015). This approach, while useful for climate change impacts, gives little indication as to what specific process should be subject to improvement.

O'Leary et al. (2015) carried out a comparison of the wheat response to CO_2 concentration, temperature, water, and nitrogen using a subset of the 26 wheat models. Results showed that the relative biomass, yield, and water use response between the simulation and observation was close and that models that simulated biomass via RUE or transpiration efficiency produced similar results (O'Leary et al., 2015). However, late planting treatments (which expose the wheat to higher air temperatures) caused the models to diverge from these observations.

Alderman et al. (2013) hosted a workshop at CIMMYT in which crop modelers, researchers, breeders, and crop physiologists worked together to understand the physiological mechanisms of temperature response in crops and how each crop model considers it. This was then followed up by a study by Wang et al. (2017), who showed that the temperature response simulation in crop models accounted for more than 50% of the uncertainties in simulated yield for a mean growing season temperature of 14–33°C. A fix to the temperature function was able to reduce such error by 19–50%, depending on the location considered, which shows how two efforts can lead to a better understanding of the underlying processes and the degree of variability in simulating growth and development processes. Webber et al. (2017) compared heat stress in irrigated wheat in a semi-arid environment. The main aim of the study was to understand if the inclusion of canopy temperature (rather than the air temperature, which is the input used by models) improves the prediction of wheat yield in their experimental settings. In their findings, the use of canopy rather than air temperature did not improve simulated yield and there was no clear indication that more complex energy balance-based models performed better.

4.4 AgMIP maize

In maize, Bassu et al. (2014) compared 23 maize models in four different locations and found that the models diverged at increasing air temperature, as with the wheat intercomparison. The authors stated that the maize models simulated LAI using biomass and specific leaf area or through a morphological leaf-number appearance module, which is a function of temperature and stress. Despite those two approaches, there was no clear better method when it came to variability at higher air temperatures. With respect to the wheat study, the maize models also diverged for the CO_2 response simulation, which could be due to the limited amount of experiments and evidence on the CO_2 effects on C4 photosynthesis or radiation use (Durand et al., 2018; Long et al., 2005).

Based on the sensitivity study by Bassu et al. (2014), a study using FACE data on maize in which models were compared against the response to water stress and CO_2 concentration was carried out (Durand et al., 2018). In that study, crop models that had an explicit mechanism for considering stomatal control on transpiration performed better than the others. However, while the overall response to CO_2 concentration had low uncertainty under well-irrigated conditions, with a water deficit, few models could satisfactorily reproduce growth and yield. The authors concluded that simulations of crop transpiration, water balance, and kernel number set required improvement. This is also limited by the quality of the dataset available for such purposes.

Kimball et al. (2019) have taken maize intercomparison a step further by using a dataset that allows comparison of maize-modeled water use. This study highlights some interesting issues with multi-model intercomparisons that could be useful for future research when it comes to improvement. The 29 maize models had a wide variability in simulating daily evapotranspiration, but among the best models there was a variety of modeling approaches (simpler to more mechanistic). A key message was that models such as the DSSAT and APSIM, having been widely used worldwide, have indirectly undergone testing in multiple environments and conditions over many years, which has helped improve them. Within those two models, a number of users ran the same version and found that choices made by users also impacted simulation outputs (Kimball et al., 2019). However, in this study, it was also concluded that most of the variability among models in predicting evapotranspiration came from the ETp, which is in contrast to what was found in wheat (Cammarano et al., 2016; Kimball et al., 2019).

4.5 AgMIP barley

While many studies on barley intercomparisons have been made, as discussed above (Rötter et al., 2012; Salo et al., 2016b; Tao et al., 2020; Tao et al., 2017), there is no formal AgMIP for barley to date. This is set to be launched relatively soon as data is compiled. An EU-based project provided a good base for barley modeling improvement (https://www.suscrop.eu/projects-first-call/barista). Depending on its use, quality is an important trait to simulate in barley. If used for malting, the quality parameter is very important. Currently, few models can simulate grain protein composition, but they are based on wheat models and so might be limited in simulating barley quality under climate change. The main challenge here is to find the right type of data for such specific improvement.

4.6 AgMIP rice

An intercomparison of 13 rice models showed that most variability to CO_2 temperature simulations come from the simulation of biomass and

HI (Li et al., 2015). They found that models that include the Farquhar-von Caemmerer-Berry biochemical model of leaf photosynthesis performed more consistently than the RUE models, which was not clearly evident in other intercomparisons. They also found that crop models with optimum response of developmental rates to temperature had a longer growth duration for high temperatures with respect to models with asymptotic response, which showed shortened durations (Li et al., 2015). While simulated changes in biomass to increased temperature were fairly consistent among models, they differ from observations, which is an area of agreement with respect to the other two intercomparisons. In a subsequent study in which simulations were compared against observations of CO_2 response, it was found that, unlike the other crop model intercomparisons, rice models had consistency in over- or under-simulating across environments (Hasegawa et al., 2017; O'Leary et al., 2015). With respect to wheat intercomparisons, the rice models tested by the authors greatly differed in LAI response to CO_2, while in wheat there was a similar LAI increase to higher CO_2 (Hasegawa et al., 2017). This means that, for rice models, identifying uncertainties and improving the models might be an easier task with respect to wheat models. As with the previous study on rice, it was found that most of the variability in simulated yield came from the biomass simulation, but the authors found that variability in simulated leaf area was the main cause for such divergence of results (Hasegawa et al., 2017; Li et al., 2015). An important conclusion of this study was that nitrogen levels make such simulation more uncertain, suggesting that another area of improvement should be CO_2 x nitrogen interactions (Hasegawa et al., 2017).

4.7 AgMIP potato

In potato, Fleisher et al. (2017) compared nine crop models with the same protocols and sensitivities as carried out by other crops described above and found that in four different agro-environmental conditions, CO_2 concentration did not impact divergence among models as much as air temperature, in which an increase of 2°C causes about 35% variability among models. The study also found that there was a divergence in simulation, for example of evapotranspiration, due to a statistically significant difference among models in the way light utilization was simulated. The authors concluded that further potato improvements should address a better sub-routine for simulating water and nutrient stresses and temperature routines. However, the need for a good dataset to address modeling shortcomings was also highlighted as one of the limiting factors for modeling improvement (Fleisher et al., 2017). Ten potato models were compared to predict potato yield from an experimental dataset with elevated CO_2 while calibrated under ambient CO_2 as a way to evaluate the modeled response to CO_2 (Fleisher et al., 2021). An interesting finding from the best-performing models in simulating tuber yield under elevated CO_2 was that it was independent of the structure (e.g. either radiation use efficiency, transpiration efficiency, or leaf-level photosynthesis).

4.8 AgMIP soybean

In soybean, a recent article by Kothari et al. (2022) compared ten soybean crop models using the same protocols shared by other crop comparisons. The uncertainty of simulated yield was high among models, and this was true for the temperature and CO_2 sensitivities. This might indicate that soybean models require, with respect to other crop models (e.g. wheat), a systematic improvement in temperature and CO_2 response functions. In their study, the simulated yields under +3°C in the coolest location showed a significant disagreement with some models simulating −24% and others +29%. However, unlike other crop comparisons, the variability due to increased atmospheric CO_2 was high - around 6–31%, as shown in Fig. 5b and reported in Kothari et al. (2022). This indicates that, for soybean modeling, additional issues due to incorrect sub-routines or their parameterization have to be addressed. The problem, as with other crops, concerns the availability of good-quality data that can be used to improve simulated processes.

4.9 Other AgMIP comparisons

Other intercomparison activities included sugarcane, livestock, and grassland. In sugarcane, Marin et al. (2015) compared two models and found that they could not accurately predict stalk development over time. One model, having only a simpler temperature response function, was not able to simulate yield in the hottest sites. The experiments were not designed to deepen the analysis of the cause of models' divergence, which again highlights the need for better experimental data.

In a simulation experiment, Ehrhardt et al. (2018) compared 24 process-based biogeochemical models on five arable crops and on grassland to evaluate their performance and uncertainty to simulated yield and N2O emissions. The different ways in which the multi-model ensemble was chosen highlighted that agricultural productivity and greenhouse gases can be estimated with a series of different models. Across models, most were within one to three standard deviations of the observations, but even in this case, the authors highlighted the need for establishing a global network of long-term experiments for measuring greenhouse gas emissions, soil organic carbon stocks, and crop/grassland productivity in order to improve model simulations.

4.10 Pest and disease model

Climate impact has mostly focused on crop growth, development, and yield, but additional biological stressors, pests, and diseases can also cause severe yield losses. Current and projected impacts can be significant and crop simulation models include such sub-routines, given their modular structure. Donatelli et al. (2017) proposed a roadmap to improve the simulations of pest diseases and optimize coupling with crop growth models. This roadmap theorizes five areas of actions, relating to improved quality and availability of data for input into models, for evaluation of the models, for integration with crop growth models, and for the subsequent evaluation. Such work also suggests the creation of a community of plant pests and disease modelers.

A study by Bregaglio et al. (2021) used a disease model (WHEATPEST) that simulated damage on wheat in terms of *light stealer* (e.g. reduction of photosynthetically active leaf area) and *assimilate sapper* (dynamic uptake of carbohydrates from daily photosynthesis prior to any partitioning) and incorporated it into four wheat growth models. The variability of simulated yield to different damages was high among the models, ranging from 1 to 3.5 t ha^{-1} across the five different diseases. This clearly highlights that better data and purposefully designed experiments are necessary. However, given the previous wheat intercomparison studies, it is likely that before integrating pest and disease models into crop growth models, additional improvement of crop growth models is necessary.

5 Conclusion and future trends in research

In summary, comparison and improvement activities can greatly contribute to improving models' responses to projected climate change, with the clear benefit of reducing uncertainty for future climate change impact studies. This will help stakeholders make an informed decision about future activities that can promote climate-smart agriculture practices and improve food security. Crop model variability in process simulations and description and response to temperature, stresses (heat, nutritional, thermal), and soil water content have all been observed in past intercomparisons (Asseng et al., 2013; Eitzinger et al., 2013; O'Leary et al., 2015; Rötter et al., 2012).

As discussed in a previous work (Palosuo et al., 2011), one of the main challenges limiting model improvement is the right amount of experimental data of a high enough quality for model development. Even when data are available to improve specific processes, some sub-routines barely improve. For example, when it comes to modeling grain yield dynamics, one element that is often overlooked is grain quality (i.e. the dynamics and mechanisms of carbohydrates and nitrogen distribution). Most crop models simulate quality through grain number, size, and N content, while other crop models only simulate grain protein composition.

Along with modeling intercomparison efforts, it is essential to develop efforts to work closely with researchers, crop physiologists, and breeders to design experiments aimed at increasing our knowledge of a given process, to provide a better understanding of how a given variable affects a process not currently simulated, and how to represent it in crop models.

Despite the models' limitations, the work highlights and focuses on areas that need improvement. CMs can still play a vital role in quantifying future climate impacts on crop productivity, whether they are used alone or as an ensemble.

6 Where to look for further information

- Agricultural Modeling Intercomparison and Improvement Project (AgMIP): www.agmip.org (verified on 10 May 2022).
- The Inter-Sectoral Impact Model Intercomparison Project (ISIMIP): www.isimip.org.
- Modelling Agriculture with Climate Change for Food Security (MACSUR): www.macsur.eu.

7 References

Ainsworth, E. A. and Long, S. P. (2020). 30 Years of free-air carbon dioxide enrichment (FACE): What have we learned about future crop productivity and its potential for adaptation? *Global Change Biology*.

Alderman, P. D., Qulligan, E., Asseng, S., Ewert, F. and Reynolds, M. P. (2013). CIMMYT, el Batan, Mexico. In: *Proceedings of the Workshop Modelling Wheat Response to High Temperature*.

Asseng, S., Ewert, F., Martre, P., Rötter, R. P., Lobell, D. B., Cammarano, D., Kimball, B. A., Ottman, M. J., Wall, G. W., White, J. W., Reynolds, M. P., Alderman, P. D., Prasad, P. V. V., Aggarwal, P. K., Anothai, J., Basso, B., Biernath, C., Challinor, A. J., De Sanctis, G., Doltra, J., Fereres, E., Garcia-Vila, M., Gayler, S., Hoogenboom, G., Hunt, L. A., Izaurralde, R. C., Jabloun, M., Jones, C. D., Kersebaum, K. C., Koehler, A.-K., Müller, C., Naresh Kumar, S., Nendel, C., O'Leary, G., Olesen, J. E., Palosuo, T., Priesack, E., Eyshi Rezaei, E., Ruane, A. C., Semenov, M. A., Shcherbak, I., Stöckle, C., Stratonovitch, P., Streck, T., Supit, I., Tao, F., Thorburn, P. J., Waha, K., Wang, E., Wallach, D., Wolf, J., Zhao, Z. and Zhu, Y. (2015). Rising temperatures reduce global wheat production. *Nature Climate Change* 5(2):143–147.

Asseng, S., Ewert, F., Rosenzweig, C., Jones, J. W., Hatfield, J. L., Ruane, A. C., Boote, K. J., Thorburn, P. J., Rötter, R. P., Cammarano, D., Brisson, N., Basso, B., Martre, P., Aggarwal, P. K., Angulo, C., Bertuzzi, P., Biernath, C., Challinor, A. J., Doltra, J., Gayler, S., Goldberg, R., Grant, R., Heng, L., Hooker, J., Hunt, L. A., Ingwersen, J., Izaurralde, R. C., Kersebaum, K. C., Müller, C., Naresh Kumar, S., Nendel, C., O'Leary, G., Olesen, J. E., Osborne, T. M., Palosuo, T., Priesack, E., Ripoche, D., Semenov, M.

A., Shcherbak, I., Steduto, P., Stöckle, C., Stratonovitch, P., Streck, T., Supit, I., Tao, F., Travasso, M., Waha, K., Wallach, D., White, J. W., Williams, J. R. and Wolf, J. (2013). Uncertainty in simulating wheat yields under climate change. *Nature Climate Change* 3(9):827–832.

Asseng, S., Martre, P., Maiorano, A., Rötter, R. P., O'Leary, G. J., Fitzgerald, G. J., Girousse, C., Motzo, R., Giunta, F., Babar, M. A., Reynolds, M. P., Kheir, A. M. S., Thorburn, P. J., Waha, K., Ruane, A. C., Aggarwal, P. K., Ahmed, M., Balkovič, J., Basso, B., Biernath, C., Bindi, M., Cammarano, D., Challinor, A. J., De Sanctis, G., Dumont, B., Eyshi Rezaei, E., Fereres, E., Ferrise, R., Garcia-Vila, M., Gayler, S., Gao, Y., Horan, H., Hoogenboom, G., Izaurralde, R. C., Jabloun, M., Jones, C. D., Kassie, B. T., Kersebaum, K.-C., Klein, C., Koehler, A.-K., Liu, B., Minoli, S., Montesino San Martin, M., Müller, C., Naresh Kumar, S., Nendel, C., Olesen, J. E., Palosuo, T., Porter, J. R., Priesack, E., Ripoche, D., Semenov, M. A., Stöckle, C., Stratonovitch, P., Streck, T., Supit, I., Tao, F., Van der Velde, M., Wallach, D., Wang, E., Webber, H., Wolf, J., Xiao, L., Zhang, Z., Zhao, Z., Zhu, Y. and Ewert, F. (2019). Climate change impact and adaptation for wheat protein. *Global Change Biology* 25:155–173.

Basso, B., Dumont, B., Maestrini, B., Shcherbak, I., Robertson, G. P., Porter, J. R., Smith, P., Paustian, K., Grace, P. R., Asseng, S., Bassu, S., Biernath, C., Boote, K. J., Cammarano, D., De Sanctis, G., Durand, J.-L., Ewert, F., Gayler, S., Hyndman, D. W., Kent, J., Martre, P., Nendel, C., Priesack, E., Ripoche, D., Ruane, A. C., Sharp, J., Thorburn, P. J., Hatfield, J. L., Jones, J. W. and Rosenzweig, C. (2018). Soil organic carbon and nitrogen feedbacks on crop yields under climate change. *Agricultural and Environmental Letters* 3(1):180026.

Bassu, S., Brisson, N., Durand, J. L., Boote, K., Lizaso, J., Jones, J. W., Rosenzweig, C., Ruane, A. C., Adam, M., Baron, C., Basso, B., Biernath, C., Boogaard, H., Conijn, S., Corbeels, M., Deryng, D., De Sanctis, G., Gayler, S., Grassini, P., Hatfield, J., Hoek, S., Izaurralde, C., Jongschaap, R., Kemanian, A. R., Kersebaum, K. C., Kim, S. H., Kumar, N. S., Makowski, D., Müller, C., Nendel, C., Priesack, E., Pravia, M. V., Sau, F., Shcherbak, I., Tao, F., Teixeira, E., Timlin, D. and Waha, K. (2014). How do various maize crop models vary in their responses to climate change factors? *Global Change Biology* 20(7):2301–2320.

Blickensdörfer, L., Schwieder, M., Pflugmacher, D., Nendel, C., Erasmi, S. and Hostert, P. (2022). Mapping of crop types and crop sequences with combined time series of Sentinel-1, Sentinel-2 and Landsat 8 data for Germany. *Remote Sensing of Environment* 269:112831.

Boote, K. J., Jones, J. W., White, J. W., Asseng, S. and Lizaso, J. I. (2013). Putting mechanisms into crop production models. *Plant, Cell and Environment* 36(9):1658–1672.

Boote, K. J., Porter, C., Jones, J. W., Thorburn, P. J., Kersebaum, K. C., Hoogenboom, G., White, J. W. and Hatfield, J. L. (2016). Sentinel site data for crop model improvement-definition and characterization. In: *Improving Modeling Tools to Assess Climate Change Effects on Crop Response*, pp. 125–158.

Bregaglio, S., Willocquet, L., Kersebaum, K. C., Ferrise, R., Stella, T., Ferreira, T. B., Pavan, W., Asseng, S. and Savary, S. (2021). Comparing process-based wheat growth models in their simulation of yield losses caused by plant diseases. *CropM, ft_macsur* 265:108108.

Bustos-Korts, D., Boer, M. P., Malosetti, M., Chapman, S., Chenu, K., Zheng, B. and van Eeuwijk, F. A. (2019). Combining crop growth modeling and statistical genetic modeling to evaluate phenotyping strategies. *Frontiers in Plant Science* 10:1491.

Cammarano, D., Basso, B., Holland, J., Gianinetti, A., Baronchelli, M. and Ronga, D. (2021). Modeling spatial and temporal optimal N fertilizer rates to reduce nitrate leaching while improving grain yield and quality in malting barley. *Computers and Electronics in Agriculture* 182:105997.

Cammarano, D., Ceccarelli, S., Grando, S., Romagosa, I., Benbelkacem, A., Akar, T., Al-Yassin, A., Pecchioni, N., Francia, E. and Ronga, D. (2019a). The impact of climate change on barley yield in the Mediterranean basin. *European Journal of Agronomy* 106:1–11.

Cammarano, D., Hawes, C., Squire, G., Holland, J., Rivington, M., Murgia, T., Roggero, P. P., Fontana, F., Casa, R. and Ronga, D. (2019b). Rainfall and temperature impacts on barley (Hordeum vulgare L.) yield and malting quality in Scotland. *CropM, ft_macsur* 241:107559.

Cammarano, D., Rötter, R. P., Asseng, S., Ewert, F., Wallach, D., Martre, P., Hatfield, J. L., Jones, J. W., Rosenzweig, C., Ruane, A. C., Boote, K. J., Thorburn, P. J., Kersebaum, K. C., Aggarwal, P. K., Angulo, C., Basso, B., Bertuzzi, P., Biernath, C., Brisson, N., Challinor, A. J., Doltra, J., Gayler, S., Goldberg, R., Heng, L., Hooker, J., Hunt, L. A., Ingwersen, J., Izaurralde, R. C., Müller, C., Kumar, S. N., Nendel, C., O'Leary, G. J., Olesen, J. E., Osborne, T. M., Palosuo, T., Priesack, E., Ripoche, D., Semenov, M. A., Shcherbak, I., Steduto, P., Stöckle, C. O., Stratonovitch, P., Streck, T., Supit, I., Tao, F., Travasso, M., Waha, K., White, J. W. and Wolf, J. (2016). Uncertainty of wheat water use: simulated patterns and sensitivity to temperature and CO_2. *CropM, ft_macsur* 198:80–92.

Cammarano, D., Valdivia, R. O., Beletse, Y. G., Durand, W., Crespo, O., Tesfuhuney, W. A., Jones, M. R., Walker, S., Mpuisang, T. N., Nhemachena, C., Ruane, A. C., Mutter, C., Rosenzweig, C. and Antle, J. (2020). Integrated assessment of climate change impacts on crop productivity and income of commercial maize farms in northeast South Africa. *Food Security* 12(3):659–678.

Connor, D. J., Loomis, R. S. and Cassman, K. G. (2011). *Crop Ecology Productivity and Management in Agricultural Systems* (2nd edn.). Cambridge: Cambridge University Press.

Davis, K. F., Downs, S. and Gephart, J. A. (2021). Towards food supply chain resilience to environmental shocks. *Nature Food* 2(1):54–65.

Doblas-Reyes, F. J., Hagedorn, R. and Palmer, T. N. (2005). The rationale behind the success of multi-model ensembles in seasonal forecasting–II. Calibration and combination. *Tellus, Series A: Dynamic Meteorology and Oceanography* 57:234–252.

Donatelli, M., Magarey, R. D., Bregaglio, S., Willocquet, L., Whish, J. P. M. and Savary, S. (2017). Modelling the impacts of pests and diseases on agricultural systems. *CropM, LiveM, TradeM, ft_macsur* 155:213–224.

Durand, J.-L., Delusca, K., Boote, K., Lizaso, J., Manderscheid, R., Weigel, H. J., Ruane, A. C., Rosenzweig, C., Jones, J., Ahuja, L., Anapalli, S., Basso, B., Baron, C., Bertuzzi, P., Biernath, C., Deryng, D., Ewert, F., Gaiser, T., Gayler, S., Heinlein, F., Kersebaum, K. C., Kim, S.-H., Müller, C., Nendel, C., Olioso, A., Priesack, E., Villegas, J. R., Ripoche, D., Rötter, R. P., Seidel, S. I., Srivastava, A., Tao, F., Timlin, D., Twine, T., Wang, E., Webber, H. and Zhao, Z. (2018). How accurately do maize crop models simulate the interactions of atmospheric CO_2 concentration levels with limited water supply on water use and yield? *European Journal of Agronomy* 100:67–75.

Ehrhardt, F., Soussana, J. F., Bellocchi, G., Grace, P., McAuliffe, R., Recous, S., Sándor, R., Smith, P., Snow, V., de Antoni Migliorati, M., Basso, B., Bhatia, A., Brilli, L., Doltra, J.,

Dorich, C. D., Doro, L., Fitton, N., Giacomini, S. J., Grant, B., Harrison, M. T., Jones, S. K., Kirschbaum, M. U. F., Klumpp, K., Laville, P., Léonard, J., Liebig, M., Lieffering, M., Martin, R., Massad, R. S., Meier, E., Merbold, L., Moore, A. D., Myrgiotis, V., Newton, P., Pattey, E., Rolinski, S., Sharp, J., Smith, W. N., Wu, L. and Zhang, Q. (2018). Assessing uncertainties in crop and pasture ensemble model simulations of productivity and N_2O emissions. *Global Change Biology* 24(2):e603–e616.

Eitzinger, J., Thaler, S., Schmid, E., Strauss, F., Ferrise, R., Moriondo, M., Bindi, M., Palosuo, T., Rötter, R., Kersebaum, K., Olesen, J., Patil, R., Şaylan, L., Çaldağ, B. and Çaylak, O. (2013). Sensitivities of crop models to extreme weather conditions during flowering period demonstrated for maize and winter wheat in Austria. *The Journal of Agricultural Science* 151(6):813–835.

Emberson, L. D., Pleijel, H., Ainsworth, E. A., van den Berg, M., Ren, W., Osborne, S., Mills, G., Pandey, D., Dentener, F., Büker, P., Ewert, F., Koeble, R. and Van Dingenen, R. (2018). Ozone effects on crops and consideration in crop models. *European Journal of Agronomy* 100:19–34.

Farquhar, G. D., von Caemmerer, S. and Berry, J. A. (1980). A biochemical model of photosynthetic CO_2 assimilation in leaves of C3 species. *Planta* 149(1):78–90.

Fleisher, D. H., Condori, B., Barreda, C., Berguijs, H., Bindi, M., Boote, K., Craigon, J., Evert, Fv., Fangmeier, A., Ferrise, R., Gayler, S., Hoogenboom, G., Merante, P., Nendel, C., Ninanya, J., Pleijel, H., Raes, D., Ramírez, D. A., Raymundo, R., Reidsma, P., Silva, J. V., Stöckle, C. O., Supit, I., Stella, T., Vandermeiren, K., van Oort, P., Vanuytrecht, E., Vorne, V. and Wolf, J. (2021). Yield response of an ensemble of potato crop models to elevated CO_2 in continental Europe. *European Journal of Agronomy* 126:126265.

Fleisher, D. H., Condori, B., Quiroz, R., Alva, A., Asseng, S., Barreda, C., Bindi, M., Boote, K. J., Ferrise, R., Franke, A. C., Govindakrishnan, P. M., Harahagazwe, D., Hoogenboom, G., Naresh Kumar, S., Merante, P., Nendel, C., Olesen, J. E., Parker, P. S., Raes, D., Raymundo, R., Ruane, A. C., Stockle, C., Supit, I., Vanuytrecht, E., Wolf, J. and Woli, P. (2017). A potato model intercomparison across varying climates and productivity levels. *Global Change Biology* 23(3):1258–1281.

French, R. and Schultz, J. (1984). Water use efficiency of wheat in a Mediterranean-type environment. I. The relation between yield, water use and climate. *Australian Journal of Agricultural Research* 35(6):743–764.

Giuliani, M. M., Gatta, G., Cappelli, G., Gagliardi, A., Donatelli, M., Fanchini, D., De Nart, D., Mongiano, G. and Bregaglio, S. (2019). Identifying the most promising agronomic adaptation strategies for the tomato growing systems in Southern Italy via simulation modeling. *European Journal of Agronomy* 111:125937.

Godfray, H. C. J., Beddington, J. R., Crute, I. R., Haddad, L., Lawrence, D., Muir, J. F., Pretty, J., Robinson, S., Thomas, S. M. and Toulmin, C. (2010). Food security: the challenge of Feeding 9 billion people. *Science* 327(5967):812–818.

Goudriaan, J. (1996). Predicting crop yields under global change. In: Walker, B. and Steffen, W. (eds.), *International Geosphere-Biosphere Programme Book Series*:260–274.

Hasegawa, T., Li, T., Yin, X., Zhu, Y., Boote, K., Baker, J., Bregaglio, S., Buis, S., Confalonieri, R., Fugice, J., Fumoto, T., Gaydon, D., Kumar, S. N., Lafarge, T., Marcaida, M., III, Masutomi, Y., Nakagawa, H., Oriol, P., Ruget, F., Singh, U., Tang, L., Tao, F., Wakatsuki, H., Wallach, D., Wang, Y., Wilson, L. T., Yang, L., Yang, Y., Yoshida, H., Zhang, Z. and

Zhu, J. (2017). Causes of variation among rice models in yield response to CO_2 examined with free-air CO_2 enrichment and growth chamber experiments. *Scientific Reports* 7(1):14858.

Hatfield, J. L., Boote, K. J., Kimball, B. A., Ziska, L. H., Izaurralde, R. C., Ort, D., Thomson, A. M. and Wolfe, D. (2011). Climate impacts on agriculture: implications for crop production. *Agronomy Journal* 103(2):351–370.

Hatfield, J. L. and Prueger, J. H. (2015). Temperature extremes: Effect on plant growth and development. *Weather and Climate Extremes* 10:4–10.

IPCC. (2021). *Climate Change: The Physical Science Basis. Contribution of Working Group I to the Sixth Assessment Report of the Intergovernmental Panel on Climate Change*. Edited by Masson-Delmotte, V., Zhai, P., Pirani, A., Connors, S. L., Péan, C., Berger, S., Caud, N., Chen, Y., Goldfarb, L., Gomis, M. I.. Cambridge: Cambridge University Press.

Jamieson, P. D., Porter, J. R., Goudriaan, J., Ritchie, J. T., van Keulen, H. and Stol, W. (1998). A comparison of the models AFRCWHEAT2, CERES-Wheat, Sirius, SUCROS2 and SWHEAT with measurements from wheat grown under drought. *CropM, ft_macsur* 55:23–44.

Kersebaum, C., Kollas, C., Bindi, M., Nendel, C., Ferrise, R., Moriondo, M., Olesen, J. E., Sharif, B., Öztürk, I., Hoffmann, H., Launay, M., Ripoche, D., Ruget, F., Bertuzzi, P., Cortazar, I. G. D., Beaudoin, N., Armas-Herrera, C., Mary, B., Müller, C., Waha, K., Ventrella, D., Palosuo, T., Rötter, R., Trnka, M., Hlavinka, P., Wu, L., Wegehenkel, M., Mirschel, W., Conradt, T., Wechsung, F., Weigel, H.-J., Manderscheid, R. and Eitzinger, J. *Modelling Complex Crop Rotations and Management Across Sites in Europe With an Ensemble of Models.*

Kersebaum, K. C., Boote, K. J., Jorgenson, J. S., Nendel, C., Bindi, M., Frühauf, C., Gaiser, T., Hoogenboom, G., Kollas, C., Olesen, J. E., Rötter, R. P., Ruget, F., Thorburn, P. J., Trnka, M. and Wegehenkel, M. (2015). Analysis and classification of data sets for calibration and validation of agro-ecosystem models. *Environmental Modelling and Software* 72:402–417.

Kim, Y., Berger, S., Kettering, J., Tenhunen, J., Haas, E. and Kiese, R. (2014). Simulation of N2O emissions and nitrate leaching from plastic mulch radish cultivation with LandscapeDNDC. *CropM, ft_macsur* 29:441–454.

Kimball, B. A., Boote, K. J., Hatfield, J. L., Ahuja, L. R., Stockle, C., Archontoulis, S., Baron, C., Basso, B., Bertuzzi, P., Constantin, J., Deryng, D., Dumont, B., Durand, J.-L., Ewert, F., Gaiser, T., Gayler, S., Hoffmann, M. P., Jiang, Q., Kim, S.-H., Lizaso, J., Moulin, S., Nendel, C., Parker, P., Palosuo, T., Priesack, E., Qi, Z., Srivastava, A., Stella, T., Tao, F., Thorp, K. R., Timlin, D., Twine, T. E., Webber, H., Willaume, M. and Williams, K. (2019). Simulation of maize evapotranspiration: an inter-comparison among 29 maize models. *Agricultural and Forest Meteorology* 271:264–284.

Kothari, K., Battisti, R., Boote, K. J., Archontoulis, S. V., Confalone, A., Constantin, J., Cuadra, S. V., Debaeke, P., Faye, B., Grant, B., Hoogenboom, G., Jing, Q., van der Laan, M., Macena da Silva, F. A., Marin, F. R., Nehbandani, A., Nendel, C., Purcell, L. C., Qian, B., Ruane, A. C., Schoving, C., Silva, E. H. F. M., Smith, W., Soltani, A., Srivastava, A., Vieira, N. A., Slone, S. and Salmerón, M. (2022). Are soybean models ready for climate change food impact assessments? *European Journal of Agronomy* 135:126482.

Li, T., Hasegawa, T., Yin, X., Zhu, Y., Boote, K., Adam, M., Bregaglio, S., Buis, S., Confalonieri, R., Fumoto, T., Gaydon, D., Marcaida III, M., Nakagawa, H., Oriol, P., Ruane, A. C.,

Ruget, F., Singh, B., Singh, U., Tang,, L., Tao, F., Wilkens, P., Yoshida, H., Zhang, Z. and Bouman, B. (2015). Uncertainties in predicting rice yield by current crop models under a wide range of climatic conditions. *Global Change Biology* 21:1328–1341.

Liu, B., Asseng, S., Müller, C., Ewert, F., Elliott, J., Lobell, D. B., Martre, P., Ruane, A. C., Wallach, D., Jones, J. W., Rosenzweig, C., Aggarwal, P. K., Alderman, P. D., Anothai, J., Basso, B., Biernath, C., Cammarano, D., Challinor, A., Deryng, D., Sanctis, G. D., Doltra, J., Fereres, E., Folberth, C., Garcia-Vila, M., Gayler, S., Hoogenboom, G., Hunt, L. A., Izaurralde, R. C., Jabloun, M., Jones, C. D., Kersebaum, K. C., Kimball, B. A., Koehler, A.-K., Kumar, S. N., Nendel, C., O'Leary, G. J., Olesen, J. E., Ottman, M. J., Palosuo, T., Prasad, P. V. V., Priesack, E., Pugh, T. A. M., Reynolds, M., Rezaei, E. E., Rötter, R. P., Schmid, E., Semenov, M. A., Shcherbak, I., Stehfest, E., Stöckle, C. O., Stratonovitch, P., Streck, T., Supit, I., Tao, F., Thorburn, P., Waha, K., Wall, G. W., Wang, E., White, J. W., Wolf, J., Zhao, Z. and Zhu, Y. (2016). Similar estimates of temperature impacts on global wheat yield by three independent methods. *Nature Climate Change* 6(12):1130–1136.

Lobell, D. B. and Asseng, S. (2017). Comparing estimates of climate change impacts from process-based and statistical crop models. *Environmental Research Letters* 12(1):13.

Lobell, D. B. and Burke, M. B. (2010). On the use of statistical models to predict crop yield responses to climate change. *Agricultural and Forest Meteorology* 150(11):1443–1452.

Long, S. P., Ainsworth, E. A., Leakey, A. D. and Morgan, P. B. (2005) Global food insecurity. treatment of major food crops with elevated carbon dioxide or ozone under large-scale fully open-air conditions suggests recent models may have overestimated future yields. *Philosophical Transactions of the Royal Society of London. Series B, Biological Sciences* 360(1463):2011–2020.

Marin, F. R., Thorburn, P. J., Nassif, D. S. P. and Costa, L. G. (2015). Sugarcane model intercomparison: structural differences and uncertainties under current and potential future climates. *Environmental Modelling and Software* 72:372–386.

Martre, P., Jamieson, P. D., Semenov, M. A., Zyskowski, R. F., Porter, J. R. and Triboi, E. (2006). Modelling protein content and composition in relation to crop nitrogen dynamics for wheat. *European Journal of Agronomy* 25(2):138–154.

Martre, P., Porter, J. R., Jamieson, P. D. and Triboï, E. (2003). Modeling grain nitrogen accumulation and protein composition to understand the sink/source regulations of nitrogen remobilization for wheat. *Plant Physiology* 133(4):1959–1967.

Martre, P., Wallach, D., Asseng, S., Ewert, F., Jones, J. W., Rötter, R. P., Boote, K. J., Ruane, A. C., Thorburn, P. J., Cammarano, D., Hatfield, J. L., Rosenzweig, C., Aggarwal, P. K., Angulo, C., Basso, B., Bertuzzi, P., Biernath, C., Brisson, N., Challinor, A. J., Doltra, J., Gayler, S., Goldberg, R., Grant, R. F., Heng, L., Hooker, J., Hunt, L. A., Ingwersen, J., Izaurralde, R. C., Kersebaum, K. C., Müller, C., Kumar, S. N., Nendel, C., O'Leary, G., Olesen, J. E., Osborne, T. M., Palosuo, T., Priesack, E., Ripoche, D., Semenov, M. A., Shcherbak, I., Steduto, P., Stöckle, C. O., Stratonovitch, P., Streck, T., Supit, I., Tao, F., Travasso, M., Waha, K., White, J. W. and Wolf, J. (2015). Multimodel ensembles of wheat growth: Many models are better than one. *Global Change Biology* 21(2):911–925.

Nendel, C., Kersebaum, K. C., Mirschel, W. and Wenkel, K. O. (2014). Testing farm management options as climate change adaptation strategies using the Monica model. *European Journal of Agronomy* 52:47–56.

Nendel, C., Melzer, D. and Thorburn, P. J. (2019). The nitrogen nutrition potential of arable soils. *Scientific Reports* 9(1):5851.
O'Leary, G. J., Christy, B., Nuttall, J., Huth, N., Cammarano, D., Stöckle, C., Basso, B., Shcherbak, I., Fitzgerald, G., Luo, Q., Farre-Codina, I., Palta, J. and Asseng, S. (2015). Response of wheat growth, grain yield and water use to elevated CO_2 under a free-air CO_2 enrichment (FACE) experiment and modelling in a semi-arid environment. *Global Change Biology* 21(7):2670–2686.
Palosuo, T., Kersebaum, K. C., Angulo, C., Hlavinka, P., Moriondo, M., Olesen, J. E., Patil, R. H., Ruget, F., Rumbaur, C., Takáč, J., Trnka, M., Bindi, M., Çaldağ, B., Ewert, F., Ferrise, R., Mirschel, W., Şaylan, L., Šiška, B. and Rötter, R. (2011). Simulation of winter wheat yield and its variability in different climates of Europe: A comparison of eight crop growth models. *European Journal of Agronomy* 35(3):103–114.
Pierce, D. W., Barnett, T. P., Santer, B. D. and Gleckler, P. J. (2009). Selecting global climate models for regional climate change studies. *Proceedings of the National Academy of Sciences of the United States of America* 106(21):8441–8446.
Porter, J. R., Jamieson, P. D. and Wilson, D. R. (1993). Comparison of the wheat simulation models Afrcwheat2, Ceres-wheat and Swheat for non-limiting conditions of crop growth. *CropM, ft_macsur* 33:131–157.
Rosenzweig, C., Elliott, J., Deryng, D., Ruane, A. C., Müller, C., Arneth, A., Boote, K. J., Folberth, C., Glotter, M., Khabarov, N., Neumann, K., Piontek, F., Pugh, T. A., Schmid, E., Stehfest, E., Yang, H. and Jones, J. W. (2014). Assessing agricultural risks of climate change in the 21st century in a global gridded crop model intercomparison. *Proceedings of the National Academy of Sciences of the United States of America* 111(9):3268–3273.
Rosenzweig, C., Jones, J. W., Hatfield, J. L., Ruane, A. C., Boote, K. J., Thorburn, P., Antle, J. M., Nelson, G. C., Porter, C., Janssen, S., Asseng, S., Basso, B., Ewert, F., Wallach, D., Baigorria, G. and Winter, J. M. (2013). The Agricultural Model Intercomparison and Improvement Project (AgMIP): Protocols and pilot studies. *Agricultural and Forest Meteorology* 170:166–182.
Rötter, R. P., Hoffmann, M. P., Koch, M. and Müller, C. (2018). Progress in modelling agricultural impacts of and adaptations to climate change. *Current Opinion in Plant Biology* 45(B):255–261.
Rötter, R. P., Palosuo, T., Kersebaum, K. C., Angulo, C., Bindi, M., Ewert, F., Ferrise, R., Hlavinka, P., Moriondo, M., Nendel, C., Olesen, J. E., Patil, R. H., Ruget, F., Takáč, J. and Trnka, M. (2012). Simulation of spring barley yield in different climatic zones of Northern and Central Europe: A comparison of nine crop models. *CropM, ft_macsur* 133:23–36.
Ruiz-Ramos, M., Ferrise, R., Rodríguez, A., Lorite, I. J., Bindi, M., Carter, T. R., Fronzek, S., Palosuo, T., Pirttioja, N., Baranowski, P., Buis, S., Cammarano, D., Chen, Y., Dumont, B., Ewert, F., Gaiser, T., Hlavinka, P., Hoffmann, H., Höhn, J. G., Jurecka, F., Kersebaum, K. C., Krzyszczak, J., Lana, M., Mechiche-Alami, A., Minet, J., Montesino, M., Nendel, C., Porter, J. R., Ruget, F., Semenov, M. A., Steinmetz, Z., Stratonovitch, P., Supit, I., Tao, F., Trnka, M., de Wit, A. and Rötter, R. P. (2018). Adaptation response surfaces for managing wheat under perturbed climate and CO_2 in a Mediterranean environment. *CropM, LiveM, TradeM, ft_macsur* 159:260–274.
Salo, T. J., Palosuo, T., Kersebaum, K. C., Nendel, C., Angulo, C., Ewert, F., Bindi, M., Calanca, P., Klein, T., Moriondo, M., Ferrise, R., Olesen, J. E., Patil, R. H., Ruget, F., Takáč, J., Hlavinka, P., Trnka, M. and Rötter, R. P. (2016) Comparing the performance

of 11 crop simulation models in predicting yield response to nitrogen fertilization. *The Journal of Agricultural Science* 154(7):1218–1240.

Tai, A. P. K., Sadiq, M., Pang, J. Y. S., Yung, D. H. Y. and Feng, Z. (2021). Impacts of surface ozone pollution on global crop yields: comparing different ozone exposure metrics and incorporating co-effects of CO_2. *Frontiers in Sustainable Food Systems* 5.

Tao, F., Palosuo, T., Rötter, R. P., Díaz-Ambrona, C. G. H., Inés Mínguez, M., Semenov, M. A., Kersebaum, K. C., Cammarano, D., Specka, X., Nendel, C., Srivastava, A. K., Ewert, F., Padovan, G., Ferrise, R., Martre, P., Rodríguez, L., Ruiz-Ramos, M., Gaiser, T., Höhn, J. G., Salo, T., Dibari, C. and Schulman, A. H. (2020). Why do crop models diverge substantially in climate impact projections? A comprehensive analysis based on eight barley crop models. *Agricultural and Forest Meteorology* 281:107851.

Tao, F., Rötter, R. P., Palosuo, T., Díaz-Ambrona, C. G. H., Mínguez, M. I., Semenov, M. A., Kersebaum, K. C., Nendel, C., Cammarano, D., Hoffmann, H., Ewert, F., Dambreville, A., Martre, P., Rodríguez, L., Ruiz-Ramos, M., Gaiser, T., Höhn, J. G., Salo, T., Ferrise, R., Bindi, M. and Schulman, A. H. (2017). Designing future barley ideotypes using a crop model ensemble. *European Journal of Agronomy* 82:144–162.

Tian, X., Engel, B. A., Qian, H., Hua, E., Sun, S. and Wang, Y. (2021). Will reaching the maximum achievable yield potential meet future global food demand? *Journal of Cleaner Production* 294:126285.

Verburg, K. Methodology in soil-water-solute balance modelling: an evaluation of the APSIM-SoilWat and SWIMv2 models. Report of an APSRU/CSIRO division of soils workshop, Brisbane, Australia, 1995, p. 95.

Wallach, D., Mearns, L. O., Asseng, S. and Rötter, R. P. *Using Ensembles of Models in Climate and Crop Modelling*.

Wallach, D., Mearns, L. O., Ruane, A. C., Rötter, R. P. and Asseng, S. (2016). Lessons from climate modeling on the design and use of ensembles for crop modeling. *Climatic Change* 139(3):551–564.

Wang, E., Martre, P., Zhao, Z., Ewert, F., Maiorano, A., Rötter, R. P., Kimball, B. A., Ottman, M. J., Wall, G. W., White, J. W., Reynolds, M. P., Alderman, P. D., Aggarwal, P. K., Anothai, J., Basso, B., Biernath, C., Cammarano, D., Challinor, A. J., De Sanctis, G., Doltra, J., Dumont, B., Fereres, E., Garcia-Vila, M., Gayler, S., Hoogenboom, G., Hunt, L. A., Izaurralde, R. C., Jabloun, M., Jones, C. D., Kersebaum, K. C., Koehler, A. K., Liu, L., Müller, C., Naresh Kumar, S., Nendel, C., O'Leary, G., Olesen, J. E., Palosuo, T., Priesack, E., Eyshi Rezaei, E., Ripoche, D., Ruane, A. C., Semenov, M. A., Shcherbak, I., Stöckle, C., Stratonovitch, P., Streck, T., Supit, I., Tao, F., Thorburn, P., Waha, K., Wallach, D., Wang, Z., Wolf, J., Zhu, Y. and Asseng, S. (2017). The uncertainty of crop yield projections is reduced by improved temperature response functions. *Nature Plants* 3:17102.

Webber, H., Martre, P., Asseng, S., Kimball, B., White, J., Ottman, M., Wall, G. W., De Sanctis, G., Doltra, J., Grant, R., Kassie, B., Maiorano, A., Olesen, J. E., Ripoche, D., Rezaei, E. E., Semenov, M. A., Stratonovitch, P. and Ewert, F. (2017). Canopy temperature for simulation of heat stress in irrigated wheat in a semi-arid environment: A multi-model comparison. *CropM, ft_macsur* 202:21–35.

White, J. W., Hoogenboom, G., Kimball, B. A. and Wall, G. W. (2011). Methodologies for simulating impacts of climate change on crop production. *CropM, ft_macsur* 124:357–368.

Wolf, J. (2002). Comparison of two potato simulation models under climate change. II. Application of climate change scenarios. *Climate Research* 21:187–198.

Xie, W., Xiong, W., Pan, J., Ali, T., Cui, Q., Guan, D., Meng, J., Mueller, N. D., Lin, E. and Davis, S. J. (2018). Decreases in global beer supply due to extreme drought and heat. *Nature Plants* 4(11):964–973.

Zheng, B., Chenu, K., Dreccer, M. F. and Chapman, S. C. (2012). Breeding for the future: What are the potential impacts of future frost and heat events on sowing and flowering time requirements for Australian bread wheat (*Triticum aestivium*) varieties? *Global Change Biology* 18(9):2899–2914.

Chapter 7

Modelling climate change impacts on livestock production

*Ryan G. McGuire, Shannon M. McLaughlin, Tanisha L. Waring, Kayley D. Barnes, Sharon A. Huws and Nigel. D. Scollan, Queen's University Belfast, UK**

1 Introduction
2 Enteric methane mitigation
3 Grassland agriculture: a net zero livestock system
4 Conclusion and future trends
5 Where to look for further information
6 References

1 Introduction

Globally, the target of net zero continues to be challenging for every sector; yet, this is most difficult for the agri-food sector, especially at the farm level. Agriculture's unique economic position, fundamental role in food security and consequential environmental footprint are a well-known trade-off. According to the Intergovernmental Panel on Climate Change (IPCC) (2019), agriculture is directly responsible for up to 8.5% of all greenhouse gases (GHG) emitted globally – with a further 14.5% coming from land-use change (LUC). For the European Union (EU), agriculture is expected to account for a third of total emissions by 2050.

To tackle this unprecedented challenge, several mitigations exist for agriculture, namely (1) strategies to reduce enteric methane (NH_4) emissions from ruminants, (2) improved on-farm carbon sequestration and storage, (3) improved manure storage and management, (4) alternative livestock feeds and (5) improved animal health and welfare standards (see Grossi et al., 2019). In this chapter, we first outline the global climate and wider sustainability challenges for agriculture – placing particular focus on ruminant production systems (RPS) and grassland agriculture. We then focus on two mitigation measures that are vital to helping global agriculture achieve the principal target of net zero by 2050: (1) enteric NH_4 reduction strategies and (2) improved on-farm carbon sequestration and storage.

* Ryan McGuire and Shannon McLaughlin are equally contributed for this chapter.

http://dx.doi.org/10.19103/AS.2022.0115.09

In addition to GHG emissions, nearly 90% of global ammonia emissions are emitted from agricultural sources, including ammonia-based fertilisers and animal manure (Ma et al., 2021). From this, in 2012, about one-fourth of global agricultural ammonia (NH_3) emissions were trade related. Globally, these emissions induce 61 000 PM2.5-related premature mortalities, with 25 000 deaths associated with crop cultivation and 36 000 deaths with livestock production (Ma et al., 2021). Regulations for agricultural NH_3 emissions are overall ineffective worldwide (Plautz, 2018). Furthermore, agriculture also accounts for ~70% of global freshwater use (Perret and Payen, 2020) and is the lead threat to 24 000 of the 28 000 (86%) species at risk of extinction.

Moreover, according to the United Nations, the majority of the world's soil resources, a vital natural carbon storage system (sink) and the source of 98.8% of global food production (Kopittke et al., 2019) are only fair, poor or very poor condition (ITPS, 2015). As a result of the above, agriculture has played a key role in anthropogenic climate change and the increasing pressure on natural capital such as the loss of natural ecosystems (e.g. forests, savannahs, natural grasslands and wetlands) and declining biodiversity (IPCC, 2019). Coupled with this, efficiency continues to be a significant issue for the global food system, with 25–30% of the total food produced becoming lost or wasted.

1.1 Economic value

From an economic viewpoint, in 2012, 19% of the global population were directly engaged in farming, but agriculture represented just 2.8% of overall income (Alston and Pardey, 2014). Yet, in 2021, agriculture represented 4.3% of global GDP (World Bank, 2022a[1]) and ~27 of total global employment (World Bank, 2022b[2]). In 2017, the value of livestock production and livestock products in the EU was equal to €170 billion, representing 40% of the total agricultural activity. Furthermore, European LPS employs around 4 million people (European Commission, 2018). The EU-28 had 131 million livestock units in 2016 and more than 50% of these units were concentrated in four countries (Peyraud and MacLeod, 2020) – with the UK identified as a leading nation.

Looking specifically at the UK, in combination, the value of farm gate products is over £12 billion per year. Whilst the UK imports £10 billion worth of meat and milk per year, it also exports £4 billion. (CIEL, 2022). Moreover, UK agriculture employed 426 000 people in 2018, 1.5% of the UK labour force. In terms of value addition, the UK farmgate value of livestock products is over £12 billion per year. Agriculture makes up 0.6% of the UK's gross value added (GVA), with the wider agri-food sector contributing 6.6% of the national GVA. Thus, livestock production systems' (LPS) unique role in one health, that is, tackling the loss of natural capital and sustaining the human population and the economy at both national and global levels is vital.

In addition to strenuous environmental challenges, the financial performance of agriculture, particularly LPS, is currently facing its greatest threat. With the increasing cost of production and the pressure The Ukraine War has added in terms of resource availability and accessibility, LPS is facing a severe challenge. This challenge also threatens to increase the existing pressure on farmers, farmer health and farm-family well-being. For example, currently, more than one agricultural worker in the UK commits suicide every week.

1.2 Food security role and challenges

Worldwide meat production has almost quadrupled from 78 Mt to 331 Mt, between 1963 and 2015. About 25% is derived from RPS (FAO, 2014). The Organisation for Economic Co-operation and Development and the UN Food and Agriculture Organisation (FAO; FAO, 2014) forecasted that global meat production will increase by 19% by 2023 (as cited by Michalk et al., 2019). Nonetheless, by 2050, the agricultural sector will face the challenge of producing 60% more than the current food, feed and fibre supply of 8.5 billion t/year to sustain a global population forecast to be 9.3 billion people (Michalk et al., 2019). From diet and public health viewpoint, despite increased publicity regarding reduced meat consumption and its role within both healthy diets and the net zero pathway, demand has gone up and, most importantly, a reduction of meat consumption could result in significant deficiencies and malnutrition (Wyness, 2016; Klurfeld, 2018, Knuppel et al., 2020; Stanton et al., 2022; Dseagu et al., 2022; McGuire et al., 2022). Coupled with this demand, global consumption shifts have contributed to about 2 billion adults becoming overweight or obese, yet approximately 821 million people remain undernourished. More specifically, it is estimated that 29.3% of the global human population - 2.3 billion people - were moderately or severely food insecure in 2021 and 11.7% (923.7 million people) faced severe food insecurity (FAO, IFAD, UNICEF, WFP and WHO, 2022). From a global nutritional perspective, Fig. 1 demonstrates a million tonnes of protein produced via different LPS in 2015 (FAO, 2023).

Despite these pressures, there is now a mandatory requirement for LPS to tackle climate change, to maximise resource use efficiency and to develop strategies that enable increased productivity from fewer inputs. Thus, policymakers, retailers and ultimately consumers want to know more about how food from LPS is produced, the impact on the planet and assurance that it was produced ethically. Globally, this is embedded within the policies of the Paris Agreement, European Green Deal, and Farm to Fork Strategy (UNFCC, 2023). For example, technologies, measurement, innovation and priority clusters deliver a changing world for a more sustainable and more inclusive economy. Thus, the global context for LPS is changing rapidly, presenting both challenges (i.e. on health, education, biodiversity loss, environmental, energy

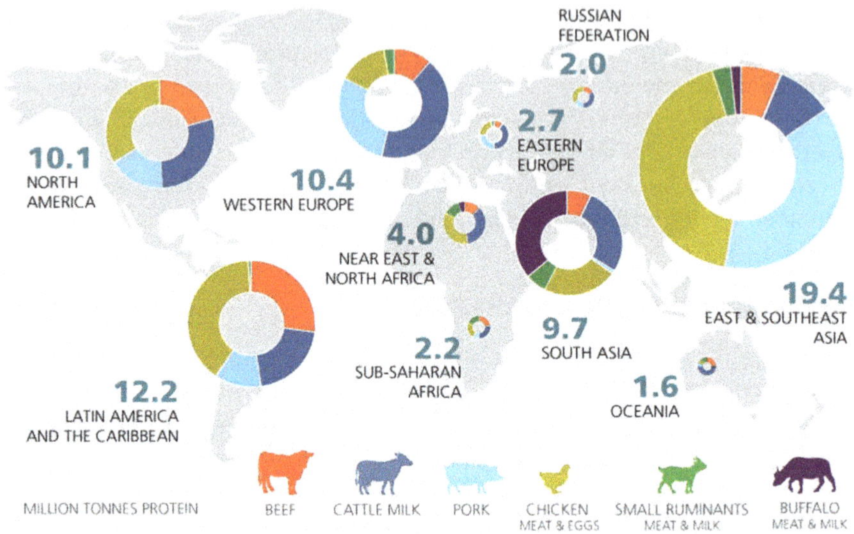

Figure 1 Regional production of protein from different types of livestock production. Source: FAO, 2023.

and food), especially to reduce GHG emissions, and opportunities (i.e. science, innovation and technology development) over the next 20–30 years.

The remainder of this chapter will be as follows. Section 2 will explore the state-of-the-art mitigation measures to address enteric NH_4 mitigation among RPS as well as models and methods to quantify NH_4 emissions. Section 3 will discuss the unique potential of grassland agriculture to sequester and store carbon to deliver a net farm carbon balance, as well as address carbon farming and accounting, measure, report and verification (MRV) methods and carbon finance. Finally, our conclusion in Section 4 outlines this chapter's key messages and the direction of future research.

2 Enteric methane mitigation

2.1 Ruminant food security and climate change

The emission of GHGs has long been recognised as causing climate change and agri-food-related GHGs are mainly produced by LPS, particularly RPS, thereby creating tension between food security, human health and environmental health. The main GHGs emanating from the agri-food sector are CH_4, carbon dioxide (CO_2) and nitrous oxide (N_2O) in descending order (FAO GLEAM – https://www.fao.org/gleam/en/). Indeed, 44% of the agri-food sector's total GHG emissions are in the form of CH_4, which is erupted by ruminants due to enteric fermentation in ruminants (FAO GLEAM – https://www.fao.org/gleam/en/). Global CH_4 emissions from ruminants have also continuously increased over the past decades (Fig. 2) and are estimated to contribute towards 17% of the global food system GHG emissions (FAO, 2020; Arndt et al., 2022; Fig. 2).

Importantly, the production of CH_4 has also been shown to divert 6–12% of energy away from productivity (Johnson and Johnson, 1995), and therefore decreasing CH_4 emissions could contribute to sustainable livestock production through enhanced growth or milk production. CH_4 also has between 27 and 30 times the global warming potential (GWP) of CO_2 (IPCC, 2021). However, recently it has been shown that the half-life of CH_4 in the atmosphere is likely approximately 12 years and that therefore its GWP is much less than that of CO_2, which stays in the atmosphere longer (Cain et al., 2019). This has resulted in the generation of a new measure, known as GWP*, to estimate GWP of CH_4 (Cain et al., 2019; Lynch et al., 2020), which accounts for the half-life of CH_4 in the atmosphere. Irrespectively, the LPS must enhance sustainable management practices resulting in lower CH_4 emissions. Indeed, many national governmental policies across the world have set a target of reaching net zero by 2050 at the latest, with some countries, such as New Zealand and Northern Ireland also setting separate CH_4 reduction targets of 10% by 2030 and 46% by 2050, respectively.

2.2 Ruminant enteric fermentation

Ruminants have a forestomach composed of four compartments – the reticulum, the rumen, the omasum and the abomasum. The rumen, the main fermentative compartment of the forestomach, is a complex, dynamic ecosystem composed of anaerobic bacteria, protozoa, anaerobic fungi, methanogenic archaea and bacteriophages. These microbes interact with each other in many ways, resulting in a symbiotic relationship with the host and provision of energy from the breakdown of plant material. The productivity and environmental impact of ruminants are mainly consequences of the biochemical process that occurs as feed enters the rumen and the process of digestion begins (Huws et al., 2018; Ungerfeld, 2020; Mizrahi et al., 2021). Essentially, dietary carbohydrates are broken down by the rumen microbes, particularly rumen bacteria (Fig. 3), and go through biochemical processes that result in the production of volatile fatty acids (VFAs), which serve as a source of energy for the animal. This process results in the generation of hydrogen, which is used by methanogenic archaea to convert CO_2 into CH_4.

Methanogenic archaea, often called methanogens, are members of the phylum Euryarcheota and sit within the orders *Methanobacteriales*, *Methanococcales*, *Methanomicrobiales*, *Methanocellales*, *Methanopyrales*, *Methanomassiliicoccales* and *Methanosarcinales*. The archaea are an ancient lineage of microbes that look phenotypically like bacteria but are phylogenetically distinct. Methanogens typically act as terminal reducers

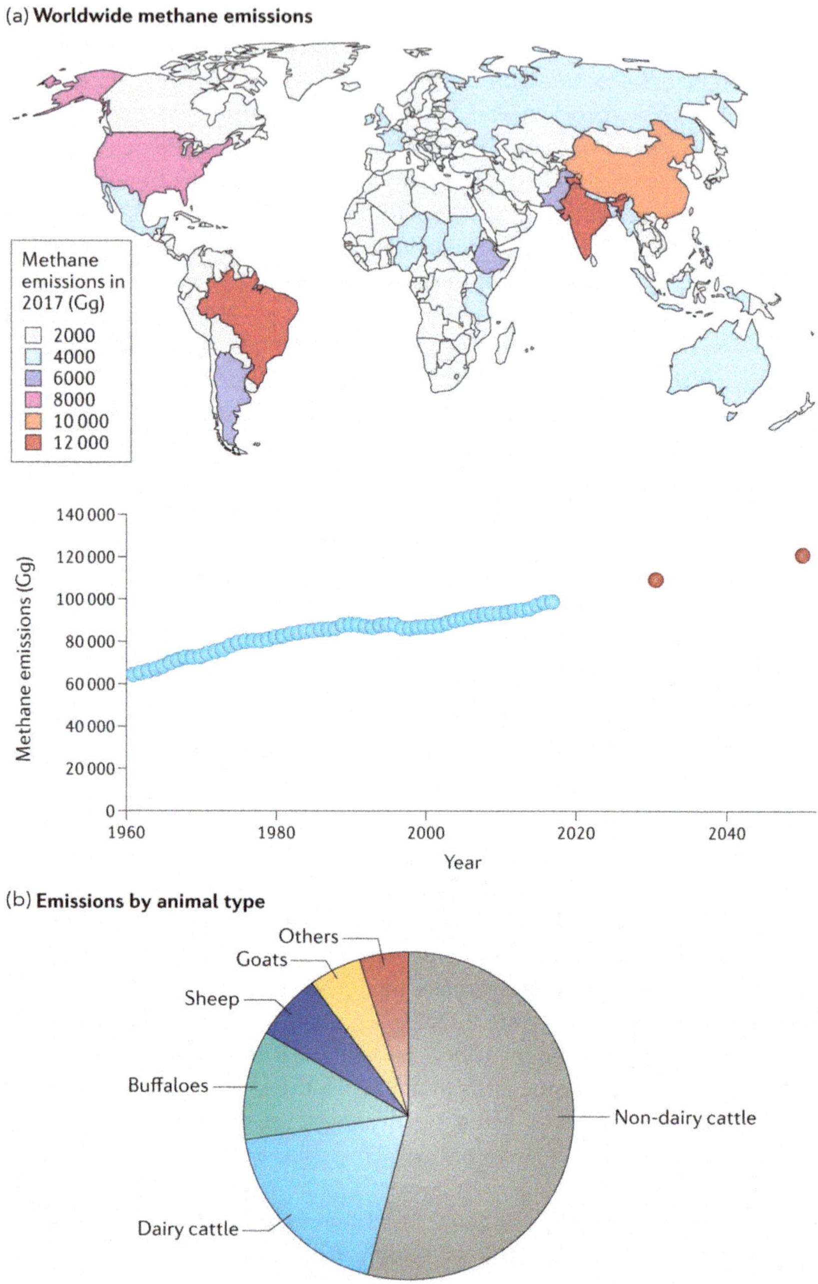

Figure 2 Global methane emissions: methane emission estimates calculated from Food and Agriculture Organization of the United Nations (FAOSTAT) enteric fermentation data, shown by (a) region (red dots on graph show predicted emissions) and (b) animal type (others mainly include non-ruminant emissions). Source: Mizrahi et al., 2021.

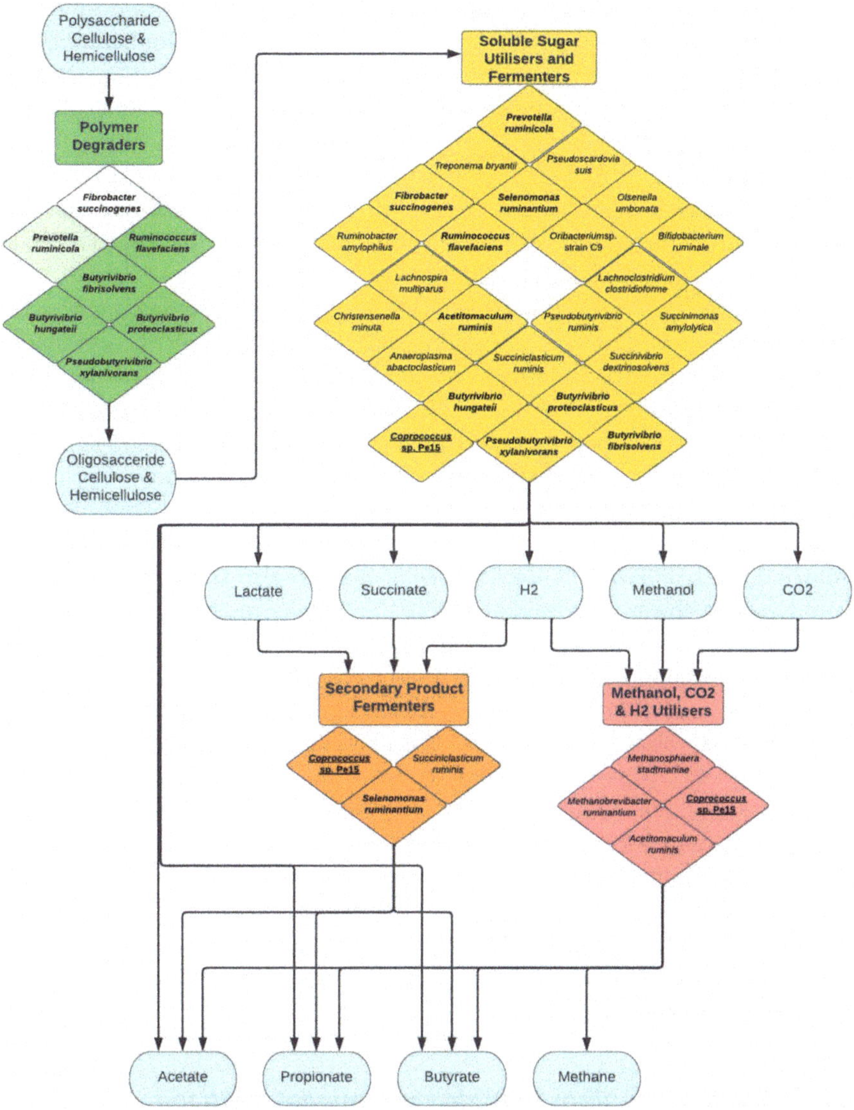

Figure 3 Examples of bacterial species that act as polymer degraders, soluble sugar utilisers and fermenters, secondary product fermenters, methanol, CO_2 and H_2 utilisers in the rumen. Many rumen organisms are specialists in specific stages of this process, but some have been identified as being active in two phases (names highlighted in bold) or even three phases (names highlighted in bold and underlined). Polymer degraders consist of primary colonizers (light green), secondary colonisers (dark green) and those with no early phase temporal pattern (white). Source: Huws et al., 2021; FAO, 2022.

of carbon in the process of dietary carbon metabolism (Attwood et al., 2020). Methane itself is produced mainly via the hydrogenotrophic pathway, which results in CH_4 being produced from hydrogen and CO_2 (Fig. 3), with a small amount being produced through the utilisation of methyl groups (methylotrophic pathway) or, even less commonly, through use of acetate as a substrate (acetoclastic pathway) (Morgavi et al., 2010; Tapio et al., 2017). More recently, it has also been shown in an *in vitro* experiment that 18% of rumen CH_4 was produced from formate and little is known regarding formate production in the rumen (Leahy et al., 2022).

Hydrogenotrophic methanogens include the genus *Methanobrevibacter* (*Mbb.*), which is subdivided into the SMT clade (*Mbb. smithii*, *Mbb. gottschalki*, *Mbb. millerae* and *Mbb. thaurei*) and the RO clade (*Mbb. ruminantium* and *Mbb. olleyae*), which are the most enumerate rumen methanogens (Tapio et al., 2017). Indeed, *Mbb. gottschalkii* and *Mbb. ruminantium* clades have been shown to be the two largest groups, accounting for 74% of all the rumen archaea globally (Henderson et al., 2015). Methylotrophic methanogens are less abundant and include *Methanosarcinales, Methanosphaera* and *Methanomassiliicoccaceae* (Attwood et al., 2020). The *Methanosarcinales* can also produce CH_4 via the acetoclastic pathway (Morgavi et al., 2010). These methanogens are strict anaerobes and very challenging to grow outside the rumen environment as they require exceptionally low redox potential (−340 mV). However, many can be grown in a laboratory, resulting in much of our understanding regarding their functional capacity. For example, the ability to grow them in the laboratory has led to 14 strains being genome sequenced, namely three unknown rumen methanogens, *Methanobacterium formicium*, two strains of *Mbb. ruminantium*, *Mbb. boviskoreani*, *Mbb. gottschalkii*, *Mbb. millerae*, *Mbb. olleyae*, *Mbb. thaueri*, *Mbb. woesei*, *Methanosarcina barkeri* and a *Mbb*. sp. The ability to grow these *in vitro* has also shown that many of the rumen methanogens associate closely with hydrogen-producing protozoa and fungi, thus enhancing their ability to produce CH_4 (Vogels et al., 1980; Bauchop and Mountfort, 1981; Cheng et al., 2009; Belanche et al., 2014).

2.3 Innovative solutions to mitigate ruminant methane emissions

Innovative solutions to reduce GHG emissions from ruminants mainly focus on the potential to breed ruminants with decreased CH_4 emissions. Breeding and dietary innovations offer a major technological advancement that can reduce CH_4 emissions by up to 30% – in terms of emissions (g/day), CH_4 yield (g/kg of dry matter intake) and CH_4 intensity 0.70 (g/kg or litre of product produced) (de Haas et al., 2017; Beauchemin et al., 2020; Smith et al., 2021; Smith et al., 2022). This reduction in CH_4 emissions through breeding is thought to be due to the influence of the host genome on the rumen microbiome (Roehe et al., 2016; Difford et al., 2018; Li et al., 2019; Wallace et al., 2019; Zhang et al., 2020; Smith et al., 2021; Martínez-Álvaro et al., 2022; Smith et al., 2022). Roehe et al. (2016, 2022) showed that some rumen microbes are heritable and that a number of rumen microbial genes are linked to low CH_4 emissions (e.g. *cofG*, *bcd*, *pccb*, *ABC.P.E.P*, *TSTA3* and *RP-L35*). Li et al. (2019) also showed some microbes to be heritable, with a modest heritability of ≥0.15, and 12 bovine chromosomes to be linked to this heritability, again indicating the possibility of breeding animals with a 'utopian' microbiome. Rumen size has also been shown to change in these breeding programmes, with sheep that emit high levels of CH_4 often having larger rumen volumes than low emitters (Goopy et al., 2014). Breeding programmes aiming to breed sheep that emit low levels of CH_4 have also unveiled differences in feeding behaviour between sheep with different CH_4 emission levels. Sheep emitting less CH_4 have been shown to eat more frequently and in smaller amounts, resulting in a smaller rumen, whereas high emitters have been shown to eat larger amounts more often, resulting in a larger rumen (Goopy et al., 2014).

Breeding-based strategies to reduce CH_4 emissions take a long time to develop and diet is the most amenable way of instigating immediate change in the rumen microbiome to enhance the ruminant phenotype. Dietary interventions enabling reductions in CH_4 emissions can be grouped into the following categories: plant-based strategies (e.g. feeding plants that are high in secondary compounds, such as tannins), targeted CH_4 inhibitors (such as 3-NOP, which is commercially known as Bovaer®), oils and oilseeds, and hydrogen sinks (e.g. chemicals or microbes that utilise hydrogen so that there is less available for methanogenesis). It has been shown that including legumes in ruminant feeding systems can reduce CH_4 emissions by approximately 18% based on CH_4 emission intensity for a given amount of milk produced (g CH_4/kg milk) due to their high tannins content (Enriquez-Hidalgo et al., 2014; Eugène et al., 2021; Arndt et al., 2022). Plant tannins are antimicrobial in nature, and a recent study showed that feeding the tannin-rich tropical plants *Calliandra calothyrsus*, *Gliricidia sepium* and *Leucaena leucocephala* reduced the densities of plant-attached *Fibrobacter succinogenes* and methanogenic archaea in the rumen, resulting in reduced CH_4 production (Rira et al., 2022).

Data on the feeding of multispecies swards containing legumes and herbs to beef cattle show that it increases average daily weight gain, which would be expected to decrease CH_4 intensity, as the animals reach slaughter weight more quickly (Boland et al., 2013). More recently, it has been found that feeding the seaweeds *Asparagopsis taxiformis*, *Alaria esculenta*, *Ascophyllum nodosum* and *Chondrus crispus* to ruminants reduces CH_4 emissions from the rumen by 10% to 80% (based on yield), with *A. taxiformis* providing the largest reductions (Wang et al., 2008; Machado et al., 2015; Kinley et al., 2016; Abbott et al., 2020;

Lean et al., 2021; Roque et al., 2021). This reduction is largely attributed to the bioactive compound bromoform, which is found in several seaweed species, especially red seaweeds such as *Asparagopsis* spp., with the mechanism of action shown to be inhibition of methanogenesis (Abbott et al., 2020). Studies have shown that the supplementation of ruminant feed with macroalgae affects the diversity of the rumen community in the short-term but that the effect lessens over time (Roque et al., 2019), although further *in vivo* studies are required to verify long-term effects.

CH_4 inhibitors offer a more targeted microbiome-targeted tools for reducing CH_4 emissions from ruminants. The best studied is 3-NOP (Beauchemin et al., 2020; Arndt et al., 2022). 3-NOP inhibits the last step in rumen methanogenesis, resulting in an approximately 35% reduction in CH_4 emissions, based on yield and intensity, without negatively affecting animal health and welfare (Hristov et al., 2015). 3-NOP is now approved for use in Australia, Brazil, Chile and the European Union and is under consideration by regulating bodies in many other countries. Encouragingly, 3-NOP seems to have minimal effects on the rumen microbiome (Romero-Perez et al., 2014; Haisan et al., 2017).

Numerous studies have shown dietary lipid supplementation (<4% of Dry Matter Intake) can decrease CH_4 production up to 20%, whilst providing increasing energy to the animal. Meta-analysis indicates that a 1–5% decrease in CH_4 (g/day) per 10 g/kg DM dietary fat can be achieved (Grainger and Beauchemin, 2011; Patra and Yu, 2013). Lipids inhibit methanogenesis by decreasing the numbers of ruminal methanogens and protozoa, and through biohydrogenation of unsaturated fatty acids (Patra and Yu, 2013). Biohydrogenation can provide an alternative hydrogen sink in the rumen to compete with methanogenesis, but this is quantitatively small (1–2%; Nagaraja et al., 1997).

The possibility of feeding chemical or microbial hydrogen sinks alone or in combination with dietary additives to reduce ruminant CH_4 emissions is also substantial. It has been shown that supplementation with the chemical phloroglucinol, together with 3-NOP, promotes capture hydrogen from reductions in methanogenesis and generates valuable VFAs as a source of energy for the host (Martinez-Fernandez et al., 2017). Likewise, the use of direct-fed microbials (DFMs), which are feed products containing a source of viable and naturally occurring microbes (Brashears et al., 2005), alone or in combination with CH_4 inhibitors, offers a mechanism for redirecting hydrogen into energy. Nollet et al. (1998) examined the addition of the cell-free supernatant of *Lactobacillus plantarum* 80 *in vitro* and noted a 30.6% reduction in CH_4 production when the supernatant was combined with an acetogenic bacterium, *Peptostreptococcus productus* ATCC 35244.

A subsequent *in vivo* study showed that the LP80 supernatant in combination with *P. productus* reduced CH_4 emissions by 80% (mmol/6h) during the initial three days, although the reduction was not persistent (Nollet et al. (1998). Cao et al. (2010a) investigated the effect of *Lactobacillus plantarum* Chikuso-1 on an ensiled total mixed ration and showed that CH_4 production decreased by 8.6 % compared with untreated silage. Further testing *in vivo* showed that silage inoculated with *L. plantarum* Chikuso-1 increased digestibility and decreased ruminal CH_4 yield by 24.7% in sheep compared with a non-inoculated control silage (Cao et al., 2010b). Jeyanathan et al. (2014) also screened 45 bacteria, including strains of LAB, Bifidobacteria and Propionibacteria, *in vitro* for their ability to reduce methanogenesis and then selected 3 strains for *in vivo* trials in sheep, with one strain, *L. pentosus* D31, resulting in a 13% reduction in CH_4 yield over 4 weeks. Although more research is required in this area, the results suggest that some DFMs have the potential to alter ruminal fermentation in a way that leads to improved production parameters and reduced CH_4 production and may have even better results in combination with dietary additives. Likewise, combining dietary interventions with different mechanisms of action may enhance CH_4 reductions whilst futureproofing such strategies in terms of decreasing the likelihood of microbial adaptation towards them.

2.4 Ruminant contribution to nitrous oxide emissions

Although CH_4 is the most important GHG emitted directly from ruminants, it should also be noted that inefficiencies in dietary nitrogen use by ruminants can also result in excess nitrogen being released to the environment, and this can be converted into N_2O by environmental microbes (e.g. in manure and soil). N_2O has approximately 273 times the GWP of CO_2 as compared over a 100-year period (IPCC, 2021). There has been a tendency for at least the last 50 years to move towards intensive farming practices, with productivity as the main goal. Enhanced productivity has been achieved through effective breeding and over-provision of protein, the latter being a strategy that is detrimental to the environment as only about 25% of the protein is utilised by the ruminant, with the rest being excreted, mainly in the urine (Huws et al., 2018). The large amounts of nitrogen accumulating in manure and deposited on the land then get converted into N_2O, creating a much larger GHG challenge. Ensuring nitrogen-use efficiency in ruminants through the provision of appropriate sources and levels of protein is thus another way of reducing global warming. Sourcing soya protein from land that has undergone land-use change, particularly deforestation, is another major contributor to global warming (CIEL, 2022). Emphasis is now being placed on evaluating local sources of feed protein obtained in a sustainable manner, feeding ruminants at levels that are optimum for their health and production, and minimising unnecessary soil nitrogen deposition.

2.5 Methods to quantify methane emissions from ruminants

Whilst developing our knowledge of the rumen microbiome and the pathways for CH_4 production is crucial to discovering mitigation methods, it is also imperative that CH_4 production from LPS and specifically enterically from ruminants is accurately quantified. Typically, prediction of CH_4 mitigation potential of dietary interventions in ruminants starts *in vitro* due to the high-throughput capacity of such approaches and their low cost in comparison to animal-based experiments, coupled with ethical obligations. *In vitro*, different substrates can be incubated with rumen fluid collected from the animal in the presence of Van Soest buffer (artificial saliva) and the end products of fermentation over-time can be assessed. The difference in CH_4 content in the headspace of the incubation vessel can be quantified using gas chromatography across treatments and compared to a negative control, and subsequently used as a guide for *in vivo* feeding recommendations. Whilst being considerably cheaper and a useful tool for screening many feed ingredients, the *in vivo* outcome often varies due to the limitations of the system (e.g. gas produced not always released in a timely manner, new substrate not added frequently or at all affecting the natural fermentative process) as well as the environmental, animal and inclusion level differences and the greater potential for the rumen microbes to find niches to avoid the effects of the intervention i.e. forming biofilms in the epithelium walls. As such, *in vivo* experimentation is required to verify the amount of CH_4 produced across ruminant LPS.

To measure ruminant enteric CH_4 emissions *in vivo* it is important to highlight the possible CH_4 release routes from the animal. Ricci et al. (2014) outlined three routes (1) the expiration of absorbed CH_4, (2) eructation of CH_4 and (3) CH_4 emitted from the hindgut in the flatus. The former two routes collectively termed the 'exhaled' gas have been deemed accountable for the greatest proportion of the released CH_4. For instance, Murray et al. (1976) documented 98% of total CH_4 production being exhaled in sheep and no recent work counters this finding. As such, the majority of the recent measurement innovations for CH_4 target the exhaled gas only, allowing for higher throughput non-confinement technologies to be developed (e.g. SF_6 tracer, open-path laser and ventilated hood system) and compared against the gold standard open-circuit respiration chamber (Fig. 4).

The open-circuit respiration chamber has been used for determining heat balance and gaseous exchange in individual animals for many years (Storm et al., 2012). Characterised by an air inlet and exhaust, animals breathe in a one-way stream of air passing through the sealed and environmentally controlled chamber space (Fig. 4a). The incoming and outgoing air is sampled for changes in CH_4 (including flatus), gaseous exchanges of oxygen (O_2) consumption and CO_2 production and monitors other trace gases. The system's ability to provide

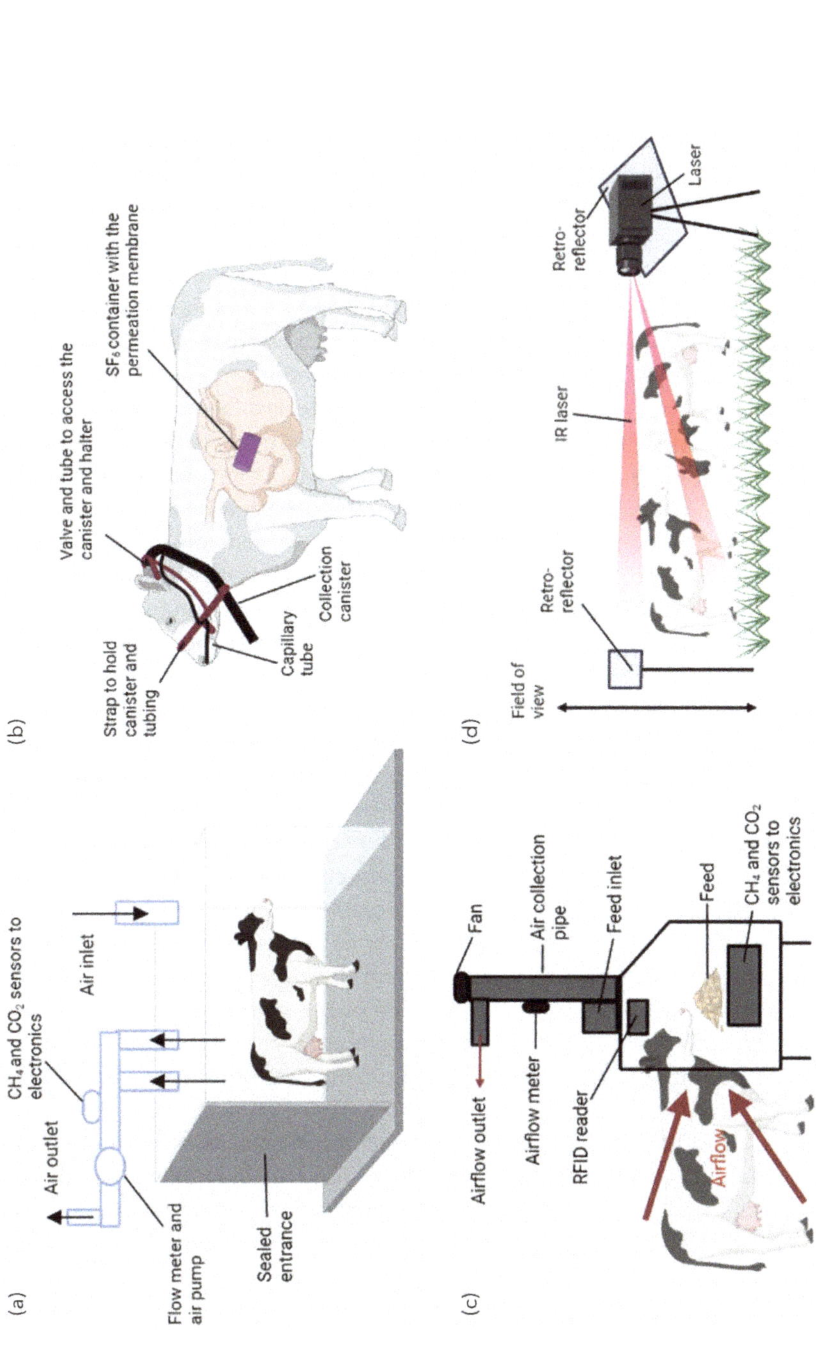

Figure 4 Schematics of (a) open-circuit respiration chamber system, (b) SF_6 tracer technique, (c) open-path laser method, (d) ventilated hood system (Greenfeed). Source: Adapted from Hill et al. (2016) by K. Barnes using BioRender.

continuous and accurate data on air composition over an extended period, alongside obtaining DMI, energy utilisation and production values allowed the development of the predictive models and equations used to estimate ruminant livestock emissions for National GHG Inventories (Hill et al., 2016). However, the system has several shortcomings. For example, the chamber limits the animal's normal behaviour, which could lead to decreased feed intake and thus an under-estimation of CH_4 production. Additionally, the system is costly to install and maintain for single-animal use especially when due to animal welfare licensing measurements can only be made over a short timeframe (Storm et al., 2012). Fortunately, other technologies have been shown to be equally effective, at lower costs and without the need for animal isolation.

The sulphur hexafluoride (SF_6) tracer technique is a favourable alternative as the animal is free to roam and their movement and thus behaviour is not limited to a chamber space. The basic idea behind the method is that the quantity of exhaled CH_4 can be measured if the emission rate of a tracer gas from the rumen is known. The method relies on the constant release of SF_6 tracer gas, delivered to the rumen via a permeation tube (demonstrated in Fig. 4b). The ratio of CH_4 to SF_6 in the exhaled breath of the animal is measured via gas chromatography and corrected to the background reference concentration. SF_6 is an ideal tracer gas because it behaves similarly to CH_4 in the rumen, is non-toxic, easily measurable, and traceable at low concentrations and due to its synthetic origin, it is not created in any biological processes in the rumen. However, SF_6 is a GHG with a GWP_{100} of 22 800 so the method is limited to research settings. In addition, the difference between measurements using the SF_6 tracer technique and respiration chambers is 10% or more (Pinares-Patiño et al., 2011) but this bias is neither consistent nor predictable (McAllister and Newbold, 2008), which increases the number of animals needed to prove feeding treatment differences in CH_4 emissions. Other difficulties include the maintenance of a constant release rate from the permeation tubes with pre- and post-animal experiments showing differences in permeation rate (Storm et al., 2012), and the measurement of background levels of SF_6 and CH_4 under field conditions due to wind direction and proximity to other animals (Williams et al., 2011).

A more recent technology for CH_4 quantification is the use of an Automated Head-Chamber System (AHCS, GreenFeed) that can monitor CH_4 and CO_2 mass fluxes from the exhaled gas from ruminants and correct readings to the background reference concentration (Fig. 4c). The AHCS identifies individual study animals via RFID tags to deliver small quantities of pelleted feed, whilst spot-sampling exhaled gases during feeding times. The basic concept is similar to the respiration chamber where gas concentration and airflow are used to determine CH_4 flux over time, but the AHCS data is extrapolated to estimate total daily emissions based on the spot measures. Whilst the AHCS allows for field measurements to be taken and multiple animals (20–25 animals/AHCS) to be measured per day, it does have some drawbacks. The difference in 'spot sampling' protocols can lead to highly variable datasets, which only provide a short-term emissions rate, and the differences in the number of times the animal visits the AHCS and the time after feed consumption (higher CH_4 emissions post-consumption of dry matter) can result in daily emission variations at an animal level. However, greater uniformity in sample frequency throughout the 24-h measurement day is attained if enough data is accumulated from a large number of animals and a representative flux can be calculated. Being less invasive in comparison to the SF_6 method and much cheaper than the respiration chamber, it is becoming increasingly popular within animal research institutes.

Whilst the former three methods are expensive, require input from specialists and are often inaccessible to producers, the open-path laser method provides a cheaper but less robust alternative. The method quantifies the dispersion of CH_4 along a defined path (distance and height) and the downwind concentration of the gas to establish the emission rate, using an 'inverse dispersion' approach (McGinn et al., 2006) (Fig. 4d). This method is often used in grazing systems or where herd scale measurements are appropriate. Measurements are made in a set area over an established period of time with a defined number of animals. However, herd level quantification poses several challenges with differences in animal CH_4 emissions being overlooked and the data being affected by high-emitting animals, herd behaviour and changes in local climatic conditions. The microclimatic effects of temperature inversions, changes in saturated vapour pressure and low wind speed are often misrepresented in calculations of total daily emissions. Several studies have documented an increase in CH_4 during the night between 23:00 h to 05:00 h, which is more likely related to these changes than biological processes (Hill et al., 2016). The use of open-path lasers to quantify CH_4 from groups of animals is new, the amount of published data is limited but in the coming years its use is likely to increase and thus accuracy may improve. Additionally, adaptations of this method such as hand-held and drone technologies have been created for wider application by the LPS sector, where the laser can be directed in-front of animals, in different areas of housing or field space for spot measures (Tedeschi et al., 2022).

There is a range of other technologies available for predicting CH_4 emissions from ruminants, some of which will increase in use as the understanding of the ruminant microbiome, CH_4 production and quantification methods improve. Emission models are a key example and are being used to predict total emissions from livestock and to evaluate the impacts of a range of mitigation strategies applied (Moraes et al., 2014). Emission models are underpinned by mechanistic models of ruminant digestion with the early works of Baldwin et al. (1977) and Mertens and Elys (1979) being notable in developing an understanding of factors that affect the nutritive value/quality of feedstuffs, fibre digestion and passage rate. The development of mechanistic models permitted the development of rationing software which utilises the above to predict performance outputs (e.g. milk yield) when animal information is supplied (e.g. parity, weight) such as the Cornell Net Carbohydrate and Protein system developed by Tylutki et al. (2008) using work from Russell et al. (1992); Fox et al. (1992). Whilst being crucial for the advancement of ruminant production, mechanistic models are considered narrow-lensed as the drivers of production tend to increase emission outputs.

As such, emissions models have been developed using existing *in vivo* CH_4 output, diet type, voluntary feed intake, digestibility of feeds, animal phenotype and animal performance data. The models can be advanced by including mathematical descriptions of biological processes such as the stoichiometric formula (Wolin, 1960) for estimating CH_4 production from rumen VFA concentrations in mmol/L: $Methane\ (mmol/L) = 0.5 * A + 0.5 * B - 0.25 * P - 0.25 * V$, where A is acetic acid (mmol/L), B is butyric acid (mmol/L), P is propionic acid (mmol/L) and V is valeric acid (mmol/L). In essence, the more predictions that can be made for the same data set the greater the robustness of the output. As such, more dynamic and mechanistic models are beginning to emerge which incorporate changes observed in the rumen microbiome in high and low-emitting animals. For example, Dai et al. (2022) reviewed the effects of ruminal protozoa abundance and taxonomy on CH_4 production. Ruminal protozoa are classified into two groups, entodiniomorphids and isotrichs, the latter having a more significant effect on ruminal methanogenesis due to having different endosymbiotic methanogens (Belanche et al., 2015, Belanche et al., 2014). The evaluation of 64 publications and 79 *in vivo* experiments in the meta-analysis by Dai et al. (2022) showed positive associations between total ruminal protozoa and isotrichids with CH_4 emissions but not between entodiniomorphids and CH_4 emissions. Concluding that both total ruminal protozoa and isotrichs can be used to model and accurately predict CH_4 emissions across different ruminant types, with 6.59% and 4.08% of the square root of the mean square prediction error in the models, respectively. In addition, to ruminal protozoa models, the composition of archaeal communities has been shown to be associated with high or low CH_4 phenotypes in ruminants. For example, ruminants with greater ruminal densities of *Methanobrevibacter gottschalkii* species tend to produce more CH_4. However, the correlation between archaeal rumen community size and CH_4 emissions is weak (Tapio et al., 2017). Nevertheless, the taxonomic identification of the microbiome is costly, requires biocomputational analysis and is not always completed for *in vivo* trials.

Other emission models include the use of gas chromatography to identify the long-chain fatty acids (LCFA) in milk (lactating ruminants only) to predict CH_4 emissions due to the de nova synthesis of both CH_4 and the precursors for milk

LCFA occurring in the rumen. Chilliard et al. (2009) predicted CH_4 (R2 = 0.95) with the milk FA 16:0, *trans*(*t*)16/*cis*(*c*) 14–18:1, *c*9-14:1, 18:2n-6, and forage intake. However, as other studies used different milk FAs, the applicability of the equations across varied feeding conditions may be limited. Mohammed et al. (2011) concluded that the best predictive equation using milk FAs also includes other rumen fermentation (e.g. rumen VFA yield and protozoal classification) or production variables (e.g. milk yield and DMI). Whilst models allow assumptions to be made for LPS enteric CH_4 outputs with similar attributes, they are often over simplified and rely on the accuracy of the *in vivo*-published data. However, projecting forward the increase in the number of animal studies that record CH_4 emissions alongside feed intake, production, fermentation and microbiome characteristics aid the validation of predictive models for CH_4 production. In addition, as our knowledge develops on the rumen microorganisms associated with high- and low-emitting ruminants, machine learning methods can be employed to build classification models. These models, for instance, could be used to classify species as high or low emitters in ruminants. What is clear is that the research space and methods used to measure and predict CH_4 from ruminants will continue to evolve over the next decades and the learnings so far will play a critical role in its progression.

3 Grassland agriculture: a net zero livestock system

3.1 Net zero

According to the IPCC, LPS on managed pastures and rangelands account for just over 50% of total N_2O emissions from agriculture (IPCC, 2019). More specifically, RPS, particularly cattle farmed in intensive systems, is the most significant source of GHG emissions within the livestock sector, with cattle alone accounting for 62% of total global emissions - primarily caused by CH_4 emissions through enteric fermentation. Evidently, such gross emissions pose significant challenges for the livestock sector and place grassland agriculture, specifically intensive RPS, at the forefront of national and global targets to reduce absolute emissions.

Although mitigations such as changes in livestock diets and improved soil and manure management can offer significant reductions in GHG emissions for RPS, the fundamental meaning behind net zero by 2050 for the agri-food sector is the *net* carbon balance of emissions and sequestration (including storage) - particularly offered within grassland agriculture. Figure 5 illustrates the most significant emission sources from agriculture, for grassland agriculture this principally includes N_2O from soils, CH_4 from enteric fermentation and CO_2 from fibre (Grossi et al., 2019). Importantly, Fig. 5 also illustrates the potential sources of carbon offsets through sequestration (CO_2 removal via above-ground biomass - see Hoque, 2022) and storage (via soil organic carbon (SOC))

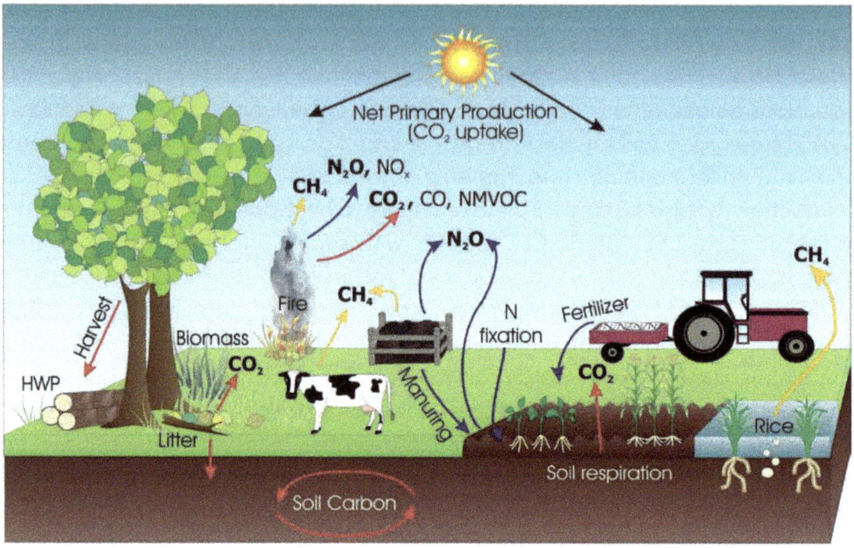

Figure 5 Emissions and carbon removals associated with farming. Source: IPCC, 2006.

that is particularly offered by grassland agriculture and recognised by both the IPCC and the UK Climate Change Committee (CCC) as integral to delivering net zero (CCC, 2020; IPCC, 2022).

3.2 Carbon farming

When considering farm carbon balance, for grassland agriculture, management practices that can improve carbon sequestration and storage include uptake of conservation, minimum or zero tillage practices, crop rotation, management of soil erosion and soil acidity, increasing the application of crop residues to soils as well as mulching to increasing SOC stocks, and improving timing and placement of nutrient (including organic and inorganic) applications using enhanced fertiliser application techniques (Hoque, 2022; McGuire et al., 2023; Paustian et al., 1997; Steinfeld et al., 2006). Improved management of pastures will also sequester carbon through the integration of plants, especially improved varieties such as increasing N_2-fixing legumes and multispecies swards (Conant et al., 2001; Skinner and Dell, 2016). More diverse pastures, such as grass-clover pastures also increase soil carbon stocks by increasing the root mass and depth (McNally et al., 2015). More specifically, 'carbon farming' within grassland agriculture can be an effective solution to combat the adverse environmental effects of production. Carbon farming refers to a collection of agricultural practices, such as silvopasture, agroforestry** and the

** Within the UK, agroforestry could sequester 6 $MtCO_2$-eq by 2050 (CCC, 2020).

use of perennial crops (Toensmeier, 2016). When such practices are combined, conservation tillage, growing cover crops, agroforestry and other measures have the potential to sequester 1.3 tonnes of carbon per hectare each year (Steinfeld et al., 2006).

Globally, grasslands represent a significant store of carbon due to the ability of grazing land to sequester CO_2 through photosynthetic assimilation by plants, carbon cycling and sequestration of carbon as SOC (Follett and Reed, 2010). Thus, SOC levels are high in grasslands compared to other ecosystems, with an estimated 650–810 gigatonnes of carbon stored globally (Gibson and Newman, 2019) - this equates to ~10–30% of global SOC (Follett and Reed, 2010, as cited by Zhou et al., 2019).

Yet, it is noteworthy that climatic conditions are vital in explaining the global differences in SOC stocks, as the biggest stores of SOC are found in more temperate zones, with low decomposition rates and high grassland productivity (Dondini et al., 2023). In contrast, arid and semi-arid climatic zones, described as having low productivity and high organic matter decomposition, have the lowest SOC stocks (Dondini et al., 2023). Thus, as temperatures continue to rise because of rising atmospheric concentrations of GHG emissions, the ability of our soils to sequester and store carbon may be reduced.

More broadly, the grassland biome is the world's largest terrestrial ecosystem, covering up to 40% of the world's ice-free land area (Hufkens et al., 2016, as cited by Zhou et al., 2019) - equating to 3.5 billion hectares of land (FAOSTAT, 2016). By region, Europe, Asia and parts of Sub-Saharan Africa, in particular, have a significant coverage of grassland, with each region contributing over 1.2 million hectare of grassland (see Fig. 6, FAOSTAT, 2022). In addition, Fig. 7 shows the ability of grassland to sequester carbon in different regions of the world - with Europe, the United Kingdom (UK) and Ireland each can remove 12.1 million gigagrams of CO_2 equivalents (CO_2-eq) from the atmosphere (FAOSTAT, 2022).

Within the literature, there is an unambiguous consensus that the grazing of pastures is a vital component of controlling global carbon processes (McSherry and Ritchie, 2013). For example, grassland areas with a reduced stocking density of grazing animals will increase both grass production and soil carbon sequestration (Holland et al., 1992; Chen et al., 2015; Gibson and Newman, 2019). More specifically, according to both Chang et al. (2021) and Reyes-Palomo et al. (2022), extensive LPS with reduced stocking rates or sparsely grazed grasslands can accelerate net climate cooling. Moreover, when this system incorporates agroforestry, GHG emissions can be offset by 54% compared to conventional farming systems and 95% of emissions from organic systems (Reyes-Palomo et al., 2022). On a global level, net sequestration from grazing lands equates to 0.3 gigatonnes of CO_2-eq per year, with improvements in grazing management and the use of legumes individually sequestering approximately 0.15 gigatonnes

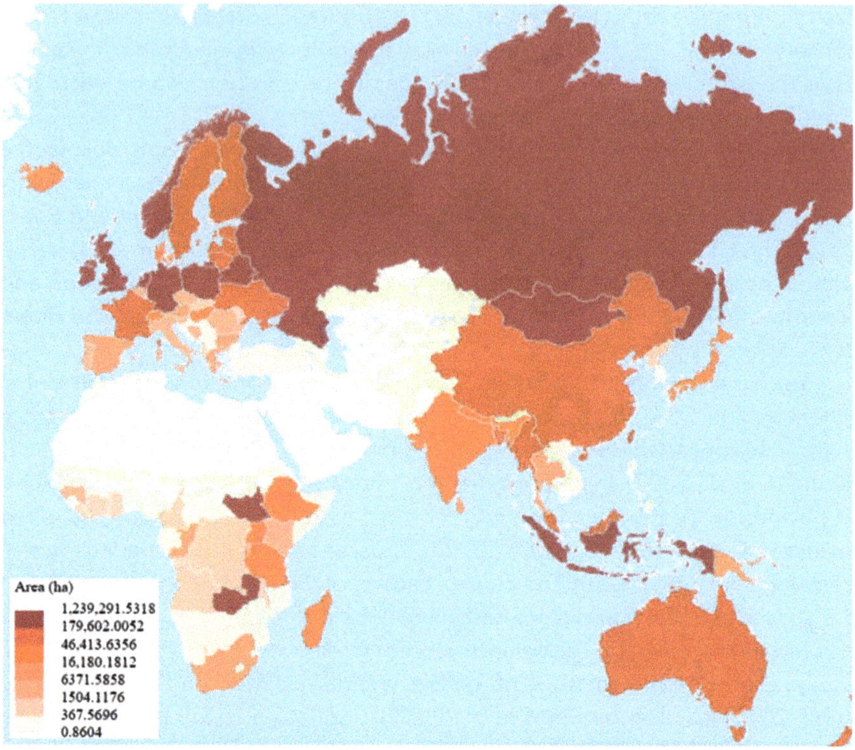

Figure 6 Area (ha) of global grassland. Source: FAOSTAT, 2022.

of CO_2-eq per year (Henderson et al., 2015). Crucially, through the adoption of practices such as agroforestry, rotational grazing and manure application, if the SOC content of grassland soils were to increase by 0.3% in the 0–30 cm depth layer, after 20 years of adoption, 0.3 tonnes of carbon per hectare could be sequestered annually (Dondini et al., 2023).

However, as temperatures continue to rise as the concentration of atmospheric GHGs increases, the ability of soils to sequester and store carbon will reduce (IPCC, 2019). Thus, it is imperative that mitigation measures to reduce CH_4 and N_2O from LPS coincide with efforts to promote and restore SOC.

3.3 Carbon accounting

3.3.1 IPCC guidelines for national greenhouse gas inventories

The IPCC has established guidelines and methodologies for measuring and reporting national inventories of GHG as well as GHG sinks (IPCC, 2006). A refinement to the global accepted methodologies was published in 2019 and is based on the most recent peer-reviewed scientific evidence. There are three

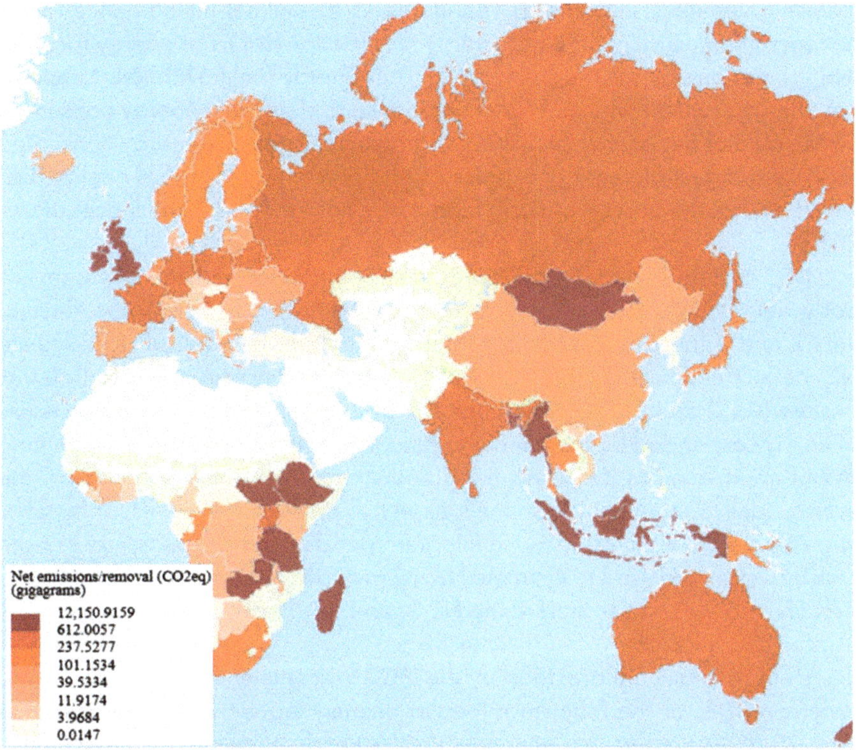

Figure 7 Net emissions/removal (CO_2-eq) from grassland (gigagrams). Source: FAOSTAT, 2022.

levels or tiers of the IPCC methodology - outputs of each are expressed as CO_2-eq. Tier 1 is a simplified method using global default values, while tier 2 is an intermediate method using calculated emissions factors (EF), and tier 3 is a more advanced and accurate model due to the data required and the complexity of the model (IPCC, 2019). For countries with large livestock populations and whose category emissions from livestock equate to a large portion of the total CH_4 emissions for the country, tier 2 models are recommended as best practice by the IPCC. Chapter 10 of the guidelines and refinement (IPCC, 2019) allows the estimation of GHGs from enteric fermentation and manure management, incorporating the GHGs CH_4 and N_2O.

To estimate total CH_4 emissions from enteric fermentation, it is necessary first to categorise livestock according to each species, age, type of production and sex (IPCC, 2019). At a minimum cattle should be grouped into mature dairy, other mature, and growing cattle. Using a tier 2 approach, emissions from enteric fermentation are estimated based on the gross energy (GE) requirements of the animal. Estimations of GE require the calculation of the following variables: net energy requirements for animal maintenance (NE_m), net energy for activity (NE_a), net energy for lactation (NE_l), net energy for work (NE_{work}), net energy for pregnancy (NE_p), net energy for growth (NE_g), ratio of net energy available in a diet for maintenance to digestible energy consumed (REM), ratio of net energy available for growth to digestible energy consumed (REG) and digestible energy expressed as a percentage of gross energy (DE) which reflects the portion of GE in feed that is not excreted in the faeces (IPCC, 2019).

NE_m is determined based on default coefficients and the average live body weight (BW) of all animals in the population (IPCC, 2019). For detailed tiers 2 and 3 models, the mean winter temperature is required. To calculate NE_a default coefficients are provided for each feeding situation; stall (0.00), pasture (0.17) and grazing large areas (0.36), which reflect the varying levels of energy expended by the animal to feed. NE_g is estimated using the mature BW of adult, animals (females, males and steers) in moderate condition, the average daily weight gain of animals, as well as default coefficients for females. The volume of milk production in kilogram per day and the fat content of the milk by weight are used to estimate NE_l, NEm and the number of hours worked per day and are used to determine NE_{work} and NE_m and a default coefficient is used to calculate NE_p.

Default values are provided by the IPCC for cattle and other ruminants to determine DE for the following: feedlot animals fed with >90% concentrate diet (75–85%), pasture-fed animals (55–75%) and animals fed on low-quality forage (45–55%). While these default values are provided, it is recommended that values for the predominate feed or forages be obtained from local or regional data where feasible and should incorporate seasonal variation as errors in estimating DE will result in overestimating emissions from the system (IPCC, 2019). For example, incorrectly measuring DE by 10% will inflate CH_4 emissions by 12–20% and emissions attributed to manure excretion by 20–45% (IPCC, 2019). Values for neutral detergent fibre (NDF), acid detergent fibre (ADF) and crude protein as well as anti-nutritional factors should also be obtained to allow for a prediction of DE (IPCC, 2019). In addition, to a range of default coefficients provided by the IPCC, DE is also used to calculate REM and REG.

In regards to feed, IPCC tier 1 and tier 2 methodologies do not account for diet composition in detail (IPCC, 2006, 2019). Estimated feed intakes are used to calculate the emissions attributed to the enteric fermentation of ruminant livestock, as well as CH_4 and N_2O from manure management (IPCC, 2019). Emissions are estimated from animals from representative animal categories.

GE is then calculated for cattle using the following equation (equation 10.16 of the IPCC methodology):

$$GE = \left\{ \frac{\left(\frac{NE_m + NE_a + NE_l + NE_{work} + NE_p}{REM} \right) + \left(\frac{NE_g + NE_{wool}}{REG} \right)}{\frac{DE\%}{100}} \right\}$$

Source: IPCC, Chapter 10, Emissions from livestock and manure management, 2019.

An EF for CH_4 emissions associated with enteric fermentation is then determined for each livestock category using GE, a CH_4 conversion factor (Y_m), and the factor 55.65, denoting the energy content of CH_4 (IPCC, 2019). Y_m reflects the rate at which feed energy is converted to CH_4. For dairy animals, default Y_m values are provided based on the productivity of the animals consuming feed with a certain DE and NDF, while values are also given for other cattle consuming a diet with more than 75% forage, 15–75% forage and feedlot diets (85–100% grain). EFs for each livestock category are multiplied by the number of animals in the category.

IPCC guidelines (IPCC, 2006, 2019) also define models to allow the estimation of CH_4 and N_2O from manure management. CH_4 emissions are calculated for each livestock species in each category. The calculation first requires the determination of the annual average volatile solids (VS) for each animal and productivity system, followed by the animal waste management system characteristics (AWMS) and the EF for CH_4 emissions from a specific manure management and productivity system (IPCC, 2019). Calculation of the annual VS per head is determined using values for the typical animal mass (TAM) for each livestock category and the VS rate, determined using default or country-specific values, derived from the estimations for GE, DE, urinary energy as a ratio of GE and ash content of the feed (IPCC, 2019). Default values are provided for AWMS and the EF; however, country-specific values can be used instead of default EFs, using daily VS, maximum CH_4-producing capacity (B0), and a CH_4 conversion factor (MCF) for the specific management system, as well as the AWMS values.

N_2O emissions from manure management fall into two categories: direct and indirect N_2O emissions. IPCC models for direct N_2O emissions at the country level are estimated again for each head of livestock in the category for a specific productivity system, the annual average nitrogen (N) excretion per head (NEX), annual nitrogen inputs, AWMS values and an EF for direct N_2O emissions. Tier 1 models use default values for many of the variables, while tier 2 models use country-specific values, and tier 3 implements an alternative country-specific methodology (IPCC, 2019). Indirect N_2O emissions are further split into emissions from the volatilisation of N and leaching associated with manure management. N_2O emissions from N volatilisation are calculated by first determining volatilisation, measured using the number of heads of livestock, the NEX, N added as co-digestates, AWMS and the fraction of managed manure N for the livestock category that volatilises as NH_3, nitric oxide (NO) and nitrogen dioxide (NO_2) ($Frac_{GasMS}$) and an EF estimating N_2O emissions from atmospheric deposition of nitrogen on soils and water surfaces (EF4) (IPCC, 2019). N losses associated with leaching are measured in the same way, except, the variable denoting $Frac_{GasMS}$ is replaced with $Frac_{LeachMS}$, which represents the fraction of managed N that is leached from the manure management system (IPCC, 2019).

IPCC methodologies also allow the estimation of N_2O emissions from managed soils and CO_2 from lime and urea application, which are important for grassland agriculture. These are also important determinants of the emissions associated with crop production for animal feed. Direct N_2O emissions are estimated by taking into account the amount of N applied to soils from synthetic fertilisers and organic fertilisers, such as manure, compost and wastewater, as well as urine and excrement from grazing animals. In addition, N from crop residues, such as N-fixing crops and forages, associated with pasture renewal are also important for the calculation of emissions. Management of soils, land use (LU) and LUC are also vital in determining N mineralisation (IPCC, 2019). To account for synthetic fertiliser, country-level statistics should be consulted. However, in their absence, data from the International Fertilizer Association (IFA) and Food and Agricultural Organization (FAO) should be used. N from urine or animal excrement from grazing animals should be taken into account based on the number of livestock and estimated average N per livestock head. Information on crop residues and forage pasture renewal should be estimated using crop yield (harvested dry matter yield per crop), default factors for above and below-ground residue, yield ratios and residue N content (IPCC, 2019). Data on crop yield can be retrieved using national sources or FAO databases if country-specific sources are available. Default values are reported for a range of crop types, such as wheat, barley, maize, legumes, forages and perennial grasses.

N mineralised due to losses in soil carbon through LU/LUC as well as management practices are also estimated using default values. Tier 1 models will use a single value for all LU and management systems, while disaggregated values by individual management systems are used for tier 2 approaches. Using a tier 1 approach, it is possible to estimate direct and indirect emissions of N_2O from soils; however, soil type, LU and detailed management systems are not considered (IPCC, 2019). In contrast, more detailed tier 2 and tier 3 methodologies should be considered. EFs are used to estimate N_2O emissions from managed soils (IPCC, 2019). Indirect N_2O emissions arise from N volatilisation or deposition (atmospheric N volatilised from managed soils) and N leaching (runoff N from synthetic and organic fertilisers and crop residues)

(IPCC, 2019). CO_2 emissions from liming and urea fertilisation are determined by multiplying annual amounts of calcic limestone or dolomite used per year by a default EF (IPCC, 2006). A similar approach is used for the estimation of CO_2 emissions from urea (IPCC, 2006).

IPCC methodologies also outline models to assess SOC stocks associated with LUC and tillage (IPCC, 1997, 2006, 2019). Using this approach, SOC stocks can be assessed for cropland, grassland and forestry using a range of LU, tillage and input factors to determine how changes in LU and management practices change SOC stocks (Sperow, 2020). Similarly, to the approach outlined above, there are three tiers of estimating SOC, tier 1 requires limited data, tier 2 required country-specific information to be provided and tier 3 requires higher resolution data (Sperow, 2020).

Using the IPCC framework and models, the UK Greenhouse Gas Inventory annual reports on emissions produced from each sector, including agriculture. The reporting of emissions by the inventory is measured at the country level; however, other tools exist to measure farm-level emissions using IPCC models. In the UK, over 60 carbon calculators are available to estimate on-farm emissions (Leinonen et al., 2019); however, most use tier 2 IPCC methodologies to estimate emissions. Calculators include the Cool Farm Tool, the Farm Carbon Toolkit, AgreCalc, Trinity AgTech Sandy, Map of AG, Intellync and AllTech E-CO_2; however, this list is not exhaustive. Both the UK inventory and carbon calculators use IPCC methodologies; however, the UK inventory does not include carbon sequestration in its assessment. Thus, the inventory assesses gross emissions from agriculture, whereas some of the carbon calculators, such as AgreCalc also measure carbon sequestration to reflect the net emissions of agriculture at the farm level.

3.3.2 Life-cycle assessment

The IPCC methodology is the globally accepted standard to measure GHG emissions; however, life-cycle assessment (LCA) is an environmental assessment method that is increasingly being used to assess GHGs, expressed as CO_2-eq (as part of the climate change impact category), as well as 15 other environmental impact areas (Guinée et al., 2011; Teillard et al., 2016; van der Werf et al., 2020). LCA uses a cradle-to-cradle approach to assess the environmental impacts of products from the raw material, processing, transportation, market, retail and consumption as well as the end of life as the system boundary (European Commission, 2021b; Muralikrishna and Manickam, 2017). The European Commission has also developed a standardised framework to allow the common measurement of environmental performance (EU Commission Recommendation 2021/2279). Based on this, updated Product Environmental

Footprint (PEF) guidelines were published in 2021 to provide guidance on LCA methodology to assess products and services.

The 16 environmental impact categories include climate change as GWP (GWP100), Kg CO_2-eq, ozone depletion potential, risk of human toxicity for cancer and non-cancers, particulate matter and the impact on human health, human exposure to ionising radiation, tropospheric ozone concentration increase, acidification, terrestrial, freshwater and marine eutrophication, freshwater ecotoxicity, LU, water use, minerals and metals resource use and fossils resource use (European Commission, 2021b).

3.3.3 Global Livestock Environmental Assessment Model

The Global Livestock Environmental Assessment Model (GLEAM) encompasses an LCA approach and the IPCC methodologies to determine the global emissions and environmental impacts of 11 livestock products: meat and milk from cattle, sheep, goats and buffalo as well as meat from pigs, and meat and eggs from poultry (FAO, 2022). GLEAM determines emissions from 'cradle-to-processing point' as the system boundary, including manure management, animal breeding and management, crop, and feed production, as well as feed processing and transport and inputs such as energy, fertilisers, pesticides and machinery use (FAO, 2022). The GLEAM model expresses emissions in terms of Kg CO_2-eq.

3.4 Measure, report and verification

The success of such sequestration and storage within grassland agriculture is fundamentally dependent on the adoption of a universal framework to monitor report and verify measure, report and verification (MRV) (Perosa et al., 2023). MRV is also acknowledged by the United Nations Framework Convention on Climate Change (UNFCCC) as a method for measuring and monitoring reductions in GHG emissions at the national level (UNFCCC, 2014). MRV is a necessary method to credit GHG reductions and obtain financial assistance as well as report on contributions to national emission reduction targets (Manzatto et al., 2018; Perosa et al., 2023).

For agriculture, one of the most significant MRV protocols available is illustrated in Fig. 8, the FAO's Global Soil Organic Carbon Map (GSOC), which sets out a protocol for measuring, reporting and verifying GHG emissions (FAO, 2020). More recently, The EU Commission's Sustainable Carbon Cycles (2021) establishes a framework to verify and provide certification of CO_2 removals in Europe, while also advocating for carbon farming to remove 310 tonnes of CO_2-eq from the land sector by 2030, as proposed by the EU Green Deal. Separately, the World Bank has also published an MRV sourcebook dedicated to SOC (World Bank, 2021). In addition, multiple SOC MRV protocols are

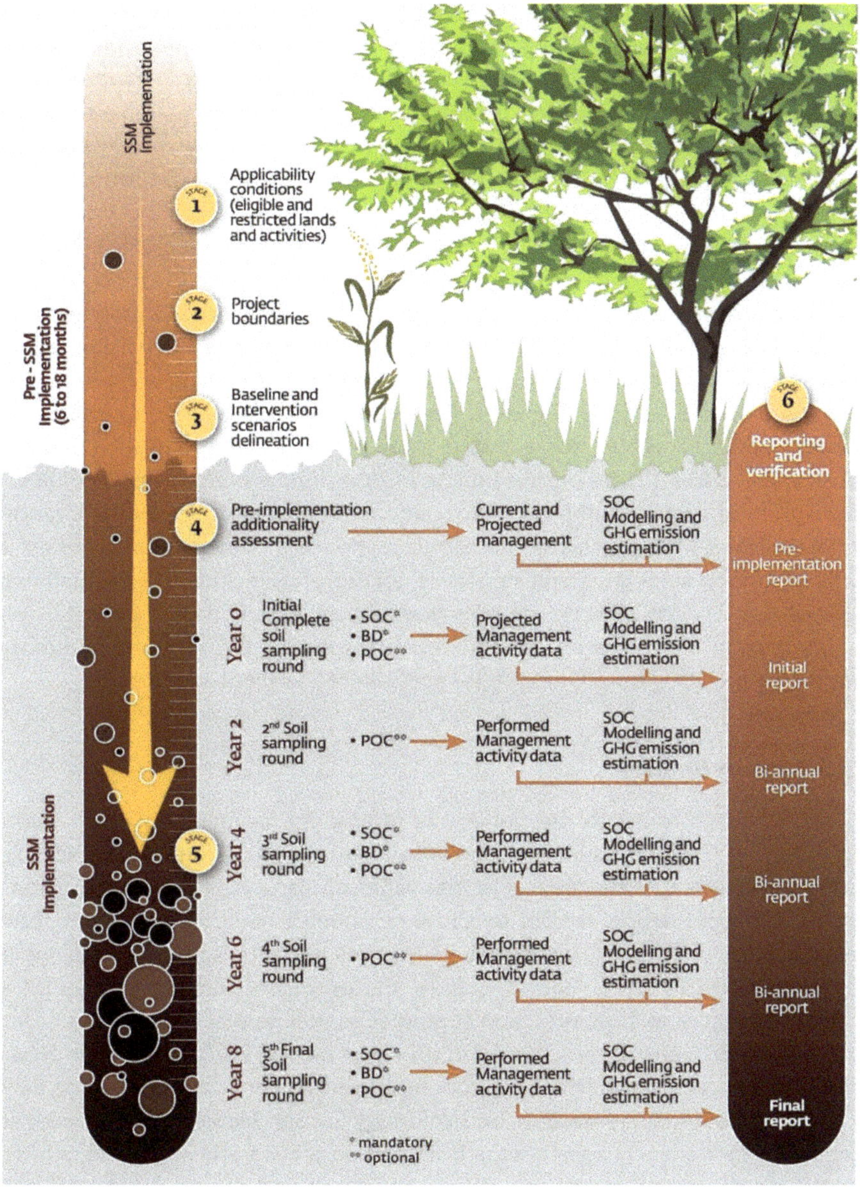

Figure 8 Measure, report and verification protocol. Source: GSOC MRV protocol (FAO, 2020).

currently in existence. These include CAR Soil Enrichment Protocol (CAR SEP); Verra Methodology for Improved Agricultural Land (VM0042); Verra Soil Carbon Quantification Methodology (VM0021); Verra Adoption of Sustainable Land Management (VM0017); Gold Standard Soil Organic Carbon Framework

Methodology (GS-SOC); Australian Carbon Credits (Carbon Farming Initiative Measurement of Soil Carbon Sequestration in Agricultural Systems and Carbon Farming Initiative-Estimating Sequestration of Carbon Using Default Values); Methodology Determination (AUS-SM, AUS-DV); Alberta Quantification Protocol for Conservation Cropping (Alberta CC); Regen Network Methodology for GHG and Co-Benefits in Grazing Systems and BCarbon Soil Carbon Credit Systems (Oldfield et al., 2021).

Such MRV protocols use a variation of methods to measure SOC and carbon stocks, ranging from physical soil sampling, and model-based approaches to remote sensing technology (Oldfield et al., 2021). Nonetheless, the inconsistencies between such protocols in how they quantify carbon stocks may lead to the creation of carbon credits within the agri-food sector that are not standardised across all systems. Consequently, this may lead to a scenario wherein on-farm carbon sequestration is not properly acknowledged (Oldfield et al., 2021) and will have significant implications for a robust carbon market for the agri-food sector. Furthermore, for nations such as the UK, despite a range of established MRV protocols and robust evidence that sequestration offers a vital mitigation strategy, sequestration is not currently included in the National GHG Inventory for agricultural emissions. Thus, at a national level, carbon sequestration and storage are not being accounted for, with the inventory reporting on gross emissions only for agriculture - and not *net* carbon.

3.5 Carbon finance

Carbon finance refers to the general term applied to financing which seeks to purchase GHG emission reductions to offset a nation's emissions (Kossoy, 2005). In essence, it places a financial value on carbon emission reductions in the form of carbon credits to allow companies to compensate for their emissions by adhering to emissions allowances or contributing to sustainable projects (UNHCR, 2014). For agriculture, the pricing of carbon will also be a necessary part of carbon reduction strategies as polluters pay for the emissions they generate within a carbon market (Burke et al., 2019). Currently, two types of carbon markets exist. These include the compliance market, such as the EU's Emissions Trading Scheme (ETS) and Voluntary Carbon Markets (VCMs), such as those within the land-based sector. Both market types trade in carbon credits, which equate to one metric ton of CO_2 or CO_2-eq and can be purchased and sold on the international market (Reichle 2020).

Using the UK as an example, VCMs for the agri-food sector are on the increase; however, according to the CCC (2022), current VCMs can produce carbon credits based on inaccurate claims and low prices, leading to limited reductions in emissions. The UK CCC suggests that if VCMs were highly regulated and reliable, carbon credits could be purchased, and this would, in turn, support the UK achieving its emission reduction targets. Yet, while VCMs may help investment reach areas that are currently underinvested, they may simultaneously dis-incentivise businesses from committing to emission reductions (CCC, 2022).

Within the UK, VCM, land-based carbon prices are calculated using Pending Issuance Unit (PIU) prices and not carbon credits verified protocols (CCC, 2022). PIUs are based on the anticipated delivery of emission reductions and as such are not guaranteed if these reductions are not met. Furthermore, PIUs range from £10 to £20 per ton of CO_2, while it is estimated that peat restoration and afforestation in the UK will restively cost £5–40 and £65–100 per ton of CO_2 in 2035 (CCC, 2022). To overcome this and to fund the negative emissions technology required to achieve net zero, according to Burke et al. (2019), a carbon price of £50 per tonne of CO_2 is recommended in 2020, £75 in 2030 and £160 in 2050. However, imposing a high carbon price will be difficult to achieve, thus, a medium-level price of £40 per tonne of CO_2 in 2020 and a price ranging from £100 to £125 in 2050 is proposed to generate £20 billion per year in revenue to help finance decarbonisation (Burke et al., 2019).

Outcome-based carbon finance is another approach that can be utilised to incentivise sustainable land use and food production at scale. This approach is dependent on outcomes being achieved. However, results-based payments would depend on consistent MRV protocols to ensure that on-farm carbon sequestration is accurately reflected in financial compensation. As mentioned previously, this is an ongoing issue as several MRV protocols exist and as such the quantification of carbon stocks is not standardised.

In relation to a compliance market, an example would be the EU's ETS. The EU ETS is governed by a legislative framework and is central to the EU'S policy directive on climate change (European Commission, 2022a). Since its installation in 2005, the EU ETS has reduced emissions from electricity, industrial production, and heat generation by over a third (European Commission, 2022a). The trading scheme operates on a cap-and-trade basis, in which a limit is placed on the amount of emissions that each operator can emit. Between July 2021 and January 2023, the price of carbon from the ETS has ranged from £42 to £97.75 per ton of CO_2 (Embe 2023). However, to date, the EU ETS does not extend to agriculture and fisheries.

3.6 Impact of climate change on milk and meat production

Agriculture and LPS can be both a contributor to GHG emissions and a solution to sequester carbon from the atmosphere. In addition, emissions from enteric fermentation can be mitigated using breeding and dietary interventions, as described in Section 2.3. However, the impacts of rising GHG emissions and the resultant increase in global temperatures (1.5°C above pre-industrial levels) will cause climate-related risks and hazards across the supply chain that will impact the production of milk and meat.

Climate change will impact the variability and availability of feed, combined with decreased productivity and nutritional value of feedstuffs. Yields of wheat, rice, soybean and temperature grasses may rise with increased atmospheric concentrations of CO_2; however, the quality of animal feed may also decline in terms of protein content (Godde et al., 2021). In addition, the yields of C4 plants such as maize, sorghum, sugarcane and tropical grasses will decline (Godde et al., 2021). The shifting of seasonal growing patterns of forages may also lead to a varying optimal growth rate (Rojas-Downing et al., 2017). Furthermore, as temperatures rise and conditions become hotter and drier, the availability of water will decline and will impact forage and crop yields and feed quality as well as increase heat stress amongst livestock (Wreford and Topp, 2020; Godde et al., 2021; Arnell and Freeman, 2021).

Increases in temperatures and disruptions to transport networks, as a result of extreme weather events will expose livestock to poorer transport conditions, such as higher humidity, limited ventilation and prolonged transport duration. That will result in heat stress and dehydration and thus reduced welfare and health outcomes (Godde et al., 2021). Heat stress will also impact forage intakes, the efficiency of feed conversion and nutritional utilisation of feeds, breeding, and changes to the lipid profile of livestock products, such as milk and ultimately body temperatures above 45°C, are fatal for many livestock species (Rojas-Downing et al., 2017; Escarcha et al., 2018).

In addition, climate change will result in an increase in vector-borne diseases from insects and endemic diseases in ruminants, increased persistence of pathogens and parasites, respiratory infections in poultry and mastitis in dairy breeds of cattle (Rojas-Downing et al., 2017; Escarcha et al., 2018; Wreford and Topp, 2020). Increased populations of agricultural pests and diseases will lead to a resultant decrease in productivity (Rojas-Downing et al., 2017; Fanzo et al., 2018; Skendžić et al., 2021). Water and land resources may also become scarce, and the quality of drinking water will decline, as well as increased competition for water access, which is compacted by the declining viability of pasture soils, as they become drier and less nutritious (Escarcha et al. 2018). Thus, leading to reduced animal welfare and life expectancy and ultimately mortality. Climate change will also cause a decline in milk yields, lower production efficiencies and reductions in growth and liveweights for ruminants (Escarcha et al., 2018). Climate change will also impact the quality of meat and milk products from livestock in terms of nutrition, eating quality, measured using carcass fatness, flavour, taste and tenderness in beef, and food safety, through the increase in carcass condemnations (Escarcha et al., 2018).

Increasing temperatures will impact weather patterns with increases in the frequency and severity of weather events, such as droughts and heat waves, as well as flooding becoming more commonplace (IPCC, 2018). For example, 80% of livestock deaths amongst pastoralist farmers in the Ethiopia-Kenya-Somalian border region are a result of extreme drought (Bogale and Erena, 2022). Climate projections for the UK suggest temperatures will increase by 2-5°C between 2080 and 2099 (Wreford and Topp, 2020). Extreme weather events, such as heat waves, flooding and storms will also impact labour availability and damage to infrastructure, such as roads and railway lines (Fanzo et al., 2018).

In terms of food distribution and storage, rising temperatures will increase food microorganisms (e.g. *E. coli* and *Salmonella*) and fungi, such as mycotoxins (e.g. aflatoxins) that result in the deterioration of food quality and shelf life, causing more food loss and waste (Mbow et al., 2017; Godde et al., 2021). Increased energy costs required to maintain the cold chain through distribution and storage (Godde et al., 2021) and a resultant decrease in food availability and increased food prices will limit the accessibility of food to consumers, especially households already food insecure (Fanzo et al., 2018). Costs, commodity prices and price volatility will also increase along the supply chain in response to climate change due to changes in costs and feed availability (Godde et al., 2021). At the consumer level, changing social norms, influenced by the role of livestock in increasing GHGs, as well as animal welfare, human health and wider environmental concerns may impact diets (Godfray et al., 2018; Godde et al., 2021). This is heightened by calls from the EAT Lancet report to reduce global meat consumption by 50% and the UK Committee on Climate Change to reduce consumption of beef, lamb and dairy products by 2050 (Willett et al., 2019; CCC, UK, 2020). Meat consumption is already on the decline in the UK, as average meat consumption has declined from 103.7 g per capita per day in 2008 to 86.3 g by 2019 (Stewart et al., 2021). An increase in the cost of meat and animal products, such as milk, may also influence alternative diets that are lower in livestock-related products (Gode et al., 2021).

3.7 Models and methods for estimating climate risks, vulnerability and impacts

Future risks from climatic change have long been estimated by the IPCC using experiments, analogies and models. These risks are then evaluated based on the interaction of these projected changes in the climate and environment with dimensions of vulnerability in societies and ecosystems. The experimental approach involves deliberately changing climatic factors affecting the subject of interest to reflect anticipated future conditions while holding all other factors constant. Spatial analogies, using examples from other parts of the world experiencing similar conditions to those anticipated in the future, and temporal analogies, using examples from the past, are used to make use of existing variations in the climate system to make inferences about the future.

Finally, models are numerical simulations of real-world systems which are then calibrated and validated by experiments and analogies like those mentioned above and run using future climate data. Scenario construction can also be used to describe possible future narratives such as the IPCC's representative concentration pathways (RCPs) for emissions scenarios and shared socio-economic pathways (SSPs). SSPs present a range of possible pathways that explore how global society, economics and demographics might change in the future and are incorporated into the IPCC's sixth assessment report as inputs for climate models. They offer five potential narratives which range from high levels of mitigation and adaptation implementation as well as lower GHG emissions to a future with no climate policy at all and unconstrained economic growth and energy use (Hausfather, 2018).

In relation to evaluating the level of risk or probability of a given impact, expert judgement is utilised by the IPCC to synthesise the diverse range of information available relating to the severity of the outcome. This is done by considering exposure and vulnerability in the context of a specific hazard and using specific criteria including but not limited to magnitude, irreversibility and timing of impacts, and limited potential of adaptation and mitigation. Risk levels are assessed as very low, low, medium, high or very high for the present timeframe, near-term future and long-term future. In relation to food security, the IPCC (2014) states that all aspects of food security are potentially affected by climate change, which includes production, access, use and price (high confidence). There is also a high level of confidence that rural areas will experience major impacts on water availability and supply, infrastructure and agricultural incomes including shifts in food production. These impacts will disproportionately affect people in poorer economic conditions such as female-headed households and those with limited access to resources and education or smallholder farms.

However, the subject of climate-related risks in the agriculture sector is only explored to a limited extent in global reports such as those produced by the IPCC (Godde et al., 2021). For instance, while it is clear that climate change will have significant impacts on the agriculture sector right across the supply chain from production to consumption, what is less clear is the nature, extent and magnitude of these risks, particularly in the livestock sector, which has been less widely explored in the literature. In addition, rates of change in global socio-economic and environmental drivers, as well as climatic variability and climatic extremes make adaptation choices difficult. Furthermore, the vulnerability and adaptive capacity of individual farming enterprises are not considered. The diversity among enterprises even within the same region in terms of resources available and changing social interactions with livestock production make it difficult to ascertain the level of risk and vulnerability to climate impacts without specific case studies. It is impossible to predict the future of a complex, socially, economically and environmentally interconnected system such as the agricultural system. Given the large uncertainties in climate projections and the range of potential future pathways outlined in the SSPs, adaptation strategies need to be carefully considered and will likely require diverse methods that cater to differences in geography, farming systems, climatic conditions, farm size and the wider socio-economic environment.

4 Conclusion and future trends

LPS play a key role in global food security (including human nutritional value), but their impact on climate change is clear. Despite the scale of this challenge, we find that there are two principle mitigations that must be implemented at national, regional and global scales: enteric CH_4 mitigation and carbon sequestration and storage.

In summary, we find that breeding and dietary innovations offer a major technological advancement that can reduce CH_4 emissions by up to 30%. A number of rumen microbial genes are linked to low CH_4 emissions, these may be inheritable, indicating the possibility of breeding animals with a 'utopian' microbiome. Dietary interventions enabling reductions in CH_4 emissions can be grouped into the following categories: plant-based strategies (e.g. tannins), multispecies swards, targeted CH_4 inhibitors (e.g. 3-NOP), lipids and hydrogen sinks. What is more, the addition of macroalgae, has the potential to offset rumen CH_4 emissions by 10–80%. The use of direct-fed DFMs may also offer an effective solution to reduce CH_4 production, however, more research is required in this area. In terms of the mitigation of nitrous oxide, emphasis is now being placed on evaluating local sources of feed protein obtained in a sustainable manner, feeding ruminants at levels that are optimum for their health and production, and minimising unnecessary soil nitrogen deposition.

Coupled with the above, LPS and RPS, particularly grassland agriculture, have a unique position within net carbon emissions. Grasslands represent a significant store of carbon, and more extensive LPS characterised by reduced stocking rates can accelerate climate cooling. Most importantly, this value will not be utilised until a universal framework for MRV carbon stocks is adopted. Therefore, agricultural sequestration is not currently included in many National GHG Inventories for scope 3 emissions. Thus, coupled with the potential of enteric CH_4 mitigations, we find that it is crucial that national inventories move beyond reporting gross emissions, and implement robust MRV protocols for carbon sequestration and storage, and to ensure net carbon emissions are calculated. Therefore, due to the predicted impacts that climate change will have on LPS and RPS, especially in regard to meat and milk production, it is imperative that the sector mitigates effectively and adapts urgently to climate change.

5 Where to look for further information

5.1 Further reading

- IPCC Assessment Reports about knowledge on climate change, its causes, potential impacts and response options.
- A good review of global ruminant agricultural challenges see Huws, S. A., Creevey, C. J., Oyama, L. B., Mizrahi, I., Denman, S. E., Popova, M., Muñoz-Tamayo, R., Forano, E., Waters, S. M., Hess, M. and Tapio, I. 2018. Addressing global ruminant agricultural challenges through understanding the rumen microbiome: past, present, and future. Frontiers in microbiology, 9, p.2161.
- For a good explanation of the current impacts of LPS on climate change, see FAO, GLEAM 2.0. 2017. Assessment of GHG emissions and mitigation potential. (Food and Agriculture Organization of the United Nations). https://www.fao.org/gleam/results/en/.
- For further reading on the role of grassland agriculture, see Follett, R. F. and Reed, D. A. (2010). Soil Carbon Sequestration in Grazing Lands: Societal Benefits and Policy Implications. *Rangeland Ecology & Management*, 63(1), 4–15. https://doi.org/10.2111/08-225.1.

5.2 Key journals and conferences

- Journal of Agricultural Economics and the *American Journal of Agricultural Economics* for peer-reviewed research relating agriculture, natural resource, and environmental economics.
- Journal of Food Security addresses the challenges of achieving food security globally.
- GGAA, the International Greenhouse Gas and Animal Agriculture Conference, is primary venue for scientists to present the latest research on the mitigation of GHGs, measurement and modelling.
- European Federation of Animal Science aims to improve the knowledge and the dissemination of advancement of animal science.

5.3 Key research projects

- Horizon Europe funded project, Re-Livestock, which focuses on understanding and mobilising the adoption of innovative practices to reduce GHG emissions from livestock farming (https://re-livestock.eu/).
- Another key Horizon Europe funded project is Pathways. Pathways for sustainable food aims to identify and increase sustainable practices along European supply and production chains (https://pathways-project.com/).

- Other notable projects include HoloRuminant (https://holoruminant.eu/), SeaSolutions (https://seasolutions.ie/), and Modelling Agriculture with Climate Change for Food Security Science-Policy Project (MACSUR Sci-Pol) (https://macsur.es/).

6 References

Abbott, D. W., Aasen, I. M., Beauchemin, K. A., Grondahl, F., Gruninger, R., Hayes, M., Huws, S., Kenny, D. A., Krizsan, S. J., Kirwan, S. F., Lind, V., Meyer, U., Ramin, M., Theodoridou, K., von Soosten, D., Walsh, P. J., Waters, S. and Xing, X. (2020). Seaweed and seaweed bioactives for mitigation of enteric methane: Challenges and opportunities. *Animals (Basel)* 10(12), 2432. https://doi.org/10.3390/ani10122432.

Adams, J., Kennedy, A., Cotton, J. and Brumby, S. (2022). Child farm-related injuries-the unique challenges of the farming workspace. *Safety and Health at Work* 13, S256.

Alston, J. M. and Pardey, P. G. (2014). Agriculture in the global economy. *Journal of Economic Perspectives* 28(1), 121–146.

Arndt, C., Hristov, A. N., Price, W. J., McClelland, S. C., Pelaez, A. M., Cueva, S. F., Oh, J., Dijkstra, J., Bannink, A., Bayat, A. R., Crompton, L. A., Eugène, M. A., Enahoro, D., Kebreab, E., Kreuzer, M., McGee, M., Martin, C., Newbold, C. J., Reynolds, C. K., Schwarm, A., Shingfield, K. J., Veneman, J. B., Yáñez-Ruiz, D. R. and Yu, Z. (2022). Full adoption of the most effective strategies to mitigate methane emissions by ruminants can help meet the 1.5°C target by 2030 but not 2050. *Proceedings of the National Academy of Sciences of the United States of America* 119(20), e2111294119. https://doi.org/10.1073/pnas.2111294119.

Arnell, N. W. and Freeman, A. (2021). The effect of climate change on agro-climatic indicators in the UK. *Climatic Change* 165(1–2), 40. https://doi.org/10.1007/s10584-021-03054-8.

Attwood, G. T., Leahy, S. C. and Kelly, W. J. (2020). The rumen Archaea. In: *Improving Rumen Function*. Burleigh Dodds Publishing, UK, pp. 133–190.

Baldwin, R. L., Koong, L. J. and Ulyatt, M. J. (1977). A dynamic model of ruminant digestion for evaluation of factors affecting nutritive value. *Agricultural Systems* 2(4), 255–288. https://doi.org/10.1016/0308-521X(77)90020-8.

Bauchop, T. and Mountfort, D. O. (1981). Cellulose fermentation by a rumen anaerobic fungus in both the absence and the presence of rumen methanogens. *Applied and Environmental Microbiology* 42(6), 1103–1110. https://doi.org/10.1128/aem.42.6.1103-1110.1981.

Beauchemin, K. A., Ungerfeld, E. M., Eckard, R. J. and Wang, M. (2020). Review: Fifty years of research on rumen methanogenesis: Lessons learned and future challenges for mitigation. *Animal* 14(S1), S2–S16. https://doi.org/10.1017/S1751731119003100.

Belanche, A., de la Fuente, G. and Newbold, C. J. (2014). Study of methanogen communities associated with different rumen protozoal populations. *FEMS Microbiology Ecology* 90(3), 663–677. https://doi.org/10.1111/1574-6941.12423.

Belanche, A., de la Fuente, G. and Newbold, C. J. (2015). Effect of progressive inoculation of fauna-free sheep with holotrich protozoa and total-fauna on rumen fermentation, microbial diversity and methane emissions. *FEMS Microbiology Ecology* 91(3), fiu026.

Bogale, G. A. and Erena, Z. B. (2022). Drought vulnerability and impacts of climate change on livestock production and productivity in different agro-Ecological zones of Ethiopia. *Journal of Applied Animal Research* 50(1), 471–489. https://doi.org/10.1080/09712119.2022.2103563.

Boland, T. M., Quinlan, C., Pierce, K. M., Lynch, M. B., Kenny, D. A., Kelly, A. K. and Purcell, P. J. (2013). The effect of pasture pregrazing herbage mass on methane emissions, ruminal fermentation, and average daily gain of grazing beef heifers. *Journal of Animal Science* 91(8), 3867–3874.

Brashears, M. M., Amezquita, A. and Jaroni, D. (2005). Lactic acid bacteria and their uses in animal feeding to improve food safety. *Advances in Food and Nutrition Research* 50, 1–31. https://doi.org/10.1016/S1043-4526(05)50001-9.

Burke, J., Byrnes, R. and Fankhauser, S. (2019). *How to Price Carbon to Reach Net-Zero Emissions in the UK*. Grantham Research Institute on Climate Change and the Environment and Centre for Climate Change Economics and Policy, London School of Economics and Political Science, London.

Cain, M., Lynch, J., Allen, M. R., Fuglestvedt, J. S., Frame, D. J. and Macey, A. H. (2019). Improved calculation of warming-equivalent emissions for short-lived climate pollutants. *NPJ Climate and Atmospheric Science* 2(1), 29. https://doi.org/10.1038/s41612-019-0086-4.

Cao, Y., Takahashi, T., Horiguchi, K. and Yoshida, N. (2010a). Effect of adding lactic acid bacteria and molasses on fermentation quality and *in vitro* ruminal digestion of total mixed ration silage prepared with whole crop rice. *Grassland Science* 56(1), 19–25. https://doi.org/10.1111/j.1744-697X.2009.00168.x.

Cao, Y., Takahash, I. T., Horiguchi, K., Yoshida, N. and Cai, Y. (2010b). Methane emissions from sheep fed fermented or non-fermented total mixed ration containing whole-crop rice and rice bran. *Animal Feed Science Technology* 157, 72–78. https://doi.org/10.1016/j.anifeedsci.2010.02.004.

Chang, J., Ciais, P., Gasser, T., Smith, P., Herrero, M., Havlík, P., Obersteiner, M., Guenet, B., Goll, D. S., Li, W., Naipal, V., Peng, S., Qiu, C., Tian, H., Viovy, N., Yue, C. and Zhu, D. (2021). Climate warming from managed grasslands cancels the cooling effect of carbon sinks in sparsely grazed and natural grasslands. *Nature Communications* 12(1), 118.

Chen, W., Huang, D., Liu, N., Zhang, Y., Badgery, W. B., Wang, X. and Shen, Y. (2015). Improved grazing management may increase soil carbon sequestration in temperate steppe. *Scientific Reports* 5(1), 10892. https://doi.org/10.1038/srep10892.

Cheng, Y. F., Edwards, J. E., Allison, G. G., Zhu, W. Y. and Theodorou, M. K. (2009). Diversity and activity of enriched ruminal cultures of anaerobic fungi and methanogens grown together on lignocellulose in consecutive batch culture. *Bioresource Technology* 100(20), 4821–4828. https://doi.org/10.1016/j.biortech.2009.04.031.

Chilliard, Y., Martin, C., Rouel, J. and Doreau, M. (2009). Milk fatty acids in dairy cows fed whole crude linseed, extruded linseed, or linseed oil, and their relationship with methane output. *Journal of Dairy Science* 92(10), 5199–5211. https://doi.org/10.3168/jds.2009-2375.

CIEL. (2022). Net zero carbon & UK livestock report. Available at: https://www.cielivestock.co.uk/expertise/net-zero-carbon-uk-livestock/report-april-2022/.

Committee on Climate Change. (2020). Land use: Policies for a net zero UK. Available at: https://www.theccc.org.uk/wp-content/uploads/2020/01/Land-use-Policies-for-a-Net-Zero-UK.pdf. UK.

Committee on Climate Change. (2022). Voluntary carbon markets and offsetting. Available at: https://www.theccc.org.uk/publication/voluntary-carbon-markets-and-offsetting/.

Conant, R. T., Paustian, K. and Elliot, E. T. (2001). Grassland management and conversion into grassland: Effects on soil carbon. *Ecological Applications* 11(2), 343–355.

Dai, X., Kalscheur, K. F., Huhtanen, P. and Faciola, A. P. (2022). Effects of ruminal protozoa on methane emissions in ruminants–A meta-analysis. *Journal of Dairy Science* 105(9), 7482–7491. https://doi.org/10.3168/jds.2021-21139.

de Haas, Y., Pszczola, M., Soyeurt, H., Wall, E. and Lassen, J. (2017). Invited review: Phenotypes to genetically reduce greenhouse gas emissions in dairying. *Journal of Dairy Science* 100(2), 855–870. https://doi.org/10.3168/jds.2016-11246.

Department of Agriculture, Environment and Rural Affairs and N. I. (2017). *Northern Ireland Expert Working Group, Delivering Our Future, Valuing Our Soils: A Sustainable Agricultural Land Management Strategy for Northern Ireland*. Department of Agriculture, Environment and Rural Affairs, Belfast.

Difford, G. F., Plichta, D. R., Løvendahl, P., Lassen, J., Noel, S. J., Højberg, O., Wright, A. G., Zhu, Z., Kristensen, L., Nielsen, H. B., Guldbrandtsen, B. and Sahana, G. (2018). Host genetics and the rumen microbiome jointly associate with methane emissions in dairy cows. *PLoS Genetics* 14(10), e1007580. https://doi.org/10.1371/journal.pgen.1007580.

Dondini, M., Martin, M., De Camillis, C., Uwizeye, A., Soussana, J.-F., Robinson, T. and Steinfeld, H. (2023). Global assessment of soil carbon in grasslands - From current stock estimates to sequestration potential. In: *FAO Animal Production and Health Paper*. Food and Agriculture Organization, Rome, p. 187. https://doi.org/10.4060/cc3981en.

Dseagu, V. L. Z., Wiseman, M. J., Allen, K., Buttriss, J. and Williams, C. (2022). Troubling assumptions behind GBD 2019 on the health risks of red meat. *The Lancet* 400(10350), 427–428.

Embe. (2023). EU carbon price tracker. Available at: https://ember-climate.org/data/data-tools/carbon-price-viewer/.

Enriquez-Hidalgo, D., Gilliland, T., Deighton, M. H., O'Donovan, M. O. and Hennessy, D. (2014). Milk production and enteric methane emissions by dairy cows grazing fertilized perennial ryegrass pasture with or without inclusion of white clover. *Journal of Dairy Science* 97(3), 1400–1412. http://doi.org/10.3168/jds.2013-7034.

Escarcha, J. F., Lassa, J. A. and Zander, K. K. (2018). Livestock under climate change: a systematic review of impacts and adaptation. *Climate* 6(3), 1–17. https://doi.org/10.3390/cli6030054.

Eugène, M., Klumpp, K. and Sauvant, D. (2021). Methane mitigating options with forages fed to ruminants. *Grass and Forage Science* 76(2), 196–204. https://doi.org/10.1111/gfs.12540.

European Commission. (2018). *Agricultural and Farm Income*. European Commission DG Agriculture and Rural Development, Brussels.

European Commission. (2021a). Carbon removal certification. Available at: https://climate.ec.europa.eu/eu-action/sustainable-carbon-cycles/carbon-removal-certification_en.

European Commission. (2021b). *Understanding Product Environmental Footprint and Organisation Environmental Footprint Methods*.

European Commission. (2022a). Report from the Commission to the European Parliament and the Council on the Functioning of the European carbon market in 2021 pursuant to Articles. Available at: https://eur-lex.europa.eu/legal-content/EN/TXT/?uri=COM:2022:516:FIN (vol. 10)(5) and 21(2) of Directive 2003/87. European Commission (as amended by Directive 2009/29. *EC and Directive (EU)* 2018/410).

European Commission. (2022b). Inspirational idea: Using data to manage environmental impact of livestock farming. Available at: https://ec.europa.eu/eip/agriculture/en/news/inspirational-idea-using-data-manage-environmental.html (accessed 27 April 2023).

Fanzo, J., Davis, C., McLaren, R. and Choufani, J. (2018). The effect of climate change across food systems: Implications for nutrition outcomes. *Global Food Security* 18, 12–19. https://doi.org/10.1016/j.gfs.2018.06.001.

FAO. (2014). Online: Accessed Food and Agriculture Organisation of the United Nations. 17/1/2018]. In: *Food Outlook: Biannual Report on Global Food Markets,* October, Rome. Available at: http://www.fao.org/3/a-i4136e.pdf.

FAO. (2020). *A Protocol for Measurement, Monitoring, Reporting and Verification of Soil Organic Carbon in Agricultural Landscapes – GSOC-MRV Protocol*. Rome. https://doi.org/10.4060/cb0509en.

FAO. (2022). Global livestock environmental assessment model: Model description version 3.0. Available at: https://www.fao.org/fileadmin/user_upload/gleam/docs/GLEAM_3.0_Model_description.pdf.

FAO. (2023). Global livestock environmental assessment model (GLEAM). Available at: https://www.fao.org/gleam/dashboard-old/en/ (accessed 20 March 2023).

FAOSTAT. (2016). *FAOSTAT Database*. Food and Agriculture Organization, Rome. Available at: www.fao.org/faostat/en/#home (accessed 27 July 2022).

FAOSTAT. (2022a). Grassland (global - national - annual - FAOSTAT). *Net Emissions/Removal*. Available at: https://data.apps.fao.org/ (accessed 17 March 2023).

FAOSTAT. (2022b). Grassland (global - national - annual - FAOSTAT). *Area (HA)*. Available at: https://data.apps.fao.org/ (accessed 17 March 2023).

FAO, G. 2.0. (2017). *Assessment of Greenhouse Gas Emissions and Mitigation Potential*. Food and Agriculture Organization of the United Nations. Available at: https://www.fao.org/gleam/results/en/.

FAO, I. F. A. D., UNICEF, WFP and WHO. (2022). *The State of Food Security and Nutrition in the World 2022. Repurposing Food and Agricultural Policies to Make Healthy Diets More Affordable*. Food and Agriculture Organization, Rome. https://doi.org/10.4060/cc0639en.

Follett, R. F. and Reed, D. A. (2010). Soil carbon sequestration in grazing lands: Societal benefits and policy implications. *Rangeland Ecology and Management* 63(1), 4–15. https://doi.org/10.2111/08-225.1.

Fox, D. G., Sniffen, C. J., O'Connor, J. D., Russell, J. B. and Van Soest, P. J. (1992). A net carbohydrate and protein system for evaluating cattle diets: III. Cattle requirements and diet adequacy. *Journal of Animal Science* 70(11), 3578–3596.

Franco, M., Roleda, M., Aasen, I. M., Hetta, M. and Steinshamn, H. (2018). *In vitro* evaluation of utilisable crude protein and methane production for a diet in which grass silage was replaced by different levels and fractions of extracted seaweed proteins. *Animal Feed Science and Technology* 255, 114225.

Gibson, D. J. and Newman, J. A. (Eds.). (2019). *Grasslands and Climate Change*. Cambridge University Press, Cambridge.

Godde, C. M., Mason-D'Croz, D., Mayberry, D. E., Thornton, P. K. and Herrero, M. (2021). Impacts of climate change on the livestock food supply chain; a review of the evidence. *Global Food Security* 28, 100488. https://doi.org/10.1016/j.gfs.2020.100488.

Godfray, H. C. J., Aveyard, P., Garnett, T., Hall, J. W., Key, T. J., Lorimer, J., Pierrehumbert, R. T., Scarborough, P., Springmann, M. and Jebb, S. A. (2018). Meat consumption, health, and the environment. *Science* 361(6399), eaam5324. https://doi.org/10.1126/science.aam5324.

Goopy, J. P., Donaldson, A., Hegarty, R., Vercoe, P. E., Haynes, F., Barnett, M. and Oddy, V. H. (2014). Low-methane yield sheep have smaller rumens and shorter rumen retention time. *British Journal of Nutrition* 111(4), 578–585. https://doi.org/10.1017/S0007114513002936.

Grainger, C. and Beauchemin, K. A. (2011). Can enteric methane emissions from ruminants be lowered without lowering their production? *Animal Feed Science and Technology* 166–167, 308–320. https://doi.org/10.1016/j.anifeedsci.2011.04.021.

Grossi, G., Goglio, P., Vitali, A. and Williams, A. G. (2019). Livestock and climate change: Impact of livestock on climate and mitigation strategies. *Animal Frontiers: The Review Magazine of Animal Agriculture* 9(1), 69–76.

Guinée, J. B., Heijungs, R., Huppes, G., Zamagni, A., Masoni, P., Buonamici, R., Ekvall, T. and Rydberg, T. (2011). Life cycle assessment: Past, present, and future. *Environmental Science and Technology* 45(1), 90–96. https://doi.org/10.1021/es101316v.

Haisan, J., Sun, Y., Guan, L., Beauchemin, K. A., Iwaasa, A., Duval, S., Kindermann, M., Barreda, D. R. and Oba, M. (2017). The effects of feeding 3-nitrooxypropanol at two doses on milk production, rumen fermentation, plasma metabolites, nutrient digestibility, and methane emissions in lactating Holstein cows. *Animal Production Science* 57(2), 282–289.

Hausfather, Z. (2018). How 'shared socio-economic pathways' explore future climate change. Available at: https://www.carbonbrief.org/explainer-how-shared-socioeconomic-pathways-explore-future-climate-change/ (accessed 27 April 2023).

Henderson, B. B., Gerber, P. J., Hilinski, T. E., Falcucci, A., Ojima, D. S., Salvatore, M. and Conant, R. T. (2015). Greenhouse gas mitigation potential of the world's grazing lands: Modeling soil carbon and nitrogen fluxes of mitigation practices. *Agriculture, Ecosystems and Environment* 207, 91–100. https://doi.org/10.1016/j.agee.2015.03.029.

Henderson, G., Cox, F., Ganesh, S., Jonker, A., Young, W., Global Rumen Census Collaborators and Janssen, P. H. (2015). Rumen microbial community composition varies with diet and host, but a core microbiome is found across a wide geographical range. *Scientific Reports* 5, 14567. https://doi.org/10.1038/srep14567.

Hill, J., McSweeney, C., Wright, A.-D. G., Bishop-Hurley, G. and Kalantar-Zadeh, K. (2016). Measuring methane production from ruminants. *Trends in Biotechnology* 34(1), 26–35. https://doi.org/10.1016/j.tibtech.2015.10.004.

Holland, E. A., Parton, W. J., Detling, J. K. and Layne, D. L. (1992). Physiological responses of plant populations to herbivory and their consequences for ecosystem nutrient flow. *The American Naturalist* 140(4), 685–706.

Hoque, M. (2022). Way forward for sustainable livestock sector. In: Akash, Mondal S. and Adusumilli, S. (Eds.), *Emerging Issues in Climate Smart Livestock Production*. Elsevier, pp. 473–488. https://doi.org/10.1016/B978-0-12-822265-2.00016-8.

Hristov, A. N., Oh, J., Giallongo, F., Frederick, T. W., Harper, M. T., Weeks, H. L., Branco, A. F., Moate, P. J., Deighton, M. H., Williams, S. R., Kindermann, M. and Duval, S. (2015). An inhibitor persistently decreased enteric methane emission from dairy cows with no negative effect on milk production. *Proceedings of the National Academy of Sciences of the United States of America* 112(34), 10663–10668. https://doi.org/10.1073/pnas.1504124112.

Hufkens, K., Keenan, T. F., Flanagan, L. B., Scott, R. L., Bernacchi, C. J., Joo, E., Brunsell, N. A., Verfaillie, J. and Richardson, A. D. (2016). Productivity of North American grasslands is increased under future climate scenarios despite rising aridity. *Nature Climate Change* 6(7), 710–714. https://doi.org/10.1038/nclimate2942.

Huws, S. A., Creevey, C. J., Oyama, L. B., Mizrahi, I., Denman, S. E., Popova, M., Muñoz-Tamayo, R., Forano, E., Waters, S. M., Hess, M., Tapio, I., Smidt, H., Krizsan, S. J., Yáñez-Ruiz, D. R., Belanche, A., Guan, L., Gruninger, R. J., McAllister, T. A., Newbold, C. J., Roehe, R., Dewhurst, R. J., Snelling, T. J., Watson, M., Suen, G., Hart, E. H., Kingston-Smith, A. H., Scollan, N. D., do Prado, R. M., Pilau, E. J., Mantovani, H. C., Attwood, G. T., Edwards, J. E., McEwan, N. R., Morrisson, S., Mayorga, O. L., Elliott, C. and Morgavi, D. P. (2018). Addressing global ruminant agricultural challenges through understanding the rumen microbiome: Past, present, and future. *Frontiers in Microbiology* 9, 2161.

Huws, S. A., Edwards, J. E., Lin, W., Rubino, F., Alston, M., Swarbreck, D., Caim, S., Stevens, P. R., Pachebat, J., Won, M. Y., Oyama, L. B., Creevey, C. J. and Kingston-Smith, A. H. (2021). Microbiomes attached to fresh perennial ryegrass are temporally resilient and adapt to changing ecological niches. *Microbiome* 9(1), 143. https://doi.org/10.1186/s40168-021-01087-w.

IPCC. (1997). Revised 1996. In: Houghton, J. T., Meria Filho, L. G., Lim, B., Treanton, K., Mamaty, I., Bonduki, Y., Griggs, D. J. and Callender, B. A. (Eds.), *IPCC Guidelines for National Greenhouse Gas Inventories Reference Manual* (Vol. 3). Intergovernmental Panel on Climate Change.

IPCC. (2006). *IPCC Guidelines for National Greenhouse Gas Inventories, Prepared by the National Greenhouse Gas Inventories Programme, Published* Eggleston, H. S., Buendia, L., Miwa, K., Ngara, T. and Tanabe, K. (Eds.). (Vol. 2006). Institute for Global Environmental Strategies, Japan.

IPCC. (2014). Synthesis report. Contribution of working groups I, II and III to the fifth assessment report of the Intergovernmental Panel on Climate Change. In: Core Writing Team, Pachauri, R. K. and Meyer, L. A. (Eds.), *Climate Change*. IPCC, Geneva, Switzerland, p. 151.

IPCC. (2018). An IPCC Special Report on the impacts of global warming of 1.5°C above pre-industrial levels and related global greenhouse gas emission pathways, in the context of strengthening the global response to summary for policymakers. Geneva.

IPCC. (2019). Summary for policymakers. In: Skea, J., Calvo Buendia, E., Masson-Delmotte, V., Pörtner, H.-O., Roberts, D. C., Zhai, P., Slade, R., Connors, S., van Diemen, R. and Ferrat, M. (Eds.), *Climate Change and Land: An IPCC Special Report on Climate*

Change, Desertification, Land Degradation, Sustainable Land Management, Food Security, and Greenhouse Gas Fluxes in Terrestrial Ecosystems [P. R. Shukla].
IPCC. (2021). The physical science basis: Sixth assessment. In: *Climate Change*. Available at: https://www.ipcc.ch/report/ar6/wg1/.
IPCC. (2022). Impacts, adaptation and vulnerability. Contribution of working Group II to the sixth assessment report of the Intergovernmental Panel on Climate Change. In: Pörtner, H.-O., Roberts, D. C., Tignor, M., Poloczanska, E. S., Mintenbeck, K., Alegría, A., Craig, M., Langsdorf, S., Löschke, S. and Möller, V. (Eds.), *Climate Change*. Cambridge University Press, Cambridge, UK and New York, NY, p. 3056. https://doi.org/10.1017/9781009325844.
IPCC, Gavrilova, O., Leip, A., Dong, H., MacDonald, J. D., Bravo, C. A. G. and Widiawati, Y. (2019). Chapter 10: Emissions from livestock and manure management. In: *Refinement to the 2006 IPCC Guidelines for National Greenhouse Gas Inventories* (Vol. 4). Agriculture, Forestry and Other Land Use, IPCC.
IPCC, Hergoualc'h, K., Akiyama, H., Bernoux, M., Chirinda, N., del Prado, A., Kasimir, A., MacDonald, J. D., Ogle, S. M. and Regina, K. (2019). Chapter 11: N_2O emissions from managed soils, and CO_2 emissions from lime and urea application. In: *Refinement to the 2006 IPCC Guidelines for National Greenhouse Gas Inventories* (Vol. 4). Agriculture, Forestry and Other Land Use, IPCC.
ITPS, F. (2015). *Status of the World's Soil Resources (SWSR)-Main Report*. Food and Agriculture Organization of the United Nations and Intergovernmental Technical Panel on Soils, p.650.
Jeyanathan, J., Martin, C. and Morgavi, D. P. (2014). The use of direct-fed microbials for mitigation of ruminant methane emissions: A review. *Animal* 8(2), 250–261. https://doi.org/10.1017/S1751731113002085.
Johnson, K. A. and Johnson, D. E. (1995). Methane emissions from cattle. *Journal of Animal Science* 73(8), 2483–2492. https://doi.org/10.2527/1995.7382483x.
Kinley, R. D., de Nys, R., Vucko, M. J., Machado, L. and Tomkins, N. W. (2016). The red macroalgae *Asparagopsis taxiformis* is a potent natural anti-methanogenic that reduces methane production during in vitro fermentation with rumen fluid. *Animal Production Science* 56(3), 282–289.
Klurfeld, D. M. (2018). What is the role of meat in a healthy diet? *Animal Frontiers: The Review Magazine of Animal Agriculture* 8(3), 5–10.
Knuppel, A., Papier, K., Fensom, G. K., Appleby, P. N., Schmidt, J. A., Tong, T. Y. N., Travis, R. C., Key, T. J. and Perez-Cornago, A. (2020). Meat intake and cancer risk: Prospective analyses in UK Biobank. *International Journal of Epidemiology* 49(5), 1540–1552.
Kopittke, P. M., Menzies, N. W., Wang, P., McKenna, B. A. and Lombi, E. (2019). Soil and the intensification of agriculture for global food security. *Environment International* 132, 105078. https://doi.org/10.1016/j.envint.2019.105078.
Kossoy, A. (2005). The role of carbon finance in project development. In: *Bioenergy-Realizing the Potential*. Elsevier, pp. 179-187.
Leahy, S. C., Janssen, P. H., Attwood, G. T., Mackie, R. I., McAllister, T. A. and Kelly, W. J. (2022). Electron flow: Key to mitigating ruminant methanogenesis. *Trends in Microbiology* 30(3), 209–212. https://doi.org/10.1016/j.tim.2021.12.005.
Lean, I. J., Golder, H. M., Grant, T. M. D. and Moate, P. J. (2021). A meta-analysis of effects of dietary seaweed on beef and dairy cattle performance and methane yield. *PLoS ONE* 16(7), e0249053. https://doi.org/10.1371/journal.pone.0249053.

Leinonen, Ilkka, Eory, Vera, Macleod, Michael, Sykes, Alasdair, Glenk, Klaus and Rees, Bob. (2019). *Comparative Analysis of Farm-Based Carbon Audits.*

Li, F., Li, C., Chen, Y., Liu, J., Zhang, C., Irving, B., Fitzsimmons, C., Plastow, G. and Guan, L. L. (2019). Host genetics influence the rumen microbiota and heritable rumen microbial features associate with feed efficiency in cattle. *Microbiome* 7(1), 92. https://doi.org/10.1186/s40168-019-0699-1.

Lynch, J., Cain, M., Pierrehumbert, R. and Allen, M. (2020). Demonstrating GWP*: A means of reporting warming-equivalent emissions that captures the contrasting impacts of short- and long-lived climate pollutants. *Environmental Research Letters: ERL [Web Site]* 15(4), 044023. https://doi.org/10.1088/1748-9326/ab6d7e.

Ma, R., Li, K., Guo, Y., Zhang, B., Zhao, X., Linder, S., Guan, C., Chen, G., Gan, Y. and Meng, J. (2021). Mitigation potential of global ammonia emissions and related health impacts in the trade network. *Nature Communications* 12(1), 6308.

Machado, L., Kinley, R. D., Magnusson, M., de Nys, R. and Tomkins, N. W. (2015). The potential of macroalgae for beef production systems in Northern Australia. *Journal of Applied Phycology* 27(5), 2001–2005.

Manzatto, C. V., Araujo, L. S., Vicente, L. E., Vicente, A. K. and Perosa, B. B. (2018). Monitoring the mitigation of carbon emissions in agriculture. In: *Agroanalysis* (3rd ed., Vol. 38, pp. 25–28). Agribusiness Center of the Getulio Vargas Foundation, Rio de Janeiro.

Martínez-Álvaro, M., Auffret, M. D., Duthie, C. A., Dewhurst, R. J., Cleveland, M. A., Watson, M. and Roehe, R. (2022). Bovine host genome acts on rumen microbiome function linked to methane emissions. *Communications Biology* 5(1), 350. https://doi.org/10.1038/s42003-022-03293-0.

Martinez-Fernandez, G., Denman, S. E., Cheung, J. and McSweeney, C. S. (2017). Phloroglucinol degradation in the rumen promotes the capture of excess hydrogen generated from methanogenesis inhibition. *Frontiers in Microbiology* 8(8), 1871. https://doi.org/10.3389/fmicb.2017.01871.

Mbow, H. O. P., Reisinger, A., Canadell, J. and O'Brien, P. (2017). *Special Report on Climate Change, Desertification, Land Degradation, Sustainable Land Management, Food Security, and Greenhouse Gas Fluxes in Terrestrial Ecosystems (SR2).* IPCC, Ginevra, p. 650.

McAllister, T. A. and Newbold, C. J. (2008). Redirecting rumen fermentation to reduce methanogenesis. *Australian Journal of Experimental Agriculture* 48(2), 7–13.

McGinn, S. M., Flesch, T. K., Harper, L. A. and Beauchemin, K. A. (2006). An approach for measuring methane emissions from whole farms. *Journal of Environmental Quality* 35(1), 14–20.

McGuire, R., Huws, S. A., Foyer, C. H., Forster, P., Welham, M., Spadavecchia, L., Curry, D. and Scollan, N. D. (2023). Agrifood and net zero. *Trends in Plant Science* 28(5), 495–497.

McGuire, R., Williams, P. N., Smith, P., McGrath, S. P., Curry, D., Donnison, I., Emmet, B. and Scollan, N. (2022). Potential co-benefits and trade-offs between improved soil management, climate change mitigation and agri-food productivity. *Food and Energy Security* 11(2), e352.

McNally, S. R., Laughlin, D. C., Rutledge, S., Dodd, M. B., Six, J. and Schipper, L. A. (2015). Root carbon inputs under moderately diverse sward and conventional ryegrass-clover pasture: Implications for soil carbon sequestration. *Plant and Soil* 392(1–2), 289–299. https://doi.org/10.1007/s11104-015-2463-z.

McSherry, M. E. and Ritchie, M. E. (2013). Effects of grazing on grassland soil carbon: A global review. *Global Change Biology* 19(5), 1347–1357.

Mertens, D. R. and Ely, L. O. (1979). A dynamic model of fiber digestion and passage in the ruminant for evaluating forage quality. *Journal of Animal Science* 49(4), 1085-1095. https://doi.org/10.2527/jas1979.4941085x.

Michalk, D. L., Kemp, D. R., Badgery, W. B., Wu, J., Zhang, Y. and Thomassin, P. J. (2019). Sustainability and future food security—A global perspective for livestock production. *Land Degradation and Development* 30(5), 561–573.

Mizrahi, I., Wallace, R. J. and Moraïs, S. (2021). The rumen microbiome: Balancing food security and environmental impacts. *Nature Reviews. Microbiology* 19(9), 553–566. https://doi.org/10.1038/s41579-021-00543-6.

Mohammed, R., McGinn, S. M. and Beauchemin, K. A. (2011). Prediction of enteric methane output from milk fatty acid concentrations and rumen fermentation parameters in dairy cows fed sunflower, flax, or canola seeds. *Journal of Dairy Science* 94(12), 6057–6068. https://doi.org/10.3168/jds.2011-4369.

Moraes, L. E., Strathe, A. B., Fadel, J. G., Casper, D. P. and Kebreab, E. (2014). Prediction of enteric methane emissions from cattle. *Global Change Biology* 20(7), 2140–2148.

Morgavi, D. P., Forano, E., Martin, C. and Newbold, C. J. (2010). Microbial ecosystem and methanogenesis in ruminants. *Animal* 4(7), 1024–1036. https://doi.org/10.1017/S1751731110000546.

Muralikrishna, I. V. and Manickam, V. (2017). Life cycle assessment. In: *Environmental Management*. Elsevier, pp. 57–68. https://doi.org/10.1016/B978-0-12-811989-1.00005-1.

Murray, R. M., Bryant, A. M. and Leng, R. A. (1976). Rates of production of methane in the rumen and large intestine of sheep. *British Journal of Nutrition* 36(1), 1–14.

Nagaraja, G., Newbold, C. J., Van Nevel, C. J. and Demeyer, D. I. (1997). Chapter 13: Manipulation of ruminal fermentation. In: Hobson, P. N. and Steward, C. S. (Eds.), *The Rumen Microbial Ecosystem* (2nd edn.). Chapman and Hall, London, UK, pp. 523–632.

Nollet, L., Mbanzamihigo, L., Demeyer, D. and Verstraete, W. (1998). Effect of the addition of *Peptostreptococcus productus* ATCC 35244 on reductive acetogenesis in the ruminal ecosystem after inhibition of methanogenesis by cell-free supernatant of *Lactobacillus plantarum* 80. *Animal Feed Science and Technology* 71(1–2), 49–66. https://doi.org/10.1016/S0377-8401(97)00135-1.

Oldfield, E. E., Eagle, A. J., Rubin, R. L., Rudek, J., Sanderman, J. and Gordon, D. R. (2021). Agricultural soil carbon credits: Making sense of protocols for carbon sequestration and net greenhouse gas removals. Environmental Defense Fund, New York, NY. Available at: Edf.org/sites/default/files/content/agricultural-soil-carbon-credits-protocolsynthesis.pdf.

Parliament House of Commons. (2016). *Soil Carbon and Climate Change: Soil Health Environmental Audit.*

Patra, A. K. and Yu, Z. (2013). Effects of coconut and fish oils on ruminal methanogenesis, fermentation, and abundance and diversity of microbial populations in vitro. *Journal of Dairy Science* 96(3), 1782–1792. https://doi.org/10.3168/jds.2012-6159.

Paustian, K., Andrén, O., Janzen, H. H., Lal, R., Smith, P., Tian, G., Tiessen, H., Noordwijk, M. and Woomer, P. L. (1997). Agricultural soils as a sink to mitigate CO_2 emissions. *Soil Use and Management* 13(s4), 230–244. https://doi.org/10.1111/j.1475-2743.1997.tb00594.x.

Perosa, B., Newton, P. and da Silva, R. F. B. (2023). A monitoring, reporting and verification system for low carbon agriculture: A case study from Brazil. *Environmental Science and Policy* 140, 286–296. https://doi.org/10.1016/j.envsci.2022.12.006.

Perret, S. R. and Payen, S. (2020). Irrigation and the environmental tragedy: Pathways towards sustainability in agricultural water use. *Irrigation and Drainage* 69(2), 263–271.

Peyraud, J. L. and MacLeod, M. (2020). *Future of EU Livestock: How to Contribute to a Sustainable Agricultural Sector?* Final Report. European Commission, Luxembourg.

Pinares-Patiño, C. S., Lassey, K. R., Martin, R. J., Molano, G., Fernandez, M., MacLean, S., Sandoval, E., Luo, D. and Clark, H. (2011). Assessment of the sulphur hexafluoride (SF6) tracer technique using respiration chambers for estimation of methane emissions from sheep. *Animal Feed Science and Technology* 166–167, 201–209. https://doi.org/10.1016/j.anifeedsci.2011.04.067.

Plautz, J. (2018). Piercing the haze. *Science* 361(6407), 1060–1063. https://doi.org/10.1126/science.361.6407.1060.

Reichle, D. E. (2020). Carbon, climate change, and public policy. In: *The Global Carbon Cycle and Climate Change*. Elsevier, pp. 253–287. https://doi.org/10.1016/B978-0-12-820244-9.00012-3.

Reyes-Palomo, C., Aguilera, E., Llorente, M., Díaz-Gaona, C., Moreno, G. and Rodríguez-Estévez, V. (2022). Carbon sequestration offsets a large share of GHG emissions in dehesa cattle production. *Journal of Cleaner Production* 358, 131918. https://doi.org/10.1016/j.jclepro.2022.131918.

Ricci, P., Chagunda, M. G. G., Rooke, J., Houdijk, M., Houdijk, J. G., Duthie, C. A., Hyslop, J., Roehe, R. and Waterhouse, A. (2014). Evaluation of the laser methane detector to estimate methane emissions from ewes and steers. *Journal of Animal Science* 92, 5239–5250. https://doi.org/10.2527/jas.2014-7676.

Rira, M., Morgavi, D. P., Popova, M., Maxin, G. and Doreau, M. (2022). Microbial colonisation of tannin-rich tropical plants: Interplay between degradability, methane production and tannin disappearance in the rumen. *Animal* 16(8), 100589.

Robinson, T. P., Wint, G. R., Conchedda, G., Van Boeckel, T. P., Ercoli, V., Palamara, E., Cinardi, G., D'Aietti, L., Hay, S. I. and Gilbert, M. (2014). Mapping the global distribution of livestock. *PLoS ONE* 9(5), e96084.

Roehe, R., Dewhurst, R. J., Duthie, C. A., Rooke, J. A., McKain, N., Ross, D. W., Hyslop, J. J., Waterhouse, A., Freeman, T. C., Watson, M. and Wallace, R. J. (2016). Bovine host genetic variation influences rumen microbial methane production with best selection criterion for low methane emitting and efficiently feed converting hosts based on metagenomic gene abundance. *PLoS Genetics* 12(2), e1005846. https://doi.org/10.1371/journal.pgen.1005846.

Rojas-Downing, M. M., Nejadhashemi, A. P., Harrigan, T. and Woznicki, S. A. (2017). Climate change and livestock: Impacts, adaptation, and mitigation. *Climate Risk Management* 16, 145–163. https://doi.org/10.1016/j.crm.2017.02.001.

Romero-Perez, A., Okine, E. K., McGinn, S. M., Guan, L. L., Oba, M., Duval, S. M., Kindermann, M. and Beauchemin, K. A. (2014). The potential of 3-nitrooxypropanol to lower enteric methane emissions from beef cattle. *Journal of Animal Science* 92(10), 4682–4693.

Roque, B. M., Brooke, C. G., Ladau, J., Polley, T., Marsh, L. J., Najafi, N., Pandey, P., Singh, L., Kinley, R., Salwen, J. K. and Eloe-Fadrosh, E. (2019). Effect of the macroalgae

Asparagopsis taxiformis on methane production and rumen microbiome assemblage. *Animal Microbiome* 1(1), 1–14.

Roque, B. M., Venegas, M., Kinley, R. D., de Nys, R., Duarte, T. L., Yang, X. and Kebreab, E. (2021). Red seaweed (*Asparagopsis taxiformis*) supplementation reduces enteric methane by over 80 percent in beef steers. *PLoS ONE* 16(3), e0247820. https://doi.org/10.1371/journal.pone.0247820.

Russell, J. B., O'connor, J. D., Fox, D. G., Van Soest, P. J. and Sniffen, C. J. (1992). A net carbohydrate and protein system for evaluating cattle diets: I. Ruminal fermentation. *Journal of Animal Science* 70(11), 3551–3561.

Skendžić, S., Zovko, M., Živković, I. P., Lešić, V. and Lemić, D. (2021). The impact of climate change on agricultural insect pests. *Insects* 12(5), 440. https://doi.org/10.3390/insects12050440.

Skinner, R. H. and Dell, C. J. (2016). Yield and soil carbon sequestration in grazed pastures sown with two or five forage species. *Crop Science* 56(4), 2035–2044. https://doi.org/10.2135/cropsci2015.11.0711.

Smith, P., Bustamante, M., Ahammad, H., Clark, H., Dong, H., Elsiddig, E. A., Haberl, H., Harper, R., House, J., Jafari, M., Masera, O., Mbow, C., Ravindranath, N. H., Rice, C. W., Robledo Abad, C., Romanovskaya, A., Sperling, F. and Tubiello, F. (2014). Mitigation of climate change. Contribution of Working Group III to the fifth assessment report of the intergovernmental panel on climate change. In: Edenhofer, O., Pichs-Madruga, R., Sokona, Y., Farahani, E., Kadner, S., Seyboth, K., Adler, A., Baum, I., Brunner, S., Eickemeier, P., Kriemann, B., Savolainen, J., Schlömer, S., von Stechow, C., Zwickel, T. and Minx, J. C. (Eds.), *Climate Change: Agriculture, Forestry and Other Land Use (AFOLU)*. Cambridge University Press, Cambridge, UK and New York, NY.

Smith, P. E., Kelly, A. K., Kenny, D. A. and Water, S. M. (2022). Differences in the composition of the rumen microbiota of finishing beef cattle divergently ranked for residual methane emissions. *Frontiers in Microbiology* 13, 855565. https://doi.org/10.3389/fmicb.2022.855565.

Smith, P. E., Waters, S. M., Kenny, D. A., Kirwan, S. F., Conroy, S. and Kelly, A. K. (2021). Effect of divergence in residual methane emissions on feed intake and efficiency, growth and carcass performance, and indices of rumen fermentation and methane emissions in finishing beef cattle. *Journal of Animal Science* 99(11), skab275. https://doi.org/10.1093/jas/skab275.

Sperow, M. (2020). Updated potential soil carbon sequestration rates on U.S. agricultural land based on the 2019 IPCC guidelines. *Soil and Tillage Research* 204, 104719. https://doi.org/10.1016/j.still.2020.104719.

Stanton, A. V., Leroy, F., Elliott, C., Mann, N., Wall, P. and De Smet, S. (2022). 36-fold higher estimate of deaths attributable to red meat intake in GBD 2019: Is this reliable? *The Lancet* 399(10332), e23–e26.

Steinfeld, H., Gerber, P., Wassenaar, T., Castel, V., Rosales, M. and de Haan, C. (2006). *Livestock's Long Shadow: Environmental Issues and Options*. Food and Agriculture Organization.

Stewart, C., Piernas, C., Cook, B. and Jebb, S. A. (2021). Trends in UK meat consumption: Analysis of data from years 1–11 (2008–09 to 2018–19) of the National Diet and Nutrition Survey rolling programme. *The Lancet. Planetary Health* 5(10), e699–e708. https://doi.org/10.1016/S2542-5196(21)00228-X.

Storm, I. M. L. D., Hellwing, A. L. F., Nielsen, N. I. and Madsen, J. (2012). Methods for measuring and estimating methane emission from ruminants. *Animals: An Open Access Journal from MDPI* 2(2), 160–183.

Tapio, I., Snelling, T. J., Strozzi, F. and Wallace, R. J. (2017). The ruminal microbiome associated with methane emissions from ruminant livestock. *Journal of Animal Science and Biotechnology* 8, 7. https://doi.org/10.1186/s40104-017-0141-0.

Tedeschi, L. O., Abdalla, A. L., Álvarez, C., Anuga, S. W., Arango, J., Beauchemin, K. A., Becquet, P., Berndt, A., Burns, R., De Camillis, C., Chará, J., Echazarreta, J. M., Hassouna, M., Kenny, D., Mathot, M., Mauricio, R. M., McClelland, S. C., Niu, M., Onyango, A. A., Parajuli, R., Pereira, L. G. R., del Prado, A., Paz Tieri, M., Uwizeye, A. and Kebreab, E. (2022). Quantification of methane emitted by ruminants: A review of methods. *Journal of Animal Science* 100(7), skac197. https://doi.org/10.1093/jas/skac197.

Teillard, F., Maia de Souza, D., Thoma, G., Gerber, P. J. and Finn, J. A. (2016). What does Life-Cycle Assessment of agricultural products need for more meaningful inclusion of biodiversity? *Journal of Applied Ecology* 53(5), 1422–1429. https://doi.org/10.1111/1365-2664.12683.

The Royal Society for the Protection of Birds. (2020). New poll shows that people in Northern Ireland want stronger protections for nature. Available at: https://www.rspb.org.uk/about-the-rspb/about-us/media-centre/press-releases/rspb-ni-poll-results-2022/ (accessed 9 January 2023).

Toensmeier, E. (2016). *The Carbon Farming Solution: A Global Toolkit of Perennial Crops and Regenerative Agriculture Practices for Climate Change Mitigation and Food Security*. Chelsea Green Publishing.

Tylutki, T. P., Fox, D. G., Durbal, V. M., Tedeschi, L. O., Russell, J. B., Van Amburgh, M. E., Overton, T. R., Chase, L. E. and Pell, A. N. (2008). Cornell net carbohydrate and protein system: A model for precision feeding of dairy cattle. *Animal Feed Science and Technology* 143(1–4), 174–202. https://doi.org/10.1016/j.anifeedsci.2007.05.010.

Ungerfeld, E. M. (2020). Metabolic hydrogen flows in rumen fermentation: Principles and possibilities of interventions. *Frontiers in Microbiology* 11, 589. https://doi.org/10.3389/fmicb.2020.00589.

UNHCR. (2014). *Carbon Financing*. United Nations, Switzerland.

United Nations Framework Convention on Climate Change (UNFCC). (2014). *Handbook on Measurement, Reporting and Verification for Developing Country Parties*.

United Nations Framework Convention on Climate Change. (2023). Paris agreement. Available at: https://legal.un.org/avl/ha/pa/pa.html (accessed 14 January 2023).

van der Werf, H. M. G., Knudsen, M. T. and Cederberg, C. (2020). Towards better representation of organic agriculture in life cycle assessment. *Nature Sustainability* 3(6), 419–425. https://doi.org/10.1038/s41893-020-0489-6.

Vogels, G. D., Hoppe, W. F. and Stumm, C. K. (1980). Association of methanogenic bacteria with rumen ciliates. *Applied and Environmental Microbiology* 40(3), 608–612.

Wallace, R. J., Sasson, G., Garnsworthy, P. C., Tapio, I., Gregson, E., Bani, P., Huhtanen, P., Bayat, A. R., Strozzi, F., Biscarini, F., Snelling, T. J., Saunders, N., Potterton, S. L., Craigon, J., Minuti, A., Trevisi, E., Callegari, M. L., Cappelli, F. P., Cabezas-Garcia, E. H., Vilkki, J., Pinares-Patino, C., Fliegerová, K. O., Mrázek, J., Sechovcová, H., Kopečný, J., Bonin, A., Boyer, F., Taberlet, P., Kokou, F., Halperin, E., Williams, J. L., Shingfield, K. J. and Mizrahi, I. (2019). A heritable subset of the core rumen microbiome dictates dairy cow productivity and emissions. *Science Advances* 5(7), eaav8391. https://doi.org/10.1126/sciadv.aav8391.

Wang, Y., Xu, Z., Bach, S. J. and McAllister, T. A. (2008). Effects of phlorotannins from *Ascophyllum nodosum* (brown seaweed) on *in vitro* ruminal digestion of mixed forage or barley grain. *Animal Feed Science and Technology* 145(1–4), 375–395.

Willett, W., Rockström, J., Loken, B., Springmann, M., Lang, T., Vermeulen, S., Garnett, T., Tilman, D., DeClerck, F., Wood, A., Jonell, M., Clark, M., Gordon, L. J., Fanzo, J., Hawkes, C., Zurayk, R., Rivera, J. A., De Vries, W., Majele Sibanda, L., Afshin, A., Chaudhary, A., Herrero, M., Agustina, R., Branca, F., Lartey, A., Fan, S., Crona, B., Fox, E., Bignet, V., Troell, M., Lindahl, T., Singh, S., Cornell, S. E., Srinath Reddy, K., Narain, S., Nishtar, S. and Murray, C. J. L. (2019). Food in the anthropocene: The EAT-Lancet Commission on healthy diets from sustainable food systems. *The Lancet* 393(10170), 447–492. https://doi.org/10.1016/S0140-6736(18)31788-4.

Williams, S. R. O., Moate, P. J., Hannah, M. C., Ribaux, B. E., Wales, W. J. and Eckard, R. J. (2011). Background matters with the SF6 tracer method for estimating enteric methane emissions from dairy cows: A critical evaluation of the SF6 procedure. *Animal Feed Science and Technology* 170(3–4), 265–276.

Wolin, M. J. (1960). A theoretical rumen fermentation balance. *Journal of Dairy Science* 43(10), 1452–1459. https://doi.org/10.3168/jds.S0022-0302(60)90348-9.

World Bank. (2021). Soil organic carbon MRV sourcebook for agricultural landscapes. Available at: https://openknowledge.worldbank.org/handle/10986/35923License:CCBY3.0IGO. World Bank, Washington, DC.

World Bank. (2022a). Agriculture, forestry, and fishing, value added (% of GDP) (2022). Available at: https://data.worldbank.org/indicator/NV.AGR.TOTL.ZS (accessed 19 July 2022).

World Bank. (2022b). Employment in agriculture (% of total employment) (2022). Available at: https://data.worldbank.org/indicator/SL.AGR.EMPL.ZS (accessed 19 July 2022).

Wreford, A. and Topp, C. F. E. (2020). Impacts of climate change on livestock and possible adaptations: A case study of the United Kingdom. *Agricultural Systems* 178, 102737. https://doi.org/10.1016/j.agsy.2019.102737.

Wyness, L. (2016). The role of red meat in the diet: Nutrition and health benefits. *Proceedings of the Nutrition Society* 75(3), 227–232.

Zhang, Q., Difford, G., Sahana, G., Løvendahl, P., Lassen, J., Lund, M. S., Guldbrandtsen, B. and Janss, L. (2020). Bayesian modeling reveals host genetics associated with rumen microbiota jointly influence methane emission in dairy cows. *ISME Journal* 14(8), 2019–2033. https://doi.org/10.1038/s41396-020-0663-x.

Zhou, G., Luo, Q., Chen, Y., He, M., Zhou, L., Frank, D., He, Y., Fu, Y., Zhang, B. and Zhou, X. (2019). Effects of livestock grazing on grassland carbon storage and release override impacts associated with global climate change. *Global Change Biology* 25(3), 1119–1132.

Chapter 8

Modeling climate change impact on low-input smallholder farming systems

Myriam Adam, Centre de coopération internationale en recherche agronomique pour le développement (CIRAD), UMR AGAP Institute, Univ Montpellier, CIRAD, INRAE, Institut Agro, Montpellier, France and National University of Battambang, Cambodia; Gatien Falconnier, CIRAD, UPR AIDA, F-34398 Montpellier and AIDA, Univ Montpellier, CIRAD, Montpellier, France and International Maize and Wheat Improvement Centre (CIMMYT), Zimbabwe; David Berre, CIRAD, UPR AIDA, F-34398 Montpellier and AIDA, Univ Montpellier, CIRAD, Montpellier, France; Katrien Descheemaeker, Wageningen University & Research, The Netherlands; Juliette Lairez, CIRAD, UPR AIDA, F-34398 Montpellier and AIDA, Univ Montpellier, CIRAD, Montpellier, France and Institut de l'Environnement et de Recherche Agricole (INERA), Burkina Faso; and Louise Leroux, CIRAD, UPR AIDA, F-34398 Montpellier and AIDA, Univ Montpellier, CIRAD, Montpellier, France, CIRAD, UPR AIDA, Nairobi and IITA, Kenya

1 Introduction
2 Modeling climate change effect on crops, livestock, grasslands and trees
3 Toward an integrative perspective
4 Conclusion
5 Where to look for further information
6 References

1 Introduction

Smallholder farming systems need to feed more than 3 billion people by 2100 (Vollset et al., 2020), while contributing to poverty alleviation, limiting agricultural expansion and reducing the environmental footprint (Foley et al., 2011). These farming systems, especially in sub-Saharan Africa (SSA), are characterized by large yields gaps (van Ittersum et al., 2016) mostly due to poor soil fertility and poor access to inputs (e.g. fertilizers, herbicides, insecticides and irrigation facilities). The vast majority of agricultural land in SSA is rainfed, and only a fifth of the area suitable for irrigation is actually irrigated (Burney et al., 2013). Average fertilizer use is 12, 2 and 3 kg ha^1 for nitrogen (N), phosphorus (P) and potassium (K), respectively (FAOSTAT, 2018), far below the

http://dx.doi.org/10.19103/AS.2022.0115.10

amounts required to alleviate nutrient limitations in most soils of the continent. Ten Berge et al. (2019) estimated that nutrient input on maize would have to increase nine to fifteen-fold to reach yields that would allow the continent's food self-sufficiency by 2050. Labor shortage further constraints the ability of these farmers to timely weed the crops (Silva et al., 2019). As a result, actual crop yields for smallholder farming systems lag far behind the yield potential. For example, yields of maize, millet and rice were between 25% and 50% of potential yield for typical smallholder farming systems of Senegal and Vietnam

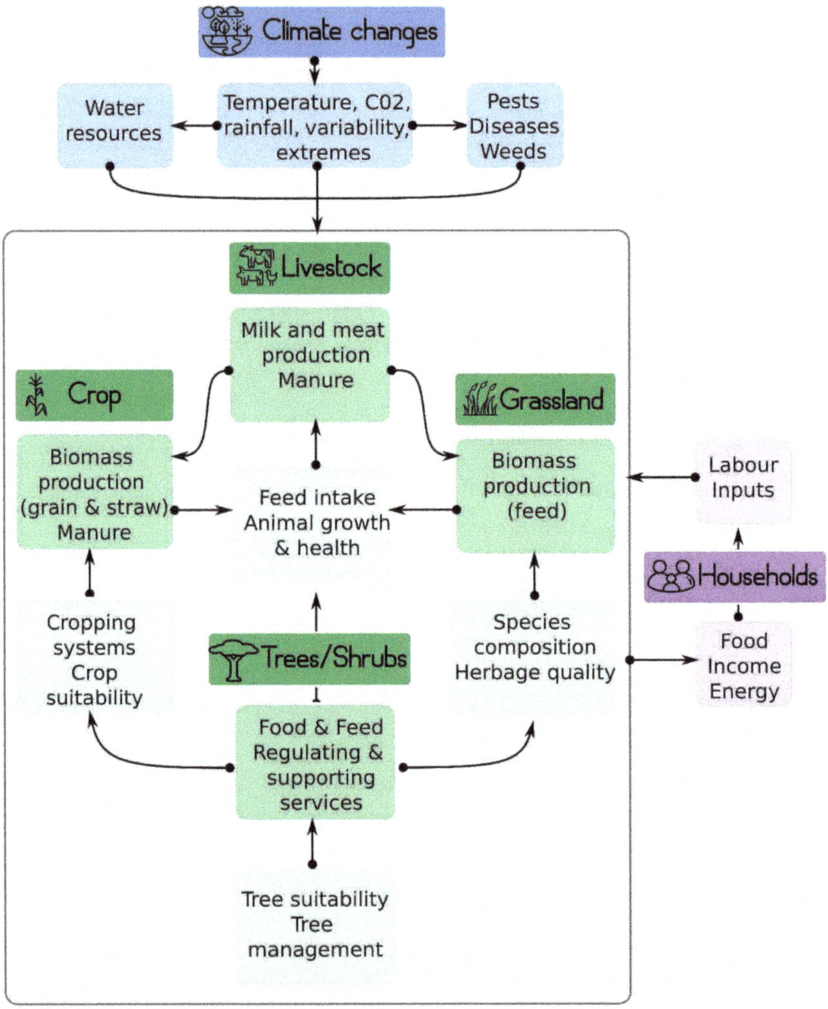

Figure 1 Representation of the effects of climate change on farming system components (dark color boxes). Intermediate color boxes are variables of the component impacted by climate change, and light color boxes are the processes affected by climate change.

(Affholder et al., 2013). On top of that, ongoing climate change is expected to exacerbate abiotic stresses on crops and poses the additional challenge of adaptation to climate change.

Climate change is undeniable: in addition to temperature, precipitation is increasing in global annual average, but with great regional and seasonal disparity. With a business as usual scenario in 20 years' time, the world will go past the global 1.5°C warming limit set by the Paris agreements. The biosphere ability to fulfill its role as a CO_2 sink will decrease as temperature increases. Climate change leads to a greater frequency and intensity of extreme events, including heat waves, intense rainfall and prolonged droughts. In addition to climate change, farmers must adapt to already existing climate variability (Huet et al., 2022). Quantitative and qualitative approaches have been developed to capture farming system's response to climate variability, and design and assess alternatives to adapt to climate change. Farming systems are complex as they integrate several components, including crops, livestock, grassland, trees and humans (Fig. 1). This chapter presents the effect of climate change on these components and highlights potential component-level strategies to adapt to climate change. We specifically focus on challenges in evaluating climate change impact and adaptation potential in the context of low-input systems. In the final section, we go beyond the component-based approach and highlight modeling and participatory approaches that help address the challenges of assessing the effect of climate change on farming systems as a whole.

2 Modeling climate change effect on crops, livestock, grasslands and trees

Climate change will impact crop, animal, trees and grazing resources of mixed farming systems. Though understanding the effect on one component alone is not enough to understand the complex effect of climate change at farm level, we first give an overview of the effect of climate change on each component.

2.1 Effect of climate change on cropping systems

2.1.1 Impact of climate change on crop productivity and adaptation strategies

Cropping systems are the most extensively studied component of the farming systems. Crops models are useful tools to analyze and unravel the impact of climate and crop management on cropping system agronomic and environmental performances. In low-input systems, studies on the impact of climate change on grain yield and potential adaptation strategies have been carried out. Faye et al. (2018) showed that cereal yields in West Africa would decrease by between 2% and 5% with a temperature increase of 1.5°C and 2°C, respectively. In Mali, Traore et al. (2017) assessed the effect of climate change on maize and pearl millet yield. They indicated a maize grain yield loss of up to 57% that could be offset by applying recommended fertilizer doses. Similar conclusions were drawn for sorghum (Adam et al., 2020), and the authors also highlighted the importance of designing improved management practices to improve crop yield at first. Further, crop production is likely to become more dependent on irrigation in many areas, but water resources are also negatively affected by, e.g. increased drought, or extreme events like floods. These extreme events are difficult to predict, because of the lack of observational data and the interactions with land-use change (Niang et al., 2014). Crop suitability may be altered, resulting in shifts in cropping patterns, shifts from cropping to livestock keeping activities (Jones and Thornton, 2009) and increased household vulnerability (e.g. Wichern et al., 2019; Ramirez-Villegas et al., 2013). These shifts can be captured with simple models like EcoCrop, originally developed by Hijmans et al. (2001).

Falconnier et al. (2020) and Faye et al. (2019) simulated that in low-input systems in SSA, intensified systems with improved yields become more sensitive to climate change impact, hence the need for appropriate adaptation strategies. Freduah et al. (2019) showed that climate change impact on grain yield ranged between 9% and 39% across sites, depending on farms. This diversity of response was mostly due to variations among farms in fertilizer applications, planting dates and soil types. Crop management and soil type diversity needs to be considered when designing farm-level adaptations strategies to climate change. The most common adaptation strategies to climate change are the adaptation of sowing dates and choice of cultivar with adapted crop cycle length. Modeling of crop growth and performance with current and future climate can help determine best sowing date and cultivar duration to escape the heat and drought stress than happen at the start, middle, and/or end of the growing season (Gérardeaux et al., 2021). More frequent heavy rainfalls (e.g. as in the Sahel; Taylor et al., 2017) will intensify nitrate leaching and other relevant adaptations will need to improve nitrogen use efficiency and minimize losses through leaching, e.g. relay intercropping with deep rooting cover crops (Baldé et al., 2011) and split applications of mineral fertilizer (Ganyo et al., 2018). In regions where drought occurrence may increase, drought-tolerant cultivars with adequate root traits and water-harvesting technologies (e.g. stone lines, tied ridging, zaï pits and contour ridging) can contribute to offset production losses. Finally, adaptation to heat stress requires specific cultivar with greater tolerance – but intercropping may also help to create a microclimate that could lower heat stress.

2.1.2 Modeling key processes of particular relevance to low-input cropping systems

For crop models to be meaningful in assessing the effect of climate change and the potential of adaptations to offset these effects, proper calibration against observed data is required, to avoid models simulating a correct output (i.e. close to the observation) for the wrong reason (Keating, 2020). Crop models are increasingly used to explore crop performance in the Global South under the effect of climate change and changes in crop management. However, experimental data from industrialized crops in the North do not necessarily reflect the reality in the fields of smallholder farmers in the Global South. There is an urgent need to fund research that would give priority to field experiments, particularly in family farmers' plots, in order to acquire the data required to document the specificities of the smallholder context in the Global South. For example, a model can accurately simulate grain yield, but this can be the result of a strong overestimation of total aboveground biomass and harvest index. There are physical boundaries to nitrogen concentration or dilution in plant tissues, and if nitrogen uptake and grain yield are not within the minimum and maximum nitrogen dilution boundaries (Fig. 2, derived from Falconnier et al., 2020), simulations are probably unrealistic.

For large-scale, intensified farming in the Global North, water and nutrient stress are often less of an issue than in low-input systems, as farmers often achieve close-to-potential crop yield. In such context, crop modelers focus is often more on yield-defining factors (radiation, temperature, CO_2) than on limiting factors (water and nitrogen), and rely on extensive datasets that have been collected in the past 40 years. For smallholder low-input farming in the Global South, water, nutrient and biotic constraints prevail: the number of processes that have to be accounted for drastically expands. Soil water dynamics (infiltration from rainfall, redistribution within the soil profile and evapotranspiration) influences soil moisture and nutrient supply from the mineralization of soil organic matter. This has a dramatic impact on the amount of nutrient (nitrogen in particular) that can be taken up by the plant in the absence of mineral fertilizer. The decomposition of previous residues buried into the soil, and the decomposition of the diversity of organic amendments (e.g. compost and manure) applied by farmers, also impact nutrient availability during the growing season. Soil water dynamics (prolonged droughts or heavy rains) also impact nitrogen leaching below the root zone and therefore the loss of nitrogen that would be otherwise available for crop uptake. On top of these issues, local crop cultivars (i.e. drought tolerant, sensitive to photoperiod), and intercropping of two or more crops in the same plot is a common feature of smallholder cropping systems (Ganeme et al., 2021). Intercropping involves competition and complementary for light interception, water and nitrogen

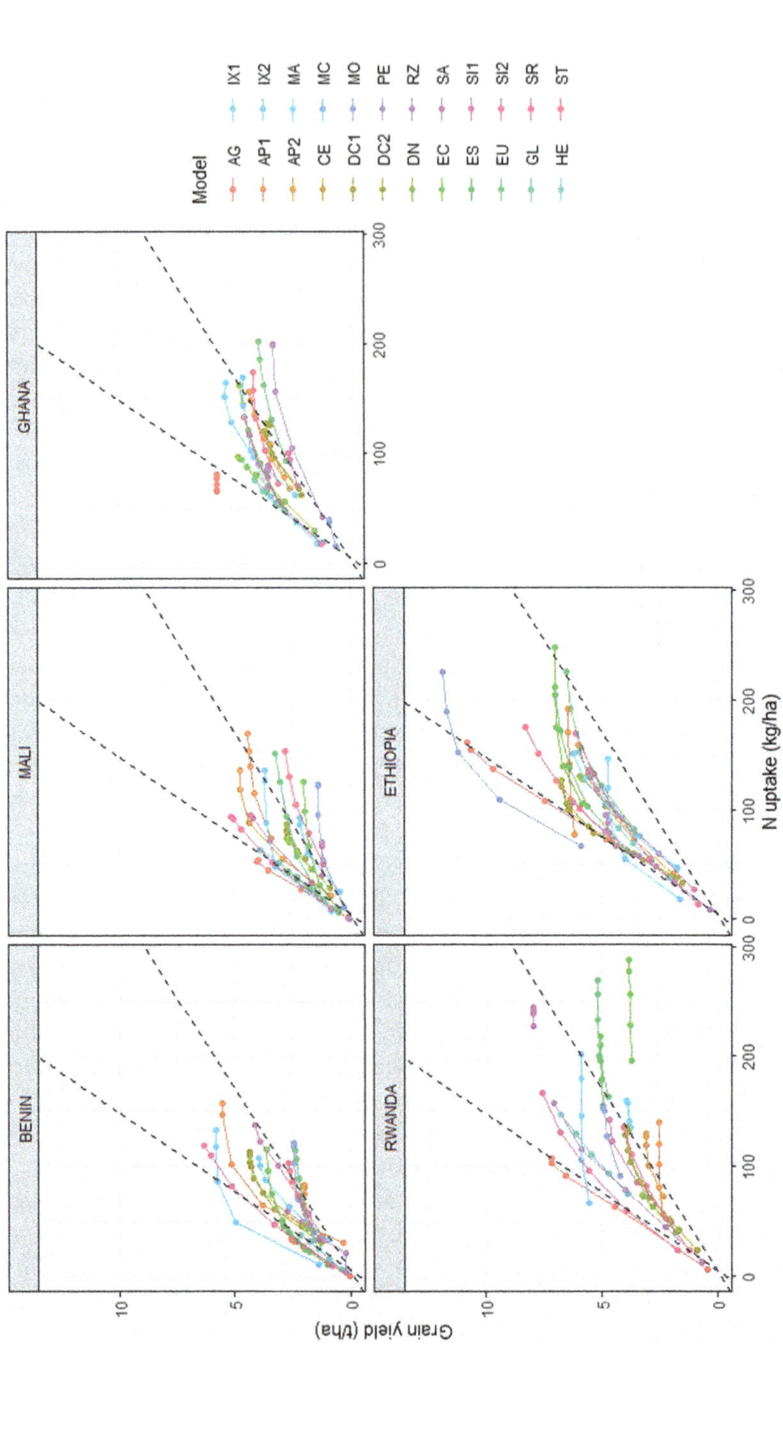

Figure 2 Simulated N uptake and grain yield for 24 AgMIP models at five sites across sub-Saharan Africa. Upper and lower dotted lines are physical boundaries corresponding to maximum and minimum N dilution in grain, respectively.

uptake between the dominant and understory plant, which adds up to the complexity of modeling low-input cropping systems. The current literature shows that the existing experiment across SSA currently lack sufficient observed data to accurately calibrate theses key processes, and keys species/varieties that characterize low-input conditions (Falconnier et al., 2020; Nendel et al., 2019).

However, a number of recent advances in calibrating crop models for the processes that matter in low-input context have been made. Soil-crop models (DSSAT, APSIM) have been calibrated to reproduce local varietal diversity (Gbegbelegbe et al., 2017; Akinseye et al., 2017; Adam et al., 2018), and the impact of complex interactions between cereal residue, soil type and climate on residue mineralization, in tropical environment in sub-humid Brazil (Maltas, 2007), sub-humid Ghana (MacCarthy et al., 2015), semi-arid Benin (Amouzou et al., 2018) and sub-humid Southern Zambia (Corbeels et al., 2016). Recently, the STICS model was also calibrated to simulate green manure decomposition and the effect on subsequent crop in the tropical context of sub-humid Madagascar (Ranaivoson et al., forthcoming). The study showed that the formalisms for residue decomposition built into the model, initially developed for temperate regions, was relevant for tropical conditions: accurate simulation of the release of nitrogen through the decomposition of the incorporated legume was achieved. STICS was also calibrated for simulating sorghum-cowpea intercropping systems in rainfed conditions in West Africa (Traoré et al., forthcoming and Ganeme et al., in prep). Competition and complementarity between crops for light, water and nitrogen uptake were well reproduced by the model: (i) cowpea and sorghum aboveground biomass was smaller in intercropping than in sole cropping, and (ii) cowpea biomass decreased more strongly than sorghum biomass, and this feature was reproduced by the calibrated model. Despite a reduction in sorghum and cowpea yield, land equivalent ratio of the intercropping for aboveground biomass was greater than one in the additive intercropping systems (Traoré et al., forthcoming).

These recent progress in modeling keys processes in low-input systems are promising, but the challenges ahead are substantial. For example, the calibration and validation of STICS-intercrop revealed prospects for improvement. The model, so far, can only simulate intercropping for interrow systems and cannot deal with more complex configurations implemented by farmers (e.g. inter- or within-hill, Ganeme et al., 2021). Traoré et al. (forthcoming) showed accurate simulation of water dynamics, but inaccurate simulation of competition for nitrogen between the cereal and the legume. Pest and disease are not yet properly considered in most crop models (Savary et al., 2019). Phosphorus deficiency, legumes, traditional crops such as pearl millet or teff are also not accurately represented.

2.2 Effect of climate change on livestock and grazing land

In low-input systems, crop-livestock integration is widespread and provides several benefits to smallholder farmers (e.g. traction, manure, saving) (Descheemaeker et al., 2016). Livestock and climate change are closely interconnected as on the one hand livestock is a significant contributor to greenhouse gas emissions (FAO, 2006), and on the other hand livestock is drastically impacted by climate change (Rojas-Downing et al., 2017). While the mitigation of greenhouse gas emissions is indeed a high priority in developed countries with intensive livestock production systems, adaptation remains the priority for smallholder farmers, as climate change strongly affects their livestock systems.

2.2.1 Impact of climate change on livestock productivity and adaptation strategies

Increase in temperature and higher frequency of drought affect feed quality and quantity when livestock systems rely on cropland and grazing land. With high temperature, feed intake and ruminant digestion are reduced (Yadav et al., 2013). Reproduction is also affected by high temperature (e.g. Wolfenson and Roth, 2019) through, e.g. disruption of oocyte development and alteration of ovarian follicular growth dynamics. Reproduction is a 'luxurious phenomenon' appropriate when animals are in a favorable environment (Chauhan and Ghosh, 2014). Fertility rate reduction is a coping strategy for animals, which can affect directly farm viability. Livestock disease outbreaks will also become more frequent under a warmer and wetter climate (Sejian et al., 2016). Climate change may favor living conditions of disease vectors and tick-borne disease, rift valley fever and bluetongue will likely affect more cattle. Parasitic diseases will also be more frequent or affect larger geographical areas, in regions where rainfall will increase.

Climate change adaptation options for low-input livestock systems can follow three overall strategies, being risk management, diversification and sustainable intensification (Descheemaeker et al., 2016). Grazing management, including seasonal herd migration, adjustment of stocking densities and rotational grazing, and weather-based index insurance schemes (Greatrex et al., 2015) are typical examples of risk management strategies that allow adapting to increased climate variability and a higher frequency of extreme events. Adapting to the increasing incidence of heat and drought stress can also be achieved through diversifying with breeds and livestock species that are better adapted to these circumstances. Tighter crop-livestock integration (e.g. improved residue and manure management, inclusion of fodder legume) and diversification of household income sources, including off-farm income, can cushion low-input systems against climate shocks. Sustainable intensification that is targeted at increasing animal productivity is a promising strategy as it also helps in reducing greenhouse gas emission, expressed per kg of animal product. In particular, feed gaps in the dry season are a major cause of livestock inefficiency, with animals loosing weight and becoming more susceptible to diseases. These feed gaps could be addressed through cultivating fodder legume crops (e.g. Mucuna pruriens) that produce high quality biomass and provide rotational benefits to other crops (Descheemaeker et al., 2018), or through using feed concentrates. If heat, drought and water stress can be managed, the introduction of more productive breeds is also an option.

For all the reasons mentioned above, milk and meat productivity, and household food security, will be affected by climate change. In addition, changes in the suitability of agro-ecological zones for keeping certain types of animals, may impact the configuration of entire farming systems, with repercussions on crops, grazing lands and herder livelihoods.

2.2.2 Modeling key processes of particular relevance for livestock and grassland low-input systems

Livestock simulation models can be used to evaluate the effect of climate change on animal performance. The LIVSIM model (LIVestock SIMulator, Rufino et al., 2009) has been used extensively for smallholder cattle systems in a range of environments, including Zimbabwe (Rufino et al., 2011; Descheemaeker et al., 2018), Kenya (Rufino et al., 2009) and Mali (de Ridder et al., 2015), and for small ruminant systems in Nigeria (Amole et al., 2017). LIVSIM calculates the performance of each individual animal in the herd on a monthly basis, according to its genetic potential, feed availability and quality, and herd management. Model outputs include milk and manure quantity and nutrient content, animal body weight and herd dynamics, including birth rate, offtake rate and mortality. LIVSIM has been used to evaluate the effects of altered feed availability due to climate change and adaptation on livestock in mixed systems in a semi-arid area of Zimbabwe (Descheemaeker et al., 2018). Taking into account differences between temperate and (sub-)tropical husbandry systems, a recent modification called LIVSIM-mod (Bateki and Dickhoefer, 2020) accurately represented voluntary feed intake capacity, growth and lactation, and metabolizable energy requirements for maintenance and weight gain of cattle in low-input tropical systems.

Studies assessing the impact of climate change on crops mostly focused on grain yield with little attention to biomass. Yet crop residues are an important feed source for livestock and for other household uses like fuel, composting, building), and as a mulch to protect the soil and recycle nutrients (Valbuena et al., 2012). Crop residues are indeed an important feed source in sub-humid

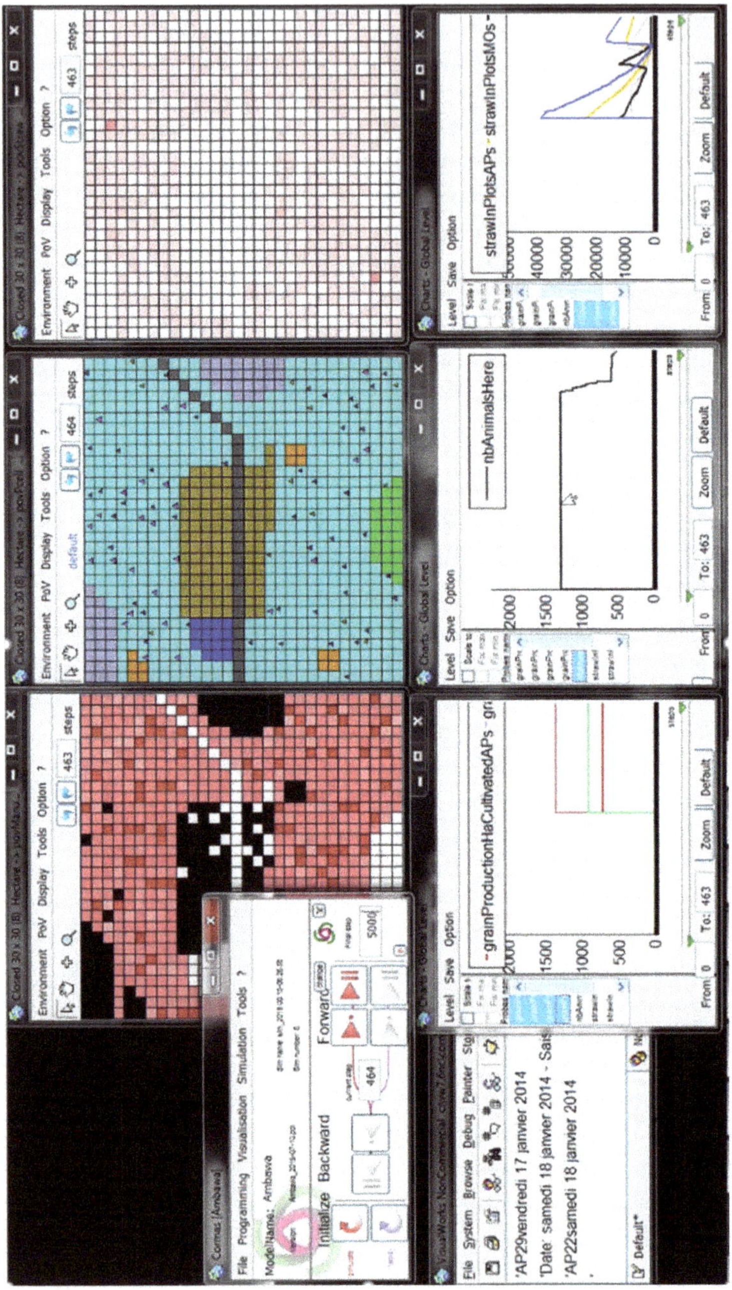

Figure 3 Example of agent-based model exploration of the effect of mulching at landscape scale (windows on the upper side are amount of manure deposited (left) during free-grazing of crop residue (right) with cattle mobility moving in the cropping land (center). Windows on the lower side are respectively crop yield (left), number of animals in the landscape (center, decreasing when cattle starts to leave the village) and crop residue decreased during the dry season).

and semi-arid areas during the dry season. Historically, livestock are authorized by common-law to graze on cropland after harvesting, which also benefits crop producers thanks to manure deposition during the livestock corralling. While the effect at field scale of crop residue mulching on yield is well-documented, in such a context of cattle mobility, spatially explicit models combining both biophysical and managerial modules are needed to understand biomass flows and correlated fertility patterns. The AMBAWA model (Berre et al., 2021) can simulate at village level the impact of an increasing share of cropland with residue mulching on cattle mobility and soil fertility (Fig. 3). Regardless of mulch effect on soil (from no effect to doubling of yield), yields are always decreasing at landscape scale if mulch is used, as cattle have to leave the landscape and manure is therefore lost for soil fertility in the village. Such complex feedbacks need to be accounted for when analyzing the impact of climate change on low-input farming systems.

In a context of cattle mobility in sub-humid and semi-arid areas, biomass inflows and outflows at farm scale are highly influenced by fodder seasonality and cattle mobility. Indeed, modeling biomass inflows and outflows in farm households allows identifying contrasted types of biomass management but does not tackle farms' interaction and inputs from grazing land at landscape scale (Assogba et al., 2022). As such, more research is needed in (i) livestock science to understand physiological processes governing feed and water intake during droughts and (ii) environmental sciences to better explore changes in feed quantity and quality in remote grazing land used during transhumance. Assouma et al. (2018) and Lo Seen et al. (1995) provide useful examples.

2.3 Effect of climate change on agroforestry

Agroforestry systems are characterized by the integration and management of trees with crop and livestock. Agroforestry has been pointed out by the Intergovernmental Panel on Climate Change as a key option to respond to climate change and land degradation while simultaneously improving global food security (IPCC, 2019). In SSA, where around 40% of people in rural areas live in landscapes with more than 10% tree cover (Zomer et al., 2014), agroforestry is a traditional and major land-use system and has long been recognized as a solution to address climate change (Mbow et al., 2014). Depending on agroclimatic situations and socio-economic environments, different types of agroforestry systems occur in SSA: e.g. Sahelian parklands systems, Kenyan rotational woodlots or intercropping systems in Malawi (Dagar et al., 2020).

2.3.1 Impact of climate change on trees and agroforestry productivity and adaptation strategies

Rainfall and temperature influence tree growth (Sanogo et al., 2016), tree density and tree species diversity (Gonzalez et al., 2012). Strong wind can increase flower dropping and hence reduce fruit production. Climate change is expected to change the current distribution of tree species in SSA. For instance, using ensemble suitability mapping, Kindt (2018) assessed the distribution of over 150 tree species with future climate. The authors showed a huge reduction of suitability areas by 2050 for most of the tree species being planted and managed today. In the Sahelian areas, a decline in tree density and tree cover is already observed, particularly in cultivated areas with high population density (Brandt et al., 2016; Hiernaux et al., 2022). This decline was attributed to climate change (Gonzalez et al., 2012).

In SSA, agroforestry systems provide an array of ecological, economic and cultural services (Sinare and Gordon, 2015; Kuyah et al., 2016; Miller et al., 2017). The expected impacts of climate change on trees in agroforestry systems can reduce their provision of ecosystem services. First, it can have a direct impact on food security. Large tree species (e.g. shea fruits, baobab fruits) have high nutritional values and households use them in coping with seasonal food shortages. Fruit yields of these trees could decrease with climate change. Second, indirect impacts of climate change on household food security and livelihoods can be expected. Trees are a source of products, income and energy (e.g. fruits, wood, charcoal) and services (e.g. microclimate amelioration, medicine) that contribute to household adaptation to climate change (Koffi et al., 2017; Bayala et al., 2014; Sida et al., 2018; Reed et al., 2017).

2.3.2 Modeling keys processes of particular relevance for agroforestry

Agroforestry models have been developed over the past decades. Due to the complexity of agroforestry systems that include interactions between crops and trees, those models have mainly been used to predict field-scale performance (Luedeling et al., 2016). None of them have been used to assess performance at farm level in SSA. Examples of such models applied at field-scale are WaNuLCAS (Van Noordwijk and Lusiana, 1998) or HyPAR (Mobbs et al., 1999). Attempts have been made to develop models to assess the economic costs and benefits of agroforestry systems at farm scale in Europe (Farm-SAFE, Graves et al., 2011). Other approaches have been proposed to assess the productivity of agroforestry systems at landscape and farm scales in SSA. Leroux et al. (2020) developed a remote sensing-based statistical model accounting for vegetation productivity and tree density to predict millet yields in Senegalese *Faidherbia albida* parklands (Fig. 4). The authors evidenced that millet yield increased with woody cover up to 35% in the surrounding landscape of agricultural fields. Then, using an integrated landscape approach relying on remote sensing data and recent advances in data analysis methods, they showed that the tree

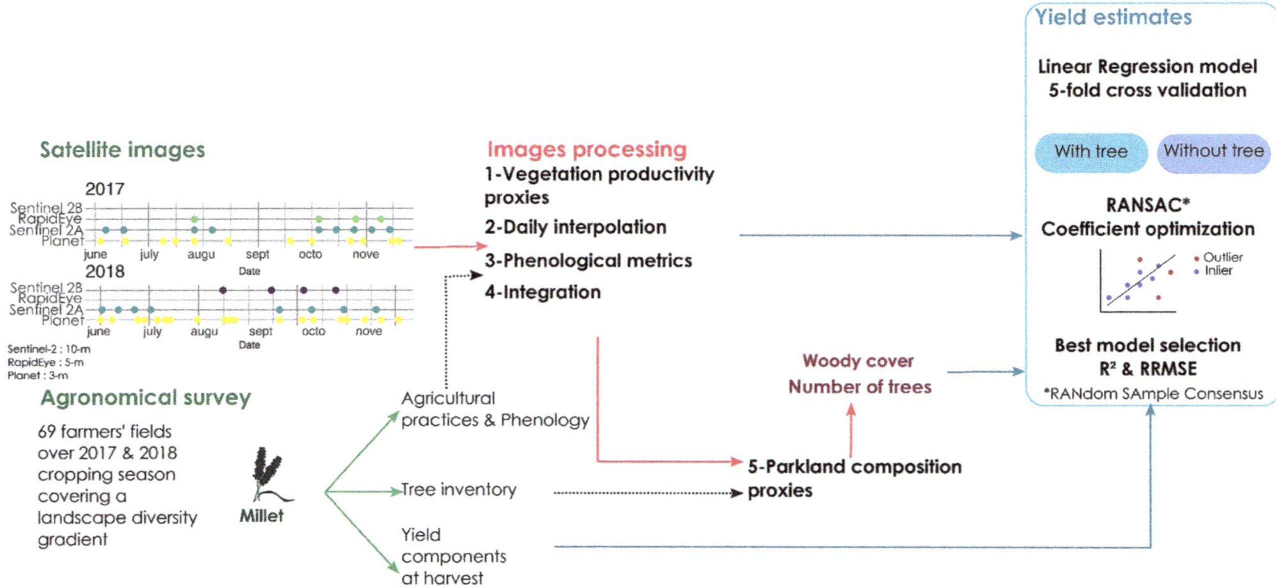

Figure 4 Methodology developed in Leroux et al. (2020) for improving millet yield estimation considering the tree component using a remote sensing-based model.

species diversity contributed mainly indirectly to household food security through the provision of ecosystem services regulating and supporting crop production (Leroux et al., 2022).

3 Toward an integrative perspective

After a review of recent progress and challenges remaining to capture the effect of climate change in the different components of low-input farming systems, it is now important to consider the integration of these different components. In this section we highlight the main approaches that address the effect of climate change at farming systems level rather than at individual component level.

3.1 Coupling models to assess climate change effect at farm scale

In order to understand climate change impacts and the effects of adaptation and mitigation options on mixed farming systems, a systems approach considering the interactions between farm components and processes at different scales is needed. The Agricultural Model Intercomparison and Improvement Project (AgMIP) Regional Integrated Assessment (RIA) approach links climate, crop, livestock and economic data and models for assessing the effect of climate change and adaptation options on heterogeneous farm populations. This modeling framework was run for a multi-farm assessment to capture the heterogeneity existing on a specific site, and to link different components of the farming systems. Results from contrasting Global Circulation Models (i.e. climate models used for future climate explorations) were used as input in two crop growth models (APSIM and DSSAT) to simulate effects on crop yield and (fodder) biomass production. The crop model results were used as input in the livestock model LIVSIM (Rufino et al., 2009), which allowed the simulation of field and herd level productivity. Both crop and livestock results fed into the TOA-MD model, which estimated economic performance at farm level (Claessens et al., 2012). This approach was carried out for different regions across the world, ranging from Africa (West, East and Southern) to Asia (Pakistan, Southern India and Indo Gangetic-Basin).

In West Africa, the impact of climate change was investigated on farmers' livelihoods under current production systems using Nioro, located in the South of the Groundnut Basin of Senegal as a case study. Nioro agriculture consists of predominately smallholder rainfed farms cultivating a range of cereals (millet, maize and sorghum) and legumes (peanut and cowpea) in rotation. Livestock plays a significant role in the functioning of the overall system through its dependence on crop residues as feed, and provision of manure to the cropping system. Fallow tends to disappear under population pressure. Very few farmers apply mineral fertilizers. As a result, average yields of cereals and peanut are low. The economic estimation of climate change impact on farming systems was done with the TOA-MD and uses inputs from two crop models (DSSAT and APSIM), which generated data through 10 GCMs (five for a 'high emissions' scenario in which little climate change mitigation happen in the future, RCP 8.5, and five for a 'medium emissions' scenario in which some climate change mitigation happens, RCP 4.5). Outcome variables such as vulnerability, defined as the proportion of households negatively affected by climate change, mean net farm returns, per capita income and poverty rates, were considered. With DSSAT simulations, the percentage of vulnerable farms varied between 24% and 59% across GCMs and RCPs. The lowest values were recorded for the cool/wet and middle scenarios. The hot/dry scenario presented the highest level of farm vulnerability. With APSIM simulations, the percent of vulnerable farms was between 42% and 49% (Fig. 5a).

Using the same approach for three farm types in semi-arid Zimbabwe, crop sensitivity to climate change in the current agricultural systems was relatively small, due mainly to poor soil fertility and low fertilizer application rates, causing severe nutrient stress. The observed small yield decline was attributed to increased temperature accelerating phenological development, and exacerbated by drought stress in the hot-dry climate. As "non-poor" farms (farmers above the poverty line) cultivated maize on better soils and with more fertilizer input, their maize was more sensitive to yield loss compared with the less-endowed farms. Livestock productivity was affected through altered production of crop residue and rangeland biomass, influencing feed intake, ranging from a slightly positive to a negative effect in the hot-wet and the hot-dry climate respectively. Non-poor farms, typically with larger stocking density, were more sensitive to feed gaps, and more strongly impacted by climate change than poor farms (Descheemaeker et al., 2018). At farm level, the simulated change in farm net returns depended on the climate and the farm type. 95% of the extremely poor households was currently below the poverty line. Their low-input production system was not sensitive to climate change, and only 40–50% of these farmers would lose (a little). In the group of best-endowed farmers, who cultivate better soils and keep more cattle, about 75% faced reduced returns in the hot-dry scenario, which increased the poverty rate from 60% to 65% of the households (Fig. 5b).

As described above, in order to assess the effect of climate change at farm scale or landscape scale, it is a common approach to first assess the effects at the component-level (climate change on crop or livestock or tree) based on general circulation models (GCM) climatic simulation, and then feed the outputs of one model into the next model in the chain. However, this raises the question of uncertainty propagation where uncertainty in data used to feed one model can lead to very high uncertainty in the outputs of subsequent models in the chain. Indeed, Corbeels et al. (2018) underlined that the variability in

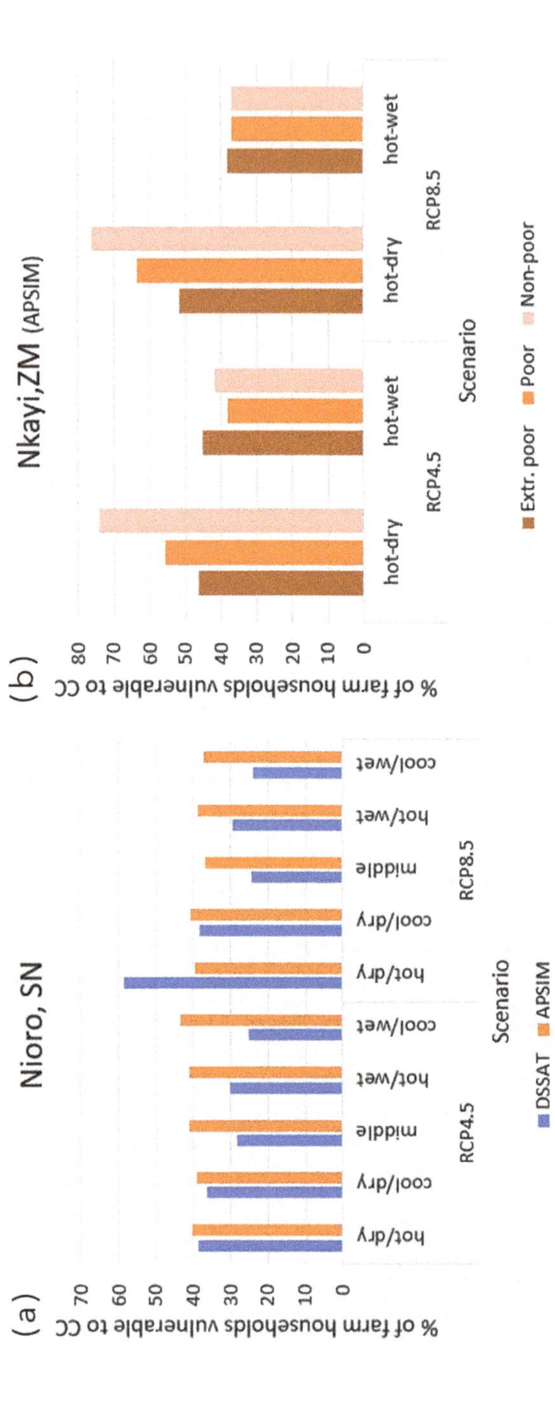

Figure 5 Percentage of farm households vulnerable to climate change (a) when considering two crop models in Nioro, Senegal and (b) according to different farmers type in Nkayi, Zimbabwe.

simulated yields arising from the use of several GCMs was stronger than the simulated yield variation that could be attributed to management adaptation. Recent ensemble modeling approaches allow to capture the uncertainty due to the GCM used (Tao et al., 2018) and crop model structure (Falconnier et al., 2020). Hence, as generated climatic data, simulated crop yield and zootechnical performance are all uncertain and the output of a farm-scale model fed with these data should be considered carefully.

3.2 Accounting for farmer decision-making in models

The AgMIP framework does not consider farmers' decision-making and farmers' preferences for different farming options nor does it account for farmers' adaptations in response to their perception of climate change. Optimization models can be used to represent farmers' decision-making. In such approaches, farms are modeled as managed by rational decision-makers (the farmers) who decide to select different farming options to achieve their goals under different constraints (cash, labor, land; Fig. 6). Using an optimization model, Wineman and Crawford (2017) found that farmers in Zambia will likely shift their choices of technologies and crops with climate change, but calorie production from crops would still decrease by 1–5%, depending on farm type. They identified that land-constrained farmers would have a high probability to fall below a minimum of calorie threshold and that autonomous adaptation would probably not be enough to offset yield losses. Boansi et al., (2021) identified with the use of an optimization model that rainfall shocks will lead to a decrease in total farm income in Ghana by 7.3–45.5%, and impact more the income and food consumption of 'asset poor farmers' who cannot cope by selling livestock.

Optimization models allow interactions between crop, tree, livestock, household and off-farm activities to be considered. Moreover, such models account for the seasonality of farming activities: by dividing a year in key periods of time, the effect of climate change on seasonal availability of farm resources (labor, forage, cash, grain for sale or self-consumption) can be analyzed and the effect on farm level indicators (e.g. income, food security) assessed.

Farmers' rationality to maximize their utility (i.e. what is important for them, e.g. profit, yield, costs, work, food security) is assumed in optimization models. Waldman et al. (2020) pinpointed that in an uncertain, changing context, these farm models represent the decision-making of farmers in an overly simplistic manner. They are used to quantify the effects of climate change in an ideal world where farmers have full knowledge of the farming options, and on the effects of climate change on these options. Assuming full rationality of farmers implies the risk to overestimate farmer adaptation to climate change. Farmers are not necessarily utility-maximizers, for example, they may not choose the

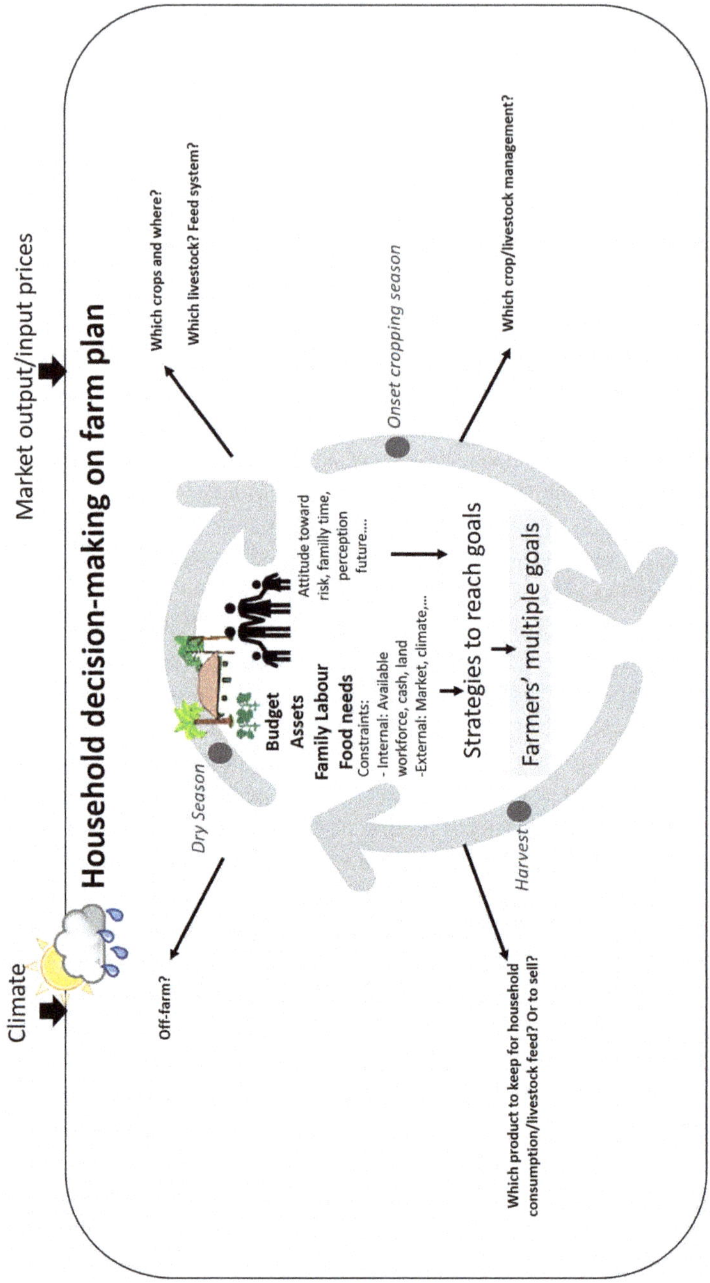

Figure 6 Representation of the key factors to consider when modeling household decision-making at farm level.

best option on every attribute, but rather make fast decisions on the option that performs best on their most important attribute (Waldman et al., 2020). Decision-making is also more complex than only utility maximization and could be linked to perceived risk, beliefs and values (Zhang et al., 2020). The perception that farmers may have of climate change may be biased, and the adaptations they may make as a result of this perception of climate change may be hard to predict. Lastly, climate change will be only one driver of change among others, such as input/output price fluctuation, policy interventions on e.g. agriculture, livestock, family planning (Falconnier et al., 2018; Assogba et al., 2022).

As climate change will generate more risks, e.g. income fluctuations (Adzawla et al., 2020), it is important to consider such risk in farmer decision-making. Accounting for risk means to assess farmer risk-aversion (see e.g. use of a lottery game in Eckel and Grossman, 2008), their perception of that risk (sometimes biased, because people tend to better remind extreme events), to identify the origins of this risk (flood, drought, delayed onset of rainy season, etc.), and the variability of performances under climate change for current systems and adaptation strategies (Huet et al., 2022).

3.3 Co-design adaptations with decision-makers

The heterogeneity of smallholder farms in large parts of SSA requires climate adaptations that are flexible and adapted to local agro-ecological and institutional conditions and to farmers' willingness and ability to invest. Models are essential tools for assessing the performance of farming systems and exploring the multidimensional effects of their adaptation, as shown in the previous sections. However, for model results to be useful in supporting decision-makers choices, modeling exercises need to be conducted in collaboration with the decision-makers so that criteria relevant to them are included in the assessment.

Decision-makers, ranging from farmers to policymakers, need science-based information to design interventions that address climate change and underpin transitions toward more sustainable farming system. AgMIP used a stakeholder-driven, integrated modeling approach to generate actionable information for policy planning processes (Homann-Kee Tui et al., 2021). Local experts and stakeholders were engaged in participatory workshops in which model results were presented and discussed, and future development pathways (called Representative Agricultural Pathways, RAPs; Valdivia et al., 2021) and adaptation options were co-designed. The process was iterative, which allowed for consecutive refinement of RAPs and adaptation options. For example, a first round of model runs for a Zimbabwean case showed that, whereas individual crop and animal productivity increased with adaptation options, whole-farm effects on food security and income were limited and would not allow to lift the majority of farmers out of poverty. Indeed, constraints related to farming system characteristics (e.g. small farm sizes, low soil fertility) and institutional barriers (e.g. absence of market incentives, poor access to inputs) were more important than climate change effects and impeded a shift toward more sustainable systems. Hence, in a second round of RAP adaptation, stakeholders and researchers co-designed future system changes that would enable farmers to earn a better living than today, even under climate change, through the adoption of relevant technologies (e.g. integrated soil fertility management, better husbandry of more productive livestock breeds) with support of appropriate institutions (e.g. functioning markets, insurance) and dedicated policies (e.g. on input and output prices, land tenure) (Homann-Kee Tui et al., 2021).

Farmers perceive strong risks from climate variability and extreme events (Huet et al., 2020), which leads to reduced investments and input use. As such, low-risk subsistence activities lower farm productivity and profit, resulting in an impact that may be stronger than that of changes in mean climate. Moreover, when a shock/hazard happens, a common coping strategy by smallholders is to sell assets (e.g. livestock, land, farming material) (Wichern et al., 2022) with prolonged impacts on income and food security and an increased likelihood of smallholders falling into poverty traps as a result. To cushion smallholder farmers against these risks, it is important to co-design relevant adaptation options with them. However, participatory processes of co-learning and co-design are challenging because farmers and researchers use different mental models of reality. Scientists rely on weather records of past meteorological data, while farmers often rely on their perception of how the climate may have changed over time. Meteorological data and farmers' perception do not necessarily align: while farmers of central Mali perceived an increase in the frequency of dry spells, the analysis of past climate data did not show such an increase (Traore et al., 2021). This investigation of the consistency between farmers' perception and weather record is crucial, as farmers' inaccurate perception of climate change can lead to maladaptation (Grothmann and Patt, 2005).

Mismatch between perception and actual weather record can be linked to the fact that perception is biased with possibly over-emphasis of recent extreme events (Marx et al., 2007), or it may also arise from the fact that scientists use indicators of climate change that do not match with the indicators used by farmers. The analysis of the link between farmers' perception of climate change and effective implementation of adaptation strategies is also crucial, as it may illustrate areas where farmers need stronger support. For example, farmers acutely perceived the increase in temperature in central Mali (an increase also shown by the analysis of historical data), yet they fell short at designing and implementing specific adaptation in their field to address this rising constraint (Traore et al., 2021). Continuous interactions between farmers and researchers are needed to develop a common understanding of the impact of climate change on crops and to design site-specific adaptation strategies.

4 Conclusion

Modeling the effect of climate change in low-input farming systems still faces some challenges. First, at component level, we notice the unbalance in studies carried out in low-input systems. In terms of modeling, much more is done for cropping systems than for livestock and let alone the agroforestry component. This can certainly be attributed to the difficulty in capturing relevant processes and acquire data for low-input systems. Even for crop models, as they have first been developed for industrialized countries, when applied to low-input cropping systems, the number of processes that have to be accounted for drastically expands (i.e. water, nutrient and biotic constraints). For livestock and agroforestry component, new approaches are need to capture the complexity of the component in low-input systems. We described a few promising attempts to address this void, but we stressed the importance of data collection in this context to account better for these constraints. With respect to modeling livestock and agroforestry components, there are novel approaches that help model important processes, i.e. agent-based modeling to represent the importance of biomass management at landscape scale, or use of remote sensing to quantify the effect of trees on crop productivity and food security. Further integration of agroforestry and livestock modeling with cropping system modeling will help account for the crop-livestock-tree integration that is so crucial for farmers of the South to adapt to climate change.

Second, there is an urgent need for farm-level models to assess climate change impacts. Such approach will permit to go beyond point-based crop modeling and coupling of input-outputs from one model to another. Impact at farm level is more than the sum of the impacts on the separate components. Unfortunately, too little is done at this level. Coupling climate, crop, livestock and economic models is a step forward. But it raises the issue of propagation of uncertainty from component to component and of integrating farm-level decision-making. It is (very) challenging to predict farmers' adaptation decisions to the effects of climate change as climate is one driver of farm change among many others (market, policies, labor supply, Mertz et al., 2009). A better understanding of farmers' climate change perception, risk-aversion and livelihood strategies can result from more interdisciplinary research involving agronomists and social scientists.

Finally, we can mention farm-level constraints (e.g. farm size) and entry points (e.g. intensification) that can easily offset climate change impact. But often family farmers have more urgent issues to solve before thinking about future climate change, and that might prevent them from acting against climate change. Hence, we argue, in our last section, on the importance of including decision-makers, including farmers, in modeling exercises to co-design adaptation strategies to climate change. Their involvement will allow first to create a constructive dialogue on the effect of climate change on farming systems, and second to better represent in farm models key factors, and finally to co-design locally relevant adaptation strategies.

5 Where to look for further information

5.1 Further reading

- For more information on effect of climate change on crop models, see the book chapter by Adam, M., Boote, K. J., Falconnier, G. N., Porter, C., Rezaei, E. E. and Webber, H. 2020. Modeling the effects of climate change on agriculture: a focus on cropping systems, in: Climate Change and Agriculture, Deryng Delphine (Ed.). Burleigh Dodds Science Publishing, Cambridge.
- For more information on a livestock model applied to tropical conditions, see Bateki, C. A. and Dickhoefer, U. 2020. Evaluation of the Modified LIVestock SIMulator for Stall-Fed Dairy Cattle in the Tropics. Animals 10, 816. https://doi.org/10.3390/ani10050816.
- For more information for agroforestry modeling, please have a look at https://worldagroforestry.org/output/wanulcas-model-water-nutrient-and-light-capture-agroforestry-systems.

5.2 Major international research projects work on integrative approaches

- AgMIP: http://www.agmip.org/.
- Big data and crop modeling groups (CGIAR initiative): https://bigdata.cgiar.org/communities-of-practice/crop-modelling/.

6 References

Adam, M., Dzotsi, K. A., Hoogenboom, G., Traoré, P. C. S., Porter, C. H., Rattunde, H. F. W., Nebie, B., Leiser, W. L., Weltzien, E. and Jones, J. W. 2018. Modelling varietal differences in response to phosphorus in West African sorghum. *Eur. J. Agron.* 100, 35–43.

Adam, M., MacCarthy, D. S., Traoré, P. C. S., Nenkam, A., Freduah, B. S., Ly, M. and Adiku, S. G. K. 2020. Which is more important to sorghum production systems in the Sudano-Sahelian zone of West Africa: Climate change or improved management practices? *Agric. Syst.* 185, 102920. https://doi.org/10.1016/j.agsy.2020.102920.

Adzawla, W., Baumüller, H., Donkoh, S. A. and Serra, R. 2020. Effects of climate change and livelihood diversification on the gendered productivity gap in Northern Ghana. *Clim. Dev.* 12(8), 743–755. https://doi.org/10.1080/17565529.2019.1689093.

Affholder, F., Poeydebat, C., Corbeels, M., Scopel, E. and Tittonell, P. 2013. The yield gap of major food crops in family agriculture in the tropics: Assessment and analysis through field surveys and modelling. *Field Crops Res.* 143, 106–118. https://doi.org/10.1016/j.fcr.2012.10.021.

Akinseye, F. M., Adam, M., Agele, S. O., Hoffmann, M. P., Traoré, P. C. S. and Whitbread, A. M. 2017. Assessing crop model improvements through comparison of sorghum (Sorghum bicolor L. Moench) simulation models: A case study of West African varieties. *Field Crops Res* 201, 19–31. http://doi.org/10.1016/j.fcr.2016.10.015.

Amole, T. A., Zijlstra, M., Descheemaeker, K., Ayantunde, A. A. and Duncan, A. J. 2017. Assessment of lifetime performance of small ruminants under different feeding systems. *Animal* 11(5), 881–889.

Amouzou, K. A., Naab, J. B., Lamers, J. P. A. and Becker, M. 2018. CERES-maize and CERES-Sorghum for modeling growth, nitrogen and phosphorus uptake, and soil moisture dynamics in the dry savanna of West Africa. *Field Crops Res.* 217, 134–149. https://doi.org/10.1016/j.fcr.2017.12.017.

Assogba, G. G. C., Adam, M., Berre, D. and Descheemaeker, K. 2022. Managing biomass in semi-arid Burkina Faso: Strategies and entry points for better production in contrasted farm systems. *Agric. Syst.* 201, 103458.

Assouma, M. H., Lecomte, P., Hiernaux, P., Ickowicz, A., Corniaux, C., Decruyenaere, V., Diarra, A. R. and Vayssières, J. 2018. How to better account for livestock diversity and fodder seasonality in assessing the fodder intake of livestock grazing semi-arid sub-Saharan Africa rangelands. *Livest. Sci.* 216, 16–23.

Baldé, A. B., Scopel, E., Affholder, F., Corbeels, M., Da Silva, F. A. M., Xavier, J. H. V. and Wery, J. 2011. Agronomic performance of no-tillage relay intercropping with maize under smallholder conditions in Central Brazil. *Field Crops Res.* 124(2), 240–251. https://doi.org/10.1016/j.fcr.2011.06.017.

Bateki, C. A. and Dickhoefer, U. 2020. Evaluation of the modified LIVestock SIMulator for stall-fed dairy cattle in the tropics. *Animals (Basel)* 10(5), 816. https://doi.org/10.3390/ani10050816.

Bayala, J., Sanou, J., Teklehaimanot, Z., Kalinganire, A. and Ouédraogo, S. 2014. Parklands for buffering climate risk and sustaining agricultural production in the Sahel of West Africa. *Curr. Opin. Environ. Sustain.* 6, 28–34. https://doi.org/10.1016/J.COSUST.2013.10.004.

Berre, D., Diarisso, T., Andrieu, N., Le Page, C. and Corbeels, M. 2021. Biomass flows in an agro-pastoral village in West-Africa: Who benefits from crop residue mulching? *Agric. Syst.* 187, 102981. https://doi.org/10.1016/j.agsy.2020.102981.

Boansi, D., Owusu, V., Akpene Tambo, J., Emmanuel, D. and Owusu, B. A. 2021. Rainfall Shocks and Household Welfare: Evidence from Northern Ghana. *Agric. Syst.* 194, 28–34. https://doi.org/10.1016/j.agsy.2021.103267.

Brandt, M., Hiernaux, P., Rasmussen, K., Mbow, C., Kergoat, L., Tagesson, T., Ibrahim, Y. Z., Wélé, A., Tucker, C. J. and Fensholt, R. 2016. Assessing woody vegetation trends in

Sahelian drylands using MODIS based seasonal metrics. *Remote Sens. Environ.* 183, 215–225. https://doi.org/10.1016/j.rse.2016.05.027.

Burney, J. A., Naylor, R. L. and Postel, S. L. 2013. The case for distributed irrigation as a development priority in sub-Saharan Africa. *Proc. Natl. Acad. Sci. U. S. A.* 110(31), 12513–12517. https://doi.org/10.1073/pnas.1203597110.

Chauhan, D. S. and Ghosh, N. 2014. Impact of climate change on livestock production:A review. *J. Anim. Res.* 4(2), 223–239.

Claessens, L., Antle, J. M., Stoorvogel, J. J., Valdivia, R. O., Thornton, P. K. and Herrero, M. 2012. A method for evaluating climate change adaptation strategies for small-scale farmers using survey, experimental and modeled data. *Agric. Syst.* 111, 85–95. https://doi.org/10.1016/j.agsy.2012.05.003.

Corbeels, M., Berre, D., Rusinamhodzi, L. and Lopez-Ridaura, S. 2018. Can we use crop modelling for identifying climate change adaptation options? *Agric. Forest Meteorol.* 256–257, 46–52.

Corbeels, M., Chirat, G., Messad, S. and Thierfelder, C. 2016. Performance and sensitivity of the DSSAT crop growth model in simulating maize yield under conservation agriculture. *Eur. J. Agron.* 76, 41–53. https://doi.org/10.1016/j.eja.2016.02.001.

Dagar, J. C., Sileshi, G. W. and Akinnifesi, F. K. 2020. Agroforestry to enhance livelihood security in Africa: Research trends and emerging challenges. In: Dagar, J. C., Gupta, S. R. and Teketay, D. (Eds.), *Agroforestry for Degraded Landscapes: Recent Advances and Emerging Challenges* (Vol.1). Singapore: Springer, pp. 71–134. https://doi.org/10.1007/978-981-15-4136-0_3.

De Ridder, N., Sanogo, O. M., Rufino, M. C., Van Keulen, H. and Giller, K. E. 2015. Milk: The new white gold? Milk production options for smallholder farmers in Southern Mali. *Animal* 9(7), 1221–1229. https://doi.org/10.1017/S1751731115000178.

Descheemaeker, K., Zijlstra, M., Masikati, P., Crespo, O. and Homann-Kee Tui, S. 2018. Effects of climate change and adaptation on the livestock component of mixed farming systems: A modelling study from semi-arid Zimbabwe. *Agric. Syst.* 159, 282–295.

Descheemaeker, K., Zijlstra, M., Ramilan, T., Senda, T., Timpong-Jones, E. C., Nenkam, A., Sajid, M., Singh, S., Baigorria, G., Adam, M., Shalander, K. and Whitbread, A. M. 2016. Impact of climate change on the livestock component of mixed farming systems: Modelling evidence from regional integrated assessments across sub-Saharan Africa and South Asia. Presented at the Global Workshop of the Agricultural Model Intercomparison and Improvement Project (AgMIP). AgMIP, Montpellier, France, pp. 30–30.

Eckel, C. C. and Grossman, P. J. 2008. Forecasting risk attitudes: An experimental study using actual and forecast gamble choices. *J. Econ. Behav. Organ.* 68(1), 1–17. https://doi.org/10.1016/j.jebo.2008.04.006.

Falconnier, G. N., Descheemaeker, K., Traore, B., Bayoko, A. and Giller, K. E. 2018. Agricultural Intensification and Policy Interventions: Exploring Plausible Futures for Smallholder Farmers in Southern Mali. *Land Use Policy*. 70, 623–34. https://doi.org/10.1016/j.landusepol.2017.10.044.

Falconnier, G. N., Corbeels, M., Boote, K. J., Affholder, F., Adam, M., MacCarthy, D. S., Ruane, A. C., Nendel, C., Whitbread, A. M., Justes, É., Ahuja, L. R., Akinseye, F. M., Alou, I. N., Amouzou, K. A., Anapalli, S. S., Baron, C., Basso, B., Baudron, F., Bertuzzi, P., Challinor, A. J., Chen, Y., Deryng, D., Elsayed, M. L., Faye, B., Gaiser, T., Galdos, M., Gayler, S., Gerardeaux, E., Giner, M., Grant, B., Hoogenboom, G., Ibrahim, E.

S., Kamali, B., Kersebaum, K. C., Kim, S. H., Laan, M. van der, Leroux, L., Lizaso, J. I., Maestrini, B., Meier, E. A., Mequanint, F., Ndoli, A., Porter, C. H., Priesack, E., Ripoche, D., Sida, T. S., Singh, U., Smith, W. N., Srivastava, A., Sinha, S., Tao, F., Thorburn, P. J., Timlin, D., Traore, B., Twine, T. and Webber, H. 2020. Modelling climate change impacts on maize yields under low nitrogen input conditions in sub-Saharan Africa. *Glob. Change Biol.* 26(10), 5942–5964. https://doi.org/10.1111/gcb.15261.

FAO. 2006. *Livestock's Long Shadow: Environmental Issues and Options*. Rome: Food and Agricultural Organization of the United Nations.

FAOSTAT. 2018. *FAOSTAT*. Rome, Italy: Food and Agriculture Organization of the United Nations. Available at: http://faostat.fao.org.

Faye, B., Webber, H., Naab, J. B., MacCarthy, D. S., Adam, M., Ewert, F., Lamers, J. P. A., Schleussner, C.-F., Ruane, A., Gessner, U., Hoogenboom, G., Boote, K., Shelia, V., Saeed, F., Wisser, D., Hadir, S., Laux, P. and Gaiser, T. 2018. Impacts of 1.5 versus 2.0°C on cereal yields in the West African Sudan Savanna. *Environ. Res. Lett.* 13(3), 034014. https://doi.org/10.1088/1748-9326/aaab40.

Faye, J. M., Maina, F., Hu, Z., Fonceka, D., Cisse, N. and Morris, G. P. 2019. Genomic signatures of adaptation to Sahelian and Soudanian climates in sorghum landraces of Senegal. *Ecol. Evol.* 9(10), 6038–6051. https://doi.org/10.1002/ece3.5187.

Foley, J. A., Ramankutty, N., Brauman, K. A., Cassidy, E. S., Gerber, J. S., Johnston, M., Mueller, N. D., O'Connell, C., Ray, D. K., West, P. C., Balzer, C., Bennett, E. M., Carpenter, S. R., Hill, J., Monfreda, C., Polasky, S., Rockström, J., Sheehan, J., Siebert, S., Tilman, D. and Zaks, D. P. M. 2011. Solutions for a cultivated planet. *Nature* 478(7369), 337–342. https://doi.org/10.1038/nature10452.

Freduah, B. S., MacCarthy, D. S., Adam, M., Ly, M., Ruane, A. C., Timpong-Jones, E. C., Traore, P. S., Boote, K. J., Porter, C. and Adiku, S. G. K. 2019. Sensitivity of maize yield in smallholder systems to climate scenarios in semi-arid regions of West Africa: Accounting for variability in farm management practices. *Agronomy* 9(10), 639. https://doi.org/10.3390/agronomy9100639.

Ganeme, A., Douzet, J.-M., Traoré, S., Dusserre, J., Kaboré, R., Tirogo, H., Nabaloum, O., Ouedraogo, N. W.-Z. S. and Adam, M. 2021. L'association sorgho/niébé au poquet, une pratique traditionnelle en zone soudano-sahélienne à faible rendement: Etat des lieux et pistes d'amélioration. *Int. J. Innov. Appl. Stud.* 31, 836–848.

Ganyo, K. K., Muller, B., Guisse, A. and Adam, M. 2018. Fertilization strategies based on climate information to enhance food security through improved dryland cereals production. In: Leal Filho, W. (Ed.), *Handbook of Climate Change Resilience*. Cham, Suisse: Springer Nature, pp. 917–934. https://doi.org/10.1007/978-3-319-93336-8_90.

Gbegbelegbe, S., Cammarano, D., Asseng, S., Robertson, R., Chung, U., Adam, M., Abdalla, O., Payne, T., Reynolds, M., Sonder, K., Shiferaw, B. and Nelson, G. 2017. Baseline simulation for global wheat production with CIMMYT mega-environment specific cultivars. *Field Crops Res.* 202, 122–135. https://doi.org/10.1016/j.fcr.2016.06.010.

Gérardeaux, E., Falconnier, G., Gozé, E., Defrance, D., Kouakou, P.-M., Loison, R., Sultan, B., Affholder, F. and Muller, B. 2021. Adapting rainfed rice to climate change: A case study in Senegal. *Agron. Sustain. Dev.* 41(4), 57. https://doi.org/10.1007/s13593-021-00710-2.

Gonzalez, P., Tucker, C. J. and Sy, H. 2012. Tree density and species decline in the African Sahel attributable to climate. *J. Arid Environ.* 78, 55–64. https://doi.org/10.1016/j.jaridenv.2011.11.001.

Graves, A. R., Burgess, P. J., Liagre, F., Terreaux, J.-P., Borrel, T., Dupraz, C., Palma, J. and Herzog, F. 2011. Farm-SAFE: The process of developing a plot- and farm-scale model of arable, forestry, and silvoarable economics. *Agrofor. Syst.* 81(2), 93–108.

Greatrex, H., Hansen, J. W., Garvin, S., Diro, R., Blakeley, S., Le Guen, M., Rao, K. N. and Osgood, D. E. 2015. *Scaling Up Index Insurance for Smallholder Farmers: Recent Evidence and Insights*. CCAFS Report No. 14. CGIAR Research Program on Climate Change, Agriculture and Food Security, Copenhagen (CCAFS).

Grothmann, T. and Patt, A. 2005. Adaptive capacity and human cognition: The process of individual adaptation to climate change. *Glob. Environ. Change* 15(3), 199–213. https://doi.org/10.1016/j.gloenvcha.2005.01.002.

Hiernaux, P., Adamou Kalilou, A., Kergoat, L., Brandt, M., Mougin, E. and Fitts, Y. 2022. Woody plant decline in the Sahel of Western Niger (1996–2017):is it driven by climate or land use changes? *J. Arid Environ.* 200, 104719. https://doi.org/10.1016/j.jaridenv.2022.104719.

Hijmans, R. J., Guarino, L., Cruz, M. and Rojas, E. 2001. Computer tools for spatial analysis of plant genetic resources data. 1. DIVA-GIS. *Plant Genet. Resour. Newsl.* 127, 15–19.

Homann-Kee Tui, S. H., Descheemaeker, K., Valdivia, R. O., Masikati, P., Sisito, G., Moyo, E. N., Crespo, O., Ruane, A. C. and Rosenzweig, C. 2021. Climate change impacts and adaptation for dryland farming systems in Zimbabwe: A stakeholder-driven integrated multi-model assessment. *Clim. Change* 168(1–2), 10. Available at: https://www.sciencedirect.com/science/article/pii/S0308521X1630186X?via%3Dihub#f0025.

Huet, E. K., Adam, M., Giller, K. E. and Descheemaeker, K. 2020. Diversity in perception and management of farming risks in southern Mali. *Agric. Syst.* 184, 102905.

Huet, E. K., Adam, M., Traore, B., Giller, K. E. and Descheemaeker, K. K. E. 2022. *Coping With Cereal Production Risks Due to the Vagaries of Weather, Labour Shortage and Input Markets Through Management in Southern Mali. Under Revision in Ag. Syst.*

IPCC. 2019. Summary for policymakers. In: *Climate Change and Land: An IPCC Special Report on Climate Change, Desertification, Land Degradation, Sustainable Land Management, Food Security, and Greenhouse Gas Fluxes in Terrestrial Ecosystems.*

Jones, P. G. and Thornton, P. K. 2009. Croppers to livestock keepers: Livelihood transitions to 2050 in Africa due to climate change. *Environ. Sci. Policy* 12(4), 427–437.

Keating, B. A. 2020. Crop, soil and farm systems models – Science, engineering or snake oil revisited. *Agric. Syst.* 184, 102903. https://doi.org/10.1016/j.agsy.2020.102903.

Kindt, R. 2018. Ensemble species distribution modelling with transformed suitability values. *Environ. Modell. Softw.* 100, 136–145. https://doi.org/10.1016/j.envsoft.2017.11.009.

Koffi, C. K., Djoudi, H. and Gautier, D. 2017. Landscape diversity and associated coping strategies during food shortage periods: Evidence from the Sudano-Sahelian region of Burkina Faso. *Reg. Environ. Change* 17(5), 1369–1380. https://doi.org/10.1007/s10113-016-0945-z.

Kuyah, S., Öborn, I., Jonsson, M., Dahlin, A. S., Barrios, E., Muthuri, C., Malmer, A., Nyaga, J., Magaju, C., Namirembe, S., Nyberg, Y. and Sinclair, F. L. 2016. Trees in agricultural landscapes enhance provision of ecosystem services in Sub-Saharan Africa. *Int. J. Biodivers. Sci. Ecosyst. Serv. Manag.* 12, 1–19. https://doi.org/10.1080/21513732.2016.1214178.

Leroux, L., Falconnier, G. N., Diouf, A. A., Ndao, B., Gbodjo, J. E., Tall, L., Balde, A. A., Clermont-Dauphin, C., Bégué, A., Affholder, F. and Roupsard, O. 2020. Using remote sensing to assess the effect of trees on millet yield in complex parklands of Central Senegal. *Agric. Syst.* 184, 1–13. https://doi.org/10.1016/j.agsy.2020.102918.

Leroux, L., Faye, N. F., Jahel, C., Falconnier, G. N., Diouf, A. A., Ndao, B., Tiaw, I., Senghor, Y., Kanfany, G., Balde, A., Dieye, M., Sirdey, N., Alobo Loison, S., Corbeels, M., Baudron, F. and Bouquet, E. 2022. Exploring the agricultural landscape diversity-food security nexus: An analysis in two contrasted parklands of Central Senegal. *Agric. Syst.* 196, 103312. https://doi.org/10.1016/j.agsy.2021.103312.

Lo Seen, D., Mougin, E., Rambal, S., Gaston, A. and Hiernaux, P. 1995. A regional Sahelian grassland model to be coupled with multispectral satellite data. II: Toward the control of its simulations by remotely sensed indices. *Remote Sens. Environ.* 52(3), 194–206.

Luedeling, E., Smethurst, P. J., Baudron, F., Bayala, J., Huth, N. I., van Noordwijk, M., Ong, C. K., Mulia, R., Lusiana, B., Muthuri, C. and Sinclair, F. L. 2016. Field-scale modeling of tree-crop interactions: Challenges and development needs. *Agric. Syst.* 142, 51–69. https://doi.org/10.1016/J.AGSY.2015.11.005.

MacCarthy, D. S., Akponikpe, P. B. I., Narh, S. and Tegbe, R. 2015. Modeling the effect of seasonal climate variability on the efficiency of mineral fertilization on maize in the coastal savannah of Ghana. *Nutr. Cycl. Agroecosyst.* 102(1), 45–64. https://doi.org/10.1007/s10705-015-9701-x.

Maltas, A. 2007. *Analyse par Expérimentation et Modélisation de la Dynamique de l'Azote dans les Systèmes Sous Semis Direct avec Couverture Végétale des Cerrados Brésiliens (These de Doctorat).* ENSA, Montpellier.

Marx, S. M., Weber, E. U., Orlove, B. S., Leiserowitz, A., Krantz, D. H., Roncoli, C. and Phillips, J. 2007. Communication and mental processes: Experiential and analytic processing of uncertain climate information. *Glob. Environ. Change* 17(1), 47–58. https://doi.org/10.1016/j.gloenvcha.2006.10.004.

Mbow, C., Van Noordwijk, M., Luedeling, E., Nuefeldt, H., Minang, P. A. and Kowero, G. 2014. Agroforestry solutions to address food security and climate change challenges in Africa. *Curr. Opin. Environ. Sustain.* 6, 61–67. https://doi.org/10.1016/j.cosust.2013.10.014.

Mertz, O., Mbow, C., Reenberg, A. and Diouf, A. 2009. Farmers' perceptions of climate change and agricultural adaptation strategies in rural Sahel. *Environ. Manag.* 43(5), 804–816. https://doi.org/10.1007/s00267-008-9197-0.

Miller, D. C., Muñoz-Mora, J. C. and Christiaensen, L. 2017. Prevalence, economic contribution, and determinants of trees on farms across Sub-Saharan Africa. *For. Policy Econ.* 84, 47–61. https://doi.org/10.1016/J.FORPOL.2016.12.005.

Mobbs, D. C., Lawson, G. J., Friend, A. D., Crout, N. M. J., Arah, J. R. M. and Hodnett, M. G. 1999. HyPAR model for Agroforestry Systems. Technical Manual. *Model Description for Version 3.0.* Institute of Terrestrial Ecology, Penicuik, UK.

Nendel, C., Melzer, D. and Thorburn, P. J. 2019. The nitrogen nutrition potential of arable soils. *Sci. Rep.* 9(1), 5851. https://doi.org/10.1038/s41598-019-42274-y.

Niang, I., Ruppel, O. C., Abdrabo, M. A., Essel, A., Lennard, C., Padgham, J. and Urquhart, P. 2014. Impacts, adaptation and vulnerability. Contribution of Working Group II to the Fifth Assessment Report of the Intergovernmental Panel on Climate Change. Africa. In *Climate Change*. Cambridge: Cambridge University Press.

Ramirez-Villegas, J., Jarvis, A. and Laederach, P. 2013. Empirical approaches for assessing impacts of climate change on agriculture: The EcoCrop model and a case study with grain sorghum. *Agric. For. Meteorol.* 170, 67–78. https://doi.org/10.1016/j.agrformet.2011.09.005.

Ranaivoson, L., Falconnier, G. N., Affholder, F., Muller, B., Autfray, P., Leroux, L., Ripoche, A. 2022. Can green manure contribute to sustainable intensification of rainfed rice production in Madagascar? *Field Crops Res.* 289 Submitted to Field Crops Research.

Reed, J., Vianen van, J., Foli, S., Clendenning, J., Yang, K., MacDonald, M., Petrokofsky, G., Padoch, C. and Sunderland, T. 2017. Trees for life: The ecosystem service contribution of trees to food production and livelihoods in the tropics. *For. Policy Econ.* 84, 62–71.

Rojas-Downing, M. M., Nejadhashemi, A. P., Harrigan, T. and Woznicki, S. A. 2017. Climate change and livestock: Impacts, adaptation, and mitigation. *Clim. Risk Manag.* 16, 145–163. https://doi.org/10.1016/j.crm.2017.02.001.

Rufino, M. C., Dury, J., Tittonell, P., van Wijk, M. T., Herrero, M., Zingore, S., Mapfumo, P. and Giller, K. E. 2011. Competing use of organic resources village-level interactions between farm types and climate variability in a communal area of NE Zimbabwe. *Agric. Syst.* 104(2), 175–190.

Rufino, M. C., Herrero, M., van Wijk, M. T., Hemerik, L., de Ridder, N. and Giller, K. E. 2009. Lifetime productivity of dairy cows in smallholder farming systems of the highlands of Central Kenya. *Animal* 3(7), 1044–1056.

Sanogo, K., Gebrekirstos, A., Bayala, J., Villamor, G. B., Kalinganire, A. and Dodiomon, S. 2016. Potential of dendrochronology in assessing carbon sequestration rates of Vitellaria paradoxa in southern Mali, West Africa. *Dendrochronologia* 40, 26–35. https://doi.org/10.1016/j.dendro.2016.05.004.

Savary, S., Willocquet, L., Pethybridge, S. J., Esker, P., McRoberts, N. and Nelson, A. 2019. The global burden of pathogens and pests on major food crops. *Nat. Ecol. Evol.* 3(3), 430–439.

Sejian, V., Gaughan, J. B., Bhatta, R. and Naqvi, S. M. K. 2016. Impact of climate change on livestock productivity. In: *Feedipedia-Animal Feed Resources Information System-INRA CIRAD AFZ and FAO*.

Sida, T. S., Baudron, F., Kim, H. and Giller, K. E. 2018. Climate-smart agroforestry: Faidherbia albida trees buffer wheat against climatic extremes in the Central Rift Valley of Ethiopia. *Agric. For. Meteorol.* 248, 339–347. https://doi.org/10.1016/J.AGRFORMET.2017.10.013.

Silva, J. V., Baudron, F., Reidsma, P. and Giller, K. E. 2019. Is labour a major determinant of yield gaps in sub-Saharan Africa? A study of cereal-based production systems in Southern Ethiopia. *Agric. Syst.* 174, 39–51. https://doi.org/10.1016/j.agsy.2019.04.009.

Sinare, H. and Gordon, L. J. 2015. Ecosystem services from woody vegetation on agricultural lands in Sudano-Sahelian West Africa. *Agric. Ecosyst. Environ.* 200, 186–199. https://doi.org/10.1016/j.agee.2014.11.009.

Tao, F., Rötter, R. P., Palosuo, T., Díaz-Ambrona, C. G. H., Mínguez, M. I., Semenov, M. A., Kersebaum, K. C., Nendel, C., Specka, X., Hoffmann, H., Ewert, F., Dambreville, A., Martre, P., Rodríguez, L., Ruiz-Ramos, M., Gaiser, T., Höhn, J. G., Salo, T., Ferrise, R., Bindi, M., Cammarano, D. and Schulman, A. H. 2018. Contribution of crop model structure, parameters and climate projections to uncertainty in climate change impact assessments. *Glob. Change Biol.* 24(3), 1291–1307. https://doi.org/10.1111/gcb.14019.

Taylor, C. M., Belušić, D., Guichard, F., Parker, D. J., Vischel, T., Bock, O., Harris, P. P., Janicot, S., Klein, C. and Panthou, G. 2017. Frequency of extreme Sahelian storms tripled since 1982 in satellite observations. *Nature* 544(7651), 475–478. https://doi.org/10.1038/nature22069.

ten Berge, H. F. M., Hijbeek, R., van Loon, M. P., Rurinda, J., Tesfaye, K., Zingore, S., Craufurd, P., van Heerwaarden, J., Brentrup, F., Schröder, J. J., Boogaard, H. L., de Groot, H. L. E. and van Ittersum, M. K. 2019. Maize crop nutrient input requirements for food security in sub-Saharan Africa. *Glob. Food Sec.* 23, 9–21. https://doi.org/10.1016/j.gfs.2019.02.001.

Thorburn, P. J., Boote, K. J., Nendel, C., Rötter, R. P. and Ewert, F. 2018. Recent advances in crop modelling to support sustainable agricultural production and food security under global change. *Eur. J. Agron.* 100, 35–43. https://doi.org/10.1016/j.eja.2018.04.001.

Traore, A., Falconnier, G. N., Mamoutou, K., Georges, S., Alassane, B. A., François, A., Michel, G. and Benjamin, S. 2021. Farmers' perception and adaptation strategies to climate change in central Mali. *Weather Clim. Soc.* 1. https://doi.org/10.1175/WCAS-D-21-0003.1.

Traoré, A., Falconnier, G. N., Sissoko, F., Sultan, B. and Affholder, F. 2022. Modelling sorghum-cowpea intercropping in semi-arid sub-Saharan Africa: Strengths and weaknesses of the Stics model. *Field Crops Res.*

Traore, B., Descheemaeker, K., van Wijk, M. T., Corbeels, M., Supit, I. and Giller, K. E. 2017. Modelling cereal crops to assess future climate risk for family food self-sufficiency in southern Mali. *Field Crops Res.* 201, 133–145. https://doi.org/10.1016/j.fcr.2016.11.002.

Valbuena, D., Erenstein, O., Homann-Kee Tui, S., Abdoulaye, T., Claessens, L., Duncan, A. J., Gérard, B., Rufino, M. C., Teufel, N., van Rooyen, A. and van Wijk, M. T. 2012. Conservation agriculture in mixed crop-livestock systems: Scoping crop residue trade-offs in Sub-Saharan Africa and South Asia. *Field Crops Res.* 132, 175–184. https://doi.org/10.1016/j.fcr.2012.02.022.

Valdivia, R. and Tui, H.-K. S., Antle, J., Subash, N., Singh, H., Nedumaram, S., Hathie, I., Geethalakshmi, V., Claessens, L. and Dickson, C. 2021. Representative agricultural pathways: A multi-scale foresight process to support transformation and resilience of farming systems. In: Rosenzweig, C., Mutter, C. Z. and Contreras, E. M. (Eds.), *Handbook of Climate Change and Agroecosystems: Climate Change and Farming System Planning in Africa and South Asia: AgMIP Stakeholder-Driven Research (In 2 Parts)* (Vol. 5). Hackensack: World Scientific Publishing.

van Ittersum, M. K., van Bussel, L. G. J., Wolf, J., Grassini, P., van Wart, J., Guilpart, N., Claessens, L., de Groot, H., Wiebe, K., Mason-D'Croz, D., Yang, H., Boogaard, H., van Oort, P. A. J., van Loon, M. P., Saito, K., Adimo, O., Adjei-Nsiah, S., Agali, A., Bala, A., Chikowo, R., Kaizzi, K., Kouressy, M., Makoi, J. H. J. R., Ouattara, K., Tesfaye, K. and Cassman, K. G. 2016. Can sub-Saharan Africa feed itself? *Proc. Natl. Acad. Sci. U. S. A.* 113(52), 14964–14969. https://doi.org/10.1073/pnas.1610359113.

Van Noordwijk, M. and Lusiana, B. 1998. WaNuLCAS, a model of water, nutrient and light capture in agroforestry systems. *Agrofor. Syst.* 43(1/3), 217–242. https://doi.org/10.1023/A:1026417120254.

Vollset, S. E., Goren, E., Yuan, C. W., Cao, J., Smith, A. E., Hsiao, T., Bisignano, C., Azhar, G. S., Castro, E., Chalek, J., Dolgert, A. J., Frank, T., Fukutaki, K., Hay, S. I., Lozano, R., Mokdad, A. H., Nandakumar, V., Pierce, M., Pletcher, M., Robalik, T., Steuben, K. M., Wunrow, H. Y., Zlavog, B. S. and Murray, C. J. L. 2020. Fertility, mortality, migration, and population scenarios for 195 countries and territories from 2017 to 2100: A forecasting analysis for the global burden of disease study. *Lancet* 396(10258), 1285–1306. https://doi.org/10.1016/S0140-6736(20)30677-2.

Waldman, K. B., Todd, P. M., Omar, S., Blekking, J. P., Giroux, S. A., Attari, S. Z., Baylis, K. and Evans, T. P. 2020. Agricultural decision making and climate uncertainty in developing countries. *Environ. Res. Lett.* 15(11), 113004. https://doi.org/10.1088/1748-9326/abb909.

Wichern, J., Descheemaeker, K., Giller, K. E., Ebanyat, P., Taulya, G. and van Wijk, M. T. 2019. Vulnerability and adaptation options to climate change for rural livelihoods – A country-wide analysis for Uganda. *Agric. Syst.* 176, 102663.

Wichern, J., Hammond, J., van Wijk, M. T., Giller, K. E. and Descheemaeker, K. 2022. Production variability and adaptation strategies of Ugandan smallholders in the face of climate variability and market shocks. *Clim. Risk Manag.*

Wineman, A. and Crawford, E. W. 2017. Climate change and crop choice in Zambia: A mathematical programming approach. *NJAS Wageningen J. Life Sci.* 81(1), 19–31. https://doi.org/10.1016/j.njas.2017.02.002.

Wolfenson, D. and Roth, Z. 2019. Impact of heat stress on cow reproduction and fertility. *Anim. Front.* 9(1), 32–38. https://doi.org/10.1093/af/vfy027.

Yadav, B., Singh, G., Verma, A. K., Dutta, N. and Sejian, V. 2013. Impact of heat stress on rumen functions. *Vet. World* 6(12), 992–996.

Zhang, L., Ruiz-Menjivar, J., Luo, B., Liang, Z. and Swisher, M. E. 2020. Predicting climate change mitigation and adaptation behaviors in agricultural production: A comparison of the theory of planned behavior and the Value-Belief-Norm theory. *J. Environ. Psychol.* 68, 101408. https://doi.org/10.1016/j.jenvp.2020.101408.

Zomer, J. R., Trabucco, A., Coe, R., Place, F., Van Noordwijk, M. and Wu, J. C. 2014. Trees on farms: An update and reanalysis of agroforestry's global extent and socio-ecological characteristics (No. 179). Bogor, Indonesia. https://doi.org/10.5716/ WP14064.PDF.

Chapter 9

Modeling climate change impact on agro-ecosystem services

Charlotte Weil, École Polytechnique Fédérale de Lausanne, Switzerland; Justin A. Johnson, University of Minnesota, USA; and Rebecca Chaplin-Kramer, University of Minnesota and SPRING, USA

1 Introduction
2 Climate change impacts on ecosystem services in agricultural systems
3 Climate change impacts on food-provisioning ecosystem services
4 From models to decision-relevant metrics: integrating demand
5 Conclusion
6 Future trends in research
7 Where to look for further information
8 References

1 Introduction

Ecosystem services, the benefits that flow from ecosystems to people, are the foundation of the world's agricultural systems. They are critical to producing food and to ensuring that agricultural landscapes produce an array of vital benefits to society beyond food, such as clean and ample water, healthy soils, hospitable corridors for wildlife and insects, and cultural values and ways of life that are meaningful to people. Over the past 10 000 years, agriculture has evolved in one of the most climatically stable periods in history, and we need to anticipate how these systems and their multiple benefits will be impacted by climate change to understand if and how they can adapt and transform to continue to meet the needs of humanity.

To prosper amidst future climate change and uncertainty, there are three alternative strategies for coping with variability and change: controlling change by minimizing its influences; adapting to change; and transforming to a new system by creating or manipulating variability (Bennett et al., 2021; Gunderson and Holling, 2002). The first strategy, controlling change, has been central to agriculture since the Green Revolution, in efforts to minimize variability in soil fertility, pest impacts, and water availability through the use of chemicals and irrigation. This has yielded major successes, but at high costs in the form of

http://dx.doi.org/10.19103/AS.2022.0115.11

unanticipated and undesirable changes to other ecosystem services provided in agricultural landscapes (Foley et al., 2011). The second strategy, adaptation, has been employed in agricultural systems since their origin, through selective breeding that has allowed domestication to track shifting environmental conditions, and in the future may involve more dramatic measures of shifting the crops grown in different regions or the types of diets supported. To some, the distinction between the second and third strategies, adaptation and transformation, may be a matter of degree. A true transformation of our agricultural systems might involve creating variability through the establishment of multi-functional agricultural landscapes that produce an array of ecosystem services with different responses to change (Hails et al., 2019). In this chapter, we will focus mainly on the second strategy, as a necessary, but perhaps more pragmatic (in the near term), approach to facing climate change.

The resilience of a system is inherent in the long-term production of a variety of ecosystem services, in the face of environmental and economic shocks resulting from variability in climate change, global trade, and other physical and socioeconomic stressors. Creating this type of stability in agricultural systems requires maintaining the quality and quantity of ecosystem services that support agriculture and that are co-produced in agricultural landscapes. Designing agricultural systems that co-produce multiple ecosystem services rather than maximize one service (such as food provision from a single commodity) at the expense of others can enhance resilience by increasing the diversity of ecological processes that maintain a regime in the face of disturbance (Gordon et al., 2008). Defining multiple societal goals for agricultural landscapes can help guide how trade-offs between ecosystem services should be assessed and their relative importance (Goldstein et al., 2012). When services considered include not only delivery of food and water – but also a larger suite of desired benefits such as climate regulation, flood mitigation, recreation opportunities, cultural heritage, livelihoods, jobs, and the provision of other goods such as timber and wild game cogenerated in the landscape – trade-offs may appear less stark and agricultural systems can better meet societal needs (Chaplin-Kramer et al., in press).

Evaluating future ecosystem service provision under future climate and other conditions requires a few key elements in a modeling framework: spatial context, appropriate scale, and endpoints relevant to human well-being. Where agriculture is located in a landscape – and how and where specific practices are employed – can have an enormous effect on the total set of services produced and how equitably they are delivered (Fischer et al., 2008; Mitchell et al., 2013; Vejre and Brandt, 2003). For example, stream buffers and low tillage regimes may matter more depending on the slope, soils, and landscape context of sources of pollution uphill or other vegetation that may provide retention downhill. Furthermore, different services operate at different scales (from a few hundred meters for pollination to a watershed for water quality regulation to a global scale for climate regulation via carbon sequestration) and must be modeled at scales appropriate to their provision in order to capture the relevant ecological processes as they play out across landscapes. Finally, the supply of an ecosystem service is not truly a 'service' until it reaches a beneficiary and the degree to which a person benefits depends on their demand or need for that service. To evaluate whether ecosystem service supply is sufficient to meet demand, the supply must also be translated into benefit-relevant metrics, put in terms of their contribution to human well-being (whether monetary, health, or other aspects).

In this chapter, we review how multiple ecosystem services may be impacted by climate change and provide an example for modeling benefit-relevant metrics that can assess the potential impacts of future global change on human well-being. Section 2 reviews the state of the art of climate impacts on supporting services for agricultural production and co-benefits of agroecosystems, Section 3 emphasizes the need for a systemic global approach for modeling food provision as an ecosystem service, in its context, and integrating climate change and societal changes cohesively, with a focus on one approach to modeling nutrition under cohesive climate and societal changes getting to benefit-relevant indicators, and Section 4 expands from the 'supply' aspect to benefit-relevant indicators, putting nutrition further in its context by integrating projections of demand.

2 Climate change impacts on ecosystem services in agricultural systems

In this section we review ecosystem services in agricultural landscapes and the measured or modeled impacts of climate change on these services. We divide our review into two categories of services: ecosystem services to agriculture, those that support the production of the primary service of interest, namely food provision; and ecosystem services from agricultural landscapes, including the cultural and regulating services that are co-produced along with food provision in these landscapes. We conclude this review with a summary of efforts to integrate modeling of these diverse ecosystem services in agricultural landscapes and research gaps that need to be filled to enable more holistic predictions of climate impacts in agricultural systems.

Past studies have reviewed the effect of climate change on ecosystem services (not specifically in agricultural systems, but covering many of the same services relevant there), and concluded that most impacts tend to be negative (Grimm et al., 2013; Scholes, 2016; Weiskopf et al., 2020). Scholes (2016) proposed three hypotheses for why this would be the case: (1) any change is disruptive before a system reaches a new equilibrium, (2) human societies evolved under the current climate and so did their demand for ecosystem services, so deviation from those conditions would necessarily be sub-optimal, and (3) humans are approaching multiple planetary boundaries that are correlated with climate change and ecosystems are being pushed past sustainable limits. All three require some human action to adapt, the first being more towards managing transition and slowing the pace of change, and the latter two involve scaling back demand for services while also scaling back other pressures for the third one. It is especially important to recognize that society's most vulnerable populations, including children, the elderly, socio-economically disadvantaged, and otherwise marginalized communities are often disproportionately impacted by climate change and the resulting impact on ecosystem services (Weiskopf et al., 2020).

2.1 Ecosystem services to agriculture

2.1.1 Maintenance of soil fertility and structure

Invertebrates and micro-organisms living in the soil maintain its fertility and structure, but are vulnerable to disruption or damage to the soil. Therefore, while agriculture relies on fertile, uncompacted soil, it can also undermine future soil productivity, depending on the management practices used and the nature of the soil (Pereira et al., 2018). Declines in soil ecosystem services have been shown to reduce both maximum yield and fertilizer-use efficiency (Brady et al., 2015), meaning that relying on synthetic substitutes for this ecosystem service can make future attempts to substitute less effective.

Climate impacts on soil fertility are likely to be complex given the complexity in the number of processes contributing to it, especially in terms of how temperature and soil moisture interact. However, rising temperatures increase rates of soil organic matter decomposition, causing losses in soil fertility that may be further compounded by climate-induced changes to soil invertebrate and microbial community composition (Weiskopf et al., 2020), both of which could compromise soil ecosystem services (Lavelle et al., 2006; Saccà et al., 2017).

Soil productivity is important not just for the current and future production of food, but also for the ways in which it interacts with other services provided in agricultural landscapes. For example, high levels of nutrients in soils (part, but not all, of soil fertility) can run off or leach out of the fields and pollute local waterways, reducing the availability of high-quality water or increasing the expense of cleaning the water to adequate quality for use. Likewise, loss of soil fertility reduces the capacity of soil to sequester carbon and avoid greenhouse gas emissions. These impacts are addressed in the section on ecosystem services from agricultural landscapes.

2.1.2 Pollination and pest regulation

Pollination and pest regulation are both insect-mediated ecosystem services, and healthy populations of such beneficial insects require non-crop habitats for floral resources, nesting, overwintering, and alternate prey (Chaplin-Kramer et al., 2011; Kennedy et al., 2013; Kremen et al., 2007; Ricketts et al., 2008; Tscharntke et al., 2002). Pollination enhances the production of 75% of the 115 major crop species grown globally (Aizen et al., 2009; Klein et al., 2007), and crop pollination provided by wild bee communities outperforms pollination by domesticated honeybees by more than two-fold (Garibaldi et al., 2013). Wild pollinators are becoming increasingly important amidst colony collapse disorder and other threats to honeybees (Potts et al., 2010). Pest regulation provided by natural enemies (predators and parasitoids of agricultural pests) is thought to provide up to 50% of the control of agricultural pest species (Pimentel et al., 2005), yet in spite of that and even with (or perhaps partly because of) the use of chemical pesticides, an estimated 20% of global crop production is lost to animal pest damage (Oerke, 2006).

Climate change is already disrupting many of these important plant-pollinators and predator-pest relationships, due to changes in the spatial distribution and seasonal timing of flowering and population growth dynamics. Climate will likely impact a pollinator and a plant it relies on differently; so if crop flowering begins before pollinators have emerged or if pollinators are active before any flowering resources are available to sustain them (Miller-Struttmann et al., 2015), fewer pollinators available for the crop. Some pollinators appear to be more plastic than others (Bartomeus et al., 2011) but we cannot count on this for every species. Meanwhile, pest dynamics are even more complex to understand or predict, because they are regulated from both the bottom up (their food source) and the top down (their enemies) and each may change in different ways (Castex et al., 2018), while at the same time pest population growth rates are often very sensitive to temperature (Sable and Rana, 2016). Global analyses have cautioned that with warming winters hundreds of pest species are expanding their ranges, an average of 2.7 km per year toward the poles (Bebber et al., 2013). Further compounding the challenges natural enemies may face in keeping up with pests' migration and growth dynamics, their own abundance and activity may be compromised through adaptive management strategies adopted by farmers to cope with climate change (Thomson et al., 2010).

2.1.3 Water provisioning

Agriculture depends innately on water provisioning and therefore impacts the water flowing through it that provides benefits to people and biodiversity downstream. We partition these entangled and overlapping ecosystem functions into water delivered to agriculture (mainly an issue of quantity) and water delivered downstream of agriculture (mainly an issue of quality and timing; discussed in the next section). Agriculture accounts for 85% of freshwater consumption globally, and currently, that use is not optimally allocated (D'Odorico et al., 2020). The volume of water available for irrigation depends primarily on climate and water storage management, rather than ecosystems per se. However, management of the agricultural ecosystem (including the types of crops grown, their need for irrigation, and the irrigation technology used) certainly affects its demand for water and the remaining water available for other agriculture (or other uses) downstream (Liu et al., 2013). The value of upstream land management to ensure dry-season flows could be considered not as the additional yield from irrigated crops compared to rainfed crops but as the additional yield and reduced risk given a managed versus unmanaged upstream watershed (Ponette-González et al., 2014).

Climate change is likely to have profound impacts on water availability for agriculture; global models suggest reversion of 20–60 million hectares of cropland from irrigated to rainfed management could be required under future climatic conditions (Elliott et al., 2014), though there is high uncertainty in magnitude and even directionality of changes in precipitation in some regions (IPCC, 2022). Extreme events like droughts have increased in frequency, duration, and severity since 1950 (Spinoni et al., 2014) and are expected to worsen under future conditions (Wang, 2005). Climate-induced changes in upstream watersheds such as snowpack melting earlier (Barnett et al., 2005) and increased wildfire frequency and severity that clog waterways with sediment and other debris (Bladon et al., 2014) will further stress water supplies to agriculture. Sobering examples have already shown these multiple stressors can combine in unpredictable ways, with droughts exacerbating wildfires and subsequent flooding causing mudslides on the fire-ravaged hillsides, ultimately shutting down water supplies (Grimm et al. 2013). As surface water supply grows unreliable, agriculture may increasingly turn to groundwater to meet its needs, but many aquifers are already overexploited and climate impacts on groundwater recharge are highly uncertain (Taylor et al., 2013).

2.2 Ecosystem services from agricultural landscapes

2.2.1 Food provision

The next section in this chapter examines climate impacts on food provisioning from agricultural landscapes in depth; however, none of the studies reviewed below have considered how the impacts on the supporting services described above will likely impact future food provision, along with direct impacts of climate on crop production. This is a key modeling gap and one that needs to be filled as we move toward a more holistic integrated assessment of climate impacts and adaptation options for our global food system (see Section 3 for further discussion). Furthermore, understanding climate impacts on the ecosystem service of food provision does not end at crop production, the supply side, but also must account for the demand side, or the nutritional needs of people which can be met in a variety of ways, although diet shifts often have cultural implications. The final section considers the climate impacts, alongside other aspects of global change, on the full ecosystem service of food provision, both supply and demand.

2.2.2 Regulation of water quality and timing of flow

Water quality, the speed of its flow, the time of year it is available, and whether it is surface water or groundwater are all influenced by upstream land cover and land management. The effects of vegetation on the downstream quality of water for drinking, the recharge of groundwater supplies for agricultural or other uses of water, or slowing the flow of potential flood waters downstream are all examples of hydrologic ecosystem services (Brauman et al., 2007) coming from agricultural landscapes. As reviewed above, the quantity of water available depends on climate and water management rather than ecosystem services per se. While large-scale changes in ecosystems can and do affect precipitation regimes (Keys et al., 2016; Swann et al., 2012), impacts occur at scales far larger than farm fields and as such are often not included in field and watershed-scale ecosystem service assessments. In contrast to water quantity, water quality is generally more dependent on land-use than climate. Fertilizer applied on fields, for example, can increase nutrient concentration downstream beyond the level acceptable for drinking water, but elements in the agricultural landscape such as riparian buffers can retain nutrients and maintain water quality for drinking water or other uses downstream (Brauman et al., 2014).

Climate change will likely increase the need for all of these services. Warming water temperatures speed up algal growth, exacerbating eutrophication resulting from nitrate loading into waterways, while decreased water flows will concentrate any pollutants that do enter streams (Whitehead et al., 2009), and therefore water quality regulation will be more critical than ever. Increased variability in precipitation and more extremes at both the wet and dry ends (Lehmann et al., 2018) make water flow regulation provided by ecosystems all the more important. Agriculture can be part of the problem or part of the solution, depending on how it is managed, and we may not have the luxury of choosing to maximize yields at the expense of water in the future.

2.2.3 Climate regulation

Agricultural landscapes can contribute to climate regulation at both the global and local scales, through the mitigation of greenhouse gasses (GHGs) and modification of the microclimate. Management practices can both increase carbon storage (in soil and vegetation) on farmlands (Hutchinson et al., 2007; Lal, 2015) and reduce major sources of GHGs such as methane from rice and livestock production (Smith et al., 2008) and nitrous oxide emissions from synthetic and organic nitrogen fertilizers (Hillier et al., 2012; Venterea et al., 2012). Agriculture can also contribute to microclimate regulation through strategic management of evapotranspiration. Integrating trees into agricultural landscapes can have substantial cooling effects (Zeppetello et al., 2022), as can irrigation (Mueller et al., 2016) and switching to perennial crops (Wang et al., 2021). However, all of these impacts could change as a result of climate change; the carbon sequestration potential or microclimate regulation potential of different management interventions is still an area of active study, and it is not well understood how those potentials could change under future climate conditions. We do know that carbon storage potential of forests is likely to decrease due to higher temperatures, increased water stress, and increased disturbance (Weiskopf et al., 2020), so similar trends may hold true for agriculture.

2.2.4 Cultural values

Agricultural landscapes are often thought of in terms of solely intensive cropland, but in reality, most of the world's agriculture operates in mosaic landscapes that incorporate a variety of natural elements alongside crops and pastures (Ellis and Ramankutty, 2008). The aesthetic quality of such landscapes is a source of identity and a sense of place for many rural residents around the world and it provides opportunities for recreation (including hiking, cycling, camping, or agritourism) for urban people as well. A growing body of research demonstrates links between experiences and exposure to such aesthetically pleasing green spaces and mental or physical health (Keeler et al., 2019; Ohly et al., 2016; van den Berg et al., 2015). These cultural services are strongly influenced by the intensity and extent of agriculture, which impact not only the aesthetic quality but the accessibility of the landscape. Cultural services can also relate to farming community heritage, manifest in some cultures that continue to grow economically disadvantageous crops in order to maintain traditional and culturally important foods (such as buckwheat grown in Yi communities in Yunnan, China) (Saunders Bulan et al., 2017).

Climate impacts on or interactions with cultural services have not received much attention. It is fairly well established that climate-driven extremes like heat waves and storms post threats to mental and physical health (Weiskopf et al., 2020); these could either counteract the effects the countryside can have or make these spaces all the more imperative to maintain. Recreational activities can also be curtailed by extreme temperatures or inclement weather. Agritourism may be more resilient to climate change than agricultural production (Valdivia and Barbieri, 2014), but if climate change makes certain types of production unviable in their most iconic landscapes (e.g. California wine country (Roehrdanz and Hannah, 2016), cultural losses could be devastating. The same could be said for heritage agriculture; climate may shift the growing ranges of certain culturally important crops beyond where the people who have been growing them for centuries, but who may be equally culturally tied to their local place, could follow.

2.3 Modeling multiple ecosystem services in agricultural landscapes

Most modeling of ecosystem services in agricultural (and other) systems tends to occur in parallel, with multiple ecosystem services impacted by common land-use or climate drivers rather than impacting each other. This provides a way of representing the multiple values of agricultural landscapes, although it may be missing some important feedback between services (Rieb et al., 2017). For example, SmartScape has been used to quantify the losses in water quality, soil, soil carbon, air quality, pollination, and biodiversity from switching from perennial to annual cropping systems (Tayyebi et al., 2016). ARIES, based on Bayesian network modeling, has been applied to explore the effects of management and climate on crop production and a suite of regulating services, demonstrating that improving air quality and water quality via reduced fertilizer application comes at a significant cost to crop yields while increasing carbon sequestration by avoiding tillage has little effect on yields (Balbi et al., 2015). However, representing feedback could reveal hidden synergies, such as through connecting carbon sequestration to longer-term maintenance and building of soil fertility, which could reduce the need for fertilizers and thereby improve yields and air and water quality simultaneously. Indeed, long-term observational studies of bundles of ecosystem services suggest that relationships between services in agricultural and surrounding landscapes are not stable over time (Renard et al., 2015), suggesting that trade-offs are not necessarily inevitable, but neither are synergies.

To better capture the interconnections and feedback between services in agricultural landscapes, an integrated, socioecological approach is needed. While frameworks for such integrated modeling have been proposed (Galic et al., 2012; Lescourret et al., 2015), no modeling systems representing connections between services are yet operational at the landscape scale, the scale over which ecosystem services operate. At coarser, regional to global scales, land-surface models have been adapted to represent ecosystem functional dynamics within agroecosystems as well as natural systems (Liu et al., 2016; Lokupitiya et al., 2009; Van den Hoof et al., 2011; Webler et al.,

2012; Wu et al., 2016). Some of these land-surface processes are captured in integrated assessment modeling (Popp et al., 2014; Stehfest et al., 2014), which has made great progress in representing the dual stressors of climate and land-use change and their effect on biodiversity, carbon, and other ecosystem services (Chaplin-Kramer et al., 2019; Leclère et al., 2020; Popp et al., 2017; Riahi et al., 2017; Rogelj et al., 2018). This has helped in identifying response options for keeping the global agricultural system within planetary boundaries (Springmann et al., 2018). However, downscaling the outputs of such models to a spatially explicit representation of landscape- or field-level processes key to agricultural ecosystems remains a challenge (Schipper et al., 2020; Suh et al., 2020), and linkages between ecosystem functions represented in global ecosystem models are not as well represented in ecosystem service models at these finer scales.

Where connections between ecosystem service models have been established at landscape scales, two-service linkages predominate, especially between water provision and crop production (Krysanova and White, 2015; Zhang et al., 2021) and between pollination and crop production (Kennedy et al., 2013; Lonsdorf et al., 2009; Ricketts and Lonsdorf, 2013). Ecological-economic modeling has been used to model the tradeoff between crop production and two services supporting it, pollination and pest control, showing that the adoption of a less toxic pesticide or better application methods by the farmers should increase the wild bees' productivity and, consequently, the total economic value of pollinators (Kleftodimos et al., 2021). Unfortunately, the complexity and context dependency of pest regulation by beneficial insects and other predators has eluded generalized modeling (Alexandridis et al., 2021; Karp et al., 2018), making it very difficult to reliably include this critical service in most crop modeling; for this particular service it may be necessary to develop different models for different archetypes of pest-enemy relationships based on their functional traits (Alexandridis et al., 2022). Other interactions between services are important to consider in agricultural landscapes, especially under climate change, but are rarely if ever included in mainstream ecosystem service and agricultural modeling tools. These include reductions in soil carbon sequestration impacting crop productivity, changes in crop nitrogen cycling (beyond the fertilizer applied) affecting downstream water quality, microclimate impacts on crop yields, atmospheric moisture recycling impacts on downwind precipitation and rainfed agriculture, and air quality impacts on labor affecting agricultural production, among others.

Furthermore, with the exception of crop-water linkages, few modeling efforts to integrate ecosystem service within agricultural systems have explored the impacts of climate change. Hydro-agronomic modeling has advanced beyond identifying potential climate impacts on water availability and crop yields (Ficklin et al., 2009; Krysanova and Arnold, 2008) toward modeling potential solutions, such as how to crop-switching can reduce irrigation requirements to mitigate expected climate impacts (Akbari et al., 2022). However, the well-established scientific evidence on how climate change impacts the diverse ecosystem services both supporting agriculture and provided by agricultural landscapes (as reviewed previously) is lacking representation in most ecosystem service modeling frameworks; this is an important frontier for further work.

Regardless of the ecosystem service or the scale at which it is modeled, the basic premise of ecosystem service modeling is that services need to be connected to human well-being in order to resonate with decision-makers and the general public - and yet ecosystem service assessments often fall short of this aim (Mandle et al., 2020). Even with modeling crop production - the primary ecosystem service from agricultural systems with clear connections to market values, livelihoods, and nutrition - many efforts predicting the impacts of climate change end at crop yields, without considering their implications for human well-being. Therefore, while Section 2 has focused on climate impacts on the variety of ecosystem services in agroecosystems, the remainder of this chapter will focus on crop production, with an intentional ecosystem service framing, linking climate impacts and related future socioecological changes to human well-being. This provides an example of how an ecosystem services approach to modeling climate impacts in agricultural systems and provide more decision-relevant information for future policy and planning.

3 Climate change impacts on food-provisioning ecosystem services

In this section, we discuss the need for a systemic approach to modeling food provision as an ecosystem service while simultaneously integrating climate change and societal changes. We emphasize aggregated and statistical approaches to this question, noting that more detailed crop-specific approaches are covered extensively in the rest of this book. The need for aggregated approaches, we argue, is that it allows us to express outcomes of the agricultural and food system in terms of benefits-relevant indicators (nutrition), discussed in Section 2. In Section 3, we review existing approaches to aggregate and statistical global crop models and discuss a new model that is able to generate insights on nutritional outcomes in the context of climate change and adaptation behaviors. Finally, we end in Section 4 with a discussion of how we can apply such crop models to cohesive scenarios of global change from the IPCC and the Intergovernmental Science-Policy Platform for Biodiversity and Ecosystem Services (IPBES) to assess the overall nutritional sufficiency of the global agricultural system under different conditions.

3.1 Global statistical model and their limits in integrating climate change impacts

Of the provisioning ecosystem services identified in the Millenium Ecosystem Assessment (2005), food provision has the highest assessed value based on market observations. Unlike many of the other ecosystem services, which typically cannot be assigned monetary value because they are not purchased on markets, agricultural production already has detailed global estimates for how much total value is produced, along with estimates of how this provisioning value is likely to be affected by climate change. Despite this abundance of evidence, there still exist questions about future scenarios of climate change, along with other scenarios including changing diets, different aggregate levels of land-use change, and different land management strategies. Many studies, including Foley (2011), Johnson et al. (2014, 2016), and Holt-Giménez (2019), analyze different strategies necessary to feed the world without undermining other sources of ecosystem services. More broadly, the tradeoff between food-provisioning and other ecosystem services has occupied a central place in the sustainability science literature (Johnson et al., 2021; Lawler et al., 2014; Nelson et al., 2008; Polasky et al., 2008).

Much of the most recent work at the global scale, however, has gone in the direction of modeling crop-specific projections of change (Müller et al., 2017; Rosenzweig et al., 2018). Relatively less attention has been paid to projecting benefit-relevant indicators of the global food system, such as those related to nutrition and overall food security. Of the models of aggregate agricultural production that do assess the food system as a whole (such as those used in the global tradeoff analyses discussed above), most are very aggregated or inaccurate compared to specific field-scale models. To improve the overall ease of conducting these analyses (and also to increase the overall comparability of results), the Natural Capital Project provides a global-scale, statistical crop model (documented in Sharp et al., 2014) to assess questions such as: how would switching from one cropping system to another affect total production? What would be the likely yield of crops grown in newly-converted agricultural land? How do different scenarios of global land-use contribute to food security goals? These are questions that are both important but not well-answered by field-scale models and which need a statistical crop model for analysis.

The crop production model from the Natural Capital Project provides global estimates of aggregate crop production via two different model approaches. First, for 175 crops and based on data from (Monfreda et al., 2008), the model can calculate how much would be produced based on a given land-use, land-cover map, and a set of assumptions on intensification. Specifically, yield increases are modeled via assuming some percentile closure of yield gaps for different climate conditions. This approach is a simplification of the work from Ray et al. (2019), and others that usefully simplified the crop production choice to land-use and partial improvement of yields to mimic the best-observed yield for given climate conditions. When more detailed input-output information is relevant to the analysis, the crop model also can calculate ten staple crops based on a regression documented in Mueller et al. (2012), based on inputted rates of nitrogen, phosphate, and potash application rates, irrigation, and other growing conditions. Although these models include great simplification of the underlying crop science, they can be quickly calculated and applied to high-resolution maps of global land-use, an area that is becoming increasingly relevant to conservation planning and ecosystem service protection (Chaplin-Kramer et al., 2014).

Both of these models can be used to inform planning under different climate change scenarios, such as in Ray et al. (2019). The underlying yield information is based on 'climate bins', discrete combinations of precipitation and growing degree days, located globally at 5 arcminute resolution. Within each climate bin, the distribution of yields is identified for each crop, creating comparable 'climate-analogues' to illustrate what yields are attainable within those climatic conditions. This framework has been especially useful in consideration of topics such as agricultural research and development or investments in yield improvements because such improvements can be seen as closing the yield gap in underperforming parts of each climate bin. In principle, it is straightforward to apply this to climate change analysis by simply recalculating where each climate bin will move as precipitation and temperature change, but to date, this has not been done and most applications, including all of those within the Natural Capital Project's tools, have been static.

Statistical models like these have significant shortcomings compared with more detailed process-based, crop-specific models. Important details, such as planting dates, growth interventions, or weather events, are not included in this type of framework. Moreover, models such as these are not able to identify *why* agricultural outputs lag behind others in similar climatic conditions. Nonetheless, statistical crop models can be useful for quickly assessing aggregated production of nutritionally relevant indicators, such as calories, protein, or micronutrients. Aggregate production statistics across all produced crops can be aggregated to different nutritional variables using, for instance, tables from (Monfreda et al., 2008), which provides a table that converts dry-tonnage production of each of the 175 crops to their nutritional makeup. Expressing aggregate nutritional outcomes expresses the production of the food system as a benefits-relevant indicator.

3.2 An integrated approach to modeling nutrition that accounts for climate change and management conditions

In this section we discuss a new approach (Weil et al., 2021) to yield modeling that focuses directly on nutrition by using the overall ability of the food system to produce calories (or other nutritional indicators), leaving unspecified exactly which crops are grown in each location. In doing so, the model is able to include the greater degree of flexibility the overall food system can have by allowing for crop-switching to occur under the combined effects of different climate, soil, and management conditions.

Leveraging machine-learning approaches and spatially explicit yield data, we project local caloric production at a global scale. To account effectively for climate change, and the inexorable adaptation that will result from it, we argue that it is necessary to innovate beyond identifying crop-specific, empirical relationships between climate and yield. Instead, our model calculates the overall ability of a location to produce calories in given growing conditions (climate, soil, management conditions, etc.), solving jointly the issues of (a) accounting effectively for a variety of climate conditions, (b) accounting for farmers' adaptation of choice of crops grown to evolving climate as a natural resilience strategy, and (c) getting to benefit-relevant indicators (to caloric sufficiency, as we will further discuss in Section 4).

This approach builds on the second resilience strategy: adaptation; by assuming adaptation of the crop mix by modeling total caloric yields, aggregating across all crop types from the outset. The response variable is the aggregated crop yields of 100 nutritionally relevant crops (excluding forage, sugar, cash crops, and oils) in calories per hectare, built from the dataset compiled by (Monfreda et al., 2008), which provides sub-national agricultural statistics of yields and area harvested for 175 crops circa the year 2000 (1997–2004). It is the most recent data available with this level of detail in terms of crop species, and at such a fine resolution. These 100 crop yields in weight are aggregated into a single aggregated caloric yield, using crops' caloric content from the FAO's food supply balance sheets (following the methods used in Johnson et al., 2014), similar to the work of West et al. (2014)).

Model inputs (predictors) are composed of climate, soil, and management information that capture fertilization, irrigation, and agriculture industrialization to the extent of known practices today. Though a technological breakthrough is not impossible, such change would result in a system-wide transformation that could not be predicted by statistical means. Even for specific crops, the combined effects of climate change and technology on future yields are not well understood. Recent studies diverge widely, with projected yields ranging from drastic decline due to climate change (Ray et al., 2019) to larger yield-enhancing impacts from technological improvements (Aggarwal et al., 2019).

In addition to the aggregation of crops translating the adaptation assumption, there are two novelties in this approach: (a) it includes a wider range of input variables than existing studies - Mueller et al. (2012) expanded on the set used by Lobell et al. (2011) by including fertilizer application and irrigation in a climate bin-based method but did not include soils data or socioeconomic variables; (b) the scale at which the regression is run is finer than existing global analyses. Specifically, we do not aggregate our data (e.g. by administrative region) and instead consider each pixel separately; this requires more computation time but captures more climate spatial variability.

The model itself uses a random forest approach with the gradient-boosting framework (Chen and Guestrin, 2016), training on a dataset of aggregated crop yields on ~1 million cropland pixels (n = 906 770) of size 5 arcmin with the 31 features describing climate, management and soil conditions. This tree-based statistical modeling approach performed well on unseen data: R^2 = 0.95; our model error is ~15% of current yield (RMSE = 1.06e9 calories per hectare), in line with previous work (Ray et al., 2019) reporting RMSE of 9 to 20% of current yields across different crops.

To ensure the validity of aggregating all crops' caloric yields and modeling directly the total caloric yield, we verify that pixels with similar inputs (i.e., similar climate, soil, and management conditions) will result in similar aggregated caloric yields. To prove this, we use the euclidean distance on standardized inputs to capture the dissimilarity between points. Sampling 4 million pixels pairs (with a set of 2000 randomly selected pixels, each associated with 2000 others), we find a strong positive Pearson correlation (+0.91, P-value 2.02e−43), on euclidean distances up to the 0.95 quantile (highest distances, thus very unsimilar points are not relevant to evaluate the impact of dissimilarity in inputs on distance in aggregated yields). In other words, the differences in caloric yields increase with the differences between inputs points, i.e. similar points have similar aggregated caloric yields.

3.3 Cohesive scenarios: Crop production projected to increase globally despite climate change

The importance of benefit-relevant indicators in the ecosystem services approach has been stressed; however such indicators are only truly relevant in context - and climate change does not happen in a vacuum. Therefore, it seems more relevant to analyze not only climate effects on yields but rather the joint effects of combined realistic changes in both society and climate. Towards this, international science bodies such as the IPCC and later the IPBES have gravitated to coupling climate change models with scenarios of other elements in the global system that change alongside. For instance, the five Shared Socioeconomic Pathways (SSPs) (Popp et al., 2017), coupled with representative concentration pathways (RCPs), describe possible futures including climate, cropland expansion, management, population, and diet in 2050. These cohesive scenarios of change help us envision the consequences of different resilience strategies.

Though existing crop models were used by the SSPs framework, projected yields do not account for climate change (Popp et al., 2017), are at a much coarser scales (from only 11 geopolitical regions in MESSAGE-GLOBIUM (Krey et al., 2016) and REMIND (Luderer et al., 2015) to around 300 agricultural land-use regions in GCAM), and are often based on demand (insofar as yield increases are defined at whatever level is necessary to meet future demands[1]). To improve on this, we apply our method for calculating total caloric yield to different SSP scenarios, innovating on past modeling approaches of crop-specific responses to climate change (Blanc, 2017; Jaggard et al., 2010; Lobell and Asseng, 2017; Lobell and Gourdji, 2012; Ostberg et al., 2017; Ray et al., 2019; Zhao et al., 2017; Zhu et al., 2019).

The inputs to our model were collected and pre-processed for each of the 5 scenarios, at a fine spatial resolution (5 arcmin). The climate input data is acquired from Worldclim v1.4 (Hijmans et al., 2005) and consists of 12 bioclimatic variables capturing annual values, seasonality, and variability of temperature and precipitation, as well as altitude and climate. These projections are available for 4 RCPs, we used the closest corresponding RCP for each SSP. To account for global circulation models (GCMs) variability, we approximated a climate ensembling method by modeling yields for each SSP with 4 different GCMs' climate data chosen based on data availability (GISS-E2-R, HadGEM2 ES, MIROC-ESM, and CCSM4), and then averaged the 4 caloric yields output maps. Topographic variables (altitude and slope) are also obtained from this source and considered unchanged. To account for management practices, we add fertilization application rates and the fraction of crop area irrigated (Hurtt et al., 2011), per crop type (perennial, annual C3, C4, nitrogen-fixing). In addition, GDP per capita (Dellink et al., 2017) captures the effects of agriculture industrialization as well as socioeconomic factors such as technology or capital investment. GDP per capita by country is categorized using quantiles to create 6 bins such that each contains a fixed amount of the data (10%, 15%, 25%, 25%, 15%, and 10%) to follow natural breaks of distribution. The reasoning behind categorizing this variable is the following: (a) we only account for adaptation here, thus no technological advances beyond the scope of what is known

[1] Several of the SSPs IAMs (Integrated Assessment Models, namely IMAGE (PBL Netherlands Environmental Assessment Agency, 2014), used by SSP1, GCAM (Calvin et al., 2019) used by SSP4 and MAgPIE (Dietrich et al., 2019) used by SSP5) rely on projected demands to infer their yields, while we are interested in assessing the biophysical capacity of land to produce sufficient food. In the GCAM models, the regional allocation of crops is considered optimal in terms of maximizing global profits, thus it does not model the optimal attainable yields for each region. The IMAGE model uses MAGNET (Woltjer & Kuiper, n.d.) as agricultural economy model where the production increases in response to the increase in demand in agricultural commodities due to demographic changes and increasing income[#]. MAgPIE uses an interplay of demand and intensification costs for its projections.

today, therefore we do not wish to extrapolate the trend of higher yields with higher GDP beyond the best technology as is available today, and (b) tree-based models such as our random forest approach tend to perform poorly to project on data points outside of the training distribution. Binning the GDP reduced the points outside of the convex hull to very few. Eight soil variables considered in the GAEZ assessment (Fischer et al., 2012) to be influencing crop performance were also included. These indices, with four to seven classes, were linearly normalized. Soil degradation is likely to further impact yields, but is complex to model and current research did not provide clear conclusions (Bindraban et al., 2012) nor could projections of soil quality for 2050 be found. Hence, soil degradation is not accounted for in this study and is left for further improvements.

Modeling crop yields under future climate and societal conditions of the SSPs, we find projected total caloric yield trends to be surprisingly similar for most geographies across these five scenarios, which are designed to cover a broad space of plausible futures (O'Neill et al., 2017). Despite climate becoming less favorable to crops in regions that are currently the most productive (Fig. 1b), we project an increase in global production for all scenarios, by 12% on average (Fig. 2). This is largely attributable to cropland expansion (Hurtt et al., 2011), in northern latitudes (e.g. Siberia) and in some parts of Africa (e.g. Botswana and Angola) (Fig. 1b), as global yields are expected to remain relatively stable (±3%). In the absence of technological breakthroughs, caloric yields are projected to decrease in the most productive regions, including Western Europe and the US Midwest, due to increasing temperature and temperature variability, and this is unlikely to be compensated for by further intensification. This is expected, given that declining yields in these regions are already observed today (Ray et al., 2019). Globally, this is compensated by increased yields in currently underperforming areas.

Our findings fall within the range of existing projections, which vary substantially: comparable forward-looking studies project changes in global yield within this timeframe of −5% to +13% (Jaggard et al., 2010), −5% to +47% (Blanc, 2017), ~+1% (Ostberg et al., 2017), and ±20% (Zhu et al., 2019). Some studies focused on climate change find consistent declines (−10% (Lobell and Asseng, 2017; Zhao et al., 2017) to almost −30% (Ray et al., 2019)). When accounting for prospective technology developments and CO_2 fertilization, most studies project increases in global yields, varying from +23% (Lobell and Gourdji, 2012) to +50% (FAO, 2009; Ray et al., 2013), though these effects vary by region and crop (INRAE, 2020; Jaggard et al., 2010; Kang et al., 2009) and include declines of -55% for maize (Kang et al., 2009). Projections extrapolating past trends fall in the same order of magnitude, with increases of +30% to +50% (Davis et al., 2019; FAO, 2009; Ray et al., 2013; van Meijl et al., 2020). Similar trends are found with sustainable intensification, focusing on irrigation

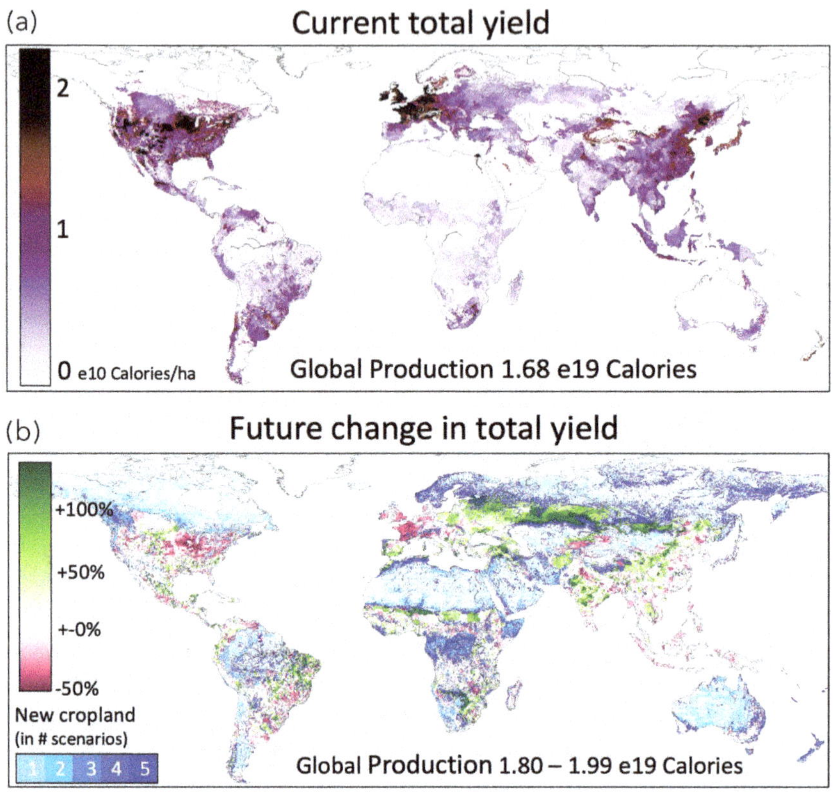

Figure 1 The maps show (a) the current status (Monfreda et al., 2008) and (b) projected changes in aggregate caloric yield for 100 crops relevant for nutrition. (a) Darker shades of purple indicate higher caloric yield for the year 2000 (the most recent year for which data is available), on which the model was trained. (b) Green denotes greater increases in yield by 2050, while pink denotes decreases. This figure shows the change in aggregated caloric yield averaged across the five SSPs. For each SSP, yields are projected with our model for each climate projection from four GCMs, then the yields are averaged (climate ensembling). Areas with new cropland are shown in blue, with the number of scenarios in which cropland is projected to increase in darker shades of blue.

(+24 to 37% (Rosa et al., 2018)), nutrient management (+45 to 70% (Mueller et al., 2012)), optimizing cropland allocation (Folberth et al., 2020) or technology (Aggarwal et al., 2019). However, the vast majority of these studies do not account for the fact that water resources are quickly becoming scarce (Turner et al., 2019), and those who do find more conservative results (e.g., Beltran-Peña et al. (2020) find similar results to ours under SSP1 but project declining production in SSP2 and SSP3). Future work is necessary on this topic to assess where current stressors to food security would potentially be compounded by lack of water availability.

4 From models to decision-relevant metrics: integrating demand

The approach outlined above merges crop-specific models to draw a fuller picture of crop-provisioning ecosystem services. While Section 3 focused on advances in crop modeling to integrate climate change impacts and nutrition and express it as ecosystem service supply, Section 4 adds the other side: demand. In the ES framework, supply only turns into a service when the human component, the need, is accounted for (Tallis et al., 2008). Modeling crop yields is just modeling a biophysical supply of ecosystem service. To model the ecosystem service of nutrition, we must account for the demand side.

4.1 Approximating diets and demand to estimate caloric sufficiency

Caloric sufficiency, defined as caloric supply relative to demand (see Equation 1), provides a coarse and imperfect but useful indicator of the ecosystem service of food availability for nutrition, and analyzing changes in caloric sufficiency (direction, magnitude, and consistency across scenarios) provides insights on the food system's ability to deliver this ecosystem service over time, a key consideration for food security. We thus interpret it in conjunction with existing national food security indicators, revealing vulnerable areas for further examination. The resilience of caloric sufficiency to climate and societal changes is assessed by the direction and approximate magnitude of change in caloric sufficiency, across different SSP scenarios:

$$\text{Caloric Sufficiency} = \frac{\text{Production}_{(\text{available})}}{\text{Demand}} = \frac{\text{Food from crops + Livestock}}{\text{ADER}_{\text{Except Fish}} * \text{Population}}$$

Equation 1: Caloric sufficiency uses SSPs' assumptions of the population (Jones and O'Neill, 2016) and dietary changes (Bodirsky et al., 2015), assuming system losses and relative amount of fish in diets (1.5%) stay constant. It accounts for changing feed/food split. ADER stands for average daily energy requirement. Our approach does not account for changes in age/gender/cultural dietary requirements.

Terrestrial caloric sufficiency is the ratio of supply over demand. Terrestrial supply corresponds to the calories available from crops and livestock, using constant feed conversion factor and system losses (Gustavsson et al., 2011) (both calculated for baseline year), taking into account changes in diets (including the proportion of livestock products by country). Terrestrial demand excludes the proportion of nutritional requirements satisfied by fish, assuming it will remain constant, i.e., that there will be more fish available overall, based on projections including aquaculture(Mente and Smaal, 2016; Searchinger et al., 2019).

As previously discussed, the SSPs framework helps us envision the consequences of different resilience strategies by drawing coherent scenarios of future changes in society and the environment. Future demand is calculated by multiplying SSPs population projections from Jones and O'Neill (2016), downscaled and adjusted to finer more recent population data (GPW (Doxsey-Whitfield et al., 2015)) with average daily caloric requirement (FAO and ESS, n.d.). The assumptions on future consumption (proportion of livestock products in diets) are reported at a national scale by Bodirsky et al. (2015). FAO Commodity Balance Sheets (FAO, Rome Food and Agriculture Organization of the United Nations, 2016) are used to complement information for livestock.

Calories are used as a proxy for diet sufficiency, informing nutritional sufficiency but do not provide a full view of it. This study does not take into account local or cultural differences in diets, nor does it calculate the demand for precise demographical nutritional requirements – though we do investigate sensitivity to the latter simplification and find little impact. In our model, diets are not optimized for health or sustainability. We include changes in future livestock consumption (Bodirsky et al., 2015), and the remainder of future diets reflect current choices, modified locally to the extent that farmers adapted crop mixes. Finally, our caloric sufficiency estimates may be overestimated because we do not account for the fact that the nutritional contents of crop decrease with CO_2 increasing, but also conversely assume no transformative innovation beyond the scope of what is implemented at baseline year.

4.2 Large decline in global caloric sufficiency: production gains are outweighed by consumption

We calculate global caloric sufficiency to be 1.96 currently, and yet about one out of nine people is undernourished (World Food Programme, 2019). At a global scale, terrestrial caloric sufficiency measures the ability of the land to produce sufficient food for the world's population, accounting for losses and feed. Mathematically, a ratio of 1 means supply matches demand, but because perfect distribution of and access to food are unrealistic, an adequate caloric sufficiency must be higher than 1. While the absolute value of sufficiency is thus challenging to interpret, *changes* in the caloric sufficiency ratio do indicate whether the overall food system is expected to become more or less strained by total production capacity.

Global caloric sufficiency is projected to drop consistently (i.e. in every scenario) because demand from a growing population (Jones and O'Neill, 2016) for eating more animal products (Bodirsky et al., 2015) outweighs increases in production (Fig. 2). The largest decline (−40%) in caloric sufficiency

	Inputs from SSPs projections						*Results from our modeling*			
Scenario	Cropland	Population	Livestock in Diet	Management $	Management (fertilizers)	Management (irrigation)	Global Caloric Sufficiency	N countries where Caloric Sufficiency < 1	Global Production	Global Yield
Current (2000)	2 bio ha	6 bio	2.7 trillion kcal (19%)	GDP	Ferti-lizers	Irriga-tion	1.96	54 (out of 166)	1.7e19 calories	7e9 cal/ha
SSP1 RCP2.6 *Sustainability*	+3%	7.9 bio	4.4 trillion kcal (24%)	high	low	med.	1.36	74	+7%	+3%
SSP2 RCP4.5 *Middle of The Road*	+12%	8.2 bio	4.4 trillion kcal (23%)	med.	med.	med.	1.38	73	+12%	≈
SSP3 RCP6.0 *Regional Rivalry*	+14%	9.9 bio	4.8 trillion kcal (21%)	low	high	med.	1.17	78	+12%	-2%
SSP4 RCP6.0 *Inequality*	+13%	8.7 bio	4.2 trillion kcal (21%)	low	med.	high	1.34	72	+13%	≈
SSP5 RCP8.5 *Fossil-fueled Development*	+16%	8.2 bio	4.0 trillion kcal (21%)	high	high	high	1.46	71	+18%	+2%

Figure 2 Quantitative, spatially-explicit projections developed for the SSPs (left side of the table) describing changes in climate (Hijmans et al., 2005), land use (Hurtt et al., 2011), management (Hurtt et al., 2011), population (Jones and O'Neill, 2016), and diet (Bodirsky et al., 2015) were used to project terrestrial food production and demand. Specifically, climate forcing and management projections (fertilizer use, irrigation, and GDP per capita (Dellink et al., 2017) as a proxy for agricultural industrialization) were used to model yields. Four different global circulation models (GCMs) were used to translate each climate-forcing scenario into gridded projections of climate variables (seasonal and annual, normal and extreme precipitation, and temperature). The SSPs have been widely used by the research and policy communities and were developed to cover a wide range of possible futures (Riahi et al., 2017), from an inclusive 'sustainability' scenario (SSP1) to a capitalist vision of geo-engineered, globalized, 'fossil-fueled development' (SSP5) (O'Neill et al., 2017). Population and diet datasets are at the national resolution, and all others are at 5 arcmin. The diet column shows the total calories provided by livestock and the proportion of livestock products in the diet, globally (though distributions by countries of consumption vary per scenario). The right side of the table presents our findings for each of the five SSPs showing that global caloric sufficiency drops in all scenarios but remains above 1. National caloric sufficiency drops below 1 in numerous countries in all scenarios. These results are explained by the following columns: production increases globally because cropland expands in all scenarios. However, this increase in the food supply is outweighed by demand, which is driven by population growth and dietary changes: people are projected to eat more animal products in the future, so the food/feed split is adjusted accordingly when calculating caloric sufficiency.

is seen in the *regional rivalry* scenario (SSP3) because of its steep growth in population, reaching nearly 10 billion by 2050, exacerbated by both increased consumption of animal products and a small decrease in our projected yields. However, even in this scenario, caloric sufficiency at the global scale remains above 1. Global caloric sufficiency is projected to be highest under *fossil-fueled development* (SSP5), closely followed by the *middle ground* (SSP2) and the *sustainability* (SSP1) pathways, though it still declines by about 30% from its current value.

4.3 Caloric sufficiency decreases consistently in most countries, including many countries facing hunger today

Caloric sufficiency is also relevant at a country level, as this is the scale at which most trade and geopolitical decisions are made. More than half of countries see their caloric sufficiency decrease in every scenario while 102 of 160 countries assessed see their caloric sufficiency decrease in most scenarios. The number of countries with caloric sufficiency under 1 in the future increases by almost 50%, regardless of the scenario (Fig. 2). National caloric sufficiency shows a significant positive correlation with the most recent information on trade independency (Kummu et al., 2020) (Pearson correlation coefficient r = +0.61), Global Food Security Index (GFSI) (Economist Intelligence Unit, 2018) (r = +0.34), and nutrition (WHO, 2019) (r = −0.25 with malnutrition). This indicates that countries with higher caloric sufficiency are less dependent on trade and less prone to malnutrition and food security issues.

We group countries empirically following eight narratives reflecting their caloric sufficiency levels, trends, and current trade situation (Table 1, Fig. 3). We identify additional potential risks by considering current national income, trade dependence (Kummu et al., 2020), and water resources (Brauman et al., 2016). Specifically:

- For *Water security*, we use a water depletion index accounting for seasonal and dry-year water scarcity (Brauman et al., 2016); water scarcity is not taken into account in our model (neither precipitation from WorldClim, nor irrigation data from the SSPs take water depletion into account). This categorical dataset at the original resolution of 5 arcmin was aggregated in a single metric by country as the percent area in any of the depleted categories (seasonally or annually).
- *Malnutrition* is defined here as the prevalence of underweight children under 5 (WHO, 2019). We use the most recent value available for each country (between 2000 and 2016) and found it negatively correlated (−0.25) with caloric sufficiency circa 2000.
- The *Global Food Security Index* (GFSI) (Economist Intelligence Unit, 2018) captures food security for 113 countries, accounting for affordability, availability, quality and safety, and natural resources, including financial, political-social, and trade/supply chain risks. It is available from 2014 to 2018; we used the most recent value and found it positively correlated (+0.34) with caloric sufficiency circa 2000.
- *Trade independency* captures the independency of a country from imports in relation to its total food supply, in terms of calories. Kummu et al. (2020)

Table 1 Categorization of countries following eight narratives reflecting their caloric sufficiency levels, trends, and trade situation. Empty cells denote variability across countries of the category. The classification is based on population projections from (Jones and O'Neill, 2016), the columns displaying the number of countries per category indicate variability when grouping countries based on caloric sufficiencies calculated with UN population projections (United Nations, Department of Economic and Social Affairs, 2019). Trade independency is calculated as the share of net imports relative to the entire food supply (WHO, 2019).

Category		N Countries (±variability)	Current trade Independency	Current Caloric Sufficiency	Future Caloric Sufficiency	Narrative	Examples
Greatest concern	Most vulnerable	35 (+4)		< 1.2	↓	*Low yields and increasing population: food demand consistently outweighs food supply.*	Afghanistan, Angola, Ethiopia, Guatemala, India (More than half of current low-income countries are in this category)
	Newly vulnerable	24 (+7)			↓ to ~1	*Large population growth or an initial low sufficiency, combined with a decline in sufficiency leads these countries to become more vulnerable.*	Benin, Burkina Faso, UK, Mexico, China, Indonesia, Sudan
Currently trade driven	Exporters	16 (=)	high	>3	↓	*World food exporters' caloric sufficiencies are declining...*	USA, France, Brazil, Denmark (w/Greenland)
	Importers	21 (−1)	low		↓ or cst	*...while these countries depend on exporters. Importers' sufficiencies are low to begin with and decrease or remain low.*	Japan, Saudi Arabia, Jordan

(*Continued*)

Table 1 (*Continued*)

Category		N Countries (±variability)	Current trade Independency	Current Caloric Sufficiency	Future Caloric Sufficiency	Narrative	Examples
Mild concern	Vulnerable but improving	4 (−3)		<1.2	↑	*Few countries with low sufficiency are improving, thanks to cropland and/or productivity increase while the population stagnates.*	Peru, Morocco
	Decreasing but not as vulnerable	28 (−10)		>2	↓	*Many (mostly European) countries with higher income and food security see their caloric sufficiency decrease. More concerning, a few countries in Africa and SE Asia have declining sufficiencies, though it is variable across scenarios.*	Spain, Austria, Turkey, Thailand, Chile, Venezuela, Chad, Mali
	Highly variable	15 (+5)				*High variability across scenarios suggests sensitivity to their differences (but finer modeling would be required to draw policy-relevant insights).*	Cuba, Liberia, Central African Republic, Guyana, Sweden
Impro-ving	Increasing sufficiency	17 (−2)		>1.2	↑	*Cropland projected to increase, leading to caloric sufficiency increase. However, SSPs' cropland expansion assumptions may be debatable and these results do not account for water scarcity.*	Botswana, Norway, Portugal, Russia

estimate it as the share of net imports of the entire food supply in a given year. Similarly to the other indices, it is scaled between 0 and 1, so that 0 represents a case when the food supply is fully dependent on imports, while 1 represents high net-exports (exports are equal to or more than the national food supply). We used the most recent values in 2013 and found them highly correlated (+0.61) with caloric sufficiency circa 2000.
- Countries' income groups from World Bank's classification (World Bank Country and Lending Groups – World Bank Data Help Desk, n.d.) based on their GNI (Gross national income) per capita captures relative wealth. We use the most recent value (2018) of GNI, and the four groups: low, lower-middle, upper-middle, and high income.

The most vulnerable countries with low yields and growing populations are located in Africa and India. Major population growth (often more than doubling) explains caloric sufficiency declines in Africa (Benin, Burkina Faso, Ivory Coast, Guinea, Sudan, Senegal, Sierra Leone, Tanzania, and Zambia). Declining sufficiency of exporters puts importers at risk: exporters (defined as countries with initial caloric sufficiency higher than 3 and trade independence higher than 0.8) are high or upper-middle income countries, with the exception of Ukraine (lower-middle income), they unsurprisingly tend to have high GFSI (>0.55), high nutrition (malnutrition <5%), and much less land area facing water depletion (water security index >0.65). The world's largest exporters (USA, Brazil, France, Denmark including Greenland, Canada, Argentina, and Uruguay) see their caloric sufficiency decrease, on average from 6 to 5 (Fig. 3a), though some see more drastic declines (e.g. by half for Australia, due to a projected doubling of the population). Although they are all still comfortable in caloric sufficiency, these declines could mean reductions in export volumes, which would likely affect the countries that import from them, especially those with lower trade independence. This is a good example of the risks associated with a highly interconnected global food system, as described by Liu et al. (2018) and Nyström et al. (2019). Most countries' importers see their caloric sufficiency decrease or remain unchanged. Though trade patterns could evolve, there is a clear trend of caloric sufficiency declining for exporters and not increasing for importers. A decline in the sufficiency of exporter countries might decrease the volume of food exported, and even though most importers are high or upper-middle income, with GFSI >0.7, they may find it difficult to meet demand, under such trade patterns. For example, Singapore is highly dependent on Malaysia for food imports, and Malaysia's sufficiency is projected to decrease by 0.84 on average. In addition, strong barriers to trade in the regional rivalry (SSP3) scenario would put importers at risk.

Across the scenarios considered, which were designed to represent a wide range of possible futures, we find the projected direction and magnitude

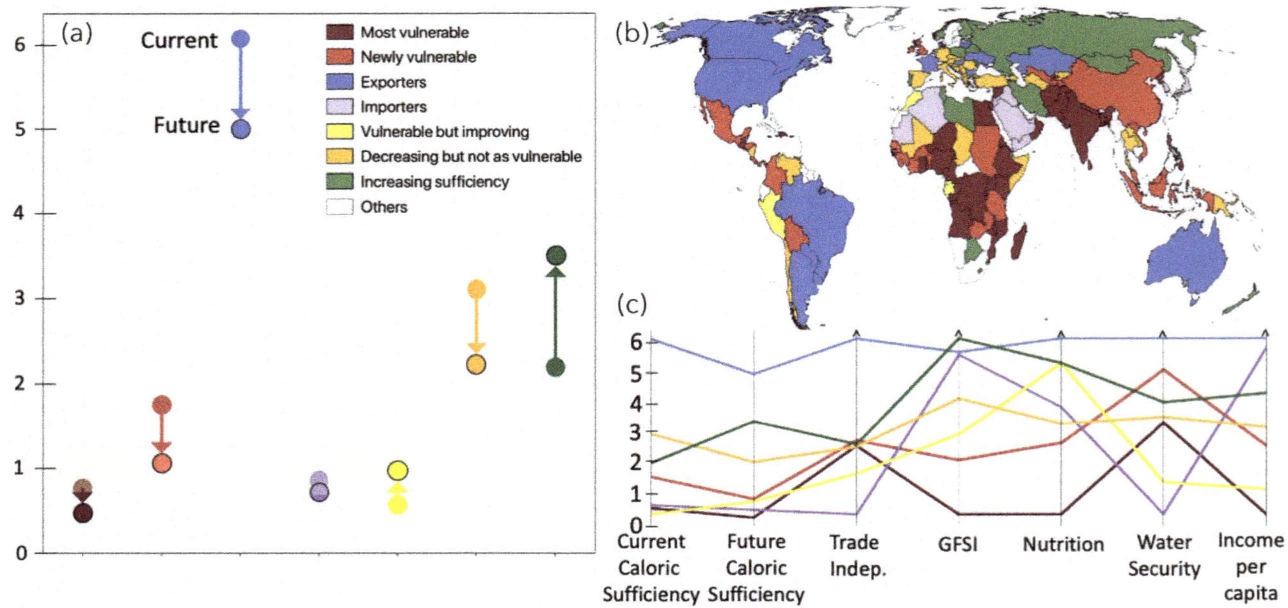

Figure 3 Trends in caloric sufficiency for 145 countries grouped into 7 categories (15 additional countries are variable across scenarios). (a) The shaded yellow bar highlights the critical caloric sufficiency threshold around 1 with a ± 0.2 tolerance. The geographic distribution of these countries is shown in (b), and the graph in (c) shows how these groups relate to the most recent information about import independency, the Global Food Security Index (GFSI), adequate nutrition (1-undernutrition), water security, and income level (GNI). The axis scale refers to caloric sufficiency, while the other metrics are scaled from 0 to 1 (details in 'Methods' section). In (c), future caloric sufficiency is averaged across SSPs and the mean values across countries in each category are displayed. The same figure with standard deviation can be found in Fig. S8 of Weil et al., 2021. Color legend applies to three subfigures.

of changes in caloric sufficiency to be surprisingly consistent, and relatively robust to uncertainty. We find that most countries already facing hunger remain vulnerable in the face of future caloric sufficiency; for 9 out of 10 countries in which more than 10% of the population is currently undernourished, caloric sufficiency is projected to decrease in most scenarios. For the remaining countries, change in caloric sufficiency is variable across scenarios, and never consistently increases. High-income countries are not spared; most (75%) see a decrease in sufficiency in at least one scenario. Half of the high-income countries actually find their sufficiency decrease in *all* scenarios, echoing previous findings projecting declining food security scores in higher-income countries.[26] We also identify 15 countries *highly variable* across SSPs, suggesting that targeted policies may have a substantial impact.

The results from this section are robust (see the sensitivity analysis in Weil et al., 2021), which replicates results using variations in the projected drivers of demand, production, and system efficiency). Regarding diets, increasing consumption of livestock products by half globally (to ~30% of diet) would reduce caloric sufficiency more than an additional billion people would, while cutting livestock consumption by half (to ~10%) would increase global caloric sufficiency by almost ~0.16. Reducing food waste could lead to comparable gains, as the sensitivity of caloric sufficiency to food waste closely mirrors diets. However, neither diets nor food waste are large enough overall drivers for reductions to maintain current levels of caloric sufficiency. A combination of all factors could theoretically lead to keeping caloric sufficiency at its current level, e.g. in SSP2, by improving yields by 34%, cutting food waste in half, and reducing livestock products in diets by 33% (see also Table S3 in Weil et al., 2021).

5 Conclusion

Reviewing the measured or modeled impacts of climate change on ecosystem service in the agricultural landscape, we find that most impacts tend to be negative (Grimm et al., 2013; Scholes, 2016) and to disproportionately affect the most vulnerable populations - children, elderly, socio-economically disadvantaged (Weiskopf et al., 2020).

Ecosystem services to agriculture supporting food provision are indeed all at risk of decline, or are already declining with climate change: Soil fertility may be compromised by rising temperature and resultant increased rates of soil organic matter decomposition; pollination and pest regulation are already affected by climate change, which disrupts important plant-pollinator and predator-pest relationships, due to changes in the spatial distribution and seasonal timing of flowering and population growth dynamics. Climate change is also likely to have profound impacts on water availability for agriculture.

Additionally, extreme events like droughts have already increased in frequency, duration, and severity since 1950 (Spinoni et al., 2014) and are expected to worsen under future conditions (Wang, 2005). As surface water supply grows unreliable, agriculture may increasingly turn to groundwater to meet its needs, but many aquifers are already overexploited and climate impacts on groundwater recharge are highly uncertain (Taylor et al., 2013).

Ecosystem services from agricultural landscapes include cultural and regulating services that are co-produced along with food provision in these landscapes. These are highly dependent on agricultural practices which can be part of the problem or the solution, conducting possible trade-offs and synergies between food production and the other services from agricultural landscapes. But for all services considered, climate change appears to lead to mostly negative impacts. Indeed, climate change will likely impact the regulation of water quality, because warming water temperatures speed up algal growth, exacerbating eutrophication resulting from agriculture fertilizers loading into waterways, while decreased water flows will concentrate any pollutants that do enter streams. Similarly, when it comes to climate regulation: agricultural management practices can increase carbon storage (in soil and vegetation) on farmlands, and contribute to microclimate regulation through strategic management of evapotranspiration. However, the potential of crops to store carbon may decrease due to higher temperatures, increased water stress, and increased disturbance, as research shows it does in forests.

For aggregate nutritional production of the food system, further work is needed to improve statistical crop modeling ability to project impacts from climate change. Implementing shifting climate-bins that move as climate conditions migrate is one potential route forward, as are other statistical methods such as quantile regression, which may be useful for identifying yield gaps and their geographic shifts over time. Modeling climate change impacts on food provisioning has been explored from several angles, but important work still needs to be done to express climate risks in a way suitable to overall ecosystem-agriculture tradeoff analysis and better able to inform food security policy. In particular, there remains a need for models that can simultaneously assess ecosystem service change in a coherent context of broader agriculture and food systems change. While most models stop at modeling the biophysical supply of the ecosystem service (food production), a full model of ecosystem service requires the human-side, i.e. the demand, to be taken into account – the supply of an ecosystem service is not truly a 'service' until it meets a need.

We attempt to explore a large view of the food-provisioning ecosystem service by modeling the biophysical supply of this service (nutrition via aggregated caloric crop yields) and estimating need (demand, as the need for nutrition) under climate change – with a global and contextual understanding of these changes, including societal changes. Building on the most recent high-resolution data available, combining 15 data sources, and leveraging machine-learning approaches, we investigated the caloric sufficiency implications of a crop mix adaptation narrative given future environmental and societal changes. We find caloric sufficiency dangerously declining globally and locally in most countries.

6 Future trends in research

The aggregated models discussed above add important detail and policy relevance to crop modeling. However, much is still missing to meet the demand from policy-makers to provide benefits-relevant indicators of food-production ecosystem services. For biophysical supply, many dependencies among agro-ecosystem services are not well studied. Regulating and supporting services will have consequent impacts on these findings, including soil degradation, water availability, pollination, and other regulating ecosystem services. Additionally, understanding CO_2 fertilization remains understudied in the ecosystem services literature. Finally, most research on the linkages of climate change and food-provisioning ecosystem services take into account only the overall evolutions in temperature and precipitation trends, ignoring relevant effects from the increased probability of extreme events, namely droughts, and floods.

On the demand side, many challenges and research gaps also remain. The evolution of diets, innovative resilience strategies, and changing consumption patterns for ag-derived products all may play large roles. Additionally, the projections we use from the SSPs do not explore the full space of different demand scenarios (Kim et al., 2021) and important simplifications must be made to allow models to reach the final decision-relevant metrics. Advances in food-provisioning modeling would thus benefit from improved fine-scale food production data, more detailed analysis of observed crop mix adaptation behaviors, improved understanding of climate change impacts on yields, and societal, environmental, and economic feedbacks such as changing trade patterns. As is the case with many machine-learning applications applied to science questions, our data-driven global modeling is also limited by data gaps (Fischer et al., 2012). Improving the availability of fine-scale food production data, as well as our understanding of climate change impacts on yields, are important to apprehend the consequences of different coping, adaptation, and transformation strategies for the food system.

Declining global and domestic caloric sufficiency suggests that adapting the choice of crops grown locally will be insufficient to feed a growing population in a changing climate. Under our scenarios, and in line with a recent comparable study (Beltran-Peña et al., 2020) assuming sustainable intensification and including water depletion, global production is projected to increase, but not as fast as the growing population's food demand. Resilience strategies beyond sole adaptation of crop mixes will be needed to avoid declines in domestic caloric sufficiency in most countries. This conclusion suggests that more than just crop mix adaptation is needed, resonating with recent work calling for a food system transformation (Díaz et al., 2019; Foley et al., 2011), and even revolution (Ehrlich and Harte, 2015). To keep global caloric sufficiency constant, major changes will be necessary, acting on demand, production, system efficiency, or, most likely, a combination of these drivers. While assessing the feasibility of such interventions is beyond the scope of this research, our results complement recent work (Gerten et al., 2020; Searchinger et al., 2019) finding that combining dietary changes (Kim et al., 2020; Springmann et al., 2018; Willett et al., 2019), food waste reduction (Kummu et al., 2012), optimizing cropland, water and nutrient management (Mueller et al., 2012) could feed the 2050 population while remaining within planetary boundaries.

We argue also that it is not sufficient to analyze the future of global agroecosystems as a single-objective optimization, which may lead to a lack of resilience and unexpected adverse effects on other ecosystem services and climate. The ecosystem service perspective presented in this chapter invites us to adopt a holistic view, re-framing as a multi-objective optimization to account for more of the interlinked ecosystem services, and environmental and societal impacts of decisions. It is important also to consider not just the impact of climate change on agriculture, but also the impact of agriculture on climate, both direct and indirect. Food-provisioning ecosystem services are not just impacted by climate change but also are an essential driver.

7 Where to look for further information

The following articles provide a good overview of the subject:

- Ruane et al. (2017).
- Reardon and Timmer (2012).
- Rosenzweig et al. (2018).
- Biggs et al. (2012).

Key research in this area can be found at the following organizations:

- The Agricultural Model Intercomparison and Improvement Project (agmip .org).
- The Natural Capital Project (https://naturalcapitalproject.stanford.edu/).
- The Global Landscapes Initiative (http://environment.umn.edu/discovery /gli/).
- The World Bank Agriculture and Food program (https://www.worldbank .org/en/topic/agriculture).

8 References

Aggarwal, P., Vyas, S., Thornton, P., Campbell, B. M. and Kropff, M. (2019). Importance of considering technology growth in impact assessments of climate change on agriculture. Global Food Security 23, 41–48. https://doi.org/10.1016/j.gfs.2019.04.002.

Aizen, M. A., Garibaldi, L. A., Cunningham, S. A. and Klein, A. M. (2009). How much does agriculture depend on pollinators? Lessons from long-term trends in crop production. Annals of Botany 103(9), 1579–1588. https://doi.org/10.1093/aob/mcp076.

Akbari, F., Shourian, M. and Moridi, A. (2022). Assessment of the climate change impacts on the watershed-scale optimal crop pattern using a surface-groundwater interaction hydro-agronomic model. *Agricultural Water Management* 265, 107508.

Alexandridis, N., Marion, G., Chaplin-Kramer, R., Dainese, M., Ekroos, J., Grab, H., Jonsson, M., Karp, D. S., Meyer, C., O'Rourke, M. E., Pontarp, M., Poveda, K., Seppelt, R., Smith, H. G., Martin, E. A. and Clough, Y. (2021). Models of natural pest control: Towards predictions across agricultural landscapes. *Biological Control* 163, 104761.

Alexandridis, N., Marion, G., Chaplin-Kramer, R., Dainese, M., Ekroos, J., Grab, H., Jonsson, M., Karp, D. S., Meyer, C., O'Rourke, M. E., Pontarp, M., Poveda, K., Seppelt, R., Smith, H. G., Walters, R. J., Clough, Y. and Martin, E. A. (2022). Archetype models upscale understanding of natural pest control response to land-use change. *Ecological Applications: A Publication of the Ecological Society of America* 32(8), e2696.

Balbi, S., del Prado, A., Gallejones, P., Geevan, C. P., Pardo, G., Pérez-Miñana, E., Manrique, R., Hernandez-Santiago, C. and Villa, F. (2015). Modeling trade-offs among ecosystem services in agricultural production systems. *Environmental Modelling and Software* 72, 314–326.

Barnett, T. P., Adam, J. C. and Lettenmaier, D. P. (2005). Potential impacts of a warming climate on water availability in snow-dominated regions. Nature 438, 7.

Bartomeus, I., Ascher, J. S., Wagner, D., Danforth, B. N., Colla, S., Kornbluth, S. and Winfree, R. (2011). Climate-associated phenological advances in bee pollinators and bee-pollinated plants. Proceedings of the National Academy of Sciences of the United States of America 108(51), 20645–20649. https://doi.org/10.1073/pnas.1115559108.

Bebber, D. P., Ramotowski, M. A. T. and Gurr, S. J. (2013). Crop pests and pathogens move polewards in a warming world. Nature Climate Change 3(11), 985–988. https://doi.org/10.1038/nclimate1990.

Beltran-Peña, A. A., Rosa, L. and D'Odorico, P. (2020). Global food self-sufficiency in the 21st century under sustainable intensification of agriculture. *Environmental Research Letters* 15(9), 095004. https://doi.org/10.1088/1748-9326/ab9388.

Bennett, E. M., Baird, J., Baulch, H., Chaplin-Kramer, R., Fraser, E., Loring, P., Morrison, P., Parrott, L., Sherren, K., Winkler, K. J., Cimon-Morin, J., Fortin, M.-J., Kurylyk, B. L., Lundholm, J., Poulin, M., Rieb, J. T., Gonzalez, A., Hickey, G. M., Humphries, M., Bahadur KC, K. and Lapen, D. (2021). Chapter 1: Ecosystem services and the resilience of agricultural landscapes. In: Bohan, D. A. and Vanbergen, A. J. (Eds.), *The Future of Agricultural Landscapes, Part II* (Vol. 64). Academic Press. https://doi.org/10.1016/bs.aecr.2021.01.001.

Bindraban, P. S., van der Velde, M., Ye, L., van den Berg, M., Materechera, S., Kiba, D. I., Tamene, L., Ragnarsdóttir, K. V., Jongschaap, R., Hoogmoed, M., Hoogmoed, W., van Beek, C. and van Lynden, G. (2012). Assessing the impact of soil degradation on food production. Current Opinion in Environmental Sustainability 4(5), 478–488. https://doi.org/10.1016/j.cosust.2012.09.015.

Bladon, K. D., Emelko, M. B., Silins, U. and Stone, M. (2014). Wildfire and the future of water supply. Environmental Science and Technology 48(16), 8936–8943. https://doi.org/10.1021/es500130g.

Blanc, É. (2017). Statistical emulators of maize, rice, soybean and wheat yields from global gridded crop models. Agricultural and Forest Meteorology 236, 145–161. https://doi.org/10.1016/j.agrformet.2016.12.022.

Bodirsky, B. L., Rolinski, S., Biewald, A., Weindl, I., Popp, A. and Lotze-Campen, H. (2015). Global food demand scenarios for the 21st century. *PLoS ONE* 10(11), e0139201. https://doi.org/10.1371/journal.pone.0139201.

Brady, M. V., Hedlund, K., Cong, R.-G., Hemerik, L., Hotes, S., Machado, S., Mattsson, L., Schulz, E. and Thomsen, I. K. (2015). Valuing supporting soil ecosystem services in agriculture: A natural capital approach. Agronomy Journal 107(5), 1809–1821. https://doi.org/10.2134/agronj14.0597.

Brauman, K. A., Daily, G. C., Duarte, T. K. and Mooney, H. A. (2007). The nature and value of ecosystem services: An overview highlighting hydrologic services. *Annual Review of Environment and Resources* 32(1), 67–98. https://doi.org/10.1146/annurev.energy.32.031306.102758.

Brauman, K. A., Richter, B. D., Postel, S., Malsy, M. and Flörke, M. (2016). Water depletion: an improved metric for incorporating seasonal and dry-year water scarcity into water risk assessments. Elementa: Science of the Anthropocene 4. https://doi.org/10.12952/journal.elementa.000083.

Brauman, K. A., van der Meulen, S. and Brils, J. (2014). Ecosystem services and river basin management. In: Risk-Informed Management of European River Basins. Berlin, Heidelberg: Springer-Verlag, 265–294.

Calvin, K., Patel, P., Clarke, L., Asrar, G., Bond-Lamberty, B., Cui, R. Y., Di Vittorio, A., Dorheim, K., Edmonds, J., Hartin, C. and Hejazi, M. (2019). GCAM v5.1: Representing the linkages between energy, water, land, climate, and economic systems. *Geoscientific Model Development*, 12(2). https://doi.org/10.5194/gmd-12-677-2019.

Castex, V., Beniston, M., Calanca, P., Fleury, D. and Moreau, J. (2018). Pest management under climate change: The importance of understanding tritrophic relations. Science of the Total Environment 616–617, 397–407. https://doi.org/10.1016/j.scitotenv.2017.11.027.

Chaplin-Kramer, R., Dombeck, E., Gerber, J., Knuth, K. A., Mueller, N. D., Mueller, M., Ziv, G. and Klein, A. M. (2014). Global malnutrition overlaps with pollinator-dependent micronutrient production. Proceedings. Biological Sciences 281(1794), 20141799. https://doi.org/10.1098/rspb.2014.1799.

Chaplin-Kramer, R., O'Rourke, M. E., Blitzer, E. J. and Kremen, C. (2011). A meta-analysis of crop pest and natural enemy response to landscape complexity. *Ecology Letters* 14(9), 922–932.

Chaplin-Kramer, R., Sharp, R. P., Weil, C., Bennett, E. M., Pascual, U., Arkema, K. K., Brauman, K. A., Bryant, B. P., Guerry, A. D., Haddad, N. M., Hamann, M., Hamel, P., Johnson, J. A., Mandle, L., Pereira, H. M., Polasky, S., Ruckelshaus, M., Shaw, M. R., Silver, J. M., Vogl, A. L. and Daily, G. C. (2019). Global modeling of nature's contributions to people. *Science* 366(6462), 255–258.

Chen, T. and Guestrin, C. (2016). XGBoost: A scalable tree boosting system. *Proceedings of the 22nd ACM SIGKDD International Conference on Knowledge Discovery and Data Mining - KDD* 16, 785–794. https://doi.org/10.1145/2939672.2939785.

Davis, K. F., Chhatre, A., Rao, N. D., Singh, D., Ghosh-Jerath, S., Mridul, A., Poblete-Cazenave, M., Pradhan, N. and DeFries, R. (2019). Assessing the sustainability of post-Green Revolution cereals in India. Proceedings of the National Academy of Sciences of the United States of America 116(50), 25034–25041. https://doi.org/10.1073/pnas.1910935116.

Dellink, R., Chateau, J., Lanzi, E. and Magné, B. (2017). Long-term economic growth projections in the shared socioeconomic pathways. Global Environmental Change 42, 200–214. https://doi.org/10.1016/j.gloenvcha.2015.06.004.

Díaz, S., Settele, J., Brondízio, E., Ngo, H. T., Guèze, M., Agard, J., Arneth, A., Balvanera, P., Brauman, K., Watson, R. T., Baste, I. A., Larigauderie, A., Leadley, P., Pascual, U., Baptiste, B., Demissew, S., Dziba, L., Erpul, G., Fazel, A., and Vilá, B. (2019). Global assessment report on biodiversity and ecosystem services, p. 45. *IPBES*. https://ipbes.net/sites/default/files/ipbes_7_10_add.1_en_1.pdf.

Dietrich, J. P., Bodirsky, B. L., Humpenöder, F., Weindl, I., Stevanović, M., Karstens, K., Kreidenweis, U., Wang, X., Mishra, A., Klein, D., Ambrósio, G., Araujo, E., Yalew, A. W., Baumstark, L., Wirth, S., Giannousakis, A., Beier, F., Chen, D. M.-C., Lotze-Campen, H. and Popp, A. (2019). MAgPIE 4 – A modular open-source framework for modeling global land systems. *Geoscientific Model Development* 12(4), 1299–1317. https://doi.org/10.5194/gmd-12-1299-2019.

D'Odorico, P., Chiarelli, D. D., Rosa, L., Bini, A., Zilberman, D. and Rulli, M. C. (2020). The global value of water in agriculture. Proceedings of the National Academy of Sciences of the United States of America 117(36), 21985–21993.

Doxsey-Whitfield, E., MacManus, K., Adamo, S. B., Pistolesi, L., Squires, J., Borkovska, O. and Baptista, S. R. (2015). Taking advantage of the improved availability of census data: A first look at the gridded population of the world version 4. *Papers in Applied Geography* 1(3), 226–234. https://doi.org/10.1080/23754931.2015.1014272.

Economist Intelligence Unit. (2018). Global Food Security Index 2018: Building Resilience in the Face of Rising Food-Security Risks. EIU.

Ehrlich, P. R. and Harte, J. (2015). Food security requires a new revolution. International Journal of Environmental Studies 72(6), 908–920. https://doi.org/10.1080/00207233.2015.1067468.

Elliott, J., Deryng, D., Müller, C., Frieler, K., Konzmann, M., Gerten, D., Glotter, M., Flörke, M., Wada, Y., Best, N., Eisner, S., Fekete, B. M., Folberth, C., Foster, I., Gosling, S. N., Haddeland, I., Khabarov, N., Ludwig, F., Masaki, Y., Olin, S., Rosenzweig, C., Ruane, A. C., Satoh, Y., Schmid, E., Stacke, T., Tang, Q. and Wisser, D. (2014). Constraints and potentials of future irrigation water availability on agricultural production under climate change. Proceedings of the National Academy of Sciences of the United States of America 111(9), 3239–3244. https://doi.org/10.1073/pnas.1222474110.

Ellis, E. C. and Ramankutty, N. (2008). Putting people in the map: anthropogenic biomes of the world. Frontiers in Ecology and the Environment 6(8), 439–447. https://doi.org/10.1890/070062.

FAO. (2009). Global agriculture towards. http://www.fao.org/fileadmin/templates/wsfs/docs/Issues_papers/HLEF2050_Global_Agriculture.pdf (vol. 2050) (*How to Feed the World in 2050*) [High Level Expert Forum].

FAO. (2016). *FAOSTAT - FAO Statistical Database for Food and Agriculture*. Rome: Food and Agriculture Organization of the United Nations. http://faostat3.fao.org/home/index.html.

FAO and ESS. (n.d.). Food security indicators. http://www.fao.org/economic/ess/ess-fs/ess-fadata/en/#.Xfz_AhdKjOQ.

Ficklin, D. L., Luo, Y., Luedeling, E. and Zhang, M. (2009). Climate change sensitivity assessment of a highly agricultural watershed using SWAT. *Journal of Hydrology* 374(1–2), 16–29.

Fischer, G., Nachtergaele, F. O., Prieler, S., Teixeira, E., Toth, G., van Velthuizen, H., Verelst, L. and Wiberg, D. (2012). Global agro-ecological zones (GAEZ v3.0)-model documentation. http://www.fao.org/soils-portal/soil-survey/soil-maps-and-databases/harmonized-world-soil-database-v12/en/.

Fischer, J., Brosi, B., Daily, G. C., Ehrlich, P. R., Goldman, R., Goldstein, J., Lindenmayer, D. B., Manning, A. D., Mooney, H. A., Pejchar, L., Ranganathan, J. and Tallis, H. (2008). Should agricultural policies encourage land sparing or wildlife-friendly farming? *Frontiers in Ecology and the Environment* 6(7), 380–385.

Folberth, C., Khabarov, N., Balkovič, J., Skalský, R., Visconti, P., Ciais, P., Janssens, I. A., Peñuelas, J. and Obersteiner, M. (2020). The global cropland-sparing potential of high-yield farming. Nature Sustainability 3(4), 281–289. https://doi.org/10.1038/s41893-020-0505-x.

Foley, J. A. (2011). Can we feed the world sustain the planet? Scientific American 305(5), 60–65.

Foley, J. A., Ramankutty, N., Brauman, K. A., Cassidy, E. S., Gerber, J. S., Johnston, M., Mueller, N. D., O'Connell, C., Ray, D. K., West, P. C., Balzer, C., Bennett, E. M., Carpenter, S. R., Hill, J., Monfreda, C., Polasky, S., Rockström, J., Sheehan, J., Siebert, S., Tilman, D. and Zaks, D. P. M. (2011). Solutions for a cultivated planet. Nature 478(7369), 337–342. https://doi.org/10.1038/nature10452.

Galic, N., Schmolke, A., Forbes, V., Baveco, H. and van den Brink, P. J. (2012). The role of ecological models in linking ecological risk assessment to ecosystem services in agroecosystems. *The Science of the Total Environment* 415, 93–100.

Garibaldi, L. A., Steffan-Dewenter, I., Winfree, R., Aizen, M. A., Bommarco, R., Cunningham, S. A., Kremen, C., Carvalheiro, L. G., Harder, L. D., Afik, O., Bartomeus, I., Benjamin, F., Boreux, V., Cariveau, D., Chacoff, N. P., Dudenhöffer, J. H., Freitas, B. M., Ghazoul, J., Greenleaf, S., Hipólito, J., Holzschuh, A., Howlett, B., Isaacs, R., Javorek, S. K., Kennedy, C. M., Krewenka, K. M., Krishnan, S., Mandelik, Y., Mayfield, M. M., Motzke, I., Munyuli, T., Nault, B. A., Otieno, M., Petersen, J., Pisanty, G., Potts, S. G., Rader, R., Ricketts, T. H., Rundlöf, M., Seymour, C. L., Schüepp, C., Szentgyörgyi, H., Taki, H., Tscharntke, T., Vergara, C. H., Viana, B. F., Wanger, T. C., Westphal, C., Williams, N. and Klein, A. M. (2013). Wild pollinators enhance fruit set of crops regardless of honey bee abundance. Science 339(6127), 1608–1611. https://doi.org/10.1126/science.1230200.

Gerten, D., Heck, V., Jägermeyr, J., Bodirsky, B. L., Fetzer, I., Jalava, M., Kummu, M., Lucht, W., Rockström, J., Schaphoff, S. and Schellnhuber, H. J. (2020). Feeding ten billion people is possible within four terrestrial planetary boundaries. Nature Sustainability 3(3), 200–208. https://doi.org/10.1038/s41893-019-0465-1.

Goldstein, J. H., Caldarone, G., Duarte, T. K., Ennaanay, D., Hannahs, N., Mendoza, G., Polasky, S., Wolny, S. and Daily, G. C. (2012). Integrating ecosystem-service tradeoffs into land-use decisions. Proceedings of the National Academy of Sciences of the United States of America 109(19), 7565–7570. https://doi.org/10.1073/pnas.1201040109.

Gordon, L. J., Peterson, G. D. and Bennett, E. M. (2008). Agricultural modifications of hydrological flows create ecological surprises. Trends in Ecology and Evolution 23(4), 211–219. https://doi.org/10.1016/j.tree.2007.11.011.

Grimm, N. B., Chapin III, F. S., Bierwagen, B., Gonzalez, P., Groffman, P. M., Luo, Y., Melton, F., Nadelhoffer, K., Pairis, A., Raymond, P. A., Schimel, J. and Williamson, C. E. (2013). The impacts of climate change on ecosystem structure and function. Frontiers in Ecology and the Environment 11(9), 474–482. https://doi.org/10.1890/120282.

Gunderson, L. H. and Holling, C. S. (2002). Panarchy: Understanding Transformations in Human and Natural Systems. Washington, DC: Island Press.

Gustavsson, J., Cederberg, C. and Sonesson, U. (2011). Global food losses and food waste: Extent, causes and prevention, 16–17 May, Düsseldorf, Germany. Food and Agriculture Organization of the United Nations (International Congress Save Food! At Interpack 2011.

Hails, R. S., Chaplin-Kramer, R., Bennett, E., Robinson, B., Daily, G., Brauman, K. and West, P. (2019). Determining the value of ecosystem services in agriculture. In: Hails, R. S., Gardner, S. M. and Ramsden, S. J. (Eds.), *Agricultural Resilience: Perspectives From Ecology and Economics*. Cambridge: Cambridge University Press, pp. 60–89. https://doi.org/10.1017/9781107705555.005.

Hijmans, R. J., Cameron, S. E., Parra, J. L., Jones, P. G. and Jarvis, A. (2005). Very high resolution interpolated climate surfaces for global land areas. International Journal of Climatology 25(15), 1965–1978. https://doi.org/10.1002/joc.1276.

Hillier, J., Brentrup, F., Wattenbach, M., Walter, C., Garcia-Suarez, T., Mila-i-Canals, L. and Smith, P. (2012). Which cropland greenhouse gas mitigation options give the greatest benefits in different world regions? Climate and soil-specific predictions from integrated empirical models. Global Change Biology 18(6), 1880–1894. https://doi.org/10.1111/j.1365-2486.2012.02671.x.

Holt-Giménez, E. (2019). Can We Feed the World Without Destroying It? New York: John Wiley & Sons.

Hurtt, G. C., Chini, L. P., Frolking, S., Betts, R. A., Feddema, J., Fischer, G., Fisk, J. P., Hibbard, K., Houghton, R. A., Janetos, A., Jones, C. D., Kindermann, G., Kinoshita, T., Klein Goldewijk, K., Riahi, K., Shevliakova, E., Smith, S., Stehfest, E., Thomson, A., Thornton, P., van Vuuren, D. P. and Wang, Y. P. (2011). Harmonization of land-use scenarios for the period 1500–2100: 600 years of global gridded annual land-use transitions, wood harvest, and resulting secondary lands. *Climatic Change* 109(1–2), 117–161. https://doi.org/10.1007/s10584-011-0153-2.

Hutchinson, J. J., Campbell, C. A. and Desjardins, R. L. (2007). Some perspectives on carbon sequestration in agriculture. Agricultural and Forest Meteorology 142(2–4), 288–302. https://doi.org/10.1016/j.agrformet.2006.03.030.

INRAE. (2020). The role of European agriculture in world trade by 2050. https://www.inrae.fr/en/actualites/agricultures-europeennes-horizon-2050.

IPCC. (2022). Impacts, adaptation, and vulnerability. Summary for policymakers. In: Pörtner, D. C., Roberts, M., Tignor, E. S., Poloczanska, K., Mintenbeck, A., Alegría, M. C., Langsdorf, S., Löschke, S., Möller, V. and Okem, A. (Eds.), Climate Change Contribution of Working Group II to the Sixth Assessment Report of the Intergovernmental Panel on Climate Change [H.-O].

Jaggard, K. W., Qi, A. and Ober, E. S. (2010). Possible changes to arable crop yields by 2050. Philosophical Transactions of the Royal Society of London. Series B, Biological Sciences 365(1554), 2835–2851. https://doi.org/10.1098/rstb.2010.0153.

Johnson, J. A., Runge, C. F., Senauer, B., Foley, J. and Polasky, S. (2014). Global agriculture and carbon trade-offs. Proceedings of the National Academy of Sciences of the United States of America 111(34), 12342–12347. https://doi.org/10.1073/pnas.1412835111.

Johnson, J. A., Runge, C. F., Senauer, B. and Polasky, S. (2016). Global food demand and carbon-preserving cropland expansion under varying levels of intensification. Land Economics 92(4), 579–592.

Johnson, J. A., Ruta, G., Baldos, U., Cervigni, R., Chonabayashi, S., Corong, E., Gavryliuk, O., Gerber, J., Hertel, T., Nootenboom, C. and Polasky, S. (2021). *The Economic Case for Nature: A Global Earth-Economy Model to Assess Development Policy Pathways*. World Bank. https://openknowledge.worldbank.org/handle/10986/35882.

Jones, B. and O'Neill, B. C. (2016). Spatially explicit global population scenarios consistent with the shared socioeconomic pathways. Environmental Research Letters 11(8), 084003. https://doi.org/10.1088/1748-9326/11/8/084003.

Kang, Y., Khan, S. and Ma, X. (2009). Climate change impacts on crop yield, crop water productivity and food security – A review. Progress in Natural Science 19(12), 1665–1674. https://doi.org/10.1016/j.pnsc.2009.08.001.

Karp, D. S., Chaplin-Kramer, R., Meehan, T. D., Martin, E. A., DeClerck, F., Grab, H., Gratton, C., Hunt, L., Larsen, A. E., Martínez-Salinas, A., O'Rourke, M. E., Rusch, A., Poveda, K., Jonsson, M., Rosenheim, J. A., Schellhorn, N. A., Tscharntke, T., Wratten, S. D., Zhang, W., Iverson, A. L., Adler, L. S., Albrecht, M., Alignier, A., Angelella, G. M., Zubair Anjum, M., Avelino, J., Batáry, P., Baveco, J. M., Bianchi, F. J. J. A., Birkhofer, K., Bohnenblust, E. W., Bommarco, R., Brewer, M. J., Caballero-López, B., Carrière, Y., Carvalheiro, L. G., Cayuela, L., Centrella, M., Ćetković, A., Henri, D. C., Chabert, A., Costamagna, A. C., De la Mora, A., de Kraker, J., Desneux, N., Diehl, E., Diekötter, T., Dormann, C. F., Eckberg, J. O., Entling, M. H., Fiedler, D., Franck, P., Frank van Veen, F. J., Frank, T., Gagic, V., Garratt, M. P. D., Getachew, A., Gonthier, D. J., Goodell, P. B., Graziosi, I., Groves, R. L., Gurr, G. M., Hajian-Forooshani, Z., Heimpel, G. E., Herrmann, J. D., Huseth, A. S., Inclán, D. J., Ingrao, A. J., Iv, P., Jacot, K., Johnson, G. A., Jones, L., Kaiser, M., Kaser, J. M., Keasar, T., Kim, T. N., Kishinevsky, M., Landis, D. A., Lavandero, B., Lavigne, C., Le Ralec, A., Lemessa, D., Letourneau, D. K., Liere, H., Lu, Y., Lubin, Y., Luttermoser, T., Maas, B., Mace, K., Madeira, F., Mader, V., Cortesero, A. M., Marini, L., Martinez, E., Martinson, H. M., Menozzi, P., Mitchell, M. G. E., Miyashita, T., Molina, G. A. R., Molina-Montenegro, M. A., O'Neal, M. E., Opatovsky, I., Ortiz-Martinez, S., Nash, M., Östman, Ö., Ouin, A., Pak, D., Paredes, D., Parsa, S., Parry, H., Perez-Alvarez, R., Perović, D. J., Peterson, J. A., Petit, S., Philpott, S. M., Plantegenest, M., Plećaš, M., Pluess, T., Pons, X., Potts, S. G., Pywell, R. F., Ragsdale, D. W., Rand, T. A., Raymond, L., Ricci, B., Sargent, C., Sarthou, J. P., Saulais, J., Schäckermann, J., Schmidt, N. P., Schneider, G., Schüepp, C., Sivakoff, F. S., Smith, H. G., Stack Whitney, K., Stutz, S., Szendrei, Z., Takada, M. B., Taki, H., Tamburini, G., Thomson, L. J., Tricault, Y., Tsafack, N., Tschumi, M., Valantin-Morison, M., Van Trinh, M., van der Werf, W., Vierling, K. T., Werling, B. P., Wickens, J. B., Wickens, V. J., Woodcock, B. A., Wyckhuys, K., Xiao, H., Yasuda, M., Yoshioka, A. and Zou, Y. (2018). Crop pests and predators exhibit inconsistent responses to surrounding landscape composition. *Proceedings of the National Academy of Sciences of the United States of America* 115(33), E7863–E7870.

Keeler, B. L., Hamel, P., McPhearson, T., Hamann, M. H., Donahue, M. L., Meza Prado, K. A., Arkema, K. K., Bratman, G. N., Brauman, K. A., Finlay, J. C., Guerry, A. D., Hobbie, S. E., Johnson, J. A., MacDonald, G. K., McDonald, R. I., Neverisky, N. and Wood, S. A. (2019). Social-ecological and technological factors moderate the value of urban nature. Nature Sustainability 2(1), 29–38. https://doi.org/10.1038/s41893-018-0202-1.

Kennedy, C. M., Lonsdorf, E., Neel, M. C., Williams, N. M., Ricketts, T. H., Winfree, R., Bommarco, R., Brittain, C., Burley, A. L., Cariveau, D., Carvalheiro, L. G., Chacoff, N. P., Cunningham, S. A., Danforth, B. N., Dudenhöffer, J. H., Elle, E., Gaines, H. R., Garibaldi, L. A., Gratton, C., Holzschuh, A., Isaacs, R., Javorek, S. K., Jha, S., Klein, A. M., Krewenka, K., Mandelik, Y., Mayfield, M. M., Morandin, L., Neame, L. A., Otieno, M., Park, M., Potts, S. G., Rundlöf, M., Saez, A., Steffan-Dewenter, I., Taki, H., Viana, B. F., Westphal, C., Wilson, J. K., Greenleaf, S. S. and Kremen, C. (2013). A global quantitative synthesis of local and landscape effects on wild bee pollinators in agroecosystems. Ecology Letters 16(5), 584–599. https://doi.org/10.1111/ele.12082.

Keys, P. W., Wang-Erlandsson, L. and Gordon, L. J. (2016). Revealing invisible water: moisture recycling as an ecosystem service. *PLoS ONE* 11(3), e0151993. https://doi.org/10.1371/journal.pone.0151993.

Kim, B. F., Santo, R. E., Scatterday, A. P., Fry, J. P., Synk, C. M., Cebron, S. R., Mekonnen, M. M., Hoekstra, A. Y., de Pee, S., Bloem, M. W., Neff, R. A. and Nachman, K. E. (2020). Country-specific dietary shifts to mitigate climate and water crises. Global Environmental Change 62, 101926. https://doi.org/10.1016/j.gloenvcha.2019.05.010.

Kim, H., Peterson, G., Cheung, W., Ferrier, S., Alkemade, R., Arneth, A., Kuiper, J., Okayasu, S., Pereira, L. M., Acosta, L. A., Chaplin-Kramer, R., Belder, E. den, Eddy, T., Johnson, J., Karlsson-Vinkhuysen, S., Kok, M., Leadley, P., Leclère, D., Lundquist, C. J., and Pereira, H. (2021). Towards a Better Future for Biodiversity and People: Modelling Nature Futures [Preprint]. https://doi.org/10.31235/osf.io/93sqp.

Kleftodimos, G., Gallai, N. and Kephaliacos, C. (2021). Ecological-economic modeling of pollination complexity and pesticide use in agricultural crops. *Journal of Bioeconomics* 23(3), 297–323.

Klein, A. M., Vaissière, B. E., Cane, J. H., Steffan-Dewenter, I., Cunningham, S. A., Kremen, C. and Tscharntke, T. (2007). Importance of pollinators in changing landscapes for world crops. *Proceedings: Biological Sciences* 274(1608), 303–313. https://doi.org/10.1098/rspb.2006.3721.

Kremen, C., Williams, N. M., Aizen, M. A., Gemmill-Herren, B., LeBuhn, G., Minckley, R., Packer, L., Potts, S. G., Roulston, T., Steffan-Dewenter, I., Vázquez, D. P., Winfree, R., Adams, L., Crone, E. E., Greenleaf, S. S., Keitt, T. H., Klein, A. M., Regetz, J. and Ricketts, T. H. (2007). Pollination and other ecosystem services produced by mobile organisms: A conceptual framework for the effects of land-use change. Ecology Letters 10(4), 299–314. https://doi.org/10.1111/j.1461-0248.2007.01018.x.

Krey, V., Havlik, P., Fricko, O., Zilliacus, J., Gidden, M., Strubegger, M., Kartasasmita, I., Ermolieva, T., Forsell, N. and Gusti, M. (2016). Message-Globiom 1.0 Documentation. Laxenburg, Austria: International Institute for Applied Systems Analysis.

Krysanova, V. and Arnold, J. G. (2008). Advances in ecohydrological modelling with SWAT–A review. *Hydrological Sciences Journal* 53(5), 939–947.

Krysanova, V. and White, M. (2015). Advances in water resources assessment with SWAT–An overview. *Hydrological Sciences Journal* 60(5), 1-13.

Kummu, M., de Moel, H., Porkka, M., Siebert, S., Varis, O. and Ward, P. J. (2012). Lost food, wasted resources: Global food supply chain losses and their impacts on freshwater, cropland, and fertiliser use. Science of the Total Environment 438, 477–489. https://doi.org/10.1016/j.scitotenv.2012.08.092.

Kummu, M., Kinnunen, P., Lehikoinen, E., Porkka, M., Queiroz, C., Röös, E., Troell, M. and Weil, C. (2020). Interplay of trade and food system resilience: gains on supply diversity over time at the cost of trade independency. Global Food Security 24, 100360. https://doi.org/10.1016/j.gfs.2020.100360.

Lal, R. (2015). Sequestering carbon and increasing productivity by conservation agriculture. Journal of Soil and Water Conservation 70(3), 55A–62A.

Lavelle, P., Decaëns, T., Aubert, M., Barot, S., Blouin, M., Bureau, F., Margerie, P., Mora, P. and Rossi, J.-P. (2006). Soil invertebrates and ecosystem services. European Journal of Soil Biology 42, S3–S15. https://doi.org/10.1016/j.ejsobi.2006.10.002.

Lawler, J. J., Lewis, D. J., Nelson, E., Plantinga, A. J., Polasky, S., Withey, J. C., Helmers, D. P., Martinuzzi, S., Pennington, D. and Radeloff, V. C. (2014). Projected land-use change impacts on ecosystem services in the United States. Proceedings of the National Academy of Sciences of the United States of America 111(20), 7492–7497. https://doi.org/10.1073/pnas.1405557111.

Leclère, D., Obersteiner, M., Barrett, M., Butchart, S. H. M., Chaudhary, A., De Palma, A., DeClerck, F. A. J., Di Marco, M., Doelman, J. C., Dürauer, M., Freeman, R., Harfoot, M., Hasegawa, T., Hellweg, S., Hilbers, J. P., Hill, S. L. L., Humpenöder, F., Jennings, N., Krisztin, T., Mace, G. M., Ohashi, H., Popp, A., Purvis, A., Schipper, A. M., Tabeau, A., Valin, H., van Meijl, H., van Zeist, W. J., Visconti, P., Alkemade, R., Almond, R., Bunting, G., Burgess, N. D., Cornell, S. E., Di Fulvio, F., Ferrier, S., Fritz, S., Fujimori, S., Grooten, M., Harwood, T., Havlík, P., Herrero, M., Hoskins, A. J., Jung, M., Kram, T., Lotze-Campen, H., Matsui, T., Meyer, C., Nel, D., Newbold, T., Schmidt-Traub, G., Stehfest, E., Strassburg, B. B. N., van Vuuren, D. P., Ware, C., Watson, J. E. M., Wu, W. and Young, L. (2020). Bending the curve of terrestrial biodiversity needs an integrated strategy. *Nature* 585(7826), 551–556.

Lehmann, J., Mempel, F. and Coumou, D. (2018). Increased occurrence of record-wet and record-dry months reflect changes in mean rainfall. Geophysical Research Letters 45(24), 468–413,476. https://doi.org/10.1029/2018GL079439.

Lescourret, F., Magda, D., Richard, G., Adam-Blondon, A.-F., Bardy, M., Baudry, J., Doussan, I., Dumont, B., Lefèvre, F., Litrico, I., Martin-Clouaire, R., Montuelle, B., Pellerin, S., Plantegenest, M., Tancoigne, E., Thomas, A., Guyomard, H. and Soussana, J.-F. (2015). A social-ecological approach to managing multiple agro-ecosystem services. *Current Opinion in Environmental Sustainability* 14, 68–75.

Liu, J., Hull, V., Godfray, H. C. J., Tilman, D., Gleick, P., Hoff, H., Pahl-Wostl, C., Xu, Z., Chung, M. G., Sun, J. and Li, S. (2018). Nexus approaches to global sustainable development. Nature Sustainability 1(9), 466–476. https://doi.org/10.1038/s41893-018-0135-8.

Liu, S., Crossman, N. D., Nolan, M. and Ghirmay, H. (2013). Bringing ecosystem services into integrated water resources management. Journal of Environmental Management 129, 92–102. https://doi.org/10.1016/j.jenvman.2013.06.047.

Liu, X., Chen, F., Barlage, M., Zhou, G. and Niyogi, D. (2016). Noah-MP-crop: Introducing dynamic crop growth in the Noah-MP land surface model. *Journal of Geophysical Research: Atmospheres* 121(23), 13953–13972.

Lobell, D. B. and Asseng, S. (2017). Comparing estimates of climate change impacts from process-based and statistical crop models. Environmental Research Letters 12(1), 015001. https://doi.org/10.1088/1748-9326/aa518a.

Lobell, D. B. and Gourdji, S. M. (2012). The influence of climate change on global crop productivity. Plant Physiology 160(4), 1686–1697. https://doi.org/10.1104/pp.112.208298.

Lobell, D. B., Schlenker, W. and Costa-Roberts, J. (2011). Climate trends and global crop production since 1980. Science 333(6042), 616–620. https://doi.org/10.1126/science.1204531.

Lokupitiya, E., Denning, S., Paustian, K., Baker, I., Schaefer, K., Verma, S., Meyers, T., Bernacchi, C. J., Suyker, A. and Fischer, M. (2009). Incorporation of crop phenology in simple biosphere model (SiBcrop) to improve land-atmosphere carbon exchanges from croplands. *Biogeosciences* 6(6), 969–986.

Lonsdorf, E., Kremen, C., Ricketts, T., Winfree, R., Williams, N. and Greenleaf, S. (2009). Modelling pollination services across agricultural landscapes. *Annals of Botany* 103(9), 1589–1600.

Luderer, G., Leimbach, M., Bauer, N., Kriegler, E., Baumstark, L., Bertram, C., Giannousakis, A., Hilaire, J., Klein, D., Levesque, A., Mouratiadou, I., Pehl, M., Pietzcker, R., Piontek, F., Roming, N., Schultes, A., Schwanitz, V. and Strefler, J. (2015). Description of the REMIND model (version 1.6). SSRN Electronic Journal. https://doi.org/10.2139/ssrn.2697070.

Mandle, L., Shields-Estrada, A., Chaplin-Kramer, R., Mitchell, M. G. E., Bremer, L. L., Gourevitch, J. D., Hawthorne, P., Johnson, J. A., Robinson, B. E., Smith, J. R., Sonter, L. J., Verutes, G. M., Vogl, A. L., Daily, G. C. and Ricketts, T. H. (2020). Increasing decision relevance of ecosystem service science. *Nature Sustainability* 4(2), 161–169. https://doi.org/10.1038/s41893-020-00625-y.

Mente, E. and Smaal, A. C. (2016). Introduction to the special issue on "European aquaculture development since 1993: The benefits of aquaculture to Europe and the perspectives of European aquaculture production." Aquaculture International 24(3), 693–698. https://doi.org/10.1007/s10499-016-0003-3.

Millennium Ecosystem Assessment (Ed.). (2005). Ecosystems and Human Well-Being: Wetlands and Water Synthesis: A Report of the Millennium Ecosystem Assessment. World Resources Institute.

Miller-Struttmann, N. E., Geib, J. C., Franklin, J. D., Kevan, P. G., Holdo, R. M., Ebert-May, D., Lynn, A. M., Kettenbach, J. A., Hedrick, E. and Galen, C. (2015). Functional mismatch in a bumble bee pollination mutualism under climate change. Science 349(6255), 1541–1544. https://doi.org/10.1126/science.aab0868.

Mitchell, M. G. E., Bennett, E. M. and Gonzalez, A. (2013). Linking landscape connectivity and ecosystem service provision: Current knowledge and research gaps. *Ecosystems* 16(5), 894–908.

Monfreda, C., Ramankutty, N. and Foley, J. A. (2008). Farming the planet: 2. Geographic distribution of crop areas, yields, physiological types, and net primary production in the year 2000. Global Biogeochemical Cycles 22(1). https://doi.org/10.1029/2007GB002947.

Mueller, N. D., Butler, E. E., McKinnon, K. A., Rhines, A., Tingley, M., Holbrook, N. M. and Huybers, P. (2016). Cooling of US Midwest summer temperature extremes from cropland intensification. Nature Climate Change 6(3), 317–322. https://doi.org/10.1038/nclimate2825.

Mueller, N. D., Gerber, J. S., Johnston, M., Ray, D. K., Ramankutty, N. and Foley, J. A. (2012). Closing yield gaps through nutrient and water management. Nature 490(7419), 254–257. https://doi.org/10.1038/nature11420.

Müller, C., Elliott, J., Chryssanthacopoulos, J., Arneth, A., Balkovic, J., Ciais, P., Deryng, D., Folberth, C., Glotter, M., Hoek, S., Iizumi, T., Izaurralde, R. C., Jones, C., Khabarov, N., Lawrence, P., Liu, W., Olin, S., Pugh, T. A. M., Ray, D. K., Reddy, A., Rosenzweig, C., Ruane, A. C., Sakurai, G., Schmid, E., Skalsky, R., Song, C. X., Wang, X., de Wit, A. and Yang, H. (2017). Global gridded crop model evaluation: Benchmarking, skills, deficiencies and implications. Geoscientific Model Development 10(4), 1403–1422. https://doi.org/10.5194/gmd-10-1403-2017.

Nelson, E., Polasky, S., Lewis, D. J., Plantinga, A. J., Lonsdorf, E., White, D., Bael, D. and Lawler, J. J. (2008). Efficiency of incentives to jointly increase carbon sequestration and species conservation on a landscape. *Proceedings of the National Academy of Sciences of the United States of America* 105(28), 9471–9476.

Nyström, M., Jouffray, J. B., Norström, A. V., Crona, B., Jørgensen, P. S., Carpenter, S. R., Bodin, Ö., Galaz, V. and Folke, C. (2019). Anatomy and resilience of the global production ecosystem. Nature 575(7781), 98–108. https://doi.org/10.1038/s41586-019-1712-3.

Oerke, E.-C. (2006). Crop losses to pests. The Journal of Agricultural Science 144(1), 31–43. https://doi.org/10.1017/S0021859605005708.

Ohly, H., White, M. P., Wheeler, B. W., Bethel, A., Ukoumunne, O. C., Nikolaou, V. and Garside, R. (2016). Attention restoration theory: A systematic review of the attention restoration potential of exposure to natural environments. Journal of Toxicology and Environmental Health. Part B, Critical Reviews 19(7), 305–343. https://doi.org/10.1080/10937404.2016.1196155.

O'Neill, B. C., Kriegler, E., Ebi, K. L., Kemp-Benedict, E., Riahi, K., Rothman, D. S., Ruijven, B. J. van, Vuuren, D. P. van, Birkmann, J., Kok, K., Levy, M. A. and Solecki, W. D. (2017). The roads ahead: Narratives for shared socioeconomic pathways describing world futures in the 21st century. Global Environmental Change 42, 169–180. https://doi.org/10.1016/j.gloenvcha.2015.01.004.

Ostberg, S., Schewe, J., Childers, K. and Frieler, K. (2017). Changes in crop yields and their variability at different levels of global warming. Earth System Dynamics 9(2), 479–496. https://doi.org/10.5194/esd-9-479-2018.

PBL Netherlands Environmental Assessment Agency. (2014). Integrated assessment of global environmental change with IMAGE 3.0–Model description and policy applications. https://www.pbl.nl/en/publications/integrated-assessment-of-global-environmental-change-with-IMAGE-3.0. PBL Netherlands: Environmental Assessment Agency.

Pereira, P., Bogunovic, I., Muñoz-Rojas, M. and Brevik, E. C. (2018). Soil ecosystem services, sustainability, valuation and management. Current Opinion in Environmental Science and Health 5, 7–13. https://doi.org/10.1016/j.coesh.2017.12.003.

Pimentel, D., Zuniga, R. and Morrison, D. (2005). Update on the environmental and economic costs associated with alien-invasive species in the United States. Ecological Economics 52(3), 273–288. https://doi.org/10.1016/j.ecolecon.2004.10.002.

Polasky, S., Nelson, E., Camm, J., Csuti, B., Fackler, P., Lonsdorf, E., Montgomery, C., White, D., Arthur, J., Garber-Yonts, B., Haight, R., Kagan, J., Starfield, A. and Tobalske, C. (2008). Where to put things? Spatial land management to sustain biodiversity and economic returns. *Biological Conservation* 141(6), 1505–1524.

Ponette-González, A. G., Marín-Spiotta, E., Brauman, K. A., Farley, K. A., Weathers, K. C. and Young, K. R. (2014). Hydrologic connectivity in the high-elevation tropics: heterogeneous responses to land change. *BioScience* 64(2), 92–104. https://doi.org/10.1093/biosci/bit013.

Popp, A., Calvin, K., Fujimori, S., Havlik, P., Humpenöder, F., Stehfest, E., Bodirsky, B. L., Dietrich, J. P., Doelmann, J. C., Gusti, M., Hasegawa, T., Kyle, P., Obersteiner, M., Tabeau, A., Takahashi, K., Valin, H., Waldhoff, S., Weindl, I., Wise, M., Kriegler, E., Lotze-Campen, H., Fricko, O., Riahi, K. and van Vuuren, D. P. (2017). Land-use futures in the shared socio-economic pathways. Global Environmental Change 42, 331–345. https://doi.org/10.1016/j.gloenvcha.2016.10.002.

Popp, A., Rose, S. K., Calvin, K., Van Vuuren, D. P., Dietrich, J. P., Wise, M., Stehfest, E., Humpenöder, F., Kyle, P., Van Vliet, J., Bauer, N., Lotze-Campen, H., Klein, D. and Kriegler, E. (2014). Land-use transition for bioenergy and climate stabilization: Model comparison of drivers, impacts and interactions with other land use based mitigation options. *Climatic Change* 123(3–4), 495–509.

Potts, S. G., Biesmeijer, J. C., Kremen, C., Neumann, P., Schweiger, O. and Kunin, W. E. (2010). Global pollinator declines: Trends, impacts and drivers. Trends in Ecology and Evolution 25(6), 345–353. https://doi.org/10.1016/j.tree.2010.01.007.

Ray, D. K., Mueller, N. D., West, P. C. and Foley, J. A. (2013). Yield trends are insufficient to double global crop production by 2050. *PLoS ONE* 8(6), e66428. https://doi.org/10.1371/journal.pone.0066428.

Ray, D. K., West, P. C., Clark, M., Gerber, J. S., Prishchepov, A. V. and Chatterjee, S. (2019). Climate change has likely already affected global food production. *PLoS ONE* 14(5), e0217148. https://doi.org/10.1371/journal.pone.0217148.

Renard, D., Rhemtulla, J. M. and Bennett, E. M. (2015). Historical dynamics in ecosystem service bundles. *Proceedings of the National Academy of Sciences of the United States of America* 112(43), 13411–13416.

Riahi, K., van Vuuren, D. P., Kriegler, E., Edmonds, J., O'Neill, B. C., Fujimori, S., Bauer, N., Calvin, K., Dellink, R., Fricko, O., Lutz, W., Popp, A., Cuaresma, J. C., Kc, S., Leimbach, M., Jiang, L., Kram, T., Rao, S., Emmerling, J., Ebi, K., Hasegawa, T., Havlik, P., Humpenöder, F., Da Silva, L. A., Smith, S., Stehfest, E., Bosetti, V., Eom, J., Gernaat, D., Masui, T., Rogelj, J., Strefler, J., Drouet, L., Krey, V., Luderer, G., Harmsen, M., Takahashi, K., Baumstark, L., Doelman, J. C., Kainuma, M., Klimont, Z., Marangoni, G., Lotze-Campen, H., Obersteiner, M., Tabeau, A. and Tavoni, M. (2017). The shared socioeconomic pathways and their energy, land use, and greenhouse gas emissions implications: An overview. *Global Environmental Change* 42, 153–168. https://doi.org/10.1016/j.gloenvcha.2016.05.009.

Ricketts, T. H. and Lonsdorf, E. (2013). Mapping the margin: Comparing marginal values of tropical forest remnants for pollination services. *Ecological Applications: A Publication of the Ecological Society of America* 23(5), 1113–1123.

Ricketts, T. H., Regetz, J., Steffan-Dewenter, I., Cunningham, S. A., Kremen, C., Bogdanski, A., Gemmill-Herren, B., Greenleaf, S. S., Klein, A. M., Mayfield, M. M., Morandin, L. A., Ochieng', A., Potts, S. G. and Viana, B. F. (2008) Landscape effects on crop pollination services: Are there general patterns? *Ecology Letters* 11(5), 499–515. https://doi.org/10.1111/j.1461-0248.2008.01157.x.

Rieb, J. T., Chaplin-Kramer, R., Daily, G. C., Armsworth, P. R., Böhning-Gaese, K., Turner, M. G. and Bennett, E. M. (2017). When, where, and how nature matters for ecosystem services: Challenges for the next generation of ecosystem service models. *Bioscience*, 1–14.

Roehrdanz, P. R. and Hannah, L. (2016). Climate change, California wine, and wildlife habitat. Journal of Wine Economics 11(1), 69–87. https://doi.org/10.1017/jwe.2014.31.

Rogelj, J., Popp, A., Calvin, K. V., Luderer, G., Emmerling, J., Gernaat, D., Fujimori, S., Strefler, J., Hasegawa, T., Marangoni, G., Krey, V., Kriegler, E., Riahi, K., van Vuuren, D. P., Doelman, J., Drouet, L., Edmonds, J., Fricko, O., Harmsen, M., Havlík, P., Humpenöder, F., Stehfest, E. and Tavoni, M. (2018). Scenarios towards limiting global mean temperature increase below 1.5°C. *Nature Climate Change* 8(4), 325–332.

Rosa, L., Rulli, M. C., Davis, K. F., Chiarelli, D. D., Passera, C. and D'Odorico, P. (2018). Closing the yield gap while ensuring water sustainability. Environmental Research Letters 13(10), 104002. https://doi.org/10.1088/1748-9326/aadeef.

Rosenzweig, C., Ruane, A. C., Antle, J., Elliott, J., Ashfaq, M., Chatta, A. A., Ewert, F., Folberth, C., Hathie, I., Havlik, P., Hoogenboom, G., Lotze-Campen, H., MacCarthy, D. S., Mason-D'Croz, D., Contreras, E. M., Müller, C., Perez-Dominguez, I., Phillips, M., Porter, C., Raymundo, R. M., Sands, R. D., Schleussner, C. F., Valdivia, R. O., Valin, H. and Wiebe, K. (2018). Coordinating AgMIP data and models across global and regional scales for 1.5°C and 2.0°C assessments. Philosophical Transactions. Series A, Mathematical, Physical, and Engineering Sciences 376(2119), 20160455. https://doi.org/10.1098/rsta.2016.0455.

Sable, M. G. and Rana, D. K. (2016). Impact of global warming on insect behavior - A review. *Agricultural Reviews* 37(1). https://doi.org/10.18805/ar.v37i1.9270.

Saccà, M., Barra Caracciolo, A., Di Lenola, M. and Grenni, P. (2017). Ecosystem services provided by soil microorganisms. In: *Soil Biological Communities and Ecosystem Resilience*, pp. 9–24. https://doi.org/10.1007/978-3-319-63336-7_2.

Saunders Bulan, M. T., Posner, J. L., Peng, D., Emshwiller, E., Wang, X., Li, J., Stoltenberg, D. and Stoecker, R. (2017). Old crop, new society: Persistence and change of Tartary buckwheat farming in Yunnan, China. Human Ecology 45(1), 37–51. https://doi.org/10.1007/s10745-016-9871-4.

Schipper, A. M., Hilbers, J. P., Meijer, J. R., Antão, L. H., Benítez-López, A., de Jonge, M. M. J., Leemans, L. H., Scheper, E., Alkemade, R., Doelman, J. C., Mylius, S., Stehfest, E., van Vuuren, D. P., van Zeist, W. J. and Huijbregts, M. A. J. (2020). Projecting terrestrial biodiversity intactness with GLOBIO 4. *Global Change Biology* 26(2), 760–771.

Scholes, R. J. (2016). Climate change and ecosystem services. WIREs Climate Change 7(4), 537–550. https://doi.org/10.1002/wcc.404.

Searchinger, T., Waite, R., Hanson, C., Ranganathan, J., Dumas, P., Matthews, E. and Klirs, C. (2019). Creating a sustainable food future: A menu of solutions to feed nearly 10 billion people by 2050. Translated by J. Ranganathan. https://apo.org.au/node/248946. World Resources Institute.

Sharp, R., Douglass, J., Wolny, S., Arkema, K., Bernhardt, J., Bierbower, W., Chaumont, N., Denu, D., Fisher, D., Kathryn, G., Griffin, R., Guannel, G., Guerry, A., Johnson, J., Hamel, P., Kennedy, C., Kim, C.-K., Layaco, M., Lonsdorf, E. and Wood, S. (2014). VEST user's guide: Integrated valuation of ecosystem services and tradeoffs, Vol. 306.

Smith, P., Martino, D., Cai, Z., Gwary, D., Janzen, H., Kumar, P., McCarl, B., Ogle, S., O'Mara, F., Rice, C., Scholes, B., Sirotenko, O., Howden, M., McAllister, T., Pan, G., Romanenkov, V., Schneider, U., Towprayoon, S., Wattenbach, M. and Smith, J. (2008). Greenhouse gas mitigation in agriculture. Philosophical Transactions of the Royal

Society of London. Series B, Biological Sciences 363(1492), 789–813. https://doi.org/10.1098/rstb.2007.2184.

Spinoni, J., Naumann, G., Carrao, H., Barbosa, P. and Vogt, J. (2014). World drought frequency, duration, and severity for 1951–2010. International Journal of Climatology 34(8), 2792–2804. https://doi.org/10.1002/joc.3875.

Springmann, M., Clark, M., Mason-D'Croz, D., Wiebe, K., Bodirsky, B. L., Lassaletta, L., de Vries, W., Vermeulen, S. J., Herrero, M., Carlson, K. M., Jonell, M., Troell, M., DeClerck, F., Gordon, L. J., Zurayk, R., Scarborough, P., Rayner, M., Loken, B., Fanzo, J., Godfray, H. C. J., Tilman, D., Rockström, J. and Willett, W. (2018). Options for keeping the food system within environmental limits. Nature 562(7728), 519–525. https://doi.org/10.1038/s41586-018-0594-0.

Stehfest, E., van Vuuren, D., Bouwman, L. and Kram, T. (2014). *Integrated Assessment of Global Environmental Change With IMAGE 3.0: Model Description and Policy Applications*. Netherlands Environmental Assessment Agency (PBL).

Suh, S., Johnson, J. A., Tambjerg, L., Sim, S., Broeckx-Smith, S., Reyes, W. and Chaplin-Kramer, R. (2020). Closing yield gap is crucial to avoid potential surge in global carbon emissions. *Global Environmental Change* 63(May), 102100.

Swann, A. L. S., Fung, I. Y. and Chiang, J. C. H. (2012). Mid-latitude afforestation shifts general circulation and tropical precipitation. Proceedings of the National Academy of Sciences of the United States of America 109(3), 712–716. https://doi.org/10.1073/pnas.1116706108.

Tallis, H., Kareiva, P., Marvier, M. and Chang, A. (2008). An ecosystem services framework to support both practical conservation and economic development. Proceedings of the National Academy of Sciences of the United States of America 105(28), 9457–9464. https://doi.org/10.1073/pnas.0705797105.

Taylor, R. G., Scanlon, B., Döll, P., Rodell, M., van Beek, R., Wada, Y., Longuevergne, L., Leblanc, M., Famiglietti, J. S., Edmunds, M., Konikow, L., Green, T. R., Chen, J., Taniguchi, M., Bierkens, M. F. P., MacDonald, A., Fan, Y., Maxwell, R. M., Yechieli, Y., Gurdak, J. J., Allen, D. M., Shamsudduha, M., Hiscock, K., Yeh, P. J.-F., Holman, I. and Treidel, H. (2013). Ground water and climate change. Nature Climate Change 3(4), 322–329. https://doi.org/10.1038/nclimate1744.

Tayyebi, A., Meehan, T. D., Dischler, J., Radloff, G., Ferris, M. and Gratton, C. (2016). SmartScape™: A web-based decision support system for assessing the tradeoffs among multiple ecosystem services under crop-change scenarios. *Computers and Electronics in Agriculture* 121, 108–121.

Thomson, L. J., Macfadyen, S. and Hoffmann, A. A. (2010). Predicting the effects of climate change on natural enemies of agricultural pests. Biological Control 52(3), 296–306. https://doi.org/10.1016/j.biocontrol.2009.01.022.

Tscharntke, T., Steffan-Dewenter, I., Kruess, A. and Thies, C. (2002). Characteristics of insect populations on habitat fragments: A mini review: Insect populations and habitat fragments. Ecological Research 17(2), 229–239. https://doi.org/10.1046/j.1440-1703.2002.00482.x.

Turner, S. W. D., Hejazi, M., Calvin, K., Kyle, P. and Kim, S. (2019). A pathway of global food supply adaptation in a world with increasingly constrained groundwater. Science of the Total Environment 673, 165–176. https://doi.org/10.1016/j.scitotenv.2019.04.070.

United Nations and Department of Economic and Social Affairs (2019). *World Population Prospects 2019, Online Edition Rev*, p. 1. https://population.un.org/wpp/DefinitionOfProjectionVariants/.

Valdivia, C. and Barbieri, C. (2014). Agritourism as a sustainable adaptation strategy to climate change in the Andean Altiplano. Tourism Management Perspectives 11, 18–25. https://doi.org/10.1016/j.tmp.2014.02.004.

van den Berg, M., Wendel-Vos, W., van Poppel, M., Kemper, H., van Mechelen, W. and Maas, J. (2015). Health benefits of green spaces in the living environment: A systematic review of epidemiological studies. Urban Forestry and Urban Greening 14(4), 806–816. https://doi.org/10.1016/j.ufug.2015.07.008.

Van den Hoof, C., Hanert, E. and Vidale, P. L. (2011). Simulating dynamic crop growth with an adapted land surface model – JULES-SUCROS: Model development and validation. *Agricultural and Forest Meteorology* 151(2), 137–153.

van Meijl, H., Shutes, L., Valin, H., Stehfest, E., van Dijk, M., Kuiper, M., Tabeau, A., van Zeist, W.-J., Hasegawa, T. and Havlik, P. (2020). Modelling alternative futures of global food security: insights from foodsecure. Global Food Security 25, 100358. https://doi.org/10.1016/j.gfs.2020.100358.

Vejre, H. and Brandt, J. (2003). *Multifunctional Landscapes: Theory, Value and History*. Southampton: WIT Press.

Venterea, R. T., Halvorson, A. D., Kitchen, N., Liebig, M. A., Cavigelli, M. A., Grosso, S. J. D., Motavalli, P. P., Nelson, K. A., Spokas, K. A., Singh, B. P., Stewart, C. E., Ranaivoson, A., Strock, J. and Collins, H. (2012). Challenges and opportunities for mitigating nitrous oxide emissions from fertilized cropping systems. Frontiers in Ecology and the Environment 10(10), 562–570. https://doi.org/10.1890/120062.

Wang, G. (2005). Agricultural drought in a future climate: Results from 15 global climate models participating in the IPCC 4th assessment. Climate Dynamics 25(7–8), 739–753. https://doi.org/10.1007/s00382-005-0057-9.

Wang, J., Li, W., Ciais, P., Li, L. Z. X., Chang, J., Goll, D., Gasser, T., Huang, X., Devaraju, N. and Boucher, O. (2021). Global cooling induced by biophysical effects of bioenergy crop cultivation. Nature Communications 12(1), 7255. https://doi.org/10.1038/s41467-021-27520-0.

Webler, G., Roberti, D. R., Cuadra, S. V., Moreira, V. S. and Costa, M. H. (2012). Evaluation of a dynamic agroecosystem model (agro-IBIS) for soybean in Southern Brazil. *Earth Interactions* 16(12), 1–15.

Weil, C., Johnson, J., Caristan, R., Brauman, K., Bodirsky, B., Jalava, M., Kummu, M., Queiroz, C., Monier, L., Popp, A., Ray, D. and Chaplin-Kramer, R. (2021). Resilience of Food Sufficiency to Future Climate and Societal Changes [Preprint]. https://doi.org/10.21203/rs.3.rs-135121/v1.

Weiskopf, S. R., Rubenstein, M. A., Crozier, L. G., Gaichas, S., Griffis, R., Halofsky, J. E., Hyde, K. J. W., Morelli, T. L., Morisette, J. T., Muñoz, R. C., Pershing, A. J., Peterson, D. L., Poudel, R., Staudinger, M. D., Sutton-Grier, A. E., Thompson, L., Vose, J., Weltzin, J. F. and Whyte, K. P. (2020). Climate change effects on biodiversity, ecosystems, ecosystem services, and natural resource management in the United States. Science of the Total Environment 733, 137782. https://doi.org/10.1016/j.scitotenv.2020.137782.

West, P. C., Gerber, J. S., Engstrom, P. M., Mueller, N. D., Brauman, K. A., Carlson, K. M., Cassidy, E. S., Johnston, M., MacDonald, G. K., Ray, D. K. and Siebert, S. (2014). Leverage points for improving global food security and the environment. Science 345(6194), 325–328. https://doi.org/10.1126/science.1246067.

Whitehead, P. G., Wilby, R. L., Battarbee, R. W., Kernan, M. and Wade, A. J. (2009). A review of the potential impacts of climate change on surface water quality. Hydrological Sciences Journal 54(1), 101–123. https://doi.org/10.1623/hysj.54.1.101.

WHO. (2019). *Joint Child Malnutrition Estimates–Levels and Trends*. WHO. http://www.who.int/nutgrowthdb/estimates2018/en/.

Willett, W., Rockström, J., Loken, B., Springmann, M., Lang, T., Vermeulen, S., Garnett, T., Tilman, D., DeClerck, F., Wood, A., Jonell, M., Clark, M., Gordon, L. J., Fanzo, J., Hawkes, C., Zurayk, R., Rivera, J. A., De Vries, W., Majele Sibanda, L., Afshin, A., Chaudhary, A., Herrero, M., Agustina, R., Branca, F., Lartey, A., Fan, S., Crona, B., Fox, E., Bignet, V., Troell, M., Lindahl, T., Singh, S., Cornell, S. E., Srinath Reddy, K., Narain, S., Nishtar, S. and Murray, C. J. L. (2019). Food in the anthropocene: The EAT-Lancet Commission on healthy diets from sustainable food systems. The Lancet 393(10170), 447–492. https://doi.org/10.1016/S0140-6736(18)31788-4.

Woltjer, G. and Kuiper, M. (n.d.). The Magnet Model: Module Description 148.

World Bank Country and Lending Groups – World Bank Data Help Desk. (n.d.). Retrieved January 24, 2020. https://datahelpdesk.worldbank.org/knowledgebase/articles/906519-world-bank-country-and-lending-groups.

World Food Programme. (2019). The State of Food Security and Nutrition in the World (SOFI): Safeguarding against economic slowdowns and downturns. https://www.wfp.org/publications/2019-state-food-security-and-nutrition-world-sofi-safeguarding-against-economic.

Wu, X., Vuichard, N., Ciais, P., Viovy, N., de Noblet-Ducoudré, N., Wang, X., Magliulo, V., Wattenbach, M., Vitale, L., Di Tommasi, P., Moors, E. J., Jans, W., Elbers, J., Ceschia, E., Tallec, T., Bernhofer, C., Grünwald, T., Moureaux, C., Manise, T. and Ripoche, D. (2016). ORCHIDEE-crop. A new process-based agro-land surface model: Model description and evaluation over Europe. *Geoscientific Model Development* 9(2), 857–873.

Zeppetello, L. R. V., Cook-Patton, S. C., Parsons, L. A., Wolff, N. H., Kroeger, T., Battisti, D. S., Bettles, J., Spector, J. T., Balakumar, A. and Masuda, Y. J. (2022). Consistent cooling benefits of silvopasture in the tropics. Nature Communications 13(1), 708. https://doi.org/10.1038/s41467-022-28388-4.

Zhang, Y., Wu, Z., Singh, V. P., Su, Q., He, H., Yin, H., Zhang, Y. and Wang, F. (2021). Simulation of crop water demand and consumption considering irrigation effects based on coupled hydrology-crop growth model. *Journal of Advances in Modeling Earth Systems* 13(11). https://doi.org/10.1029/2020MS002360.

Zhao, C., Liu, B., Piao, S., Wang, X., Lobell, D. B., Huang, Y., Huang, M., Yao, Y., Bassu, S., Ciais, P., Durand, J. L., Elliott, J., Ewert, F., Janssens, I. A., Li, T., Lin, E., Liu, Q., Martre, P., Müller, C., Peng, S., Peñuelas, J., Ruane, A. C., Wallach, D., Wang, T., Wu, D., Liu, Z., Zhu, Y., Zhu, Z. and Asseng, S. (2017). Temperature increase reduces global yields of major crops in four independent estimates. Proceedings of the National Academy of Sciences of the United States of America 114(35), 9326–9331.

Zhu, X., Troy, T. J. and Devineni, N. (2019). Stochastically modeling the projected impacts of climate change on rainfed and irrigated US crop yields. Environmental Research Letters 14(7), 074021. https://doi.org/10.1088/1748-9326/ab25a1.

Chapter 10

Modelling climate change impacts on agricultural commodity markets

Ignacio Pérez Domínguez, Jordan Hristov, Christian Elleby and Thomas Fellmann, European Commission, Joint Research Centre, Spain; Andrea Toreti, European Commission, Joint Research Centre, Italy; Thomas Chatzopoulos and Ana Luisa Barbosa, European Commission, Joint Research Centre, Spain; and Frank Dentener, European Commission, Joint Research Centre, Italy

1 Introduction
2 Measuring direct impacts of climate change on agriculture
3 Economic adaptation potential
4 The role of technologies for mitigation and adaptation
5 Conclusion
6 Future trends in research
7 Where to look for further information
8 References

1 Introduction

Climate change is increasingly perceived as a source of concern for the agricultural sector, especially regarding its medium- to long-term impacts on the global agri-food system (Von Lampe et al., 2014). This is due to the prevailing adverse biophysical effects of changes in mean temperature, precipitation, and the impacts of climate extremes on the quantity and quality of crop yields (IPCC, 2014; IPCC, 2022; EEA, 2019). Climate change, including increases in the frequency and intensity of extremes, has reduced food and water security, hindering efforts to meet Sustainable Development Goals (*high confidence*). Although overall agricultural productivity has increased, climate change has slowed this growth over the past 50 years globally (*medium confidence*). Related negative impacts were mainly in mid- and low-latitude regions, but positive impacts occurred in some high-latitude regions (*high confidence*). Climate change will increasingly put pressure on food production and access, especially in vulnerable regions, undermining Food and nutrition security (*high confidence*). Increases in frequency, intensity, and severity of droughts, floods, and heatwaves and

http://dx.doi.org/10.19103/AS.2022.0115.12

continued sea level rise will increase risks to food security (*high confidence*) in vulnerable regions from moderate to high between 1.5°C and 2°C global warming level (IPCC, 2022).

This chapter provides insights into how climate change can affect agricultural commodity markets in the medium to long term. It includes some methodological aspects on how various quantitative modelling tools, ranging from earth system models to farm-level production decisions, can be combined in the analysis of climate change impacts on agricultural markets. The analysis presented here is supported by several agro-economic studies carried out by the Joint Research Centre of the European Commission, including *PESETA* (Projection of Economic impacts of climate change in Sectors of the European Union based on bottom-up Analysis)[1]; *AgCLIM50* (Challenges of Global Agriculture in a Climate Change Context by 2050; Van Meijl et al., 2017); *C2ESAM* (Concurrent Climate Extremes and Shocks on Agricultural Markets)[2]; and *EcAMPA* (Economic Assessment of Greenhouse Gas Mitigation Policy Options for EU Agriculture; Pérez Domínguez et al., 2020).

The above-mentioned studies focus on the potential for the (European) agricultural sector to adapt to climate change. Farmers have always been adapting to changing biophysical conditions as well as market forces influencing supply and demand (e.g. land use policies and consumer preferences). However, effective adaptation to climate change cannot be entirely autonomous and needs to be accompanied by changes in technology (e.g. better irrigation systems) and crop management practices (e.g. fallowing of histosols). Climate change is not the only factor to be taken into account as the agricultural sector is subject to different policies and regulations with multiple, and partly diverging, objectives. For example, the objectives of the Farm to Fork Strategy aiming at increasing the sustainability of EU production, processing, distribution, and consumption will have to be achieved simultaneously with the EU's transition to cleaner energy and climate neutrality (European Green Deal).

Most of the studies mentioned above incorporate climate change as a global dimension and make use of global climate, biophysical, and economic models. They largely rely on Integrated Assessment Models (IAMs) and model inter-comparison exercises (e.g. Rosenzweig et al., 2014). While this chapter is focusing on EU agricultural markets, similar modelling efforts have been done for instance in the US, where a multi-model framework for quantitative sectoral impact analysis was used to estimate climate change impacts and economic damages to multiple US sectors under different scenarios (EPA,

[1] URL: https://joint-research-centre.ec.europa.eu/peseta-projects_en
[2] URL: https://knowledge4policy.ec.europa.eu/node/36086_de

2017). Other global studies have recently proposed integrating a geospatially resolved physical representation of impacts into a coupled human-Earth system modelling framework when conducting a multi-sectoral assessment of climate impacts under a range of consistent and integrated economic and climate scenarios that are responsive to new policies and business expectations (Monier et al., 2018). Wang and Tang (2022) also estimate the climate damage for China by using different open-source IAMs under coupled shared socioeconomic pathways and representative concentration pathways (RCPs).

2 Measuring direct impacts of climate change on agriculture

Concerning the analysis of climate change impacts on agriculture, we use a combination of climate, crop growth, and economic models, an approach that is frequently used in the literature (Hasegawa et al., 2018; Van Meijl et al., 2018) (see Fig. 1). Climate-driven agricultural productivity changes across the globe are obtained from the Global Gridded Crop Models (GGCMs), ISI-MIP FastTrack Database (Hempel et al., 2013; Rosenzweig et al., 2014). Yield data are available for four crops (wheat, maize, rice, and soybeans) from five GGCMs and 20 different bias-adjusted climate scenarios under four representative concentration pathways (RCPs: RCP2.6, RCP4.5, RCP6.0, RCP8.5) and for five general circulation models (GCMs: HadGEM2-ES, GFDL-ESM2M, IPSL-CM5A-LR, MIROC-ESM-CHEM, NorESM1-M). Data are extracted for different time slices of 30 years (e.g. 2016–2045 and 2070–2099), which can be compared against the historic baseline (1986–2015)[3]. For each baseline, a 30-year average growth

[3] The historical CMIP5 simulations end in 2005. Many impact studies have extended to 2015 by selecting a specific RCP.

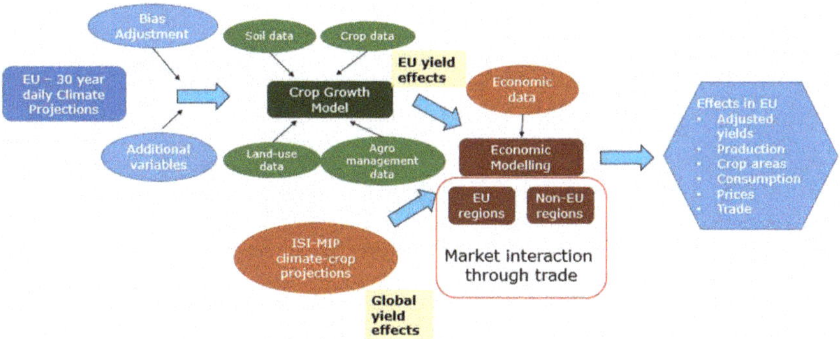

Figure 1 Information flow between biophysical and agro-economic models in the PESETA IV study. Source: (Hristov et al., 2020).

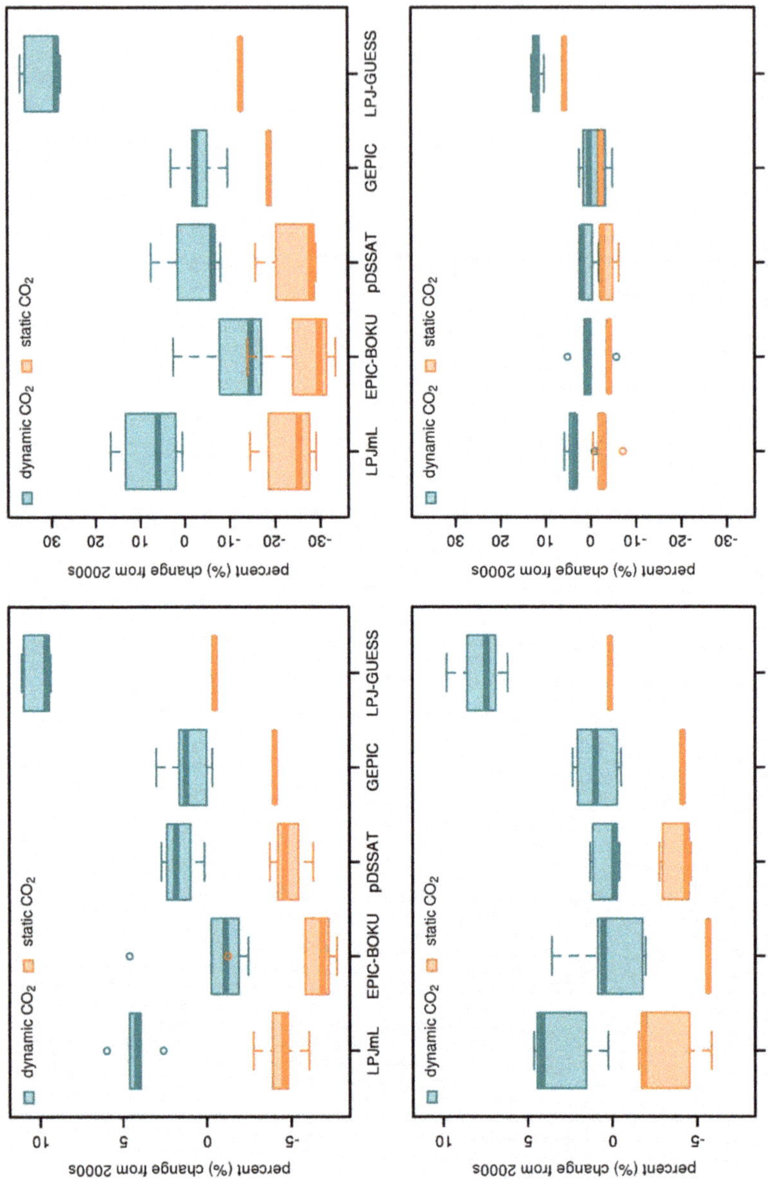

Figure 2 Relative changes in total global crop productivity under RCP8.5 (upper row) and RCP2.6 (lower row) for the five different crop models for the periods 2015–2045 (left column) and 2070–2100 (right column). Source: AgCLIM50-3 report (unpublished). Note: Boxes (interquartile range) and whiskers display the range of projected changes. The solid horizontal bar represents the median case. Outliers (>1.5 the interquartile range) are depicted as dots.

rate is computed for each crop and country to describe the climate-induced change in productivity on average conditions (van Meijl et al., 2017, 2018).

The ISI-MIP initiative has provided a set of combinations of climate projections, climate models, and crop models that allow for a robust quantitative analysis of crop productivity changes linked to different climate change potential pathways. Figure 2 shows that global mean productivity changes are larger over a time horizon closer to the end of the century (30-year period around 2085 vs. 2030). Moreover, yearly crop productivity changes are more variable in scenarios having a larger climate change signal linked to the specific emission pathway (e.g. RCP8.5 vs. RCP2.6). As expected (Toreti et al., 2020), the impact of CO_2 fertilization[4] on crop productivity (blue-coloured box-whiskers vs. orange-coloured ones in Fig. 2) is a relevant and important factor to consider.

3 Economic adaptation potential

There are still many uncertainties in fully understanding and modelling climate change impacts on agriculture. The choice of a crop model is one source of uncertainty since crop models may be very different in their responses (Müller et al., 2017). In addition, there is also an inconsistency in yield estimates from regional biophysical modelling as they usually do not consider global agro-economic changes that may influence agriculture in the region of interest. Economic aspects and related market adjustments are often not taken into account in regional studies of climate impacts. Thus, there is a need for comprehensive global analysis, because market feedback may be an important adaptation measure to climate change impacts on agriculture (Nelson et al., 2014). However, the evidence on market feedback is still not clear due to often contradictory results from economic models (Fernández and Blanco, 2015).

To better quantify the uncertainty of the crop model choice, two different modelling approaches and respective scenario ensembles were selected. An innovative modelling approach based on a new version of the World Food Studies (WOFOST) crop model (driven by 10 bias-adjusted regional climate models) and the ISI-MIP fast track simulations, consisting of results from 35 global scenarios, was recently used (Hristov et al., 2020). Using these two approaches allows to calculate a range of impacts in Europe with relatively high accuracy, while uncertainty resulting from diverse models is more comprehensively calculated by the ISI-MIP model ensemble. To understand better the feedback of agricultural markets, the biophysical crop yield effects were introduced in the global agro-economic CAPRI (Common Agricultural Policy Regional Impact Analysis) model. CAPRI captures in detail agricultural production at the regional

[4] Enhanced atmospheric CO_2 concentration stimulates photosynthesis and growth while reducing leaf transpiration. This so-called fertilization effect varies by plant species and growing conditions, and it is more pronounced in C_3 crops.

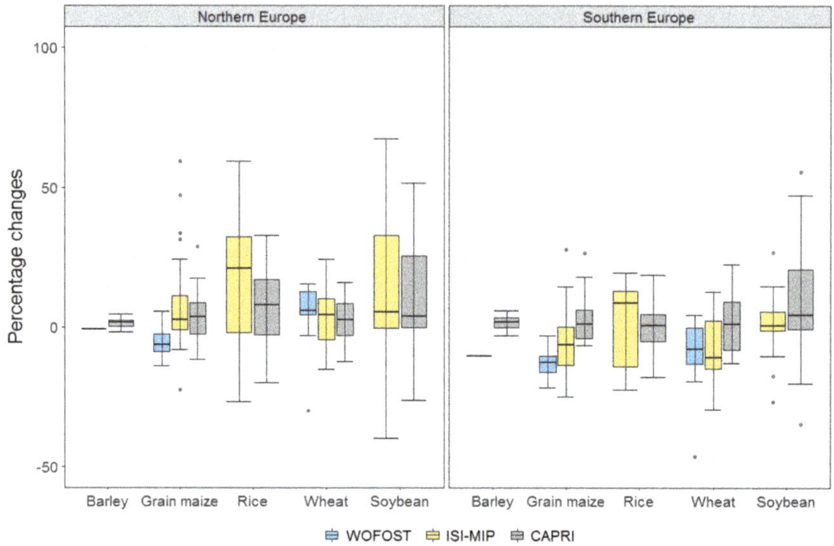

Figure 3 Northern and Southern Europe crop yield changes (exogenous yield shocks (WOFOST/ISI-MIP) and endogenous response (CAPRI)) in 2050 relative to the baseline. Source: Hristov et al. (2020). Note: The boxplot displays the median, two hinges, and two whiskers of a continuous variable distribution. The lower and upper hinges (coloured bars) correspond to the first and third quartiles (the 25th and 75th percentiles). The upper whisker extends from the hinge to the largest value no further than 1.5 than the interquartile range (IQR), or distance between the first and third quartiles. The lower whisker extends from the hinge to the smallest value at most 1.5 * IQR of the hinge. Data beyond the end of the whiskers are 'outlier' points and are plotted individually as dots.

level in the EU considering global market interaction through bilateral trade for major agricultural commodities (Britz and Witzke, 2014). However, since the WOFOST yield projections were limited to Europe, only the estimates from the ISI-MIP simulations were further considered in the economic analysis.

The comparison of the yield changes ensemble shows a partial overlap between the WOFOST and ISI-MIP approaches (Fig. 3). WOFOST projections for wheat and grain maize in 2050 in Southern Europe as well as for wheat in Northern Europe are in line with the ISI-MIP yield projections. However, the magnitude differs due to the different underlying assumptions. Significant yield differences are noticeable for grain maize in Northern Europe with negative median yield changes projected by WOFOST and positive changes by the ISI-MIP ensemble[5]. Both WOFOST and ISI-MIP approaches consider irrigation and water availability as constraints. In ISI-MIP, they are fixed to historical averages; whereas scenarios with current agro-management as well as with simplified adaptation strategies were run with WOFOST (Hristov et al., 2020).

[5] CO_2 fertilization is not included in these estimates.

Differences are also noticeable when comparing the biophysical yield effects implemented as exogenous shocks in CAPRI and the endogenous yield responses caused by economic considerations that drive farmers' management decisions. Results show that some of the negative effects of climate change on agricultural productivity may be partially compensated by the re-adjustment of commodity markets through farmers' production decisions and trade. Farmers may indeed respond to changes in market prices with changes in crop choices and input intensities, allowing them to offset the negative impacts of climate change. Trade plays here the role of rebalancing global markets as a response to climate change-induced productivity shocks.

Trade feedback and the related market adjustments are the reasons for the projected positive net effect on wheat or maize yields (median effect) in CAPRI. The economic reasoning behind this effect is that farmers initially will prioritize the crops with positive yield effects and produce fewer crops with more negative yield effects. As a consequence, producer prices will decrease for the prioritized crops, and increase for the crops that are produced less. This in turn, together with the stronger negative yield changes in other non-EU regions, further will influence farmers' decisions to produce alternative crops in a more intensive way, which in the end will result in the net effect of positive yield changes (despite the initially negative yield shock induced by climate change).

As a result of the positive yield adjustments, agricultural production indicators (i.e. area, gross/net farm income, production level, net trade) show that the market-adjusted yield developments lead to positive production effects in the EU. These positive effects may of course come with an environmental cost, e.g. higher greenhouse gas (GHG) emissions from the EU agriculture sector, or higher water use. However, this effect to some extent may be offset by the area reduction of other crops with lower relative profitability at expense of the area expansion driven by the market effect.

4 The role of technologies for mitigation and adaptation

Apart from being directly affected by climate change due to altered weather conditions and resulting biophysical effects, the agricultural sector will need to change to reduce its GHG emissions and contribute to mitigating climate change. Food system emissions represent about a third of total GHG emissions, with two-thirds corresponding to agriculture and land use/land use change activities (Crippa et al., 2021). Thus, agriculture, forestry, and other land use (AFOLU) are important strategic factors to reach global climate change goals (IPCC, 2022). Mitigation efforts in the agricultural sector raise concerns about adverse impacts on food security (Hasegawa et al., 2018; van Meijl et al., 2018). Technological (i.e. technical and management-based) mitigation and adaptation options can help diminish adverse effects on food production, which is why their development, transfer, and adoption play an important role to achieve AFOLU emission reductions and removals (Smith et al., 2014; Frank et al., 2019).

For instance, the EcAMPA 3 study (Pérez Domínguez et al., 2020) provides a technical framework to analyse the potential contribution of different technological mitigation options in EU agriculture to achieve GHG emissions mitigation targets. The study focuses on methodological developments for a better representation of technological mitigation options in EU agriculture with the CAPRI model, testing and analysing their mitigation potential and related costs.

Emissions in agriculture can be mitigated by three main approaches (Himics et al., 2020): implementation of technological options (e.g. precision farming); reduction of production levels (e.g. fewer cows); and changes in the composition or intensity of farming activities based on current management practices. Figure 4 shows the total mitigation that could be achieved in the EU by the application of technological options and changes in the production mix and levels under different climate change mitigation scenarios, expressed in terms of different carbon prices. Four major characteristics related to technological options can be derived. First, the contribution of technologies to total mitigation increases with increasing carbon prices but at a decreasing rate.

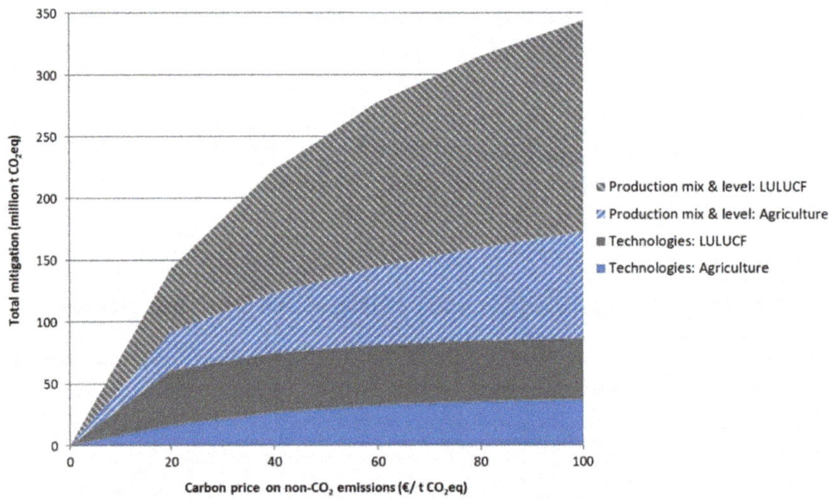

Figure 4 Total mitigation under different carbon price scenarios, EU (compared to the baseline, 2030). Source: Fellmann et al. (2020). Note: 'Production mix' refers to changes in the composition of farming activities based on current management practices (e.g. more pulses and less wheat) and 'level' to the reduction of production quantity (e.g. decrease in the harvested area and production intensity). Emissions under the common reporting sector 'agriculture': methane and nitrous oxide, under 'LULUCF': carbon dioxide.

Second, the adoption of mitigation technologies increases considerably under lower carbon prices, but further uptake is increasingly limited with carbon prices beyond 60 EUR/t CO_2eq. Third, due to the previous two characteristics, the relative contribution of technology adoption to total mitigation is decreasing (from 40% at a carbon price of €20 to 25% in CP 100). Last but not least, the relative contribution of 'agriculture' (i.e. mainly non-CO_2) emissions mitigated by technological options remains at about 11% in all carbon price scenarios, whereas the relative contribution of technology-related 'LULUCF' (Land Use, Land Use Change and Forestry) emissions decreases from 28% to 14% with rising carbon prices.

An important contribution of this study is that it provides a consistent framework to explore disparities between technology-specific marginal abatement cost curves (MACCs) derived from two different commonly used approaches. The modelling approach also explicitly accounts for factors beyond profitability and profit maximization that influence farmers' adoption behaviour of mitigation technologies. For the two approaches, a large set of technological mitigation measures was tested in simulations individually and together, and MACCs specific to mitigation technologies and regions were calculated. First, in a 'standalone measures approach', each mitigation technology was tested in isolation to assess its theoretical maximum mitigation potential and the related costs. Second, in a 'combined measures approach', all mitigation options together were made available to farmers at the same time and could be adopted cumulatively. Uptake in the combined measures approach is triggered through the introduction of different carbon prices on non-CO_2 emissions, taking interactions between the options into account.

The analysis highlights the importance of assessing emission mitigation from different perspectives, clearly demonstrating that technological options need to be considered in combination and from a regional perspective. For instance, there might be different possible feed additives in the market, but only the cost-effective ones will be adopted. Moreover, abandoning crop production on organic-rich soils could be a promising management option to mitigate carbon emissions, but can only be applied where those soils are available. Regarding a possible ranking of mitigation technologies, the analysis allows classification in terms of their mitigation potential and attached costs. This is valuable information when considering policy incentives for the adoption of mitigation technologies, e.g. within the future common agricultural policy of the EU. However, the analysis underlines the need to consider mitigation technologies as 'a bundle', avoiding the simple aggregation of mitigation potentials by individual measures without taking into account their interactions from both biophysical and economic perspectives. Considering the theoretical maximum potential of single mitigation technologies does not seem to give a robust picture of the real-world overall mitigation potential and related costs, as this does not consider the interplay with other technologies. Moreover, the spatial analysis of how mitigation measures might influence differently the agricultural sector in different EU Member States underlines that there is no 'one-fits-all' rule that could be followed. Thus, recommendations on optimal mitigation technologies should be implemented at the regional level (Pérez Domínguez et al., 2020; Fellmann et al., 2021).

The framework and the analysis of the potential of technological mitigation options also provide valuable insights for the assessment of technologies and management-based options for the sectors' adaptation to climate change. Technological options, like seed varieties, precision farming, and innovative irrigation systems, can play an important role to reduce the severe impacts of climate change and extremes on food production (EEA, 2019). Similar to mitigation, also adaptation options interact and influence each other, and their real overall effectiveness will only be revealed when analysed in a combined manner and will also depend on the region of interest. Accordingly, an assessment of adaptation technologies requires their consideration in a combined and regional approach. Furthermore, when analysing the application of technologies for adaptation, their synergies, trade-offs, and co-benefits with mitigation should also be taken into account.

5 Conclusion

This chapter provides some methodological aspects on how different quantitative tools can be combined in the analysis of climate change impacts on agricultural production. Applied to EU agriculture, results show that without adaptation and market adjustments, climate change will substantially lower crop productivity in southern Europe, and to a lesser extent in northern Europe. However, total EU production could still increase due to the interplay of different market forces and, hence, economic adaptation. On the one hand, this is due to farmers adjusting to the climate change-induced yield impacts and changing their production mix. On the other hand, this is due to lower projected negative yield effects in Europe compared to other important world regions in terms of production/exports (e.g. the US). Therefore, the EU might have a comparative advantage in terms of climate change impacts on agricultural productivity compared to other world regions, which could positively affect EU competitiveness. Our results underline the importance of taking market feedback and adjustments into account when analysing the impacts of climate change on agricultural production. Yield effects as projected by the biophysical models can be altered considerably as a result of such market adjustments. In this respect, it is also important to concentrate not only on climate change impacts in one region in isolation, but take global effects and interactions into account in order to capture the full market feedback and effects.

Methodological limitations and uncertainties still pose challenges in terms of developing and testing sustainable adaptation strategies. Among them, there are the use of different climate and crop models, varying emission and warming scenarios, different modelling assumptions around irrigated and rain-fed yields, and the effects of elevated atmospheric CO_2 concentrations in certain growing conditions. A particular challenge is the accuracy of modelling the impacts of weather and climate extreme events. While recent scientific development in the detection and attribution of extreme events helps assess the impacts of heatwaves, droughts, cold spells, and floods on productivity, their full consideration and analysis also in agro-economic models (typically focusing on the medium- and long term) remains a challenge. It is important to acknowledge that the analysis of mean temperature changes over time will fall short of capturing the potential adverse impacts of climate change on regional agricultural markets.

As regards the incorporation of technological options to mitigate climate change, a developed framework for mitigation technologies shows that specific options should not be assessed in isolation. Interactions between technologies can influence their individual cost-effectiveness, and the latter can also vary considerably at the regional level. These issues also have to be taken into account when analysing technological options for climate change adaptation.

It is important to critically reflect on the validity of the economic models used for the analysis of climate change impacts on agricultural commodity markets. For instance, regional aggregation might become an issue for a very granular spatial analysis, since agricultural markets in global economic models are represented only for large countries and smaller ones are usually regionally aggregated. Moreover, agro-economic models are calibrated against a reference historical period (i.e. real-world data from different statistical sources) and projected into a medium- to long-term future based on historical trends and/or expert judgement. Important assumptions regarding the development of policies, economic and population growth, and other macroeconomic indicators need to be made (and harmonized across models). The temporal scale is also annual, so seasonal market adjustments (e.g. in fruits and vegetable markets) cannot be depicted.

6 Future trends in research

A comprehensive assessment of climate change impacts on agriculture requires a combined analysis of direct and indirect impacts on production, the assessment of climate change mitigation efforts, and the potential of technological options to facilitate both mitigation and adaptation. With respect to the application of technologies, synergies, trade-offs, and co-benefits between mitigation and adaptation measures need to be taken into account in future research. In addition, comprehensive assessments need to consider the broader market environment that affects food and feed prices while climate and climate extremes evolve. Recent simulations of single-year/single-country as well as concurrent and recurrent climate shocks (including extremes), for example, have shown that climate anomalies can significantly distort expected market developments while trade and stockholding can operate as important alleviative mechanisms (Chatzopoulos et al., 2020).

The risk of food-supply instability is expected to increase along with the frequency and intensity of extreme climatic events such as heatwaves, droughts, and floods. Therefore, future impact assessments should focus not only on estimating the probability of occurrence of such events but also on linking them with supply and demand indicators beyond crop yields. In a recent example of addressing this, a global simulation model of agricultural markets and policies was calibrated with an indicator of compound climate stress (Chatzopoulos et al., 2021). Then, 1000 sets of partially stochastic (in space and time) climate shocks were simulated while agricultural markets endogenously and dynamically adjusted to a new equilibrium. By treating the frequency values of simulated climate (input) and market (output) data as approximations of their probabilities, the authors synthesized food market balances attributable to extreme events into formal and coherent market risk and vulnerability indicators. A promising avenue, therefore, is the analysis and monitoring of such (or similar) indicators in order to identify commodities prone to climate hazards, anticipate potential consequences on local food systems, and develop contingency mechanisms to evaluate the prospective risks (e.g. persistent dependence on imports, alarmingly low stock-to-use).

Last but not least, two frequently cited ways to scale up the climate resilience of global food systems pertain to coordinated shifts towards fewer trade barriers (e.g. lower import tariffs) and targeted redistribution of food reserves to meet projected demand. Therefore, future agro-economic modelling frameworks need to address these market challenges *better* and *simultaneously* with adaptation, mitigation, and risk management measures.

To this respect, it is important to acknowledge that for a robust analysis of the impacts of extreme climatic events, the set of model results offered by the ISI-MIP database is not representative of all relevant processes and is affected by some issues, due to non-stationarity (e.g. induced by trends in climate variables and atmospheric CO_2 concentrations) and various biases between models, crops, and regions (Müller et al., 2017). Moreover, it is not possible to represent the full range of plausible climate change scenarios within a small set of climate models (McSweeney and Jones, 2016) as supplied for the ISI-MIP Fast Track Database. Also, the differences in climate patterns between GCMs and yield responses between crop models (Rosenzweig et al., 2014; Müller et al., 2015) complicate the comparison of individual cases, where regional differences can mask the scenario-induced differences (Van Meijl et al., 2018).

Future research also aims at improving the representation of land use change and forestry sources and sinks taking the advantage of the latest developments for land monitoring systems (e.g. the EU Copernicus Atmosphere and climate services) to allow for an integrated analysis of climate change mitigation policies affecting both the agricultural and forestry sectors. Modelling improvements also require implementing new farm practices likely to be addressed by the new EU agricultural policy architecture (e.g. crop cover, no tillage, and mandatory rotation) as a way to reflect a higher level of environmental ambition and to contribute to the achievement of the EU environmental objectives (e.g. SDGs).

Equally important is assessing the economic impacts of climate change in combination with land degradation in the global agricultural sector. Future assessments require considering (i) different land management measures and change in land uses by worsening land degradation (declining yields due to land degradation); (ii) reinforcing the policies for protecting land areas key to prevent land degradation and maintain ecosystem services, and (iii) different levels of climate change impacts to assess the interaction between land degradation/restoration and climate change. This is essential to implement known, proven actions to combat land degradation and accelerate the implementation of restoration measures.

Future impact assessments need to focus more on climate extremes (e.g. heatwaves, droughts, and floods). Currently, impact models do not fully take climate and weather-related processes into account or do not include them at all (e.g. the impacts of heavy precipitation on production, quality, and harvest conditions). The expected increase in frequency and intensity of climate extremes as well as the projected recurrent and concurrent events in key producing regions of the world (Toreti et al., 2019a,b) may trigger yield and production losses, inducing higher price variability and altering global food markets (Chatzopoulos et al., 2020), which can be a serious threat to food security. Effects of extreme events can be offset or at least smoothened by stockholding of agricultural commodities or tailored trade policies. Current agro-economic modelling frameworks need to better quantify these effects and related adaptation measures.

7 Where to look for further information

Key research in this area can be found at the following organizations:

- Intergovernmental Panel on Climate Change (IPCC), Sixth Assessment report (https://www.ipcc.ch/assessment-report/ar6/).
- The Agricultural Model Intercomparison and Improvement Project (AGMIP) (https://agmip.org/).
- The European Green Deal (https://ec.europa.eu/info/strategy/priorities-2019-2024/european-green-deal_en).

8 References

Britz, W. and Witzke, P. (Eds.). (2014). *"Capri Model Documentation 2014*. Available at: https://www.capri-model.org/.

Chatzopoulos, T., Perez-Dominguez, I., Toreti, A., Adenäuer, M. and Zampieri, M. (2021). Potential impacts of concurrent and recurrent climate extremes on the global food system by 2030. *Environmental Research Letters* 16(12), 124021.

Chatzopoulos, T., Perez-Dominguez, I., Zampieri, M. and Toreti, A. (2020). Climate extremes and agricultural commodity markets: A global economic analysis of regionally simulated events. *Weather and Climate Extremes* 27, 100193.

Crippa, M., Solazzo, E., Guizzardi, D., Monforti-Ferrario, F., Tubiello, F. N. and Leip, A. (2021). Food systems are responsible for a third of global anthropogenic GHG emissions. *Nature Food* 2(3), 198–209. https://doi.org/10.1038/s43016-021-00225-9.

EPA. (2017). *Multi-Model Framework for Quantitative Sectoral Impacts Analysis: A Technical Report for the Fourth National Climate Assessment*. United States Environmental Protection Agency, EPA 430-R-17-001.

European Environment Agency. (2019). *Climate Change Adaptation in the Agriculture Sector in Europe*. EEA Report 04/2019. Luxembourg: Publications Office of the European Union. https://doi.org/10.2800/537176.

Fellmann, T., Pérez-Domínguez, I. P., Witzke, P., Weiss, F., Hristov, J., Barreiro-Hurle, J., Leip, A. and Himics, M. (2021). Greenhouse gas mitigation technologies in agriculture: Regional circumstances and interactions determine cost-effectiveness. *Journal of Cleaner Production* 317(1), 128406.

Fernández, F. J. and Blanco, M. (2015). Modelling the economic impacts of climate change on global and European agriculture: Review of economic structural approaches. *Economics* 9(1), 1–53.

Frank, S., Havlík, P., Stehfest, E., van Meijl, H., Witzke, P., Pérez-Domínguez, I., van Dijk, M., Doelman, J. C., Fellmann, T., Koopman, J. F. L., Tabeau, A. and Valin, H. (2019). Agricultural non-CO_2 emission reduction potential in the context of the 1.5°C target. *Nature Climate Change* 9(1), 66–72. https://doi.org/10.1038/s41558-018-0358-8.

Hasegawa, T., Fujimori, S., Havlík, P., Valin, H., Bodirsky, B. L., Doelman, J. C., Fellmann, T., Kyle, P., Koopman, J. F. L., Lotze-Campen, H., Mason-D'Croz, D., Ochi, Y., Pérez Domínguez, I., Stehfest, E., Sulser, T. B., Tabeau, A., Takahashi, K., Takakura, J., van Meijl, H., van Zeist, W., Wiebe, K. and Witzke, P. (2018). Risk of increased food insecurity under stringent global climate change mitigation policy. *Nature Climate Change* 8(8), 699–703. https://doi.org/10.1038/s41558-018-0230-x.

Hempel, S., Frieler, K., Warszawski, L., Schewe, J. and Piontek, F. (2013). *A Trend-Preserving Bias Correction - The ISI-MIP Approach*. München: European Geopyhsical Union.

Himics, M., Fellmann, T. and Barreiro-Hurle, J. (2020). Setting climate action as the priority for the common agricultural policy: A simulation experiment. *Journal of Agricultural Economics* 71(1), 50–69.

Hristov, J., Toreti, A., Pérez Domínguez, I., Dentener, F., Fellmann, T., Elleby, C., Ceglar, A., Fumagalli, D., Niemeyer, S., Cerrani, I., Panarello, L. and Bratu, M. (2020). *Analysis of Climate Change Impacts on EU Agriculture by 2050, EUR 30078*. Luxembourg: EN, Publications Office of the European Union. ISBN 978-92-76-10617-3. https://doi.org/10.2760/121115, JRC119632.

IPCC. (2014). Climate Change 2014: Synthesis Report. [Core Writing Team, R. K. Pachauri and L. A. Meyer (eds.)]. *Contribution of Working Groups I, II and III to the Fifth Assessment Report of the Intergovernmental Panel on Climate Change*. IPCC, Geneva, Switzerland. 151.

IPCC. (2022). Climate change. 2022: Impacts, adaptation, and vulnerability. In: H.-O. Pörtner (Ed.), *Contribution of Working Group II to the Sixth Assessment Report of the Intergovernmental Panel on Climate Change*. Cambridge: Cambridge University Press.

McSweeney, C. F. and Jones, R. G. (2016). How representative is the spread of climate projections from the 5 CMIP5 GCMs used in ISI-MIP? *Climate Services* 1, 24–29.

Monier, E., Paltsev, S., Sokolov, A., Chen, Y. H., Gao, X., Ejaz, Q., Couzo, E., Schlosser, C. A., Dutkiewicz, S., Fant, C., Scott, J., Kicklighter, D., Morris, J., Jacoby, H., Prinn, R. and Haigh, M. (2018). Toward a consistent modeling framework to assess multi-sectoral climate impacts. *Nature Communications* 9(1), 660. https://doi.org/10.1038/s41467-018-02984-9.

Müller, C., Elliott, J., Chryssanthacopoulos, J., Arneth, A., Balkovic, J., Ciais, P., Deryng, D., Folberth, C., Glotter, M., Hoek, S., Iizumi, T., Izaurralde, R. C., Jones, C., Khabarov, N., Lawrence, P., Liu, W., Olin, S., Pugh, T. A. M., Ray, D. K., Reddy, A., Rosenzweig, C., Ruane, A. C., Sakurai, G., Schmid, E., Skalsky, R., Song, C. X., Wang, X., de Wit, A. and Yang, H. (2017). Global gridded crop model evaluation: Benchmarking, skills, deficiencies and implications. *Geoscientific Model Development* 10(4), 1403–1422, https://doi.org/10.5194/gmd-10-1403-2017.

Müller, C., Elliott, J., Chryssanthacopoulos, J., Deryng, D., Folberth, C., Pugh, T. A. M and Schmid, E. (2015). Implications of climate mitigation for future agricultural production. *Environmental Research Letters* 10, 125004, doi: 10.1088/1748-9326/10/12/125004.

Nelson, G. C., Valin, H., Sands, R. D., Havlík, P., Ahammad, H., Deryng, D., Elliott, J., Fujimori, S., Hasegawa, T., Heyhoe, E., Kyle, P., Von Lampe, M., Lotze-Campen, H., Mason d'Croz, D., van Meijl, H., van der Mensbrugghe, D., Müller, C., Popp, A., Robertson, R., Robinson, S., Schmid, E., Schmitz, C., Tabeau, A. and Willenbockel, D. (2014). Climate change effects on agriculture: Economic responses to biophysical shocks. *Proceedings of the National Academy of Sciences of the United States of America* 111(9), 3274–3279.

Pérez Domínguez, I., Fellmann, T., Witzke, P., Weiss, F., Hristov, J., Himics, M., Barreiro Hurlé, J., Gómez Barbero, M. and Leip, A. (2020). *Economic Assessment of GHG Mitigation Policy Options for EU Agriculture: A Closer Look at Mitigation Options and Regional Mitigation Costs (EcAMPA 3), EUR 30164*. Luxembourg: EN, Publications Office of the European Union. ISBN 978-92-76-17854-5, https://doi.org/10.2760/4668, JRC120355.

Rosenzweig, C., Elliott, J., Deryng, D., Ruane, A. C., Müller, C., Arneth, A., Boote, K. J., Folberth, C., Glotter, M., Khabarov, N., Neumann, K., Piontek, F., Pugh, T. A., Schmid, E., Stehfest, E., Yang, H. and Jones, J. W. (2014). Assessing agricultural risks of climate change in the 21st century in a global gridded crop model intercomparison. *Proceedings of the National Academy of Sciences of the United States of America* 111(9), 3268–3273.

Smith, P. (2014). Mitigation of climate change. Contribution of working group III to the fifth assessment report of the IPCC. In: *Climate Change: Agriculture, Forestry and Other Land Use (AFOLU)*, edited by Edenhofer. Cambridge: Cambridge University Press.

Toreti, A., Belward, A., Perez-Dominguez, I., Naumann, G., Luterbacher, J., Cronie, O., Seguini, L., Manfron, G., Lopez-Lozano, R., Baruth, B., van den Berg, M., Dentener, F., Ceglar, A., Chatzopoulos, T. and Zampieri, M. (2019a). The exceptional 2018 European water seesaw calls for action on adaptation. *Earth's Future* 7(6), 652–663.

Toreti, A., Cronie, O. and Zampieri, M. (2019b). Concurrent climate extremes in the key wheat producing regions of the world. *Scientific Reports* 9(1), 5493.

Toreti, A., Deryng, D., Tubiello, F. N., Müller, C., Kimball, B. A., Moser, G., Boote, K., Asseng, S., Pugh, T. A. M., Vanuytrecht, E., Pleijel, H., Webber, H., Durand, J., Dentener, F., Ceglar, A., Wang, X., Badeck, F., Lecerf, R., Wall, G. W., van den Berg, M., Hoegy, P., Lopez-Lozano, R., Zampieri, M., Galmarini, S., O'Leary, G. J., Manderscheid, R., Mencos Contreras, E. and Rosenzweig, C. (2020). Narrowing uncertainties in the effects of elevated CO_2 on crops. *Nature Food* 1(12), 775–782.

Van Meijl, H., Havlik, P., Lotze-Campen, H., Stehfest, E., Witzke, P., Pérez Domínguez, I. P., Bodirsky, B. L., van Dijk, M., Doelman, J., Fellmann, T., Humpenöder, F., Koopman, J. F. L., Müller, C., Popp, A., Tabeau, A., Valin, H. and van Zeist, W. (2018). Comparing impacts of climate change and mitigation on global agriculture by 2050. *Environmental Research Letters* 13(6), 064021.

Van Meijl, H., Havlik, P., Lotze-Campen, H., Stehfest, E., Witzke, P., Pérez Domínguez, I., Bodirsky, B., van Dijk, M., Doelman, J., Fellmann, T., Humpenoeder, F., Levin-Koopman, J., Mueller, C., Popp, A., Tabeau, A. and Valin, H. (2017). Challenges of global agriculture in a climate change context by 2050 (AgCLIM50). JRC Science for Policy Report. En: EUR 28649. https://doi.org/10.2760/772445.

von Lampe, M., Willenbockel, D., Ahammad, H., Blanc, E., Cai, Y., Calvin, K., Fujimori, S., Hasegawa, T., Havlik, P., Heyhoe, E., Kyle, P., Lotze-Campen, H., Mason d'Croz, D., Nelson, G. C., Sands, R. D., Schmitz, C., Tabeau, A., Valin, H., van der Mensbrugghe, D. and van Meijl, H. (2014). Why do global long-term scenarios for agriculture differ? An overview of the AgMIP global economic model inter-comparison. *Agricultural Economics* 45(1), 3–20.

Wang, T. P. and Teng, F. (2022). A multi-model assessment of climate change damage in China and the world. *Advances in Climate Change Research* 13(3), 385–396. ISSN 1674-9278, https://doi.org/10.1016/j.accre.2022.04.005.

Chapter 11

Modelling transition of agricultural systems in response to climate change

Arnout van Soesbergen and Fiona Jones, King's College London, UK

1 Introduction
2 The transition of agricultural systems in response to climate change
3 The role of models in informing adaptation responses
4 Capturing the transition in modelling approaches and climate change impact studies
5 Overarching limitations of current modelling approaches
6 Future trends in research
7 Where to look for further information
8 References

1 Introduction

The world faces pressing challenges in the face of climate change and worldwide ecological degradation. A particularly relevant challenge is to achieve global food security for a growing population under these increasing strains. It is estimated that agricultural production will have to increase by between 25% and 70% to feed the world in 2050 (Hunter et al., 2017) and this growth in production is constrained by land and water availability for crops and livestock and declining wild fishery stocks (Wheeler and von Braun, 2013). Agriculture and consequently food security are already affected by climate change (Ray et al., 2019) and impacts are expected to continue or worsen e.g. it is projected that for each degree of global temperature increase, wheat yields are expected to reduce by 4–6% (Liu et al., 2016). At the same time, however, agriculture and food value chains are significant emitters of GHGs and thus contribute to climate change. Therefore agriculture is a key focus for adaptation and mitigation initiatives to address climate change and ensure food security, e.g. through climate-smart agriculture (CSA).

Recent advances in agriculture and climate modelling have demonstrated that approaches can be applied to not only explore the potential impacts and risks that agricultural systems must overcome but have also been shown

http://dx.doi.org/10.19103/AS.2022.0115.13

to shed light on transitions and adaptations that may help to make global systems more resilient and sustainable. For example, the development and application of integrated assessment modelling approaches for agri-climate models have successfully investigated the impacts of climate change on various types of crop, livestock and combined crop–livestock systems around the world and in doing so contextually assess the sustainability of agricultural systems; how they are anticipated to respond to the future impacts of climate change, their resilience to more extreme weather events, what natural adaptive responses might occur within the ecosystems that underpin such systems, and what this translates to in terms of risk if no additional action is taken to mitigate it.

This chapter will examine how agricultural systems modelling can support decision-making to identify and implement suitable 'climate-smart' production practices, thus helping to mitigate climate risk in agricultural systems and contributing to enhanced and more sustainable crop and livestock production. Current key challenges will be outlined and synthesised, pointing to areas in need of further research and development. Finally, the chapter will discuss some of the most promising future research trends.

2 The transition of agricultural systems in response to climate change

2.1 Natural responses

Agricultural systems are fundamentally shifting in response to climate change, as the very ecosystems that underpin them are having to adapt in order to survive changes in precipitation, temperatures, humidity and seasonal weather patterns, as well as more intense and frequent extreme weather events. Globally, spatial shifts in the agricultural suitability of land due to an array of factors, including climate change, are expected to continue. In some cases, agroecological systems have the innate adaptive capacity to naturally tolerate moderate climatic changes and might even benefit from the milder initial impacts of climate change, such as warmer temperatures and increased carbon dioxide concentrations that enhance the yield production of more drought-tolerant crops. For example, a recent modelling study found that maize crops will become increasingly difficult to grow in higher latitudes such as the tropics, with yield steadily decreasing by up to 24%, while wheat-growing range could initially expand until around mid-century, increasing global yield up to an estimated 17%, before net losses in production incur over the latter half of the century (Jägermeyr et al., 2021).

However, in the majority of cases, changes in climatic suitabilities impact crop production negatively. Progressive climatic change can render historically thriving agroecological zones as arid and unable to support production. Furthermore, rising carbon dioxide concentrations will impact the nutritional value of crops (Myers et al., 2014), and other biotic stress factors brought on by climate change will also have serious impacts, such as water resource constraints or surplus moisture, as well as new regional incidences of pest and disease. It is important to note that local patterns of agroecological responses to climate change drivers may differ greatly from the mean global, long-term climate signals, which are uncertain in themselves. This is because the interactions between and within biotic and abiotic systems are complex and cascading. Furthermore, regional geographical characteristics (such as latitude, land cover and topography) as well as growing season exposure also have a weighty influence on how these systems interact (Ruane and Rosenzweig, 2018).

Beyond the farm and purely agroecological impacts, the same physical risks of climate change will have ramifications on the entire production system's value chain, from transportation to storage and processing facilities. With a growing population and an increasing reliance on fragile, intertwined global supply chains, food security in such uncertain environmental settings is gearing up to be a major risk globally, with weather shocks further exacerbating the problem particularly for developing import-dependent countries, which are often the most vulnerable and most exposed to impacts (Ruane and Rosenzweig, 2018). Where there is an opportunity to diagnose climate vulnerabilities and engineer risk mitigation, man-made adaptation responses in agricultural systems are also being seen to contribute towards their transition in response to climate change.

2.2 Man-made responses

Broadly speaking, agricultural adaptation responses to climate change fall over two core levels (Holzkämper, 2017):

1) Incremental short-term, which farmers tend to tactfully select autonomously, is based on their previous experience and local knowledge, in reaction to observed changes.
2) Transformative long-term is usually implemented on a larger spatial scale (either regional, national or international) in anticipation of future climate impacts and requires strategic planning.

The first level of response works within the realms of existing technological and governance systems, however, the second level likely changes fundamental attributes of them. Typically, the observed ratio of these responses will depend on the balance between the adaptive capacity of the socio-ecological system in question and the climatic risk it faces.

Mild/moderate climate impacts occurring at a slower pace often do not push an existing agriculture system beyond its boundaries. So, if adaptive capacity is sufficient, risks and challenges may usually be overcome through short-term responses only, without much requirement for strategic planning or wide stakeholder involvement. This type of response typically includes local, contained changes to farm management practices. For example, adapting management of nutrients, water, soil and canopy, as well as utilising different crops or cultivars that have a stronger resilience to stress factors driven by climate change. Another simple adaptation is to shift sowing dates or crop rotations, such that exposure to climatic stresses is minimised or avoided during sensitive phenological periods. Double- or triple-cropping may also enhance land productivity if climate change prolongs the annual growing season. Adaptation options for livestock systems may entail flexible herd management, adoption of alternative species or breeds, including genetically modified breeds, pest management innovation and indoor climate control systems. Under highly uncertain and variable climates, diversifying production systems through the use of different crops and staggering sowing dates may increase stability as yields that are less negatively affected, or even positively affected, by a climate event, then compensate for other yield losses.

When the impacts of climate change ramp up beyond a certain threshold, Level 1 responses will no longer sufficiently alleviate the ramifications, and more transformative adaptations are needed to do so. Level 2 responses may entail more fundamental, structural changes to production systems as well as to farming activities. Higher level responses might also include spatial shifts in agroecological zones, as well as the breeding or genetic modification of crops and cultivars to produce new, more climate-resistant species with phenology tailored to certain climate conditions (Rosenzweig et al., 2013). An especially sought-after feature is phenotypic plasticity, enabling crops to respond and adapt to a changing environment (Bloomfield et al., 2014; Nicotra et al., 2010). Typical breeding goals include improved nutrient- and water-use efficiencies, as well as heightened tolerance to drought, flooding or increased salinity. Such transformative adaptation, especially when at scale, requires strategic coordination and foresight. Compared to the incremental adaptation routes outlined in Level 1, Level 2 responses involve greater uncertainty, risk and more complex challenges to overcome. Transformative adaptation hence requires robust risk management.

3 The role of models in informing adaptation responses

Holzkämper (2017) proposed that models facilitate risk management and decision-making across both levels in three core ways; however, different features prove more or less useful when considering different types of adaptive responses:

i) Models can increase understanding of biophysical and socio-economic systems.
ii) Predictive capabilities can be used to forecast climate risk under different scenarios, hence identifying vulnerabilities in agricultural production systems.
iii) Models can be employed to test climate adaptation measures in terms of environmental and socio-economic impacts, hence identifying the optimal response to mitigate climate risk.

Uncertainty and limitations in models may lead to decision-making under incorrect or misinformed assumptions, which ultimately results in undesirable outcomes, or 'maladaptation'. Maladaptation is especially prominent when short-term benefits are favoured over long-term costs, when system interlinkages such as biophysical feedback loops are neglected from the model, and when negative externalities of adaptive responses are not properly accounted for in the decision-making process. Advancing modelling techniques by reducing uncertainty and improving integration may help to overcome maladaptation and avoid such outcomes. In addition, once an adaptation route has been selected and implemented, the portfolio of alternative routes for the future is often then reduced due to financial constraints or overuse of resources. Having then shifted away from the original system, complex biophysical cascades may have been sparked which could upset delicate feedback loops and cause further negative externalities, all of which take time to materialise and rectify.

4 Capturing the transition in modelling approaches and climate change impact studies

The transition of agricultural systems in response to climate change, either naturally or agent-driven, is captured in an array of different modelling approaches such as empirical crop models, mechanistic models, Integrated Assessment Models (IAMs), land use suitability models and other models that describe other elements of the agricultural system, e.g. soil organic matter turnover, grazing behaviour and farm economy models. Whilst not aiming to be comprehensive (see Jones et al., 2017, for a review of agricultural system models), this section aims to summarise this broad spectrum of available models for capturing agricultural transitions under climate change. Finally, the section concludes with an overview of how some of these models have been used to project future transitions.

4.1 Empirical crop models

Empirical crop models are predominantly utilised for agricultural–climate impact studies as they typically simulate climate–yield relationships through employing data panels, which capture the temporal and spatial changes in climate and yield alongside empirical time series (Holzkämper, 2017). Crop yield responses can be applied to fit aggregated (e.g. monthly) or relatively scarce climate data, which gives this approach a degree of flexibility. Empirical modelling in this capacity allows for statistical testing of regional, causal agricultural–climate relationships, which enhances an analytical understanding of the system's most prominent drivers and main feedback loops. However, hypotheses of causal relationships based on observed conditions may not sufficiently account for key, underlying mechanisms at play, such as plant functioning, ecosystem dynamics and more complex biophysical and biogeochemical feedbacks. The causal associations quantified may obscure the very process relationships that give rise to them. Empirical models tend to also use a reductionist approach towards representing the complex dimensions of the climate, with extreme cases focusing on the impact of a single climate variable whilst others are omitted from the study. This becomes increasingly problematic if a non-stationary correlation between the measured and omitted variables exists.

In turn, these shortcomings impact this approach's ability to simulate accurate results beyond observations. Generally, empirical crop models are used to analyse short-term, annual or sub-annual fluctuations, such as weather shocks, to quantify the relationship between the climate and agricultural outcomes selected for the study. Under this application, the approach has limited capacity to model and predict the transition of agricultural systems in response to progressive climate change unless longer-term data is used. This would implicitly account for autonomous adaptation over time, such as the shifting of sowing dates. Nevertheless, this approach is still limited by the inability to validate projections beyond the observational data used to fit the model. Therefore the approach is ill-suited for applications outside of those utilised for model development or parameterisation (Pasquel et al., 2022). This hinders its capacity to make robust climate risk predictions at scale and explore theoretical adaptation scenarios to inform decision-making (Jones et al., 2017).

4.2 Mechanistic models

Mechanistic models simulate processes mathematically, based on an understanding of the underlying workings involved in the agricultural system. These models are system based and simulate relationships between two or more of the 'atmosphere-soil-plant-management' components. This often requires multidisciplinary inputs to blend in ecological, biophysical, physical and economical approaches (Pasquel et al., 2022). The majority of mechanistic models are biophysical and focus on crop growth and are known as global gridded crop growth models (GGCMs); however, others focus more broadly on physiological responses to environmental variability and change. This may include simulating water and nutrient cycling, phenology, carbon dynamics and photosynthetic pathways. Some of these models are described in Section 4.3.

Mechanistic models may be stochastic, including random effects, or deterministic, where there are no random variations such that the result will always be the same for a set of inputs. Most agricultural models are an amalgam of process-based, system models and empirical models to create mechanistic deterministic models (Pasquel et al., 2022). Modelling techniques and ensemble designs have become more robust as research into this field has progressed. An experimental approach can be adopted when using mechanistic models to explore their sensitivity to certain parameters, which can provide useful information about the system under study. Various calibration techniques exist that enable an assessment of adequacy for each simulated model output, in contrast to observational data (Luo et al., 2022). These include history matching methods, as well as calculations of distance and implausibility. Subsequent judgements can then be made regarding any necessary refinement of the parameter space, which in turn enables researchers to fine-tune the model and select the most probabilistic projections. Furthermore, automated parameter sampling techniques enable modellers to conduct high volumes of simulations which might not be possible manually due to time constraints. This facilitates a more exploratory approach and arguably a more detailed analysis of agricultural systems.

Mechanistic models can be integrated with general circulation models (GCMs) to examine the historic transition of agricultural systems under various climate change scenarios and simulate future projections under different forcings. However, more recently, researchers have interfaced biophysical crop models with land-surface models, as well as economic models, to create a more complete representation of the interacting elements that contribute to the transition of agricultural systems when forced under climate change scenarios (Peng et al., 2020).

4.3 Models of elements of the agricultural system

4.3.1 Crop rotation models

Diversification of crop rotation is a key option to increase the resilience of crop production under climate change. Cropping plans and crop rotation decisions have been modelled using many approaches based on different objectives and handled at very different scales (Dury et al., 2012). However, in many of these studies, the focus was on isolated monocultures, whereas, in many agricultural regions around the world, multi-crop rotations better represent local production systems. For example, Kollas et al. (2015) compared the capability of 15 crop growth simulation models to predict yields in crop rotations at five sites across Europe under minimal calibration. They found that the continuous simulation of multi-year crop rotations yielded results of slightly higher quality compared to the simulation of single years and single crops. Teixeira et al (2018) modelled crop rotation for Kaituna catchment in New Zealand and showed that direction and magnitude of climate change impacts, and the response to adaptation, varied spatially and were affected by rotation carryover effects due to agronomical (e.g. timing of sowing and harvesting) and soil (e.g. residual nitrogen, N) aspects.

4.3.2 Ruminant digestion

Livestock agriculture is a main contributor to GHG emissions. Therefore, it is increasingly important to have models capable of simulating GHG emissions from livestock and assess options to reduce them. A number of studies have reviewed models for grassland-based ruminant production (Bateki et al., 2019; Bryant and Snow, 2008; Snow et al., 2014). To effectively simulate ruminant livestock systems, models must consider the interplay of various components, including animal genetics (breed), nutrition (forage), management practices and their environmental impacts (emissions to air and water). Modelling ruminant systems remains a challenge due to the complex symbiotic relationship between rumen microbial anaerobic fermentation and mammalian metabolism. The digestion process involves the breakdown of microbial biomass, fermentation by-products and dietary components that bypass rumen fermentation. To overcome these challenges, a systems approach is recommended to investigate ruminant production through modelling and simulation (Dougherty et al., 2019)

4.3.3 Soil organic matter turnover

Models of soil organic carbon (SOC) storage and turnover can be useful tools to analyse the effects of soil and crop management practices and climate change on soil organic carbon stocks. Brilli et al. (2017) provide a comprehensive review of the state-of-the-art in modelling the carbon cycle in agricultural systems. Soil organic carbon models tend to be based on data for soil structure and soil physical properties. An important shortcoming of the current generation of models is their poor representation of plant-soil interactions, in particular, the responses of soil organic matter (SOM) decomposition and nutrient release to altered plant inputs and ultimately plant uptake of mineral nutrients (Hinsinger et al., 2011; Zaehle et al., 2014). Brilli et al. (2017) propose the explicit representation of soil microbial biomass to drive SOM turnover. A recent model that does this is the JSM model (Yu et al., 2020).

4.3.4 Nitrate leaching

Nitrogen losses from agricultural lands reduce soil fertility and plant yield but also have adverse impacts on the environment. Nitrate leaching from soils has a great impact on ground- and surface waters and can cause eutrophication of these waters. Nitrate leaching is governed by a number of factors that affect accumulation and movement of residual NO_3 in the soil such as plant characteristics, soil properties, seasonal fluctuations and climate changes. There are many models that capture nitrate leaching, but often these are created for specific contexts, e.g. Spijker et al. (2021) developed a predictive modelling framework to create annual maps with full national coverage of nitrate concentrations leaching from the root zone of Dutch agricultural soils. Matteau et al. (2019) coupled a soil surface empirical nitrate production model adapted to the regional agro-climatic conditions of the study area in Quebec, Canada, with a physical soil water dynamics model to determine the most suitable combination in order to predict nitrate leaching into subsurface drains

4.4 Agro-economic models

Agro-economic models are mathematical models that are used to analyse and understand the economic behaviour of agricultural systems. These models are designed to simulate the interactions between the various components of agricultural systems, such as producers, consumers and markets. Broadly, there are two types of agro-economic models: partial equilibrium (PE) and computable general equilibrium (CGE) models. Partial equilibrium models analyse the agricultural sector in isolation and assume no or little impact from other markets or sectors whereas CGE models analyse interactions between all markets and sectors of an economy. The latter models are thus more complex but can be used to analyse the overall effects of policy changes on the entire economy, taking into account feedback effects and interdependencies between markets. Both types of models have been used to model impacts of climate change on the agricultural sector (e.g. Nelson et al., 2014) and climate change adaptation scenarios (e.g. Janssens et al., 2020). Most of these studies integrate the economic model with GCMs and other models. For example, Leclère et al. (2014) combined the EPIC GGCM and the GLOBIOM high-resolution Global Economic Model of the agricultural sector to extend previous literature with global insights into the transformations required of agricultural systems to face climate change. De Pinto et al. (2020) used data from the Spatial Production Allocation Model (SPAM), DSSAT (Jones et al., 2003) and the International Model for Policy Analysis of Agricultural Commodities and Trade (IMPACT v3.3; Robinson et al., 2015) to simulate the effects of CSA adoption on yields, production, harvested areas and world commodity prices of maize, wheat and rice.

4.5 Integrated assessment models

IAMs model the impact of various societal changes, such as global policies, population growth, socio-economic development, GHG emissions and technological advancements on land, air and water resources (Ruane et al., 2017). They typically cover larger spatial and temporal scales than the previously described models. While they integrate crop models with market models, they often lack connections with high-resolution models that can represent complex decision-making such as farm-level models. However, the IAM approach is considered the most advanced for climate impact research, as it enables researchers to more accurately resolve the complex feedback loops that dictate the state of agricultural systems under progressive climate change. In the process, this approach uncovers important information about the transition of agricultural systems over spatial and temporal scales in response to climate variability. Here there is also the capacity to include technological innovation and agent-driven adaptation within the IAM framework. For example, by including sustainability pathways in scenarios, e.g. the effects of changing diets on the amount of grassland and cropland that could be freed up through the substitution of animal-based products (Stehfest et al., 2009).

The underlying paradigm of IAMs is mostly the coupling of mechanistic systems approaches, however, due to the complexity of each constituent system and how they interlink, it can be difficult to give detailed descriptions of IAMs. Typically the framework description will include some specific system elements, feedbacks between the elements, organisational detail, borders, scale and resolution and inputs and outputs that are most relevant to the modelling goals. Table 1 summarises some of the most prominent IAMs for simulating climate change responses of agricultural systems. There are many other models that can be considered IAMs, e.g. HAPPI & CMIP5 (Mitchell et al., 2017), GRACE (Aaheim et al., 2018), C31AM (Wei et al., 2018) and COFFEE-TEA (Cunha et al., 2020; Rochedo, 2016), some of which have a specific focus or are less useful for agricultural systems modelling, are developed for a specific geography (e.g. CAPRI (Britz and Witzke, 2008) for the EU) or are not open-source or part of intercomparison projects.

In addition to the individual IAMs, there have been agricultural model intercomparison projects such as the Agricultural Model Intercomparison and Improvement Project (AgMIP; Rosenzweig et al., 2013, 2021) and the Inter-Sectoral Impact Model Intercomparison Project (ISI-MIP; Warszawski et al., 2014) that aim to provide a common framework for analysis of agricultural challenges,

Table 1 Summary of prominent agriculture integrated assessment models for simulating climate change responses

Model	Resolution and domain	Description	Benefits/Limitations
IMAGE-MAGNET (Doelman et al., 2018)	Global, using world regions	Agro-economic model 'computable general equilibrium (CGE) model that is connected via a soft link to the core model of IMAGE'	Demographic changes and shifts in incomes are primary model drivers. Agricultural yields as well as the carbon content of crops and natural vegetation are affected by climatic change
IMAGE 3.0 (Stehfest et al., 2014)	Large-scale (Global) and long-term assessments up to 2100	Partial equilibrium model for simulating land cover and land use scenarios includes - Agro-ecological zone models	IMAGE provides - regional land availability assessment based on crop yields, soil profiles etc. - climate change Technological change treated exogenously
MESSAGE-GLOBIOM (Messner and Strubegger, 1995)	Global using 11 world regions	General equilibrium model. Also includes the air pollution and GHG model GAINS, the aggregated macro-economic model MACRO and the simple climate model MAGICC	Detailed inclusion of land use sector through GLOBIOM model
REMIND-MAgPIE (Luderer et al., 2015)	Global, using world regions	An optimisation model that integrates the macro-economic, agriculture and land-use, energy, water and climate systems	Pasture areas remain constant Technological change treated endogenously
GCAM (Wise et al., 2014)	Global and regional	Represents the behaviour of and complex interactions between five major systems - energy, water, land, climate and the economy - at global and regional scales	Pasture areas driven by animal products demand Technological change treated exogenously

using multi-model ensembles and common simulation approaches. A key recommendation from the AgMIP project regarding IAMs is that they should better incorporate crop responses to changes in carbon dioxide, temperature, water, nitrogen and adaptation, thereby combining strengths of network and gridded approaches (Ruane et al., 2017).

4.6 Suitability models

Land use potential for crops under future climate is commonly quantified using land suitability approaches which have the potential to highlight regions of increasing or decreasing suitability and provide indications of adaptation potential in areas where cultivation zones change. These types of models often come under the umbrella of land-use models. Land use modelling plays a critical role in major environmental assessments. It enables the evaluation of the drivers of global environmental change and aids in designing measures to mitigate or adapt to such changes (Verburg et al., 2019). These approaches can be instrumental in the identification and targeting of adaptation measures. Suitability approaches aim to estimate the occurrence of particular crops rather than yields and thus are most commonly modelled using statistical species distribution modelling that use biophysical factors such as climate and soil conditions as predictors, e.g. MaxEnt (Phillips and Dudík, 2008), Bioclim (Booth et al., 2014) or related approaches such as the CLIMEX bioclimate modelling tool (Kriticos et al., 2015). There are many examples of applications of these approaches, e.g. for rapeseed in Europe (Jaime et al., 2018).

Another crop suitability approach is the FAO Global Agro-Ecological Zones (GAEZ v4; Fischer et al., 2021), which determines crop cultivation potentials for around 50 crops globally, modelled on over 300 generic production systems. Crop cultivation potentials are defined as the agronomically possible upper limit to produce crops given climatic, soil and terrain conditions for specific levels of agricultural inputs and management. In addition to the analysis of current climatic conditions, the database includes projections based on future climate conditions from the AR5 Earth System Model outputs for four Representative Concentration Pathways (RCPs). The database is made available through a dedicated website at https://gaez.fao.org/ These data are used in numerous assessments of crop-specific suitability under changing climatic conditions, e.g. for potatoes in South Korea (Kang et al., 2022), cereal crop yield (Tatsumi et al., 2011) and wheat yields (Raimondo et al., 2021).

An example of an application of a crop suitability approach under climate change is described in Box 1. This study (Hannah et al., 2020) employed three modelling methods: (1) EcoCrop (Hijmans et al., 2001); (2) Maxent (Phillips and Dudík, 2008) and (3) frequency of daily critical maximum and minimum temperatures (Gourdji et al., 2013) to assess the climatic suitability of 12 globally important crops to identify Agricultural Frontiers (areas that transition from zero crops suitable in the current climate to one or more crops suitable in the future climate scenario) and the potential impacts of such a transition on water, biodiversity and carbon storage.

Box 1

Due to shifting climatic suitabilities, multiple crops will be moving. A study by Hannah et al. (2020) assessed crop suitabilities for 12 major global commodity crops due to climate change and the potential impacts of such a transition on water, biodiversity and carbon storage. Areas that become newly suitable for one or more crops are called Climate-driven Agricultural Frontiers. The study used downscaled (Worldclim; Hijmans et al., 2005) climate outputs for 17 GCMs for the RCP 4.5 and 8.5 projections and three crop suitability modelling methods (EcoCrop, Maxent and Critical temperatures). Suitability for each crop was then determined for current climate and in each of 34 future scenarios (17 GCM, 2 radiative forcing) as the agreement of the three suitability methods. Under the high emission pathway (8.5), the agricultural frontiers cover between 10.3 and 24.1 million km^2 of the earth's surface by 2060–2080 (Fig. 1) with a median of 15.1 million km^2 which is equivalent to 59% of current global cultivated and managed vegetation area. These frontiers comprise mainly suitabilities for cold-tolerant temperate crops such as potatoes, wheat, maize and soy. The frontiers have major potential impacts on biodiversity in tropical mountains, on water resources downstream and on carbon storage in high latitude lands, with frontier soils containing up to 177 Gt of C.

Areas that transition from no current suitability for major commodity crops to suitability for one or more crops are depicted in blue, while currently uncultivated areas that transition to suitability for multiple major commodity crops are shown in red. The intensity of colour indicates the level of agreement between simulations driven by different GCMs for the RCP 8.5 radiative concentration pathway. Terrestrial areas in white are either currently suitable for at least one modelled crop or not suitable for any modelled crops in the projected climatic conditions. Suitability under current and projected climates is defined as a universal agreement of suitability methods (EcoCrop, Maxent, Frequency of Extreme Temperatures). https://doi.org/10.1371/journal.pone.0228305.g001

4.7 Modelling future pathways of agriculture

While there are many studies that explore the potential impacts of climate change on agriculture and food security, very few focus on modelling the transition of agriculture to adapt to a changing climate. The studies that do tend to focus on elements of transition but do not necessarily offer a holistic

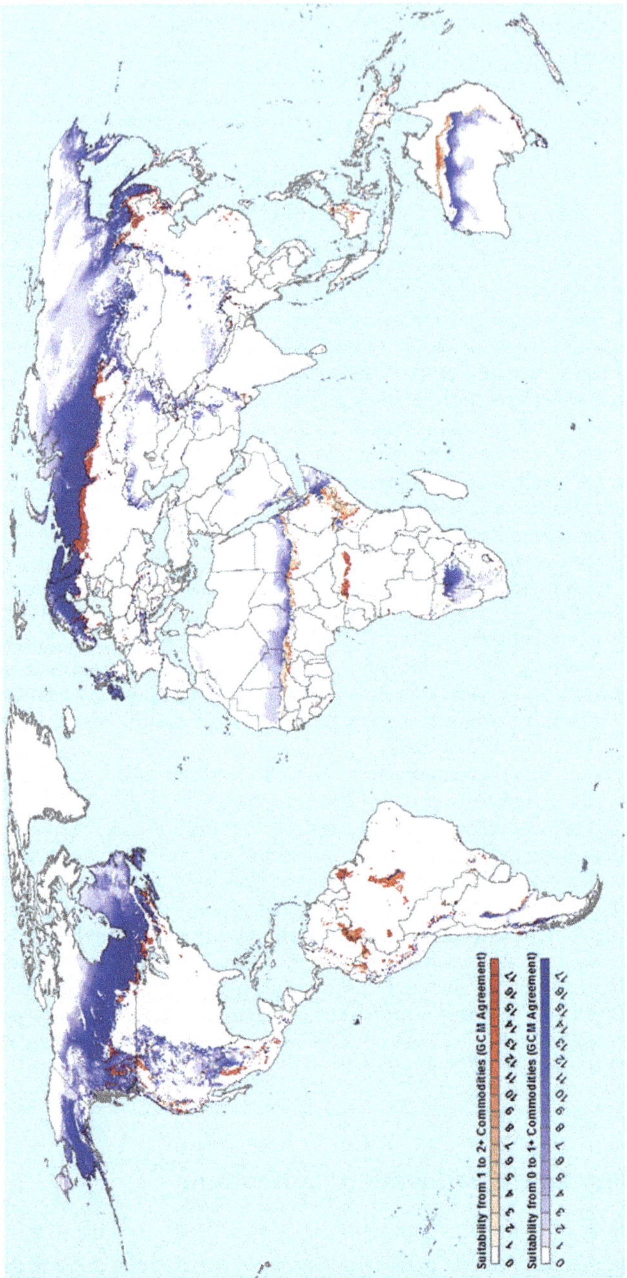

Figure 1 Global climate-driven agricultural frontiers for representative concentration pathways 8.5 2060–2080. Source: Hannah et al. (2020).

perspective across economic, technological and environmental drivers. Most global modelling approaches focus on food security and utilise the Shared Socioeconomic Pathways (SSP) scenarios to project alternative futures. For example, Van Meijl et al. (2018), applied five global climate and agro-economic models to assess the range of potential impacts of climate change on the agricultural sector by 2050 using SSP scenarios. The study also provided a comparison of climate change impacts relative to the impact of mitigation measures. Doelman et al. (2018) use the IMAGE model to explore land use dynamics under SSP scenarios, noting that key drivers for diverging trends are population growth, changes in food consumption and agricultural efficiency. Transitions in livestock systems have been modelled by Weindl et al. (2015) who explore the effectiveness of livestock system transitions as an adaptation strategy using an impact modelling chain, finding that a shift towards mixed crop-livestock systems represent a resource- and cost-efficient adaptation option. Röös et al. (2017) model four different livestock futures combined with dietary scenarios showing that without improvements in crop productivity or reductions in today's waste levels, available cropland will only suffice if production of all protein currently supplied by animal foods is replaced by (hypothetical) artificial variants not requiring any land.

5 Overarching limitations of current modelling approaches

Each modelling approach discussed in the previous section has its merits and shortcomings as described in the individual model sections. However, there are a number of key challenges that all agricultural modelling approaches to climate transitions face. The broad diversity of studies on agricultural modelling of climate change impacts and adaptive responses in terms of scope and focus make it difficult to fully understand the nature of the modelling challenges (Kipling et al., 2019), e.g. whether the modelling of climate change impacts on agricultural systems in general is sufficient for decision-making by stakeholders and policymakers or if there are agricultural modelling challenges specific to climate change adaptation that need to be addressed by the research community. In the following sections we outline some of the key challenges with current modelling approaches.

5.1 Uncertainty and uncertainty propagation

There are many well-known uncertainties in climate model projections due to the range of emission trajectories and internal variability of the climate system. In most cases, it is necessary to downscale climate simulations to a more relevant spatial scale before they can be used in crop models which adds bias and uncertainty. Agriculture itself is also inherently risky and yields tend to be affected by an array of interacting variables. Process-based crop models require a number of steps (parameterisation, bias correction, model stacking that can all add to overall uncertainty). Farmer decision-making under climate risk introduces an additional layer of uncertainty, e.g. in studies on adapting agriculture to climate change, there was often an assumption of 'full capacity' in implementing adaptations, even though this may not necessarily be the case, particularly in regions with subsistence agriculture (Howden et al., 2007; Morton, 2007). The combination of uncertainty in climate change projections and farmer decision-making under risk should therefore be a focus of approaches that combine statistical economic farming systems modelling with crop models in risk assessment (Moriondo et al., 2010). While uncertainty in crop modelling under climate change is inevitable, there are a number of best practices that can help in addressing and communicating this uncertainty (Challinor et al., 2018). These include the use of multiple methods to critically assess the use of climate model outputs and increasing transparency and inter-comparability by making modelling assumptions explicit.

5.2 Interactions and feedbacks

A common challenge in agricultural modelling is that of interactions and feedbacks. There are many interacting factors and mechanisms that are still not well enough understood to be modelled, e.g. confounding effects of heat stress on animal productivity or the application of fertiliser and soil interactions. Crop models tend to treat climate, soils and management factors as exogenous inputs into the system and thus do not account for any dynamic feedbacks between crop productivity and input variables. Integrated modelling approaches often link single models with static outputs and thus do not include any representation of feedbacks in e.g. adaptation choices and are therefore limited in their utility of assessment of costs and benefits of such adaptation strategies (Kipling et al., 2019). The same applies to market feedbacks, where models rely on exogenous price assumptions for agricultural inputs and outputs unrelated to management for adaptation. Other biophysical climate-related feedbacks such as pests, disease, pollution, water availability for irrigation systems, land degradation and resource availability are also very difficult to include in agricultural model simulations. While some of these factors may be accounted for in statistical modelling approaches through their linkage to climate parameters (Rötter et al., 2018), they are largely ignored in process-based models.

IAMs are better suited to capturing the dynamic feedbacks between crop growth, soils, water and management as well as more explicitly accounting for markets and policies. However, it remains challenging to include human decision-making and relationships between humans and the biophysical environment in such models, particularly describing these linkages in mathematical form (Calvin and Bond-Lamberty, 2018).

5.3 Data limitations

A key limitation to agricultural modelling, and modelling in general, is the lack of reliable data for model parameterisation, validation and calibration. Jones et al. (2017) argue that limitations of agricultural system models are more around data limitations than knowledge gaps. Lack of data is especially prevalent in developing countries while in data-rich regions, often data does exist but is not accessible, e.g. when it is owned by the private sector.

To successfully calibrate a model, high-quality and compatible datasets are required (i.e. same units, scale and resolution). This is particularly relevant for modelling future crop performance under climate change where a wide range of historical data is required. In addition to slow onset climate change, modelling climate risk in crop models is challenging as these models tend to perform well under expected conditions but less so under extreme weather events due to a lack of data reflecting such conditions (Benami et al., 2021). Climate change projections are also difficult to downscale to finer resolutions and suffer from deep uncertainty. For climate impact studies on crop production, it is necessary to focus on longer time scales for which climate projections are highly uncertain which may lead to maladaptive management options (Nissan et al., 2019). Spatial data on soil properties and land management decisions (e.g. sowing dates and planting densities) used in models is often too coarse to appropriately reflect heterogeneity of farming conditions to predict changes in yields (Pongratz et al., 2018). Pasquel et al. (2022) note that most spatial crop models are evaluated using aspatial indicators which may lead to spatial autocorrelation and thus result in erroneous model evaluation. Therefore, new indicators and coefficients that consider spatial autocorrelation should be developed by crop modelling and statistical communities. Given the current trend of upscaling crop simulation models from the field to large regions, a main challenge is the calibration of model inputs beyond the field scale (Pearl, 2019; Jung et al., 2021).

Other crop management practices such as tillage, crop rotation and cover crops may not be included at all in crop models as spatiotemporal data on these does not exist (Peng et al., 2020). These are critical datasets for the assessment of adaptation outcomes. New approaches in remote sensing are being developed to overcome some of these limitations (Kubitza et al., 2020) although thus far they are of limited use in mapping management practices or crop types in heterogeneous smallholder plots, particularly in cloud-rich environments (Ibrahim et al., 2021). Nevertheless, remote-sensing data of higher temporal and spatial resolution is increasingly becoming available which can be used to develop new types of crop models, e.g. Huang et al. (2019) combined remote-sensing data with crop models to improve estimates of crop yield, water use and nitrogen uptake across large scales. Remote sensing data itself however also requires ground-truthed data for calibration and learning which limits the development and reliability of these datasets.

6 Future trends in research

Agricultural modelling under climate change is a rapidly developing field and advances are made for all types of modelling described in this chapter. There are however some key advancements and developments that will shape the discipline in the coming years. This section sets out future trends in research by describing new visions for model development, data availability for modelling, scaling and new and emerging concepts for capturing agricultural transition under climate change in modelling approaches.

6.1 Model development

A good overview of research priorities for agricultural system models is provided by Antle et al. (2017) who describe a vision for 'NextGen' data and models that are able to accelerate the rate of agricultural innovation and meet the global need for food and fibre. A key pillar of this vision is the increase in computational agricultural research and development. For example, carrying out simulation experiments with models greatly reduce the need for trial-and-error learning in research on new varieties and management practices. A second pillar is the need to connect people with models by providing useful information to different types of users which will increase relevance and credibility of models.

To advance the modelling of agricultural impacts of and adaptation to climate change, Rötter et al. (2018) identify two main avenues of research. First, the need for a more mechanistic understanding of climate impacts and management options for adaptation and mitigation, and second, a focus on cropping systems and integrative multi-scale assessments, particularly in complex tropical environments rather than focusing on single season and crops.

Silva and Giller (2020) highlight a number of research priorities specific to crop models, including the development of crop models for crops other than cereals (e.g. root and tuber crops and tropical perennials). To expand the current capabilities of crop models (i.e. simulation of potential, water-limited and nitrogen-limited yields) and to simulate limitations of phosphorus and potassium and yield reductions due to pests and diseases. Finally, they also highlight the need to simulate processes at cropping systems level and to contextualise model applications within broader farm(ing) systems and food systems aspects.

Another area of research that is receiving increased attention is that of agro-ecology as it is considered a challenging model for agricultural systems that can play a key role in a transition to sustainable agriculture and adaptation to climate change. Agro-ecology has its roots in smallholder agriculture but urgently needs to be mainstreamed in large-scale farming (Tittonell et al., 2020). Gascuel-Odoux et al. (2022) set out five lines of research for better modelling of agro-ecology. The first is to address knowledge gaps on interactions among structural or functional components of agroecosystems, e.g. better representation of microbial communities and their functions in models of soil biogeochemical cycles (Moorhead et al., 2014). A second line of research centres around developing different modeling approaches to complexity, e.g. incorporating data from complex interactions, such as remote sensing data and network theory. Third, better consideration of spatial and temporal dimensions, e.g. models that capture transient dynamics are essential to study systems not at equilibrium, as is the case for agroecological transitions. The fourth line of research should be around including more phenotypic dimensions to study mechanisms involved in mixture performances and finally, research should focus on a better representation of the cascade connecting cropping practices to biodiversity, ecosystem functioning and ultimately benefits and values to people. This should include the design of socio-ecological models that include findings of actors' behaviours from the human and social sciences.

Land-use change models are important models that play a critical role in environmental assessments as they can evaluate drivers of global change but also help to design measures to mitigate or adapt to climate change (Verburg et al., 2019). However, these models often do not sufficiently represent land system dynamics beyond the local scale and are frequently insufficiently evaluated. Verburg et al. (2019) therefore argue that land system modelling needs to work on new model structures and more focus on multi-scale interactions, diversity in human agency and the links between demand and supply. This will inevitably lead to more complex models that may require more interpretation and improved narratives for use in policy but with greater realism and rigour in the modelling outputs.

Machine learning is a powerful approach that is increasingly used in agricultural modelling as it enables the extraction of information from a complex interplay of drivers in large datasets. It is therefore able to detect responses that may not be reflected yet in process-based crop models. There are now a wide variety of machine learning algorithms applied in e.g. yield prediction under climate change (e.g. Crane-Droesch, 2018; Feng et al., 2019). In many cases, machine learning shows better performance compared to conventional linear regression models (Feng et al., 2018; Leng and Hall, 2020). However, to apply machine learning approaches appropriately, a high level of expert judgement is required to select the best model as selecting these based only on statistical criteria like model performance may lead to targeting irrelevant predictors. In addition, machine learning outcomes can vary depending on model assumptions (Lischeid et al., 2022). A recent and novel development in the field of machine learning approaches is the so-called theory-guided or physics-guided machine learning, where models integrate some specific theoretical knowledge, laws or principles from process-based models with machine learning models. These approaches have been applied in hydrology and hydrogeology (e.g. Adombi et al., 2021) and are starting to be applied in agricultural modelling (e.g. Liu et al., 2022) as well. Machine learning approaches hold great promise in adaptation research, e.g. by coupling with crop models as shown by Tao et al. (2022) who applied this approach to design wheat cultivar adaptation to future climate change across China.

In recent years, there has been a lot of research on incorporating genetics and genomics into modelling of crop and livestock systems. For crop models incorporating genetics allows for better predictions of yield based on genetic regulation. This approach is therefore able to improve simulations of adaptive responses to climate change for crops (Wang et al., 2019). Crop models that have incorporated functional relationships between environmental variables and traits for genetic variation have already been used in research on drought-tolerant maize hybrids. There are also highly detailed plant systems models that simulate gene network regulation, metabolic reactions and metabolite transport. These highly detailed models could be integrated with crop models in order to target for example crop genetic improvements but can also help in improving both types of models. There is a lot of promise in using this integrated approach to better represent genetic traits in crop models. However, this research domain is still very underdeveloped (Peng et al., 2020).

6.2 Improvements in data

As described previously, a big challenge in modelling agricultural systems' response to climate change is data availability to parameterise, validate and calibrate models that accurately represent biophysical conditions and land management decisions. Driven by advances in remote sensing and machine learning, more and more datasets at higher temporal and spatial resolutions have become available in recent years. An example of this is the development of microclimate data with hourly simulations of future climate at high spatial resolution (e.g. Maclean, 2020) that can be used in assessing climate suitability for crops (Gardner et al., 2021). There have also been advances in using remote sensing approaches to improve data on land management practices such as grassland management (Reinermann et al., 2020), cropping intensities, irrigation, tillage etc. (Kubitza et al., 2020). However, these studies have mostly been applied to small scales and for single time periods, although progress is being made to apply this to larger (national) scales (Blickensdörfer et al., 2022). Large-scale automated continuous monitoring would significantly advance the utility of these data and their use in agricultural models. In addition, while remote-sensing-based data and analysis techniques are improving, there is an opportunity to start connecting these with field data and data on secondary data on socio-economic characteristics that would improve data on land management practices and factors that drive adoption patterns of agricultural practices such as CSA (Kubitza et al., 2020). Machine learning approaches are key to developing new datasets as they are better able to cope with large volumes of data, e.g. remote sensing data but can also be applied to data collected through citizen science. For example, Hampf et al. (2021) applied machine learning to data collected through a smartphone app to detect crop diseases and associated yield losses in the Southern Amazon. Machine learning is also better able to allow for non-linearity and interactions across variables.

6.3 Scaling

In order to reliably model agricultural adaptation to climate change, models need to operate at a large scale as adaptation decisions and policy tends to be made at broad scales. However, crop models have traditionally been developed for plot and field scales and are thus less useful to assess the impacts of adaptation decisions. Larger scale Earth Systems Models and IAMs are better suited to assess national and global policies for climate change adaptation but tend to be limited in their approach to crop modelling. Therefore, there is an urgent need to integrate large-scale models with highly specific crop models. Such integration would enable a consistent and systematic assessment of impacts of adaptation on multiple sectors of the Earth system at different scales (Peng et al., 2020). Such integration can only be achieved through large-scale projects that can capture all aspects and dimensions of future agricultural systems which would require networks of researchers from different countries to more easily exchange and synthesise knowledge, methods and data in order to improve modelling from local to global scales (Müller et al., 2020). An example of such a project that brings together the global and local modelling communities is the Glassnet project, led by Purdue university: https://mygeohub.org/groups/glassnet which aims to develop a network of networks that facilitates integration across research teams through the creation of a common data typology and infrastructure and workshops and seminars. Such trans-disciplinary collaboration is critical to advance agricultural systems modelling under climate change and address wider global sustainability challenges.

6.4 New concepts in modelling agricultural transition under climate change

Modelling the transition of the agricultural system under climate change is inherently complex due to the multifaceted nature of the system and the various factors that influence it which require a multidisciplinary approach that integrates knowledge from agronomy, economics, social science and climate science to capture the complex interactions and feedbacks between the various components of the system and to identify the most effective adaptation strategies.

Moreover, modelling such a transformative change is challenging as it can radically change the conceptualisation of the system, with changes in state variables and structure, e.g. different classes, processes and relationships need to be included to adequately represent the dynamics of the new system (Müller et al., 2020).

To successfully model adaptation pathways, it is therefore critical that models and model systems capture all of this complexity and incorporate all elements that make up the agricultural system. This will include the development of new models and model approaches as well as the development of new projections of alternative futures that consider all aspects of agricultural transition. For example, global modelling approaches such as those described in Section 4.5 in most cases include assumptions about technological improvements. However, technological changes and innovations are often not linear and can be more important than captured in modelling approaches. For example, Aggarwal et al. (2019), found, through a meta-analysis that observed yield changes for maize, rice and wheat were considerably higher than projected yield changes under climate change because of the large yield-enhancing impact of technological improvements compared to the negative impacts of climate change. Their study thus suggested that technological growth tends to be underestimated in global impact assessments and needs to be appropriately considered. Given that new technology, such as robotics, nanotechnology and artificial intelligence are expected to revolutionise agriculture (Shepherd et al., 2020) and potentially lead to what is dubbed the fourth agricultural revolution, or 'Agriculture 4.0' (Rose and Chilvers, 2018) it is essential that such game-changing technologies are included in scenario assessments that aim to explore transition pathways towards sustainable agricultural systems (Klerkx and Rose, 2020). There are some studies that have started to develop new scenarios based on these technological changes, e.g. Ehlers et al. (2022) recently developed explorative scenarios of future digitalisation of agricultural systems in Europe to assess potential implications for agricultural policy. However, these scenarios have not yet been quantitatively assessed with models.

Similarly, emerging land uses such as agri-photovoltaic (PV) which can involve agricultural activities around pre-existing PV facilities or as designed co-production of agricultural crops and PV have considerable potential to simultaneously produce food and sustainable energy as well as reduce land use competition. While many countries have set targets for PV power production (e.g. India: Giri and Mohanty, 2022), agri-PV is not often included in future agriculture modelling studies, although there are some studies emerging that assess the feasibility of agri-PV. For example, Choi et al. (2021) used life-cycle analysis to analyse the land use, energy, GHG emissions, economic feasibility and the environmental co-benefits of low-density PV over existing crops for Indonesia and found that small-scale dual land-use systems have the potential for multiple co-benefits (e.g. rural electrification) and are economically viable.

Another element of transition that is currently not adequately captured in modelling approaches is a move to more organic production as an adaptation option. There are some studies that focus on this transition, e.g. Barbieri et al. (2021) analyse different scenarios of organic conversion and found that smaller areas planted with wheat, rice and maize must be offset by more nitrogen-fixing crops, such as beans, alfalfa and clover and that caloric energy would fall by about 27% from current production. Similarly, Kuczuk and Widera (2021) analyse scenarios for the transition to organic farming in Poland by the year 2030. However, while useful, these studies did not quantitatively model these scenarios and thus may not be realistic in terms of land area required, trade-offs with other land uses, nutritional requirements and other limiting factors.

There are many other processes and elements of agricultural adaptation under climate change that are currently not adequately captured in modelling approaches, e.g. farm composition such as intercropping, agroforestry and crop diversification which is typical for small-holdings but also changes in diets and consumer behaviour which has the potential to drastically change the agricultural system. These are all challenges that the agricultural modelling community needs to address. It is not completely clear though if these are all novel modelling challenges or if adaptation adds new motivations to old challenges (Kipling et al., 2019).

7 Where to look for further information

In addition to all cited literature in this chapter, there are a number of useful resources to learn more about this topic. A wealth of information is provided through the CGIAR (Consultative Group for International Agricultural Research) that comprises 15 research centres that focus on specific elements of agriculture and crops. Of particular relevance is their climate adaptation and mitigation impact area. All publications and research data are made available through their website: https://www.cgiar.org/. To find out more about the various IAMs, there is a good overview available at the IAM Wiki available here: https://www.iamcdocumentation.eu/index.php/IAMC_wiki. A really useful website that provides an overview of plant and crop models aimed at researchers is Quantitative Plant (https://www.quantitative-plant.org/), which aims to help plant researchers to find the right tool or model for their research (Lobet et al., 2013). The previously mentioned Glassnet community welcomes new researchers and provides tools, data, training and some funding opportunities: https://mygeohub.org/groups/glassnet. Another community of interest is the Global Land Programme (https://glp.earth/) which is an interdisciplinary community of science and practice on the study of land systems and sustainability solutions that help linking scientists, post jobs and organise open science meetings every couple of years. Data for agricultural modelling can be found through the Global Open Data for Agriculture and Nutrition initiative (GODAN, www.godan.info).

8 References

Aaheim, H. A., Orlov, A., Wei, T. and Glomsrød, S. 2018. GRACE model and applications. CICERO Report.

Adombi, A. V. D. P., Chesnaux, R. and Boucher, M. A. 2021. Review: Theory-guided machine learning applied to hydrogeology–State of the art, opportunities and future challenges. *Hydrogeology Journal* 29(8), 2671-2683.

Aggarwal, P., Vyas, S., Thornton, P., Campbell, B. M. and Kropff, M. 2019. Importance of considering technology growth in impact assessments of climate change on agriculture. *Global Food Security* 23, 41-48.

Antle, J. M., Jones, J. W. and Rosenzweig, C. 2017. Next generation agricultural system models and knowledge products: Synthesis and strategy. *Agricultural Systems* 155, 179-185.

Barbieri, P., Pellerin, S., Seufert, V., Smith, L., Ramankutty, N. and Nesme, T. 2021. Global option space for organic agriculture is delimited by nitrogen availability. *Nature Food* 2(5), 363-372.

Bateki, C. A., Cadisch, G. and Dickhoefer, U. 2019. Modelling sustainable intensification of grassland-based ruminant production systems: A review. *Global Food Security* 23, 85-92.

Benami, E., Jin, Z., Carter, M. R., Ghosh, A., Hijmans, R. J., Hobbs, A., Kenduiywo, B. and Lobell, D. B. 2021. Uniting remote sensing, crop modelling and economics for agricultural risk management. *Nature Reviews Earth and Environment* 2(2), 140-159.

Blickensdörfer, L., Schwieder, M., Pflugmacher, D., Nendel, C., Erasmi, S. and Hostert, P. 2022. Mapping of crop types and crop sequences with combined time series of Sentinel-1, Sentinel-2 and Landsat 8 data for Germany. *Remote Sensing of Environment* 269, 112831.

Bloomfield, J. A., Rose, T. J. and King, G. J. 2014. Sustainable harvest: Managing plasticity for resilient crops. *Plant Biotechnology Journal* 12(5), 517-533.

Booth, T. H., Nix, H. A., Busby, J. R. and Hutchinson, M. F. 2014. Bioclim: The first species distribution modelling package, its early applications and relevance to most current MAXENT studies. *Diversity and Distributions* 20(1), 1-9.

Brilli, L., Bechini, L., Bindi, M., Carozzi, M., Cavalli, D., Conant, R., Dorich, C. D., Doro, L., Ehrhardt, F., Farina, R., Ferrise, R., Fitton, N., Francaviglia, R., Grace, P., Iocola, I., Klumpp, K., Léonard, J., Martin, R., Massad, R. S., Recous, S., Seddaiu, G., Sharp, J., Smith, P., Smith, W. N., Soussana, J. F. and Bellocchi, G. 2017. Review and analysis of strengths and weaknesses of agro-ecosystem models for simulating C and N fluxes. *Science of the Total Environment* 598, 445–470.

Britz, W. and Witzke, P. 2008. *Capri Model Documentation 2008 Version 2*. Institute for Food and Resource Economics, University of Bonn, Bonn.

Bryant, J. R. and Snow, V. O. 2008. Modelling pastoral farm agro-ecosystems: A review. *New Zealand Journal of Agricultural Research* 51(3), 349–363.

Calvin, K. and Bond-Lamberty, B. 2018. Integrated human-earth system modeling–State of the science and future directions. *Environmental Research Letters* 13(6), 063006.

Challinor, A. J., Müller, C., Asseng, S., Deva, C., Nicklin, K. J., Wallach, D., Vanuytrecht, E., Whitfield, S., Ramirez-Villegas, J. and Koehler, A. K. 2018. Improving the use of crop models for risk assessment and climate change adaptation. *Agricultural Systems* 159, 296–306.

Choi, C. S., Ravi, S., Siregar, I. Z., Dwiyanti, F. G., Macknick, J., Elchinger, M. and Davatzes, N. C. 2021. Combined land use of solar infrastructure and agriculture for socioeconomic and environmental co-benefits in the tropics. *Renewable and Sustainable Energy Reviews* 151, 111610.

Crane-Droesch, A. 2018. Machine learning methods for crop yield prediction and climate change impact assessment in agriculture. *Environmental Research Letters* 13(11), 114003.

Cunha, B. S., Garaffa, R. and Gurgel, Â. C. 2020. *TEA Model Documentation*.

De Pinto, A., Cenacchi, N., Kwon, H. Y., Koo, J. and Dunston, S. 2020. Climate smart agriculture and global food-crop production. *PLoS ONE* 15(4), e0231764.

Doelman, J. C., Stehfest, E., Tabeau, A., van Meijl, H., Lassaletta, L., Gernaat, D. E. H. J., Hermans, K., Harmsen, M., Daioglou, V., Biemans, H., van der Sluis, S. and van Vuuren, D. P. 2018. Exploring SSP land-use dynamics using the IMAGE model: Regional and gridded scenarios of land-use change and land-based climate change mitigation. *Global Environmental Change* 48, 119–135.

Dougherty, H. C., Ahmadi, A., Oltjen, J. W., Mitloehner, F. M. and Kebreab, E. 2019. Review: Modeling production and environmental impacts of small ruminants–Incorporation of existing ruminant modeling techniques, and future directions for research and extension. *Applied Animal Science* 35(1), 114–129.

Dury, J., Schaller, N., Garcia, F., Reynaud, A. and Bergez, J. E. 2012. Models to support cropping plan and crop rotation decisions: A review. *Agronomy for Sustainable Development* 32(2), 567–580.

Ehlers, M. H., Finger, R., El Benni, N., Gocht, A., Sørensen, C. A. G., Gusset, M., Pfeifer, C., Poppe, K., Regan, Á., Rose, D. C., Wolfert, S. and Huber, R. 2022. Scenarios for European agricultural policymaking in the era of digitalisation. *Agricultural Systems* 196, 103318.

Feng, P., Wang, B., Liu, D. L., Xing, H., Ji, F., Macadam, I., Ruan, H. and Yu, Q. 2018. Impacts of rainfall extremes on wheat yield in semi-arid cropping systems in eastern Australia. *Climatic Change* 147(3–4), 555–569.

Feng, P., Wang, B., Li Liu, D. L., Waters, C. and Yu, Q. 2019. Incorporating machine learning with biophysical model can improve the evaluation of climate extremes impacts on

wheat yield in south-eastern Australia. *Agricultural and Forest Meteorology* 275, 100–113.
Fischer, G., Nachtergaele, F. O., van Velthuizen, H., Chiozza, F., Francheschini, G., Henry, M., Muchoney, D. and Tramberend, S. 2021. *Global Agro-Ecological Zones (Gaez v4)-Model Documentation*.
Gardner, A. S., Maclean, I. M. D., Gaston, K. J. and Bütikofer, L. 2021. Forecasting future crop suitability with microclimate data. *Agricultural Systems* 190, 103084.
Gascuel-Odoux, C., Lescourret, F., Dedieu, B., Detang-Dessendre, C., Faverdin, P., Hazard, L., Litrico-Chiarelli, I., Petit, S., Roques, L., Reboud, X., Tixier-Boichard, M., de Vries, H. and Caquet, T. 2022. A research agenda for scaling up agroecology in European countries. *Agronomy for Sustainable Development* 42(3), 53.
Giri, N. C. and Mohanty, R. C. 2022. Design of agrivoltaic system to optimize land use for clean energy-food production: A socio-economic and environmental assessment. *Clean Technologies and Environmental Policy* 24(8), 2595–2606.
Gourdji, S. M., Sibley, A. M. and Lobell, D. B. 2013. Global crop exposure to critical high temperatures in the reproductive period: Historical trends and future projections. *Environmental Research Letters* 8(2), 024041.
Hampf, A. C., Nendel, C., Strey, S. and Strey, R. 2021. Biotic yield losses in the Southern Amazon, Brazil: Making use of smartphone-assisted plant disease diagnosis data. *Frontiers in Plant Science* 12, 621168.
Hannah, L., Roehrdanz, P. R., K C, K. B., Fraser, E. D. G., Donatti, C. I., Saenz, L., Wright, T. M., Hijmans, R. J., Mulligan, M., Berg, A. and van Soesbergen, A. 2020. The environmental consequences of climate-driven agricultural frontiers. *PLoS ONE* 15(2), e0228305.
Hijmans, R. J., Guarino, L., Cruz, M. and Rojas, E. 2001. Computer tools for spatial analysis of plant genetic resources data: 1. DIVA-GIS. *Plant Genetic Resources Newsletter*, pp. 15–19.
Hinsinger, P., Brauman, A., Devau, N., Gérard, F., Jourdan, C., Laclau, J. P., Le Cadre, E., Jaillard, B. and Plassard, C. 2011. Acquisition of phosphorus and other poorly mobile nutrients by roots. Where do plant nutrition models fail? *Plant and Soil* 348(1–2), 29–61.
Holzkämper, A. 2017. Adapting agricultural production systems to climate change– What's the use of models? *Agriculture* 7(10), 86.
Howden, S. M., Soussana, J. F., Tubiello, F. N., Chhetri, N., Dunlop, M. and Meinke, H. 2007. Adapting agriculture to climate change. *Proceedings of the National Academy of Sciences of the United States of America* 104(50), 19691–19696.
Huang, J., Gómez-Dans, J. L., Huang, H., Ma, H., Wu, Q., Lewis, P. E., Liang, S., Chen, Z., Xue, J. H., Wu, Y. and Zhao, F. 2019. Assimilation of remote sensing into crop growth models: Current status and perspectives. *Agricultural and Forest Meteorology* 276, 107609.
Hunter, M. C., Smith, R. G., Schipanski, M. E., Atwood, L. W. and Mortensen, D. A. 2017. Agriculture in 2050: Recalibrating targets for sustainable intensification. *BioScience* 67(4), 386–391.
Ibrahim, E. S., Rufin, P., Nill, L., Kamali, B., Nendel, C. and Hostert, P. 2021. Mapping crop types and cropping systems in Nigeria with sentinel-2 imagery. *Remote Sensing* 13(17), 3523.
Jägermeyr, J., Müller, C., Ruane, A. C., Elliott, J., Balkovic, J., Castillo, O., Faye, B., Foster, I., Folberth, C., Franke, J. A., Fuchs, K., Guarin, J. R., Heinke, J., Hoogenboom, G., Iizumi, T., Jain, A. K., Kelly, D., Khabarov, N., Lange, S., Lin, T., Liu, W., Mialyk,

O., Minoli, S., Moyer, E. J., Okada, M., Phillips, M., Porter, C., Rabin, S. S., Scheer, C., Schneider, J. M., Schyns, J. F., Skalsky, R., Smerald, A., Stella, T., Stephens, H., Webber, H., Zabel, F. and Rosenzweig, C. 2021. Climate impacts on global agriculture emerge earlier in new generation of climate and crop models. *Nature Food* 2(11), 873-885.

Jaime, R., Alcántara, J. M., Manzaneda, A. J. and Rey, P. J. 2018. Climate change decreases suitable areas for rapeseed cultivation in Europe but provides new opportunities for white mustard as an alternative oilseed for biofuel production. *PLoS ONE* 13(11), e0207124.

Janssens, C., Havlík, P., Krisztin, T., Baker, J., Frank, S., Hasegawa, T., Leclère, D., Ohrel, S., Ragnauth, S., Schmid, E., Valin, H., Van Lipzig, N. and Maertens, M. 2020. Global hunger and climate change adaptation through international trade. *Nature Climate Change* 10(9), 829-835.

Jones, J. W., Hoogenboom, G., Porter, C. H., Boote, K. J., Batchelor, W. D., Hunt, L. A., Wilkens, P. W., Singh, U., Gijsman, A. J. and Ritchie, J. T. 2003. The DSSAT cropping system model. *European Journal of Agronomy* 18(3-4), 235-265.

Jones, J. W., Antle, J. M., Basso, B., Boote, K. J., Conant, R. T., Foster, I., Godfray, H. C. J., Herrero, M., Howitt, R. E., Janssen, S., Keating, B. A., Munoz-Carpena, R., Porter, C. H., Rosenzweig, C. and Wheeler, T. R. 2017. Brief history of agricultural systems modeling. *Agricultural Systems* 155, 240-254.

Jones, J. W., Antle, J. M., Basso, B., Boote, K. J., Conant, R. T., Foster, I., Godfray, H. C. J., Herrero, M., Howitt, R. E., Janssen, S., Keating, B. A., Munoz-Carpena, R., Porter, C. H., Rosenzweig, C. and Wheeler, T. R. 2017. Toward a new generation of agricultural system data, models, and knowledge products: State of agricultural systems science. *Agricultural Systems* 155, 269-288.

Jung, J., Maeda, M., Chang, A., Bhandari, M., Ashapure, A. and Landivar-Bowles, J. 2021. The potential of remote sensing and artificial intelligence as tools to improve the resilience of agriculture production systems. *Current Opinion in Biotechnology* 70, 15-22.

Kang, M., Hyun, S. and Kim, K. S. 2022. Spatial assessment of climate suitability for summer cultivation of potato in North Korea. *Korean Journal of Agricultural and Forest Meteorology* 24(1), 35-47.

Kipling, R. P., Topp, C. F. E., Bannink, A., Bartley, D. J., Blanco-Penedo, I., Cortignani, R., del Prado, A., Dono, G., Faverdin, P., Graux, A. I., Hutchings, N. J., Lauwers, L., Özkan Gülzari, Ş., Reidsma, P., Rolinski, S., Ruiz-Ramos, M., Sandars, D. L., Sándor, R., Schönhart, M., Seddaiu, G., van Middelkoop, J., Shrestha, S., Weindl, I. and Eory, V. 2019. To what extent is climate change adaptation a novel challenge for agricultural modellers? *Environmental Modelling and Software: With Environment Data News* 120, 104492.

Klerkx, L. and Rose, D. 2020. Dealing with the game-changing technologies of Agriculture 4.0: How do we manage diversity and responsibility in food system transition pathways? *Global Food Security* 24, 100347.

Kollas, C., Kersebaum, K. C., Nendel, C., Manevski, K., Müller, C., Palosuo, T., Armas-Herrera, C. M., Beaudoin, N., Bindi, M., Charfeddine, M., Conradt, T., Constantin, J., Eitzinger, J., Ewert, F., Ferrise, R., Gaiser, T., Cortazar-Atauri, I. Gd., Giglio, L., Hlavinka, P., Hoffmann, H., Hoffmann, M. P., Launay, M., Manderscheid, R., Mary, B., Mirschel, W., Moriondo, M., Olesen, J. E., Öztürk, I., Pacholski, A., Ripoche-Wachter, D., Roggero, P. P., Roncossek, S., Rötter, R. P., Ruget, F., Sharif, B., Trnka, M., Ventrella, D., Waha, K.,

Wegehenkel, M., Weigel, H. and Wu, L. 2015. Crop rotation modelling–A European model intercomparison. *European Journal of Agronomy* 70, 98–111.

Kriticos, D. J., Maywald, G. F., Yonow, T., Zurcher, E. J., Herrmann, N. I. and Sutherst, R. 2015. Exploring the effects of climate on plants, animals and diseases. *CLIMEX Version* 4, 184.

Kubitza, C., Krishna, V. V., Schulthess, U. and Jain, M. 2020. Estimating adoption and impacts of agricultural management practices in developing countries using satellite data: A scoping review. *Agronomy for Sustainable Development* 40(3), 1-21.

Kuczuk, A. and Widera, K. 2021. A greater share of organic agriculture in relation to food security resulting from the energy demand obtained from food–Scenarios for Poland until 2030. *Energies* 14(21), 6959.

Leclère, D., Havlík, P., Fuss, S., Schmid, E., Mosnier, A., Walsh, B., Valin, H., Herrero, M., Khabarov, N. and Obersteiner, M. 2014. Climate change induced transformations of agricultural systems: Insights from a global model. *Environmental Research Letters* 9(12), 124018.

Leng, G. and Hall, J. W. 2020. Predicting spatial and temporal variability in crop yields: An inter-comparison of machine learning, regression and process-based models. *Environmental Research Letters: ERL [Web Site]* 15(4), 044027.

Lischeid, G., Webber, H., Sommer, M., Nendel, C. and Ewert, F. 2022. Machine learning in crop yield modelling: A powerful tool, but no surrogate for science. *Agricultural and Forest Meteorology* 312, 108698.

Liu, B., Asseng, S., Müller, C., Ewert, F., Elliott, J., Lobell, D. B., Martre, P., Ruane, A. C., Wallach, D., Jones, J. W., Rosenzweig, C., Aggarwal, P., Alderman, P., Anothai, J., Basso, B., Biernath, C., Cammarano, D., Challinor, A., Deryng, D., Sanctis, G., Doltra, J., Fereres, E., Folberth, C., Garcia-Vila, M., Gayler, S., Hoogenboom, G., Hunt, L., Izaurralde, R., Jabloun, M., Jones, C., Kersebaum, K., Kimball, B., Koehler, A., Kumar, S., Nendel, C., O'Leary, G., Olesen, J., Ottman, M., Palosuo, T., Prasad, P., Priesack, E., Pugh, T., Reynolds, M., Rezaei, E., Rötter, R., Schmid, E., Semenov, M., Shcherbak, I., Stehfest, E., Stöckle, C., Stratonovitch, P., Streck, T., Supit, I., Tao, F., Thorburn, P., Waha, K., Wall, G., Wang, E., White, J., Wolf, J., Zhao, Z. and Zhu, Y. 2016. Similar estimates of temperature impacts on global wheat yield by three independent methods. *Nature Climate Change* 6(12), 1130–1136.

Liu, L., Xu, S., Tang, J., Guan, K., Griffis, T. J., Erickson, M. D., Frie, A. L., Jia, X., Kim, T., Miller, L. T. and Peng, B. 2022. KGML-ag: A modeling framework of knowledge-guided machine learning to simulate agroecosystems: A case study of estimating N_2O emission using data from mesocosm experiments. *Geoscientific Model Development* 15(7), 2839–2858.

Lobet, G., Draye, X. and Périlleux, C. 2013. An online database for plant image analysis software tools. *Plant Methods* 9(1), 1-8.

Luderer, G., Leimbach, M., Bauer, N., Kriegler, E., Baumstark, L., Bertram, C., Giannousakis, A., Hilaire, J., Klein, D., Levesque, A. and Mouratiadou, I. 2015. *Description of the REMIND Model (Version 1.6).*

Luo, Q., Hoogenboom, G. and Yang, H. 2022. Uncertainties in assessing climate change impacts and adaptation options with wheat crop models. *Theoretical and Applied Climatology* 149(1-2), 805–816.

Maclean, I. M. D. 2020. Predicting future climate at high spatial and temporal resolution. *Global Change Biology* 26(2), 1003-1011.

Matteau, J.-P., Gumiere, S. J., Gallichand, J., Létourneau, G., Khiari, L., Gasser, M.-O. and Michaud, A. 2019. Coupling of a nitrate production model with HYDRUS to predict nitrate leaching. *Agricultural Water Management* 213, 616–626.

Messner, S. and Strubegger, M. 1995. *User's Guide for Message* (Vol. III).

Mitchell, D., AchutaRao, K., Allen, M., Bethke, I., Beyerle, U., Ciavarella, A., Forster, P. M., Fuglestvedt, J., Gillett, N., Haustein, K., Ingram, W., Iversen, T., Kharin, V., Klingaman, N., Massey, N., Fischer, E., Schleussner, C., Scinocca, J., Seland, Ø., Shiogama, H., Shuckburgh, E., Sparrow, S., Stone, D., Uhe, P., Wallom, D., Wehner, M. and Zaaboul, R. 2017. Half a degree additional warming, prognosis and projected impacts (HAPPI): Background and experimental design. *Geoscientific Model Development* 10(2), 571–583.

Moorhead, D., Lashermes, G., Recous, S. and Bertrand, I. 2014. Interacting microbe and litter quality controls on litter decomposition: A modeling analysis. *PLoS ONE* 9(9), e108769.

Moriondo, M., Pacini, C., Trombi, G., Vazzana, C. and Bindi, M. 2010. Sustainability of dairy farming system in Tuscany in a changing climate. *European Journal of Agronomy* 32(1), 80–90.

Morton, J. F. 2007. The impact of climate change on smallholder and subsistence agriculture. *Proceedings of the National Academy of Sciences of the United States of America* 104(50), 19680–19685.

Müller, B., Hoffmann, F., Heckelei, T., Müller, C., Hertel, T. W., Polhill, J. G., Van Wijk, M., Achterbosch, T., Alexander, P., Brown, C., Kreuer, D., Ewert, F., Ge, J., Millington, J. D. A., Seppelt, R., Verburg, P. H. and Webber, H. 2020. Modelling food security: Bridging the gap between the micro and the macro scale. *Global Environmental Change* 63, 102085.

Myers, S. S., Zanobetti, A., Kloog, I., Huybers, P., Leakey, A. D., Bloom, A. J., Carlisle, E., Dietterich, L. H., Fitzgerald, G., Hasegawa, T., Holbrook, N. M., Nelson, R. L., Ottman, M. J., Raboy, V., Sakai, H., Sartor, K. A., Schwartz, J., Seneweera, S., Tausz, M. and Usui, Y. 2014. Increasing CO_2 threatens human nutrition. *Nature* 510(7503), 139–142.

Nelson, G. C., Valin, H., Sands, R. D., Havlík, P., Ahammad, H., Deryng, D., Elliott, J., Fujimori, S., Hasegawa, T., Heyhoe, E., Kyle, P., Von Lampe, M., Lotze-Campen, H., Mason d'Croz, D., van Meijl, H., van der Mensbrugghe, D., Müller, C., Popp, A., Robertson, R., Robinson, S., Schmid, E., Schmitz, C., Tabeau, A. and Willenbockel, D. 2014. Climate change effects on agriculture: Economic responses to biophysical shocks. *Proceedings of the National Academy of Sciences of the United States of America* 111(9), 3274–3279.

Nicotra, A. B., Atkin, O. K., Bonser, S. P., Davidson, A. M., Finnegan, E. J., Mathesius, U., Poot, P., Purugganan, M. D., Richards, C. L., Valladares, F. and van Kleunen, M. 2010. Plant phenotypic plasticity in a changing climate. *Trends in Plant Science* 15(12), 684–692.

Nissan, H., Goddard, L., de Perez, E. C., Furlow, J., Baethgen, W., Thomson, M. C. and Mason, S. J. 2019. On the use and misuse of climate change projections in international development. *WIREs Climate Change* 10(3), e579.

Pasquel, D., Roux, S., Richetti, J., Cammarano, D., Tisseyre, B. and Taylor, J. A. 2022. A review of methods to evaluate crop model performance at multiple and changing spatial scales. *Precision Agriculture* 23(4), 1489–1513.

Pearl, J. 2019. The seven tools of causal inference, with reflections on machine learning. *Communications of the ACM* 62(3), 54–60.

Peng, B., Guan, K., Tang, J., Ainsworth, E. A., Asseng, S., Bernacchi, C. J., Cooper, M., Delucia, E. H., Elliott, J. W., Ewert, F., Grant, R. F., Gustafson, D. I., Hammer, G. L., Jin, Z., Jones, J. W., Kimm, H., Lawrence, D. M., Li, Y., Lombardozzi, D. L., Marshall-Colon, A., Messina, C. D., Ort, D. R., Schnable, J. C., Vallejos, C. E., Wu, A., Yin, X. and Zhou, W. 2020. Towards a multiscale crop modelling framework for climate change adaptation assessment. *Nature Plants* 6(4), 338–348.

Phillips, S. J. and Dudík, M. 2008. Modeling of species distributions with Maxent: New extensions and a comprehensive evaluation. *Ecography* 31(2), 161–175.

Pongratz, J., Dolman, H., Don, A., Erb, K. H., Fuchs, R., Herold, M., Jones, C., Kuemmerle, T., Luyssaert, S., Meyfroidt, P. and Naudts, K. 2018. Models meet data: Challenges and opportunities in implementing land management in Earth system models. *Global Change Biology* 24(4), 1470–1487.

Raimondo, M., Nazzaro, C., Marotta, G. and Caracciolo, F. 2021. Land degradation and climate change: Global impact on wheat yields. *Land Degradation and Development* 32(1), 387–398.

Ray, D. K., West, P. C., Clark, M., Gerber, J. S., Prishchepov, A. V. and Chatterjee, S. 2019. Climate change has likely already affected global food production. *PLoS ONE* 14(5), e0217148.

Reinermann, S., Asam, S. and Kuenzer, C. 2020. Remote sensing of grassland production and management–A review. *Remote Sensing* 12(12), 1949.

Robinson, S., Mason-D'Croz, D., Sulser, T., Islam, S., Robertson, R., Zhu, T., Gueneau, A., Pitois, G. and Rosegrant, M. W. 2015. The international model for policy analysis of agricultural commodities and trade (Impact): Model description for version 3. *SSRN Electronic Journal*.

Rochedo, P. R. R. 2016. *Development of a global integrated energy model to evaluate the Brazilian role in climate change mitigation scenarios*. PhD Thesis, PPE-COPPE/UFRJ, 08.

Röös, E., Bajželj, B., Smith, P., Patel, M., Little, D. and Garnett, T. 2017. Greedy or needy? Land use and climate impacts of food in 2050 under different livestock futures. *Global Environmental Change* 47, 1–12.

Rose, D. C. and Chilvers, J. 2018. Agriculture 4.0: Broadening responsible innovation in an era of smart farming. *Frontiers in Sustainable Food Systems* 2, 87.

Rosenzweig, C. and Hillel, D. 2012. Conclusion agricultural solutions for climate change at global and regional scales. In: *Handbook of Climate Change and Agroecosystems: Global and Regional Aspects and Implications–Joint Publication with the American Society of Agronomy*. Crop Science Society of America, and Soil Science Society of America, pp. 281–292.

Rosenzweig, C., Jones, J. W., Hatfield, J. L., Ruane, A. C., Boote, K. J., Thorburn, P., Antle, J. M., Nelson, G. C., Porter, C., Janssen, S., Asseng, S., Basso, B., Ewert, F., Wallach, D., Baigorria, G. and Winter, J. M. 2013. The Agricultural Model Intercomparison and Improvement Project (AgMIP): Protocols and pilot studies. *Agricultural and Forest Meteorology*. The Agricultural Model Intercomparison and Improvement Project 170, 166–182.

Rosenzweig, C., Mutter, C. Z., Ruane, A. C., Contreras, E. M., Boote, K. J., Valdivia, R. O., Houtkamp, J. M., MacCarthy, D. S., Claessens, L. F. G., Adhikari, R., Durand, W., Homann-Kee Tui, S., Ahmad, A., Subash, N., Vellingiri, G. and Nedumaran, S. 2021. AgMIP regional integrated assessments: High-level findings, methods, tools, and studies (2012–2017). In: *Handbook of Climate Change and Agroecosystems* (Vol. 5). World Scientific Publishing.

Rötter, R. P., Hoffmann, M. P., Koch, M. and Müller, C. 2018. Progress in modelling agricultural impacts of and adaptations to climate change. *Current Opinion in Plant Biology* 45(B), 255–261.

Ruane, A. C., Rosenzweig, C., Asseng, S., Boote, K. J., Elliott, J., Ewert, F., Jones, J. W., Martre, P., McDermid, S. P., Müller, C., Snyder, A. and Thorburn, P. J. 2017. An AgMIP framework for improved agricultural representation in integrated assessment models. *Environmental Research Letters* 12(12), 125003.

Ruane, A. C. and Rosenzweig, C. 2018. Climate change impacts on agriculture. In: *Agriculture & Food Systems to 2050: Global Trends, Challenges and Opportunities*, pp. 161–191.

Shepherd, M., Turner, J. A., Small, B. and Wheeler, D. 2020. Priorities for science to overcome hurdles thwarting the full promise of the 'digital agriculture' revolution. *Journal of the Science of Food and Agriculture* 100(14), 5083–5092.

Silva, J. V. and Giller, K. E. 2020. Grand challenges for the 21st century: What crop models can and can't (yet) do. *The Journal of Agricultural Science* 158(10), 794–805.

Snow, V. O., Rotz, C. A., Moore, A. D., Martin-Clouaire, R., Johnson, I. R., Hutchings, N. J. and Eckard, R. J. 2014. The challenges–and some solutions–to process-based modelling of grazed agricultural systems. *Environmental Modelling and Software* 62, 420–436.

Spijker, J., Fraters, D. and Vrijhoef, A. 2021. A machine learning based modelling framework to predict nitrate leaching from agricultural soils across the Netherlands. *Environmental Research Communications* 3(4), 045002.

Stehfest, E., Bouwman, L., Van Vuuren, D. P., Den Elzen, M. G. J., Eickhout, B. and Kabat, P. 2009. Climate benefits of changing diet. *Climatic Change* 95(1–2), 83–102.

Stehfest, E., van Vuuren, D., Bouwman, L. and Kram, T. 2014. *Integrated Assessment of Global Environmental Change With IMAGE 3.0: Model Description and Policy Applications*. Netherlands Environmental Assessment Agency (PBL).

Tao, F., Zhang, L., Zhang, Z. and Chen, Y. 2022. Designing wheat cultivar adaptation to future climate change across China by coupling biophysical modelling and machine learning. *European Journal of Agronomy* 136, 126500.

Tatsumi, K., Yamashiki, Y., Valmir da Silva, R., Takara, K., Matsuoka, Y., Takahashi, K., Maruyama, K. and Kawahara, N. 2011. Estimation of potential changes in cereals production under climate change scenarios. *Hydrological Processes* 25(17), 2715-2725.

Teixeira, E. I., de Ruiter, J., Ausseil, A. G., Daigneault, A., Johnstone, P., Holmes, A., Tait, A. and Ewert, F. 2018. Adapting crop rotations to climate change in regional impact modelling assessments. *Science of the Total Environment* 616–617, 785–795.

Tittonell, P., Piñeiro, G., Garibaldi, L. A., Dogliotti, S., Olff, H. and Jobbagy, E. G. 2020. Agroecology in large scale farming—A research agenda. *Frontiers in Sustainable Food Systems* 4, 584605.

Van Meijl, H., Havlik, P., Lotze-Campen, H., Stehfest, E., Witzke, P., Domínguez, I. P., Bodirsky, B. L., van Dijk, M., Doelman, J., Fellmann, T., Humpenöder, F., Koopman, J. F. L., Müller, C., Popp, A., Tabeau, A., Valin, H. and van Zeist, W. 2018. Comparing impacts of climate change and mitigation on global agriculture by 2050. *Environmental Research Letters* 13(6), 064021.

Verburg, P. H., Alexander, P., Evans, T., Magliocca, N. R., Malek, Z., Rounsevell, M. D. and van Vliet, J. 2019. Beyond land cover change: Towards a new generation of land use models. *Current Opinion in Environmental Sustainability* 38, 77–85.

Wang, E., Brown, H. E., Rebetzke, G. J., Zhao, Z., Zheng, B. and Chapman, S. C. 2019. Improving process-based crop models to better capture genotype× environment× management interactions. *Journal of Experimental Botany* 70(9), 2389–2401.

Warszawski, L., Frieler, K., Huber, V., Piontek, F., Serdeczny, O. and Schewe, J. 2014. The inter-sectoral impact model intercomparison project (ISI-MIP): Project framework. *Proceedings of the National Academy of Sciences of the United States of America* 111(9), 3228–3232.

Wei, Y. M., Han, R., Liang, Q. M., Yu, B. Y., Yao, Y. F., Xue, M. M., Zhang, K., Liu, L. J., Peng, J., Yang, P., Mi, Z. F., Du, Y., Wang, C., Chang, J., Yang, Q., Yang, Z., Shi, X., Xie, W., Liu, C., Ma, Z., Tan, J., Wang, W., Tang, B., Cao, Y., Wang, M., Wang, J., Kang, J., Wang, K. and Liao, H. 2018. An integrated assessment of INDCs under shared socioeconomic pathways: An implementation of C3IAM. *Natural Hazards* 92(2), 585–618.

Weindl, I., Lotze-Campen, H., Popp, A., Müller, C., Havlík, P., Herrero, M., Schmitz, C. and Rolinski, S. 2015. Livestock in a changing climate: Production system transitions as an adaptation strategy for agriculture. *Environmental Research Letters* 10(9), 094021.

Wheeler, T. and Von Braun, J. 2013. Climate change impacts on global food security. *Science* 341(6145), 508–513.

Wise, M., Calvin, K., Kyle, P., Luckow, P. and Edmonds, J. A. E. 2014. Economic and physical modeling of land use in GCAM 3.0 and an application to agricultural productivity, land, and terrestrial carbon. *Climate Change Economics* 5(2), 1450003.

Yu, L., Ahrens, B., Wutzler, T., Schrumpf, M. and Zaehle, S. 2020. Jena soil model (JSM v1.0; revision 1934): A microbial soil organic carbon model integrated with nitrogen and phosphorus processes. *Geoscientific Model Development* 13(2), 783–803.

Zaehle, S., Medlyn, B. E., De Kauwe, M. G., Walker, A. P., Dietze, M. C., Hickler, T., Luo, Y., Wang, Y. P., El-Masri, B., Thornton, P., Jain, A., Wang, S., Warlind, D., Weng, E., Parton, W., Iversen, C. M., Gallet-Budynek, A., McCarthy, H., Finzi, A., Hanson, P. J., Prentice, I. C., Oren, R. and Norby, R. J. 2014. Evaluation of 11 terrestrial carbon–nitrogen cycle models against observations from two temperate free-air CO_2 enrichment studies. *New Phytologist* 202(3), 803–822.

Part 3

Modelling climate change impacts on regional agricultural systems

Chapter 12

Modelling the impact of climate change on agriculture in Europe

Mirek Trnka and Rudolf Brázdil, Global Change Research Institute CAS, Czech Republic; Lorenzo Brilli, Institute for BioEconomy (IBE) – National Research Council (CNR), Italy; Hannah C. von Czettritz und Neuhaus, Leibniz Centre for Agricultural Landscape Research (ZALF), Germany; Sergi Costafreda-Aumedes, Institute for BioEconomy (IBE) – National Research Council (CNR), Italy; Petr Dobrovolný, Petr Holub and Karel Klem, Global Change Research Institute CAS, Czech Republic; Matthias Kuhnert, University of Aberdeen, UK; Luisa Leolini, University of Florence, Italy; Marco Moriondo, Institute for BioEconomy (IBE) – National Research Council (CNR), Italy; Nina Muntean, Czech University of Life Sciences Prague, Czech Republic; Jørgen Eivind Olesen, Aarhus University, Denmark; Markéta Poděbradská, Vera Potopová, Petr Štěpánek and Otmar Urban, Global Change Research Institute CAS, Czech Republic; and Peter Zander and Claas Nendel, Leibniz Centre for Agricultural Landscape Research (ZALF), Germany

1 Introduction
2 Past climate and future projections for Europe
3 Agricultural systems across Europe
4 Agricultural system modelling: models and platforms
5 Modelling and experimental work on impacts of climate change: case studies
6 Future trends and conclusion
7 Where to look for further information
8 Acknowledgements
9 References

1 Introduction

1.1 Europe

Europe is recognised as the western part of Eurasia, situated on the Afro-Eurasia plate. It is bordered by the Arctic Ocean to the north, the Atlantic Ocean to the west, the Mediterranean Sea to the south and Asia to the east. The eastern border is composed of the watershed of the Ural Mountains, the Ural River, the

http://dx.doi.org/10.19103/AS.2022.0115.14

northern shoreline of the Caspian Sea, the Caucasus Mountains, the southern shoreline of the Black Sea and the Turkish Straits. Considered as a continent of its own, Europe covers about 10.2 million km^2 or 6.8% of the global land area. Europe is divided into 47 sovereign states, although concepts exist in which Georgia, Azerbaijan and even Armenia are fully accounted to Europe. Europe has a total population of about 744 million (UN, 2022), with Russia contributing the majority (104 million) followed by Germany (82 million), UK (66 million) and France (65 million). Eight smaller regions have a politically autonomous status, while four other regions are currently a subject of legal disputes.

1.2 Climate

The European climate is classified as mostly temperate. Western Europe is under the influence of the Atlantic Ocean and the Gulf Stream and mostly eastward-moving weather patterns, bringing warm summers, cool winters and often cloudy conditions with ample rainfall to the coastal areas and up to ~500 km further inlands. Southern Europe has a distinctively Mediterranean climate (Köppen climate classification Cs a/b), which features often hot and dry summers and rather mild winters, with considerably less precipitation when compared to winters in Western Europe. Towards the east, continental climate takes over, featuring hot summers, but cold winters, with a high annual temperature amplitude. A shift of these zones has already been observed in recent decades as a result of ongoing climate change (Fig. 1). Most parts of Europe experience four distinct seasons, while in the Mediterranean Basin, a wet and dry season pattern prevails, with a rainy season spanning roughly from October to February. These seasons govern agricultural practices and are the most fundamental reason for the current pattern of cultivation and the traditions, rules and techniques that have developed here over the course of ~9000 years, when early farmers settled in the Thessalia lowlands (Greece).

Climate change is about to influence the geographical distribution of the most frequent adverse weather events relevant for agriculture (Fig. 2). Excessive rainfall is expected to remain the dominant limitation for crop and fodder production over much of the northwestern coastal area of the continent and the British Isles, while snow continuously limits production in the northeastern and eastern part of the domain and in Alpine areas. Drought is projected to become the major limiting factor for crop and grass production over most of the Mediterranean region, with heat stress dominating only small areas in Turkey and Central Europe. Low temperatures are the leading cause of adverse weather events in parts of Scandinavia and Central Europe. According to our estimates of future climatic conditions, overly wet conditions will become an increasing problem across northwestern Europe, whereas snow-related

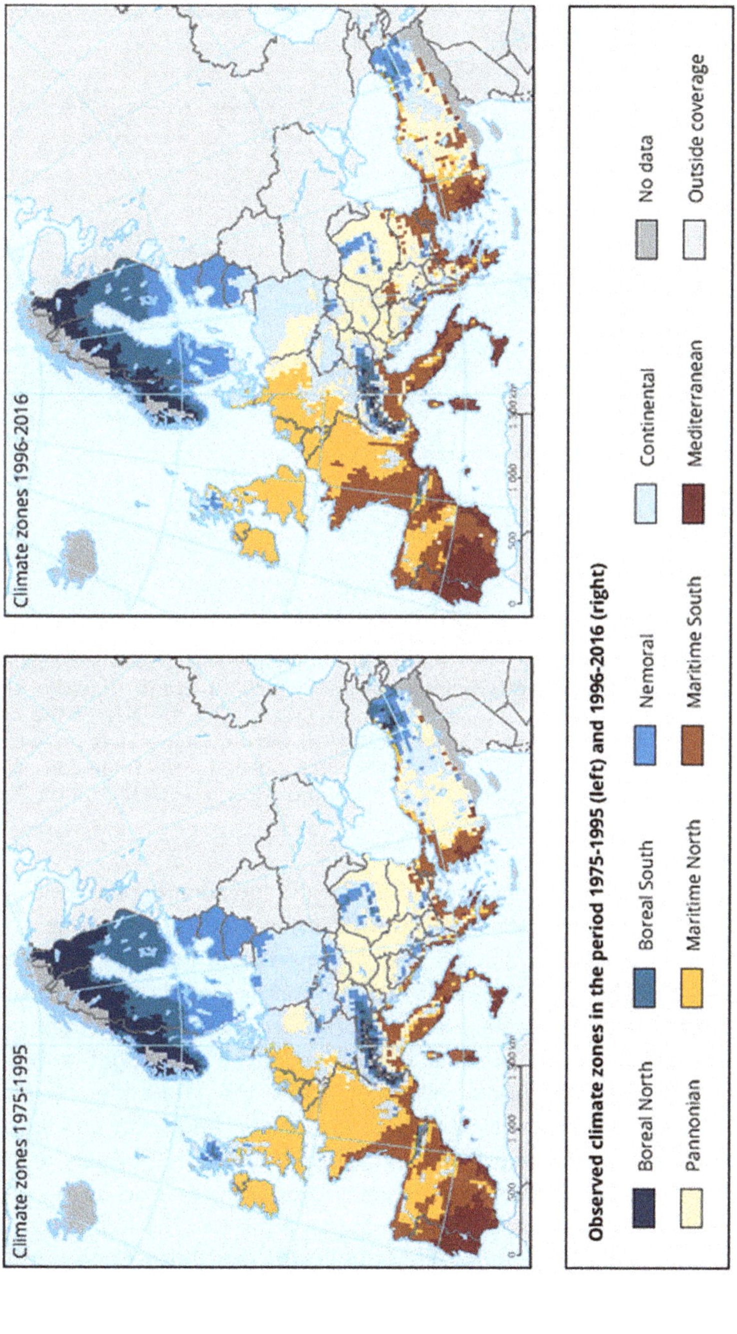

Figure 1 Observed climate zones in the period 1975–95 (left) and 1996–2016 (right). Source: European Environmental Agency, based on Ceglar et al. (2019). Different from the Köppen approach, this zonation is derived from season length and active temperature sum.

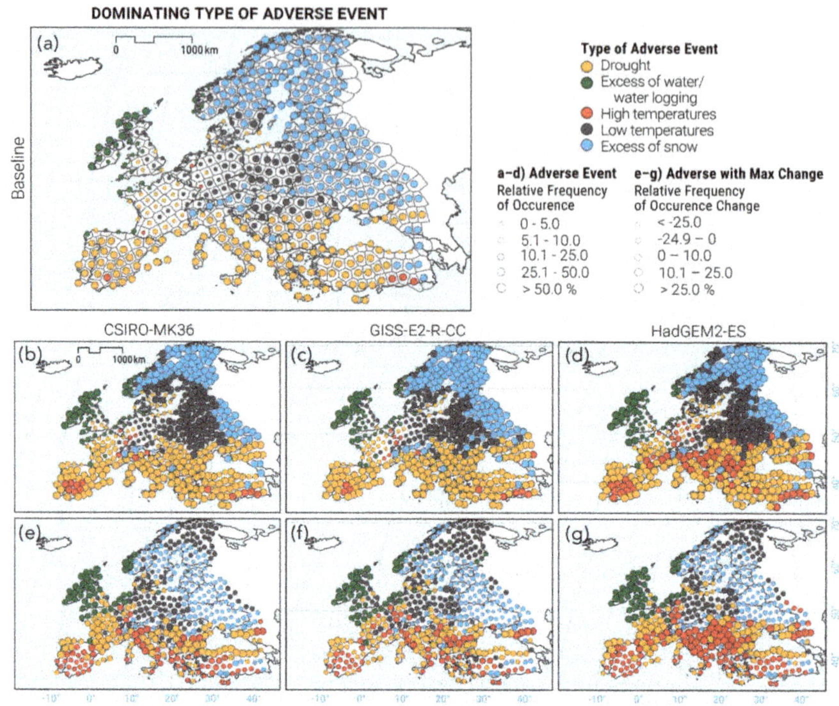

Figure 2 Dominant type of adverse events for croplands and grasslands and their expected frequency for (a) the baseline period (1981–2010) and (b–d) under future (2041–60) climate conditions according to three GCMs, with the size of the circle corresponding to the event frequency. Panels (e–g) show the type of adverse event that exhibited the largest change compared to the baseline for the three considered GCMs: (b, e) CSIRO-RCP4.5, (c, f) GISS-RCP4.5 and (d, g) HadGEM-RCP4.5 (Trnka et al., 2020).

issues together with the occurrence of damaging low temperatures will likely be a major concern in eastern and northeastern Europe as well as in the Alpine region (Fig. 2). Even for the relatively modest RCP4.5 (Representative Concentration Pathway) emission scenario, three global climate models (GCMs) agree on a sharp increase in the probability of heat and drought stress (Trnka et al., 2020). A major increase in the dominance of drought- and heat-related adverse weather events will occur across the whole Mediterranean region, but these events are also likely to increase in frequency and severity for much of France, Germany, Denmark, southern Sweden, Hungary and the Czech Republic. There is also an increase in drought projected as the single most important adverse event in eastern parts of the UK. The high degree of correspondence among all three GCMs indicates that these projections are quite robust.

1.3 Importance of agriculture

Europe and North America are the two regions of the world, with the lowest rates of food insecurity, with approximately 6.5% of the European population being moderately and 1.5% being severely food insecure (FAO, 2022). Despite the fact that agriculture is not the driving engine of Europe's economy, agricultural land represents the most important form of land use, accounting for almost 48% of the European land area. Europe, despite its relatively small share of global agricultural land total area (9.8%), has been one of the world's largest and most productive suppliers of food and fibre. Europe accounted for 17.6% of global cereal production during 2014–8, and average yields in European Union (EU) countries were 6.8% higher than the world average in the same period, particularly in Western Europe (FAOSTAT, 2021). In 2017, the EU's agricultural industry produced a total output value of €427 billion; 50% of this value came from crops, 40% from livestock and the remainder from secondary activities. Within the EU, the share of agriculture in overall employment is approximately 4%, with the relative share being higher for most East and South European countries. However, in the EU alone agriculture provides an estimated 20.5 million full- and part-time jobs, but the workforce demographics looms as a considerable challenge over the entire sector, with only 5.6% of all European farms being run by farmers younger than 35 years, while more than 31% of all farmers are older than 65 years.

1.4 The focus of this chapter

With agriculture and its primary production being seen as a sector of waning importance over the past centuries, it has been rising back to prominence over the last decade. First of all, the huge influence of agriculture on the European landscape means that the quality of many key ecosystem services beyond food production (e.g. water provisioning, biodiversity, recreational and cultural services) is intertwined with the status, financial means and motivation of farmers. At the same time, agriculture may provide, at least to some extent, a means of mitigating the looming climate crisis through its ability to influence net greenhouse gases (GHG) emissions on a large scale through so-called carbon farming (sequestering atmospheric carbon) while contributing significantly to lowering N_2O emissions from crop and livestock production. The following sections describe the current state of European agricultural systems, the expected climate change projections for the region and how various modelling approaches use these projections to estimate climate change impacts on agricultural production in Europe.

2 Past climate and future projections for Europe

Long-term climate variability in Europe can be characterised by different pan-European or regional climate reconstructions based on various proxies (e.g. tree rings, pollen and lake sediments), documentary and instrumental data. Particularly valuable among them are those combining different data sources, also referred to as multi-proxy reconstructions. An example of multi-proxy reconstruction is the reconstruction of June, July and August (JJA) temperatures across Europe by Luterbacher et al. (2016), presented here for the 1501–2003 period and expressed as temperature anomalies with respect to the 1961-90 reference period (Fig. 3a). JJA temperatures show high interannual variability, with the absolute temperature range reaching 4°C with the coldest year 1821 (−1.7K below the mean value of the period 1961-90) and the warmest year 2003 (+2.3K above the mean). The coldest climate defined for the running 30-year periods occurred during the last decades of the sixteenth century and beginning of the seventeenth century (1579–1608) during the Little Ice Age period with mean anomaly of -0.8K. JJA temperatures demonstrate generally increasing tendency during the whole analysed period. The period of recent global warming is partly reflected by the higher temperatures in the 1940s and particularly by the steep increase from the late 1970s (1974–2003 being the warmest 30-year interval according to this reconstruction with a mean anomaly of +0.4K above the mean value of the period 1961-90). This increase has continued in recent years, as documented, for example, by the Central European JJA series reconstructed from documentary evidence (Dobrovolný et al., 2010). The Central European series also demonstrates that fluctuations in the regional series compared to the pan-European scale show higher variability and confirms dramatic temperature increases in the past decades, Compared to the coldest and warmest 30-year periods, the higher variability of JJA temperatures in the warmer period (Fig. 3b) is remarkable. Looking at the spatial distribution of JJA temperature anomalies over the European territory in 1579–1608, the most intense cooling appeared particularly in the southwest of Central Europe and in Scandinavia, while the weakest cooling occurred in a broad belt extending from southeastern Europe and the Black Sea area to the north up to Belarus (Fig. 3c). In the warmest summers of 1974–2003, the highest warming occurred in the belt between latitudes of 45°N and 55°N (Fig. 3d).

In order to provide a general overview of the expected temperature and precipitation shift, two periods are shown: 2041-60 and 2071-90. Air temperature and precipitation simulated for future climate were compared to the reference period 1981-2010, using emission scenarios SSP2-45 and SSP8-85 (Riahi et al., 2017). For further processing, only models with 100-km resolution were selected (CMIP6 HighResMIP cover only the 2015-50 period and scenario SSP5-8.5), with CMIP6 being described by O'Neill et al. (2016).

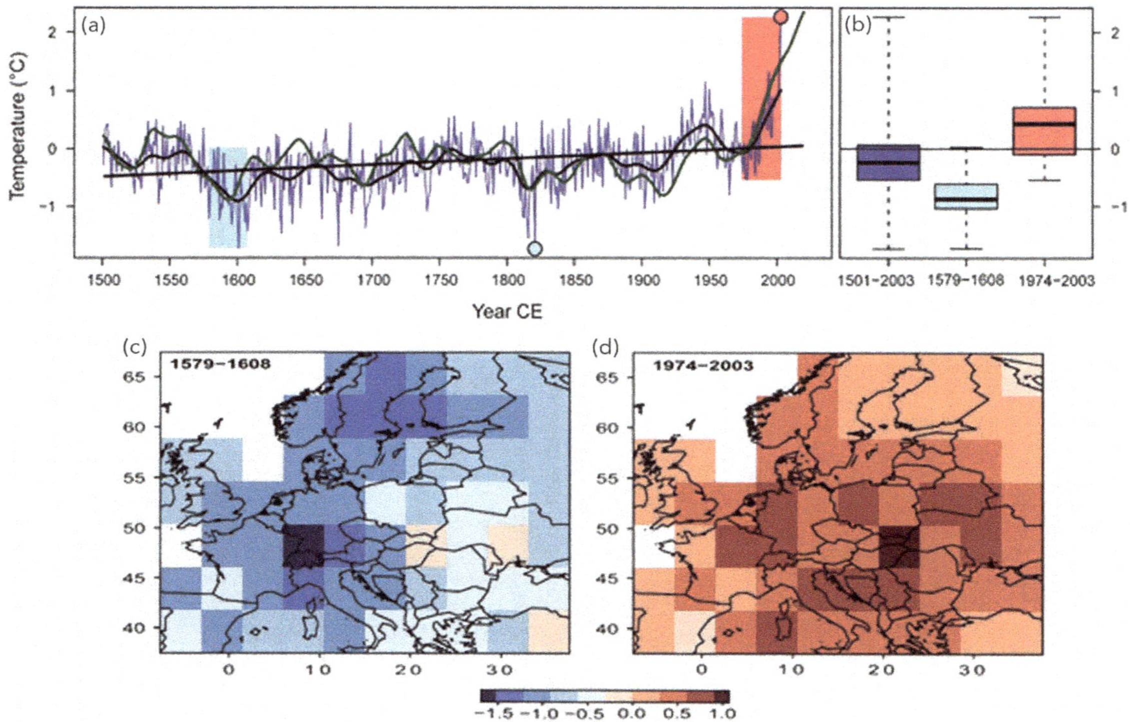

Figure 3 European summer temperatures in the 1501–2003 CE period, expressed as anomalies with respect to the 1961–90 reference period (Luterbacher et al., 2016): (a) annual fluctuations smoothed by 30-year Gaussian filter, complemented by coloured bands representing extreme 30-year periods (in blue and red) with the lowest and highest values portrayed by small circles (green line – summer temperatures for Central Europe smoothed by 30-year Gaussian filter in the 1501–2020 CE period – Dobrovolný et al., 2010, complemented); (b) box plots (median, lower and upper quartile, minimum and maximum) for the whole series and the most extreme cool and warm 30-year periods represented by blue and red colours, respectively; (c) spatial patterns of summer temperature anomalies in the coldest 1579–1608 period; (d) spatial patterns of summer temperature anomalies in the warmest 1974–2003 period.

In this particular analysis, the processing area was continental Europe (including the British Isles), with the area defined to the east by 45th longitude and including the whole of Turkey. The results for both Shared Socioeconomic Pathway (SSP) scenarios and two time periods are shown in Tables 1 and 2 and in Fig. 4 (temperature) and Fig. 5 (precipitation).

Regarding averaged values over the whole of Europe (area limited by the 45th longitude), compared to the reference period 1981–2010, air temperature is expected to increase in the near future period 2041–60 for the whole year by 2.3°C for SSP2-45 and by 2.9°C for SSP5-85 while, for the distant future period (2071-2090), air temperature is expected to increase by 3.1°C (SSP2-45) and 4.8°C (SSP5-85). The highest increase of temperature is expected in the June to Sugust JJA season (5.4°C for SSP5-85 and far future), followed by September to November (SON) and December to February (DJF) and the lowest is expected in the March to May (MAM) season (4.4°C for the same scenario and far future).

There are pronounced differences in regional climate change, as both Figs 4 and 5 indicate. In the case of annual average air temperature, the lowest values (in terms of increase in air temperature) are found in the British Isles but, in general, they are increasing from the Iberian Peninsula to Finland (from southwest to northeast) from about 2°C (in average of used CMIP6 models, 1°C as the lower limit of uncertainty) to 4°C for SSP2-45 and far future (2071–90) compared to the reference period 1981–2010. For winter (DJF months), the climate change also varies in the case of annual average temperatures, increasing from southwest to northeast but with a wider range of values (spanning from 1.5°C to 4.5°C, with uncertainty from 0.5°C in British Isles to almost 6.5°C in Kola Peninsula in northern Russia). In contrast, increases in air temperature in summer (JJA months) change with latitude, together with notable effects of cooling seas along coasts. For inland regions, increases in summer temperatures vary, depending on averages between models, from 5°C in Southern Europe to 3°C in Northern Europe (some models indicate even 6.5°C as maxima on the South and 2°C as minima on the North).

Compared to the reference period 1981–2010, precipitation will increase (Table 2), on average for the whole of Europe (area limited by the 45th longitude), in the near future period 2041–60 for the whole year by 5% for SSP2-45 and by 6% for SSP5-85, while in the distant future the increase will be by 4% (both SSP2-45 and SSP5-85). The highest increase in precipitation is expected in the DJF season by 17% for SSP5-85 and far future, followed by MAM and SON, and the decrease is expected in the JJA season (decrease by 16% for the same scenario and far future). In Table 1, values of uncertainty in the form of minimum and maximum values, taken from the processed models, are indicated in parenthesis.

For SSP5-85 and the 2071–90 time period, the annual precipitation is expected to change with latitude, from 20% decrease in the South (some

Table 1 Climate change for air temperature, average values of differences over whole Europe for a whole year and individual seasons [values of uncertainty (minimum and maximum of all models) are indicated in parenthesis]

SCENARIO	PERIOD	YEAR	DJF	MAM	JJA	SON
SSP245	2041-60 vs. 1981-2010	2.3 (1.6-3.3)	2.3 (1.5-3.4)	2.2 (1.4-3.4)	2.5 (1.5-3.5)	2.3 (1.3-3.4)
SSP245	2071-90 vs. 1981-2010	3.1 (2.1-4.3)	3.1 (1.7-4.2)	2.8 (1.7-4.0)	3.5 (2.2-5.1)	3.2 (2.1-4.4)
SSP585	2041-60 vs. 1981-2010	2.9 (1.9-4.2)	2.8 (1.7-4.6)	2.5 (1.5-3.8)	3.2 (2.0-4.6)	2.9 (2.0-4.4)
SSP585	2071-90 vs. 1981-2010	4.8 (3.7-6.6)	4.7 (3.2-7.2)	4.4 (3.1-6.3)	5.4 (3.9-7.2)	4.9 (3.5-6.6)

Table 2 Climate change for precipitation, average values of ratios over whole Europe for a whole year and individual seasons [values of uncertainty (minimum and maximum of all models) are indicated in parenthesis]

SCENARIO	PERIOD	YEAR	DJF	MAM	JJA	SON
SSP245	2041–60 vs. 1981–2010	1.05 (0.98–1.12)	1.09 (0.98–1.19)	1.09 (0.98–1.21)	0.99 (0.83–1.16)	1.03 (0.93–1.17)
SSP245	2071–90 vs. 1981–2010	1.06 (0.99–1.16)	1.13 (1.01–1.26)	1.10 (0.98–1.25)	0.95 (0.76–1.16)	1.05 (0.91–1.20)
SSP585	2041–60 vs. 1981–2010	1.04 (0.96–1.15)	1.10 (0.98–1.24)	1.09 (0.97–1.23)	0.95 (0.76–1.15)	1.03 (0.89–1.16)
SSP585	2071–90 vs. 1981–2010	1.04 (0.95–1.16)	1.17 (1.01–1.36)	1.11 (0.97–1.30)	0.84 (0.65–1.07)	1.02 (0.89–1.15)

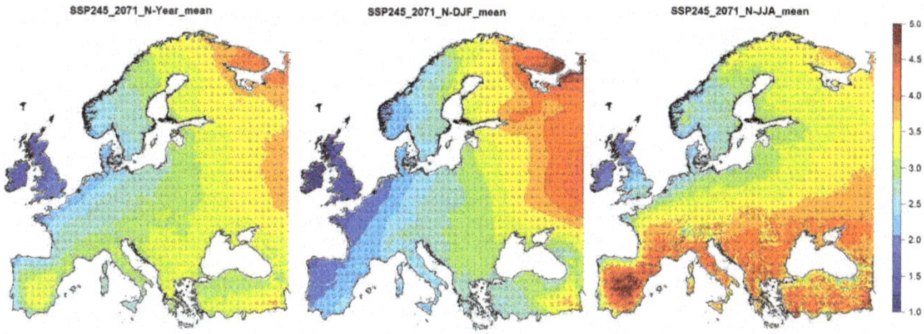

Figure 4 Average over CMIP6 GCM simulations for Europe, SSP2-45 scenario, differences between periods 2071-90 and the reference period 1981-2010 (for a whole year and DJF and JJA seasons).

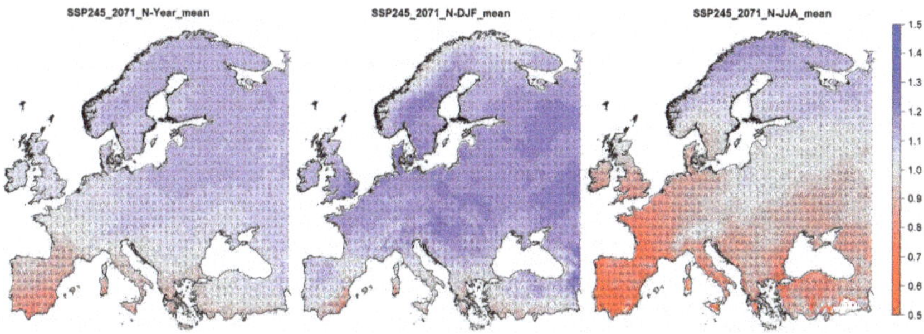

Figure 5 Average over CMIP6 GCM simulations for Europe, SSP5-85 scenario, ratios between periods 2071-90 and the reference period 1981-2010 (for a whole year and DJF and JJA seasons).

models indicating even -30% in southern Portugal) to +10% on the North (upper uncertainty limit +30%), with a transient zone between latitudes 44 and 50 on the west (France), dropping to latitudes 36 and 41 on the east (Turkey). For the DJF season, more precipitation is expected (compared to the reference period 1981-2010), with transient zone being situated on latitude 42, the most southern parts of Europe with –10% of precipitation in far future and for SSP2-45, and +20% in the northern part of Europe (with, according to some models, up to +40% precipitation in some areas). In contrast, for the JJA season, most of Europe is projected to experience a decrease in precipitation, with a transient zone around 57th latitude, the most northern parts of Fennoscandia having +10% of precipitation (+30% upper limit of uncertainty), dropping to -20% in the Pyrenean peninsula (-40% according to some models).

3 Agricultural systems across Europe

European farmers produce a wide range of food products in response to Europe's diverse climatic and geographic conditions. Therefore, agricultural systems across Europe vary widely in terms of their characteristics, production, intensity, actors involved and the challenges they face. Agricultural systems range from extensive ruminant to feed-reliant intensive systems and from systems characterised by large-field (>1000 ha) industrial crop production, including monocultures, to integrated crop-livestock systems and small-scale organic farming. The area in Europe that is used for organic agriculture has been increasing over the past decades, and in 2017, it covered more than 14 million hectares (European Environmental Agency, 2023). The most specialised agricultural systems are found in lowland Western Europe, while more extensive systems are located in southern, eastern and mountainous regions. Agriculture represents around 48% of the European land area, with about 62% of the area used as croplands, shaping the character of the landscape in rural areas (FAOSTAT, 2021). Grasslands cover about 304 million hectares of Europe's land area. The proportion of arable land and permanent grassland as a percentage of the total utilised agricultural area is highly region specific, ranging from 99% and 89% of arable land in Finland and Denmark to 98% and 89% of permanent grassland in Iceland and Ireland, respectively (Eurostat, 2016).

In the last two decades, the amount of land used for agricultural production has remained relatively stable. However, there has been a sharp reduction in farm numbers, indicating a trend to land consolidation and a bigger proportion of larger holdings. In the EU, a small number (3% of all farms) of very large farms (over 100 ha) operate on more than half of the utilised agricultural area, while very small farms represent 69% of all holdings (Eurostat, 2016). Over 22 million people work regularly in the agricultural sector (numbers are representative of the EU in 2013), but the full-time work equivalent is much smaller as the sector is often characterised by seasonal and part-time workers. The demographics of European farmers is highly skewed, 31% of farmers are 65 years or older, while only 6% of farmers are 35 years or younger. The lack of successors and the insufficient availability of qualified labour (both seasonal and permanent) present one of the challenges the European agricultural sector faces. This challenge is exacerbated by many other uncertainties, such as changing climate and impact of extreme events, controversies about current, mainstream agricultural practices and interruptions to trade chains (caused, for example, by coronavirus disease-19, Brexit and the war in Ukraine).

Provision of food remains the primary function of Europe's agricultural systems, although they also play an important role in rural development, landscape management, and achieving the objectives of EU's Biodiversity Strategy and the UN Sustainable Development Goals. The agricultural sector has a complex relationship with the environment, and it has various direct and indirect impacts on it and is a major user of natural resources (OECD, 2017). Even though the overall contribution of agriculture to Europe's GDP is fairly modest, it plays an important role in rural development and management of natural resources. Post-war agricultural policies aimed to boost productivity through technological advancements and resulted in increased yields and more intensive production methods. This had a positive impact on Europe's food self-sufficiency but it has also led to increased pressure on biodiversity (Tilman et al., 2001), which is expected to continue and even intensify in the future (Petit et al., 2001). Since 1962, the EU agricultural sector has been operating under the Common Agricultural Policy (CAP). CAP-specific objectives change over time, but generally allow farmers to fulfil their functions in society in the areas of food production, rural community development and environmentally sustainable farming. Since January 2023, a new CAP aims to secure the future of European agriculture and forestry while also achieving the objectives of the European Green Deal.

Europe is the world's largest producer of wine, olive oil and tomatoes. The major European producers of wine are Italy, France and Spain; however, wine production also has a regional importance in other European countries. The EU member countries alone account for nearly half of the global wine-growing area, 64% of global production and 48% consumption. Europe is also the leading producer (nearly 70% of global production) and consumer (over 50% of global consumption) of olive oil, mostly clustered in the Mediterranean region. The main producers, while also being the main consumers, are Spain, Italy and Greece. Major tomato production areas are also located in the Mediterranean region. The main European producers are Turkey (12 Mt), Italy (6 Mt) and Spain (5 Mt, FAOSTAT, 2021). Besides these products, the EU is also the world's biggest producer of beet sugar (around 50% of global production). However, the majority of global sugar is produced from sugarcane (around 80%).

In the world context, Europe is also an important producer of cereals. Around 18% of global cereal is produced in Europe (FAOSTAT, 2021). The most important cereal is wheat, representing about 50% of the total cereal production. The most important wheat-producing countries are the Russian Federation (28% of Europe's production), France (13.5%) and Ukraine (12%). Maize and barley represent about one-third of the total cereal production. The major producers of maize are Ukraine (30%), France (11%), Italy (11%), the Russian Federation (11%) and Romania (10%), and the major producers of barley are the Russian Federation (20%), France (13%) and Germany (12%; FAOSTAT, 2021). The remaining cereal production is a mix of smaller quantities of rye, oats and spelt. Rye is the primary candidate crop for marginal soils and climates, as it tolerates conditions where other cereals fail to produce. Cereal crops are primarily used for animal feed (approximately two-thirds) and about one-third of the production is for human consumption. Only small amounts of cereal production are used as biofuels. However, over the last decade, energy and climate policies have driven an increase in energy crop production.

Cereals, together with oilseeds and legumes, are the main constituents of European crop rotations. Well-designed crop rotation systems, compared to continuous monoculture production, general reduce system reliance on external inputs (e.g. fertilisers and disease- and pest-controlling chemicals) through the benefits of growing different crops in the same field year after year which improve soil physical, chemical, and biological properties (e.g. reducing soil erosion, improving soil structure and enhancing soil water retention capacity and organic matter content). The CAP recognises crop rotation as one of the key measures for achieving sustainability goals in European agriculture. Rotation systems are usually designed around one to three main crops that are followed by legumes or other crops. In European conventional agriculture, crop rotations typically last from 3 to 5 years, while longer rotations (5–10 years) are typically used in organic agriculture. In Europe, the type of crop rotation is dependent on the farming system as well as on biogeophysical properties, such as climate (Jänicke et al., 2022; Reumaux et al., 2023). In conventional arable farming systems crop rotations focus primarily on market/cash crops, especially cereals which are present in a high number of rotation sequences. In Europe, other crops may include root crops, oilseeds or maize but rarely legumes that are mainly used in humid or oceanic climate zones. Legumes are a more characteristic rotation crop in organic farming and in other farming systems, such as conventional cattle rearing systems or dairy farming systems (European Commission, 2020). In the Mediterranean region, where there are limited options for the selection of rotating crops, fallowing is often used in combination with crop rotation. An increasing demand for renewable energy sources can lead to more frequent addition of energy crops into the crop rotation systems in the near future (Zegada-Lizarazu and Monti, 2011). Intermediate crops for energy production, a typical feature of European cropping systems, are an important component of pesticide-free agriculture which enhances crop diversification. Capturing the spatiotemporal complexity and variability of crop production caused by crop rotation and other crop diversification practices has been the subject of current agricultural modelling research (e.g. Blickensdörfer et al., 2022; Kollas et al., 2015).

Dairy products are also an important production asset in Europe. Almost all milk that is produced in Europe is also processed there (very little is imported or exported). Livestock production occurs mostly in large-scale specialised livestock holdings, and since the 1980s there has been an increase in poultry, pig and veal production, while beef, sheep and goat production has been decreasing.

4 Agricultural system modelling: models and platforms

The history of using mathematical models to describe ecosystem behaviour has been largely influenced by early developments in the Netherlands and the USA.

Picking up differential equation systems that describe the movement of water (Luthin and Gaskell, 1950) and solutes (Biggar and Nielsen, 1962) in porous media, Wageningen researchers combined them with similar approaches to represent simple plant physiological processes (De Wit, 1970) to produce the first dynamic models to describe the interaction of crop growth to water availability in soils (Seligman and van Keulen, 1981; van Keulen et al., 1982). The growing awareness of nitrogen being a threat to groundwater quality led to the development of the first agro-ecosystem models that considered the turnover of organic matter in soils (Beek and Frissel, 1973) and management by the farmer (sowing, harvesting and fertilisation; Dutt et al., 1972; Hagin and Amberger, 1974). Parallel to developments in Eastern Europe (Unger, 1968) and in the USA (Jones et al., 2017), these model developments built the stock of algorithms that we still find functioning in many contemporary models (Nendel, 2010).

Some groups developing a particular model have been active in promoting them more widely, which is why some models are used by many scientists globally, including in Europe. Many groups in Europe today use the DSSAT (Jones et al., 2003) and APSIM (Keating et al., 2003) but also CropSyst (Stöckle et al., 2003), EPIC (Wang et al., 2012), STICS (Brisson et al., 2003), WOFOST (van Diepen et al., 1989), LINTUL (Spitters and Schapendonk, 1990), AquaCrop (Vanuytrecht et al., 2014) and MONICA (Nendel et al., 2011), of which the last five were developed in Europe. Through grassroots activities starting in the 1990s, modellers benefited from intercomparisons of their models for their own development, starting with single-community workshops (Diekkrüger et al., 1995; Kabat et al., 1995), European network groups (Maracchi et al., 2006; Eitzinger et al., 2009; Orlandini and Nejedlik, 2012) and later continuing with larger projects that drew largely on intrinsically motivated scientists rather than funding, such as AgMIP (Rosenzweig et al., 2013) and MACSUR (Rötter et al., 2013). While the efforts to improve individual models were not always successful, these activities introduced a new culture of sharing code and data benefitting the research community.

Today, agricultural modelling in Europe is developing along multiple different lines. Crop modelling on one hand focuses on plant physiological processes. Currently, crop modelling concentrates on simulating more complex cropping systems (mixed cropping, Malezieux et al., 2009; agroforestry, Kraft et al., 2021; crop-livestock systems, Le Gal et al., 2022) and the link between genetic traits of main staple crops and their response to environmental stresses (Hammer et al., 2019). On the other hand, agro-ecosystem modelling aims at representing the crop as part of a larger environmental system. Agro-ecosystem modellers incorporate the above-mentioned crop model enhancements for a better representation of water, C and N cycles in the context of climate change mitigation (GHG emissions and soil C sequestration) and large-scale assessments of climate impact and adaptation options. Farm economy models are conceptually different and use a range of different modelling approaches (further discussed in Section 5.3). These models simulate strategic decisions by farmers strategic decisions in the context of environmental, social and economic drivers. As farm economy models use crop yields as an input, this has led to couplings between farm and agro-ecosystem models, in which the latter also provide N fertiliser and irrigation water inputs to the economic models (e.g. Hampf et al., 2018; Kuhn et al., 2020; Schuler et al., 2020).

Coupling models to achieve representation of agricultural systems at higher spatial and organisational scales (farm and region) led to the development of model frameworks, which include a set of crop or agro-ecosystem models, farm economy models and market models. To be able to provide a useful framing for the farm-level economy and for decisions to be taken at a farm level, a coupling between the global-scale trade models, mostly implemented as computable general equilibrium models, and specialised agricultural sector models has been promoted (Delzeit et al., 2020). Economic sector models are addressed in detail elsewhere in this book. The SEAMLESS project (van Ittersum et al., 2008) pioneered such work, inspiring later developments of the RECORD platform (Bergez et al., 2013) in France, the MESSAGE*ix*-GLOBIOM framework (Fricko et al., 2017) in Austria and the SIMPLACE (Enders et al., 2010) and CASSIS (Berg-Mohnicke and Nendel, 2022) frameworks in Germany.

5 Modelling and experimental work on impacts of climate change: case studies

5.1 Crop yields

Among the variables that describe the effect of changing climate conditions on agriculture, crop yields have always been of crucial importance and interest. The prediction of crop yields under different management scenarios has evolved as an accepted alternative to field experiments. Such *in-silico* experiments inform more basic science on crop physiology and soil processes through hypothesis testing and increase the understanding of yield formation under variable environmental and management conditions. On the other hand, the prediction of yields under hypothetical conditions, such as the future under climate change, forms today an important pillar in policy support and decision-making. In this chapter, we present some examples, ranging from experimental work informing model development to the application of models to help answer key societal questions relating to agriculture.

5.1.1 On-the-ground experiments to inform agro-ecosystem models

Models can only be as good as the data they have been constructed with. A growing number of experiments deliver evidence for the various interactions of biophysical processes in the soil-plant-atmosphere continuum. However, the more factors are involved, the more complex and costly such experiments become. Here, we present one example, in which the CO_2 effect on crop growth is investigated under the influence of extreme weather conditions such as drought, high temperatures, UV radiation or limited nutrient availability. Such compound effects have been little studied so far, and experimental data to inform models are scarce.

One example of such an approach and part of the Analysis and Experimentation on Ecosystems (AnaEE) project is a multi-year experiment in automatic lamellar chambers (Fig. 6) which was established in the Moravian Highlands, Czech Republic, with winter wheat and spring barley. Elevated [CO_2] (EC, 700 ppm) stimulated wheat yield by more than 40% under optimal nitrogen and water supply. However, this stimulatory effect of EC gradually decreased when plant growth was limited by other factors (Fig. 7). Reduced water availability in the second half of the stem elongation phase decreased the positive CO_2 effect to below 40%. UV radiation reduced the positive CO_2

Figure 6 Lamellar open top chambers with cultivated winter wheat and spring barley under elevated CO_2 concentration, variable nitrogen and water availability and UV radiation, established in the Moravian Highlands, Czech Republic.

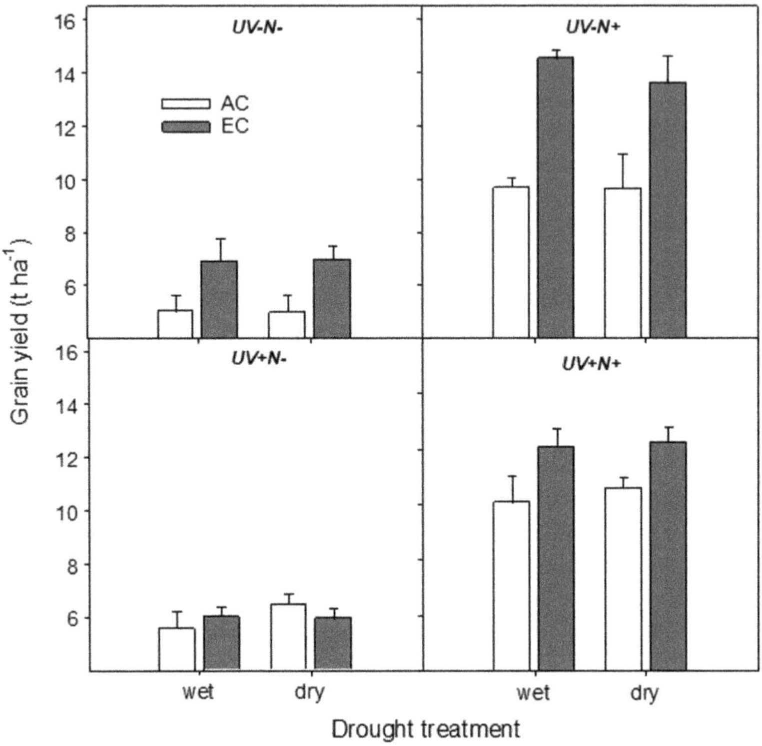

Figure 7 Effect of elevated CO_2 concentration on grain yield of winter wheat (variety Bohemia) and interactions with nitrogen and water availability and UV radiation. Means (columns) and standard deviations (error bars) are shown. AC denotes ambient CO_2 concentration (ca. 400 ppm) and EC denotes elevated CO_2 concentration (ca. 700 ppm). Two water availability treatments were applied during the second half of stem elongation: one well-watered (wet) and one reduced (dry). Optimum nitrogen dose (N+) was dimensioned to 200 kg N/ha and reduced nitrogen availability (N–) to 0 kg N/ha. UV+ denotes an ambient UV radiation treatment and UV– excluded UV-B and UV-A radiation.

effect to below 20%, and, in combination with reduced water availability, to below 15%. However, UV radiation alone did not have a negative impact on wheat yield in the control variant with an ambient [CO_2] of 400 ppm (AC). In contrast, limited nitrogen reduced both the yield under the AC and also the positive effect of EC. When drought, reduced nitrogen availability and UV exposure were combined, the effect of EC on wheat grain yield was even slightly negative. Similar results were also found for barley. EC reduced the grain protein content by about 10% under optimal growth conditions in both crops, and this effect was alleviated in the presence of other limiting factors. Reduced water availability had a rather variable effect on protein content, while in some cases it also enhanced the effect of EC on the protein content.

This experimental evidence of the complex behaviour of grain cereals to multiple stressors is a welcome source for further improving crop and agro-ecosystem models, especially for yield projections under climate change and establishing crop breeding targets and adaptation measures in farming.

5.1.2 Heat and drought stress outlook for wheat and maize

An understanding of what determines yields in the face of climate change is necessary for adaptation planning and responding to changing production risks. Webber et al. (2018) used an ensemble of crop models applied on a spatial grid to quantify contributions of various climatic drivers to past yield variability in grain maize and winter wheat of European cropping systems and drivers of climate change impacts for 2050. Their study reveals that for current genotypes and mix of irrigated and rainfed production, climate change would lead to yield losses for grain maize and gains for winter wheat. Across Europe, on average, heat stress does not increase for either crop in rainfed systems, while drought stress intensifies for maize only. In low-yielding years, drought stress persists as the main driver of losses for both crops, with EC offering no yield benefit in these years.

5.1.3 Crop rotations as basis for simulations in Central Europe

Crop rotation, fertilisation and residue management affect the water balance and productivity of cropping systems and are, in turn, sensitive to climate change. Pohanková et al. (2022) employed a crop model ensemble on two rotations with the same set of crops (winter wheat, silage maize, spring barley and winter rape) with different use of residues and cover crops. After initial performance tests (Kostková et al., 2021), simulations were performed on two virtual soil types within three sites in the Czech Republic representing different temperature and precipitation gradients. In general, the ensemble predicted higher productivity for C_3 cereals in the future (2051-80) when part of rotations prioritizing maintenance of soil organic matter, with ensemble models agreeing better for a Chernozem compared to a Cambisol soil type. For silage maize, simulated yields varied more across the sites and pointed to likely yield losses in the future, especially in the Czech lowlands. The results demonstrated the potential for higher biomass production from cover crops, but intensive rotation was associated with much higher water losses through evapotranspiration. Stella et al. (2019) modelled the effects of the management of crop residues on sequestration of carbon in soils in Germany. They found that a significant capture of carbon would only be possible on clayey soils with low organic matter levels. Faye et al. (2023) addressed a similar research question at the European scale but focused more on the sensitivity of climate change impact simulations to assumptions about annual model re-initialisation, specification of crop rotations and the amount of residue retained in fields.

5.1.4 The future of soybean in Europe

Rising global temperatures have painted an image of a bright future for soybean production in Europe, but emerging climatic risks such as a more frequent and severe drought occurrence have so far not been included in any of those outlooks. Nendel et al. (2023) presented a simulation study on future soybean production and the most prominent risk factors across Europe. Using an ensemble of five GCMs and four soybean growth models, they investigated how the distribution of recommendations for growing specific soybean maturity groups would shift northwards on the European continent, and what this would mean for the productivity and the potential production area. Exploiting the full potential of process-based simulation models, they also evaluated how the different weather-related risks that currently impede soybean production would change, especially in Central and Northern Europe. Their projections suggest a substantial increase in potential soybean production area and productivity in Central Europe, while Southern European production would become increasingly dependent on supplementary irrigation. Average productivity would increase by 8.3% (RCP4.5) to 8.7% (RCP8.5) as a result of improved growing conditions and farmers adapting to them by using cultivars with longer phenological cycles (Fig. 8). The suitable production area would increase by 31.4% (RCP4.5) to 37.7% (RCP8.5) by mid-century, contributing significantly more than increases in productivity potential, narrowing the protein gap in Europe. The models and climate data analysis anticipated that drought and heat will become the dominant limitations in the future. The authors

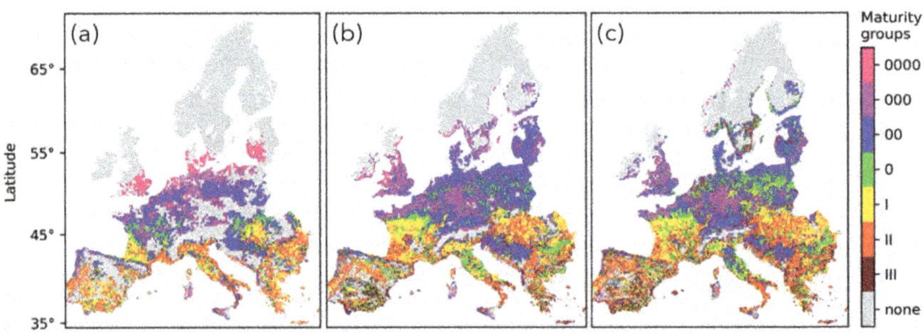

Figure 8 Soybean maturity group (0000 - extremely early; III - middle) that produced the highest yield in individual pixels in a simulation of an ensemble of four crop models (MONICA, HERMES, APSIM and STICS). (a) 1981-2010 baseline, (b) 2040-69 RCP4.5 scenario ensemble mean of five global circulation models and (c) 2040-69 RCP8.5 scenario ensemble mean of five global circulation models. Source: Reproduced from: Nendel et al. (2023).

specifically pointed to the fact that heat stress effects on soybean growth were not yet sufficiently covered by experimental evidence and contemporary agro-ecosystem models.

5.2 Changes in soil organic carbon in Europe

Soil organic carbon (SOC) is a crucial soil variable, which is used as an indicator for soil health and soil quality (Bünemann et al., 2018), affects soil physical properties (Wösten et al., 1999) and shows potential to mitigate climate change by storage of atmospheric CO_2 (Paustian et al., 2016). Climate change will impact SOC through all carbon fluxes associated with changes in the biomass production and changes in the environmental conditions that affect the decomposition and leaching (e.g. temperature, soil moisture and $[CO_2]$). The main fluxes affecting the SOC changes are the organic carbon input (mainly plant material in the form of remaining roots and residues left on the field or applied organic material by management activities, e.g. manure), soil respiration (CO_2 emissions to the atmosphere due to decomposition) and leaching (transport of organic carbon by water percolation in the soil).

Stergiadi et al. (2016) analysed all three fluxes for sandy soils in northwestern Europe using the model CENTURY. For the period 2013–2100, they projected a decrease of SOC for croplands, with the increased decomposition rate exceeding the impacts of the higher biomass production. The main driver for these changes was temperature, while precipitation modulated the dissolved organic carbon (DOC) leaching towards the groundwater. CENTURY predicted increasing decomposition but a decrease of DOC.

In contrast, Lugato et al. (2014) focused on analysing the impacts of different management options. Here, SOC changes were mainly driven by crop and soil management options, indicating some potential to maximise SOC through optimised management (Paustian et al., 2016). Lugato et al. (2014) used CENTURY to simulate the impacts by reduced tillage, optimised residue management, introducing ley cropping systems and cover crops to increase carbon sequestration. In addition, different land-use options (replacing arable with grassland and *vice versa*) were tested. The study showed that changes applied on 12–28 % of arable land area can sequester between 101 and 336 Mt CO_2-eq by 2020 and between 549 and 2141 Mt CO_2-eq by 2100. However, land-use change from arable to grassland showed the largest SOC change but also impaired food production. As climate change mitigation must balance with food security targets, such management changes require a wider analysis beyond SOC sequestration (Frank et al., 2017).

Spatial simulations of SOC suffer from a wide range of methodological problems. Lugato et al. (2014) highlighted that the variability of key variables and feedback relations were not considered in their study. This includes changes of bulk density, hydraulic properties or soil aggregation during the simulation run. Besides the model internal limitations, there are two major challenges in SOC modelling: first, the uncertainties connected to the initial SOC content for the simulation and second the equilibrium assumption of the ecosystem. The initial SOC content is a key variable for carbon dynamic simulations, which for spatial simulations is often derived from soil maps. However, soil maps do not sufficiently differentiate between forest, grassland or cropland soils. As a consequence, an average value would always overestimate the initial SOC for croplands. The second challenge is associated with the complex chemical structure of SOC, their interaction with the mineral matrix of the soil and thus their accessibility to microbes for decomposition. This is reflected by the use of different carbon pools in the models. Each of these pools comes with a different decomposition rate that reflects differences in their chemical structure or accessibility to microbes. The initial distribution of the carbon in the different carbon pools of the model has a strong impact on the carbon fluxes and all associated processes. A standardised approach for setting the initial conditions is to let the model run until equilibrium (carbon input equals carbon output) is reached, before the targeted simulation period begins (spin up). For this, however, modellers often assume constant management over the spin-up period, which has often not been the case in the real world.

There is a wide range of SOC models available. The most common approach is based on carbon pools and uses first-order kinetics to describe the fluxes that leave these pools. In contrast, Wieder et al. (2013) argued that first-order kinetics are not appropriate to capture impacts of climate change on microbial processes. Therefore, they introduced a different approach that is based on Michaelis–Menten kinetics. While the pool structure reflects the complexity of the different organic compounds, the new approach represents the functional process of decomposition by enzymes. The pool approach proved its functionality in a wide range of studies, while Campbell and Paustian (2015) found it difficult to determine the quality of the new approach due to a lack of studies. More recently, approaches have been introduced that reflect more the interaction with the soil matrix through physical protection mechanisms against microbial decomposition (Robertson et al., 2019), which seems to be necessary for the representation of soil organic matter processes in iron- and aluminium-dominated soils (Nendel et al., 2019). Currently, for studies covering Europe, the existing simulation models for SOC turnover show satisfactory performance when tested against long-term experiments (Smith et al 1997; Farina et al., 2021) and in rare cases also against high-resolution carbon flux data (Specka et al., 2016). However, most of the SOC models do not consider the crop and its management explicitly. Only few agro-ecosystem models have participated in such studies (e.g. DAISY, DNDC and MONICA), so up to now there is little evidence that agro-ecosystem models simulate both carbon (and nitrogen) fluxes and crop performance well simultaneously across Europe (Aiteew et al., 2023). This has implications for model-based assessments of crop management targeting climate change mitigation.

5.3 Greenhouse gas emissions

The main GHGs resultng from agricultural activity are carbon dioxide (CO_2) and nitrous oxide (N_2O). The non-CO_2 GHGs are especially important to consider because of their much higher warming potential and due to their specific chemical behaviour in the stratosphere. Agriculture is the main source of N_2O emissions, contributing 79% of global N_2O emissions (IPCC, 2018). In addition, livestock production contributes to 39% of global CH_4 emissions (IPCC, 2018). On the other hand, there is potential to offset a fraction of these emissions by absorbing CO_2 in agricultural soils for permanent storage, also referred to as negative emissions (Paustian et al., 2016).

The capability to model GHG emissions is crucial to quantify emissions, predict future changes and to develop and evaluate strategies for their reduction. Modelling GHG emissions for croplands has different objectives, including the support of national inventories (Smith, 2013), change predictions (Carozzi et al., 2022), the development of emission reduction strategies (Frank et al., 2015) and analyses for trade-offs with other mitigation strategies (Lugato et al., 2018). The simulation of GHG emissions in Europe requires a detailed set of input variables and parameters. The availability of input data (Lugato et al., 2014) and data aggregation (Hoffmann et al., 2016) are major challenges for performing such spatial simulations.

Models and tools of different complexity are used for simulations of GHG emissions. Most recently, Carozzi et al. (2022) simulated N_2O, CH_4 and CO_2 emissions for crop- and grassland systems across Europe. This approach combined the cropping system model CERES-EGC (Gabrielle et al., 2006) with the grassland model PaSim (Riedo et al., 1998). Besides productivity, the main objective was to estimate impacts of climate change on GHG emissions. They simulated an N_2O emission rate of 2.2 g N_2O-N ha^{-1} a^{-1} for the period 1978–2004 and an exponential increase for simulations under future climate scenario RCP8.5 (up to 10 g N_2O-N ha^{-1} a^{-1}), but a moderate increase (3.6 g N_2O-N ha^{-1} a^{-1}) using scenario RCP4.5 for the period 2050–99 with a decrease of emissions in the first half of the century. Carozzi et al. (2022) identified temperature as the main driver for the changes in N_2O, which is in line with findings by Abdalla et al. (2009). Besides the sensitivity to air temperature, studies also showed high sensitivity to the amount of applied nitrogen. As this is an uncertain input variable, this can affect generic factors (Gerber et al., 2016) or the model performance (Abdalla et al., 2009).

A crucial European dataset is the land use/cover area frame statistical survey. Initially set up as a European land use and cover database obtained through the COPERNICUS land monitoring service, it was extended in 2009 with a regular topsoil survey that is carried out across the EU to derive policy-relevant statistics on the effect of land management on soil characteristics (Orgiazzi et al., 2018). Quantification is based on combining on-site measurements with the process-based model DAYCENT (Lugato et al., 2017). Lugato and colleagues estimated the average N_2O emissions as 2.27 kg N-N_2O for the EU. Reinds et al. (2012) used the model INTEGRATOR (Reinds et al., 2011) to estimate the N_2O emissions of croplands, extrapolating from 450 individual sites to the European scale. In this study emissions were quantified within a range of 1.1–2.4 kg N-N_2O ha^{-1} a^{-1} for the year 2000.

Hillier et al. (2012) use the Cool Farm Tool (Hillier et al., 2011) to estimate different mitigation options to reduce GHG emissions from croplands globally. The Cool Farm Tool is a decision support tool that is based on empirical model approaches. The authors concluded that this model was most suitable to advise policy decisions, but it was not possible to determine absolute values without extra validation for different regions.

There are some local studies investigating the development of GHG emissions in the future and deriving recommendations on how to reduce them. The TIMES model (Ireland, Chiodi et al., 2013) and a system dynamics model (Latvia, Dace et al., 2015) were used for these future projections. None of these models is process based; they utilise historical database analyses to arrive at future projections.

Nitrogen fluxes are most challenging to simulate as they are highly sensitive to management (e.g. fertiliser application) and environmental impacts (e.g. precipitation events). Models have been developed to capture the dynamics of nitrogen fluxes, but they often fail to reproduce the magnitude of emissions (Gu et al., 2014). Measurements of nitrogen fluxes, especially chamber measurements, provide very local information and do not always represent the heterogeneity of conditions that prevail in the respective region (Laville et al., 1999). Despite these challenges, models are still considered the most suitable tools to integrate assessments and to predict future emissions. However, the number of spatial modelling studies addressing climate change impacts on European cropland emissions (e.g. Haas et al., 2022) is still very limited.

5.4 Farm economics

Farm models are important tools for scientists and policymakers to assess the economic and ecological impacts of drivers such as markets, policies or climate change on farm-level decisions. Political challenges such as increasing inequality, biodiversity loss, climate change and increasing complexity in agricultural policies are accelerating the development of new models and indicators adapted to specific research questions.

Agent-based models (ABMs) aim to model the behaviour of autonomous agents (e.g. farmers, consumers and market providers). Agricultural systems evolve through their interactions, limiting the need for *a priori* assumptions on properties of the system as a whole. They are therefore a useful tool when assessing structural changes (Huber et al., 2018). In ABMs like the Agricultural Policy Simulator (Happe et al., 2006) and mathematical programming-based multiagent systems (MP-MAS), agents are represented by single farms creating a strong link between ABM and bio-economic farm models (BEFMs). MP-MAS combines microeconomic modelling and a range of optional biophysical modules. The central model repeatedly solves a constrained optimisation problem. Despite the optimisation, the model is descriptive ('positive') aiming to simulate *actual* decision-making of heterogeneous farms (Schreinemachers and Berger, 2011).

BEFMs simulate or optimise farmers' management decisions regarding production possibilities and the associated externalities. They therefore combine biophysical components, e.g. crop growth and soil quality, with economic factors (Janssen and van Ittersum, 2007). Main advantages are a detailed representation of agricultural systems, inclusion of the interaction between different farm activities (e.g. crop, biogas and livestock) and the inclusion of parameter uncertainty allowing for sensitivity analysis (Wossink and Renkema, 1994). BEFMs can use positive or normative approaches. Positive approaches use positive mathematical programming (PMP) methods to calibrate models to empirical data, while normative approaches use economic theory-based rules to find optimal production patterns. A review by Reidsma et al. (2018) showed that model-based assessment studies have covered all EU countries and all agricultural sectors throughout the last decade, with the largest number of articles focusing on arable farming. Most models used mathematical programming (MP) approaches. Farms in MP are represented by a set of 'activities' that link inputs and outputs through technical coefficients. Outputs can be exported or used internally, e.g. crop yields as well as environmental goods or externalities. Constraints define limitations (e.g. resource, legal, financial or environmental) depending on the orientation of the model. The set is optimised for an objective function in which coefficients show the contribution of each activity. Results are highly dependent on the chosen activities and constraints which are based on coherent assumptions. To compare and validate results, it is therefore important to make a detailed model description accessible (Reidsma et al., 2018). Even though research in recent years has focused increasingly on strong, multifaceted indicators, there is still a lack of good societal indicators. The CAP reform in 2013 shifted the focus towards environmental aspects and increased the need for detailed policy assessments at farm level. The number of studies using BEFMs has been increasing since, and improvements have been made regarding environmental indicators (Reidsma et al., 2018). In the following sections, we illustrate the use of farm models with three exemplary studies.

5.4.1 Assessing climate change impacts and adaptation strategies in Southern France

The models FarmDesign (Groot et al., 2012) and MODAM (Uthes et al., 2010) are two static whole farm models. FarmDesign couples a BEFM with a multi-objective optimisation algorithm. Evolutionary optimization makes it possible to evaluate multiple objectives simultaneously resulting in a set of Pareto-optimal solutions. MODAM uses linear programming allowing for multiple indicators for constraining the solution while optimising for a single objective. The inclusion of ecological indicators for each crop production activity allows a quick generation of trade-off functions between single economic and ecological objectives. The two models have different advantages and weaknesses and can be used together. FarmDesign was developed to support participatory processes and works best in interaction with farmers, as it provides a range of near-to-optimal solutions with different results for the economic and ecological indicators to be discussed. Schuler et al. (2020) applied both models on farms in Southern France assessing climate change impacts and short-term on-farm adaptation strategies. Baseline and future scenarios were compared based on differing climate variables, cropping patterns and levels of water availability. Indicators in both models cover a range of economic, ecological and social dimensions, allowing for a holistic sustainability assessment. Results showed that a lack of water for irrigation is one of the strongest drivers for farmers to implement adaptation strategies, while cropping pattern adjustments played only a minor role. The study demonstrates how model-based climate change outlooks can merge with participatory studies for the derivation of recommendations for adaptation.

5.4.2 Model-aided life cycle sustainability assessment of European beef production

The single-farm model FARMDYN differs from other models mainly through the consideration of the highly-detailed management options. It includes modules for risk behaviour, different farm activities or management options (Britz et al., 2021). Due to its flexibility, the model is particularly valuable when analysing sectors with heterogeneous farm types. The European beef sector is among the largest globally and characterised by highly diverse farm structures. Beef production causes unwanted environmental externalities. Therefore, Kokemohr et al. (2022) used FARMDYN on European beef production for a life cycle sustainability assessment assessing farm-level environmental, social and economic performances on three different farm types in three European countries. The use of BEFM facilitates a sensitivity analysis that considers the effects of prices, yields and animal traits. The study revealed major differences between the systems caused by different beef/calf prices and emission allocation. FARMDYN integrates all indicators in a consistent framework, considering price effects and farmers' behaviour for a heterogeneous population of farms and practices. Results enable a more informed debate on sustainable beef production (Kokemohr et al., 2022).

5.4.3 Assessing impacts of Common Agricultural Practice diversification measures

The Individual Farm Model for Common Agricultural Policy Analysis (IFM-CAP) is a static PMP model. It assesses the CAP impact on farm-level economy and environment using economic and environmental indicators (Louhichi et al., 2015) although with a strong focus on the economic components (Britz et al., 2021). Louhichi et al. (2017) analysed impacts of the CAP 2013 diversification measures and found that the measures have only minor effects on the European farming sector. At the EU level, <1% was reallocated due to those measures, while the crop production and income decreased by <0.5%. Impacts on the farm level were in some cases more significant; however, the share of affected farms is relatively low (around 5%). The 75% threshold for growing main crops on European fields, as the CAP prescribes, emerged as the most constraining element of the diversification measures. As the model differentiates by farm size, it could be demonstrated that large farms were affected less than medium-sized or small farms. Using IFM-CAP also allowed assessment of distributional effects across EU farms (Louhichi et al., 2017).

As shown by these examples, farm models can support farmers' decision-making as well as the development of new technologies and farm type-targeted policies and their evaluation based on their effect on local agro-environmental conditions. Currently, many studies focus on climate change and related risks, water management, biodiversity and ecosystem services or distributional impacts of different policies. Most applications show that models are valuable for complementing qualitative studies with quantitative simulations.

5.5 Animal production systems

Livestock are affected by climate change directly through impacts on their well-being and health and indirectly through effects on feed production, quality and availability. There are large differences in the management of livestock across Europe depending on climatic conditions, the type of livestock and production systems. There is both globally and in Europe a lack of systematic studies of impacts and adaptation of livestock systems to climate change, including lack of linkages between impacts and adaptation, little emphasis on mixed crop–livestock systems and monogastric livestock and underrepresentation of quantitative methods including models for predicting productivity (Escarcha et al., 2018). The need to reduce GHG emissions from livestock further stresses the need to integrate modelling of livestock with respect to climate change impact and adaptation with mitigation.

Monogastric animals (e.g. pigs and poultry) are mostly kept in confined conditions (animal houses), where conditioning of the housing temperature is a prerequisite for optimal production. However, heat stress during summer in warm climates has several documented negative effects on pig production, with low farrowing rate of sows mated in summer, increased carcass fatness of progeny born to the sows mated in summer and slower growth rate of finisher pigs in summer (Liu et al., 2022). Heat stress also affects animal health. Impacts and adaptations therefore depend on complex interactions between heat stress events and the housing environment. Modelling provides an important way to assess impacts and adaptation strategies (Özkan et al., 2016).

The projected increase in frequency of heat waves calls for novel approaches to environmental control in livestock facilities, and implementation of cooling technologies is a key factor for alleviating these negative impacts on animal health and productivity (Godýn et al., 2020). Air temperature may be decreased by using adiabatic cooling technologies such as high-pressure fogging system or evaporative pads. There is a need to develop energy- and water-saving cooling methods, using for example an earth–air or earth-to-water heat exchanger. Modelling adaptation needs therefore links to modelling future energy systems.

Dairy cattle respond negatively to heat stress, which is aggravated by high relative humidity (Hill and Wall, 2015). Heat stress causes many negative consequences in terms of physiological and behavioural disturbances and significant losses in milk production and quality. The effort to reduce the risk of the occurrence of heat stress among dairy cows involves several aspects, including methods for predicting heat stress (Herbut et al., 2018). There is some variation among equations to predict effects on heat stress, which may relate to genetic differences between cattle breeds and to seasonal and lagged effects of heat stress (Herbut et al., 2018). A heat stress model was used to assess the effects of heat stress on milk quality in Spanish organic dairy farms (Blanco-Penedo et al., 2020). The effect of heat stress was deemed low on biological costs through milk yield. However, milk quality (protein and fat content) was negatively impacted. This shows that multiple indicators are needed to assess impacts of climate change on livestock production.

Ruminant animals depend to a large extent on grassland for feed production. Ruminants convert plant materials indigestible to humans into nutritious food. Much of the effects of climate change on cattle production thus depend on the effects of climate change on grassland productivity. Therefore, grassland models are important in predicting and understanding the impacts of climate change on grassland systems and cattle production and assessing the efficacy of potential adaptation strategies (Kipling et al., 2016). There is a particular need to better predict the interannual and seasonal variation in grassland productivity, as affected by climate change and increased climatic variability (Graux et al., 2011).

Farm-level models that represent both livestock and associated feed production (grassland and other fodder crops) are essential for quantifying the farm-scale components affecting GHG emissions (Schils et al., 2007; del Prado et al., 2013). Livestock systems are major sources of GHG emissions, and models have been used to identify measures to reduce these emissions (Weiske et al., 2006). Important mitigation measures include livestock feeding, manure management and crop and grassland management.

5.6 Grapevine and olives

Agoecosystems based on growing grapes and olives are important globally in both economic and cultural terms. Europe has the highest area of grape cultivation in the world (38%, Fraga et al., 2013), while the Mediterranean Basin concentrates while 90% of globally-cultivated olives are concentrated in the Mediterranean Basin (Vossen, 2007). Besides purely economic aspects, these agroecosystems provide other ecosystem services, including protection against soil erosion, carbon storage in soil and a typical landscape with high cultural identification for the people living within (Loumou and Giourga, 2003). Experimental evidence indicates that both systems could contribute to climate mitigation, thanks to their long-term carbon sequestration capacity in soil and woody tissues in vines and olive trees (Nieto et al., 2010). Specifically, olive orchards may overall be considered a considerable carbon pool, contributing to sequestration of atmospheric CO_2 with a net ecosystem exchange comparable to several Italian forestry systems (Valentini et al., 2000). Similarly, grapevine cultivation was found to contribute to carbon sequestration, even though with a lower efficiency due to inter-row management practices (Brunori et al., 2016).

Despite the multiple benefits provided by these systems, a changing climate is expected to substantially alter this picture. For instance, at global scale, Hannah et al. (2013) indicated a substantial decrease of land viable for grapevine cultivation due to warmer and drier conditions, altering the climate metrics of most wine regions. Moriondo et al. (2011) emphasised that the Mediterranean Basin will likely suffer the most negative impacts, with its southern regions becoming increasingly unsuitable for grapevine cultivation if no adaptation is considered. Cultivation of grapevines will progressively shift to the Northern European regions. This trend, assessed through the combination of bioclimatic indices as predictors of the presence of a wine region, is corroborated by results obtained using process-based growth models. Fraga et al. (2015) indicated that on the southern shores of the Mediterranean region, a warmer climate will result in such a high reduction in yield that grapevine cultivation will no longer be economically viable. A similar trend may be outlined for olive trees whose cultivation is expected to shift from its original range to the northwest European regions, while olive productivity in Southern Europe will likely decrease in the western regions, particularly in the Iberian Peninsula (Fraga et al., 2022; Ponti et al., 2014).

The impact of climate change, in addition to production aspects, will affect the ecosystem services provided by these agroecosystems. Future warming is expected to modify the seasonal carbon sequestration capacity, with a higher carbon uptake in the winter and a prolonged period of photosynthesis inhibition during summer compared to the baseline. As a result of this pattern, carbon sequestration capacity of olive orchards will substantially decrease across all Mediterranean areas with respect to the present, especially in extensively managed olive groves (Brilli et al., 2019). These negative impacts may be counterbalanced by the implementation of irrigation as an adaptation option. However, this strategy should take into account the future availability of water resources in the region.

Although future climate change has high spatial variability across Europe (Giorgi and Lionello, 2008), the use of crop growth models makes it possible to identify some promising adaptation practices, especially in areas characterised by a trend towards drier and warmer climates such as the Mediterranean region. Deficit irrigation strategies are currently considered an effective measure for maintaining both olive tree and grapevine agroecosystems by increasing water use efficiency and crop performance while limiting water usage (Fraga et al., 2018; Santos et al., 2020). Crop models indicate that rainfed low-density olive tree orchards will be the most affected, especially in driest areas such as in the Iberian Peninsula and Southern Italy. Deficit irrigation will reduce the expected impacts on yield while also improving the carbon sequestration capacity of the system. At the same time, new intensive orchards are expected to become more productive with respect to the present, even considering the same irrigation regime (Mairech et al., 2021). Fraga et al. (2013) indicated that deficit irrigation strategies are an effective tool to alleviate the effect of a warmer and drier climate. However, modelling by the same author indicated that this approach, indicating that this approach is less effective in drier regions of Portugal (Fraga et al., 2022). This suggests that additional adaptation measures aiming at reducing evapotranspiration such as grass cutting and mulching might be required in the warmer regions of the Mediterranean. The use of these practices may allow one to maintain a suitable level of production and, in turn, a mitigation capacity. These practices may make it possible to maintain both levels of production and mitigation capacity, should be considered interacting rather than alternatives to each other, resulting in the most useful approach to maintain a suitable level of production and at the same time reducing GHG emissions (Smith and Olesen, 2010).

5.7 Vegetables

Field-grown vegetable production generates high economic returns per unit of land and thus offers promising income prospects across Europe. By 2031, the EU consumption of fresh vegetables is expected to increase as a result of increasing consumer awareness of the benefits of adopting a vegetable-rich diet (FreshPlaza, 2021). Field vegetable production often relied on high volumes on nitrogen and water to maintain profitable yields and the produce quality standards consumers demand. Early simulation models for field vegetables, therefore, concentrated on nitrogen and water use efficiency within crop rotations (Nendel, 2009). Generic crop growth models often do not cover parameterisations for vegetables, because their phenology, growth patterns and habitus are extremely diverse. The DAISY model is one of the few exceptions (Mollerup et al., 2005), while the EUROTATE_N model (Rahn et al., 2010) avoids using a fully-fledged growth algorithm to simulate water and nitrogen dynamics in field vegetable rotations. For many other vegetables, single model solutions are available, for example the CROPGRO family (Jones et al., 2003).

Some thermophilic vegetables that currently grow mostly in Southern Europe (eggplants, tomatoes and peppers) may become suitable for cultivation in lowland areas in Central Europe in the future (Potopová et al., 2017). A reduction in cold stress in a future climate, an increase in the length of growing season and an overall warmer climate, will see a substantial increase of areas suitable for open-field cultivation of tomatoes and thermophilic vegetables in Europe, especially in the more northern latitudes (Karkanis et al., 2018; Litskas et al., 2019; Ide et al., 2022; Potopová et al., 2016). Conversely, current production areas will soon become marginal or unsuitable for open-field cultivation of tomatoes due to progressively increasing heat and drought stress (Silva et al., 2017; dos Santos et al., 2022; Cammarano et al., 2022). Simulations have shown that climate change may cause an advance of the date of last spring-frost, yet the benefits of a reduced frost risk derive primarily from the delay in the date of first frost in autumn (Potop et al., 2014a,b). However, the increasing likelihood for extreme weather conditions may counteract a production increase.

Research has recently addressed the vulnerability of vegetables to adverse weather events during the growing period due to characteristics such as low tolerance to frost, waterlogging, heat waves and droughts (e.g. Potopová et al., 2017). Research on tomatoes has suggested that drought stress during fruit formation reduces fruit size by 30% (Boote et al., 2012) and that higher temperature and more frequent drought reduces fruit formation (Muntean et al., 2021). Ventrella et al. (2011) studied agronomic adaptation strategies in Italy under climate change and found that tomato phenology may alter by 2050 due to warmer temperatures and may reduce tomato yields. Saadi et al. (2015) showed that air temperature is projected to have a dominant role in the shortening and anticipation of the tomato-growing cycle over the Mediterranean region, as a consequence of which the major tomato-growing areas in Italy may lose their productivity (Cammarano et al., 2022). Potopová et al. (2017) showed that droughts, heavy precipitation, heat waves and days with heat stress can significantly affect cultivation of at last 12 vegetable crops in Central Europe. They found positive long-term impacts of the recent warming on tomatoes (from +4.9°C^{-1} to +12.2%°C^{-1}) but a decrease in the yield stability for root vegetables in the warmest areas. Bisbis et al. (2018) provided a comprehensive overview of the state of the art in climate change impact research on the quality of vegetables in Europe. Little is known about how climate change will affect vegetable production in the next 10–30 years, especially because much open-field cultivation is irrigated. However, dynamic crop simulation models can be a useful tool to simulate climate effects, including compound adverse weather, on vegetable production.

Irrigation is widely applied to ensure stable yields of vegetables (Potopová et al., 2022). The amount of irrigation water applied often depends on how much water is available from local sources. A significant increase in irrigation requirements can be expected in all vegetable-producing regions in the EU for 2031–50. Vertical farming is a way to optimise vegetable supply chains and holds significant potential. Indeed, when compared to traditional technology, cultivation technologies using multi-layer stacking can reduce water used for growing lettuce by 96%. However, vertical farming systems are energy intensive, since light, temperature and humidity need to be controlled (de Carbonnel et al., 2022), which will have major influence on uptake of the technology.

Climate change also affects the product quality of field-grown vegetables and may alter their nutritional quality. Heat stress causes decreases in the lycopene concentrations in tomatoes (antioxidants with anticancer properties). Fungal diseases (potato blight, brown spot and leaf spot) cause a loss of tomato quality and hence reduce the yield worldwide (Panthee and Chen, 2010). Because prevailing weather conditions determine the severity of the disease, crop loss may range from a mild loss of productivity to a complete loss of a particular crop, resulting in great economic losses for the food industry. Litskas et al. (2019) examined the impact of climate change on tomatoes and their main pests for 29 countries in Europe, South Asia and Africa. They projected that 30–100% of the tomato-producing area under irrigation will become unsuitable for tomato production, in part due to the increasing probability of mite outbreaks.

Vegetable grafting, aeroponics and hydroponic growing systems provide potential adaptation tools for boosting adverse climate change effects. Grafted vegetables tend to produce higher yields of high-quality fruit, making field cultivation financially rewarding. Spain utilises ~50–70 million grafted plants per year, accounting for about 40% of the country's tomato production (Grieneisen et al., 2018). The main advantages of modern hydroponic systems is lower water use and higher productivity per unit area.

In summary, field vegetables are affected differently by climate change and increasing adverse weather conditions during their main productive development stages. This suggests that adaptation strategies need to be tailored based on region-specific characteristics of each vegetable crop.

6 Future trends and conclusion

The role of European agriculture has considerably changed and shifted from the focus on provisioning of food and other materials as an ecosystem service to a greater focus on sustainability. Therefore, strengthening other ecosystem services beyond production, such as nutrient cycling, water protection and biodiversity, has become more important, together with supporting the cultural and recreational potential of the iconic agricultural landscapes of Europe. This shift of emphasis has not only affected EU agricultural policy but also modelling initiatives in European agriculture. Modelling has moved from approaches used for the optimisation of production to methods to estimate the best use of arable land, to optimise fertilisation and irrigation to ensure a balance between maximum production and minimum impact on water quality and availability and to estimate the agricultural soil's carbon sequestration potential. This required moving away from single-crop simulations, using simple crop models and ignoring carry-over effects, to comprehensive agro-ecosystem models that represent and improve crop rotations. This has increased the demand for less common crops being parameterised and introduced into the crop models, including intermediate crops. Consequently, the effect of tillage practices and their effect on soil characteristics and nutrient distribution have also become increasingly important. Putting emphasis on the sustainability of particular practices has made consecutive multiannual runs even more necessary. In the same way, it has become more important to look beyond the impact of particular weather patterns in a particular season and, instead, assess long-term impacts of particular crop rotations and even land-use types. This then facilitates the proper accounting of crop rotation effects on soil carbon and nitrogen dynamics over the long term. The growing area of organic agriculture (another 16% of European farmland is to be turned to organic farming by 2030 according to the Green Deal's Farm to Fork strategy of the European Commission; European Commission 2020) has also resulted in demand for developing models or modifying existing ones to handle organic farming techniques, including the use of organic amendments (composts, digestates and biochar) or mulching. The efforts of many European nations to lower the GHG emissions in agriculture has led to agro-ecosystem models being applied for GHG emission assessments and consequently to the efforts to improve these models in this regard. With these different agendas, agro-ecosystem modelling has become excessively complex and difficult to understand and handle. The issue has been addressed through creative visualisation and by aggregating the outputs through common denominators, e.g. costs or profits. As the financial aspect makes the complex system easier to comprehend for many stakeholders, agro-ecosystem models have been linked to economic models (or vice versa) and used in combination.

These trends are likely to continue in the future. The demand for agriculture to contribute to climate change mitigation will mean that more attention will be paid to the ability of the crop models to reliably estimate GHG fluxes. This will lead to much more frequent use of agro-ecosystem models as an important constituent of GHG inventories, identifying avoidable GHG emissions. Sustainable agriculture requires as-low-as-possible emissions of any kind (GHG, nitrates and pesticides) and thus the as-efficient-as-possible use of mineral fertilisers and pesticides. To identify such optima, agro-ecosystem models have emerged as very useful tools. In combination with extended-range weather forecasts, even tactical decisions in day-to-day farming will soon be more reliably based on simulations. Even more valuable will be the introduction of pest and disease models, potentially providing more precise estimates of the pest and disease epidemiology within the season and accurate decision support for spraying much reduced amounts of pesticides and fungicides.

Remotely (and proximally) sensed data are becoming more accessible at very high resolution and will greatly increase the value of process-based models. In Europe, the European Commission seeks to replace the current agricultural surveillance system to control the subsidy pay-out (IACS) with satellite remote sensing, as soon as the sensor platform performance and the method toolbox are mature enough to detect the crops grown in the current season accurately and early enough in the season (d'Andrimont et al., 2021; Blickensdörfer et al., 2022). However, remote sensing can also provide spatially explicit input data for models and thus allow for a very precise representation of the entire landscape. Having the models recalibrated frequently during the season with remotely sensed soil moisture, biomass N contents and leaf area is a computationally intensive process. Therefore, artificial intelligence (AI) using big data will soon compete with traditional mechanistic approaches and, more likely, use them as a physical constraint to deliver more powerful predictions where process monitoring for explanatory purposes is not necessarily required.

Agro-ecosystem models are treasure troves of agronomy knowledge and containers of fine-tuned and generalised functions and data collected by multiple generations of researchers. Yet, they remain easily accessible and open sourced, which makes it difficult to build a successful business model around such software. From the business standpoint, it seems far more efficient to invest in simple-to-use tools, e.g. based on AI, and develop proprietary software that can be licensed and sold. However, both approaches will find their place in agricultural research and in emerging agricultural service provided by companies of all sizes across Europe.

7 Where to look for further information

The European Commission's Green Deal Farm-to-Form Strategy:

- https://food.ec.europa.eu/horizontal-topics/farm-fork-strategy_en.

The European Drought Observatory:

- https://edo.jrc.ec.europa.eu/edov2/php/index.php?id=1000.

CzechGlobe's Drought Monitor for Central Europe:

- https://www.intersucho.cz/en/.

Agri4Cast toolbox of the Joint Research Centre:

- https://agri4cast.jrc.ec.europa.eu.

EO4Agri – European capacity for improving operational agriculture monitoring:

- https://eo4agri.eu.

Sen4CAP – Sentinels for Common Agriculture Policy:

- http://esa-sen4cap.org.

8 Acknowledgements

The contribution of the Global Change Research Institute co-authors was supported by the Adaptation strategies for sustainable ecosystem services and food security under adverse environmental conditions CZ.02.1.01/0.0/0.0/16_019/0000797 (http://sustes.czechglobe.cz/).

9 References

Abdalla, M., Jones, M., Smith, P. and Williams, M. (2009). Nitrous oxide fluxes and denitrification sensitivity to temperature in Irish pasture soils. *Soil Use and Management* 25(4), 376–388.

Aiteew, K., Rouhiainen, J. and Dechow, R. (2023). Evaluation and optimization of the soil carbon turnover routine in the Monica model. *Geoscientific Model Development*.

Beek, J. and Frissel, M. (1973). *Simulation of Nitrogen Behaviour in Soils.* Wageningen: PUDOC, p. 67.

Berg-Mohnicke, M. and Nendel, C. (2022). A case for object capabilities as the foundation of an environmental model and simulation infrastructure. *Environmental Modelling and Software* 156, 105471.

Bergez, J.-E., Chabrier, P., Gary, C., Jeuffroy, M. H., Makowski, D., Quesnel, G., Ramat, E., Raynal, H., Rousse, N., Wallach, D., Debaeke, P., Durand, P., Duru, M., Dury, J., Faverdin, P., Gascuel-Odoux, C. and Garcia, F. (2013). An open platform to build, evaluate and simulate integrated models of farming and agro-ecosystems. *Environmental Modelling and Software* 39, 39–49.

Biggar, J. W. and Nielsen, D. R. (1962). Miscible displacement. II. Behavior of tracers. *Soil Science Society of America Journal* 26(2), 125–128.

Bisbis, M. B., Gruda, N. and Blanke, M. (2018). Potential impacts of climate change on vegetable production and product quality – A review. *Journal of Cleaner Production* 170, 1602–1620.

Blanco-Penedo, I., Velarde, A., Kipling, R. P. and Ruete, A. (2020). Modeling heat stress under organic dairy farming conditions in warm temperate climates within the Mediterranean basin. *Climatic Change* 162(3), 1269–1285.

Blickensdörfer, L., Schwieder, M., Pflugmacher, D., Nendel, C., Erasmi, S. and Hostert, P. (2022). Mapping of crop types and crop sequences with combined time series of Sentinel-1, Sentinel-2 and Landsat 8 data for Germany. *Remote Sensing of Environment* 269, 112831.

Boote, K. J., Rybak, M. R., Scholberg, J. M. S. and Jones, J. W. (2012). Improving the CROPGRO-tomato model for predicting growth and yield response to temperature. *HortScience* 47(8), 1038–1049.

Brilli, L., Lugato, E., Moriondo, M., Gioli, B., Toscano, P., Zaldei, A., Leolini, L., Cantini, C., Caruso, G., Gucci, R., Merante, P., Dibari, C., Ferrise, R., Bindi, M. and Costafreda-Aumedes, S. (2019). Carbon sequestration capacity and productivity responses of Mediterranean olive groves under future climates and management options. *Mitigation and Adaptation Strategies for Global Change* 24(3), 467–491.

Brisson, N., Gary, C., Justes, E., Roche, R., Mary, B., Ripoche, D., Zimmer, D., Sierra, J., Bertuzzi, P., Burger, P., Bussière, F., Cabidoche, Y. M., Cellier, P., Debaeke, P., Gaudillère, J. P., Hénault, C., Maraux, F., Seguin, B. and Sinoquet, H. (2003). An overview of the crop model STICS. *European Journal of Agronomy* 18(3–4), 309–332.

Britz, W., Ciaian, P., Gocht, A., Kanellopoulos, A., Kremmydas, D., Müller, M., Petsakos, A. and Reidsma, P. (2021). A design for a generic and modular bio-economic farm model. *Agricultural Systems* 191, 103133.

Britz, W., Lengers, B., Kuhn, T. and Schäfer, D. (2016). *A Highly Detailed Template Model for Dynamic Optimization of Farms - FARMDYN*. University of Bonn, Institute for Food and Resource Economics, p. 147.

Brunori, E., Farina, R. and Biasi, R. (2016). Sustainable viticulture: The carbon-sink function of the vineyard agro-ecosystem. *Agric. Ecosyst. Environ.* 223, 1–21.

Bünemann, E. K., Bongiorno, G., Bai, Z. G., Creamer, R. E., De Deyn, G., de Goede, R., Fleskens, L., Geissen, V., Kuyper, T. W., Mäder, P., Pulleman, M., Sukkel, W., van Groenigen, J. W. and Brussaard, L. (2018). Soil quality – A critical review. *Soil Biology and Biochemistry* 120, 105–125.

Cammarano, D., Jamshidi, S., Hoogenboom, G., Ruane, A. C., Niyogi, D. and Ronga, D. (2022). Processing tomato production is expected to decrease by 2050 due to the projected increase in temperature. *Nature Food* 3(6), 437–444.

Campbell, E. E. and Paustian, K. (2015). Current developments in soil organic matter modeling and the expansion of model applications: A review. *Environmental Research Letters* 10(12), 123004.

Carozzi, M., Martin, R., Klumpp, K. and Massad, R. S. (2022). Effects of climate change in European croplands and grasslands: Productivity, greenhouse gas balance and soil carbon storage. *Biogeosciences* 19(12), 3021–3050.

Ceglar, A., Zampieri, M., Toreti, A.and Dentener, F. (2019). Observed northward migration of agro-climate zones in Europe will further accelerate under climate change. *Earth's Future* 7(9), 1088–1101.

Chiodi, A., Gargiulo, M., Rogan, F., Deane, J. P., Lavigne, D., Rout, U. K. and Gallachóir, B. P. Ó. (2013). Modelling the impacts of challenging 2050 European climate mitigation targets on Ireland's energy system. *Energy Policy* 53, 169–189.

Dace, E., Muizniece, I., Blumberga, A. and Kaczala, F. (2015). Searching for solutions to mitigate greenhouse gas emissions by agricultural policy decisions – Application of system dynamics modeling for the case of Latvia. *Science of the Total Environment* 527–528, 80–90.

d'Andrimont, R., Verhegghen, A., Lemoine, G., Kempeneers, P., Meroni, M. and van der Velde, M. (2021) From parcel to continental scale – A first European crop type map based on Sentinel-1 and LUCAS Copernicus in-situ observations. *Remote Sensing of Environment* 266, 112708.

de Carbonnel, M., Stormonth-Darling, J. M., Liu, W., Kuziak, D. and Jones, M. A. (2022). Realising the environmental potential of vertical farming systems through advances in plant photobiology. *Biology* 11(6), Article 922.

De Wit, C. T. (1970). Dynamic concepts in biology. Prediction and measurement of photosynthetic productivity. In: Meet, T. (Ed.), *Proceedings of the IBP/PP Tech*. Wageningen: PUDOC.

del Prado, A., Crosson, P., Olesen, J. E. and Rotz, C. A. (2013). Whole-farm models to quantify greenhouse gas emissions and their potential use for linking climate change mitigation in temperate grassland and ruminant-based farming systems. *Animal* 7(S2), 373–385.

Delzeit, R., Beach, R., Bibas, R., Britz, W., Chateau, J., Freund, F., Lefevre, J., Schuenemann, F., Sulser, T., Valin, H., van Ruijven, B., Weitzel, M., Willenbockel, D. and Wojtowicz, K. (2020). Linking global CGE models with sectoral models to generate baseline scenarios: approaches, challenges, and opportunities. *Journal of Global Economic Analysis* 5(1), 162–195.

Diekkrüger, B., Söndgerath, D., Kersebaum, K. C. and McVoy, C. W. (1995). Validity of agroecosystem models - A comparison of results of different models applied to the same data set. *Ecological Modelling* 81(1–3), 3–29.

Diepen, C. A., Wolf, J., Keulen, H. and Rappoldt, C. (1989). WOFOST: A simulation model of crop production. *Soil Use and Management* 5(1), 16–24.

Dobrovolný, P., Moberg, A., Brázdil, R., Pfister, C., Glaser, R., Wilson, R., van Engelen, A., Limanówka, D., Kiss, A., Halíčková, M., Macková, J., Riemann, D., Luterbacher, J. and Böhm, R. (2010). Monthly and seasonal temperature reconstructions for Central Europe derived from documentary evidence and instrumental records since AD 1500. *Climatic Change* 101(1–2), 69–107.

Dong, J., Gruda, N., Li, X., Tang, Y., Zhang, P. and Duan, Z. (2020). Sustainable vegetable production under changing climate: The impact of elevated CO_2 on yield of vegetables and the interactions with environments - A review. *Journal of Cleaner Production* 253, 119920.

dos Santos, M. M., da Silva, R. S., Soares, M. A., Lopes, É. M. G. and Shabani, F. (2022). Global risks of Bedellia somnulentella (Lepidoptera: Bedelliidae) invasion: A modeling exercise using a mechanistic model, CLIMEX. *Theoretical and Applied Climatology* 149(1–2), 401–411.

Dutt, G. R., Shaffer, M. J. and Moore, W. J. (1972). *Computer Simulation Model of Dynamic Bio-Physicochemical Processes in Soils*. Technical Bulletin. University of Arizona, Agricultural Experimental Station.

Eitzinger, J., Kubu, G., Thaler, S., Alexandrov, V., Utset, A., Mihailovic, D. T., Lalic, B., Trnka, M., Zalud, Z., Semeradova, D., Ventrella, D., Anastasiou, D. P., Medany, M., Altaher, S., Olejnik, J., Lesny, J., Nemeshko, N., Nikolaev, M., Simota, C. and Cojocaru, G. (2009). Adaptation of agriculture in European regions at environmental risk under climate change - Final report; Specific Support Action, FP6-2005-SSP-5-A, Proj. No.044210. Vienna: Institute of Meteorology, University of Natural Resources and Applied Life Sciences (BOKU), p. 450.

Enders, A., Diekkrüger, B., Laudien, R., Gaiser, T. and Bareth, G. (2010). The IMPETUS spatial decision support systems. In: Speth, P., Christoph, M. and Diekkrüger, B. (Eds.), *Impacts of Global Change on the Hydrological Cycle in West and Northwest Africa*. Berlin, Heidelberg: Springer, pp. 360–393.

Escarcha, J. F., Lassa, J. A. and Zander, K. K. (2018). Livestock under climate change: A systematic review of impacts and adaptation. *Climate* 6(3), 54.

European Commission. (2020). *Farm to Fork Strategy. COM/2020/381 Final*. Brussels.

European Environmental Agency. (2023). https://www.eea.europa.eu/data-and-maps/figures/organic-agricultural-land-coverage-in.

Eurostat. (2016). https://ec.europa.eu/eurostat/de/data/database.

FAO, I. F. A. D., UNICEF, WFP and WHO. (2022). *The State of Food Security and Nutrition in the World 2022. Repurposing Food and Agricultural Policies to Make Healthy Diets More Affordable*. Rome: Food and Agriculture Organization.

FAOSTAT. (2021). https://www.fao.org/faostat/.

Farina, R., Sándor, R., Abdalla, M., Álvaro-Fuentes, J., Bechini, L., Bolinder, M. A., Brilli, L., Chenu, C., Clivot, H., De Antoni Migliorati, M., Di Bene, C., Dorich, C. D., Ehrhardt, F., Ferchaud, F., Fitton, N., Francaviglia, R., Franko, U., Giltrap, D. L., Grant, B. B., Guenet, B., Harrison, M. T., Kirschbaum, M. U. F., Kuka, K., Kulmala, L., Liski, J., McGrath, M. J., Meier, E., Menichetti, L., Moyano, F., Nendel, C., Recous, S., Reibold, N., Shepherd, A., Smith, W. N., Smith, P., Soussana, J. F., Stella, T., Taghizadeh-Toosi, A., Tsutskikh, E. and Bellocchi, G. (2021). Ensemble modelling, uncertainty and robust predictions of organic carbon in long-term bare fallow soils. *Global Change Biology* 27(4), 904–928.

Faye, B., Webber, H., Gaiser, T., Müller, C., Zhang, Y., Stella, T., Latka, C., Reckling, M., Heckelei, T., Helming, K. and Ewert, F. (2023). Climate change impacts on European arable crop yields: Sensitivity to assumptions about rotations and residue management. *European Journal of Agronomy* 142, 126670.

Flichman, G., Belhouchette, H., Komarek, A. M., Drogue, S., Hawkins, J., Chenoune, R., Haile, B., Roberts, C. and Msangi, S. (2016). *Dynamic Agricultural Household Bio-Economic Simulator (DAHBSIM) Model Description*. Washington, DC: International Food Policy Research Institute.

Fraga, H., de Cortázar Atauri, I. G. and Santos, J. A. (2018). Viticultural irrigation demands under climate change scenarios in Portugal. *Agricultural Water Management* 196, 66–74.

Fraga, H., García de Cortázar Atauri, I., Malheiro, A. C. and Santos, J. A. (2016). Modelling climate change impacts on viticultural yield, phenology and stress conditions in Europe. *Glob. Change Biol.* 22(11), 3774–3788.

Fraga, H., Guimarães, N., Freitas, T. R., Malheiro, A. C. and Santos, J. A. (2022). Future scenarios for olive tree and grapevine potential yields in the World Heritage Côa Region, Portugal. *Agronomy* 12(2), 350.

Fraga, H., Malheiro, A. C., Moutinho-Pereira, J. and Santos, J. A. (2013). Future scenarios for viticultural zoning in Europe: Ensemble projections and uncertainties. *International Journal of Biometeorology* 57(6), 909–925.

Frank, S., Havlík, P., Soussana, J.-F., Levesque, A., Valin, H., Wollenberg, E., Kleinwechter, U., Fricko, O., Gusti, M., Herrero, M., Smith, P., Hasegawa, T., Kraxner, F. and Obersteiner, M. (2017). Reducing greenhouse gas emissions in agriculture without compromising food security? *Environmental Research Letters* 12(10), 105004.

Frank, S., Schmid, E., Havlík, P., Schneider, U. A., Böttcher, H., Balkovič, J. and Obersteiner, M. (2015). The dynamic soil organic carbon mitigation potential of European cropland. *Global Environmental Change* 35, 269–278.

FreshPlaza, E. C. (2021). *EU Agricultural Outlook for Markets, Income and Environment, 2021–2031. European Commission, DG Agriculture and Rural Development, Brussels*.

Fricko, O., Havlik, P., Rogelj, J., Klimont, Z., Gusti, M., Johnson, N., Kolp, P., Strubegger, M., Valin, H., Amann, M., Ermolieva, T., Forsell, N., Herrero, M., Heyes, C., Kindermann, G., Krey, V., McCollum, D. L., Obersteiner, M., Pachauri, S., Rao, S., Schmid, E., Schoepp, W. and Riahi, K. (2017). The marker quantification of the shared socioeconomic pathway 2: A middle-of-the-road scenario for the 21st century. *Global Environmental Change* 42, 251–267.

Gabrielle, B. P., Laville, O. Duval, B., Nicoullaud, J. C. and Germon, C. (2006). Process-based modeling of nitrous oxide emissions from wheat-cropped soils at the subregional scale Hénault. *Global Biogeochemical Cycles* 20, GB4018.

Gerber, J. S., Carlson, K. M., Makowski, D., Mueller, N. D., Garcia de Cortazar-Atauri, I., Havlík, P., Herrero, M., Launay, M., O'Connell, C. S., Smith, P. and West, P. C. (2016) Spatially explicit estimates of N_2O emissions from croplands suggest climate mitigation opportunities from improved fertilizer management. *Global Change Biology* 22(10), 3383–3394.

Giorgi, F. and Lionello, P. (2008). Climate change projections for the Mediterranean region. *Global and Planetary Change* 63(2–3), 90–104.

Giuliani, M. M., Gatta, G., Cappelli, G., Gagliardi, A., Donatelli, M., Fanchini, D., De Nart, D., Mongiano, G. and Bregaglio, S. (2019). Identifying the most promising agronomic

adaptation strategies for the tomato growing systems in Southern Italy via simulation modeling. *European Journal of Agronomy* 111.
Godýn, D., Herbut, P., Angrecka, S. and Corrêa Vieira, F. M. (2020). Use of different cooling methods in pig facilities to alleviate the effects of heat stress - A review. *Animals: An Open Access Journal from MDPI* 10(9), 1459.
Graux, A.-I., Gaurut, M., Agabriel, J., Baumont, R., Delagarde, R., Delaby, L. and Soussana, J.-F. (2011). Development of the pasture simulation model for assessing livestock production under climate change. *Agriculture, Ecosystems and Environment* 144(1), 69–91.
Grieneisen, M. L., Aegerter, B. J., Scott Stoddard, C. and Zhang, M. (2018). Yield and fruit quality of grafted tomatoes, and their potential for soil fumigant use reduction. A meta-analysis. *Agronomy for Sustainable Development* 38(3), 29.
Groot, J. C. J., Oomen, G. J. M. and Rossing, W. A. H. (2012). Multi-objective optimization and design of farming systems. *Agricultural Systems* 110, 63–77.
Gruda, N., Bisbis, M. and Tanny, J. (2019). Influence of climate change on protected cultivation: Impacts and sustainable adaptation strategies - A review. *Journal of Cleaner Production* 225, 481–495.
Gu, J., Loustau, D., Hénault, C., Rochette, P., Cellier, P., Nicoullaud, B., Grossel, A. and Richard, G. (2014). Modeling nitrous oxide emissions from tile-drained winter wheat fields in Central France. *Nutrient Cycling in Agroecosystems* 98(1), 27–40.
Haas, E., Carozzi, M., Massad, R. S., Butterbach-Bahl, K. and Scheer, C. (2022). Long term impact of residue management on soil organic carbon stocks and nitrous oxide emissions from European croplands. *Science of the Total Environment* 836, 154932.
Hagin, J. and Amberger, A. (1974). *Contribution of Fertilizers and Manures to the N and P Load of Waters: A Computer Simulation*. Technische Universität München, p. 123.
Hammer, G. C., Messina, A. W. and Cooper, M. (2019). Biological reality and parsimony in crop models—Why we need both in crop improvement! In Silico Plants 1(1), diz010.
Hampf, A. C., Carauta, M. M. M., Latynski, E., Libera, A. A. D., Monteiro, L. A., Sentelhas, P. C., Troost, C., Berger, T. and Nendel, C. (2018). The biophysical and socio-economic dimension of yield gaps in the Southern Amazon – A bio-economic modelling approach. *Agricultural Systems* 165, 1–13.
Hannah, L., Roehrdanz, P. R., Ikegami, M., Shepard, A. V., Shaw, M. R., Tabor, G., Zhi, L., Marquet, P. A. and Hijmans, R. J. (2013). Climate change, wine and conservation. *Proc. Natl Acad. Sci. U. S. A.* 110(17), 6907–6912.
Happe, K., Kellermann, K. and Balmann, A. (2006). Agent-based analysis of agricultural policies: An illustration of the agricultural policy simulator AgriPoliS, its adaptation and behavior. *Ecology and Society* 11(1).
Herbut, P., Angrecka, S. and Walczak, J. (2018). Environmental parameters to assessing of heat stress in dairy cattle—A review. *International Journal of Biometeorology* 62(12), 2089–2097.
Hill, D. L. and Wall, E. (2015). Dairy cattle in a temperate climate: The effects of weather on milk yield and composition depend on management. *Animal* 9(1), 138–149.
Hillier, J., Brentrup, F., Wattenbach, M., Walter, C., Garcia-Suarez, T., Mila-i-Canals, L. and Smith, P. (2012). Which cropland greenhouse gas mitigation options give the greatest benefits in different world regions? Climate and soil-specific predictions from integrated empirical models. *Global Change Biology* 18(6), 1880–1894.

Hillier, J., Walter, C., Malin, D., Garcia-Suarez, T., Mila-i-Canals, L. and Smith, P. (2011). A farm-focused calculator for emissions from crop and livestock production. *Environmental Modelling and Software* 26(9), 1070–1078.

Hoffmann, H., Zhao, G., Asseng, S., Bindi, M., Biernath, C., Constantin, J., Coucheney, E., Dechow, R., Doro, L., Eckersten, H., Gaiser, T., Grosz, B., Heinlein, F., Kassie, B. T., Kersebaum, K. C., Klein, C., Kuhnert, M., Lewan, E., Moriondo, M., Nendel, C., Priesack, E., Raynal, H., Roggero, P. P., Rötter, R. P., Siebert, S., Specka, X., Tao, F. L., Teixeira, E., Trombi, G., Wallach, D., Weihermüller, L., Yeluripati, J. and Ewert, F. (2016). Impact of spatial soil and climate input data aggregation on regional yield simulations. *PLoS ONE* 11(4), Article e0151782.

Huber, R., Bakker, M., Balmann, A., Berger, T., Bithell, M., Brown, C., Grêt-Regamey, A., Xiong, H., Le, Q. B., Mack, G., Meyfroidt, P., Millington, J., Müller, B., Polhill, J. G., Sun, Z., Seidl, R., Troost, C. and Finger, R. (2018). Representation of decision-making in European agricultural agent-based models. *Agricultural Systems* 167, 143–160.

Ide, R., Ichiki, A., Suzuki, T. and Jitsuyama, Y. (2022). Analysis of yield reduction factors in processing tomatoes under waterlogging conditions. *Scientia Horticulturae* 295, 110840.

IPCC. (2018). Summary for policymakers. In: *Global Warming of an IPCC Special Report on the Impacts of Global Warming of 1.5°C Above Pre-Industrial Levels and Related Global Greenhouse Gas Emission Pathways, in the Context of Strengthening the Global Response to the Threat of Climate Change, Sustainable Development, and Efforts to Eradicate Poverty.* Cambridge, UK and New York, NY: Cambridge University Press, pp. 3–24.

Jänicke, C., Goddard, A., Stein, S., Steinmann, H. H., Lakes, T., Nendel, C. and Müller, D. (2022). Field-level land-use data reveal heterogeneous crop sequences with distinct regional differences in Germany. *European Journal of Agronomy* 141, Article 126632.

Janssen, S. and van Ittersum, M. K. (2007). Assessing farm innovations and responses to policies: A review of bio-economic farm models. *Agricultural Systems* 94(3), 622–636.

Jones, J. W., Antle, J. M., Basso, B., Boote, K. J., Conant, R. T., Foster, I., Godfray, H. C. J., Herrero, M., Howitt, R. E., Janssen, S., Keating, B. A., Munoz-Carpena, R., Porter, C. H., Rosenzweig, C. and Wheeler, T. R. (2017). Brief history of agricultural systems modeling. *Agricultural Systems* 155, 240–254.

Jones, J. W., Hoogenboom, G., Porter, C. H., Boote, K. J., Batchelor, W. D., Hunt, L. A., Wilkens, P. W., Singh, U., Gijsman, A. J. and Ritchie, J. T. (2003). The DSSAT cropping system model. *European Journal of Agronomy* 18(3–4), 235–265.

Kabat, P., Marshall, B., van den Broek, B. J., Vos, J. and van Keulen, H. (1995). *Modelling and Parametrization of the Soil-Plant-Atmosphere System – A Comparison of Potato Growth Models*. Wageningen, the Netherlands: Wageningen Pers, p. 513.

Karkanis, A., Ntatsi, G., Alemardan, A., Petropoulos, S. and Bilalis, D. (2018). Interference of weeds in vegetable crop cultivation, in the changing climate of Southern Europe with emphasis on drought and elevated temperatures: A review. *The Journal of Agricultural Science* 156(10), 1175–1185.

Keating, B. A., Carberry, P. S., Hammer, G. L., Probert, M. E., Robertson, M. J., Holzworth, D., Huth, N. I., Hargreaves, J. N. G., Meinke, H., Hochman, Z., McLean, G., Verburg, K., Snow, V., Dimes, J. P., Silburn, M., Wang, E., Brown, S., Bristow, K. L., Asseng, S., Chapman, S., McCown, R. L., Freebairn, D. M. and Smith, C. J. (2003). An overview of APSIM, a model designed for farming systems simulation. *European Journal of Agronomy* 18(3–4), 267–288.

Kipling, R. P., Virkajärvi, P., Breitsameter, L., Curnel, Y., De Swaef, T., Gustavsson, A. M., Hennart, S., Höglind, M., Järvenranta, K., Minet, J., Nendel, C., Persson, T., Picon-Cochard, C., Rolinski, S., Sandars, D. L., Scollan, N. D., Sebek, L., Seddaiu, G., Topp, C. F. E., Twardy, S., Van Middelkoop, J., Wu, L. and Bellocchi, G. (2016). Key challenges and priorities for modelling European grasslands under climate change. *Science of the Total Environment* 566–567, 851–864.

Kokemohr, L., Escobar, N., Mertens, A., Mosnier, C., Pirlo, G., Veysset, P. and Kuhn, T. (2022). Life cycle sustainability assessment of European beef production systems based on a farm-level optimization model. *Journal of Cleaner Production* 379, 134552.

Kollas, C., Kersebaum, K. C., Nendel, C., Manevski, K., Müller, C., Palosuo, T., Armas-Herrera, C. M., Beaudoin, N., Bindi, M., Charfeddine, M., Conradt, T., Constantin, J., Eitzinger, J., Ewert, F., Ferrise, R., Gaiser, T., Cortazar-Atauri, I. Gd., Giglio, L., Hlavinka, P., Hoffmann, H., Hoffmann, M. P., Launay, M., Manderscheid, R., Mary, B., Mirschel, W., Moriondo, M., Olesen, J. E., Öztürk, I., Pacholski, A., Ripoche-Wachter, D., Roggero, P. P., Roncossek, S., Rötter, R. P., Ruget, F., Sharif, B., Trnka, M., Ventrella, D., Waha, K., Wegehenkel, M., Weigel, H. and Wu, L. (2015). Crop rotation modelling–A European model intercomparison. *European Journal of Agronomy* 70, 98–111.

Kostková, M., Hlavinka, P., Pohanková, E., Kersebaum, K. C., Nendel, C., Gobin, A., Olesen, J. E., Ferrise, R., Dibari, C., Takáč, J., Topaj, A., Medvedev, S., Hoffmann, M. P., Stella, T., Balek, J., Ruiz Ramos, M., Rodrígues, A., Hoogenboom, G., Shelia, V., Ventrela, D., Giglio, L., Sharif, B., Oztürk, I., Rötter, R. P., Balkovič, J., Skalský, R., Moriondo, M., Thaler, S., Žalud, Z. and Trnka, M. (2021). Performance of 13 crop simulation models and their ensemble for simulating four field crops in Central Europe. *Journal of Agricultural Science* 159(1–2), 69–89.

Kraft, P., Eyshi Rezaei, E. E., Breuer, L., Ewert, F., Große-Stoltenberg, A., Kleinebecker, T., Seserman, D. M. and Nendel, C. (2021). Modelling agroforestry's contributions to people–A review of available models. *Agronomy* 11(11), 2106.

Kuhn, T., Enders, A., Gaiser, T., Schäfer, D., Srivastava, A. K. and Britz, W. (2020). Coupling crop and bio-economic farm modelling to evaluate the revised fertilization regulations in Germany. *Agricultural Systems* 177, 102687.

Laville, P., Jambert, C., Cellier, P. and Delmas, R. (1999). Nitrous oxide fluxes from a fertilised maize crop using micrometeorological and chamber methods. *Agricultural and Forest Meteorology* 96(1–3), 19–38.

Le Gal, P.-Y., Andrieu, N., Bruelle, G., Dugué, P., Monteil, C., Moulin, C.-H., Penot, E. and Ryschawy, J. (2022). Modelling mixed crop-livestock farms for supporting farmers' strategic reflections: The CLIFS approach. *Computers and Electronics in Agriculture* 192, 106570.

Litskas, V. D., Migeon, A., Navajas, M., Tixier, M. and Stavrinides, M. C. (2019). Impacts of climate change on tomato, a notorious pest and its natural enemy: Small scale agriculture at higher risk. *Environmental Research Letters* 14(8), 084041.

Liu, F., Zhao, W., Le, H. H., Cottrell, J. J., Green, M. P., Leury, B. J., Dunshea, F. R. and Bell, A. W. (2022). Review: What have we learned about the effects of heat stress on the pig industry? *Animal* 16(Suppl. 2), 100349.

Louhichi, K., Ciaian, P., Espinosa, M., Colen, L., Perni, A., Gomez, Y. and Paloma, S. (2015). *Individual Farm Model for Common Agricultural Policy Analysis (IFM-CAP). JRC Research Reports JRC92574.*

Louhichi, K., Ciaian, P., Espinosa, M., Colen, L., Perni, A. and Paloma, S. (2017). Does the crop diversification measure impact EU farmers' decisions? An assessment using an Individual Farm Model for CAP Analysis (IFM-CAP). *Land Use Policy* 66, 250–264.

Louhichi, K., Kanellopoulos, A., Janssen, S., Flichman, G., Blanco, M., Hengsdijk, H., Heckelei, T., Berentsen, P., Lansink, A. O. and Ittersum, M. V. (2010). FSSIM, a bio-economic farm model for simulating the response of EU farming systems to agricultural and environmental policies. *Agricultural Systems* 103(8), 585–597.

Loumou, A. and Giourga, C. (2003). Olive groves: The life and identity of the Mediterranean. *Agriculture and Human Values* 20(1), 87–95.

Lugato, E., Bampa, F., Panagos, P., Montanarella, L. and Jones, A. (2014). Potential carbon sequestration of European arable soils estimated by modelling a comprehensive set of management practices. *Global Change Biology* 20(11), 3557–3567.

Lugato, E., Leip, A. and Jones, A. (2018). Mitigation potential of soil carbon management overestimated by neglecting N_2O emissions. *Nature Climate Change* 8(3), 219–223.

Lugato, E., Paniagua, L., Jones, A., de Vries, W. and Leip, A. (2017). Complementing the topsoil information of the land use/land cover area frame survey (LUCAS) with modelled N_2O emissions. *PLoS ONE* 12(4), e0176111.

Luterbacher, J., Werner, J. P., Smerdon, J. E., Fernández-Donado, L., González-Rouco, F. J., Barriopedro, D., Ljungqvist, F. C., Büntgen, U., Zorita, E., Wagner, S., Esper, J., McCarroll, D., Toreti, A., Frank, D., Jungclaus, J. H., Barriendos, M., Bertolin, C., Bothe, O., Brázdil, R., Camuffo, D., Dobrovolný, P., Gagen, M., García-Bustamante, E., Ge, Q., Gómez-Navarro, J. J., Guiot, J., Hao, Z., Hegerl, G. C., Holmgren, K., Klimenko, V. V., Martín-Chivelet, J., Pfister, C., Roberts, N., Schindler, A., Schurer, A., Solomina, O., von Gunten, L., Wahl, E., Wanner, H., Wetter, O., Xoplaki, E., Yuan, N., Zanchettin, D., Zhang, H. and Zerefos, C. (2016). European summer temperatures since Roman times. *Environmental Research Letters* 11(2), 024001.

Luthin, J. N. and Gaskell, R. A. (1950). Numerical solution for tile drainage of layered soils. *Transactions of the American Geophysical Union* 31, 595–602.

Mairech, H., López-Bernal, Á., Moriondo, M., Dibari, C., Regni, L., Proietti, P., Villalobos, F. J. and Testi, L. (2021). Sustainability of olive growing in the Mediterranean area under future climate scenarios: Exploring the effects of intensification and deficit irrigation. *European Journal of Agronomy* 129, 126319.

Malezieux, E., Crozat, Y., Dupraz, C., Laurans, M., Makowski, D., Ozier-Lafontaine, H., Rapidel, B., de Tourdonnet, S. and Valantin-Morison, M. (2009). Mixing plant species in cropping systems: Concepts, tools and models - A review. *Agronomy for Sustainable Development* 29(1), 43–62.

Maracchi, G., Mestre, A., Toulios, L., Kajfež-Bogataj, L. and Hočevar, A. (2006). *COST Action 718: Meteorology Applications for Agriculture 22255*. Office for Official Publications of the European Communities, p. 197.

Mollerup, M., Hansen, S. and Abrahamsen, P. (2005). *Parameteriseringer Af Grøntsager Til Daisy. Tåstrup: den Kgl. Veterinær- og Landbohøjskole, Institut for Jordbrugsvidenskab*, p. 200.

Moriondo, M., Jones, G. V., Bois, B., Dibari, C., Ferrise, R., Trombi, G. and Bindi, M. (2013). Projected shifts of wine regions in response to climate change. *Climatic Change* 119(3–4), 825–839.

Muntean, N., Chawdhery, R. A., Potopová, V. and Tűrkott, L. (2021). The ability of CROPGRO-Tomato model to simulate the growth characteristics of Thomas F1 tomato cultivar grown under open field conditions. *J. Agric. Sci.* 159(7–8), 473–487.

Nendel, C. (2009). Evaluation of best management practices for N fertilisation in regional field vegetable production with a small-scale simulation model. *European Journal of Agronomy* 30(2), 110–118.

Nendel, C. (2010). Modelle zum Stickstoffhaushalt. In: Fränzle, O., Müller, F. and Schröder, W. (Eds.), *Handbuch der Umweltwissenschaften - 19. Erg. Lfg* (Vols. 11/2010). Weinheim, Germany: Wiley-VCH Press.

Nendel, C., Melzer, D. and Thorburn, P. J. (2019). The nitrogen nutrition potential of arable soils. *Scientific Reports* 9(1), 5851.

Nendel, C., Reckling, M., Debaeke, P., Schulz, S., Berg-Mohnicke, M., Constantin, J., Fronzek, S., Hoffmann, M., Jakšić, S., Kersebaum, K. C., Klimek-Kopyra, A., Raynal, H., Schoving, C., Stella, T. and Battisti, R. (2023). Future area expansion outweighs increasing drought risk for soybean in Europe. *Global Change Biology* 29(5), 1340–1358.

Nieto, O. M., Castro, J., Fernández, E. and Smith, P. (2010). Simulation of soil organic carbon stocks in a Mediterranean olive grove under different soil-management systems using the RothC model. *Soil Use and Management* 26(2), 118–125.

OECD 2018. Organization for economic co-operation and development (OECD) work on environment 2017–2018. Available at: https://www.oecd.org/env/OECD-work-on-environment-2017-2018.pdf.

O'Neill, B. C., Tebaldi, C., van Vuuren, D. P., Eyring, V., Friedlingstein, P., Hurtt, G., Knutti, R., Kriegler, E., Lamarque, J., Lowe, J., Meehl, G. A., Moss, R., Riahi, K. and Sanderson, B. M. (2016). The scenario model intercomparison project (ScenarioMIP) for CMIP6. *Geoscientific Model Development* 9(9), 3461–3482. https://doi.org/10.5194/gmd-9-3461-2016.

Orgiazzi, A. ,Ballabio, C., Panagos, P., Jones, A. and Fernández-Ugalde, O. (2018). LUCAS soil, the largest expandable soil dataset for Europe: A review. *European Journal of Soil Science* 69(1), 140–153.

Orlandini, S. and Nejedlik, P. (2012). *Climate Change Impacts on Agriculture in Europe - Final Report of COST Action 734*. Firenze University Press, p. 206.

Özkan, Ş., Vitali, A., Lacetera, N., Amon, B., Bannink, A., Bartley, D. J., Blanco-Penedo, I., Haas, Y., Dufrasne, I., Elliott, J., Eory, V., Fox, N. J., Garnsworthy, P. C., Gengler, N., Hammami, H., Kyriazakis, I., Leclère, D., Lessire, F., Macleod, M., Robinson, T. P., Ruete, A., Sandars, D. L., Shrestha, S., Stott, A. W., Twardy, S., Vanrobays, M. L., Ahmadi, B. V., Weindl, I., Wheelhouse, N., Williams, A. G., Williams, H. W., Wilson, A. J., Østergaard, S. and Kipling, R. P. (2016). Challenges and priorities for modelling livestock health and pathogens in the context of climate change. *Environmental Research* 151, 130–144.

Panthee, D. R. and Chen, F. (2010). Genomics of fungal disease resistance in tomato. *Current Genomics* 11(1), 30–39.

Paustian, K., Lehmann, J., Ogle, S., Reay, D., Robertson, G. P. and Smith, P. (2016). Climate-smart soils. *Nature* 532(7597), 49–57.

Petit, S., Firbank, L., Wyatt, B. and Howard, D. (2001). MIRABEL: Models for integrated review and assessment of biodiversity in European landscapes. *Ambio* 30(2), 81–88.

Pohanková, E., Hlavinka, P., Kersebaum, K. C., Rodríguez, A., Balek, J., Bednařík, M., Dubrovský, M., Gobin, A., Hoogenboom, G., Moriondo, M., Nendel, C., Olesen, J.

E., Rötter, R. P., Ruiz-Ramos, M., Shelia, V., Stella, T., Hoffmann, M. P., Eitzinger, J., Eitzinger, J., Dibari, C., Ferrise, R., Bláhová, M. and Trnka, M. (2022). Expected effects of climate change on the production and water use of crop rotation management reproduced by crop model ensemble for Czech Republic sites. *European Journal of Agronomy* 134, 126446.

Potop, V., Zahradníček, P., Türkott, L., Štěpánek, P. and Soukup, J. (2014a). Risk occurrences of damaging frosts during the growing season of vegetables in the Elbe River lowland, the Czech Republic. *Natural Hazards* 71(1), 1–19.

Potop, V., Zahradníček, P., Türkott, L., Štěpánek, P. and Soukup, J. (2014b). Potential impacts of climate change on damaging frost during growing season of vegetables. *Scientia Agriculturae Bohemica* 45(1), 26–35.

Potopová, V., Štěpánek, P., Farda, A., Türkott, L., Zahradníček, P. and Soukup, J. (2016). Impacto de las sequías en la producción de los cultivos en el bajo Río Elba entre 1961 y 2014. *Cuadernos de Investigacion Geografica* 42(1), 127–143.

Potopová, V., Trnka, M., Vizina, A., Semerádová, D., Balek, J., Chawdhery, M. R. A., Musiolková, M., Pavlík, P., Možný, M., Štěpánek, P. and Clothier, B. (2022). Projection of 21st century irrigation water requirements for sensitive agricultural crop commodities across the Czech Republic. *Agricultural Water Management* 262, 107337.

Potopová, V., Zahradníček, P., Štěpánek, P., Türkott, L., Farda, A. and Soukup, J. (2017). The impacts of key adverse weather events on the field-grown vegetable yield variability in the Czech Republic from 1961 to 2014. *International Journal of Climatology* 37(3), 1648–1664.

Rahn, C. R., Zhang, K. F., Lillywhite, R. D., Ramos, C., de Paz, J. M., Doltra, J., Riley, H., Fink, M., Nendel, C., Thorup-Kristensen, K., Pedersen, A., Piro, F., Venezia, A., Firth, C., Schmutz, U., Rayns, F. and Strohmeyer, K. (2010). EU-Rotate_N – A decision support system – To predict environmental and economic consequences of the management of nitrogen fertiliser in crop rotations. *European Journal of Horticultural Science* 75(1), 20–32.

Reidsma, P., Janssen, S., Jansen, J. and van Ittersum, M. K. (2018). On the development and use of farm models for policy impact assessment in the European Union – A review. *Agricultural Systems* 159, 111–125.

Reinds, G., Kros, J., Uiterwijk, M. and de Vries, W. (2011). The integrator model: a tool to assess greenhouse gas emissions and nitrogen fluxes in Europe. In: *Nitrogen and Global Change Conference*, Edinburgh, UK.

Reinds, G. J., Heuvelink, G. B. M., Hoogland, T., Kros, J. and de Vries, W. (2012). Estimating nitrogen fluxes at the European scale by upscaling INTEGRATOR model outputs from selected sites. *Biogeosciences* 9(11), 4527–4536.

Reumaux, R., Chopin, P., Bergkvist, G., Watson, C. A. and Öborn, I. (2023). Land Parcel Identification System (LPIS) data allows identification of crop sequence patterns and diversity in organic and conventional farming systems. *SSRN Electronic Journal. European Journal of Agronomy, Under Revision.*

Riahi, K., van Vuuren, D. P., Kriegler, E., Edmonds, J., O'Neill, B. C., Fujimori, S., Bauer, N., Calvin, K., Dellink, R., Fricko, O., Lutz, W., Popp, A., Cuaresma, J. C., Kc, S., Leimbach, M., Jiang, L., Kram, T., Rao, S., Emmerling, J., Ebi, K., Hasegawa, T., Havlik, P., Humpenöder, F., Da Silva, L. A., Smith, S., Stehfest, E., Bosetti, V., Eom, J., Gernaat, D., Masui, T., Rogelj, J., Strefler, J., Drouet, L., Krey, V., Luderer, G., Harmsen, M., Takahashi, K., Baumstark, L., Doelman, J. C., Kainuma, M., Klimont, Z., Marangoni, G., Lotze-Campen, H., Obersteiner, M., Tabeau, A. and Tavoni, M. (2017). The shared socioeconomic pathways and their energy, land use, and greenhouse gas emissions implications: an overview. *Global Environmental Change* 42, 153–168. https://doi.org/10.1016/j.gloenvcha.2016.05.009.

Riedo, M., Grub, A., Rosset, M. and Fuhrer, J. (1998). A pasture simulation model for dry matter production, and fluxes of carbon, nitrogen, water and energy. *Ecological Modelling* 105(2–3), 141–183.

Robertson, A. D., Paustian, K., Ogle, S., Wallenstein, M. D., Lugato, E. and Cotrufo, M. F. (2019). Unifying soil organic matter formation and persistence frameworks: The MEMS model. *Biogeosciences* 16(6), 1225–1248.

Rojas-Downing, M. M., Nejadhashemi, A. P., Harrigan, T. and Woznicki, S. A. (2017). Climate change and livestock: Impacts, adaptation, and mitigation. *Climate Risk Management* 16, 145–163.

Rosenzweig, C., Jones, J. W., Hatfield, J. L., Ruane, A. C., Boote, K. J., Thorburn, P., Antle, J. M., Nelson, G. C., Porter, C., Janssen, S., Asseng, S., Basso, B., Ewert, F., Wallach, D., Baigorria, G. and Winter, J. M. (2013). The agricultural model intercomparison and improvement project (AgMIP): Protocols and pilot studies. *Agricultural and Forest Meteorology* 170, 166–182. The Agricultural Model Intercomparison and Improvement Project.

Rötter, R. P., Ewert, F., Palosuo, T., Bindi, M., Kersebaum, K. C., Olesen, J. E., Trnka, M., van Ittersum, M. K., Janssen, S., Rivington, M., Semenov, M., Wallach, D., Porter, J. R., Stewart, D., Verhagen, J., Angulo, C., Gaiser, T., Nendel, C., Martre, P. and de Wit, A. (2013). Challenges for agro-ecosystem modelling in climate change risk assessment for major European crops and farming systems. In: *Impacts World Conference,* Potsdam, Germany.

Saadi, S., Todorovic, M., Tanasijevic, L., Pereira, L. S., Pizzigalli, C. and Lionello, P. (2015). Climate change and Mediterranean agriculture: Impacts on winter wheat and tomato crop evapotranspiration, irrigation requirements and yield. *Agricultural Water Management* 147, 103–115.

Santos, J. A., Fraga, H., Malheiro, A. C., Moutinho-Pereira, J., Dinis, L. T., Correia, C., Moriondo, M., Leolini, L., Dibari, C., Costafreda-Aumedes, S., Kartschall, T., Menz, C., Molitor, D., Junk, J., Beyer, M. and Schultz, H. R. (2020). A review of the potential climate change impacts and adaptation options for European viticulture. *Applied Sciences* 10(9), 3092.

Schils, R. L. M., Olesen, J. E., del Prado, A. and Soussana, J. F. (2007). A review of farm level modelling approaches for mitigating greenhouse gas emissions from ruminant livestock systems. *Livestock Science* 112(3), 240–251.

Schreinemachers, P. and Berger, T. (2011). An agent-based simula-tion model of human-environment interactions in agricultural systems. *Environmental Modelling and Software* 26(7), 845–859.

Schuler, J., Adelhart Toorop, R., Willaume, M., Vermue, A., Schläfke, N., Uthes, S., Zander, P. and Rossing, W. (2020). Assessing climate change impacts and adaptation options for farm performance using bio-economic models in Southwestern France. *Sustainability* 12(18), 7528.

Seligman, N. C. and van Keulen, H. (1981). PAPRAN: A simulation model of annual pasture production limited by rainfall and nitrogen. In: Frissel, M. J. and van Veen, J. A. (Eds.), *Simulation of Nitrogen Behaviour of Soil-Plant Systems*. Wageningen: PUDOC, pp. 192–221.

Silva, R. S., Kumar, L., Shabani, F. and Picanço, M. C. (2017). Assessing the impact of global warming on worldwide open field tomato cultivation through CSIRO-Mk3·0 global climate model. *The Journal of Agricultural Science* 155(3), 407–420.

Smith, J. E., Heath, L. S. and Hoover, C. M. (2013). Carbon factors and models for forest carbon estimates: national greenhouse gas emission inventories of the United States. *Forest Ecol. Manag.* 307, 7–19.

Smith, P. and Olesen, J. E. (2010). Synergies between the mitigation of, and adaptation to, climate change in agriculture. *Journal of Agricultural Science* 148(5), 543–552.

Smith, P., Smith, J. U., Powlson, D. S., McGill, W. B., Arah, J. R. M., Chertov, O. G., Coleman, K., Franko, U., Frolking, S., Jenkinson, D. S., Jensen, L. S., Kelly, R. H., Klein-Gunnewiek, H., Komarov, A. S., Li, C., Molina, J. A. E., Mueller, T., Parton, W. J., Thornley, J. H. M. and Whitmore, A. P. (1997). A comparison of the performance of nine soil organic matter models using datasets from seven long-term experiments. *Geoderma* 81(1–2), 153–225.

Specka, X., Nendel, C., Hagemann, U., Pohl, M., Hoffmann, M., Barkusky, D., Augustin, J., Sommer, M. and van Oost, K. (2016). Reproducing CO_2 exchange rates of a crop rotation at contrasting terrain positions using two different modelling approaches. *Soil and Tillage Research* 156, 219–229.

Spitters, C. J. T. and Schapendonk, A. H. C. M. (1990). Evaluation of breeding strategies for drought tolerance in potato by means of crop growth simulation. *Plant and Soil* 123(2), 193–203.

Stella, T., Mouratiadou, I., Gaiser, T., Berg-Mohnicke, M., Wallor, E., Ewert, F. and Nendel, C. (2019). Estimating the contribution of crop residues to soil organic carbon conservation. *Environ. Res. Lett.* 14(9), 094008.

Stergiadi, M., van der Perk, M., de Nijs, T. C. M. and Bierkens, M. F. P. (2016). Effects of climate change and land management on soil organic carbon dynamics and carbon leaching in northwestern Europe. *Biogeosciences* 13(5), 1519–1536.

Stöckle, C. O. M., Donatelli, M. and Nelson, R. (2003). CropSyst, a cropping systems simulation model. *European Journal of Agronomy* 18(3–4), 289–307.

Tilman, D., Fargione, J., Wolff, B., D'Antonio, C., Dobson, A., Howarth, R., Schindler, D., Schlesinger, W. H., Simberloff, D. and Swackhamer, D. (2001). Forecasting agriculturally driven global environmental change. *Science* 292(5515), 281–284.

Trnka, M., Balek, J., Semenov, M. A., Semerádová, D., Bělínová, M., Hlavinka, P., Olesen, J. E., Eitzinger, J., Schaumberger, A. and Zahradníček, P. (2020). Future agroclimatic conditions and implications for European grasslands. *Biologia Plantarum* 64, Article 865–880.

Unger, K. (1968). *Biophysik Pflanzlicher Systeme, Studia Biophysica* (Vol. 11). Quedlinburg.

United Nations, Department of Economic and Social Affairs, Population Division. (2022). *World Population Prospects 2022, Online Edition*.

Uthes, S., Sattler, C., Zander, P., Piorr, A., Matzdorf, B., Damgaard, M., Sahrbacher, A., Schuler, J., Kjeldsen, C., Heinrich, U. and Fischer, H. (2010). Modeling a farm population to estimate on-farm compliance costs and environmental effects of a grassland extensification scheme at the regional scale. *Agricultural Systems* 103(5), 282–293.

Valcárcel, M., Lahoz, I., Campillo, C., Martí, R., Leiva-Brondo, M., Roselló, S. and Cebolla-Cornejo, J. (2020). Controlled deficit irrigation as a water-saving strategy for processing tomato. *Scientia Horticulturae* 261, 108972.

Valentini, R., Matteucci, G., Dolman, A. J., Schulze, E. D., Rebmann, C., Moors, E. J., Granier, A., Gross, P., Jensen, N. O., Pilegaard, K., Lindroth, A., Grelle, A., Bernhofer, C., Grünwald, T., Aubinet, M., Ceulemans, R., Kowalski, A. S., Vesala, T., Rannik, U., Berbigier, P., Loustau, D., Guethmundsson, J., Thorgeirsson, H., Ibrom, A., Morgenstern, K., Clement, R., Moncrieff, J., Montagnani, L., Minerbi, S. and Jarvis, P. G. (2000). Respiration as the main determinant of carbon balance in European forests. Nature 404(6780), 861-865.

van Ittersum, M. K., Ewert, F., Heckelei, T., Wery, J., Alkan Olsson, J., Andersen, E., Bezlepkina, I., Brouwer, F., Donatelli, M., Flichmann, G., Olsson, L., Rizzoli, A. E., van der Wal, T., Wien, J. E. and Wolf, J. (2008). Integrated assessment of agricultural systems - A component-based framework for the European Union (SEAMLESS). *Agricultural Systems* 96(1-3), 150-165.

van Keulen, H., Penning de Vries, F. W. T. and Drees, E. M. (1982). A summary model for crop growth. In: Penning de Vries, F. W. T. and van Laar, H. H. (Eds.), *Simulation of Plant Growth and Crop Production*. Wageningen: PUDOC, pp. 87-97.

Vanuytrecht, E., Raes, D., Steduto, P., Hsiao, T. C., Fereres, E., Heng, L. K., Garcia Vila, M. and Mejias Moreno, P. (2014). AquaCrop: FAO's crop water productivity and yield response model. *Environmental Modelling and Software* 62, 351-360.

Ventrella, D., Charfeddie, M., Moriondo, M., Rinaldi, M. and Bindi, M. (2011). Agronomic adaptation strategies under climate change for winter durum wheat and tomato in Southern Italy. *Reg. Envir. Change* 12, 407-419.

Vossen, P. (2007). Olive oil: History, production, and characteristics of the world's classic oils. *HortScience* 42(5), 1093-1100.

Wang, X., Williams, J. R., Gassman, P. W., Baffaut, C., Izaurralde, R. C., Jeong, J. and Kiniry, J. R. (2012). EPIC and APEX: Model use, calibration and validation. *Transactions of the ASABE* 55(4), 1447-1462.

Webber, H., Ewert, F., Olesen, J. E., Müller, C., Fronzek, S., Ruane, A. C., Bourgault, M., Martre, P., Ababaei, B., Bindi, M., Ferrise, R., Finger, R., Fodor, N., Gabaldón-Leal, C., Gaiser, T., Jabloun, M., Kersebaum, K. C., Lizaso, J. I., Lorite, I. J., Manceau, L., Moriondo, M., Nendel, C., Rodríguez, A., Ruiz-Ramos, M., Semenov, M. A., Siebert, S., Stella, T., Stratonovitch, P., Trombi, G. and Wallach, D. (2018). Diverging importance of drought stress for maize and winter wheat in Europe. *Nature Communications* 9(1), 4249.

Weiske, A., Vabitsch, A., Olesen, J. E., Schelde, K., Michel, J., Friedrich, R. and Kaltschmitt, M. (2006). Mitigation of greenhouse gas emissions in European conventional and organic dairy farming. *Agric. Ecosys. Environ.* 112(2-3), 221-232.

Wieder, W. R., Bonan, G. B. and Allison, S. D. (2013). Global soil carbon projections are improved by modelling microbial processes. *Nature Climate Change* 3(10), 909-912.

Wossink, A. and Renkema, J. (1994). Analysis of future change in Dutch arable farming: A farm economics approach. *European Review of Agricultural Economics* 21(1), 95-112.

Wösten, J. H. M., Lilly, A., Nemes, A. and Le Bas, C. (1999). Development and use of a database of hydraulic properties of European soils. *Geoderma* 90(3-4), 169-185.

Zander, P. and Kächele, H. (1999). Modelling multiple objectives of land use for sustainable development. *Agricultural Systems* 59(3), 311-325.

Zegada-Lizarazu, W. and Monti, A. (2011). Energy crops in rotation. A review. *Biomass and Bioenergy* 35(1), 12-25.

Chapter 13

Modeling the impact of climate change on agriculture in the United States

Bruno Basso, Neville Millar and Lydia Price, Michigan State University, USA

1 Introduction

2 Climate change impact on crop-based agriculture

3 Can new genetics mitigate climate change impacts on yields?

4 Overview of incentives and payment mechanisms for carbon markets

5 References

1 Introduction

Climate change is widespread, rapid, and intensifying and refers to long-term (several decades or longer) shifts in temperature and weather patterns. Driven by increasing concentrations of greenhouse gases (GHGs) in the atmosphere, the resulting rise in average temperature is responsible for shrinking mountain glaciers, rising sea levels, shifts in flower, and plant blooming times and changes in atmospheric processes. As a consequence of the latter, weather patterns change to more extreme and enduring weather situations, including severe droughts, heat spells, and thunderstorms.

Climate change presents real threats to U.S. agricultural production with significant implications not just for producers but for the wider population. Farmers and other land managers continue to feel the pressures of a changing climate and its effects on weather. As these ongoing risks amplify, they will be faced with the challenges of adaptation and mitigation.

We divide our chapter into three major sections: 1) an overview of the recent past and predicted near-future changes in climate at the national and regional scale in the United States, and the impacts of this on crop-based agriculture, in particular the direct effect on soil, water, and major cash crops; 2) a case study that uses a crop and soil biogeochemical process-based model (SALUS; Basso et al., 2006; Basso and Ritchie, 2015) to investigate potential responses of corn yield to continued trends in temperature and precipitation for the near future using adapted cultivar; and 3) a summary of some of the incentives and payment mechanisms related to carbon markets, especially those that focus on agricultural product supply chains.

http://dx.doi.org/10.19103/AS.2022.0115.15

2 Climate change impact on crop-based agriculture

This section draws from prior comprehensive syntheses, reviews, and data presentation, including the U.S. Department of Agriculture (USDA; Walthall et al., 2012), the U.S. Environmental Protection Agency (EPA; https://www.epa.gov/climate-indicators), and U.S. Global Change Research Program (GCRP) National Climate Assessments (NCA; https://www.globalchange.gov/browse/reports). Our focus is primarily on the direct impacts of climate change on crop-based agriculture due to altered temperature and precipitation and frequency of extreme events. The reader is encouraged to access the above sources for further information.

2.1 Overview

Crops grown in the United States are critical globally; about a quarter of all grains on the global market are supplied by U.S.-based farms. Agricultural productivity in the United States will continue to be impacted by increasing atmospheric GHG concentrations, rising air temperatures, altered precipitation patterns, and higher incidences of extreme weather events. These environmental stressors compromise the ability of farmland to provide key ecosystem services that include the provision of food, fiber and fuel, carbon sequestration, biodiversity conservation, soil health and water provision, pollination habitat, and pest regulation, and may exacerbate disservices, such as water pollution, health risks from pests and diseases, and biodiversity loss (Swinton et al., 2007).

Here, we do not provide an overview of adaptation and mitigation measures to counter these climate change stressors, but note that they can and should take place with changes in behavior and management throughout the agricultural sector. Specific farmer responses can include changing crop type and rotation, adjusting planting and harvesting time – due to, for example, planting new cultivars – and altering chemical inputs and soil management practices. These responses will likely see and require an increased demand for current and new payment-based incentive programs.

2.2 U.S. national and regional climate: past, present and future

Climate varies considerably across the diverse regions of the United States. Regional climates have been relatively stable for centuries, but with significant inter-annual variations due to large-scale patterns of variability, such as the El Niño–Southern Oscillation (ENSO). Extended droughts, temperature changes, and shifts in precipitation were relatively short-lived (decadal at most), and afterward, conditions reverted to those more long-term typical for the region.

More recently, clearly observable trends in the United States, consistent with many other areas in the world, show that these shorter-term fluctuations are superimposed on longer-term changes in temperature and precipitation. U.S. climate will continue to change by magnitudes larger and rates of change more rapid than in the past.

2.2.1 U.S. greenhouse gas emissions

The burning of fossil fuels, deforestation, and a variety of industrial processes and agricultural practices release GHG emissions into the atmosphere (Figs 1 and 2) and have increased the atmospheric concentrations of carbon dioxide (CO_2) and other GHGs. Within the agriculture sector, the main GHGs emitted are methane (CH_4) and nitrous oxide (N_2O), of which the latter is most important in crop-based agriculture due to emissions that occur primarily because of soil management and the application of nitrogen (N)-based fertilizers (Figs 3 and 4).

2.2.2 Temperature

Across the contiguous U.S. states, average surface temperature has risen at an average rate of nearly 0.1°C per decade since the start of the twentieth century. Since the late 1970s, this rate has increased up to threefold with the U.S. warming at a faster rate than the global average. Relatedly, nine of the ten warmest years in the United States have occurred in the last 25 years.

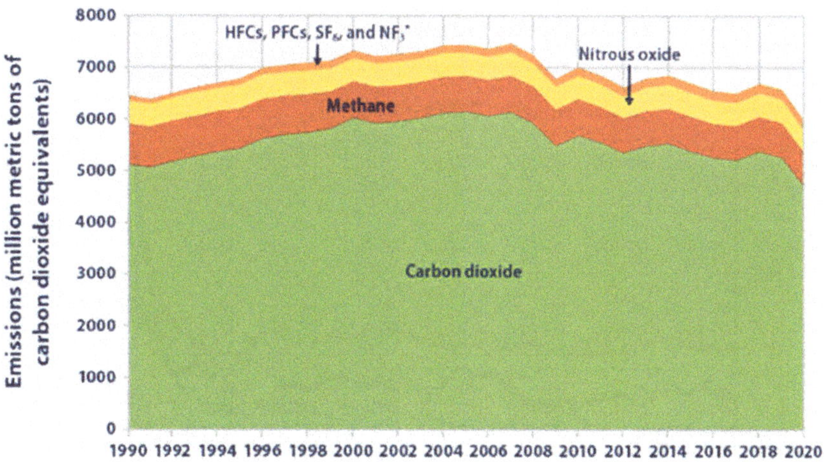

Figure 1 U.S. greenhouse gas (GHG) emissions by gas (million metric tons [MMT] of carbon dioxide equivalent [CO2e]). Carbon dioxide, methane, and nitrous oxide are the major GHGs associated with agricultural activities. Data source: U.S. Environmental Protection Agency (EPA), 2022. www.epa.gov/ghgemissions/inventory-us-greenhouse-gas-emissions-and-sinks.

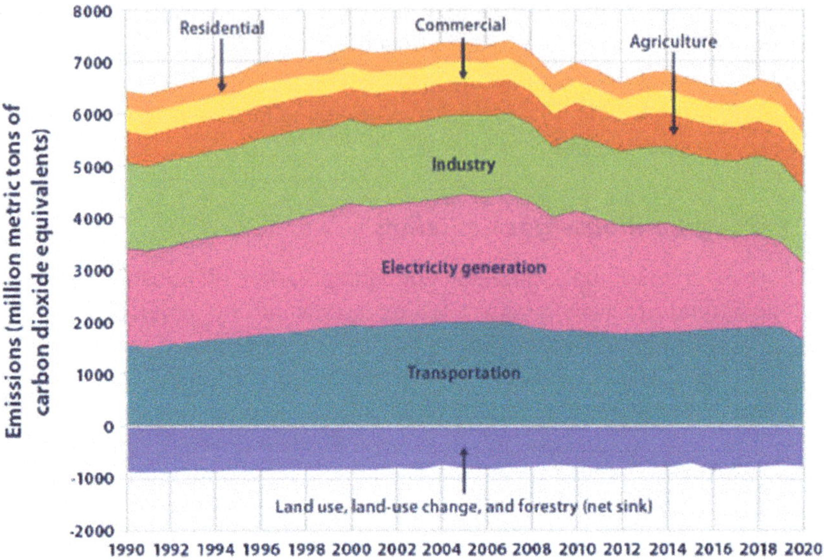

Figure 2 U.S. greenhouse gas (GHG) emissions and sinks (negative values) by economic sector (million metric tons [MMT] of carbon dioxide equivalent [CO2e]) in the United States from 1990 to 2020. Data source: U.S. Environmental Protection Agency (EPA), 2022. www.epa.gov/ghgemissions/inventory-us-greenhouse-gas-emissions-and-sinks.

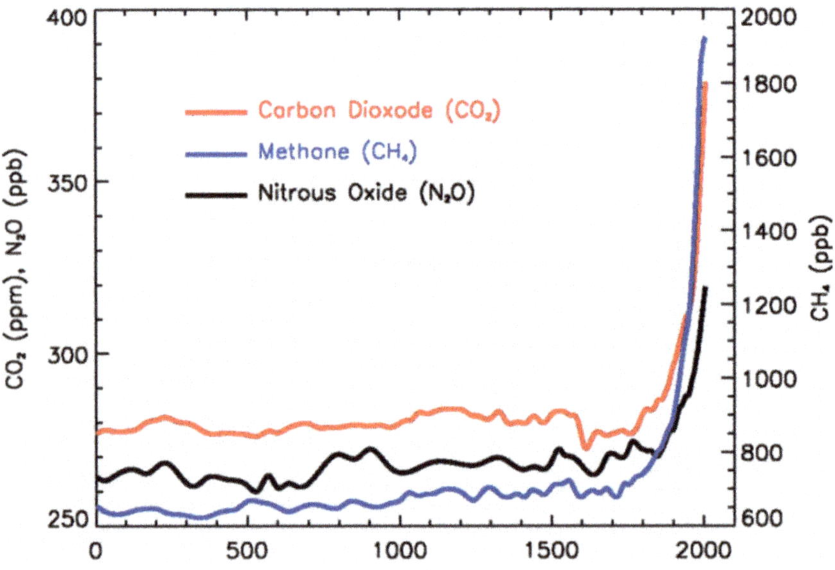

Figure 3 Atmospheric concentrations of agriculturally important long-lived GHGs over the last 2,000 years. Concentration units are parts per million (ppm) or parts per billion (ppb). Data combined and simplified from IPCC Climate Change 2007: Working Group I: The Physical Science Basis.

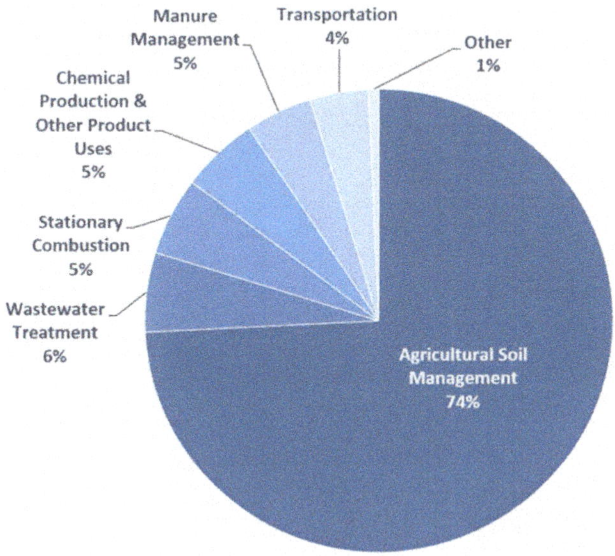

Figure 4 Nitrous oxide emissions, by source in 2020. Data source: U.S. Environmental Protection Agency (EPA), 2022.

In most U.S. regions, mean annual temperatures have increased significantly, but with considerable variability (Fig. 5). Alaska, the North and West (including the northern Midwest) and the Southwest have warmed significantly, whereas parts of the Southeast have not seen much change. The United States will continue to warm substantially until the middle of this century (Karl et al., 2009), with the interior experiencing increases of 2-3°C, and the southeastern and western coastal regions about 1-2°C.

Changes in U.S. temperature have also varied by season (Figs 6 and 7). Since the late 1800s, average winter, spring, summer, and fall temperatures in the contiguous states have increased by 1.7, 1.1, 0.8, and 0.8°C, respectively. The trend toward warmer winters is consistent with reductions in snow and shorter ice seasons and along with warmer springs and falls, is compatible with longer growing seasons in nearly all states. Increases in growing season are greater in the West than in the East and are most pronounced in the Southwest (Fig. 8). Growing season patterns are largely mirrored in the overall timing of spring and fall frost, which now occur about 3 days earlier and later, respectively, than they did in 1980. Similarly, plant growing degree days (GDDs; using 10°C as a standard baseline) that are based on temperature have increased on average by 9% across all long-term measurement stations throughout the United States, with the West showing increases of 20% or more (Fig. 9).

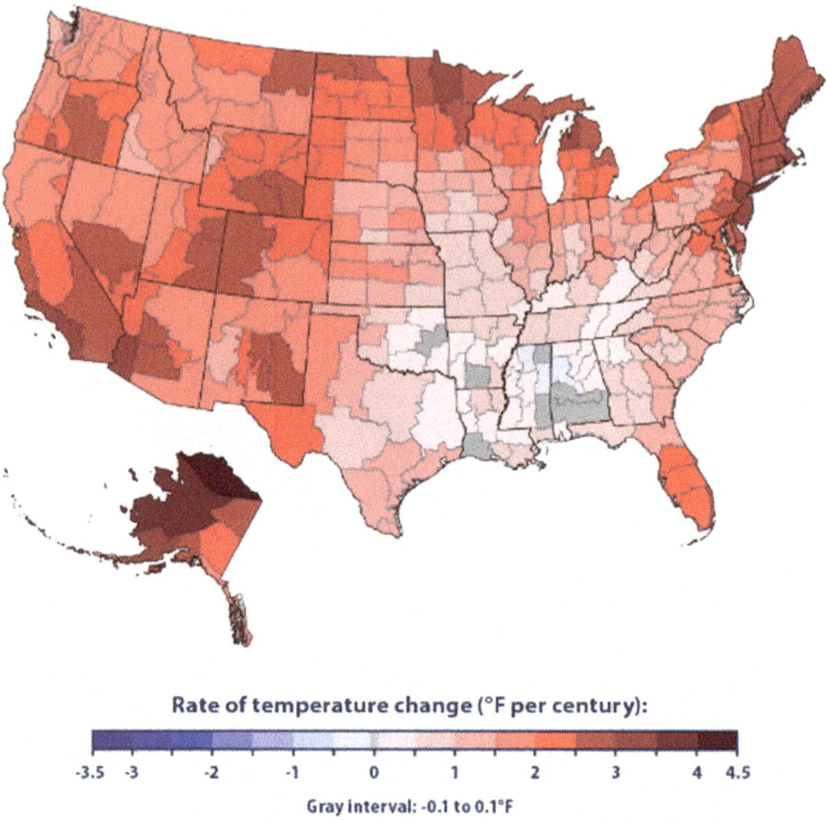

Figure 5 Changes in annual average air temperature in the United States between 1901 and 2021 (since 1925 for Alaska). Data shown are for climate divisions, as defined by the National Oceanic and Atmospheric Administration (NOAA). Data source: NOAA, 2022. Climate at a glance. www.ncdc.noaa.gov/cag.

2.2.3 Extreme temperature events and droughts

Abnormally hot summer days (i.e. highs) are now much more common in the United States than they were only a few decades ago. The occurrence of unusually hot summer nights (i.e. lows) has increased at an even faster rate (Fig. 10), indicating less 'cooling off' during the night. Since the 1980s, unusually cold winter temperatures have become less common. From 2000 to 2009, in the United States, there were twice as many record high temperatures as record low temperatures. As hotter than usual days and nights become more common, heat waves are expected to become more frequent and intense (USGCRP, https://science2017.globalchange.gov/).

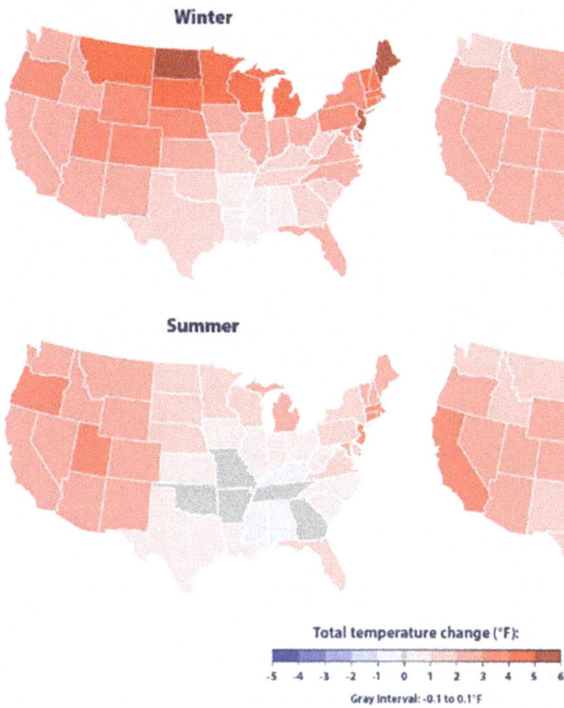

Figure 6 Maps showing the total change in the average seasonal temperature for the contiguous 48 states from 1896 to 2021, in degrees Fahrenheit. Total change in temperature was calculated from the long-term average rate of change. The 3-month seasons are defined as follows: winter (December, January, and February), spring (March, April, and May), summer (June, July, and August), and fall (September, October, and November). Data source: NOAA (National Oceanic and Atmospheric Administration). National Centers for Environmental Information. www.ncei.noaa.gov.

While the last few decades have generally been wetter than average across the contiguous states, large and consistent decreases in the Standardized Precipitation Evapotranspiration Index (SPEI; balance between water supply and demand relevant for impacts on agriculture) have been observed throughout the western United States. Decreases have been especially prominent in the Southwest, particularly California, Arizona, and New Mexico (Fig. 11).

2.2.4 Precipitation

On average, total annual precipitation over the United States has increased. Average decadal rates of increase across the contiguous states since the early 1900s (5 mm) are four times greater than the global value. While average precipitation trends across time and over large areas can mask the variability that occurs over smaller areas or during specific time periods, regional U.S.

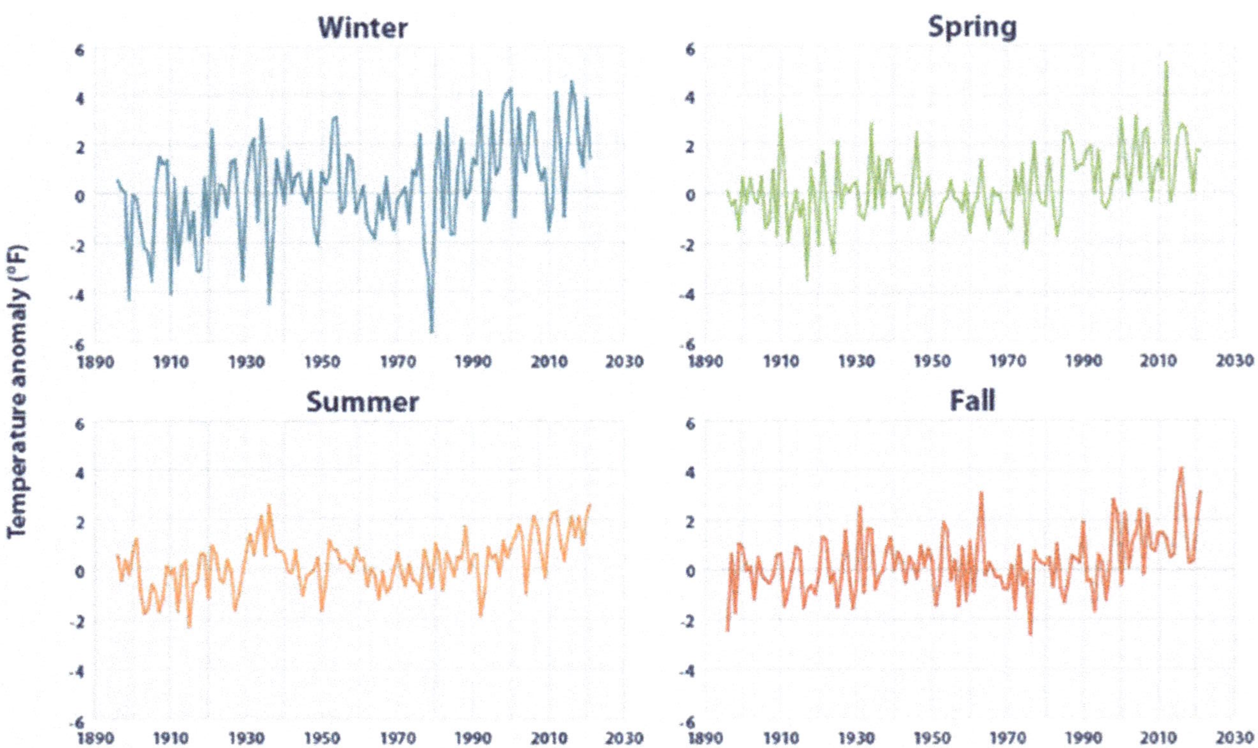

Figure 7 Changes in the average temperature for each season across the contiguous 48 states from 1896 to 2021. Seasons are defined as follows: winter (December, January, and February), spring (March, April, and May), summer (June, July, and August), and fall (September, October, and November). The 1901–2000 average is used as a baseline for depicting change. Data source: NOAA (National Oceanic and Atmospheric Administration), 2022. National Centers for Environmental Information. www.ncei.noaa.gov.

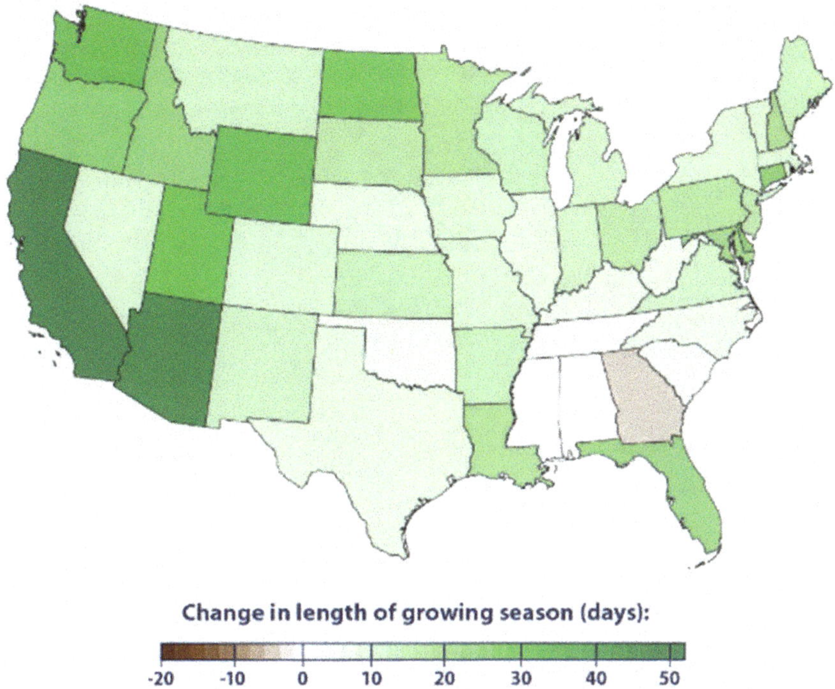

Figure 8 Total change in length of the growing season from 1895 to 2020 for the contiguous 48 states. https://www.epa.gov/climate-indicators/climate-change-indicators-length-growing-season. Data source: Kunkel, 2021.

patterns, while not always significant at the state level, exist. For example, large areas in the Northwest, Central, and Southern United States now receive more precipitation when compared to the early twentieth century, while parts of the Eastern Seaboard and the Rocky Mountains and much of the Southwest receive less (Fig. 12).

Projected changes in precipitation are sensitive to local conditions and shifts in large-scale atmospheric circulation. They are therefore more uncertain and heterogeneous than temperature projections. However, climate models are in agreement that until the middle of this century, certain northern central and eastern regions will see a 5–15% increase in total annual precipitation, whereas the Northwest will likely experience reductions of 15–25% in summertime precipitation, with much of the central South will probably see decreases in annual precipitation of about 5%.

Changes in the amount and timing of winter precipitation impact snow availability and the timing of melt runoff, particularly in the higher latitudes and western regions which are dependent on snow accumulation and a gradual release of snowpack water during the spring and summer. The Northwest and Northeast regions are projected to continue to see increases of 5–15% in

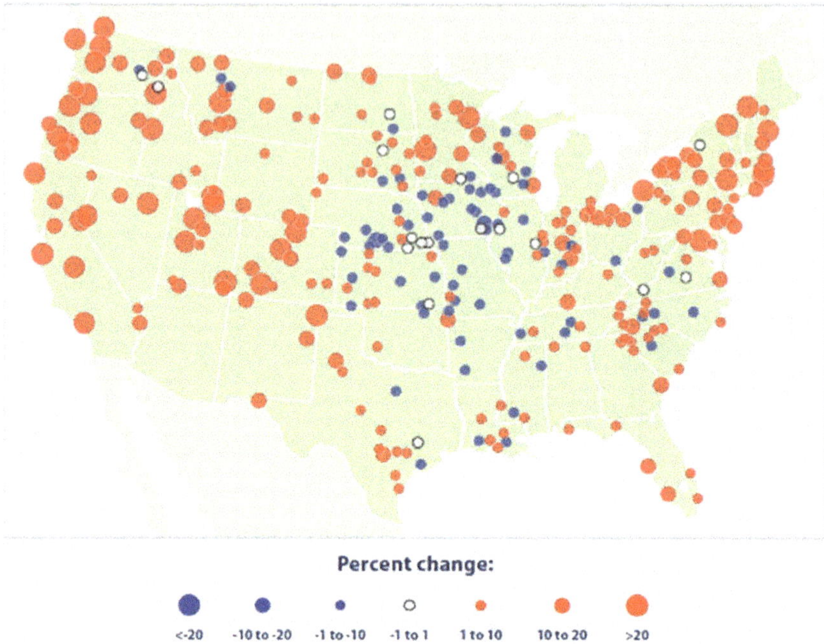

Figure 9 Trends in the total number of growing degree days per year at 305 weather stations across the contiguous U.S. states. The color and size of the symbols represent percent change between 1948 and 2020, based on the long-term average rate of change. Data source: NOAA (National Oceanic and Atmospheric Administration), 2021. Global Historical Climatology Network Daily: Data access. https://www.ncei.noaa.gov/products/land-based-station/global-historical-climatology-network-daily.

winter precipitation, whereas areas along the southern U.S. border will likely experience decreases of 5–10%.

2.2.5 Extreme precipitation events

Extreme or 'heavy precipitation' refers to events (e.g. within a single day) during which the amount of rain or snow substantially exceeds the normal for that location. This 'definition' varies according to the location and the time of year. Over the last century, a larger percentage of precipitation has come in the form of intense single-day events, with occurrences increasing substantially since the 1980s (Fig. 13).

The intensity of precipitation has increased in most areas, even those where there is less total precipitation. This is consistent with a general warming, as the water vapor carrying capacity of the atmosphere increases due to higher average temperature. In future, more precipitation is projected to fall in shorter, more intense storms. This will lead to more rapid runoff, and along with

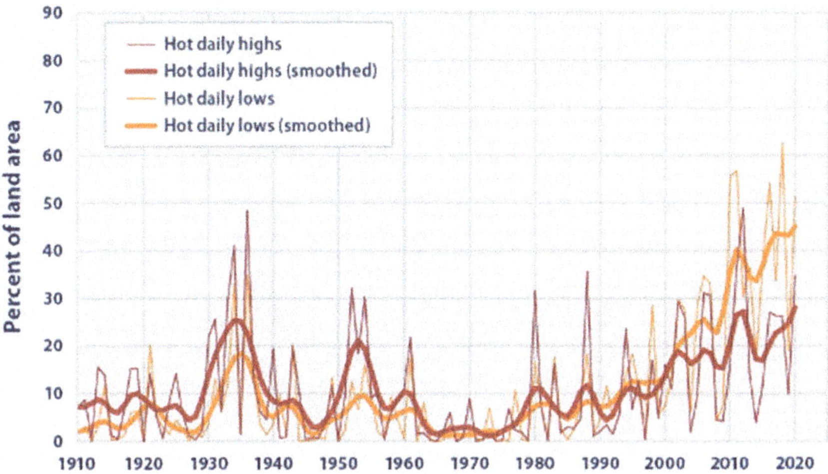

Figure 10 Percentage of the land area of the contiguous U.S. states with unusually hot daily high and low temperatures during the months of June, July, and August. Thin lines represent individual years, while the thick lines show a 9-year weighted average. Red lines represent daily highs, and orange lines represent daily lows. 'Unusual' term based on the long-term average conditions at each location. Data source: NOAA (National Oceanic and Atmospheric Administration), 2021. U.S. Climate Extremes Index. www.ncdc.noaa.gov/extremes/cei.

enhanced evapotranspiration, losses due to higher temperatures may offset any increases in mean precipitation.

2.3 Effects of climate change on soil and water

Soil processes that are sensitive to climate change include erosion and loss of organic matter. Climate change can impact erosion via changes in rainfall, snowmelt, and wind patterns, while the rate of soil organic matter accumulation or loss may be affected due to changes in crop biomass and type, and soil temperature and water availability, that alter organic matter input. Water available for crop growth may be impacted by altered timing and amount of precipitation.

Soil erosion by water is a major environmental threat to sustainable crop production (Pimentel and Burgess, 2013). As extreme precipitation events increase in number, essential soil and water resources will be further and more widely degraded, with crop growth harmed through excessive runoff, leaching, and flooding. Increased erosion can accelerate climate change effects through the loss of carbon and nutrient stocks stored in the soil as well as increase sedimentation in rivers and lakes (Frank et al., 2015).

Extreme precipitation events, high streamflow events, and large nutrient and sediment loadings in river systems are highly correlated (e.g. NRC, 2009).

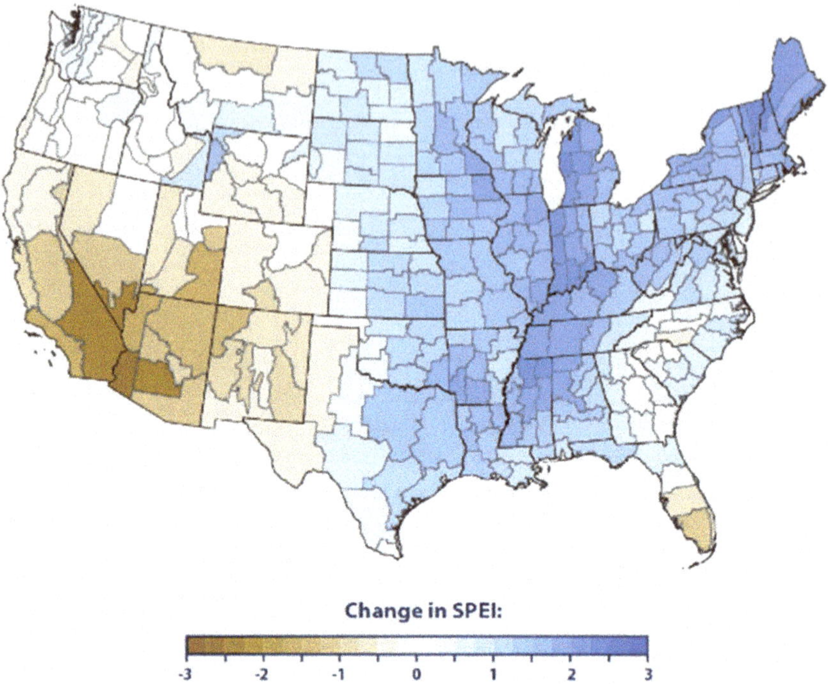

Figure 11 Total change in drought conditions across the contiguous states, based on the long-term average rate of change in the 5-year Standardized Precipitation Evapotranspiration Index (SPEI) from 1900 to 2020. Data are displayed for small regions called climate divisions. Blue areas represent increased moisture; brown areas represent decreased moisture or drier conditions. Data sources: WestWide Drought Tracker, 2021; PRISM, 2021.

Along with increased temperatures, these loadings have caused increases in the duration, intensity, and extent of low-oxygen conditions (hypoxia) in U.S. coastal and freshwater systems over many decades (e.g. Testa et al., 2014).

The effects of changing climate on precipitation, streamflow, and soil water evaporation will impact surface and groundwater systems. Rising temperatures and shifting precipitation patterns will alter crop water requirements, crop water availability, crop productivity, and costs of water access across the agricultural landscape. Higher temperatures will restrict the snow storage season leading to reduced snow accumulation and earlier spring snowmelt, as well as increased evaporative losses from land and water surfaces and transpiration losses from non-crop land cover, both potentially resulting in reduced runoff and streamflow.

2.4 Crop response to climate change

Crop response to climate change is determined by complex interactions of a set of factors that include atmospheric CO_2, temperature, and precipitation.

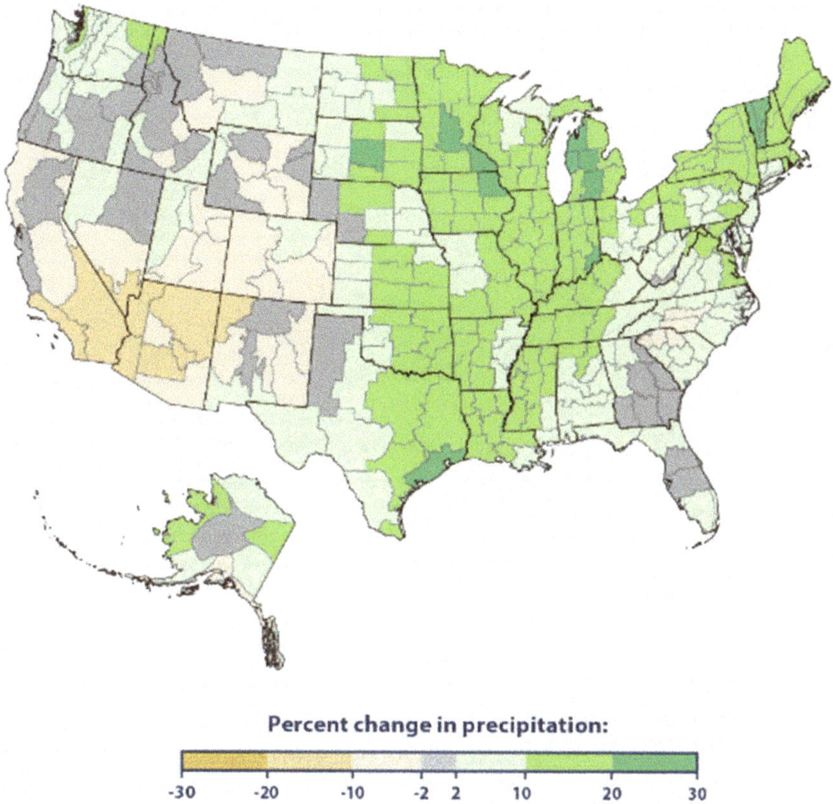

Figure 12 Rate of change in total annual precipitation in the United States (since 1901 for the contiguous states and 1925 for Alaska). The data are shown for climate divisions, as defined by the National Oceanic and Atmospheric Administration (NOAA). Data source: NOAA 2022. Climate at a glance.

Increasing CO_2 concentrations in the atmosphere stimulate photosynthesis typically increasing plant growth, while reducing crop water loss due to partial stomatal closure (e.g. Bunce, 2004). However, elevated CO_2 (leading to higher air temperatures) also increases crop tissue temperatures that can increase damage to plant reproductive processes. Reduced plant nitrogen (Ainsworth and Long, 2005) and protein content (Erbs et al., 2010) can also occur leading to lower grain and forage quality. Higher CO_2 concentrations can also promote increased weed growth relative to crop growth, contributing to a higher risk of crop failure from weed pressure. Reduced crop water loss and increased water-use efficiency (WUE) under elevated CO_2 may be advantageous where precipitation is low or projected to decline, but other changing conditions may offset, e.g. higher temperatures and larger plants that increase crop water

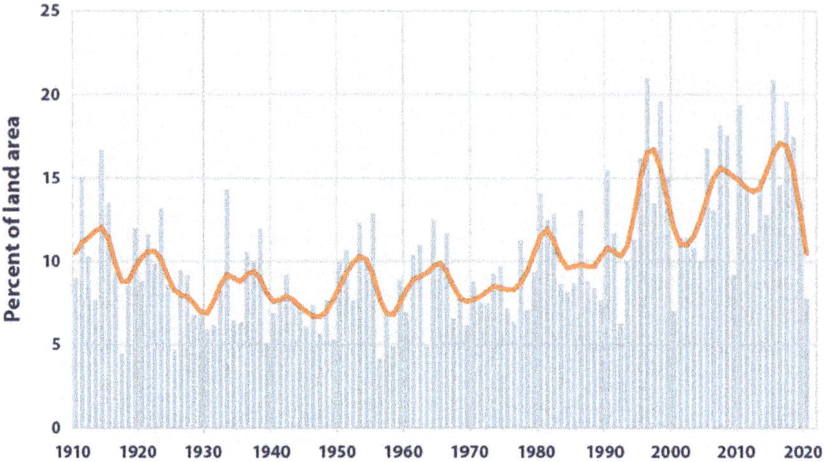

Figure 13 Percentage of the land area of the contiguous states where a much greater than normal portion of total annual precipitation has come from extreme single-day precipitation events between 1910 and 2020. Bars represent individual years, while the line is a 9-year weighted average. Data source: NOAA (National Oceanic and Atmospheric Administration) 2021. U.S. Climate Extremes Index. www.ncdc.noaa.gov/extremes/cei.

demand (Betts et al., 2007), or complement, e.g. declining near-surface winds that may decrease evapotranspiration, this effect.

The effect of increased temperature on a particular crop will depend on its optimal temperature for growth and reproduction (USGCRP, 2014). High temperatures can decrease yields of the major U.S. grain crops corn, soybean, and wheat, and climate change is predicted to decrease yields of other important crops including rice, sorghum, cotton, and oats (Marshall et al., 2015). At a certain point, higher air temperatures will negatively affect plant growth, pollination, and reproductive processes (Klein et al., 2007; Sacks and Kucharik, 2011). In some areas, warming may help crops typically planted in a particular area or allow a shift to crops normally grown in warmer areas through earlier planting dates, faster plant maturation, and longer growing seasons. Increases in minimum temperatures may have more impact as they are more likely to increase over broad geographic scales (Knowles et al., 2006). Higher minimum temperatures affect nighttime plant respiration and can reduce biomass accumulation. Higher maximum temperatures are modulated by local conditions, especially soil water content and evaporative heat loss and their peaks (heat spells) add damage potential to already drought-affected crops through the denaturation of pollen proteins during the flowering phase, leading to spikelet sterility and the inability to develop grains (Alfaro et al., 2006).

2.5 Indirect climate change effects

As well as the direct impacts on major cash crops, the indirect effects of climate change, defined here as the effects on non-crop species such as weeds, insects, pathogens, and invasive species, are of great importance. Altered conditions arising from climate change can impact insect population, pathogen occurrence, weed infestation, and their timing and distribution. Below we briefly overview some indirect effects on weeds and the potential consequences for crops.

Agronomic and invasive weeds (i.e. undesired plant species in crop fields) compete with crops for light, nutrients, and water. Increases in the number of frost-free days associated with warmer temperature may lead to an expansion northward of (sub)tropical weeds associated with current large grain crop losses in southern U.S. states. Conversely, for the same reason, cold-tolerant weeds may become less problematic in the northern states (Mcdonald et al., 2009). Warming may support weed invasion in wetter conditions and inhibit or have no effect in drier environments. Warming may also likely favor invading C4 weeds over C3 crops while inhibiting invading C3 weeds competing with C4 crops. While reduced precipitation may inhibit invasion, droughts may decrease native species resistance and allow for increased invasion once wetter conditions return.

Agronomic weeds typically produce relatively more biomass than crops in response to higher concentrations of CO_2, particularly when both plant species have the same photosynthetic pathway (e.g. Ziska, 2003). Many invasive plants also respond strongly under elevated CO_2 (e.g. Song et al., 2009) and may be particularly favored in water-limited conditions.

2.6 Spatial scales and their disparity

There is a divergence between the typical spatial and temporal scales of climate science and the data requirements for practical management change in crop-based agriculture. Farmers require information at the field to local scale to help them understand and adopt practice change, whereas climate change science and projections typically operate at the regional to national scale.

Similarly, while decadal climate change projections are useful, they are still not well suited for informing annual agricultural decisions. Improved seasonal to annual forecasting is critical for providing information useful for farmers to make in-season management decisions and near-term planning.

Therefore, there is a need for higher spatial and temporal resolution simulations and projections of change to help determine the local-scale effects of climate on agriculture and to help inform real-world management decisions.

3 Can new genetics mitigate climate change impacts on yields?

3.1 Overview

In this section, we present a case study that uses a biogeochemical process-based model (SALUS; Basso et al., 2006; Basso and Ritchie, 2015) to investigate the response of corn yield to projected, near-future climate change trends of rising temperature (due to increases in the atmospheric concentration of CO_2) when compared to a business-as-usual (baseline) scenario using an adapted cultivar as results of improved genetic traits.

3.2 Methods

The SALUS model was used to examine the potential effects of future climate conditions on corn yield in the U.S. Midwest. Model simulations were conducted at 1354 stations enrolled in the National Weather Service Cooperative Observer Program (NWS-COOP). Quality-controlled, daily weather data (minimum and maximum air temperatures and precipitation) for the years 1984 to 2019 were provided for each station by the Iowa Environmental Mesonet. Summer (June–August) averages for maximum and minimum temperatures and total precipitation were calculated. Sen's slopes (Sen, 1968), a nonparametric method to obtain an upper and lower confidence interval, and median of all slopes in climate data, were computed for a 30-year period from 1990 to 2019 and extrapolated to 2050. Projected values for 2050 were used to calculate an additive factor for temperature and a multiplicative factor for precipitation. The delta method was applied to the summer months for the baseline period 1984 to 2019 to represent the future climate scenario at each station. Additional information can be found in Basso et al. (2021).

To fulfill the SALUS model input requirements, daily solar radiation data from the closest NASA Power grid cell to each station were downloaded. These values were not adjusted for the future climate. Required soil inputs for the SALUS model were extracted from the Soil Survey Geographic Database (SSURGO) using characteristics of the predominant soil within a 40-km radius of each station to best represent surrounding cropland.

Field management was simulated by using a continuous corn cropping system with sufficient nitrogen (N) inputs (200 kg N ha^{-1}) applied annually with a split application; at planting and 45 days after planting. Conservation soil management was simulated with no-till and an overwintering rye cover crop. Weekly planting progress reports for corn at the state level for 1981–2019 were downloaded from USDA National Agricultural Statistics Service (NASS) and interpolated spatially based on county-level corn production. Annual dates when at least 50% of the state was planted were used for the historical baseline scenario. For the future climate scenario, planting dates were adjusted to 10 days earlier than the baseline to account for projected warmer temperatures and extended growing seasons allowing farmers to plant earlier in the spring and harvest later in the fall.

We varied the corn cultivar spatially, at the county-level when available, and otherwise at the state level. Associated changes in the timing and length of grain fill, timing of maturity, and daily rate of kernel fill were implemented. For the future scenarios, we compared simulations using the historical cultivar and an adapted cultivar. The scenario with the adapted cultivar assumed genetic advances will allow for an earlier onset and longer period of grain fill. In addition, the daily rate of kernel fill was optimized under historical weather conditions. The solar radiation use efficiency of the adapted corn cultivar was increased from 4.5 to 5.0 g MJ^{-1}. To account for increasing atmospheric CO_2 levels, a value of 375 ppm was used in the historical baseline scenario and 500 ppm was used in the future scenarios based on the intermediate representative concentration pathway (RCP) 7.0 described in the IPCC 6 Technical Report.

Simulation results were processed by removing 138 sites in the historical baseline scenario where yield was too low, or soil organic carbon (SOC) was too high due to soil inputs uncharacteristic of cropland. Mean yields across the time periods were calculated at each of the remaining 1216 stations. The values were spatially interpolated using ordinary kriging through ArcGIS software (version 10.8, ESRI). The historical baseline map was subtracted from the future maps to show the difference between the scenarios. The differences in yield between the scenarios were calculated at the station-level for statistical analysis. Changes in SOC were calculated in the top 40 cm of the soil profile by subtracting the value in the first year from the last year of the simulated time period. The difference in SOC was divided by the number of years to calculate the annual rate of SOC change for each station and scenario.

3.3 Results

For the future scenario using the historical cultivar, varying changes in yield were found when compared to the historical baseline, ranging from a decrease of 2.9 Mg ha^{-1} to an increase of 2.3 Mg ha^{-1} (Fig. 14). The average change in yield was negligible (a decrease of 42 kg ha^{-1} with a standard deviation of 503 kg ha^{-1}). Many of the stations (729; 60%) showed a decrease in yield. The others (487; 40%) with an increased yield were typically located in northernmost parts of the region, including northern Michigan, Wisconsin, Minnesota, North Dakota, South Dakota, and northeast Nebraska (Fig. 15). The spatial pattern of the yield in the future scenario was consistent with the historical spatial pattern,

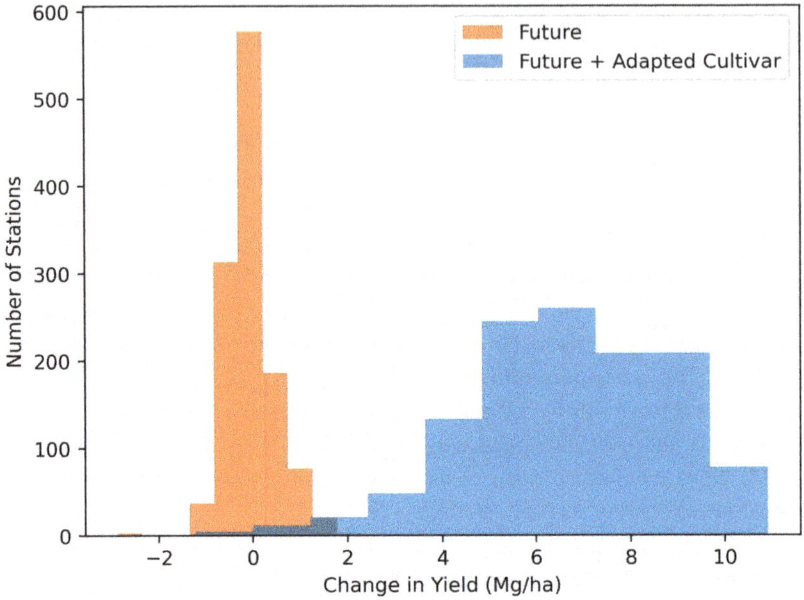

Figure 14 Histogram depicting the change in yield in Mg ha^{-1} from the historical to the future scenario and the future with the adapted cultivar scenario for 1216 stations.

with the highest yielding areas occurring in the main corn-producing areas including Iowa, Illinois, and Indiana.

For the future scenario with the adapted cultivar, a large increase in average yield (6.7 Mg ha^{-1} with a standard deviation of 2.1 Mg ha^{-1}) was found when compared to the historical baseline. The change in yield ranged from −1.2 Mg ha^{-1} to 10.9 Mg ha^{-1}. Only four stations decreased in yield with the adapted cultivar compared to the historical baseline. The spatial pattern of the yields for the future scenario with the adapted cultivar displayed higher yields in the southern part of the region and lower yields in the north. Correspondingly, the largest increases in yield from the historical baseline were in the southern part of the region.

Both historical and future scenarios resulted in an increase in SOC over time with the conservation management practices (Fig. 16). The historical baseline scenario increased by an average of 425 kg ha^{-1} yr^{-1} with a standard deviation of 227 kg ha^{-1} yr^{-1}. The future scenarios showed a slightly larger increase in SOC with an average of 469 kg ha^{-1} yr^{-1} with a standard deviation of 240 kg ha^{-1} yr^{-1} for the future scenario using the historical cultivar, and an average of 480 kg ha^{-1} yr^{-1} with a standard deviation of 232 kg ha^{-1} yr^{-1} for the future scenario using the adapted cultivar.

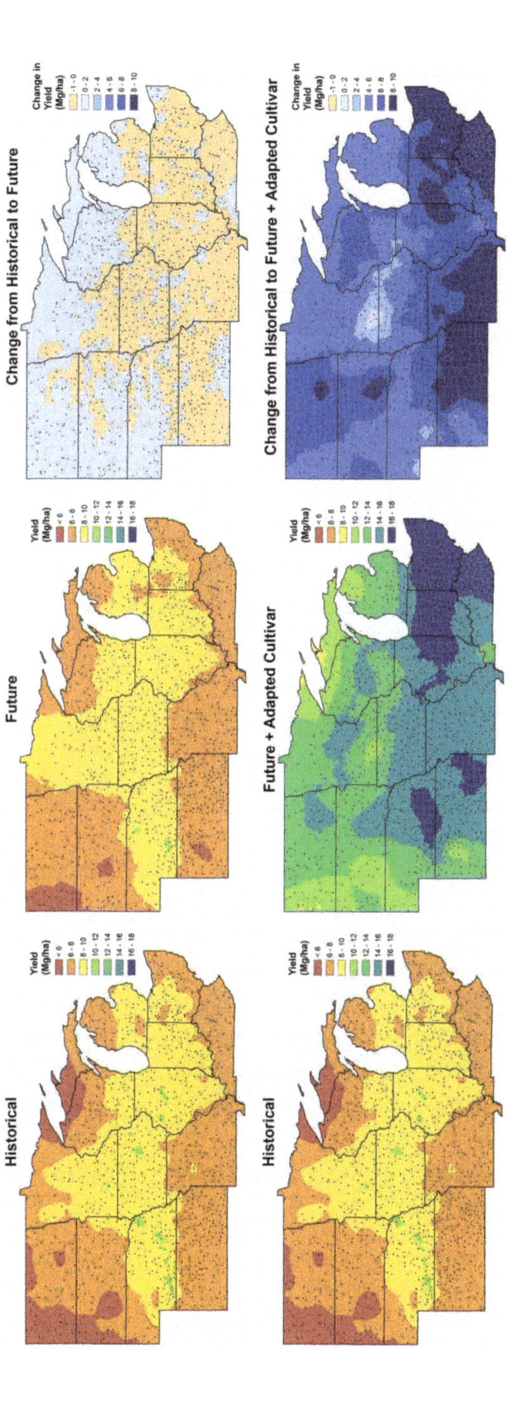

Figure 15 Maps of the mean yields in Mg ha^{-1} for the historical baseline (left), future scenario with the historical cultivar (middle, top), and future scenario with the adapted cultivar (middle, bottom). The historical baseline was subtracted from the future scenarios to show the changes in yield in Mg ha^{-1} for each future scenario accordingly (right).

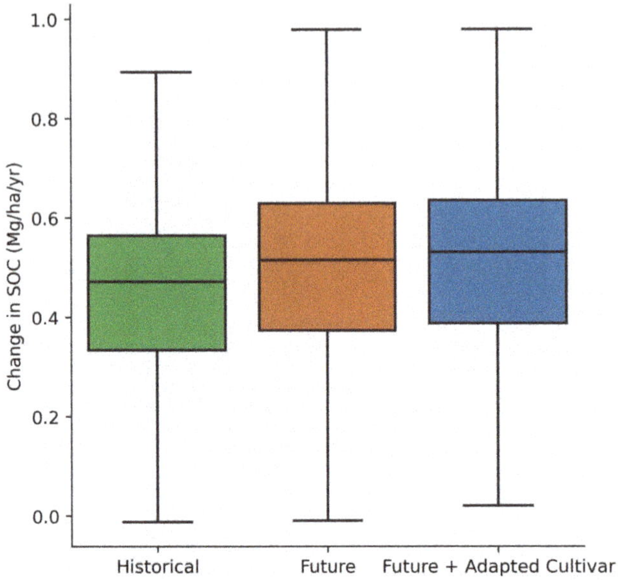

Figure 16 Box-and-whisker plot of the change in SOC in Mg ha-1 yr-1 for historical, future, and future with adapted cultivar scenarios. The box shows the first and third quartiles, with the line at the median, and the whiskers at 1.5 times the interquartile range. Outliers are not shown.

3.4 Implications

Comparison of these projected future scenarios indicates a yield decrease for much of the U.S. Midwest unless the genetics of corn cultivars are improved. Advancements in corn breeding, as represented by the scenario with the adapted cultivar, could cause a significant increase in future yield. If new hybrids can take advantage of increasing temperatures and a longer growing season, southern areas of the U.S. Midwest in particular would, from a corn yield perspective, see an adaptation to mid-century climate conditions. In addition, the conservation soil management practices used in the simulations demonstrate that no-till and cover cropping leads to an increase in SOC, which in combination with improved cultivars, may be more pronounced in the future.

4 Overview of incentives and payment mechanisms for carbon markets

Climate change adds to the increasing pressure on agricultural production systems to help accomplish food security and prevent environmental degradation. Incentivizing farmers to improve soil health and provide other environmental benefits through the adoption of conservation practices could include changes in land use, cropping practice, tillage, nutrient and manure management, and other mitigation and adaptation practices that reduce GHG emissions, sequester carbon, and limit environmental damage. In this section, we describe carbon markets and some options in the United States that may help farmers adopt and maintain conservation management practices on their cropland. We focus on relatively recent opportunities that involve agricultural product supply chains.

4.1 Carbon markets as an opportunities to sequester carbon and reduce greenhouse gas emissions

Carbon markets treat emissions reductions and removals as a commodity and are a policy tool that can be used as part of a broader climate strategy to mitigate GHG emissions. The commodity trading unit is one metric ton (tonne) of carbon dioxide equivalent (CO_2e) emissions. In the agriculture sector, the most important GHG emissions other than CO_2 are N_2O and CH_4, the former being the most important in cropping systems due to its emissions associated with soil management and the application of nitrogen (N)-based fertilizers. Non-CO_2 gases are converted to CO_2e's using global warming potential (GWP) values that are based on how much energy the gas absorbs in the atmosphere when compared to an equivalent mass of emitted CO_2, typically over a 100-year duration. The larger the GWP, the more it warms the globe compared to CO_2 (IPCC, 2021).

Carbon markets generally fall into two broad categories: compliance markets and voluntary markets. Compliance markets often support a program that requires certain sectors and corporations to reduce (cap) their GHG emissions, whereas voluntary markets provide businesses, and other organizations and individuals that are not regulated in terms of their GHG emissions the ability to purchase carbon offsets to support their environmental or other objectives.

A traded carbon unit or commodity can be referred to as a credit, offset, or inset. Credit and offset terms are often used interchangeably but can be distinguished; a credit (or allowance) is a 'permission' from a regulator (e.g. government) to generate GHG emissions and is the basis for cap-and-trade schemes and emissions trading systems (ETS), whereas an offset is a measurable reduction, avoidance, or sequestration of GHG emissions that is used to compensate for emissions elsewhere. From a timing perspective, in an offsetting mechanism, the traded reduction in emissions has already occurred, while in an ETS for example, organizations trade credits to emit in the future.

With insets, companies can implement methods to reduce their internal (indirect) emissions, as opposed to offsets where external emissions are reduced. With insetting, companies reduce indirect emissions that are associated with products within their own supply chain. These emissions can include those generated in the production and transport of agricultural raw materials or ingredients such as grains that the company purchases. Purchasing these insets helps a company with their (commonly termed) Scope 3 emissions (e.g. https://www.epa.gov/climateleadership/scope-3-inventory-guidance#description). These are distinct from Scope 1 (direct emissions controlled by the company) and Scope 2 (indirect emissions from the energy that the company buys) emissions. In the agriculture sector, food and beverage companies can work with farmers that they purchase goods from to lower their Scope 3 supply chain emissions and make claims about these reductions in their annual environmental, social, and governance (ESG) reports submitted to the U.S. Securities and Exchange Commission (SEC). Benefits to Scope 3 insets include the ability for multiple companies to co-finance and co-claim these improvements to their shared supply chains. New rules around Scope 3 mandatory reporting (e.g. https://www.sec.gov/news/press-release/2022-46) are not aimed at farms and farmers, but rather larger publicly traded companies (e.g. https://www.americanprogress.org/article/the-secs-proposed-scope-3-emissions-disclosure-will-not-affect-farms-and-ranches/). Many businesses including those from the food and beverage sector have set near-future voluntary targets to reduce their GHG emissions. Many of these organizations use the Science-Based Targets Initiative (SBTi; (https://sciencebasedtargets.org/faqs) to help set these targets and guide them through this process. The Greenhouse Gas Protocol (https://ghgprotocol.org/standards) is the commonly accepted standard used by food and beverage companies to determine their inset inventories. Since very few of these companies produce their own raw materials, and many agricultural products have large GHG footprints, many food and beverage companies have very high Scope 3 supply chain emissions, up to nearly 90% of their total GHG inventory (Reavis et al., 2022).

Key concerns regarding the quality of agricultural (and other sector) carbon units include additionality, baselines, leakage, and permanence. These issues and others are well known and have been described in detail elsewhere (e.g. Oldfield et al., 2021).

4.2 U.S. carbon markets for agriculture

In the United States, there are two current compliance markets: the California cap-and-trade program (https://ww2.arb.ca.gov/our-work/programs/cap-and-trade-program) and the Regional Greenhouse Gas Initiative (RGGI; https://www.rggi.org/). Both these regulatory systems allow companies to reduce a small portion of their annual required GHG reductions through the purchase of offsets generated by other non-capped sectors, such as agriculture. The agricultural offsets that can be purchased from these programs can currently only be generated from livestock manure management and rice cultivation projects, both of which focus on avoiding emissions of CH_4. Therefore, cropland (other than rice) farmers are not currently catered for in U.S. compliance markets.

There are several carbon registries that operate in the United States and that work with voluntary markets. These registries track offset ownership and issue credits for verified and certified units of emissions reductions or removals, and include Verra, American Carbon Registry, Climate Action Reserve, and The Gold Standard. They also provide additional services such as developing and approving protocols and operating standards and accrediting third-party verifiers. Reviews and comparisons of these registries and their role in agricultural carbon markets are available (e.g. Oldfield et al., 2021). The agriculture protocols and projects that have been approved for offset generation in the voluntary market are more varied and numerous than in the compliance market. They include those for N_2O mitigation from reduced N fertilizer application, sustainable grassland management, biochar utilization in soil, reduced enteric CH_4 through feed alteration, avoided conversion of grasslands and shrublands, and adoption of sustainable agricultural land management practices.

Farmers in the United States can register carbon offsets they have generated on their farms via conservation practices using the relevant methodology on the registry. They can then sell that carbon offset to a potential buyer. However, as a percentage of the total number of offsets generated in the compliance and voluntary markets by various sectors, agriculture's contribution is very low, about 1%, of which the vast majority is from projects that digest manure CH_4 (https://agfundernews.com/carbon-credits-just-one-percent-from-agriculture).

In recent years, the number of private market initiatives or programs in the United States that pay farmers and landowners to generate insets (and offsets) for use in carbon markets has grown substantially. More detailed overviews and comparisons of these entities are available (e.g. Plastina and Wongpiyabovorn, 2023; Plastina, 2022, and Stubbs et al., 2021). Briefly, the programs that operate currently include carbon and ecosystems credit entities, carbon credit entities, input suppliers, and data platforms, with differing eligibility based on, for example, location, acreage, crops, and practices, as well as requirements in contract length and data sharing, the process and rigor of certification and verification, and the degree and method of compensation.

There are therefore current and future opportunities for farmers to engage with food and beverage companies as part of these market initiatives, in particular to generate carbon insets through changes to their management practices. Typical, eligible practices include reduced or no-till operations, cover crops, intercropping, improved nutrient management, and new or diversified crops and rotations.

5 References

Ainsworth, E. A. and Long, S. P. 2005. What have we learned from 15 years of free-air CO2 enrichment (FACE)? A meta-analytic review of the responses of photosynthesis, canopy properties and plant production to rising CO_2. *New Phytologist* 165(2): 351–371.

Alfaro, E. J., Gershunov, A. and Cayan, D. 2006. Prediction of summer maximum and minimum temperature over the central and western United States: The roles of soil moisture and sea surface temperature. *Journal of Climate* 19(8): 1407–1421.

Basso, B., Martinez-Feria, R. A., Rill, L. and Ritchie, J. T. 2021. Contrasting long-term temperature trends reveal minor changes in projected potential evapotranspiration in the US Midwest. *Nature Communications* 12(1): 1476. doi:10.1038/s41467-021-21763-7.

Basso, B. and Ritchie, J. T. 2015. Simulating crop growth and biogeochemical fluxes in response to land management using the SALUS model. In: Hamilton, S., et al. (Eds), *The Ecology of Agricultural Landscapes: Long-Term Research on the Path to Sustainability*. Oxford University Press, 252–274.

Basso, B., Ritchie, J. T., Grace, P. R. and Sartori, L. 2006. Simulation of tillage systems impact on soil biophysical properties using the SALUS model. *Italian Journal of Agronomy* 1(4): 667–688.

Betts, R. A., Falloon, P. D., Goldewijk, K. K. and Ramankutty, N. 2007. Biogeophysical effects of land use on climate: Model simulations of radiative forcing and large-scale temperature change. *Agricultural and Forest Meteorology* 142(2–4): 216–233.

Bunce, J. A. 2004. Carbon dioxide effects on stomatal responses to the environment and water use by crops under field conditions. *Oecologia* 140(1): 1–10.

Erbs, M., Manderscheid, R., Jansen, G. and Seddig, S., Pacholski, A. and Weigel, H-J. 2010. Effects of free-air CO_2 enrichment and nitrogen supply on grain quality parameters and elemental composition of wheat and barley grown in crop rotation. *Agriculture, Ecosystems and Environment* 136: 59–68.

Frank, D., Reichstein, M., Bahn, M., Thonicke, K., Frank, D., Mahecha, M. D., Smith, P., van der Velde, M., Vicca, S., Babst, F., Beer, C., Buchmann, N., Canadell, J. G., Ciais, P., Cramer, W., Ibrom, A., Miglietta, F., Poulter, B., Rammig, A., Seneviratne, S. I., Walz, A., Wattenbach, M., Zavala, M. A. and Zscheischler, J. 2015. Effects of climate extremes on the terrestrial carbon cycle: Concepts, processes and potential future impacts. *Global Change Biology* 21(8): 2861–2880. doi:10.1111/gcb.12916.

Iowa environmental mesonet 2019. *NWS COOP Network: Quality Controlled and Estimated Data*. Iowa State University. Available at: http://mesonet.agron.iastate.edu/request/coop/fe.phtml.

IPCC 2021. Climate change 2021: the physical science basis. In: Masson-Delmotte, V., Zhai, P., Pirani, A., Connors, S. L., Péan, C., Berger, S., Caud, N., Chen, Y., Goldfarb, L., Gomis, M. I., et al. (Eds). Contribution of Working Group I to the Sixth Assessment Report of the Intergovernmental Panel on Climate Change. Cambridge University Press, Cambridge, United Kingdom and New York, NY, 2391 Available at: https://www.ipcc.ch/report/ar6/wg1/downloads/report/IPCC_AR6_WGI_FullReport_small.pdf. doi:10.1017/9781009157896.

Karl, T., Melillo, J., Peterson, T. and Hassol, S. J. (Eds) 2009. *Global Climate Change Impacts in the United States*. Cambridge University Press.

Klein, A. M., Vaissière, B. E., Cane, J. H., Steffan-Dewenter, I., Cunningham, S. A., Kremen, C. and Tscharntke, T. 2007. Importance of pollinators in changing landscapes for world crops. *Proceedings. Biological Sciences* 274(1608): 303–313.

Knowles, N., Dettinger, M. D. and Cayan, D. R. 2006. Trends in snowfall versus rainfall in the western United States. *Journal of Climate* 19(18): 4545–4559.

Kunkel, K. E. 2004. 2021 update to data originally published in: Kunkel, K.E., D.R. Easterling, K. Hubbard, and K. Redmond. Temporal variations in frost-free season in the United States: 1895–2000. *Geophysical Research Letters* 31(3): L03201.

Lewis, T. K. Maycock and Stewart, B. C. (Eds). *U.S. Global Change Research Program*, Washington, DC, 391–437. doi:10.7930/NCA4.2018.CH10.

Marshall, E., Aillery, M., Malcolm, S. and Williams, R. 2015. Climate Change, Water Scarcity, and Adaptation in the U.S. Fieldcrop Sector, ERR-201, U.S. Department of Agriculture, Economic Research Service, November 2015.

McDonald, A., Riha, S., DiTommaso, A. and DeGaetano, A. 2009. Climate change and the geography of weed damage: analysis of U.S. maize systems suggests the potential for significant range transformations. *Agriculture, Ecosystems and Environment* 130: 131–140. doi:10.1016/j.agee.2008.12.007.

NASA. *Prediction of Worldwide Energy Resources (Power)* (Data Access Viewer 2018).

National Research Council (NRC) 2009. *Nutrient Control Actions for Improving Water Quality in the Mississippi River Basin and Northern Gulf of Mexico*. The National Academies Press, Washington, DC, 90 pp. doi:10.17226/12544.

Oldfield, E. E., Eagle, A. J., Rubin, R. L., Rudek, J., Sanderman, J. and Gordon, D. R. 2021. Agricultural soil carbon credits: Making sense of protocols for carbon sequestration and net greenhouse gas removals. Environmental Defense Fund, New York, NY. Available at: https://www.edf.org/sites/default/files/content/agricultural-soil-carbon-credits-protocol-synthesis.pdf.

Pimentel, D. and Burgess, M. 2013. Soil erosion threatens food production. *Agriculture* 3(3): 443–463. doi:10.3390/agriculture3030443.

Plastina, A. 2022. The U.S. voluntary agricultural carbon market: Where to from here? Farm Foundation Issue Report, May 2022. Available at: https://www.farmfoundation.org/wp-content/uploads/2022/06/Farm-Foundation-Issue-Report-on-Agricultural-Carbon-Market.pdf.

Plastina, A. and Wongpiyabovorn, O. 2023. How to grow and sell carbon credits in U.S. agriculture. Iowa State Extension, Ag Decision Maker, File A1-76. Updated Feb. 2023. https://www.extension.iastate.edu/agdm/crops/pdf/a1-76.pdf.

Reavis, M., Ahlen, J., Rudek, J. and Naithani, K. 2022. Evaluating greenhouse gas emissions and climate mitigation goals of the global food and beverage sector. *Frontiers in Sustainable Food Systems* 5: 789499. doi:10.3389/fsufs.2021.789499.

Sacks, W. J. and Kucharik, C. J. 2011. Crop management and phenology trends in the U.S. corn belt: Impacts on yields, evapotranspiration and energy balance. *Agricultural and Forest Meteorology* 151(7): 882–894.

Sen, P. K. 1968. Estimates of the regression coefficient based on Kendall's tau. *Journal of the American Statistical Association* 63(324): 1379–1389.

Soil Survey Staff. Soil Survey Geographic (SSURGO) Database, 2019. Available at: https://sdmdataaccess.sc.egov.usda.gov.

Song, L., Wu, J., Li, C., Li, F., Peng, S. and Chen, B. 2009. Different responses of invasive and native species to elevated CO_2 concentration. *Acta Oecologica* 35(1): 128–135.

Stubbs, M., Hoover, K. and Ramseur, J. L. 2021. Agriculture and forestry offsets in carbon markets: background and selected issues. Congressional Research Service, R46956. Available at: https://crsreports.congress.gov/product/pdf/R/R46956.

Swinton, S. M., Lupia, F., Robertson, G. P. and Hamilton, S. K. 2007. Ecosystem services and agriculture: Cultivating agricultural ecosystems for diverse benefits. *Ecological Economics* 64(2): 245–252. doi:10.1016/j.ecolecon.2007.09.020.

Testa, J. M., Li, Y., Lee, Y. J., Li, M., Brady, D. C., Di Toro, D. M., Kemp, W. M. and Fitzpatrick, J. J. 2014. Quantifying the effects of nutrient loading on dissolved O_2 cycling and hypoxia in Chesapeake Bay using a coupled hydrodynamic–biogeochemical model. *Journal of Marine Systems* 139: 139–158. doi:10.1016/j.jmarsys.2014.05.018.

USDA-NASS. Quick stats. http://quickstats.nass.usda.gov/(2019).

U.S. Environmental Protection Agency 2022. Inventory of U.S. greenhouse gas emissions and sinks: 1990–2020. EPA 430-R-22-003. Available at: www.epa.gov/ghgemissions/inventory-us-greenhouse-gas-emissions-and-sinks.

United States Global Change Research Program (USGCRP), Hatfield, J., Takle, G., Grotjahn, R., Holden, P., Izaurralde, R. C., Mader, T., Marshall, E., and Liverman, D. 2014. Ch. 6: Agriculture. In: Melillo, J. M., Richmond, Terese (T.C.), and Yohe, G. W. (Eds), *Climate Change Impacts in the United States: The Third National Climate Assessment*. U.S. Global Change Research Program, 150–174.

United States Global Change Research Program (USGCRP), Gowda, P., Steiner, J. L., Olson, C., Boggess, M., Farrigan, T. and Grusak, M. A. 2018. Agriculture and rural communities. In: Reidmiller, D. R., Avery, C. W., Easterling, D. R. and Kunkel, K. E. (Eds.), *Impacts, Risks, and Adaptation in the United States: Fourth National Climate Assessment, Volume II*.

Walthall, C., Backlund, P., Hatfield, J., Lengnick, L., Marshall, E., Walsh, M., Adkins, S., Aillery, M., Ainsworth, E. A., Amman, C., Anderson, C. J., Bartomeus, I., Baumgard, L. H., Booker, F., Bradley, B., Blumenthal, D. M., Bunce, J., Burkey, K., Dabney, S. M., Delgado, J. A., Dukes, J., Funk, A., Garrett, K., Glenn, M., Grantz, D. A., Goodrich, D., Hu, S., Izaurralde, R. C., Jones, R. A. C., Kim, S.-H., Leaky, A. D. B., Lewers, K., Mader, T. L., McClung, A., Morgan, J., Muth, D. J., Nearing, M., Oosterhuis, D. M., Ort, D., Parmesan, C., Pettigrew, W. T., Polley, W., Rader, R., Rice, C., Rivington, M., Rosskopf, E., Salas, W. A., Sollenberger, L. E., Srygley, R., Stöckle, C., Takle, E. S., Timlin, D., White, J. W., Winfree, R., Wright-Morton, L. and Ziska, L. H. 2012. Climate change and agriculture in the United States: Effects and adaptation. USDA Technical Bulletin 1935. U.S. Department of Agriculture, Washington, DC, 186 pp. http://www.usda.gov/oce/climate_change/effects_2012/CC%20and%20Agriculture%20Report%20(02-04-2013)b.pdf.

WestWide Drought Tracker. 2021. SPEI 60 month. https://wrcc.dri.edu/wwdt.

Wongpiyabovorn, O., Plastina, A. and Crespi, J. M. 2022. Challenges to voluntary Ag carbon markets. *Applied Economic Perspectives and Policy* 2022: 1–14. doi:10.1002/aepp.13254.

Ziska, L. H. 2003. Evaluation of the growth response of six invasive species to past, present and future atmospheric carbon dioxide. *Journal of Experimental Botany* 54(381): 395–404.

Chapter 14

Modeling the impact of climate change on agriculture in Latin America

Fábio R. Marin and Ivo Zution Gonçalves, University of São Paulo, Brazil; Francisco J Meza, Forestal. Pontificia Universidad Católica de Chile, Chile; Néstor M. Riaño H., Ministry of Agriculture and Rural Development, Colombia; Andrés J. Peña Q, Colombian Corporation for Agricultural Research (AGROSAVIA), Colombia; Alexandre B. Heinneman, EMBRAPA, Brazil; Nereu A. Streck and Alencar Jr Zanon, Universidade Federal de Santa Maria, Brazil; Murilo S. Vianna, University of Bonn, Germany; and Evandro H. F. Silva and Nilson A. Vieira Jr, University of São Paulo, Brazil*

1 Introduction
2 Developments in crop modeling and climate change in Chile
3 Developments in crop modeling and climate change in Colombia
4 Developments in crop modeling in Brazil: upland rice production and climate change
5 Developments in crop modeling in Brazil: irrigated rice production and climate change
6 Developments in crop modeling in Brazil: common bean production and climate change
7 Developments in crop modeling in Brazil: soybean production and climate change
8 Developments in crop modeling in Brazil: sugarcane production and climate change
9 Developments in crop modeling in Brazil: grass forage production for livestock and climate change
10 Conclusion
11 Where to look for further information
12 References

1 Introduction

Between 2017 and 2050, the world's population is projected to rise from 7.7 billion to 9.8 billion, with Latin America and the Caribbean region potentially having an extra 57 million people by 2031 (United Nations, 2017). The projected

* fabio.marin@usp.br.

http://dx.doi.org/10.19103/AS.2022.0115.16

increase in income across the developing world will also potentially increase demand for meat and dairy products by 50–100% in the next few decades (Thornton, 2010; Foley et al., 2011; Tilman et al., 2011; Searchinger et al., 2019). This may result in grain price increases (Marchand et al., 2016), which might increase food insecurity in several Latin American countries (Sternberg, 2012; D'Amour et al., 2016).

Around 38% of available land in Latin America and the Caribbean is used for agriculture (about 10% for crops and 28% for pasture) with 46% covered with forests. Latin America is an important global producer and exporter of a variety of agricultural commodities. According to OECD-FAO (2019) trade data, the region accounts for 14% of global food production. Brazil is the largest agricultural and food exporter (USD 79.3 billion in 2017) in the region, followed by Argentina (USD 35.0 billion), Mexico (USD 32.5 billion), Chile (USD 17 billion), Ecuador (USD 10.4 billion), and Peru (USD 8.8 billion). South America is the main global supplier of agricultural produce such as corn, soybeans, cotton, sugar, chicken meat, and beef as well as products derived from crops such as ethanol. The region is expected to remain the largest soybean producer in the world, accounting for 53% of global production with a predicted growth of 12.4 Mha in harvested area by 2031. Overall agricultural and livestock production in the region is expected to grow 14% over the next decade, 64% of this growth coming from crop production, 28% from the livestock sector, and the rest coming from the fishery. Productivity gains have contributed greatly to agricultural production growth in recent years with estimated average yield gains of around 10% by 2031 for key agricultural commodities (OECD-FAO, 2022).

However, rising temperatures, more frequent weather extreme events, and changes in rainfall patterns are some of the effects climate change is having on Latin America. Climate change is projected to significantly reduce crop yields (Morris et al., 2020). The Intergovernmental Panel on Climate Change (IPCC) special report on climate change and land predicts reductions of 6% in the production of 11 staple crops in the Latin America and the Caribbean region by 2046–2055 (IPCC, 2019). Climate change is expected to affect agriculture worldwide in the next decade (Zabel et al., 2019).

Models are a consolidated set of assumptions, commonly presented as interrelated algorithms that seek to represent complex systems. Decision-makers and others rely on models to estimate the response of a system under specific conditions. In the case of crop models, key inputs are time series of meteorological data, soil physical properties (chemical and biological parameters are also sometimes needed for specific functions), and agricultural decisions or interventions that can be classified as general agronomic management. Global circulation models (GCMs) are central to climate change research (Field, 2014) and can be coupled to process-based crop models

(PBCM) to investigate the impacts of climate change on agriculture (Rosenzweig et al., 2013; He et al., 2017). Different approaches include niche models focused on impacts on particular corps and potential mitigation strategies (e.g. Castro et al., 2019; Läderach et al., 2013; Läderach et al., 2017) and more general models (e.g. Sotelo et al., 2020; Eitzinger et al., 2014) assessing the effect of climate on the agroecosystem.

This chapter provides an overview of crop modeling in three Latin American countries: Chile, Colombia, and Brazil. It assesses model characteristics, strengths and weaknesses, and areas for future development.

2 Developments in crop modeling and climate change in Chile

Chile occupies a long, relatively narrow strip of land in South America between the Andes to the east and the Pacific Ocean to the west. Chile is the southernmost country in the world and the closest to Antarctica. It covers an area of 756 096 km^2, with a population of 19.8 million (INE, 2022). Chile is among the top ten agricultural exporters in the world. Key exports include wine and high-value fruits such as grape, cherry, cranberry, apple, kiwi, peach, and plum. The country is still an important producer of dairy, meat, and fishery products. In 2020, the agriculture industry contributed $27 billion or 11% of Chile's GDP, accounting for 28% of overall trade and 105 of the country's workforce (ITA, 2022). The country's mountainous landscape of Chile means that arable land represents only 2.62% (22.4 km^2) of the total territory.

Crop modeling can be traced back to the Netherlands in the 1960s, spreading to the US in the 1970s (Pinter et al., 2003; Jones et al., 2016). It was introduced in Chile from the US in the 1970s via the Rockefeller Foundation Scholarship program (Quesada, 2013). Some of the pioneers and local leaders in this field were Aldo Norero from Pontificia Universidad Católica de Chile, and Edmundo Acevedo and Fernando Santibañez from Universidad de Chile. Early challenges included limited predictive capabilities in a highly heterogeneous soil and management environment, difficulties in accessing reliable data for local parameterization and validation, and low literacy in computer programming. One formidable exception was the work of Claudio Stöckle, a key figure in the development of CROPSYST (Stöckle et al., 2003). This platform for cropping systems modeling synthesizes soil, water, and atmospheric impacts on crop ecophysiology and has been extensively used and discussed.

Around the late 1990s, crop simulation models were used more extensively to answer initial questions about vulnerability, impacts, and adaptation of agricultural systems to climate change. These were presented as part of the National Communications, a report presented to the United Framework Convention on Climate Change as part of the commitments of Non-Annex

I listed parties. Second, the Third and Fourth Communications contained detailed maps and information on crop yield changes as functions of future climate scenarios, produced using the SIMPROC crop simulation model (Santibáñez, 1997). These were also used to generate estimates of economic losses associated with climate change (CEPAL, 2012).

In the early 2000s, crop simulation models such as EPIC (Williams et al., 1989) were used to analyze the main impacts of seasonal climate variations, particularly those associated with the El Niño-Southern Oscillation phenomenon (Meza et al., 2003; Meza and Wilks, 2003) and to assess tactical interventions to adapt to good or bad years (Meza and Wilks, 2004). There were also modeling developments in livestock management (Castellaro et al., 2006) to improve existing model platforms such as the Crop Environment Resource Synthesis (CERES) (Cazanga et al., 2019), to evaluate their applicability to predict crop performance under limiting conditions (Artacho et al., 2011), and to study the effect of nutrients on crop growth (Valdés-Gomez et al., 2014). There has also been an increase in research on using models to predict the impact of climate change on major crops such as wheat and maize (Meza and Silva, 2009), analyzing optimum grid resolution (Orrego et al., 2014) and exploring non-traditional adaptation strategies such as double cropping (Meza et al., 2008). Other more recent uses of models include assessing biofuel crop potential (Morales-Salinas et al., 2015), the impact of irrigation management under climate change scenarios (Pizarro et al., 2022), and optimization of harvest dates (Herrera-Cáceres et al., 2017), as well the use of STICS (Simulateur mulTIdisciplinaire pour les Cultures Standard) model to study the performance of grapevines (Valdés-gomez et al., 2014).

The first approximations on crop yield impacts were done at very coarse resolution, which limited accuracy (Jones and Thornton, 2000, 2003). A breakthrough in crop modeling in Chile occurred in 2020 with the development of the Climate Change Risk Atlas for Chile (ARCLIM, www.arclim.mma.goc.cl). The agricultural component of this project was developed using a combination of CROPSYST and a model inspired by the Simple Model (Zhao et al., 2019). It has improved assessments of impacts on crop yields for what, maize, beans, cherries, walnut, and wine grapes as well as on pasture growth affecting meat and milk productivity. Figure 1 shows an example of digital maps generated using weather data at 5 km resolution as well as more detailed soil databases using information from the Center of Soils at Universidad de Talca and FAO databases (https://www.fao.org/soils-portal/data-hub/soil-maps-and-databases/regional-and-national-soil-maps-and-databases/en/). Parallel computing was used to handle larger databases and minimize simulation time.

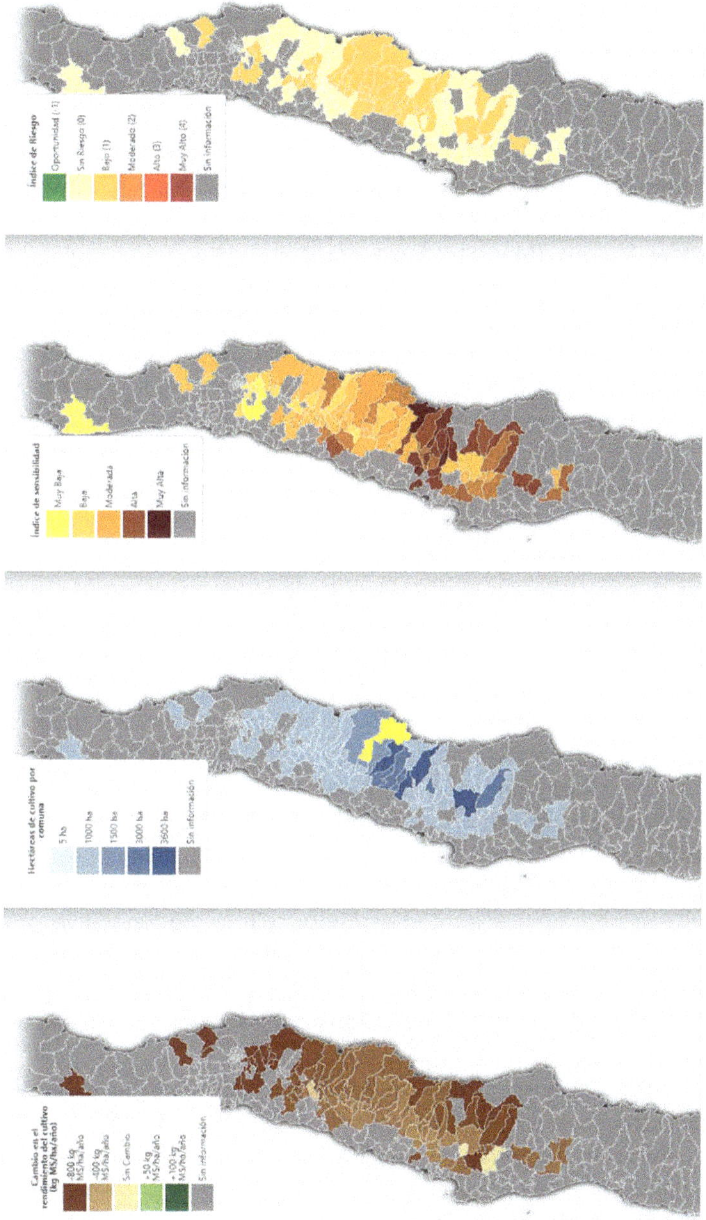

Figure 1 Example of digital maps generated under the ARCLIM project for the agricultural sector. From left to right, maps present data on crop yield impacts, number of hectares of the specific crop exposed, proxy data of vulnerability consolidated as an index, and risk comparison.

There are a number of priorities for future research. Agriculture in Chile is a mixture of irrigated and rainfed systems. There is therefore an ongoing need to couple crop simulation models with hydrological models to better represent the dynamics of water availability, study the effects of major land-use changes on the local hydrological cycle, and analyze the trade-offs between multiple water users under climate change and drought conditions Meza et al. (2014).

Another promising area is the integration of remote sensing data and crop simulation models to obtain more robust regional estimates of crop yields and irrigation water demands. Estimation of crop water requirements has used models such as METRIC (Mapping EvapoTranspiration at high Resolution with Internalized Calibration) (see an example in Ortega-Salazar et al., 2019) or SEBAL (Surface Energy Balance Algorithms for Land) (García-Gutiérrez et al., 2021). Increasing the availability of eddy covariance data under different conditions will produce more precise calibration of models looking at net ecosystem exchange and water fluxes. Climate-driven models of crop phenology need to be integrated in simple computation platforms for regional analysis of crop yield performance and to generate reliable early warning systems of threats to crop production.

A third area of research is the development of crop simulation models for perennial horticultural crops (grapevine, almonds, blueberries, cherries), annual industrial crops (tomato, sweet corn), and specialty crops with interesting nutritional properties and bioproducts such as quinoa. These models are underrepresented in the literature. Chile is a major producer of such crops and has the potential to lead in modeling. This requires moving from simple evapotranspiration (ET) and photosynthesis model calibration to incorporate phenological development, root, and rootstock–variety interactions. This requires the establishment of coordinated flux networks with free access to information

Finally, there is a need for integrated approaches to evaluate the co-benefits of mitigation and adaptation strategies in agriculture. Regenerative grazing, for instance, relies on scheduling livestock feeding to allow for major root development and carbon capture. Models can be used to assess the impact of management on yearly biomass produced and the increase in organic carbon stock over time.

There is also a general need for more investment in crop modeling, improving access to environmental information (soils and climate), taking advantage of high-throughput phenotyping, incorporating artificial intelligence algorithms, and, above all, training the future generation of researchers and practitioners in programming skills.

3 Developments in crop modeling and climate change in Colombia

Colombia is a mostly equatorial country located in the north of South America. It covers 111.5 million (M) hectares (ha), of which 56.7% is natural forest and 38.6% (43 M ha) is used for agriculture and commercial forestry. After pasture and forage for livestock production, which occupy 34.4 M ha, the crops with the largest planted area are coffee (737 thousand ha), oil palm (609 thousand ha), sugarcane (480 thousand ha), rice (702 thousand ha), corn (667 thousand ha), avocado (79 thousand ha), and banana (49 thousand ha). About 529000 ha are dedicated to commercial forestry. Agricultural activity was traditionally concentrated in the Andean and Caribbean regions but has more recently expanded into the Orinoquia region, which has diversified from meat and milk production into cultivation of permanent crops (e.g. oil palm), semi-permanent crops (e.g. sugar cane), and annual crops (e.g. corn, rice, and soybeans). This means Colombian agriculture is now practiced in very varied climatic, soil, and socioeconomic conditions. The country is considered highly vulnerable to variations in weather patterns associated with climate change, which also makes adaptation challenging (Ramírez et al., 2012; Ramirez and Khoury, 2013; Feola et al., 2015; Perez et al., 2020).

With few exceptions (e.g. Sotelo et al., 2020), there are currently few examples of using crop models to inform decision-making in agriculture. Decisions about adapting agricultural practices in the face of climatic variability are still primarily made via monthly regional meetings of experts (Loboguerrero et al., 2018).

There have been developments to better parameterize, calibrate, and refine models to assess the behavior of various crops under different climatic and soil conditions. Some examples include CREFT, MPPC, and CASUPRO. CREFT is a platform that allows the simulation of production of biomass, its redistribution, and extraction of nutrients for several species in intertropical equatorial climatic conditions (Riaño et al., 2005). MPPC models coffee production (Montoya et al., 2009), integrating five sub-models that assess the effect of management (planting date) and climate on the crop. CASUPRO, which was designed to be part of DSSAT (Decision Support System for Agrotechnology Transfer), simulates the volume of sugarcane stems and sugar produced per surface unit depending on climate, soil, and management (Villegas et al., 2005).

The late 1990s and early 2000s saw developments in parameterization and validation of models within the DSSAT platform (Jones et al., 2003), including pastures (Giraldo et al., 1998) and potato (Olarte, 2000). There have also been efforts to parameterize, calibrate, and validate the AquaCrop model (Steduto et al., 2009); for example, ryegrass and fodder oats have parameterized for high Andean conditions (Terán, 2015). The AquaCrop Corn model has been calibrated and validated for regions in Colombia (FAO, 2013; García et al., 2017):

AquaCrop sugarcane for the Cauca River Valley (FAO, 2013b) and AquaCrop potato for the Savanna of Bogotá (Quintero and Díaz, 2020). AquaCrop has also been to simulate specialty coffee production in local conditions (Vega et al., 2018). Models that simulate the dynamics of the Musaceae population as a function of air temperature have also been calibrated and validated (Guarín and Ochoa, 2012).

Models have also been developed for different crops such as peas (Galindo, 2006; Galindo and Clavijo, 2007; Galindo and Clavijo, 2009); plantain (Chaves et al., 2009); broccoli and cabbage (Carranza et al., 2008); and potatoes (Soto et al., 2018). Models have also been developed to simulate fruit growth for different species such as 'Lulo' (Cruz et al., 2007; Pulido et al., 2008); Granadilla (Rodríguez et al., 2015); Rosas (Rodríguez and Florez, 2006); Carnations (López et al., 2010); Feijoa (Parra et al., 2016a); Gulupa (Franco et al., 2014); Curuba (Mayorga et al., 2017; Mayorga et al., 2020); and Cacao (Romero Vergel et al., 2022). Research has also focused on the definition of development threshold temperatures (Acero Camelo et al., 2021; Parra et al., 2016b), crop coefficients (Chaali et al., 2020); and other ecophysiological parameters. However, with the exception of the CREFT model, which is used by companies in the timber sector, other models have had a limited application in decision-making by producers, suggesting a continuation between research and practice.

4 Developments in crop modeling in Brazil: upland rice production and climate change

Brazil is a mostly tropical country located in the center of South America, covering 8.5 million km^2, of which 58.5% is natural forest. In the rest of the country, pasture (for livestock), soybean, sugarcane, maize, rice, and common beans are the main crops cultivated. Brazil is a major global food producer. Brazil is the fourth largest producer of grains (rice, barley, soy, corn, and wheat), accounting for 7.8% of world production. In 2020, it produced 239 million tons and exported 123 million tons of grains, making it the second largest grain exporter in the world with 19% of the international market. Brazilian grain exports totaled US$30 billion in 2020 and US$346 billion in the last two decades. Brazil also currently accounts for 50% of the world's soybean trade. Additionally, Brazil is the leading exporter of beef, chicken meat, sugarcane, and coffee (EMBRAPA, 2021).

Brazil is the largest rice producer outside Asia with rice playing a strategic role in its economy and society (United States Department of Agriculture, 2018). There are two main rice production systems:

- irrigated rice (IR); and
- upland (rainfed) rice (UR).

IR accounts for 85% of Brazilian production and 78% of the rice-cropped area (CONAB, 2021). UR is grown at latitudes lower than 20° South in Brazil, with the largest production areas concentrated mainly in the states of Mato Grosso (MT), Tocantins (TO), Rondônia (RO), and Goiás (GO), accounting for around 90% of UR (Heinemann et al., 2015). The UR area, in the 2018/2019 growing season in these states, was 346.6 thousand ha, representing 20% of the rice area in Brazil, with an average yield of 2.35 Mg ha^{-1}. The climate is characterized as tropical, with a characteristic wet and dry period, and corresponds to the Köppen climate classification category 'Aw' (i.e. tropical wet and dry or Savanna climate), with mean annual precipitation of 1000–1500 mm (mono-modal summer rains) and elevation ranging from 85 to 1190 m above sea level.

There are growing concerns about the vulnerability of IR production, which is concentrated in the subtropical region in the extreme south of Brazil and is critically dependent on water availability (ANA, 2020). An example of this vulnerability is the damaging effect of the 'ENSO' phenomenon on subtropical IR yield due to extreme rainfall or drought events (Grimm et al., 2020; Cai et al., 2020). In this context, rice supply diversification by improving UR yield in the Midwest and Northern regions of Brazil is a strategic priority of the country. UR is also key to the livelihoods and food security of farming and non-farming populations in the Midwest and North regions (Martínez et al., 2014; Oliveira Neto, 2015).

UR can be incorporated into existing crop rotation systems (e.g. soybean-maize, cotton-maize) in the Brazilian Savanna (Heinemann et al., 2017). Crop rotation is a key practice in sustainable agricultural systems (Garnett et al., 2013). UR intercropping with tropical forage grass (with adequate management of N fertilizer) can also improve land-use efficiency and revenue for the long-term sustainability of tropical agriculture (Crusciol et al., 2021a). UR is also suitable for the no-till system, widely used in the Savanna region (Crusciol et al., 2021b). UR systems also use 66% less water than lowland rice to produce 1 kg of rice (Bouman et al., 2006). More broadly, UR has a lower global warming potential (GWP) (~1 Mg CO_2-eq) than the IR in Brazil, ~1 and ~6 Mg CO_2-eq, respectively (Yuan et al., 2021).

However, in the last two decades, the UR cropped area has reduced by 70% mainly due to high levels of agro-climatic risk (Heinemann et al., 2017). Drought is a key limiting factor with 30–40% probability of occurrence, resulting in up to 30% yield reduction (Heinemann et al., 2008; Rabello et al., 2008), particularly during the reproductive phase in the states of GO, MT, TO, and Rondônia, during the growing season from November to February (Heinemann et al., 2015; Ramirez-Villegas et al., 2018).

The crop model ORYZA2000 (Bouman et al., 2001) has been used to assess the effects of UR on changes in the prevalence and intensity of drought stress resulting from climate change. Given that environment and drought-stress pattern impact may vary, depending on CO_2 levels, simulations were conducted with two sets of crop model parameters (initial and maximum rates of assimilation) to account for this variation (Ramirez-Villegas et al., 2018). The study used historical 1981–2005 weather data from 51 weather stations within the study region (Heinemann et al., 2015). Climate scenarios were from the CMIP5 ensemble (Taylor et al., 2012) for all four representative concentration pathways (RCPs 2.6, 4.5, 6.0, and 8.5). Two bias correction (BC) methods were applied (Hawkins et al., 2013):

- The delta method that corrects mean outputs; and
- The change factor method that corrects both mean outputs and variability in GCM output.

Simulation involved a combination of 12 [GCMs] × 4 [RCPs] × 2 [BC methods] × 2 (CO_2 parameterization), making a total of 10,752 simulations per weather station region over 25 years. This totaled ca. 13.7 million simulations for the entire UR produce region. A typical short-cycle cultivar named BRS Primavera, which is a common reference cultivar in the UR breeding trials and thus representative of materials that breeders are currently selecting, was used for model calibration and validation. The sowing dates were based on the Brazilian Government risk zoning for the UR (Heinemann et al., 2015; http://www.agricultura.gov.br).

Ensemble mean temperature increases are substantial, ranging from 1.5°C (minimum for RCP 2.6) to 3.1°C (maximum for RCP 8.5). The largest temperature increases are projected to occur in the state of MT, whereas the lowest temperature increases are projected for the state of TO (Northeast).

In contrast to temperature projections, expected precipitation changes were relatively small (mean regional changes between −2% and −5%). Decreases in precipitation of up to 5% are projected in the state of MT for all RCPs. Particularly in the northern part of MT, precipitation projections showed substantial (>70%) agreement in the direction of the projected change. Elsewhere, however, uncertainty was large, with percentage agreement rarely reaching 60%. For TO, climate change models indicated decreased precipitation. For RO, precipitation gains were projected mostly across the north-western areas. Precipitation changes projected for GO differed across RCPs, with RCP 2.6 and RCP 8.5 showing precipitation gains in the south of the state, and RCP 4.5 and RCP 6.0 showing precipitation decreases across all the state. GO is also a state where GCM agreement is low (around 50% in most weather station regions).

Current mean yield levels are in the range 500–4500 kg ha^{-1}. The ensemble of simulations indicated that mean yield is projected to reduce across most of the western part of the production region, and increase across the east and southeast, with some differences among RCPs. Mean yield changes ranged from -600 to 600 kg ha^{-1}, with the largest reductions (400–600 kg ha^{-1}) projected in the central part of MT, followed by north-western and south-western MT (between 200 and 400 kg ha^{-1}). Model agreement can be measured as the percentage of model simulations out of the 384 simulations per soil and weather station combination (i.e. 8 [sowing dates] × 12 [GCMs] × 2 [BC methods] × 2 [CO_2 parameterization]) that predicted similar median yield change. In this simulation, agreement was generally above 60% (i.e. roughly two-thirds of the model simulations) and for RCP 8.5 specifically, it was above 80% (Ramirez-Villegas et al., 2018). Yield gains were projected across the south-eastern part of GO, as well as across south-eastern and northern TO. Model agreement in these regions was, as in the areas of yield decline, above 60% and sometimes above 80% for both RCPs. Only in specific pockets within MT and RO (<10% of total study area) was model agreement close to 50% (i.e., no agreement). In these areas, median-projected yield changes were small, likely because of uncertainty in the direction of yield changes across model projections.

Yield stability is projected to decrease across virtually the entire study region. Projections of yield coefficient of variation (CV) indicated increases in yield variability in all weather station and soil combinations within the study region, except for south-eastern GO, where decreases in yield CV are projected. For central MT, eastern TO, and northern RO, yield CV increases were above 10% and often above 20%, with high agreement (>80%) in model projections (Ramirez-Villegas et al., 2018). Simulations suggested a substantial shift in growing conditions for UR and thus of breeding priorities.

There was a general trend towards reproductive (52% – overall probability of occurrence) and terminal (29%) stress across the entire UR study region, with only 13% probability of stress-free conditions. The probability of occurrence of joint reproductive and grain-filling stress (i.e. reproductive-to-grain-filling stress) ranged between 25% and 28% (depending on the RCP chosen), making it the most important stress after terminal stress (29–40% overall probability). The probability of reproductive stress reduced to less than half depending on RCP (to 17–21%), whereas the probability of stress-free conditions remained the lowest (12–13%). Climate change could enhance the contrast between high- and low-yielding environments. In general, there has been a homogenization of drought-stress patterns and intensity within the region. Projected changes in crop yield and loss in yield stability will thus bring numerous challenges for UR cropping in Brazil, highlighting the need for adaptation. Due to the diversity of stresses (current and future climate) found, a revised breeding strategy is suggested for UR in Brazil.

These climate change scenarios require a weighted selection strategy in breeding. Although economic and technical considerations remain key to breeding, the objectives of future projected climatic conditions are an important consideration. Improving the adaptive traits of germplasm to respond better under drought stress will ultimately facilitate UR system adaptation to climate change, improving food security and farmer livelihoods (Ramirez-Villegas et al., 2018).

5 Developments in crop modeling in Brazil: irrigated rice production and climate change

In Latin America, countries such as Peru and Mexico consume between 36.5 kg and 91 kg of rice per person^{-1} year^{-1}, while countries like Uruguay, Chile, Argentina, and Paraguay consume less than 36.5 kg per person year^{-1} of rice. Brazil is the largest producer and consumer of rice outside the Asian continent, with an annual consumption of around 12 Mt and average consumption per person of 55 kg year^{-1} (FIESP, 2018). According to MAPA (2018) and FIESP (2018), future Brazilian demand for rice by 2028 has been estimated at 12.2 Mt and 13 Mt, respectively. The production of rice in Brazil is small compared to the Asian countries, but the potential to increase its production is great, making Brazil a potential rice exporter. Three key models used in Brazil are the SimulArroz model, the GD Arroz program, and the PlanejArroz app.

SimulArroz is a dynamic ecophysiological process-based model, developed by the FieldCrops Team at the Federal University of Santa Maria, for simulating the growth, development, and yield of IR in the subtropical region of Brazil. SimulArroz is the result of two rice models, ORYZA2000 (Bouman et al., 2001), which is the rice crop simulation model of the International Rice Research Institute (IRRI), and InfoCrop (Aggarwal et al., 2006). Both rice models were developed and tested in rice ecosystems in Asia. What mainly differentiates SimulArroz from these older models is that it has a sub-model for calculating leaf appearance and the final leaf number on the main stem. SimulArroz calculates the main ecophysiological processes of a rice ecosystem, such as the biomass accumulation of roots, leaves, stems, and panicles, the development of the crop (phenology by the scale of Counce et al., 2000) and the appearance of leaves (scale of Haun, 1973) on a daily basis. In version 1.1, SimulArroz simulates the yield potential and the yield at three technological levels of the crop (high, medium, and low) (Duarte Júnior et al., 2021).

An application of the SimulArroz model is its use in Agricultural Climate Risk Zoning for IR in Southern Brazil (Steinmetz, 2018). The model was used to define sowing periods, taking into account yield levels and climatic risk (Walter et al., 2014; Duarte Júnior et al., 2021). Details on how to download the software for use on desktops and laptops on Windows and IOS operating systems, as well as for instructions and manuals for using the software, are available at: www.ufsm.br/simularroz. On the website, version 1.1 of the SimulArroz model is also available for free download. For more information about the SimulArroz model, please consult Meus et al. (2021).

Another example of rice modeling is the GD Arroz software (Steinmetz et al., 2015), which assists farmers to carry out irrigation and fertilization management practices according to plant developmental stage (PDS). PDS varies substantially depending on the temperature. The R1 stage and the other PDS are expressed in degree-days (DD) (Streck et al., 2006b; Steinmetz, 2010).

The PlanejArroz software was developed in partnership with several research and extension institutions of Brazil, and is intended for management planning and estimating grain yield and development stages in IR in the Rio Grande do Sul state. It contains two modules:

- A management module; and
- A yield module.

The management module supports planning of key crop management practices based on the prediction of PDSs of rice plants as a function of the accumulation of DD throughout the cycle (since temperature is key to PDSs) (Streck et al., 2006b; Steinmetz, 2010). This module estimates dates of six stages of development for 41 cultivars recommended for Southern Brazil. It can be used, for example, to plan sowing dates, which can have a significant impact on yield (Menezes et al., 2013; Steinmetz, 2009). The Web version and Application for the Android platform were launched in February 2020. The PlanejArroz software is available at: http://planejarroz.cpact.embrapa.br.

The second yield module estimates grain yield both 'in season' and over 30 years for the main cultivars used in South America: IRGA 424 RI, GURI INTA CL, and PUITÁ INTA CL. Process-based models such as SimulArroz can be used to estimate rice grain yield at different time and spatial conditions (Rosa, 2015; Ribas et al., 2016), supporting public policies such as those originating from Agricultural Climate Risk Zoning (Steinmetz, 2018).

In an attempt to understand the impact of climate changes on food production, the United Nations (UN) created the IPCC, which has published a series of reports on climate change. In the southern region of Brazil, future projections show a trend for greater amounts of rain, which may result in a higher frequency of floods, especially in areas near to rivers. However, this does not preclude years of drought (Pinheiro et al., 2013), as occurred in the 2004/2005, 2019/2020, and 2021/2022 in South Brazil, Uruguay, and Northeast Argentina. An increase mainly in minimum temperatures will result in warmer nights, less intense winters, and more frequent and more severe heat waves in the summer. These projections with a tendency of increased rainfall together with the temperature rise increase evaporation and evapotranspiration of crops, which implies a greater need of water for agriculture.

An increase in CO_2 can enhance plant growth since it promotes photosynthesis. Rice is a C_3 plant, which particularly benefits from increasing concentration of atmospheric CO_2. However, an increase in temperature accompanied by an increase in CO_2 could decrease the physiological benefits of CO_2 due to a higher level of respiration and shortening of the cycle (Masutomi et al., 2009).

A study by Walter et al. (2014) has assessed the impact of future climate scenarios on rice cultivation in the Rio Grande do Sul state. This suggests that there is potential for increasing grain yield in most of the state, except for the West, where future climatic scenarios may increase the sterility of spikelets and crop respiration due to high daytime temperatures. Projections indicate that regions further south of Southern Brazil and Uruguay would have a greater increase in yield per year. In Southern Brazil, Northeast Argentina, and Paraguay, there may be a decrease in future yield until the end of the 21st century. Results suggest that months currently not recommended for sowing due to low temperatures may become favorable for planting rice crops, such as in early spring and in fall months.

6 Developments in crop modeling in Brazil: common bean production and climate change

Common bean (*Phaseolus vulgaris* L.) is the main legume consumed in many developing countries (Pachico, 1989; Broughton et al., 2003; Fageria et al., 2013; CIAT, 2016). The wide range of genetic material growing across the globe (Burle et al., 2010; Rodriguez et al., 2015) allows its cultivation in a large range of environments and seasons. Brazil is ranked the third major global common bean producer (ca. 3 million metric tons of grain; FAOSTAT, 2018). The common bean cropping system has three major sowing seasons in a year: the wet season (spring sowing), dry season (summer sowing), and winter season (fall sowing) (CONAB, 2021) (Fuscaldi and Prado, 2005).

There have been two studies assessing the impact of climate change on common bean cultivation in Brazil:

- One study targeting the near future (2020–2045, target ~2030) (Heinemann et al., 2017); and
- One study targeting the mid-century (2040–2070, target 2055) (Antolin et al., 2021).

Both studies were based on the CMIP5 protocol and applied the CSM-CROPGRO-DRYBEAN model (Boote et al., 1998), based on two common cultivars but with distinct regional scales.

The near-future study gathered daily weather data for 26 weather stations within the study region (GO state) from the Brazilian Meteorological Institute

(INMET, http://www.inmet.gov.br) from 1980 to 2005. The study applied an ensemble of 12 GCMs, but with two methods of BC, delta and change factor, and four RCPs (2.6, 4.5, 6.0, and 8.5). This made a total of 12 [GCMs] × 4 [RCPs] × 2 [BC methods] = 96 individual climate scenarios for the period 2020–2045.

The mid-century study encompassed all common bean production regions in Brazil and considered three major sowing seasons in a year (wet, dry, and winter), focusing only on GO state and rainfed seasons (wet and dry). The CSM-CROPGRO-DRYBEAN model has been widely used in Brazil in common bean studies for irrigation management optimization (Heinemann et al., 2000, 2002; Teixeira et al., 2017; Justino et al., 2019, 2022), environment characterization (Heinemann et al., 2016), climatic change prediction (Heinemann et al., 2017; Antolin et al., 2021), and yield prediction (Oliveira et al., 2012). The study defined 24 agroclimatic homogeneous zones (HZ) (Van Wart et al., 2013), which encompasses 81% of common bean production region in Brazil, using baseline daily weather data from NASAPOWER API Client (Sparks, 2018) from 1988 to 2019. Future daily weather for the mid-century study was generated from downscaled datasets of 20 GCMs using a tool provided by the Agricultural Model Intercomparison and Improvement Project (AgMIP, www.agmip.org) and two RCPs for atmospheric CO_2: RCP 4.5 and 8.5.

In the near-future study (2020–2045), the ensemble mean temperature increases for the wet and dry sowing seasons ranged from 1.10°C to 1.66°C and 1.03°C to 1.59°C, respectively, across all RCPs, with RCP 8.5 showing the largest warming. The largest temperature increases are projected to occur in the southwest region of GO (RCP 8.5), the largest grain production region for both growing seasons (wet and dry). In contrast, the lowest temperature increases are projected for the Northeast state, the hottest region for both seasons. Projected precipitation changes varied in both direction and magnitude across the GO for both growing seasons. Mean regional changes were between −1.8% (-2.0) and 3.2% (6.7%) for the wet (dry) seasons. The uncertainty among models was, however, high in most of the study area (model agreement around 50%), indicating low robustness in precipitation projections. In the wet growing season, model agreement for precipitation projections was greater for RCP 8.5 (mean 60.7%; SD 7.4%), whereas for the dry TPE the greatest model agreement was found for RCP 2.6 (mean 59.7%; SD 7.9%). There were also differences in the direction of the projected precipitation change across RCPs, with RCP 2.6 and RCP 4.5 showing precipitation gains in the south of the state, and RCP 8.5 and RCP 6.0 showing gains in the Northeast. These differences between RCPs indicate that future global emissions strongly condition the future precipitation conditions in the state.

For the mid-century study, there is an expected increase in maximum temperatures (2.05 ± 0.99°C) compared to minimum temperature (1.96 ±

0.66°C) at 2055 for both RCPs, 4.5 and 8.5 and all HZs. The HZs in the Brazilian Northeast showed higher temperature variations when compared with those in other regions. Rainfall will increase where rainfall is well distributed throughout the year (South and Southwest) and decrease in places where rainfall follows a different regime, such as in the Brazil Midwest and Northeast. Locations in South Brazil are expected to see increases in rainfall up to 7.5% when compared to the current annual accumulation. On the other hand, the Northeast and northern Southeast may see precipitation levels of 6.9% lower than the 1988-2019 average.

The near-future study, in highly favorable environments during the wet growing season in high favorable environments (HFEw) predicted overall higher yields (3904–4025 kg ha^{-1}, depending on cultivar), but higher probabilities of occurrence (34.3–53.2%). In the dry growing season in high favorable environments (HFEd), higher yields were between 3059 kg ha^{-1} and 3129 kg ha^{-1} depending on the cultivar, with a probability of occurrence between 47.3% and 69.4%. Under RCP 8.5, median yields for cv. Pérola increased by 2% for HFEw. This increase was 6% for HFEd. Albeit minor, the change would suggest a net positive effect on bean yield from climate change, likely as a result of CO_2 stimulation, overall precipitation gains, and moderate (<1.5°C) increases in temperature by the 2030s.

In general, there was an increase in the frequency of FE, and a decrease in the frequency of HFE, except for cv. Pérola, in the wet growing season. This means that the probabilities of occurrence of both environments (HFE and FE), regardless of cultivar or growing season (wet or dry), will increase. The future homogenization of environments (growing seasons) is clearer and depends on the sowing date. This is particularly evident for the wet growing season, for which early sowing has ca. 30% less benefit as compared to now. These variations could potentially have important breeding and management implications

For the mid-century study, the central region of Brazil (Midwest, Northeast, and northern Southeast) showed the worst scenarios for future grain yield, mainly in RCP 4.5. Among the seasons in this region, the third season experienced the highest impact, considering that, in the first and second seasons, beans are mainly planted alongside soybean crops. In addition, the region is home to the highest producing areas (e.g. Cristalina/GO and Unaí/MG), with an average increase ranging from 2.5% to 5% for RCP 8.5. The Brazilian South region had the highest average yield for both scenarios. It is expected that with the future projected climate, the region will have better environmental conditions for crop growth and development. Projections of black bean production in Brazil, which is mainly concentrated in the southern states, showed increases from 5% to 15% in comparison with current yield levels. In general, the Midwest and Northeast of Brazil showed the highest level of uncertainty relative to the Southern areas.

The near-future analyses suggest that climate change will make drought more frequent across GO state. For the dry growing season, the overall probability of occurrence of drought stress (reproductive and/or terminal) changes from 29.6% (baseline) to ca. 70% (2030, RCP8.5), whereas for the wet growing season, it increases from 16% (baseline) to ca. 43% (2030, RCP8.5). These changes suggest that future common bean breeding programs will need to account for drought tolerance. Additionally, changes in climate also have a variety of management implications. In particular, the importance is of early sowing and of cultivars that can escape drought, as seen in the generally higher yields in both growing seasons (wet and dry). The shorter cycle characteristic helps to overcome drought (by drought escape), and this characteristic becomes yet more important under climate change following temperature-driven reductions in growing cycle duration.

In the mid-century study, although the scenarios showed increases in average crop yield, there is a projected risk increase of not achieving the level of production needed to meet future demands relative to current conditions. In the South region, the projected risk increases ranging from 5% to 22%. For the Brazilian Midwest region, where GO state is situated, the projected risk increases by almost 25%. The main cause of higher demand and lower yield, despite an increase in yield, is related to air temperature effects on different crop stages, which limit how much CO_2 increase can contribute to yield gains. The increasing risk of production below expected demand may cause recurrent price fluctuations, which, in extreme cases, may force producers to replace 'winter growing season' common bean cultivation areas with more profitable crops.

Risk projections suggest that Brazilian production could rely heavily on area expansion where suitable areas for agriculture will be a major limitation. Protecting natural ecosystems from conversion to agriculture is currently the subject of intense discussion in Brazil and elsewhere. Improving crop management to overcome yield limitations in the field and studies aimed at improving genetic adaptations for climate change must be taken into account by local governments if there is still interest in keeping common beans as a reliable protein source for poor populations with minimum additional expansion of new areas.

7 Developments in crop modeling in Brazil: soybean production and climate change

In the past 20 years, Latin America has experienced one of the most significant expansions of agricultural land use (Zalles et al., 2019). It has established itself as the world's largest soybean producer with over 38.5 million hectares of soybean production area (USDA, 2021). Soybean has a harvested area of

123 million ha and total production of 337 million Mg in 2020 (USDA, 2020), accounting for 53% of total global oilseed production (Beta and Isaak, 2016). As well as being a food crop, it is also one of the main sources of protein for livestock (Smárason et al., 2019; Wajid et al., 2020; Parisi et al., 2020).

Early agrometeorological models sought to forecast crop yields as a function of meteorological conditions during crop development (e.g. Mota, 1983; Pedro Júnior et al., 1983; Callis and Sakamoto, 1984; Camargo, 1984, 1986, 1988). These models were based in part on soil water balance and regression equations.

The use of dynamic soybean crop simulation models in Brazil began in the 1990s (e.g. Siqueira et al., 1994; Klosowski, 1997; Dourado-Neto et al., 1998). These were built on dynamic models developed in the 1980s such as SOYGRO (Wilkerson et al., 1983; Hoogenboom et al., 1992). The model version described by Boote et al. (1998) was improved with the incorporation of a soil water balance (Ritchie, 1985, 1998) and crop phenological modules (Grimm et al., 1993, 1994). After various improvements (Jones et al., 1989), the SOYGRO model was the origin of the DSSAT/CROPGRO model, with new components such as leaf-level, hedgerow canopy photosynthesis (Boote and Pickering, 1994; Boote et al., 1998), an N_2 fixation module (Boote et al., 2009), and a soil N balance and N uptake (Godwin and Jones, 1991; Godwin and Singh, 1998).

There have been several applications of agricultural modeling to soybean cultivation in Brazil:

(i) Improvement of soybean resilience to drought and soil water dynamics (e.g. Battisti and Sentelhas, 2017);
(ii) Soybean irrigation requirements and soil water dynamics (e.g. Silva et al., 2021a);
(iii) Intercomparison of performance of soybean crop models and their ensemble (e.g. Battisti and Sentelhas 2017, Kothari et al., 2022);
(iv) Assessments of soybean yield gaps (e.g. Sentelhas et al., 2015; Di Mauro et al., 2018; Nóia and Sentelhas, 2020; Santos et al., 2021; Tagliapietra et al., 2021; Silva, 2021b; Marin et al., 2022);
(v) Soybean evapotranspiration and crop water productivity (e.g. Silva, 2021b);
(vi) Soybean response to N-fertilization (e.g. Silva, 2021b; Silva et al., 2023);
(vii) Simulation of grain oil and protein concentration (e.g. Silva, 2021b);
(viii) Soybean production and profitability (e.g. Battisti et al., 2020); and
(ix) Soybean sowing dates and plant densities (e.g. Sampaio et al., 2020).

These studies have used state-of-the-art agricultural models to propose alternatives for sustainable management of soybeans in Latin America in the face of climate change.

Battisti et al. (2018) used a crop model ensemble (DSSAT/DSSAT/CROPGRO, APSIM (Agricultural Production Systems slMulator) and MONICA

(Model for Nitrogen and Carbon in Agro-ecosystems)) and two climate scenarios projected to mid-century (2041–2070) with low (+2.2°C, A1BLs) and high (+3.2°C, A1BHs) parameters for air temperature and with atmospheric (CO_2) of 600 ppm. The authors evaluated several management practices such as irrigation, sowing date, cultivar maturity group, and planting density to increase the resilience of soybean under climate change. Their main findings showed that supplementary irrigation, a sowing date on 15th October with cultivar maturity group 7.8, and a plant density of 50 plants/m^2 resulted in higher yields in all climate scenarios in Southern Brazil.

Rio et al. (2016) combined ETA and PRECIS (Providing REgional Climates for Impacts Studies) models with DSSAT/CROPGRO and found a potential decrease in yield of up to 35% for the Southern Brazilian region depending on climate change scenarios. In the Pampas region of Argentina, Rolla et al. (2018) projected soybean yield increase exceeding 50% based on CSM4, RCP 4.5 and 8.5 and DSSAT/CROPGRO.

Justino et al. (2013) used DSSAT/CROPGRO, HadCRM3 for GCM, atmospheric CO_2 concentrations of 550 ppm and 770 ppm projected over 2070–20100. They predicted an overall soybean yield increase of up to 60% for the Midwest and Northern Brazilian regions, but with high variability in annual yields. In line with results from Justino et al. (2013), Silva et al. (2021c) simulated soybean yields in Midwest and Northern Brazilian regions, predicting an increase of up to 60% increase in yield production in a warmer future climate. They used DSSAT/CROPGRO and 20 GCM combined with RCP 4.5 and 8.5. The authors incorporated variables such as future CO_2 levels, several maturity groups, sowing date windows, and soil water content (Silva et al., 2021c). The authors also documented a high variability among future climate scenarios from 20 GCM, with soybean yield increases ranging between −40% to +60%, due to variability in amount and distribution of rainfall.

Reasons for an increase in yield include the impact of increased CO_2 on C3 plant photosynthesis, yield, and stomatal regulation (Allen et al., 1991; Ainsworth et al., 2002; Allen et al., 2020). Drag et al. (2020) found that a CO_2 level of 900 ppm promoted photosynthesis and dry matter accumulation of soybean. Previous free-air CO_2 enrichment (FACE) soybean experiments showed yield increases due to CO_2 increases in a range from 2.2 kg ha^{-1} ppm^{-1} to 13.3 kg ha^{-1} ppm^{-1} (Bunce, 2014; Sakurai et al., 2014).

Silva et al. (2021c) reported that soybean yield increase due to elevated CO_2 is also strongly related to water conservation in rainfed crops, reducing stomatal conductance (Tubiello and Ewert, 2002; Alagarswamy et al., 2006; Bunce, 2016). With the reduction of stomatal conductance, dry matter accumulation increases (Ainsworth et al., 2002). Bernacchi et al. (2007) reported a reduction of up to 16% in soybean evapotranspiration at 550 ppm CO_2, which agreed with results from Silva et al. (2021c), which showed that water use efficiency in soybean will improve with increased CO_2. The increase in water productivity is a key factor in the increase in soybean yield, potentially enabling countries such as Brazil and Argentina to continue as major soybean producers. However, this could be offset by a decrease in rainfall. Silva et al. (2019) found water deficit as the main limiting factor for soybean production in current Brazilian weather conditions. The increase in rainfall instability, however, is expected to be aggravated in some tropical regions (Duffy, O'Gorman, and Back, 2020; Saint-Lu et al., 2020).

Hampf et al. (2020) simulated future yields of soybean, maize, and cotton in the Southern Amazon using MONICA. They concluded that a potentially sustainable solution for agricultural production in the face of climate change could be a combination of preventing further deforestation and more climate-friendly production systems such as integrated crop–livestock–forest systems, drought-tolerant crop varieties, or improved crop rotations.

Rio et al. (2016) and Silva et al. (2021) documented reduction of crop life cycle length with climate change. Silva et al. (2021) found that crop phenology was affected by the reduction of the crop cycle, mainly in the phase R1 (first flower) and R5 (first seed) (Fehr and Caviness, 1977). They found that, on average, there would be a future reduction of 3 days for RCP 4.5 and 5 days for RCP 8.5. They suggested that air temperature was the factor in reducing crop life cycle length. In the DSSAT/CROPGRO model, the main traits affected were EM-FL (time between plant emergence and flower appearance); FL-SH (time between first flower and first pod); FL-SD (time between first flower and first seed); SD-PM (time between first R5 seed and physiological maturity).

Kothari et al. (2022) have shown that an ensemble of crop models focused on grain legumes could reduce uncertainty from individual models. Assessments of ensemble models could therefore increase the accuracy of crop management and yield forecasts. They highlighted the need for more experiments on elevated CO_2 and temperature conditions to reduce uncertainty from individual models.

8 Developments in crop modeling in Brazil: sugarcane production and climate change

Sugarcane is the main feedstock for sugar production in the world and has emerged as a major source of bioethanol (Goldemberg et al., 2014; Marin et al., 2016). Seventy-five per cent of global production is concentrated in Brazil, India, China, Thailand, and Pakistan (FAO, 2019). Brazil is the largest producer (ca. 38% of global production), with approximately 655 million metric tons (MMT) of harvested stalk fresh mass grown in 8.6 million ha (CONAB, 2020). The crop is also cultivated in other Latin American countries such as Mexico, Colombia, Guatemala, Argentina, and Cuba (ca. 8% of global production). However, given its importance, many climate impact studies are focused on Brazilian sugarcane production (Karp et al., 2021).

As a large C_4 perennial grass, sugarcane (*Saccharum officinarum*) is well-adapted to warm and wet tropical conditions. Crop development is significantly reduced below 10°C, and it does not tolerate frosts (Dias and Inman-Bamber, 2020). This long growing cycle (9–18 months) is the main barrier for the introduction of the crop in high latitudes. Although adapted to elevated temperatures, a cool and dry period is necessary during the last 3–4 months before harvest to promote the accumulation of sucrose within the plant's stalk (Inman-Bamber, 2004). Very high air temperatures (>40°C) also negatively affect sugarcane germination and shoot emergence, leaf phenology, and increased plant respiration (Bonnett, 2013; Merwe et al., 2013). In Brazil, sugarcane is typically cultivated under rainfed conditions in the Southeast and Central regions, with less than 7% of plantations under fully irrigated systems (ANA, 2019). The re-growth characteristic of sugarcane plantations (ratooning) also plays an important role as growers typically harvest the same field over five to eight seasons until re-planting the crop (Marin et al., 2019).

The long growing cycle and re-growth characteristics expose sugarcane plants to changing environmental conditions throughout the year, with drought stress and temperature oscillations being the major factors affecting yield variability (Marin et al., 2016; Ukkola et al., 2020). When exposed to high atmospheric concentrations of CO_2, sugarcane plants generally do not show significantly higher photosynthetic capacity. As a C_4 plant, it only benefits indirectly from elevated CO_2 because of improved water use efficiency (Stokes et al., 2016; Marin et al., 2017; Marin et al., 2014).

PBCMs integrate soil–plant–atmosphere and management conditions to understand crop response to Genotype × Environment × Management interactions (Jones et al., 2017).

Several models have been developed for sugarcane (Dias and Inman-Bamber, 2020). In Brazil, they have been widely applied, for example, to assess:

- Climate risks (Vianna and Sentelhas, 2014);
- Irrigation and water use (Dias and Sentelhas, 2018a; Vianna et al., 2020; Vianna and Sentelhas, 2016);
- Yield efficiency (Dias and Sentelhas, 2018b; Marin et al., 2008, 2016; Marin and Carvalho, 2012);
- Soil organic carbon dynamics (Galdos et al., 2009); and
- Climate change impact (Dias et al., 2021; Marin et al., 2013; Pinto et al., 2018; Singels et al., 2014).

Two key models are DSSAT-CANEGRO (Jones and Singels, 2018) and APSIM-Sugar (Keating et al., 1999). A third (SAMUCA) initially developed by Marin and Jones (2014) has been updated and incorporated into the DSSAT model by Vianna et al. (2020). These models are able to replicate the response of sugarcane to changes in carbon dioxide levels, temperature, and water availability (CTW-response) (Ruane et al., 2017). The effect of nitrogen availability on sugarcane growth (N-response) is only accounted for in the APSIM-Sugar model (Dias and Inman-Bamber, 2020). With the introduction of the maintenance of a 'trash blanket' (mulch) in sugarcane plantations, models that can simulate soil nutrient dynamics have become important to optimize residue management (Marin et al., 2014; Vianna et al., 2020).

Process-based models for assessing the impact of climate change on sugarcane production generally predict increased yields. Marin et al. (2013) found that stalk fresh yields of sugarcane could increase +15 to +24% by 2050 in the São Paulo state. The DSSAT-CANEGRO model has been used with projections from two GCMs, PRECIS and CSIRO (Commonwealth Scientific and Industrial Research Organisation), and two scenarios (A2 and B2). Singels et al. (2014) found a +9 to +20% increase in sugarcane yields to 2050 in Piracicaba. They also found a potential decrease in sucrose content of about 18%, due to a large increase in maintenance respiration when total accumulated biomass was high (>75 ton ha^{-1}). A study of sugarcane production in Mexico found similar potential yield increases (Baez-Gonzalez et al., 2018). In contrast, ensemble projections from Dias et al. (2021) using the APSIM-Sugar model suggested that cane yields might decline slightly by the end of the century across ten sites in Brazil. They concluded that water deficit is the main driver for the decline, but acknowledged high variability in climate variables considered in an ensemble of 29 GCMs.

Jaiswal et al. (2017) employed the generic crop model BioCro and found significant yield losses in production in the north, smaller losses in the south, and small increases in the most southerly and in coastal areas of Brazil. Their simulations show that yield losses were mainly associated with rainfall decline offset partly by positive effects of elevated CO_2 and warmer temperatures in the south. Using the Century model, Carvalho et al. (2015) found reduced potential yields in Northeast Brazil. Flack-Prain et al., (2021) also projected a decrease in yields in São Paulo, Brazil (−4% to −12%).

Although there have been many studies modeling the impact of climate change on sugarcane, it is difficult to compare results due to the disparity in methodology, scale, and datasets utilized among studies. The adoption of multi-model ensembles has proven to significantly improve the simulation performance and reduce the uncertainty of PBCMs (Asseng et al., 2013). Marin et al. (2015) provided the first intercomparison analysis between two sugarcane process-based models to understand their structural differences and uncertainties as a pilot study for the AgMIP (Rosenzweig et al., 2013). Pinto et al. (2018) evaluated the effect of climate change on sugarcane yields in São Paulo, Brazil, using both DSSAT-CANEGRO and APSIM-Sugar. Their simulations using five GCMs, two scenarios (RCP 4.5 and RCP 8.5) in 38 locations indicated a positive yield trend of +16% using APSIM-Sugar and +4% using DSSAT-CANEGRO.

Physiological processes considered in these models include crop development, carbon assimilation and respiration, biomass partitioning, senescence, and transpiration. More complex mechanisms modeled include sugarcane tillering, aging effect, sucrose accumulation at stalk and phytomer level, and source–sink relations (Dias et al., 2019; Jones and Singles, 2018; Vianna et al., 2020). Crop lodging is also accounted for in some models (e.g. DSSAT-CANEGRO, APSIM-Sugar). Flowering is not simulated by any current model even though it triggers a substantial remobilization of stalk sucrose to inflorescence organs, reducing crop value (Moore and Botha, 2013).

The role of root plasticity and deeper roots in offsetting drought stress still needs to be quantified and added in dynamic crop models. Laclau and Laclau (2009) found that the root length density of sugarcane was 49% higher in rainfed than in irrigated conditions, suggesting that the plasticity of the sugarcane root system can offset drought stress. One difficulty in simulating sugarcane root growth is uncertainty about the fraction of roots that remains active between ratooning seasons (Smith et al., 2005). Sugarcane models simulate variable root depth only for the first plant-cane season (Jones and Singels, 2018; Vianna et al., 2020). Improving root growth modeling could help not only reduce simulation uncertainties but also allow the in-silico identification of roots ideotypes that better adapts to agronomic constraints (e.g. soil compaction, dry spells) (Pierre et al., 2019).

Sugarcane yields are generally expressed as stalk biomass, but sucrose content is a key quality and production indicator (e.g. when to harvest or to apply chemical maturing agents). In sugarcane models like DSSAT-CANEGRO, APSIM-Sugar, and SAMUCA, sucrose accumulation is mainly considered a function of air temperature and soil moisture (Marin et al., 2015; Vianna et al., 2020). When soil moisture and/or temperature drops below optimum thresholds for vigorous plant expansion, the proportion of carbohydrates allocated in stalk reserves increases. This mechanism mirrors sucrose content observed at field trials (Vianna et al., 2020).

However, complex interactions of carbon translocation, maintenance respiration, and source-to-sink modulation may play an important role in sucrose dynamics (Moore and Botha, 2013; O'Leary, 2000; van Heerden et al., 2010; Wang et al., 2013). This is linked with the Reduced Growth Phenomena observed in high yielding sugarcane fields, which are still not fully understood (van Heerden et al., 2010). Advances in kinetic models and advection-diffusion theory may help investigate these mechanisms at plant level (Uys et al., 2021).

Capturing the bidirectional feedback between croplands and the atmosphere is difficult (Hanssen et al., 2020), and therefore was not included in any climate impact study discussed in this chapter. Improving the representation of sugarcane growth in large-scale models such as land surface models may offer an opportunity to further understand the interplay between bioenergy production and climate (Lapola et al., 2009). This could help understand how to balance bioenergy with carbon capture and storage (van Vuuren et al., 2018; Köberle et al., 2019).

Adding the effect of climate on sucrose content or the integration of pests and diseases could improve model accuracy (Fattori et al., 2021). The improvement of sugarcane models depends on high-quality datasets covering different environmental conditions. Exploring plant-level modeling, new techniques for phenotyping, and data acquisition from remote sensing can also increase the amount and quality of data for crop models.

9 Developments in crop modeling in Brazil: grass forage production for livestock and climate change

Livestock production based on extensive pastures occurs across diverse types of environmental conditions in tropical ecosystems (Roser and Ritchie, 2019). Grass forage production in Brazil covers 196 million ha (Goldewijk et al., 2017). 90% of Brazilian pasturelands are cultivated with *Brachiaria* and *Megathyrsus* grasses due to their great productivity and adaptability (da Silva et al., 2015). Cultivated tropical pastures in Brazil are managed in many different ways. These include extensive and intensive areas managed under rotational stocking, continuous stocking, deferred stocking, and cut-and-carry systems (Jank et al., 2014).

Crop models can help improve forage production and pasture management (Sage and Kubien, 2003; Lara et al., 2012). Several models have been used in tropical pasture simulation. These include ALMANAC (Agricultural Land Management Alternative with Numerical Assessment Criteria), Century, Orchidee-GM (Orchidee Grassland Management), CROPGRO–Perennial Forage of DSSAT, APSIM-Tropical Pasture of APSIM platform, and STICS, and empirical models based on climatic variables (Andrade et al., 2016).

The CROPGRO–Perennial Forage model has been adapted to simulate the growth of the tropical species for Brazil, using experimental data on growth, development, dry matter accumulation, and partitioning for Xaraes palisadegrass (*Brachiaria brizantha* (A. Rich.) Stapf. cv. Xaraes) (Pedreira et al., 2011). The CROPGRO–Perennial Forage model has also been calibrated for forage species, such as *Panicum maximum* Jacq. cv. 'Tanzânia' (Lara et al., 2012) and Marandu palisadegrass (*Brachiaria brizantha* (A. Rich.) Stapf. cv. Marandu; Pequeno et al., 2014). This has, for example, led to the evaluation of forage production and soil moisture using Marandu palisadegrass (Santos et al., 2019). Bosi et al. (2020a) calibrated the CROPGRO-Perennial Forage model for Piatã palisadegrass (*Brachiaria brizantha* (syn. Urochloa) cv. BRS Piatã) growth in monoculture and a silvopastoral system with eucalyptus.

The APSIM-Tropical Pasture model was first calibrated for Brazil by Araujo et al. (2013). The performance of the model in simulating forage production of Mombaça Guinea grass (*Panicum maximum* Jacq.) was compared to three empirical models. The APSIM-Tropical Pasture model showed greater flexibility in representing different management, soil, and climate conditions. The APSIM-Tropical Pasture model has been parameterized for Piatã palisadegrass (*Brachiaria brizantha* (syn. Urochloa) brizantha (Hochst. ex A. Rich.) Stapf 'BRS Piatã') in monoculture (Bosi et al., 2020b) and silvopastoral system (Bosi et al., 2022). It has also been parameterized for Marandu palisadegrass (*Brachiaria brizantha* (A. Rich.) Stapf. cv. Marandu) under continuous stocking in both systems (Gomes et al., 2020).

Pezzopane et al. (2018) has proposed a mechanistic agrometeorological model based on solar radiation, temperature, and soil water availability to estimate Marandu palisadegrass production. This kind of approach can be useful in regions where data availability is scarce. Empirical models also can be developed to estimate forage production based on field measurements through the correlation between variables measured and the final biomass production. Domiciano et al. (2021) evaluated the physiological responses and forage accumulation of Marandu palisadegrass and Mombaça guineagrass to nitrogen fertilizer. The authors concluded that CO_2 assimilation and canopy photosynthesis rates correlate well with forage accumulation and could be used to estimate forage production in livestock systems.

Mechanistic models still do not effectively model processes such as tillering, leaf growth, senescence, and particularly photosynthesis in determining growth of forage species (Andrade et al., 2016). Issues needing further investigation include:

(i) The lack of phenology simulations, partitioning changes and senescence between phenological phases (CROPGRO-Perennial Forage and APSIM-Tropical Pasture) (Bosi et al., 2020b);
(ii) Remobilization of reserves (carbon and nutrients) and translocation of C and N (Brunetti et al., 2021);
(iii) Poor simulation of plant nutrition that considers only effects of nitrogen (N) rather than other nutrients on growth, especially phosphorus (APSIM-Tropical Pasture) (Bosi et al., 2022, 2020b, Gomes et al., 2020);
(iv) The effects of grazing on pasture structure and productivity (for both models) (Brunetti et al., 2021; Bosi et al., 2020b); and
(v) Damage caused by pests and diseases.

Brunetti et al. (2021) have developed the CROPGRO-Perennial Forage Model for simulation of growth and biomass partitioning of Guinea grass, which could be used to develop pasture-grazing strategies. Parameterization of APSIM-Tropical Pasture carried out by Bosi et al. (2020b) has shown the limitations of the model in simulating pasture growth in tropical environments.

Studies combining different approaches could improve the simulation of livestock production. These include:

(i) Evaluating animal production based on simulated amount of forage (Kowal et al., 2019);
(ii) Estimating profitability of different systems (Vogel et al., 2021);
(iii) Simulating scenarios for climate changes and their impacts on pasture growth (Brunetti et al., 2020; Pezzopane et al., 2016) and animal thermal comfort (Magalhães et al., 2020; Vieira Junior et al., 2019); and
(iv) Evaluating crop-livestock and silvopastoral systems (Vieira Junior et al., 2022; Peterson et al., 2020).

The main challenges to livestock production posed by climate change relate to the effects of greenhouse gas emissions (Pezzopane et al., 2016). At the same time, the sector is under global pressure to reduce deforestation and to increase productivity to help reduce these emissions (Pedreira et al., 2018). Pasture areas in Brazil are largely cultivated under rainfed conditions, increasing the potential effect of weather conditions on forage production (Pezzopane et al., 2016). Simulations of new genotypes better adapted to these conditions could, for example, guide future breeding programs (Andrade et al., 2016).

Another area of research is alternative, more sustainable systems. Peterson et al. (2020) calibrated the modules Soybean and AusFarm Pasture from APSIM to simulate productivity dynamics in an integrated cover crop grazing agroecosystem typical of Southern Brazil, making it easier to adapt to different climate change scenarios. Vieira Junior et al. (2022) have simulated forage production in silvopastoral systems. The authors coupled a mechanistic model (CROPGRO-Perennial Forage) to a 3D model developed to represent the structure of trees and light interception, addressing a weakness in other mechanistic models (Bosi et al., 2022; Bosi et al., 2020a; Gomes et al., 2020).

10 Conclusion

Modeling has been used to assess the impact of climate change (e.g. changes in temperature, CO_2 concentrations, and rainfall patterns) on agricultural production and assess adaptation options for policymakers in Latin America. Modeling studies have been used to investigate potential solutions such as genotypes resistant to drought and heat stress, changes to management practices such as sowing dates, the use of more sustainable farming practices such as crop rotation and integrated crop-livestock systems, as well as preservation of riparian forests, water resources, and natural biomass. There is a need for more investment in modeling activities to improve accuracy in predicting climate change impacts on particular crops in specific regional contexts.

11 Where to look for further information

- New Report Details Dire Climate Impacts in Latin America and the Caribbean (https://unfccc.int/news/new-report-details-dire-climate-impacts-in-latin-america-and-the-caribbean).
- Climate Change 2022: Mitigation and Climate Change 2022: Impacts, Adaptation and Vulnerability (https://www.ipcc.ch/report/ar6/wg3/).
- United in Science 2022: A multi-organization high-level compilation of the most recent science related to climate change, impacts and responses (https://library.wmo.int/index.php?lvl=notice_display&id=22128#.Yzc0dXbMKiN).
- WMO Greenhouse Gas Bulletin: The State of Greenhouse Gases in the Atmosphere Based on Global Observations through 2020 (https://library.wmo.int/doc_num.php?explnum_id=10904).
- WMO issues report State of Climate in Latin America and Caribbean (https://public.wmo.int/en/media/press-release/wmo-issues-report-state-of-climate-latin-america-and-caribbean).

12 References

Acero Camelo, R. A., Molina, M. R. E., Parra Coronado, A., Fischer, G. and Carulla Fornaguera, J. E. 2021. Base growth temperature and phyllochron for kikuyu grass (Cenchrus clandestinus; Poaceae). *Acta Biológica Colombiana* 26(2), 160–169. http://doi.org/10.15446/abc.v26n2.83199.

Aggarwal, P. K., Kalra, N., Chander, S. and Pathak, H. 2006. InfoCrop: A dynamic simulation model for the assessment of crop yields, losses due to pests, and environmental impact of agro-ecosystems in tropical environments. I. Model description. *Agricultural Systems* 89(1), 1–25. https://doi.org/10.1016/j.agsy.2005.08.001.

Ainsworth, E. A., Davey, P. A., Bernacchi, C. J., Dermody, O. C., Heaton, E. A., Moore, D. J., Morgan, P. B., Naidu, S. L., Yoo Ra, H., Zhu, X., Curtis, P. S. and Long, S. P. 2002. A meta-analysis of elevated [CO_2] effects on soybean (Glycine max) physiology, growth and yield. *Global Change Biology* 8(8), 695–709.

Alagarswamy, G., Boote, K. J., Allen Jr, L. H. and Jones, J. W. 2006. Evaluating the CROPGRO-soybean model ability to simulate photosynthesis response to carbon dioxide levels. *Agronomy Journal* 98(1), 34–42.

Allen, L. H., Bisbal, E. C., Boote, K. J. and Jones, P. H. 1991. Soybean dry matter allocation under subambient and superambient levels of carbon dioxide. *Agronomy Journal* 83(5), 875–883.

Allen, L. H., Kimball, B. A., Bunce, J. A., Yoshimoto, M., Harazono, Y., Baker, J. T., Boote, K. J. and White, J. W. 2020. Fluctuations of CO_2 in free-air CO_2 enrichment (FACE) depress plant photosynthesis, growth, and yield. *Agricultural and Forest Meteorology* 284, 107899.

ANA. 2019. *Levantamento da Cana-de-Açúcar Irrigada e Fertirrigada no Brasil. Agência Nacional de ÁGUAS.*

ANA (Agência Nacional de Águas e Saneamento Básico. Brasil). 2020. *Mapeamento do arroz irrigado no Brasil/Agência Nacional de Águas e Saneamento Básico. Companhia Nacional de Abastecimento*. Brasília: American Nurses Association, p. 40.

Andrade, A. S., Santos, P. M., Pezzopane, J. R. M., de Araujo, L. C., Pedreira, B. C., Pedreira, C. G. S., Marin, F. R. and Lara, M. A. S. 2016. Simulating tropical forage growth and biomass accumulation: An overview of model development and application. *Grass and Forage Science* 71(1), 54–65. https://doi.org/10.1111/gfs.12177.

Antolin, L. A. S., Heinemann, A. B. and Marin, F. R. 2021. Impact assessment of common bean availability in Brazil under climate change scenarios. *Agricultural Systems* 191, 103174. https://doi.org/10.1016/j.agsy.2021.103174.

Araujo, L. C., Santos, P. M., Rodriguez, D., Pezzopane, J. R. M., Oliveira, P. P. A. and Cruz, P. G. 2013. Simulating guinea grass production: Empirical and mechanistic approaches. *Agronomy Journal* 105(1), 61–69. https://doi.org/10.2134/agronj2012.0245.

Artacho, P., Meza, F. and Alcalde, J. A. 2011. Evaluation of the ORYZA2000 rice growth model under nitrogen-limitted conditions in an irrigated Mediterranean environment. *Chilean JAR* 71(1), 23–33.

Asseng, S., Ewert, F., Rosenzweig, C., Jones, J. W., Hatfield, J. L., Ruane, A. C., Boote, K. J., Thorburn, P. J., Rötter, R. P., Cammarano, D., Brisson, N., Basso, B., Martre, P., Aggarwal, P. K., Angulo, C., Bertuzzi, P., Biernath, C., Challinor, A. J., Doltra, J., Gayler, S., Goldberg, R., Grant, R., Heng, L., Hooker, J., Hunt, L. A., Ingwersen, J., Izaurralde, R. C., Kersebaum, K. C., Müller, C., Naresh Kumar, S., Nendel, C., O'Leary, G., Olesen, J. E., Osborne, T. M., Palosuo, T., Priesack, E., Ripoche, D., Semenov, M. A., Shcherbak, I., Steduto, P., Stöckle, C., Stratonovitch, P., Streck, T., Supit, I., Tao, F., Travasso, M., Waha, K., Wallach, D., White, J. W., Williams, J. R. and Wolf, J. 2013. Uncertainty in simulating wheat yields under climate change. *Nature Climate Change* 3(9), 827–832.

Baez-Gonzalez, A. D., Kiniry, J. R., Meki, M. N., Williams, J. R., Alvarez Cilva, M., Ramos Gonzalez, J. L. and Magallanes Estala, A. 2018. Potential impact of future climate change on sugarcane under dryland conditions in Mexico. *Journal of Agronomy and Crop Science* 204(5), 515–528.

Battisti, R., Ferreira, M. D. P., Tavares, É. B., Knapp, F. M., Bender, F. D., Casaroli, D. and Júnior, J. A. 2020. Rules for grown soybean-maize cropping system in Midwestern Brazil: Food production and economic profits. *Agricultural Systems* 182, 102850.

Battisti, R. and Sentelhas, P. C. 2017. Improvement of soybean resilience to drought through deep root system in Brazil. *Agronomy Journal* 109(4), 1612–1622.

Battisti, R., Sentelhas, P. C. and Boote, K. J. 2017. Inter-comparison of performance of soybean crop simulation models and their ensemble in southern Brazil. *Field Crops Research* 200, 28–37.

Battisti, R., Sentelhas, P. C., Parker, P. S., Nendel, C., Gil, G. M. D. S., Farias, J. R. B. and Basso, C. J. 2018. Assessment of crop-management strategies to improve soybean resilience to climate change in Southern Brazil. *Crop and Pasture Science* 69(2), 154–162.

Bernacchi, C. J., Kimball, B. A., Quarles, D. R., Long, S. P. and Ort, D. R. 2007. Decreases in stomatal conductance of soybean under open-air elevation of [CO_2] are closely coupled with decreases in ecosystem evapotranspiration. *Plant Physiology* 143(1), 134–144.

Beta, T. and Isaak, C. 2016. *Grains Around the World: Encyclopedia of Food Grains*, pp. 349–428.

Bezerra, L. M. C. and Hasparyk, R. G. 2017. Dinâmica da produção de feijão no Brasil: Progresso técnico e fragilidades. *Inf. Agropecuário, Belo Horiz.* 38, 84–91.

Bonnett, G. D. 2013. Developmental stages (phenology). In: *Sugarcane: Physiology, Biochemistry, and Functional Biology*. John Wiley & Sons Ltd, pp. 35–53.

Boote, K. J. and Pickering, N. B. 1994. Modeling photosynthesis of row crop canopies. *Hortscience* 29(12), 1423–1434.

Boote, K. J., Jones, J. W. and Hoogenboom, G. 1998. Simulation of crop growth: CROPGRO model. In: Peart, R. M. and Curry, R. B. (Eds.), *Agricultural Systems Modeling and Simulation*. New York: Marcel Dekker, pp. 651–692.

Bosi, C., Huth, N. I., Sentelhas, P. C. and Pezzopane, J. R. M. 2022. APSIM model performance in simulating Piatã palisade grass growth and soil water in different positions of a silvopastoral system with eucalyptus. *Agricultural Systems* 195. https://doi.org/10.1016/j.agsy.2021.103302.

Bosi, C., Sentelhas, P. C., Huth, N. I., Pezzopane, J. R. M., Andreucci, M. P. and Santos, P. M. 2020a. APSIM-tropical pasture: A model for simulating perennial tropical grass growth and its parameterisation for palisade grass (*Brachiaria brizantha*). *Agricultural Systems* 184. https://doi.org/10.1016/j.agsy.2020.102917.

Bosi, C., Sentelhas, P. C., Pezzopane, J. R. M. and Santos, P. M. 2020b. CROPGRO-perennial forage model parameterization for simulating Piatã palisade grass growth in monoculture and in a silvopastoral system. *Agricultural Systems* 177, 102724. https://doi.org/10.1016/j.agsy.2019.102724.

Bouman, B. A. M. 2001. *ORYZA2000: Modelling Lowland Rice*. Los Baños: IRRI e Wageningen University and Research Centre, p. 235.

Bouman, B. A. M., Kropff, M. J., Tuong, T. P., Wopereis, M. C. S., Berge, H. F. M. ten and van Laar, H. H. 2001. *ORYZA2000: Modeling Lowland Rice*. Los Baños, Philippines: International Rice Research Institute, p. 235.

Broughton, W. J., Hernández, G., Lair, M., Beebe, S., Gepts, P. and Venderleyden, J. 2003. Bean (Phaseolus spp) - Model food legumes. *Plant and Soil* 252(1), 55–128. https://doi.org/10.1023/A:1024146710611.

Brunetti, H. B., Boote, K. J., Santos, P. M., Pezzopane, J. R. M., Pedreira, C. G. S., Lara, M. A. S., Moreno, L. S. B. and Hoogenboom, G. 2021. Improving the CROPGRO perennial forage model for simulating growth and biomass partitioning of guineagrass. *Agronomy Journal* 113(4), 3299–3314. https://doi.org/10.1002/agj2.20766.

Brunetti, H. B., Cavalcanti, P. P., Dias, C. T. S., Pezzopane, J. R. M. and Santos, P. M. 2020. Climate risk and seasonal forage production of marandu palisadegrass in Brazil. *Anais da Academia Brasileira de Ciencias* 92(3), e20190046. https://doi.org/10.1590/0001-3765202020190046.

Bunce, J. A. 2014. Limitations to soybean photosynthesis at elevated carbon dioxide in free-air enrichment and open top chamber systems. *Plant Science: An International Journal of Experimental Plant Biology* 226, 131–135.

Bunce, J. A. 2016. Responses of soybeans and wheat to elevated CO_2 in free-air and open top chamber systems. *Field Crops Research* 186, 78–85.

Cai, W., McPhaden, M. J., Grimm, A. M., Rodrigues, R. R., Taschetto, A. S., Garreaud, R. D., Dewitte, B., Poveda, G., Ham, Y., Santoso, A., Ng, B., Anderson, W., Wang, G., Geng, T., Jo, H., Marengo, J. A., Alves, L. M., Osman, M., Li, S., Wu, L., Karamperidou, C., Takahashi, K. and Vera, C. 2020. Climate impacts of the El Niño–Southern oscillation on South America. *Nature Reviews Earth and Environment* 1(4), 215–231. https://doi.org/10.1038/s43017-020-0040-3.

Callis, S. L. and Sakamoto, C. 1984. *Brazil Soybean Yield Covariance Model* (No. E84-10104).

Camargo, M. B. P. 1984. Exigências bioclimáticas e estimativa da produtividade para Quatro cultivares de soja no Estado de São Paulo. Tese (Mestrado). Piracicaba, ESALQ.

Camargo, M. B. P., Brunini, O. and Miranda, M. A. C. D. 1986. Agrometeorological model for soybean yield estimation in São Paulo State, Brazil. *Bragantia* 45(2), 279–292.

Camargo, M. B. P., Miranda, M. A. C. D., Pedro Júnior, M. J., Pereira, J. C. V. N. A. and Mascarenhas, H. A. A. 1988. Yield potential of soybean cultivars under the climatic conditions of Ribeirão Preto, São Paulo State, Brazil. *Bragantia* 47(2), 277–288.

Carranza, C., Lanchero, O., Miranda, D., Salazar, M. and Chaves, B. 2008. Modelo simple de simulación de distribución de masa seca en brócoli (Brassica sp.) variedad Coronado y repollo (Brassica oleracea) híbrido Delus cultivados en la Sabana de Bogotá. *Agronomía Colombiana* 26(1), 23–31.

Carvalho, A. L. de, Menezes, R. S. C., Nóbrega, R. S., Pinto, A. de S., Ometto, J. P. H. B., von Randow, C. and Giarolla, A. 2015. Impact of climate changes on potential sugarcane yield in Pernambuco, northeastern region of Brazil. *Renewable Energy* 78, 26–34.

Castellaro, G., Squella, F. and Simple, A. Simulation model to estimate pasture growth, phenology and water balance of annual pastures of Mediterranean climate. *Agricultura Técnica (Chile)* 66(3), 271–282.

Castro, F., Hyman, G., Rubiano, J., Ramirez-Villegas, J. and Achicanoy, H. 2019. Climate change favors rice production at higher elevations in Colombia. *Mitigation Adaptation Strategy Global Change* 24(8), 1401–1430. https://doi.org/10.1007/s11027-019-09852-x.

Cazanga, R., Ledent, J. F. and von Benewitz, E. 2019. Adaptation of CERES-Maize model for a better simulation of plant population density effects. *Field Crops Research* 238, 27–34. https://doi.org/10.1016/j.fcr.2019.04.017.

CEPAL. 2012. La economía del cambio climático en Chile. http://repositorio.cepal.org/bitstream/handle/11362/35372/1/S2012058_es.pdf. Accessed June 2022.

Chaali, N., Ouazaa, S., Jaramillo-Barrios, C. I., Araujo Carrillo, G. A. and Ávila Pedraza, E. Á. 2020. Edaphoclimatic characterization and crop water requirement of arracacha (Arracacia xanthorrhiza Bancroft) roots in upland production areas. *Scientia Horticulturae* 272, 109533. https://doi.org/10.1016/j.scienta.2020.109533.

Chaves, B., Cayón, G. and Jones, J. 2009. Modeling plantain (Musa A A B Simmonds) potential yield. *Agronomía Colombiana* 27(3), 359–366.

CIAT. 2016. Beans. https://ciat.cgiar.org/what-we-do/breeding-better-crops/beans/. Accessed 25 September 2018. Inter-American Center of Tax Administrators.

Companhia Nacional de Abastecimento, C. O. N. A. B. 2021. Safras: Séries históricas. *Feijão 3ª safra, 494 Março/2021*. https://www.conab.gov.br/info-agro/safras/serie-historica-das-495 safras?start=20. Accessed 20 February 2022.

CONAB. 2020. Acompanhamento da safra brasileira: grãos 1o. https://www.conab.gov.br/info-agro/safras/graos/boletim-da-safra-de-graos. In: 1o Levantamento - Safra 2020/21. *Levantamento - Safra* (vol. 2020/1) [Data set].

CONAB. 2021. *Companhia Nacional de Abastecimento. Acompanhamento da Safra Brasileira de Grãos – Brasília: Conab*. http://www.conab.gov.br (Vol. 2013).

Counce, P. A., Keisling, T. C. and Mitchell, A. J. 2000. *A Uniform, Objective, and Adaptive System for Expressing Rice Development. Crop Science Department* (Vol. 40), pp. 436–443.

CRUSCIOL, C. A. C., Momesso, L., Portugal, J. R., Costa, C. H. M., Bossolani, J. W., Costa, N. R., Pariz, C. M., Castilhos, A. M., Rodrigues, V. A., Costa, C., Franzluebbers, A. J. and Cantarella, H. 2021a. Upland rice intercropped with forage grasses in an integrated crop-livestock system: optimizing nitrogen management and food production. *Field Crops Research* 261, 108008.

CRUSCIOL, C. A. C., Momesso, L., de Campos, M., Bossolani, J. W., Portugal, J. R., Moretti, L. G., Volf, M. R., Calonego, J. C. and Rosolem, C. A. 2021b. Liming optimizes nitrogen fertilization in a maize-upland rice rotation under no-till conditions. *Agronomy* 11(10), 2005.

Cruz, P., Acosta, K., Cure, R. and Rodríguez, D. 2007. Desarrollo y fenología del lulo Solanum quitoense var. septentrionale bajo polisombra desde siembra hasta primera fructificación. *Agronomía Colombiana* 25(2), 288–298.

da Silva, E. H. F. M., Gonçalves, A. O., Pereira, R. A., Fattori Júnior, I. M., Sobenko, L. R. and Marin, F. R. 2019. Soybean irrigation requirements and canopy-atmosphere coupling in Southern Brazil. *Agricultural Water Management* 218, 1–7.

da Silva, S., Sbrissia, A. and Pereira, L. 2015. Ecophysiology of C4 forage grasses—Understanding plant growth for optimising their use and management. *Agriculture* 5(3), 598–625. https://doi.org/10.3390/agriculture5030598.

Departamento Administrativo Nacional de Estadística (DANE). 2014. Censo Nacional agropecuario. https://www.dane.gov.co/index.php/estadisticas-por-tema/agropecuario/censo-nacional-agropecuario-2014#1.

Di Mauro, G., Cipriotti, P. A., Gallo, S. and Rotundo, J. L. 2018. Environmental and management variables explain soybean yield gap variability in Central Argentina. *European Journal of Agronomy* 99, 186–194.

Dias, H. B. and Inman-Bamber, G. 2020. Sugarcane: Contribution of process-based models for understanding and mitigating impacts of climate variability and change on production. In: Ahmed, M. (Ed.), *Systems Modeling*. Singapore: Springer, pp. 217–260.

Dias, H. B., Inman-Bamber, G., Bermejo, R., Sentelhas, P. C. and Christodoulou, D. 2019. New APSIM-sugar features and parameters required to account for high sugarcane yields in tropical environments. *Field Crops Research* 235, 38–53.

Dias, H. B. and Sentelhas, P. C. 2019. Dimensioning the impact of irrigation on sugarcane yield in Brazil. *Sugar Tech* 21(1), 29–37. https://doi.org/10.1007/s12355-018-0619-x.

Dias, H. B. and Sentelhas, P. C. 2018. Sugarcane yield gap analysis in Brazil - A multi-model approach for determining magnitudes and causes. *The Science of the Total Environment* 637–638, 1127–1136.

Dias, H. B., Sentelhas, P. C., Inman-Bamber, G. and Everingham, Y. 2021. Sugarcane yield future scenarios in Brazil as projected by the APSIM-sugar model. *Industrial Crops and Products* 171, 113918.

Domiciano, L. F., Santos, M. L., Boote, K. J., Santos, P. M., Pereira, D. H. and Pedreira, B. C. 2021. Physiological responses and forage accumulation of Marandu palisadegrass and Mombaça guineagrass to nitrogen fertilizer in the Brazilian forage-based systems. *Grassland Science* 67(2), 93–101. https://doi.org/10.1111/grs.12291.

Dourado-Neto, D., Teruel, D. A., Reichardt, K., Nielsen, D. R., Frizzone, J. A. and Bacchi, O. O. S. 1998. Principles of crop modelling and simulation: II. The implications of the objective in model development. *Scientia Agricola* 55, 51–57.

Drag, D. W., Slattery, R., Siebers, M., DeLucia, E. H., Ort, D. R. and Bernacchi, C. J. 2020. Soybean photosynthetic and biomass responses to carbon dioxide concentrations ranging from pre-industrial to the distant future. *Journal of Experimental Botany* 71(12), 3690–3700.

Duarte Junior, A. J., Streck, N. A., Zanon, A. J., Ribas, G. G., Silva, M. R., Cera, J. C., Pilecco, I. B., Nascimento, M. F. and Puntel, S. 2021. Rice yield potential as a function of sowing date in Southern Brazil. *Agronomy Journal* 206, 1–10.

Eitzinger, A., Läderach, P., Bunn, C., Quiroga, A., Benedikter, A., Pantoja, A., Gordon, J. and Bruni, M. 2014. Implications of a changing climate on food security and smallholders' livelihoods in Bogotá, Colombia. *Mitigation Adaption Strategy Global Change* 19(2), 161–176. https://doi.org/10.1007/s11027-012-9432-0.

EMBRAPA - Brazilian Agricultural Research Corporation. 2021. Estudos socioeconômicos e ambientais. https://www.embrapa.br/documents/10180/62618376/O+AGRO+NO+BRASIL+E+NO+MUNDO.pdf.

Fageria, N. K., Melo, L. C. and Oliveira, J. 2013. Nitrogen use efficiency in dry bean genotypes. *Journal of Plant Nutrition* 36(14), 2179–2190. https://doi.org/10.1080/01904167.2013.836225.

FAO. 2013a. *Uso del Modelo AquaCrop para Estimar Rendimientos para el Cultivo de Maíz en los Departamentos de Cordoba, Tolima, Meta y Valle del Cauca*. ISBN I3429S/1/08.13, p. 62.

FAO. 2013b. *Uso del Modelo AquaCrop para Estimar Rendimientos para el Cultivo de Caña de Azúcar en el Departamento del Valle del Cauca*. ISBN I3427S/1/08.13, p. 66.

FAO. 2019. Food and Agriculture Organization corporate statistical database. FAOSTAT. http://www.fao.org/faostat/en/#home.

FAOSTAT. 2018. Production quantity - Dry beans. http://www.fao.org/faostat/en/#data/QC.

Fattori, I. M., Sentelhas, P. C. and Marin, F. R. 2022. Assessing the impact of climate variability on Asian rust severity and soybean yields in different Brazilian mega-regions. *International Journal of Plant Production* 16(1), 17–28. https://doi.org/10.1007/s42106-021-00169-x.

Fehr, W. R. and Caviness, C. E. 1977. Stages of soybean development. Special Report (Vol. 87).

Feola, G., Agudelo Vanegas, L. A. and Contesse Bamón, B. P. 2015. Colombian agriculture under multiple exposures: A review and research agenda. *Climate and Development* 7(3), 278–292. https://doi.org/10.1080/17565529.2014.934776.

FIESP and DAS INDÚSTRIAS DO ESTADO DE SÃO PAULO, F. 2018. *Outlook Fiesp 2028: Projeções para o Agronegócio Brasileiro*. São Paulo, FIESP. ISBN: 978-85-7201035-1, p. 86.

Figueiredo Moura da Silva, E. H., Silva Antolin, L. A., Zanon, A. J., Soares Andrade, A., Antunes de Souza, H., dos Santos Carvalho, K., Aparecido Vieira, N. and Marin, F. R. 2021. Impact assessment of soybean yield and water productivity in Brazil due to climate change. *European Journal of Agronomy* 129, 126329.

Flack-Prain, S., Shi, L., Zhu, P., Rocha, H. R., Cabral, O., Hu, S. and Williams, M. 2021. The impact of climate change and climate extremes on sugarcane production. *GCB Bioenergy* 13(3), 408–424.

Franco, G., Cartagena, R., J. and Correa, G. 2014. Análisis de crecimiento del fruto de gulupa (Passiflora edulis Sims), en las condiciones ecológicas del bosque húmedo montano bajo de Colombia. *Revista U.D.C.A Actualidad and Divulgación Científica* 17(2), 391–400.

Fuscaldi, K. and Prado, G. 2005. *Análise econômica da cultura do feijão (Rev. Política Agrícola).*

Galdos, M. V., Cerri, C. C., Cerri, C. E. P., Paustian, K. and Van Antwerpen, R. 2009. Simulation of soil carbon dynamics under sugarcane with the century model. *Soil Science Society of America Journal* 73(3), 802–811.

Galindo, J. R. and Clavijo, J. 2007. Modelos alométricos para estimar el área de los foliolos de arveja (*Pisum sativum* L.). *Ciencia and Tecnología Agropecuaria* 8(1), 37–43.

Galindo, R. 2006. *Desarrollo de un modelo explicativo de la productividad potencial del cultivo de arveja (Pisum sativum L.) var. Santa Isabel bajo condiciones de la sabana de Bogotá.* Disertación Doctoral. Facultad de Agronomía. Bogotá: Universidad Nacional de Colombia.

García, J., Fischer, G. and Riaño, N. 2017. Effect of fertilization level on water use and production of corn (*Zea mays* L.) in a cereal producing area in Colombia - A modeling exercise using AquaCrop-FAO. *Agronomía Colombiana* 35(1), 68–74.

García-Gutiérrez, V., Stöckle, C., Gil, P. M. and Meza, F. J. 2021. Evaluation of Penman-Monteith model based on Sentinel-2 data for the estimation of Actual evapotranspiration in vine-yards. *Remote Sensing* 13(3), 478; https://doi.org/10.3390/rs13030478.

Giraldo, L., Lizcano, L. , Gijsman, A., Rivera, B. and Franco, L. 1998. Adaptación del Modelo DSSAT para simular la producción de Brachiaria decumbens. *Pasturas Tropicales* 20(2), 2–12.

Goldemberg, J., Mello, F. F. C., Cerri, C. E. P., Davies, C. A. and Cerri, C. C. 2014. Meeting the global demand for biofuels in 2021 through sustainable land use change policy. *Energy Policy* 69, 14–18.

Gomes, F. J., Bosi, C., Pedreira, B. C., Santos, P. M. and Pedreira, C. G. S. 2020. Parameterization of the APSIM model for simulating palisadegrass growth under continuous stocking in monoculture and in a silvopastoral system. *Agricultural Systems* 184. https://doi.org/10.1016/j.agsy.2020.102876.

Grimm, A. M., Almeid, A. S., Beneti, C. A. A. and Leite, E. A. 2020. The combined effect of climate oscillations in producing extremes: The 2020 drought in southern Brazil. *RBRH* 25. https://doi.org/10.1590/2318-0331.252020200116.

Guarín, G. and Ochoa, A. 2012. Calibración y validación del modelo SIMBA-POP en la zona bananera del Urabá Antioqueño, Colombia. *Rev.Fac.Nal. Agriculturists Medellín* 65(1), 6419–6429.

Hampf, A. C., Stella, T., Berg-Mohnicke, M., Kawohl, T., Kilian, M. and Nendel, C. 2020. Future yields of double-cropping systems in the Southern Amazon, Brazil, under climate change and technological development. *Agricultural Systems* 177, 102707. https://doi.org/10.1016/j.agsy.2019.102707.

Hanssen, S. V., Daioglou, V., Steinmann, Z. J. N., Doelman, J. C., Van Vuuren, D. P. and Huijbregts, M. A. J. 2020. The climate change mitigation potential of bioenergy with carbon capture and storage. *Nature Climate Change* 10(11), 1023–1029.

Haun, J. R. 1973. Visual quantification of wheat development 1. *Agronomy Journal* 65(1), 116–119. https://doi.org/10.2134/agronj1973.00021962006500010035x.

Hawkins, E., Osborne, T. M., Ho, C. K. and Challinor, A. J. 2013. Calibration and bias correction of climate projections for crop modelling: An idealised case study over Europe. *Agricultural and Forest Meteorology* 170, 19–31.

Heinemann, A. B., Barrios-Perez, C., Ramirez-Villegas, J., Arango-Londoño, D., Bonilla-Findji, O., Medeiros, J. C. and Jarvis, A. 2015. Variation and impact of drought-stress patterns across upland rice target population of environments in Brazil. *Journal of Experimental Botany* 66(12), 3625–3638. https://doi.org/10.1093/jxb/erv126.

Heinemann, A. B., Dingkuhn, M., Luquet, D., Combres, J. C. and Chapman, S. 2008. Characterization of drought stress environments for upland rice and maize in central Brazil. *Euphytica* 162(3), 395–410. https://doi.org/10.1007/s10681-007-9579-z.

Heinemann, A. B., Hoogenboom, G. and Faria, R. T. 2002. Determination of spatial water requirements at county and regional levels using crop models and GIS. *Agricultural Water Management* 52(3), 177–196. https://doi.org/10.1016/S0378-3774(01)00137-8.

Heinemann, A. B., Hoogenboom, G., Georgiev, G. A. and Faria, R. T. 2000. Center pivot irrigation management optimization of dry beans in humid areas. *Transactions of the ASAE. American Society of Agricultural Engineers* 43, 1507–1516. https://doi.org/10.13031/2013.3050.

Heinemann, A. B., Ramirez-Villegas, J., Nascente, A. S., Zeviani, W. M., Stone, L. F. and Sentelhas, P. C. 2017. Upland rice cultivar responses to row spacing and water stress across multiple environments. *Experimental Agriculture* 53(4), 609–626. https://doi.org/10.1017/S0014479716000612.

Heinemann, A. B., Ramirez-Villegas, J., Stone, L. F. and Didonet, A. D. 2017. Climate change determined drought stress profiles in rainfed common bean production systems in Brazil. *Agricultural and Forest Meteorology* 246, 64–77, https://doi.org/10.1016/j.agrformet.2017.06.005.

Herrera-Cáceres, C., Pérez-Galarce, F., Alvarez-Miranda, E. and Candía-Véjar, A. 2017. Optimization of the harvest planning in the olive oil production: A case study in Chile. *Computers and Electronics in Agriculture* 141, 147–159.

INE and National Institute of Statistics. 2022. https://www.ine.cl/.

Inman-Bamber, N. G. 2004. Sugarcane water stress criteria for irrigation and drying off. *Field Crops Research* 89(1), 107–122.

IPCC. 2019. Climate change and land: An IPCC special report on climate change, desertification, land degradation, sustainable land management, food security, and greenhouse gas fluxes in terrestrial ecosystems. https://www.ipcc.ch/srccl/.

ITA - International Trade Administration. 2022. Chile - Agricultural sector. https://www.trade.gov/country-commercial-guides/chile-agricultural-sector.

Jaiswal, D., De Souza, A. P., Larsen, S., LeBauer, D. S., Miguez, F. E., Sparovek, G., Bollero, G., Buckeridge, M. S. and Long, S. P. 2017. Brazilian sugarcane ethanol as an expandable green alternative to crude oil use. *Nature Climate Change* 7(11), 788–792.

Jank, L., Barrios, S. C., do Valle, C. B., Simeão, R. M. and Alves, G. F. 2014. The value of improved pastures to Brazilian beef production. *Crop and Pasture Science* 65(11), 1132–1137. https://doi.org/10.1071/CP13319.

Jones, J. W., Antle, J. M., Basso, B., Boote, K. J., Conant, R. T., Foster, I., Godfray, H. C. J., Herrero, M., Howitt, R. E., Janssen, S., Keating, B. A., Munoz-Carpena, R., Porter, C. H., Rosenzweig, C. and Wheeler, T. R. 2017. Brief history of agricultural systems modeling. *Agricultural Systems* 155, 240–254.

Jones, J. W., Boote, K. J., Hoogenboom, G., Jagtap, S. S. and Wilkerson, G. G. (1989). SOYGRO V5.42, soybean crop growth simulation model. User's guide. Florida Agricultural Experiment Station Journal No. 8304, University of Florida, Gainesville, FL, USA.

Jones, J. W., Hoogenboom, G., Porter, C. H., Boote, K. J., Batchelor, W. D., Hunt, L. A., Wilkens, P. W., Singh, U., Gijsman, A. J. and Ritchie, J. T. 2003. The DSSAT cropping system model. *European Journal of Agronomy* 18(3–4), 235–265.

Jones, M. R. and Singels, A. 2018. Refining the Canegro model for improved simulation of climate change impacts on sugarcane. European journal of agronomy. *The Journal of the European Society for Agronomy* 100, 76–86.

Jones, P. G. and Thornton, P. K. 2000. MarkSim: Software to generate daily weather data for Latin America and Africa. *Agronomy Journal* 92(3), 445–453.

Jones, P. G. and Thornton, P. K. 2003. The potential impacts of climate change on maize production in Africa and Latin America in 2055. *Global Environmental Change* 13(1), 51–59.

Justino, F., Oliveira, E. C., de Ávila Rodrigues, R., Gonçalves, P. H. L., Souza, P. J. O. P., Stordal, F. and Silva Lindemann, D. 2013. Mean and interannual variability of maize and soybean in Brazil under global warming conditions. *American Journal of Climate Change* 2(4), 237–253.

Justino, L. F., Alves Júnior, J., Battisti, R., Heinemann, A. B., Leite, C. V., Evangelista, A. W. P. and Casaroli, D. 2019. Assessment of economic returns by using a central pivot system to irrigate common beans during the rainfed season in Central Brazil. *Agricultural Water Management* 224, 105749. https://doi.org/10.1016/j.agwat.2019.105749.

Justino, L. F., Battisti, R., Stone, L. F. and Heinemann, A. B. 2022. In silico assessment of sowing dates and nitrogen management in common bean crops. *European Journal of Agronomy* 133, 126434. https://doi.org/10.1016/j.eja.2021.126434.

Karp, S. G., Medina, J. D. C., Letti, L. A. J., Woiciechowski, A. L., Carvalho, J. C., Schmitt, C. C., Oliveira Penha, R., Kumlehn, G. S. and Soccol, C. R. 2021. Bioeconomy and biofuels: The case of sugarcane ethanol in Brazil. *Biofuels, Bioproducts and Biorefining* 15(3), 899–912. https://doi.org/10.1002/bbb.2195.

Keating, B. A., Robertson, M. J., Muchow, R. C. and Huth, N. I. 1999. Modelling sugarcane production systems I. Development and performance of the sugarcane module. *Field Crops Research* 61(3), 253–271.

Klosowski, E. S. 1997. Estimativa da produtividade de cultivares de soja por meio do modelo SOYGRO para Londrina, estado do Paraná. *Revista UNIMAR* 19(3), 751–765.

Köberle, A. C., Portugal-Pereira, J., Cunha, B., Garaffa, R., Lucena, A. F. P., Szklo, A. and Schaeffer, R. 2019. Brazilian ethanol expansion subject to limitations. *Nature Climate Change* 9(3), 209–210. https://doi.org/10.1038/s41558-019-0422-z.

Kothari, K., Battisti, R., Boote, K. J., Archontoulis, S. V., Confalone, A., Constantin, J., Cuadra, S. V., Debaeke, P., Faye, B., Grant, B., Hoogenboom, G., Jing, Q., van der Laan, M.,

Macena da Silva, F. A., Marin, F. R., Nehbandani, A., Nendel, C., Purcell, L. C., Qian, B., Ruane, A. C., Schoving, C., Silva, E. H. F. M., Smith, W., Soltani, A., Srivastava, A., Vieira, N. A., Slone, S. and Salmerón, M. 2022. Are soybean models ready for climate change food impact assessments? *European Journal of Agronomy* 135, 126482.

Kowal, V. A., Jones, S. M., Keesing, F., Allan, B. F., Schieltz, J. M. and Chaplin-Kramer, R. 2019. A coupled forage-grazer model predicts viability of livestock production and wildlife habitat at the regional scale. *Scientific Reports* 9(1), 19957. https://doi.org/10.1038/s41598-019-56470-3.

Laclau, P. B. and Laclau, J. P. 2009. Growth of the whole root system for a plant crop of sugarcane under rainfed and irrigated environments in Brazil. *Field Crops Research*. https://www.sciencedirect.com/science/article/pii/S0378429009002391.

Läderach, P., Martinez-Valle, A., Schroth, G. and Castro, N. 2013. Predicting the future climatic suitability for cocoa farming of the world's leading producer countries, Ghana and Côte d'Ivoire. *Climatic Change* 119(3–4), 841–854. https://doi.org/10.1007/s10584-013-0774-8.

Läderach, P., Ramirez-Villegas, J., Navarro-Racines, C., Zelaya, C., Martinez-Valle, A. and Jarvis, A. 2017. Climate change adaptation of coffee production in space and time. *Climatic Change* 141(1), 47–62. https://doi.org/10.1007/s10584-016-1788-9.

Lapola, D. M., Priess, J. A. and Bondeau, A. 2009. Modeling the land requirements and potential productivity of sugarcane and jatropha in Brazil and India using the LPJmL dynamic global vegetation model. *Biomass and Bioenergy* 33(8), 1087–1095. https://doi.org/10.1016/j.biombioe.2009.04.005.

Lara, M. A. S., Pedreira, C. G. S., Boote, K. J., Pedreira, B. C., Moreno, L. S. B. and Alderman, P. D. 2012. Predicting growth of Panicum maximum: an adaptation of the CROPGRO-perennial forage model. *Agronomy Journal* 104(3), 600–611. https://doi.org/10.2134/agronj2011.0272.

Loboguerrero, A. M., Boshell, F., León, G., Martinez-Baron, D., Giraldo, D., Recaman Mejía, L., Díaz, E. and Cock, J. 2018. Bridging the gap between climate science and farmers in Colombia. *Climate Risk Management* 22, 67–81. https://doi.org/10.1016/j.crm.2018.08.001.

López, M., Chaves, B., Flórez, V. and Salazar, M. 2010. Modelo de aparición de nudos en clavel (*Dianthus caryophyllus* L.) cv. Delphi cultivado en sustratos. *Agronomía Colombiana* 28(1), 47–54.

Magalhães, C. A. S., Zolin, C. A., Lulu, J., Lopes, L. B., Furtini, I. V. v., Vendrusculo, L. G., Zaiatz, A. P. S. R., Pedreira, B. C. and Pezzopane, J. R. M. 2020. Improvement of thermal comfort indices in agroforestry systems in the southern Brazilian Amazon. *Journal of Thermal Biology* 91, 102636. https://doi.org/10.1016/j.jtherbio.2020.102636.

MAPA and MINISTÉRIO, D. A. Agricultura, pecuária E abastecimento. Projeções do Agronegócio Brasil 2017/18 a 2027/8, projeções de longo prazo. Brasília, MAPA – Secretaria de Política Agrícola 2018, p. 112. ISBN 978- 85-7991-116-3.

Marchand, P., Carr, J. A., Dell'Angelo, J., Fader, M., Gephart, J. A., Kummu, M., Magliocca, N. R., Porkka, M., Puma, M. J., Ratajczak, Z., Rulli, M. C., Seekell, D. A., Suweis, S., Tavoni, A. and D'Odorico, P. 2016. Reserves and trade jointly determine exposure to food supply shocks. *Environmental Research Letters* 11(9), 095009.

Marin, F. R. and de Carvalho, G. L. 2012. Spatio-temporal variability of sugarcane yield efficiency in the state of São Paulo, Brazil. *Pesquisa Agropecuária Brasileira* 47(2), 149–156.

Marin, F. R. and Jones, J. W. 2014. Process-based simple model for simulating sugarcane growth and production. *Scientia Agricola* 71(1), 1–16.

Marin, F. R., Jones, J. W., Singels, A., Royce, F., Assad, E. D., Pellegrino, G. Q. and Justino, F. 2013. Climate change impacts on sugarcane attainable yield in southern Brazil. *Climatic Change* 117(1–2), 227–239.

Marin, F. R., Lopes-Assad, M. L., Assad, E. D., Vian, C. E. and Santos, M. C. 2008. Sugarcane crop efficiency in two growing seasons in São Paulo State, Brazil. *Pesquisa Agropecuária Brasileira* 43(11), 1449–1455.

Marin, F. R., Martha, G. B., Jr., Cassman, K. G. and Grassini, P. 2016. Prospects for increasing sugarcane and bioethanol production on existing crop area in Brazil. *BioScience* 66(4), 307–316.

Marin, F. R., Nassif, D. S. P., Costa, L. G., Vianna, M. S., Carvalho, K. and Pereira, P. R. 2017. Predicting the effect of climate change on sugarcane cultivation. *Achieving Sustainable Cultivation of Sugarcane* 1, 277–290. https://doi.org/10.19103/AS.2017.0035.30.

Marin, F. R., Rattalino Edreira, J. I., Andrade, J. and Grassini, P. 2019. On-farm sugarcane yield and yield components as influenced by number of harvests. *Field Crops Research* 240, 134–142.

Marin, F. R., Ribeiro, R. V. and Marchiori, P. E. R. 2014. How can crop modeling and plant physiology help to understand the plant responses to climate change? A case study with sugarcane. *Theoretical and Experimental Plant Physiology* 26(1), 49–63.

Marin, F. R., Thorburn, P. J., da Costa, L. G. and Otto, R. 2014. Simulating long-term effects of trash management on sugarcane yield for Brazilian cropping systems. *Sugar Tech* 16(2), 164–173.

Marin, F. R., Thorburn, P. J., Nassif, D. S. P. and Costa, L. G. 2015. Sugarcane model intercomparison: Structural differences and uncertainties under current and potential future climates. *Environmental Modelling and Software* 72, 372–386.

Marin, F. R., Zanon, A. J., Monzon, J. P., Andrade, J. F., Silva, E. H. F. M., Richter, G. L., Antolin, L. A. S., Ribeiro, B. S. M. R., Ribas, G. G., Battisti, R., Heinemann, A. B. and Grassini, P. 2022. Protecting the Amazon forest and reducing global warming via agricultural intensification. *Nat Sustain* 5, 1018–1026. https://doi.org/10.1038/s41893-022-00968-8.

Martínez, C. P., Torres, E. A., Chatel, M., Mosquera, G., Duitama, J., Ishitani, M., Selvaraj, M., Dedicova, B., Tohme, J., Grenier, C., Lorieux, M., Cruz, M., Berrío, L., Corredor, E., Zorrilla De San Martín, G., Breseghello, F., Peixoto, O., Colombari Filho, J. M., Castro, A. P., Lopes, S. I. G., Barbosa, M., Funck, G. R. D., Blanco, P. H., Pérez De Vida, F., Molina, F., Rosas, J. E., Martínez, S., Bonnecarrere, V., Carracelas, G., Marin, A., Correa-Victoria, F., Camargo, I. and Bruzzone, C. B. 2014. Rice breeding in Latin America. *Plant Breeding Reviews* 38, 187–277. https://doi.org/10.1002/9781118916865.

Masutomi, Y., Takahashi, K., Harasawa, H. and Matsuoka, Y. 2009. Impact assessment of climate change on rice production in Asiain comprehensive consideration of process/parameter uncertainty in general circulation models. *Agriculture, Ecosystems and Environment* 131(3–4), 281–291. https://doi.org/10.1016/j.agee.2009.02.004.

Mayorga, M., Fisher, G. and Melgarejo, L. 2017. Fotosíntesis, uso eficiente del agua y desarrollo de curuba (Passiflora tripartita var. mollissima) en dos zonas de Pasca (Cundinamarca - Colombia). III. Congreso Latinoamericano y I Mundial de Pasifloras. Neiva, Huila, Colombia.

Mayorga, M., Fisher, G., Melgarejo, L. and Parra, A. 2020. Growth, development and quality of *Passifora tripartita* var. mollissima fruits under two environmental tropical conditions. *Journal of Applied Botany and Food Quality* 93, 66-75.

Menezes, V. G., Anghinoni, I., Silva, P. D., Macedo, V. R. M., Petry, C., Grohs, D. S., Freitas, T. D. and Valente, L. D. L. 2013. *Projeto 10 - Management Strategies to Increase Productivity and Sustainability of Irrigated Rice Growth in the State of Rio Grande do Sul, Brasil: Developments and New Challenges* (Vol. 100). Cachoeirinha: IRGA/Estação Experimental do Arroz.

Meus, L. D., Silva, M. R., Ribas, G. G., Zanon, A. J., Rossato, I. G., Fogliato, V., Pilecco, I. B., Ribeiro, B. S. M. R., Souza, P. M., Nascimento, M. F. and Poersch, A. H. 2021. *Ecophysiology of Rice for Reaching High Yields.* 1st ed., Vol. 1. Santa Maria: DR, p. 312.

Meza, F. J. and Silva, D. 2009. Dynamic adaptation of maize and wheat production to climate change. *Climatic Change* 94(1-2), 143-156.

Meza, F. J., Silva, D. and Vigil, H. 2008. Climate change impacts on irrigated maize in Mediterranean climates. Evaluation of double cropping as an emerging adaptation alternative. *Agricultural Systems* 98(1), 21-30.

Meza, F. J., Vicuña, S., Jelinek, M., Bustos, E. and Bonelli, S. 2014. Assessing water demands and coverage sensitivity to climate change in the urban and rural sectors in Central Chile. *Journal of Water and Climate Change* 5(2), 192-203*.

Meza, F. J. and Wilks, D. S. 2003. Value of operational forecasts of seasonal average sea surface temperature anomalies, for selected rain-fed agricultural locations of Chile. *Agricultural and Forest Meteorology* 116(3-4), 137-158.

Meza, F. J. and Wilks, D. S. 2004. Use of seasonal forecasts of sea surface temperature anomalies for potato fertilization management. Theoretical study considering EPIC model results at Valdivia, Chile. *Agricultural Systems* 82(2), 161-180.

Meza, F. J., Wilks, D. S., Riha, S. J. and Stedinger, J. R. 2003. Value of perfect forecasts of sea surface temperature anomalies, for selected rain-fed agricultural locations of Chile. *Agricultural and Forest Meteorology* 116(3-4), 117-135.

Ministerio de Agricultura y Desarrollo Rural. *Boletín Estadístico Forestal Marzo 2022*.

Ministerio de Agricultura y Desarrollo Rural. Estadísticas agropecuarias. https://www.agronet.gov.co/estadistica/Paginas/home.aspx?cod=1. ND, AGRONET.

Ministerio de agricultura y Desarrollo Rural and Federación Nacional de Cafeteros de Colombia, Corporación Colombiana de Investigación Forestal. 51 p.

Moore, P. H. and Botha, F. C. 2013. *Sugarcane: Physiology, Biochemistry and Functional Biology*. Chichester: John Wiley & Sons.

Morales-Salinas, L., Acevedo, E., Castellaro, G., Román, L., Morales-Inostroza, A. and Alonso, M. 2015. A simple method for estimating suitable territory for bioenergy species in Chile. Cien. *Investigación Agraria* 42(2), 227-242.

Morris, M., Sebastian, A. R. and Perego, V. M. E. 2020. *Future Foodscapes: Re-Imagining Agriculture in Latin America and the Caribbean*. Washington, DC: International Bank for Reconstruction and Development/The World Bank.

Mota, F. S. 1983. Weather-technology models for corn and soybeans in the south of Brazil. *Agricultural Meteorology* 28(1), 49-64.

Nóia, R. S. and Sentelhas, P. C. 2020. Yield gap of the double-crop system of main-season soybean with off-season maize in Brazil. *Crop and Pasture Science* 71(5), 445-458.

OECD. 2019-2028. Latin American agriculture: Prospects and challenges. https://www.fao.org/3/CA4076EN/CA4076EN_Chapter2_Latin_American_Agriculture.pdf. FAO agricultural outlook 2019. *Agricultural Outlook*.

OECD. 2022-2031. Regional outlook: Latin America and the Caribbean. https://www.oecd-ilibrary.org/docserver/f1b0b29c-en.pdf?expires=1658859748&id=id&accname=guest&checksum=B06DBF50008DFF34A56CD2A1833C8B82. FAO agricultural outlook 2022. *Agricultural Outlook*.

Olarte, S. 2000. *Validación del Modelo de Simulación Substor de la Plataforma DSSAT 3.5 en dos Variedades Nativas de Papa. Trabajo de Grado. Facultad de Agronomía*. Bogotá: Universidad Nacional de Colombia.

O'Leary, G. J. 2000. A review of three sugarcane simulation models with respect to their prediction of sucrose yield. *Field Crops Research* 68(2), 97–111.

Oliveira, E. C. D., Costa, J. M. Nd., Paula Júnior, T. J. D., Ferreira, W. P. M., Justino, F. B. and Neves, L. D. O. 2012. The performance of the CROPGRO model for bean (*Phaseolus vulgaris* L.) yield simulation. *Acta Scientiarum. Agronomy* 34(3), 239–246.

Oliveira Neto, A. A. (Org.) 2015. *A cultura do arroz*. Brasília, DF, CONAB. http://www.conab.gov.br/OlalaCMS/uploads/arquivos/16_03_01_16_56_00_a_cultura_do_arroz_-_conab.pdf.

Orrego, R., Avila, A., Meza, F. J. and Matus, F. 2014. Using a crop simulation model to select the optimal climate grid cell resolution. *Journal of Soil Science and Plant Nutrition* 14(2), 407–420.

Ortega-Salazar, S., Ortega-Farías, S., Kilic, A. and Allen, R. 2021. Performance of the METRIC model for mapping energy balance components and actual evapotranspiration over a superintensive drip-irrigated olive orchard. *Agricultural Water Management* 251 https://doi.org/10.1016/j.agwat.2021.106861.

Pacheco, J. R. G. and Porras, J. C. 2009. Fenología del cultivo de Arveja (Pisum sativum L. var. Santa Isabel) en la sabana de Bogotá en campo abierto y bajo cubierta plástica. *Ciencia and Tecnología Agropecuaria* 10(1), 5–15.

Pachico, D. 1989. Trends in world common bean production. Bean Production Problems in the Tropics 10(1), 1–33.

Parisi, G., Tulli, F., Fortina, R., Marino, R., Bani, P., Dalle Zotte, A., Angelis, A., Piccolo, G., Pinotti, L., Schiavone, A., Terova, G., Prandini, A., Gasco, L., Roncarati, A. and Danieli, P. P. 2020. Protein hunger of the feed sector: The alternatives offered by the plant world. *Italian Journal of Animal Science* 19(1), 1204–1225.

Parra, A., Fischer, G. and Chaves, B. 2016. Tiempo térmico para estados fenológicos reproductivos de la feijoa (Acca sellowiana (O. Berg) Burret). *Acta Biológica Colombiana* 20(1), 163–173. http://doi.org/10.15446/abc.v20n1.43390.

Pedreira, B. C., Domiciano, L. F., Vilela, L., Salton, J. C., Marchió, W., Wruck, F. J. and Zolin, C. A. 2018. O estado da arte e estudos de caso em sistemas integrados de produção agropecuária no Centro-Oeste do Brasil. In: de Souza, E. D., da Silva, F. D., Assmann, T. S., Carneiro, M. A. C., Carvalho, P. C. F. and Paulino, H. B. (Eds.), *Sistemas Integrados de Produção Agropecuária no Brasil* (1st ed., pp. 277–300). Tubarão, SC, Brazil: Copiart.

Pedreira, B. C., Pedreira, C. G. S., Boote, K. J., Lara, M. A. S. and Alderman, P. D. 2011. Adapting the CROPGRO perennial forage model to predict growth of Brachiaria brizantha. *Field Crops Research* 120(3), 370–379. https://doi.org/10.1016/j.fcr.2010.11.010.

Pedro Júnior, M. J., Camargo, M. B. P., Brunini, O., Alfonsi, R. R., Orto Lani, A. A. and Miranda, M. A. C. 1983. Teste de um modelo para estimativa de produtividade da soja. In: *Campinas. Resumos Congresso Brasileiro de Agrometeorologia, 3.* Campinas: Instituto Agronômico, pp. 138–139.

Pequeno, D. N. L., Pedreira, C. G. S. and Boote, K. J. 2014. Simulating forage production of Marandu palisade grass (*Brachiaria brizantha*) with the CROPGRO-perennial forage model. *Crop and Pasture Science* 65(12), 1335–1348. https://doi.org/10.1071/CP14058.

Perez, L., Rios, D. A., Giraldo, D. C., Twyman, J., Blundo-Canto, G., Prager, S. D. and Ramirez-Villegas, J. 2020. Determinants of vulnerability of bean growing households to climate variability in Colombia. *Climate and Development* 12(8), 730–742, https://doi.org/10.1080/17565529.2019.1685931.

Peterson, C. A., Bell, L. W., Carvalho, P. C. de F. and Gaudin, A. C. M. 2020. Resilience of an integrated crop-livestock system to climate change: A simulation analysis of cover crop grazing in Southern Brazil. *Frontiers in Sustainable Food Systems* 4. https://doi.org/10.3389/fsufs.2020.604099.

Pezzopane, J. R. M., Santos, P. M., Cruz, P. G. da, Bosi, C. and Sentelhas, P. C. 2018. An integrated agrometeorological model to simulate Marandu palisade grass productivity. *Field Crops Research* 224, 13–21. https://doi.org/10.1016/j.fcr.2018.04.015.

Pezzopane, J. R. M., Santos, P. M., Evangelista, S. R. M., Bosi, C., Cavalcante, A. C. R., Bettiol, G. M., de Miranda Gomide, C. A. and Pellegrino, G. Q. 2017. Panicum maximum cv. Tanzânia: Climate trends and regional pasture production in Brazil. *Grass and Forage Science* 72(1), 104–117. https://doi.org/10.1111/gfs.12229.

Pierre, J. S., Perroux, J. M. and Rae, A. L. 2019. Screening for sugarcane root phenes reveals that reducing tillering does not lead to an increased root mass fraction. *Frontiers in Plant Science* 10, 119.

Pinheiro, A., Graciano, R. L. G. and Severo, D. L. 2013. Tendência das séries temporais de precipitação da região sul do Brasil. *Revista Brasileira de Meteorologia* 28(3), 281–290.

Pinter, P. J., Ritchie, J. C., Hatfield, J. L. and Hart, G. F. 2003. The agricultural research service's remote sensing program. *Photogrammetric Engineering and Remote Sensing* 69(6), 615–618.

Pinto, H. M. S., Viana, M. D. S., Costa, L. G. and Marin, F. R. 2018. Produtividade de cana-de-açúcar no Estado de São Paulo baseada em simulações multimodelos e mudanças climáticas. *Agrometeoros* 26(1), 11–20.

Pizarro, E., Galleguillos, M., Barría, P. and Callejas, R. 2022. Irrigation management or climate change? Which is more important to cope with water shortage in the production of table grape in a Mediterranean context. *Agricultural Water Management* 263, 107467. https://doi.org/10.1016/j.agwat.2022.107467.

Pulido, S., Bojacá, C. R., Salazar, M. and Chaves, B. 2008. Node appearance model for Lulo (*Solanum quitoense* Lam.) in the high altitude tropics. *Biosystems Engineering* 101(4), 383–387. https://doi.org/10.1016/j.biosystemseng.2008.09.009.

Quesada, F. 2013. The Rockefeller Foundation fellowship program in Chile (1938–1970). https://rockarch.issuelab.org/resource/the-rockefeller-foundation-fellowship-program-in-chile-1938-1970.html. Accessed June 2022.

Quintero, D. and Díaz, E. 2020. A comparison of two open-source crop simulation models for a potato crop. *Agronomía Colombiana* 38(3), 382–387.

Rabello, A. R., Guimaraes, C. M., Rangel, P. H. N., da Silva, F. R., Seixas, D., de Souza, E., Brasileiro, A. C., Spehar, C. R., Ferreira, M. E. and Mehta, A. 2008. Identification of drought-responsive genes in roots of upland rice (*Oryza sativa* L.). *BMC Genomics* 9, 485.

Ramirez, J. and Khoury, C. K. 2013. Reconciling approaches to climate change adaptation for Colombian agriculture. *Climatic Change* 119(3–4), 575–583. https://doi.org/10.1007/s10584-013-0792-6.

Ramírez, J., Salazar, M., Jarvis, A. and Navarro-Racines, C. E. 2012. A way forward on adaptation to climate change in Colombian agriculture: Perspectives towards 2050. *Climatic Change* 115(3–4), 611–628. https://doi.org/10.1007/s10584-012-0500-y.

Ramirez-Villegas, J., Heinemann, A. B., Castro, A., Breseghello, F., Navarro-Racines, C., Li, T., Rebolledo, M. C. and Challinor, A. J. 2018. Breeding implications of drought stress under future climate for upland rice in Brazil. *Global Change Biology* 24(5), 2035–2050. https://doi.org/10.1111/gcb.14071.

Riaño, N., Tangarife, G., Osorio, O., Giraldo, J., Ospina, C., Obando, D., Gómez, L. and Jaramillo, F. 2005. *Modelo de Crecimiento y Captura de Carbono para Especies Forestales en el Trópico*.

Ribas, G. G., Streck, N. A., Lago, I., Zanon, A. J., Waldow, D. A. G., Duarte Junior, A. J., Nascimento, Md. Fd. and Fontana, V. 2016. Acúmulo de matéria seca e produtividade em híbridos de arroz irrigado simulados com o modelo SimulArroz. *Pesquisa Agropecuária Brasileira* 51(12), 1907–1917.

Rio, A., Sentelhas, P. C., Farias, J. R. B., Sibaldelli, R. N. R. and Ferreira, R. C. 2016. Alternative sowing dates as a mitigation measure to reduce climate change impacts on soybean yields in southern Brazil. *International Journal of Climatology* 36(11), 3664–3672.

Ritchie, J. T. (1985). A user-orientated model of the soil water balance in wheat. In: Day, W. and Atkin, R. K. (Eds.), *Wheat Growth and Modelling. NATO ASI Science* (Vol. 86). Springer, Boston, MA. https://doi.org/10.1007/978-1-4899-3665-3_27.

Rodriguez, A. K., Rodriguez, A., Melgarejo, L., Miranda, D., Fischer, G. and Martínez, O. 2015. Caracterización fenológica de granadilla (Passiflora ligularis Juss) crecida a diferentes altitudes en el departamento del Huila. En: *Libro: Granadilla (Passiflora Ligularis Juss): Caracterización Ecofisiológica del Cultivo de Granadilla*.

Romero Vergel, A. P., Camargo Rodriguez, A. V., Ramirez, O. D., Arenas Velilla, P. A. and Gallego, A. M. 2022. A crop modelling strategy to improve cacao quality and productivity. *Plants* 11(2), 157. https://doi.org/10.3390/plants11020157.

Rosa, H. T. 2015. Simulação do crescimento e produtividade de arroz no Rio Grande do Sul pelo modelo SimulArroz. *Revista Brasileira de Engenharia Agrícola e Ambiental* 19, 1159–1165. https://doi.org/10.1590/1807.

Rosenzweig, C., Jones, J. W., Hatfield, J. L., Ruane, A. C., Boote, K. J., Thorburn, P., Antle, J. M., Nelson, G. C., Porter, C., Janssen, S., Asseng, S., Basso, B., Ewert, F., Wallach, D., Baigorria, G. and Winter, J. M. 2013. The agricultural model intercomparison and improvement project (AgMIP): Protocols and pilot studies. *Agricultural and Forest Meteorology* 170, 166–182.

Roser, M. and Ritchie, H. 2019. Yields and land use in agriculture. https://ourworldindata.org/yields-and-land-use-in-agriculture.

Ruane, A. C., Rosenzweig, C., Asseng, S., Boote, K. J., Elliott, J., Ewert, F., Jones, J. W., Martre, P., McDermid, S. P., Müller, C., Snyder, A. and Thorburn, P. J. 2017. An AgMIP framework for improved agricultural representation in integrated assessment models. *Environmental Research Letters* 12(12), 125003.

Sage, R. F. and Kubien, D. S. 2003. An ecophysiological perspective on global change and the future of C4 plants. *Photosynthesis Research.*

Saint-Lu, M., Chadwick, R., Lambert, F. H., Collins, M., Boutle, I., Whitall, M. and Daleu, C. 2020. Influences of local and remote conditions on tropical precipitation and its response to climate change. *Journal of Climate* 33(10), 4045–4063.

Sakurai, G., Iizumi, T., Nishimori, M. and Yokozawa, M. 2014. How much has the increase in atmospheric CO_2 directly affected past soybean production? *Scientific Reports* 4, 4978.

Sampaio, L. S., Battisti, R., Lana, M. A. and Boote, K. J. 2020. Assessment of sowing dates and plant densities using CSM-CROPGRO-Soybean for soybean maturity groups in low latitude. *The Journal of Agricultural Science* 158(10), 819–832.

Santibáñez, F. 1997. Possible impact on agriculture due to climatic change and variability in South America. In: Yoshino, M., Domrös, M., Douguédroit, A., Paszyński, J. and Nkemdirim, L. C. (Eds.), *Climates and Societies – A Climatological Perspective. The GeoJournal Library* (Vol. 36). Dordrecht: Springer, pp. 255–277. https://doi.org/10.1007/978-94-017-1055-8_12.

Santos, M. G., Boote, K. J., Faria, R. T. and Hoogenboom, G. 2019. Simulation of productivity and soil moisture under Marandu palisade grass using the CSM-CROPGRO-perennial forage model. *Crop and Pasture Science* 70(2), 159–168. https://doi.org/10.1071/CP18258.

Santos, T. G., Battisti, R., Casaroli, D., Alves Jr., J. and Evangelista, A. W. P. 2021. Assessment of agricultural efficiency and yield gap for soybean in the Brazilian Central Cerrado biome. *Bragantia* 80.

Searchinger, T., Waite, R., Hanson, C., Ranganathan, J., Dumas, P., Matthews, E. and Klirs, C. 2019. Creating a sustainable food future: A menu of solutions to feed nearly 10 billion people by 2050. Final report. Available at: https://agritrop.cirad.fr/593176/1/WRR_Food_Full_Report_0.pdf. Accessed in 4/4/2023.

Sentelhas, P. C., Battisti, R., Câmara, G. M. S., Farias, J. R. B., Hampf, A. C. and Nendel, C. 2015. The soybean yield gap in Brazil–magnitude, causes and possible solutions for sustainable production. *The Journal of Agricultural Science* 153(8), 1394–1411.

Silva, E. H. F. M., Boote, K. J., Hoogenboom, G., Gonçalves, A. O., Junior, A. S. A. and Marin, F. R. 2021. Performance of the CSM-CROPGRO-soybean in simulating soybean growth and development and the soil water balance for a tropical environment. *Agricultural Water Management* 252, 106929.

Silva, E. H. F. M., Boote, K. J., Hoogenboom, G., Gonçalves, A. O., Junior, A. S. A. and Marin, F. R. 2021a. Performance of the CSM-CROPGRO-Soybean in simulating soybean growth and development and the soil water balance for a tropical environment. *Agricultural Water Management* 252, 106929.

Silva, E. H. F. M. 2021b. *Crop modeling for understanding yield-gap causes and the potential for sustainable intensification of soybean in Brazil.* PhD Thesis, ESALQ, Piracicaba.

Silva, E. H. F. M., Cafaro La Menza, N., Munareto, G., Zanon, A., Santos Carvalho, K. and Marin, F. 2023. Soybean seed protein concentration is limited by nitrogen supply in tropical and subtropical environments in Brazil. *The Journal of Agricultural Science* 1–12. doi:10.1017/S0021859623000199.

Singels, A., Jones, M., Marin, F. R., Ruane, A. and Thorburn, P. 2014. Predicting climate change impacts on sugarcane production at sites in Australia, Brazil and South Africa using the Canegro model. *Sugar Tech* 16(4), 347–355.

Siqueira, O. J. F., de Farias, J. R. B. and Sans, L. M. A. 1994. Potential effects of global climate change for Brazilian agriculture and adaptive strategies for wheat, maize and soybean. *Revista Brasileira de Agrometeorologia* 2, 115–129.

Smárason, B. Ö., Alriksson, B. and Jóhannsson, R. 2019. Safe and sustainable protein sources from the forest industry-The case of fish feed. *Trends in Food Science and Technology* 84, 12–14.

Smith, D. M., Inman-Bamber, N. G. and Thorburn, P. J. 2005. Growth and function of the sugarcane root system. *Field Crops Research* 92(2–3), 169–183.

Sotelo, S., Guevara, E., Llanos-Herrera, L., Agudelo, D., Esquivel, A., Rodriguez, J., Ordoñez, L., Mesa, J., Muñoz Borja, L. A., Howland, F., Amariles, S., Rojas, A., Valencia, J. J., Segura, C. C., Grajales, F., Hernández, F., Cote, F., Saavedra, E., Ruiz, F., Serna, J., Jimenez, D., Tapasco, J., Prager, S. D., Epanchin, P. and Ramirez-Villegas, J. 2020. Pronosticos a climate Colombia: A system for the provision of information for climate risk reduction in Colombia. *Computers and Electronics in Agriculture* 174.

Soto, A., Cotes, J. and Rodríguez, D. 2018. Modelo de simulación y crecimiento de la papa criolla. *Ciencia en Desarrollo* 9(1), 9–20.

Sparks, A. H. 2018. Nasapower: A NASA power global meteorology, surface solar energy and climatology data client for R. *Journal of Open Source Software* 3(30). Available at: https://doi. https://doi.org/10.21105/joss.01035.

Steduto, P., Hsiao, T. C., Raes, D. and Fereres, E. 2009. AquaCrop–The FAO crop model to simulate yield response to water: I. Concepts and underlying principles. *Agronomy Journal* 101(3), 426–437. https://doi.org/10.2134/agronj2008.0139s.

Steinmetz, S. 2009. *Produtividade de cultivares de arroz irrigado em função da época de semeadura*. Pelotas: Embrapa Clima Temperado, p. 22. (Embrapa Clima Temperado. *Boletim de Pesquisa e Desenvolvimento* (Vol. 111)).

Steinmetz, S. 2010. *Uso de graus-dia para estimar a data de diferenciação da panícula (DP) de seis subgrupos de cultivares de arroz irrigado visando à adubação nitrogenada em cobertura no Rio Grande do Sul*. Pelotas: Embrapa Clima Temperado. 75p. (Embrapa Clima Temperado. *Boletim de Pesquisa e Desenvolvimento* (Vol. 121)).

Steinmetz, S. 2018. *Períodos de Semeadura para o Arroz Irrigado no Rio Grande do Sul Com Base na Estimativa de Produtividade de Grãos e em Níveis de Risco Climático* (Vol. 57). Pelotas: Embrapa Clima Temperado.

Steinmetz, S., Cuadra, S. V., Pereira, C. B., dos Santos, E. L. and de Almeida, I. R. 2015. *GD Arroz: Programa Baseado em graus-dia Como Suporte Ao Planejamento e à Tomada De Decisão no Manejo do Arroz Irrigado*. Pelotas, RS: Embrapa Clima Temperado: Circular Técnica 162.

Sternberg, T. 2012. Chinese drought, bread and the Arab Spring. *Applied Geography* 34, 519–524.

Stöckle, C. O., Donatelli, M. and Nelson, R. 2003. CropSyst, a cropping systems simulation model. *European Journal of Agronomy* 18(3–4), 289–307. http://doi.org/10.1016/S1161-0301(02)00109-0.

Stokes, C. J., Inman-Bamber, N. G., Everingham, Y. L. and Sexton, J. 2016. Measuring and modelling CO_2 effects on sugarcane. *Environmental Modelling and Software* 78, 68–78.

Streck, N. A., Bosco, L. C., Michelon, S., Rosa, H. T., Walter, L. C., Paula, G. Md., Camera, C., Lago, I. and Marcolin, E. 2006. Avaliação da resposta ao fotoperíodo em genótipos de arroz irrigado. *Bragantia* 65(4), 533–541.

Tagliapietra, E. L., Zanon, A. J., Streck, N. A., Balest, D. S., da Rosa, S. L., Bexaira, K. P., Richter, G. L., Ribas, G. G. and Silva, M. R. 2021. Biophysical and management factors causing yield gap in soybean in the subtropics of Brazil. *Agronomy Journal* 113(2), 1882-1894.

Taylor, K. E., Stouffer, R. J. and Meehl, G. A. 2012. An overview of CMIP5 and the experiment design. *Bulletin of the American Meteorological Society* 93(4), 485-498.

Teixeira, G. C. D. S., Stone, L. F., Santos, A. B. D., Silva, S. C. D. and Heinemann, A. B. 2017. Early sowing can improve irrigation water use efficiency and yield of common bean. *Pesquisa Agropecuaria Tropical* 47(1), 118-126.

Terán, C. 2015. *Determinación de la huella hídrica y modelación de la producción de biomasa de cultivos forrajeros a partir del agua en la Sabana de Bogotá (Colombia).* Disertación Doctoral. Departamento de Ingeniería Hidraúlica y Ambiental. Universitat Politécnica de Valencia, p. 397.

Thornton, P. K. 2010. Livestock production: recent trends, future prospects. *Philosophical Transactions of the Royal Society of London. Series B, Biological Sciences* 365(1554), 2853-2867.

Tubiello, F. N. and Ewert, F. 2002. Simulating the effects of elevated CO_2 on crops: Approaches and applications for climate change. *European Journal of Agronomy* 18(1-2), 57-74.

Ukkola, A. M., De Kauwe, M. G., Roderick, M. L., Abramowitz, G. and Pitman, A. J. 2020. Robust future changes in meteorological drought in CMIP6 projections despite uncertainty in precipitation. *Geophysical Research Letters* 47(11), 1029.

United States Department of Agriculture. 2018. Grain: World markets and trade. https://apps.fas.usda.gov/psdonline/circulars/grain-rice.pdf.

USDA. 2020. *World Agricultural Production*. WASDE Report. Washington, DC: United States Department of Agriculture.

USDA. 2021. *World Agricultural Production*. WASDE Report. Washington, DC: United States Department of Agriculture.

Uys, L., Hofmeyr, J.-H. S. and Rohwer, J. M. 2021. Coupling kinetic models and advection-diffusion equations. 1. Framework development and application to sucrose translocation and metabolism in sugarcane. *In Silico Plants* 3(1). https://doi.org/10.1093/insilicoplants/diab013.

Valdés-Gomez, H., Gary, C., Brisson, N. and Matus, F. 2014. Modelling indeterminate development, dry matter partitioning and the effect of nitrogen supply in tomato with the generic STICS crop-soil model. *Scientia Horticulturae* 175, 44-56.

van der Merwe, M. J. and Botha, F. C. 2013. Respiration as a competitive sink for sucrose accumulation in sugarcane culm: Perspectives and open questions. *Sugarcane: Physiology, Biochemistry, and Functional Biology* 155-168. https://doi.org/10.1002/9781118771280.ch7.

van Heerden, P. D. R., Donaldson, R. A., Watt, D. A. and Singels, A. 2010. Biomass accumulation in sugarcane: Unravelling the factors underpinning reduced growth phenomena. *Journal of Experimental Botany* 61(11), 2877-2887.

van Vuuren, D. P., Stehfest, E., Gernaat, D. E. H. J., van den Berg, M., Bijl, D. L., de Boer, H. S., Daioglou, V., Doelman, J. C., Edelenbosch, O. Y., Harmsen, M., Hof, A. F. and van Sluisveld, M. A. E. 2018. Alternative pathways to the 1.5°C target reduce the need for negative emission technologies. *Nature Climate Change* 8(5), 391-397.

Van Wart, J., van Bussel, L. G. J., Wolf, J., Licker, R., Grassini, P., Nelson, A., Boogaard, H., Gerber, J., Mueller, N. D., Claessens, L., van Ittersum, M. K. and Cassman, K. G. 2013. Use of agro-climatic zones to upscale simulated crop yield potential. *Field Crops Research* 143, 44–55. https://doi.org/10.1016/j.fcr.2012.11.023.

Vega, A., Paola Pertuz, V., Perez, A., Isabel Ort, M. and M. 2018. Calibration of the AquaCrop model in special coffee (Coffea arabica) crops in the Sierra Nevada of Santa Marta, Colombia. *Journal of Agronomy* 17(4), 241–250. https://doi.org/10.3923/ja.2018.241.250.

Vianna, M. dos S., Nassif, D. S. P., dos Santos Carvalho, K. and Marin, F. R. 2020. Modelling the trash blanket effect on sugarcane growth and water use. *Computers and Electronics in Agriculture* 172, 105361.

Vianna, M. dos S. and Sentelhas, P. C. 2014. Simulação do risco de deficit hídrico em regiões de expansão do cultivo de cana-de-açúcar no Brasil. *Pesquisa Agropecuaria Brasileira* 49(4), 237–246.

Vianna, M. dos S. and Sentelhas, P. C. 2016. Performance of DSSAT CSM-CANEGRO under operational conditions and its use in determining the 'saving irrigation' impact on sugarcane crop. *Sugar Tech* 18(1), 75–86.

Vieira, N. A., de Aguiar, E., Caramori, P. H., Nitsche, P. R., Corrêa, K. A. B., Alves, D. S. and Alves, D. S. 2019. Temperature, thermal comfort, and animal ingestion behavior in a silvopastoral system. *Semina: Ciencias Agrarias* 40(1), 403–416. https://doi.org/10.5433/1679-0359.2019v40n1p403.

Vieira, Jr., N. A., Evers, J., dos Santos Vianna, M., Pedreira, B. C., Pezzopane, J. R. M. and Marin, F. R. 2022. Understanding the arrangement of Eucalyptus-Marandu palisade grass silvopastoral systems in Brazil. *Agricultural Systems* 196. https://doi.org/10.1016/j.agsy.2021.103316.

Villegas, F., Daza, O., Jones, J. and Royce, F. 2005. CASUPRO: An industrial-driven sugarcane model. https://www.researchgate.net/publication/271420619_CASUPRO_An_industrydriven_sugarcane_model.

Vogel, E., Martinelli, G. and Artuzo, F. D. 2021. Environmental and economic performance of paddy field-based crop-livestock systems in Southern Brazil. *Agricultural Systems* 190, 103109. https://doi.org/10.1016/j.agsy.2021.103109.

Wajid, A. L. I., Ahmad, M. M., İftakhar, F., Qureshi, M. and Ceyhan, A. 2020. Nutritive potentials of Soybean and its significance for humans health and animal production: A review. *Eurasian Journal of Food Science and Technology* 4(1), 41–53.

Walter, L. C., Streck, N. A., Rosa, H. T., Ferraz, S. E. T. and Cera, J. C. 2014. Climate change and its effects on irrigated rice yield in the state of Rio Grande do Sul, Brazil. *Pesquisa Agropecuária Brasileira* 49(12), 915–924.

Wang, J., Nayak, S., Koch, K. and Ming, R. 2013. Carbon partitioning in sugarcane (Saccharum species). *Frontiers in Plant Science* 4, 201.

Wilkerson, G. G., Jones, J. W., Boote, K. J., Ingram, K. T. and Mishoe, J. W. 1983. Modeling soybean growth for crop management. *Transactions of the ASAE* 26(1), 63.

Williams, J. R., Jones, C. A., Kiniry, J. R. and Spanel, D. A. 1989. The EPIC crop growth model. *Transactions of the ASAE: American Society of Agricultural Engineers* 32(2), 497–511.

Yuan, S., Linquist, B. A., Wilson, L. T., Cassman, K. G., Stuart, A. M., Pede, V., Miro, B., Saito, K., Agustiani, N., Aristya, V. E., Krisnadi, L. Y., Zanon, A. J., Heinemann, A. B., Carracelas,

G., Subash, N., Brahmanand, P. S., Li, T., Peng, S. and Grassini, P. 2021. Sustainable intensification for a larger global rice bowl. *Nature Communications* 12(1), 7163.

Zabel, F., Delzeit, R., Schneider, J. M., Seppelt, R., Mauser, W. and Václavík, T. 2019. Global impacts of future cropland expansion and intensification on agricultural markets and biodiversity. *Nature Communications* 10(1), 2844.

Zalles, V., Hansen, M. C., Potapov, P. V., Stehman, S. V., Tyukavina, A., Pickens, A., Song, X. P., Adusei, B., Okpa, C., Aguilar, R., John, N. and Chavez, S. 2019. Near doubling of Brazil's intensive row crop area since 2000. *Proceedings of the National Academy of Sciences of the United States of America* 116(2), 428–435.

Zhao, C., Liu, B., Xiao, L. J., Hoogenboom, G., Boote, K. J., Kassie, B. T., Pavan, W., Shelia, V., Kim, K. S., Hernandez-Ochoa, I. M., Wallach, D., Porter, C. H., Stöckle, C. O., Zhu, Y. and Asseng, S. 2019. A SIMPLE crop model. *European Journal of Agronomy* 104, 97–106 https://doi.org/10.1016/j.eja.2019.01.009.

Chapter 15

Modelling the impact of climate change on agriculture in Australia and Oceania

Enli Wang, CSIRO Agriculture and Food, Australia; Edmar Teixeira, The New Zealand Institute for Plant & Food Research Limited, New Zealand; Bangyou Zheng, CSIRO Agriculture and Food, Australia; Neal Hughes, Australian Bureau of Agricultural and Resource Economics and Sciences (ABARES), Australia; Karine Chenu, The University of Queensland, Queensland Alliance for Agriculture and Food Innovation (QAAFI), Australia; James Hunt, University of Melbourne, Australia; Afshin Ghahramani, University of Southern Queensland, Australia; Andries B. Potgieter, The University of Queensland, Queensland Alliance for Agriculture and Food Innovation (QAAFI), Australia; Junqi Zhu and Rogerio Cichota, The New Zealand Institute for Plant & Food Research Limited, New Zealand; and Neil Huth, CSIRO Agriculture and Food, Australia

1 Introduction
2 Past climate change and future projections
3 Challenges to agriculture
4 Agricultural systems modelling
5 Modelling the impact of climate change: case studies
6 Future trends in research
7 Conclusion
8 Where to look for further information
9 References

1 Introduction

1.1 The Oceania region

The Oceania region (Fig. 1) includes Australia, Papua New Guinea (PNG), New Zealand, and 22 countries and territories of the Pacific Islands (i.e. American Samoa, Cook Islands, Federated States of Micronesia, Fiji, French Polynesia, Guam, Kiribati, Marshall Islands, Nauru, New Caledonia, Niue, Northern Mariana Islands, Palau, Pitcairn Islands, Samoa, Solomon Islands, Tokelau, Tonga, Tuvalu, Vanuatu, Wallis, and Futuna). The region covers over 100 million km^2 of the

http://dx.doi.org/10.19103/AS.2022.0115.17

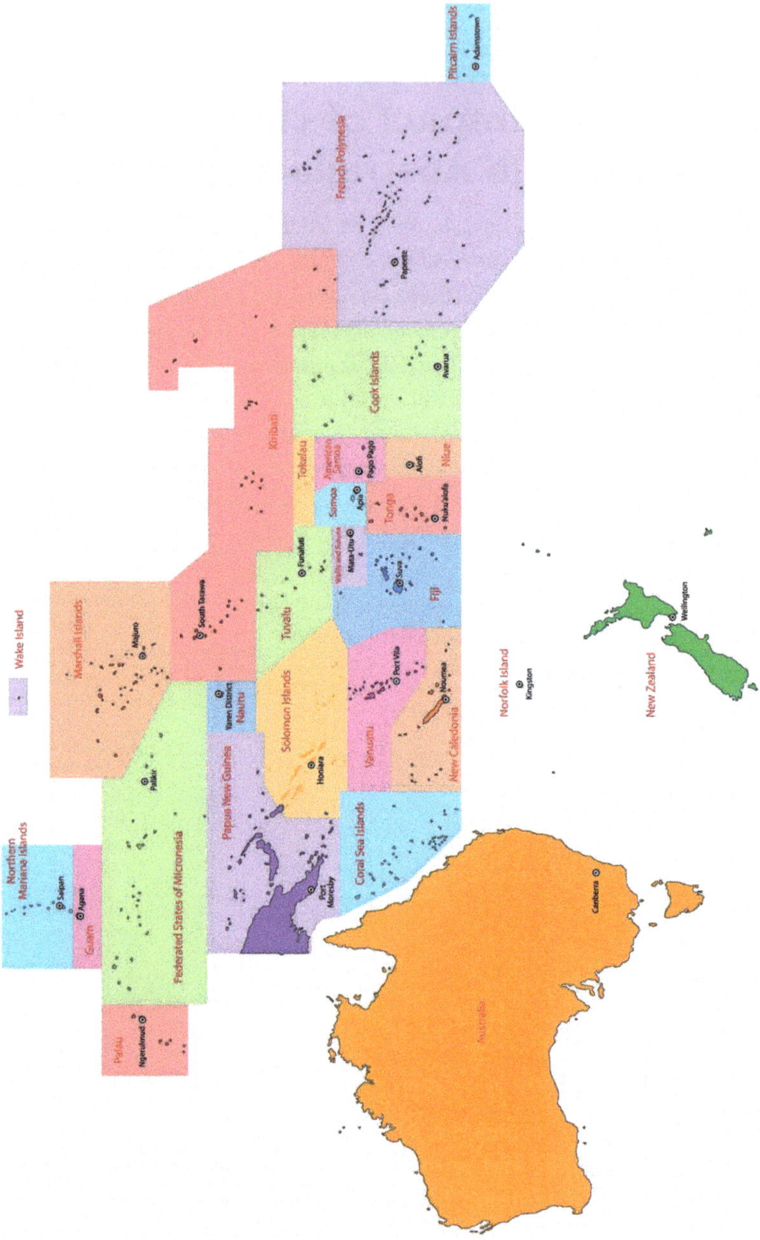

Figure 1 Countries and territories of Oceania. Source: www.canstockphoto.com.

Pacific Ocean, only 8.5% of which being land surface area. Australia has the largest land surface area (7 million km^2), followed by PNG (463 000 km^2) and New Zealand (268 000 km^2). These three countries make up around 93% of the total land area in the region.

The Oceania region is the least populated region in the world. It had a total population of 41 million in 2018, of which 60% was in Australia, 21% in PNG, 11% in New Zealand, and the rest in the 22 countries and territories of the Pacific Islands. It has a diverse mix of economies, ranging from highly developed countries (Australia, New Zealand) to much less developed ones such as PNG and Vanuatu.

The region is unique in its relative proportion of ocean to land mass and the significant dominance of ocean impact on climate and other environmental features. There are diverse terrestrial and marine systems that are responsive to climate change. If not managed properly, they can lead to significant environmental and ecological problems, with impact beyond the region.

This chapter will focus on agricultural production systems in Australia, New Zealand, and PNG. These three countries dominate land mass, population, and production.

1.2 Climate

Australia's climate is largely dominated by the dry and sinking air of the sub-tropical high-pressure belt (sub-tropical ridge) that moves north and south with seasons. In winter, the high-pressure systems move north, and westerly and southwesterly; winds and rain-bearing cold fronts prevail in southern Australia, resulting in a winter rainfall-dominant climate suited to winter crop and pasture production. In summer, the high-pressure systems move south, and north-westerly wind associated with Australian monsoon bring moist air to tropical regions of northern Australia, causing showers and thunderstorms, where rainfall is summer dominant and the region suited to both summer and winter crop production.

Due to its large land mass and the surrounding ocean, the climate in Australia is highly variable both inter-annually and spatially, with frequent droughts and sporadic flooding rainfall. The high spatiotemporal variability is linked to large-scale circulation systems, including El Niño-Southern Oscillation (ENSO), the Indian Ocean Dipole, the Australian monsoon and the Madden-Julian Oscillation, and the Southern Annular Mode (http://www.bom.gov.au/climate/about/). In general, most parts of inland Australia are characterised by desert or semi-arid climate, while the areas near the coast are more humid and have more rainfall (Peel et al., 2007) (Fig. 2). For the near-coastal regions, the climate changes from tropical to temperate climate from north to south accompanied with a decline in both rainfall fraction in summer months and temperatures. The southeast and southwest corners are characterised by temperate climate with winter-dominant rainfall patterns. The mean annual

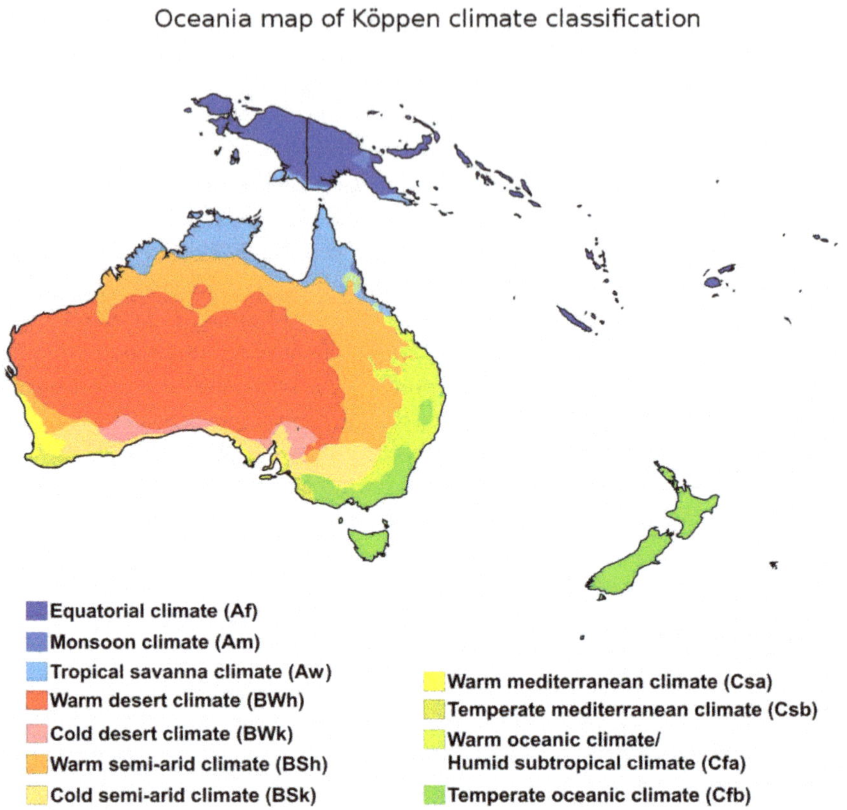

Figure 2 Oceania map of Köppen climate classification. Source: https://en.wikipedia.org/wiki/Oceania#/media/File:Oceania_map_of_K%C3%B6ppen_climate_classification.svg.

rainfall declines from above 1500 mm in coastal areas to less than 100 mm in the central inland, with 80% of Australia having a mean annual rainfall of less than 60 mm (www.bom.gov.au/climate/extremes/). Mean annual temperature can range from less than 5°C in the alpine environments of the southeast to more than 35°C in northern tropical areas. Frost mostly occurs in southeast and southwest inland regions.

New Zealand has a temperate maritime climate with well-defined seasons in most of its regions (Sturman and Tapper, 2005; NIWA, 2021). The country's long latitudinal range and contrasting landscapes create diverse climate zones across the country; from sub-tropical in far north to cool temperate in far south and from semi-arid in central southern regions to extremely wet on the west coastal areas of the South Island. In particular, the country's Southern Alps block prevailing westerly winds and define the wettest and driest of New Zealand's environments, with annual rainfall varying from less than 600 mm to more than 1600 mm with different regional seasonal distributions. Similarly, mean annual temperatures range from around 10°C to 16°C from the southern to northern agricultural areas of the country. Frosts can occur across most agricultural regions in New Zealand, but snow fall is mostly restricted to mountainous areas.

All Pacific Islands are characterised by tropical climates (Fig. 2), though temperature and precipitation change among them. ENSO is the major cause of year-to-year variability of climate. In many of the island countries, the rainfall in El Niño years is often much lower than in other years. ENSO also influences temperature, sea level, and the risk of tropical cyclones. Changes in sea-surface temperature and winds associated with El Niño and La Niña cause large-scale shifts in rainfall patterns. The Interdecadal Pacific Oscillation influences climate variability over a longer time scale. The monsoon and the related wet season during the summer months contrast with dry winds in the winter which blow over the ocean from the Asian landmass, resulting in only two distinct seasons - a humid, warmer period from November to April and a drier, cooler period from May to October (CSIRO et al., 2015). Most of the island countries have an annual rainfall of more than 1000 mm, with some reaching more than 7000 mm. Temperature seasonal changes vary between countries, with some experiencing little change, such as Kiribati and Tuvalu, with monthly averages staying around 27.5°C to 28.5°C, while others, such as Vanuatu, having moderate changes in average temperature from 21°C to 27°C.

PNG is characterised by a hot, humid tropical climate with two distinctive seasons: wet (December–March) and dry (June–September). In many parts of PNG, the annual rainfall exceeds 2500mm, with an average monthly rainfall range of 250–350 mm and an average temperature of 26–28°C (https://climateknowledgeportal.worldbank.org/country/papua-new-guinea/climate-data-historical).

1.3 Importance of agriculture

Agriculture is a major form of land use in Oceania and is important for regional economies, particularly in countries with larger land areas. In Australia, agriculture accounts for 55% of land use, 25% of water use, 11% of goods and services exports, 1.9% of gross domestic product (GDP), and 2.6% of employment in 2019–20 (Weragoda and Duver, 2021). Around 70% of agricultural products are exported. In terms of production value, cropping, horticulture, livestock, and fishery account for 23%, 18%, 51%, and 8%, respectively.

The types of agricultural systems are largely determined by climate, water availability, and soils and are also influenced by markets. While extensive livestock grazing is widespread, grain cropping and horticulture are generally concentrated in areas relatively close to the coast. In the southern part of Australia where rainfall is either uniform or winter-dominant, winter cropping is dominant. In northeast and central east regions where rainfall is concentrated in summer, farmers have adopted opportunity cropping systems. Wheat, barley, and canola are the three largest grain crops in Australia, grown mainly in the temperate southern part of Australia. Summer crops like cotton, sorghum, and sunflower are predominately grown in summer rainfall regions.

Australian agriculture is more exposed to climate change than agriculture in most other countries due to the highly variable climate. Farmers in Australia receive low levels of financial government subsidy, in comparison to other members of the Organisation for Economic Co-operation and Development, with limited support provided through investment in research and development, e.g. via the Research and Development Corporations. Government-facilitated financial tools such as farm management deposits and income tax smoothing help farms manage risks that can arise from Australia's variable climate (Weragoda and Duver, 2021) .

In New Zealand, agriculture is a key component of the economy. Primary industries currently account for more than 5% of the GDP and 6% of the country's employment (Sturman and Tapper, 2005; Ministry for the Environment and Stats NZ, 2020; NIWA, 2021). Being the largest component of the country's tradable economy, agricultural land use and exports are currently dominated by the livestock (dairy and meat), forestry, and horticultural sectors. The value of the horticulture industry has increased by more than 60% in the past 10 years to $6.6 billion, representing more than half of the total gross domestic product value of the agriculture, forestry, and fishing industry (Horticulture New Zealand 2020). Kiwifruit, wine grapes, and apples are the three largest horticulture sectors in New Zealand.

For the various sectors, agricultural production systems, species choice, and management largely differ depending on local climatic conditions and market drivers (Ministry for the Environment and Stats NZ, 2017). For example, although livestock farming is largely based on mixed pasture production (e.g. ryegrass and white clover) with animals supplemented during the winter (e.g. silage maize, brassicas, and fodder beet), these can range from less intensive dryland sheep farming in the hilly areas of the country to highly intensified irrigated dairy production systems in lowland regions. Broadacre cropping is mostly intensively managed for high yields and relies on multi-species crop rotations, adapted to local climate and market demands. A wide range of C_3 broadacre crop species such as wheat, barley, potatoes, and peas are cultivated across New Zealand's agricultural regions while C_4 crop species, such as maize for silage or grains, are confined to the warmer climate zones.

For the Pacific Islands, the importance of agriculture varies between the countries. For countries with larger land mass, such as PNG, Fiji, Solomon Islands, Samoa, and Vanuatu, agriculture plays a more significant role than in other island countries. The major commercial crops include oil palm and coffee in PNG; sugarcane in Fiji; oil palm, copra, and cocoa in Solomon Island; taro in

Samoa; and kava and coffee in Vanuatu. In those countries, agriculture accounts for 10–24% of the GDP. In smaller island countries, agricultural faces challenges with limited land areas and poor soils. Farming is mainly for household consumption and cultural practices (Iese et al., 2020).

1.4 The focus of this chapter

In the past decades, the Oceania region has experienced a general trend of rising temperature, increased rainfall variability, declining rainfall in parts of Australia and New Zealand, and increased rainfall, particularly extreme events in other parts of the Oceanian region. These changes in climate pose challenges to agriculture, forcing agricultural management to adapt.

Systems modelling has played a key role in understanding how climate variability and change impact different aspects of agricultural systems, ranging from growth, development, and yield of different crops, management practices (such as types of crops and cropping systems, input levels etc.), and farm profits. Such studies have enabled an understanding of future changes and the development of adaptation strategies for climate change mitigation. Great efforts have been made in Australia and New Zealand to develop the agricultural systems modelling platform of the Agricultural Production SIMulator (APSIM (Holzworth et al., 2014)), which has also been applied in PNG, other Pacific Island countries, and other countries worldwide.

This chapter summarises the climate change in the recent past and future climate projections for the Oceanian region, briefly reviews how agriculture has responded to past climate changes, and how APSIM and other related models have been developed to assist in the development of adaptation strategies. We then present case studies in Australia, New Zealand, and PNG to illustrate how systems modelling has helped to quantify the impact, evaluate adaptation strategies, and influence policy. We also discuss future research needs and provide links for further information on relevant research and applications.

2 Past climate change and future projections

Climate change in the recent past of Oceania has been governed by global warming due to anthropogenic greenhouse gas emissions, and so too are future changes in climate. The newest Intergovernmental Panel on Climate Change (IPCC) assessment has projected a continued increase in global surface temperature until at least the mid-century under all emissions scenarios considered. Global warming of 1.5°C and 2°C will be exceeded during the twenty-first century unless deep reductions in CO_2 and other greenhouse gas emissions occur in the coming decades. This will be accompanied by an increase in the frequency and intensity of heat extremes, marine heat waves, heavy precipitation, and agricultural and ecological droughts in some regions (IPCC, 2021), including Australia.

2.1 Australia

Australia's climate in recent decades has been characterised by warming and increased frequency of extreme weather events (Fig. 3) (http://www.bom.gov.au/state-of-the-climate/). Air temperature increased on average by 1.44 ± 0.24°C since 1910, which has been accompanied by an increased frequency of extreme heat events. Rainfall in the main crop and pasture growing season from April to October has declined around 12% since 1990 in southeast Australia and by 16% since 1970 in southwest Australia (with the largest decrease in May–July by 20%). The decline in rainfall has resulted in a decline in streamflow across southern Australia, affecting irrigation agriculture. However, in northern Australia, there has been an increase in rainfall and streamflow since the 1970s.

Rainfall decline in southern Australia has been due to a more southerly position of the sub-tropical high-pressure belt during the cool half of the year, a synoptic pattern associated with higher sea-surface temperatures (Cai et al., 2012; Cai and Cowan, 2013). This has notably reduced the frequency and size of rain events at the start of the growing season which is essential for the timely establishment of winter crops and pastures (Pook et al., 2009; Flohr et al., 2021). High-pressure systems are also associated with an increase in the frequency and severity of frost during critical reproductive phases of grain and horticultural crop development (Crimp et al., 2016a,b).

In the coming decades, Australia's climate is projected (Fig. 3) to have continued increases in air temperatures, with more heat extremes and fewer cold extremes. The decline in growing season rainfall in southern and eastern Australia will continue, likely leading to prolonged periods of drought, more intense, short-duration heavy rainfall events. The frequency of dangerous fire weather days will increase with a longer fire season for southern and eastern Australia. There will be further sea level rise and continued warming and acidification of the oceans around Australia, leading to increased and longer-lasting marine heat waves. It is also projected to have fewer tropical cyclones, but a greater proportion projected to be of high intensity, with large variations from year to year.

2.2 New Zealand

Land areas in New Zealand have warmed by around 1.1°C between ~1910 and 2020 (IPCC, 2021) with different intensity patterns across regions (Mullan et al., 2016) (Fig. 4). Recent downscaled CMIP5 climate change projections for New Zealand show large regional differences both for average conditions

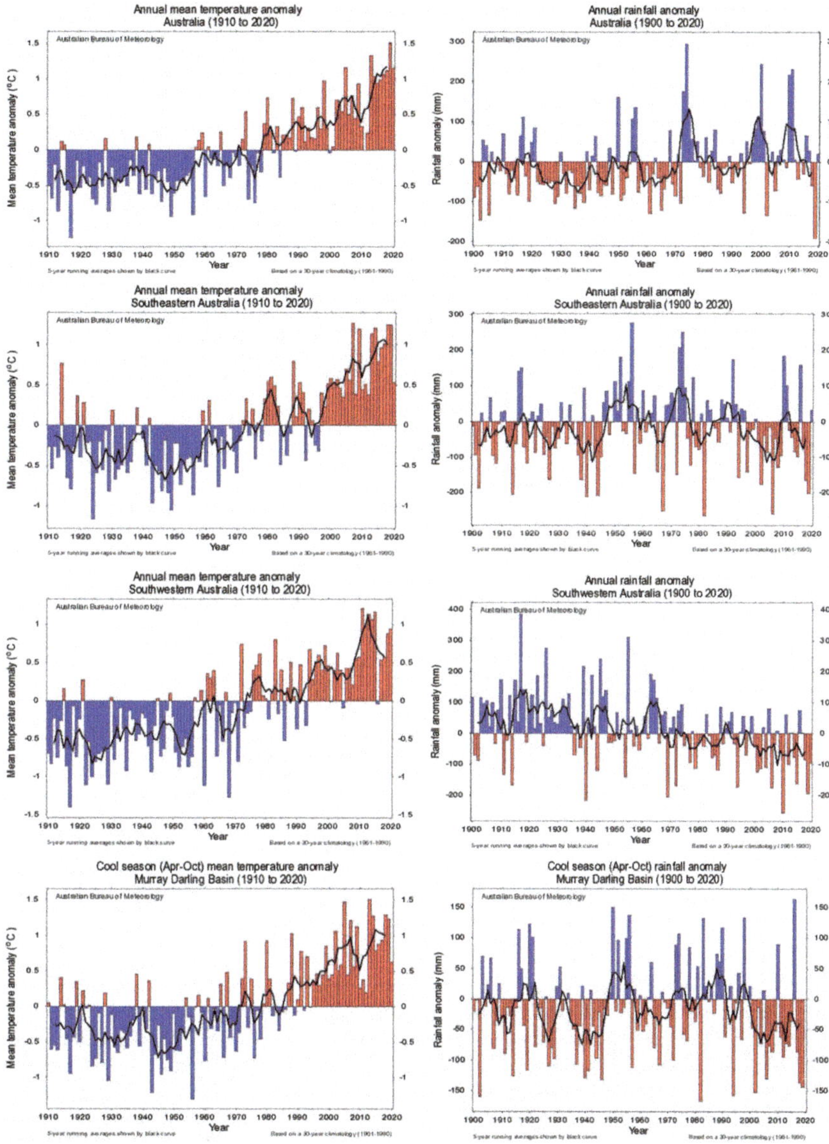

Figure 3 Annual temperature and rainfall anomalies in Australia, southeast and southwest Australia, and cool season temperature and rainfall anomalies in the Murray Darling Basin of Australia. Source: http://www.bom.gov.au/climate/change/#tabs=Tracker&tracker=timeseries.

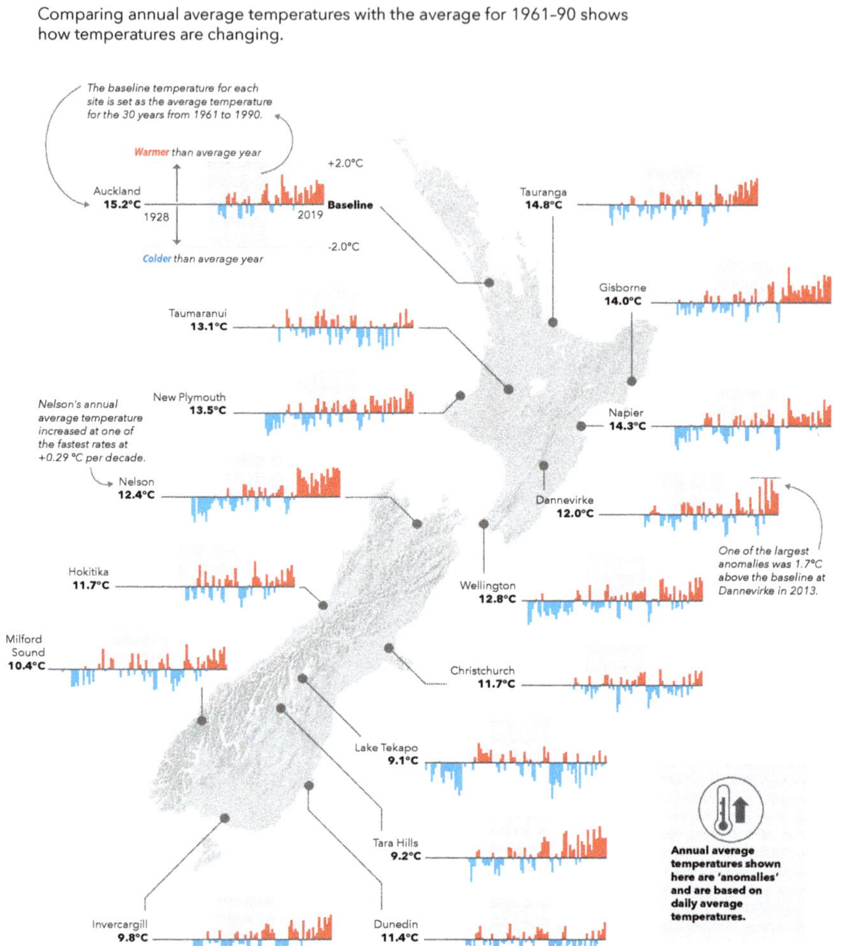

Figure 4 Average temperature anomalies from a 1961–1990 reference across different New Zealand regions. Note: Temperature records have different start dates. Only sites with enough data to calculate a baseline temperature are shown here. Image: Map data by Land Information New Zealand (CC BY 4.0) adapted from Ministry for the Environment & Stats NZ (2020). New Zealand's Environmental Reporting Series: Our atmosphere and climate 2020. Available from www.mfe.govt.nz and www.stats.govt.nz (https://environment.govt.nz/assets/Publications/Files/our-atmosphere-and-climate-2020.pdf).

and the intensity and frequency of extreme climatic events (Mullan et al., 2016). For example, depending on the Representative Concentration Pathway (RCP) scenario considered, average temperature increases can range from 0.7°C to 1.0°C by 2040s and 0.7°C to 3.0°C by 2090s across the country, with greater warming estimated for the summer/autumn periods and also at higher elevations, where snow days are projected to decline (Ministry for the

Environment, 2018). Similar ranges of warming projections for New Zealand were suggested by the most recent IPCC reports (IPCC, 2021).

In contrast with the consistent increase in temperature, the direction of changes in projected rainfall patterns largely differs with location and across seasons in New Zealand. For example, higher rainfall amounts are mostly projected on the west coast and south of the South Island for the winter/spring period (IPCC, 2021 report). In contrast, drier conditions are projected for the east and north of the country during this period. During summer, more rainfall is projected in the east of both islands and less in the west and central North Island.

A key aspect relevant to New Zealand agricultural systems is the projected changes in the frequency and intensity of extreme climatic events. From 1972 to 2019, the number of heat wave days has increased in many sites monitored across the country (IPCC, 2021). Future projections using the most recent downscaled climate data show an overall reduction in the frequency of frost days (i.e. minimum temperature <0°C) and an increase in the frequency of 'hot days' (i.e. maximum temperature > 25°C), although with varied intensity across the country's regions (Mullan et al., 2016). Such climatic extremes influence agricultural productivity by affecting, for example, vernalisation of horticultural crops (Clothier et al., 2012) and heat stress risks in both agricultural (Teixeira et al., 2013) and livestock systems (Polsky and von Keyserlingk, 2017). Similarly, the risk of extreme precipitation events is projected to increase under future emission scenarios, particularly for the southwest of the South Island. Finally, stronger winds are projected for the South Island, although with a low degree of certainty on the magnitude of changes. The already observed retreat of New Zealand glaciers is projected to continue in the coming decades due to warming trends (IPCC, 2021).

2.3 Pacific Islands

Global warming led to region-wide temperature rise in Pacific Islands, with mean annual temperatures increased by 0.18°C per decade since 1961, accompanied by the increased number of hot days. Since 1950, islands in the west pacific become wetter, with increased inter-annual rainfall variability. Sea levels have risen 10–15 cm region-wide, 2-4 times the global average. Sea-surface temperatures have increased by 0.07–0.23°C per decade.

For the future, the temperature is projected to increase by 0.6–1.4°C by 2050, with more hot days. Rainfall is expected to increase, with more extreme rainfall events and floods. Sea levels are likely to rise by 17–38 cm by 2050, and at least 1 m by 2100. Sea-surface temperature will further increase by 0.9–1.4°C by 2050 (USAID, 2018; https://www.climatelinks.org/resources/climate-risk-profile-pacific-islands).

For PNG, the 'most likely' scenario indicates that the mean monthly air temperature will increase by about 0.9°C by 2030 (the range is 0.4–1.0°C) (Hennessy et al., 2011), which is higher than recorded temperature increases over the past 30 years (0.2°C per decade). The 'most likely' scenario indicates that mean annual rainfall in PNG will increase by about 8% by 2030. The scenario showing the biggest change indicates that PNG will become much wetter, with a 34% increase in rainfall in October to March (the northwest monsoon season), and an increase of 25% in mean annual rainfall (Bourke, 2018).

3 Challenges to agriculture

3.1 Australia

The rise in temperature across Australia and the decline in winter season (April to October) rainfall in southwestern and southeastern Australia have significantly impacted broadacre agriculture in Australia. The projected continuation of future warming and drying poses even greater challenges. For dryland cropping, rainfall amount and distribution, together with temperature, determine the start and end of the crop growing season as well as the water and temperature stress experienced by crops during critical periods of yield determination. The rise in temperature and reduction in seasonal rainfall has increased the frequency of water stress in major Australian grain producing regions. They have also impacted the timing of germination, hastened the time from germination to flowering and maturity, and subsequently reduced yield.

An increase in atmospheric CO_2 concentration could have a positive effect by increasing the rate of photosynthesis in some plants growing under adequate water and nutrient supply (Wang et al., 2011). However, such a positive impact is restricted by increasing temperatures and declining rainfall. It was estimated that a 10% reduction in rainfall would be likely to remove the CO_2 fertilisation benefit for wheat in Australia (Howden et al., 1999; Crimp et al., 2002). Hochman et al. (2017) estimated that reduced rainfall and increased temperatures have reduced national water-limited potential wheat yield by 27% since 1990, but that CO_2 fertilisation only prevented a 4% further loss.

Past warming has reduced the number of frost events in some regions while increasing risk in others, in particular in southwestern and southeastern Australia. However, due to higher mean temperature accelerating crop development, frost impact has increased on crops like wheat even in regions where no significant trend of frost occurrence was observed (Zheng et al., 2015; Crimp et al., 2016b). In a few regions of western and southern Australia, the number of frost days has increased due to the lower minimum temperature (Zheng et al., 2016). Over recent decades, the frequency of extreme high temperature and heat stress on major crops in Australia has also increased, inducing substantial productivity losses (Ababaei and Chenu, 2020). Heat stress impacts Australian crops mostly after flowering, during the grain filling period of crops, e.g. wheat (Ababaei and Chenu, 2020), field pea (Sadras et al., 2012), chickpea (Lake et al., 2016).

Australian-rainfed crops frequently experience drought stress during grain fill (e.g. chickpea (Lake et al., 2016; Chauhan et al., 2017)), as well as around flowering (e.g. wheat (Chenu et al., 2011, 2013)), sorghum (Hammer et al., 2014), field pea (Sadras et al., 2012)). The past and projected future climate change means that the farmers need to continuously adapt through new management practices such as early sowing and use of new cultivars. Earlier sowing and early flowering can typically help to reduce or avoid the impact of late water and heat stress, but they also increase the risks of frost damage.

The yield of major summer grain crops like sorghum in the dryland cropping region of northeast Australia is also heavily influenced by climate change and the high climatic variability associated with the ENSO phenomenon (Stone et al., 1996; Allan, 2000).

Changes in rainfall lead to insubstantial decline inflows to river systems, reducing the water availability for irrigated agriculture in the Murray Darling Basin of Australia. This impacts irrigated production groups, including livestock (e.g. beef and sheep for meat, dairy), broadacre (cotton, rice, and other grains), and horticulture (grapes, stone fruit, and vegetables). Without mitigation (CO_2 level of 450 ppm), economic production was projected to fall by 12%, 49%, and 92% by 2030, 2050, and 2100, respectively (450 ppm CO_2). With mitigation (CO_2 at 450 ppm), these losses would be 3%, 6%, and 6%) (Garnaut, 2008).

In addition, the more intense, short-duration heavy rainfall events may lead to more frequent flooding and devastating damage to agricultural crops across Australia.

3.2 New Zealand

New Zealand agriculture is currently impacted by climate change in multiple forms. The overall ~1.1°C warming observed for the 1909 to 2016 period (IPCC, 2021 report) is reflected in a range of relevant agronomic indicators measured across multiple sites in the past 30 years (Fig. 5). The faster accumulation of thermal-time (growing degree days) observed in recent decades (Ministry for the Environment and Stats NZ, 2020) causes shortening of crop cycles through faster phenological development, leading to earlier maturity in both annual and perennial species (Salinger et al., 2020). Warmer weather has an impact on the species composition of pastures, affecting feed quality and seasonal growth patterns (Baars et al., 1990; Dodd et al., 2009). Similarly, the increase in warm days and frequency of heat wave years have been associated with

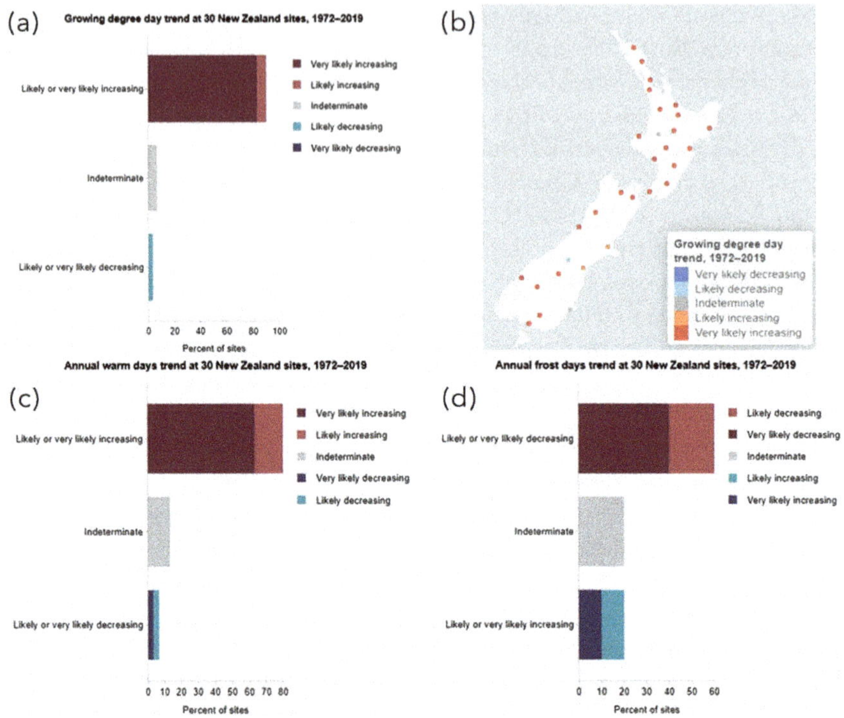

Figure 5 Observed trends in metrics relevant to agricultural systems (a, c, and d) for 30 monitored sites across New Zealand (b) for the period from 1972 to 2019. Growing degree days (GDD in panels a and b) calculated above a 10°C average temperature, warm days (c) considering maximum temperature above 25°C and frost days (d) considering a minimum temperature below 0°C (Source: Adapted from: www.stats.govt.nz/indicators/).

negative impacts on both productivity and quality of potato crops (Salinger et al., 2020). The number of frost days in the past 30 years has also declined in most sites monitored across the country (Fig. 5).

Recent surveys to assess farmers' perceptions of climate change show contrasting flexibility to adapt to the different primary sectors in New Zealand (Cradock-Henry et al., 2020). For example, a range of tactical adaptations already used by the broadacre cropping sector to adjust to short-term inter-annual weather variability and market conditions (e.g. shift of crop species, varietals and sowing time) can be partially extended to also minimise negative impacts of climate change. In contrast, other sectors such as perennial horticulture (apples, kiwifruit and grapevine) show less short-term flexibility to adjust due to long-term commitments with orchard establishment, infrastructure, and market shares of specific varietals. The pastoral sector fits in-between, with some adaptation already needed, especially in some regions where drought and heat waves are already being felt, while solutions for long-term challenges, affecting either forage production or animal welfare, are still being investigated (Keller et al., 2014; Lieffering et al., 2016; Kalaugher et al., 2017). Long-term strategies such as breeding for crop performance under warmer target environments are already being considered by industry, such as in the kiwifruit sector (Cradock-Henry, 2017) and new forages for the pastoral sector (Chapman et al., 2011; DairyNZ, 2018).

3.3 Papua New Guinea

Climate change challenge to agriculture in PNG has been studied in less detail compared to Australia and New Zealand. Take oil palm as an example, which is a major agricultural crop grown in a wide range of geographical locations in PNG. Historically, spatial and temporal climate variability has required tailored management of important issues for productivity and environmental sustainability. Climate change will lead to changes in extreme events, drought, flooding, heat wave, as well as infestation of pests and diseases, forcing the management to adapt. In areas and periods of high rainfall and extreme rainfall events, palm yield will be suppressed due to the clouding effect and risk of soil nutrient loss. However, rainfall increase may lesson water stress in relatively drier areas.

The impact of climate change on tropical crops such as oil palm is mostly unknown (Murphy et al., 2021); however, a detailed review of the major oil palm growing regions in Malaysia (Abubakar et al., 2021) suggests that climate change predictions show some similarity with those predicted for PNG. Adaptation options need to be identified, such as breeding of heat-tolerant hybrid varieties; sustainable management of soil, water, fertilisers, and pesticides; zero burning; and minimum tillage.

4 Agricultural systems modelling

Agricultural systems modelling has become an effective means to integrate knowledge and data and disentangle complex processes that interact and determine productivity and environmental outcomes of agricultural systems. It has been used as a key tool to study both past and future impacts of climate change on crops and cropping systems, and to explore how management intervention can help to adapt and mitigate the negative impact of climate change. For most of the modelling studies in the Oceania region, the APSIM farming systems model has been used to quantify the impact of climate variability and change on crop growth, development, and yield. Some APSIM derivatives and other models are also used to study various aspects of agricultural systems.

4.1 The Agricultural Production Systems Simulator (APSIM)

The APSIM model was developed to simulate the biophysical processes in farming systems, in particular where there is interest in the economic and ecological outcomes of management practice in the face of climatic risk (Keating et al., 2003). It contains a suite of modules that enable the simulation of systems for a diverse range of plant, animal, soil, climate, and management interactions (Fig. 6). APSIM is undergoing continual development, evolving into a framework with capability ranging from simulation of gene expression to multi-field farms and beyond (Holzworth et al., 2014), with new capability being added to APSIM Next Generation (Holzworth et al., 2018). Its development and maintenance are underpinned by rigorous science and software engineering standards.

APSIM can simulate a diverse range of crops, pastures, and trees. A complete list of crops simulated in APSIM can be found at https://www.apsim.info/apsim-model/models-in-release/. Great efforts have been made to capture the common physiological processes and organ structures among different crop species so that a generic framework is achieved for modelling various crops (Wang et al., 2002; Brown et al., 2014). The Plant Modelling

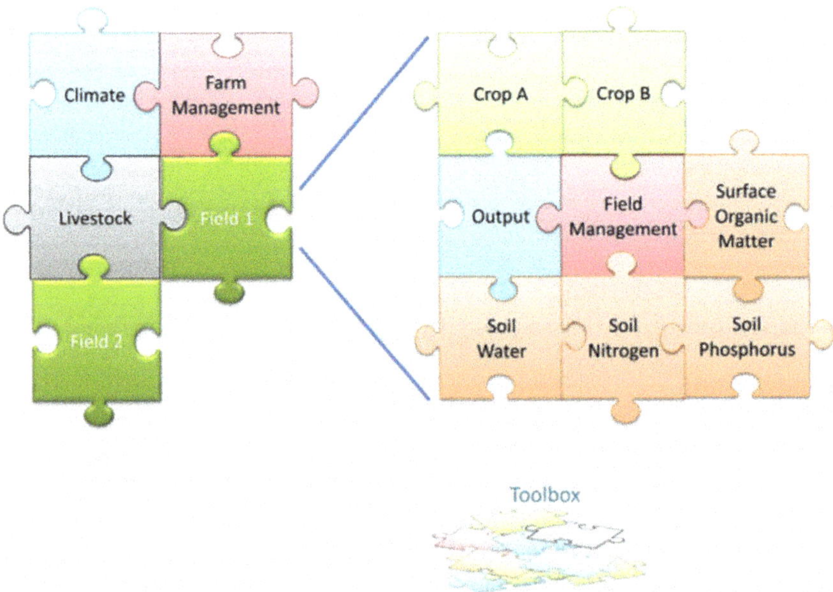

Figure 6 A conceptual representation of an APSIM simulation shows a 'top level' farm (with climate, farm management, and livestock) and two fields. The farm and each field are built from a combination of models found in the toolbox. The APSIM infrastructure connects all selected model pieces together to form a coherent simulation. Source: Adopted from: Holzworth et al. (2014).

Framework (Brown et al., 2014) is currently used in APSIM NextGen to simulate key processes including crop phenology, canopy development, resource (light, water, N) capture and use efficiency, and arbitration for assimilate and nitrogen (N) allocation and redistribution among organs. The impact of climate variability and change is simulated through the response of various physiological processes to changes in temperature, rainfall, and other aspects of climate, including the rising CO_2 concentration in the air (Vanuytrecht and Thorburn, 2017).

Recent development includes a focus on improved modelling of genotype × environment × management (G × E × M) interactions (Hammer et al., 2006, 2016; Wang et al., 2019) so that variations at genetic and trait levels can be captured to assist in crop design for adaptation to drought and high-temperature risks in future climate (Hammer et al., 2020; Collins et al., 2021) .

Key soil processes simulated in APSIM include soil temperature dynamics, water movement and balance in soil layers, surface organic matter dynamics, soil carbon formation and decomposition, N and P transformations, soil pH, and soil erosion. Recent interests have been in soil carbon sequestration and greenhouse gas emissions (CO_2, N_2O) from cropping systems, as well as other environmental outcomes in response to management changes.

One strength of APSIM is that it offers a set of management modules that allow the user to flexibly specify the intended management rules that characterise the scenario being simulated and that control the simulation, e.g. different sowing rules, rotations, and other management interventions.

It is important to note that APSIM does not mechanistically simulate the effect of frost or several facets of heat damage on yield. In studies that use APSIM and consider frost, effect on yield are applied *post hoc* to simulation output assuming a percentage decrease based on the occurrence of threshold temperatures during sensitive stages of development. These are derived either from expert opinion (e.g. Bell et al., 2015; Zheng et al., 2015; Lilley et al., 2019; Cann et al., 2020), published field studies (e.g. Cann et al., 2020), or a combination of both including fitting to observed data (e.g. Lilley et al., 2019). Increased temperatures affect crop yield through diverse mechanisms (Hunt et al., 2018), and APSIM captures some of these (e.g. accelerated crop development, increased soil evaporation, reduced transpiration efficiency) but not others (e.g. heat-induced sterility, loss of green leaf area).

4.2 The farmpredict *micro-simulation model in Australia*

farmpredict is a statistical model of Australian broadacre farming businesses (Hughes et al., 2022b) based on the Australian Agricultural and Grazing Industry Survey (AAGIS). The AAGIS collects detailed physical and financial information for around 1600 broadacre farms across Australia each financial year to provide representative coverage of all Australian broadacre farming regions and industries, including extensive cropping, livestock (beef and sheep) and mixed farming types.

A sample of around 40 000 farms observations (drawn from the AAGIS over the period 1988-9 to 2018-9) is used to fit the model, with each farm linked via point location geocoding to spatial climate data. Variations and changes in climate relevant to crop seasons and crop growth are captured, including those of rainfall amount, average maximum and minimum temperatures, and exposure to upper and lower temperature extremes (Hughes et al., 2022b). The statistical model is fit via a non-parametric machine learning-based method, involving the *xgboost* regression algorithm (Chen and Guestrin, 2016) with multi-target 'stacking' (Spyromitros-Xioufis et al., 2016), and verified with out-of-sample validation tests.

farmpredict predicts - at a farm level - six crop outputs, four livestock outputs, and seven stock (inventory) holdings including livestock numbers and on-farm crop and wool storage (Fig. 7). These production outcomes are then used to simulate farm financial results including various measures of profit. The model can help identify a wide range of relationships with climate influencing: crop planting and storage decisions; input usage (particularly fertiliser and fodder); livestock turn-off, birth and death rates; and farm prices received (via quality effects on livestock and crop outputs). Growing season rainfall via its

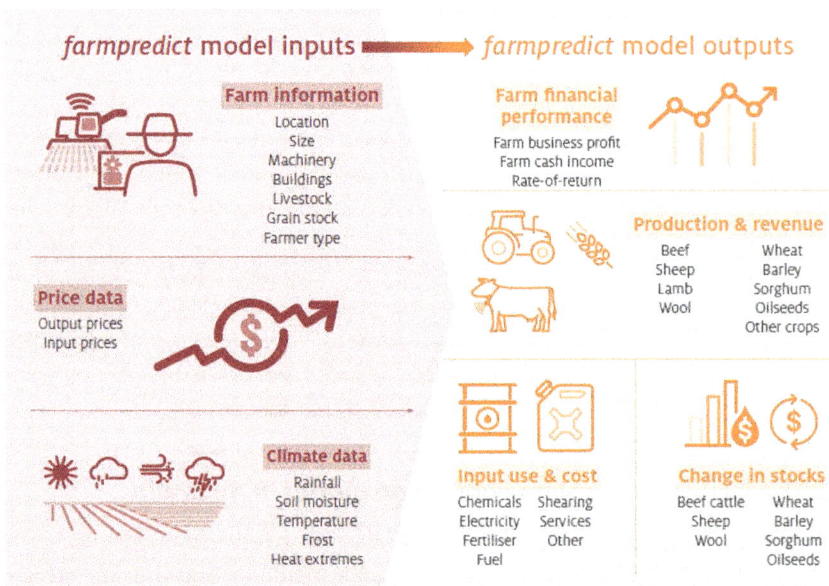

Figure 7 An overview of the *farmpredict* model, showing model inputs (left) and model outputs (right).

effect on crop yields is one of the key drivers of climate effects, temperatures play an important role in many of these responses, proving particularly important for livestock birth and death rates (Hughes et al., 2022b).

4.3 Other models

In addition, several other models have been used to investigate the impacts of climate change and action to mitigate or adapt to these changes. The majority of studies employ process-based, farm systems models, or modelling frameworks. Those other models working at a similar level of complexity include DSSAT (Jones et al., 2003), DairyMod (Johnson et al., 2008), the WholeFarmModel (Beukes et al., 2008), and Biome-BGC (Thornton et al., 2021).

For grazing systems in Australia, several other models have been used, including GRAZPLAN (Freer et al., 1997; Moore et al., 1997), GRASP (Rickert et al., 2000; for savanna rangelands not temperate pasture), SGS pasture model (Cullen et al., 2009), and DairyMod (Johnson et al., 2008). APSIM was coupled with GRAZPLAN in AusFarm (Moore, 2014; Ghahramani et al., 2015) to provide GRAZPLAN with greater soil water and nutrient balance capability.

5 Modelling the impact of climate change: case studies

5.1 Dryland cropping: wheat and sorghum production in Australia

Numerous modelling studies have been conducted in Australia to investigate the impact of past and future climate change on dryland crops, with the majority focusing on wheat and sorghum, as representative winter and summer crops, respectively. This section summarises case studies of how modelling is used to detect the impact of climate change on crop yield, quantify crop responses, and evaluate improved practices in both agronomy and breeding.

5.1.1 Impact of recent climate change on crop yield and production

Climate change has already had a substantial impact on Australian wheat production. Hochman et al. (2017) used APSIM to derive water-limited potential yield (Yw) defined as the yield that can be achieved with best-adapted commercial cultivars managed such that crops experience no biotic or abiotic stresses other than water. They showed that Yw declined by 27% from 1990–2015 relative to 1900–89 (Fig. 8). They attributed 83% of the decline to reduced growing season rainfall during the period and 17% to increased temperatures reducing crop life cycle. In addition, Ababaei and Chenu (2020) estimated an

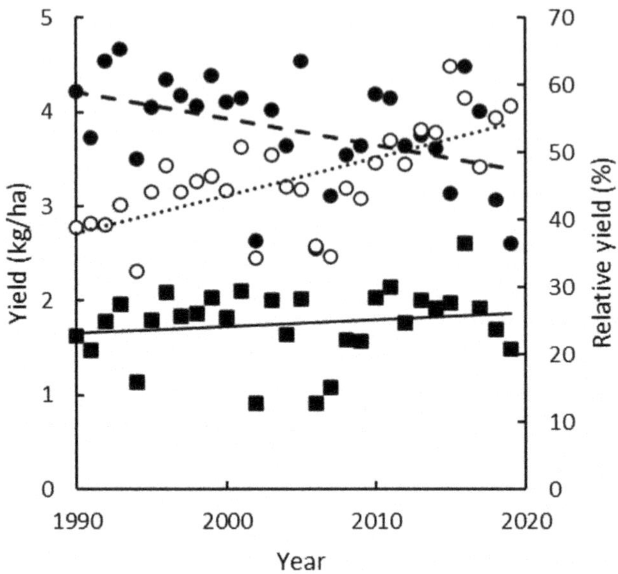

Figure 8 Actual Australian farm wheat yield (■, solid regression line is of form $y = 0.007x - 12.953$, $R^2 = 0.03$), water limited potential yield derived from crop simulation (•, long dashed regression line is of form $y = -0.028x + 60$, $R^2 = 0.18$), and relative yield which is farm yield expressed as a percentage of water limited potential yield (o, short dashed regression line is of form $y = 0.556x - 1068$, $R^2 = 0.45$). See Hochman et al. (2017) for further details. Source: Figure reprinted from: Hunt et al. (2021) by permission of the Society for Experimental Biology.

increase of yield loss of 4.6% per decade between 1985 and 2017. It is likely that the observed reductions in Yw identified in wheat also apply to major cool-season grain crops including barley, oats, canola and grain legumes although no studies have specifically quantified this.

At the national level, both dry land wheat and sorghum production fluctuate from year-to-year and have been strongly impacted by the change in accumulated effective rainfall during the fallow (FRain) and during the crop growth (ICRain) periods (Potgieter et al., 2016). During the past four decades (1982–2020), the area harvested has decreased by −0.4%/year for sorghum and increased by 0.5%/year for wheat. At the same time, there has been about a 1%/year increase in total production for both crops at the national scale. Rainfall during the fallow period for sorghum declined by −0.3%/year and in-crop rainfall increased by 0.3%/year. The pattern was the opposite for wheat as expected since the fallow rain period of sorghum coincides with the in-crop rainfall period for wheat (i.e. April to October) (Fig. 9).

The modelled decline in wheat Yw was not entirely reflected in national actual farm yields (Ya) due to technology-driven gains closing the gap between Ya and modelled Yw by 25 kg/ha/year enabling relative yields (Ya as a percentage of

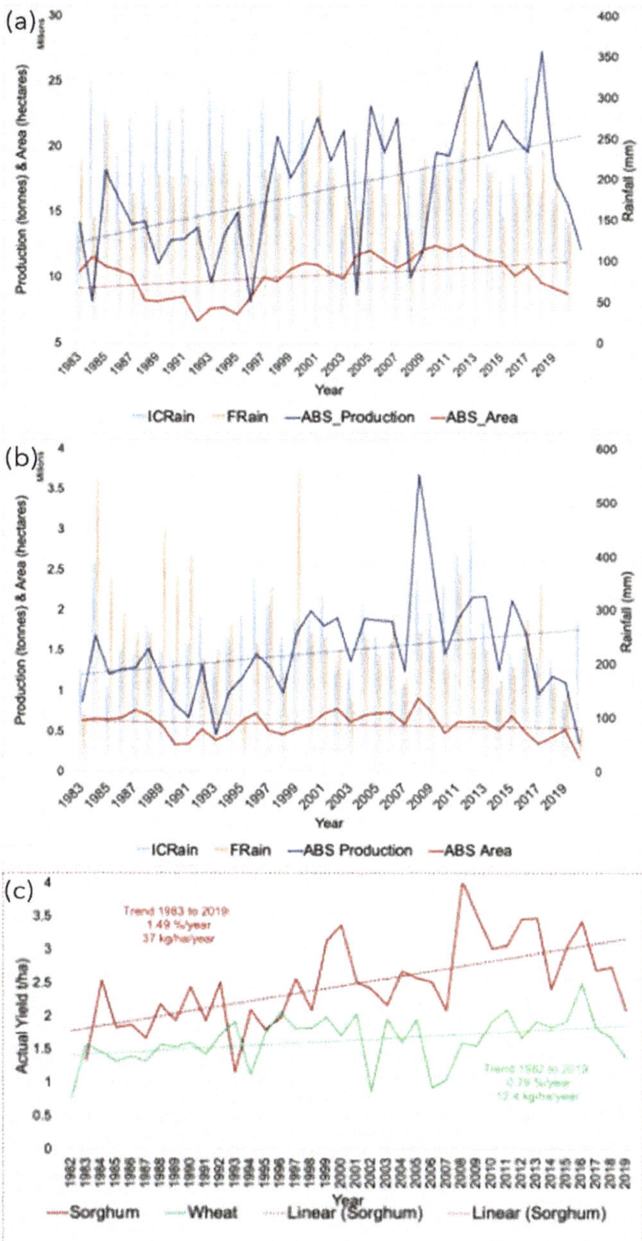

Figure 9 Annual variability in national crop production, planted area, in-crop (ICRain) and fallow rainfall (FRain) for (a) wheat and (b) sorghum in Australia since 1982–1983 (Source: ABS Commodity Statistics 2021). (c) Seasonally adjusted yield trends derived through the integration of observed production data and biophysical modelling approaches for sorghum (Potgieter et al., 2005) and wheat (Potgieter et al., 2006) (Source: Analysis updated from: Potgieter et al., 2016).

Yw) to increase from 39% in 1990 to 55% in 2015. This increase has continued in the years since the study of Hochman et al. (2017) was published with a relative yield of 57% achieved in 2019 (Hunt et al., 2021). Likewise, Potgieter et al. (2016) adjusted and updated yield trends while accounting for the likely effect of climate (1982–3 to 2019–20). Their results clearly show that the advances in technologies (e.g. breeding, agronomic practices) for both sorghum and wheat have been steadily on the increase at a rate of 1.67%/year and 1.54%/year, respectively (Fig. 9). The observed overall advance in technology was 0.8%/year and 1.5%/year for wheat and sorghum, respectively. This suggests that the advances in technologies for both sorghum and wheat have been able to overcome the overall effect of climate up to now. However, it is evident that more will need to be done in terms of enhancing management practices and breeding for heat and drought-tolerant cultivars to ensure technology advances continue to keep up with the projected climate trends, especially in drier and hotter environments (Potgieter et al., 2016; Hammer et al., 2020; Collins et al., 2021).

5.1.2 Impact of technology advances and adaptation on crop yield

The technology-driven gains that have offset decline in Yw have resulted from both genetic and management improvements and the interaction between the two. These are extensively described by Hunt et al. (2021) for wheat, but a summary follows. Australian wheat breeders have consistently achieved yield gains of 0.5% per year during the period of recent change (Sadras and Lawson, 2011), although there is evidence of this slowing in some environments (Flohr et al., 2018b). Notable management changes have included the near-universal adoption of no-tillage farming since 1990 (Llewellyn and Ouzman, 2019), which has allowed earlier (Anderson et al., 2016) and dry sowing (Fletcher et al., 2015), reducing losses due to drought and heat. Wheat has increasingly been grown following broad-leaf break crops, particularly canola and grain legumes (Collins and Norton, 2019), which increases yield and water-use efficiency relative to wheat grown after other cereals through a variety of mechanisms (Angus et al., 2015). Retention of crop residues over summer and stringent control of summer fallow weeds in southern and western Australia has allowed more efficient capture and use of summer fallow rain, which on average contributes one-third of wheat yield in these environments (Hunt and Kirkegaard, 2011). This practice was already well established in northern Australia. Finally, rates of N fertiliser applied to crops have increased rapidly (Angus et al., 2017), allowing crops to capitalise on the additional yield potential provided by the abovementioned changes.

Hochman and Horan (2018) used simulation to identify the potential for further yield gains in the current climate, and whilst restricted to farming system factors simulated by APSIM found that N deficiency was the single biggest cause of low relative yields. This was corroborated by field surveys (Armstrong et al., 2019). Hunt et al. (2019)also found that Yw could be increased under the current climate through earlier sowing of long-cycle cultivars.

5.1.3 Impact on growing season and crop phenology (flowering and maturity time)

An increase in temperature accelerates crop development, thus reducing time for resource acquisition, mismatching between key sensitive growth stages and extreme climate events, and resulting in potential yield loss (Zheng et al., 2016; Hunt et al., 2018). Wheat flowering time is determined by the interaction of temperature response, vernalisation requirement and photoperiod sensitivity. An increase in average temperature has resulted in shortening the wheat crop cycle by 0.74 day per decade since 1957 (Zheng et al., 2016) or 1.6 days per decade since 1985 (Ababaei and Chenu, 2020) with significant spatial variations (Fig. 10). The shift in flowering date had a coastward-eastern and coastward-southern direction in the eastern part of Australia, and a weak centre-ward direction in West Australia as a result of spatial variation of climate changes (Zheng et al., 2016), showing a similar pattern to the observed bird distributions in Australia (VanDerWal et al., 2013).

Further increase in temperature is projected to shorten the wheat season by up to 2 weeks by 2030 and 6 weeks at some sites by 2050 (Zheng et al., 2012, 2016). This acceleration is occurring at a higher rate for the vegetative

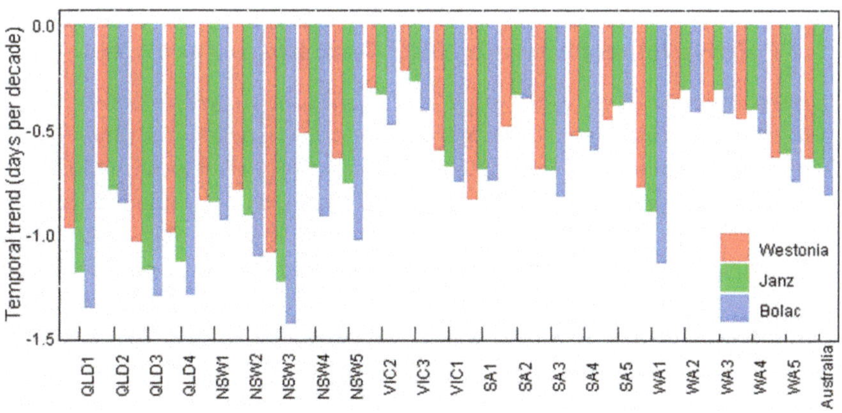

Figure 10 Temporal trends of flowering (day per decade) for wheat cultivar Westonia (early-season cultivar, sown at 22 June), Janz (mid-season cultivar, sown at 1 June), and Bolac (long-season cultivar, sown at 11 May) at 22 representative regions in Australia from 1957 to 2010. The temporal trend for the whole wheatbelt is presented on the right hand side of the bar chart ('Australia'). Source: Redrawn from: Fig 5 in Zheng et al. (2016).

period than the post-flowering period, leading to earlier flowering and changes in the timing of sensitive stages to abiotic stresses, and reduced yield potential as a result of shortened duration to capture resources (e.g. radiation, water and nutrition). To allow current/earlier sowing to avoid stresses (i.e. frost, heat and terminal drought) and to maintain the current growing period in future climates, breeders will need to develop and/or introduce new genetic sources for slower-developing cultivars.

Changes in rainfall timing have also affected the length of the growing season in southern and western Australia. Dryland wheat crops in these environments are reliant on the first significant rainfall in autumn to germinate seeds and commence the crop life cycle, and this is referred to as the 'autumn break'. The median date of the autumn break has become between 3 and 11 days later in southeastern Australia, and 2 and 17 days earlier in southwestern Australia (Flohr et al., 2021). Pook et al. (2009) used APSIM to quantify the yield reduction due to delayed germination at a single location in southeastern Australia. The earlier breaks being experienced in western Australia present an opportunity for growers to capitalise on early rains with slower-developing cultivars suited to early sowing as proposed by Hunt et al. (2019), and this adaptation is in the early stages of adoption.

For sorghum, the climate has a similar but varying effect on flowering and thus final yield outcomes across the dry land cropping systems of northeastern Australia (Wang et al., 2020). The projected increase in the frequencies of high temperatures will expose the summer crops to damaging temperatures around flowering causing an increased incidence of yield loss in the current and future climates (Hammer et al., 2014, 2020; Lobell et al., 2015).

5.1.4 Relative impact on water, frost, heat stresses and yield

Changes in wheat phenology and climate variables are projected to impact abiotic stresses. Over recent decades, the negative impact of frost (Zheng et al., 2015), heat (Ababaei and Chenu, 2020), and drought (Collins and Chenu, 2022) has increased in major production regions across Australia, while the increase in CO_2 had a small positive impact on wheat productivity (Collins and Chenu, 2022). While general trends usually remain, estimated impacts can highly vary depending on the management practices considered and the climate scenarios studied. It is important to keep in mind that impacts reported from different studies were estimated using different hypothesis (e.g. sowing date, cultivar, climate scenarios). Hence, figures presented here may not always be directly comparable.

Frost occurrence has significantly decreased over the last few decades in almost half of the northeastern part of the Australian wheatbelt, while it has increased in other regions with, in addition, an extension of the frost season (Zheng et al., 2015; Crimp et al., 2016b). Increased temperatures from global warming are expected to accelerate crop development, which could counterintuitively increase the chance of frost at sensitive post-heading stages (Zheng et al., 2012). Between 1981 and 2018, frost-induced yield loss has increased by 6.4% per decade in Australia, for a mid-maturing cultivar and common management practices (Collins and Chenu, 2022). Without adaptation, Australian yield losses due to frost are projected to decrease by 6% by 2050, with frost causing 2% of overall yield loss for historic best sowing and cultivar practices (Fig. 12; (Collins et al., 2021)).

The frequency of daytime heat shocks has significantly increased across the wheatbelt over the past decades. From 1985 to 2017, an extra 0.6 and 1.2 hot days (maximum temperature > 26°C) per decade have occurred for sensitive stages around anthesis and during early-mid grain filling, respectively, for common management practices (Ababaei and Chenu, 2020). By contrast, the occurrence of warm nights (minimum temperature > 17°C) has not significantly changed in most regions (Ababaei and Chenu, 2020). Heat-induced yield loss has increased by 4.6% per decade for Australian wheat, increasingly affecting grain size (4.0% per decade) more than grain number (1.0% per decade). By 2050, without agronomic and breeding adaptation, yield losses due to heat stress are expected to increase by 2.2%, thus causing 13.5% yield loss of crops grown with optimal historical sowing × cultivar combinations (Fig. 11; (Collins et al., 2021)).

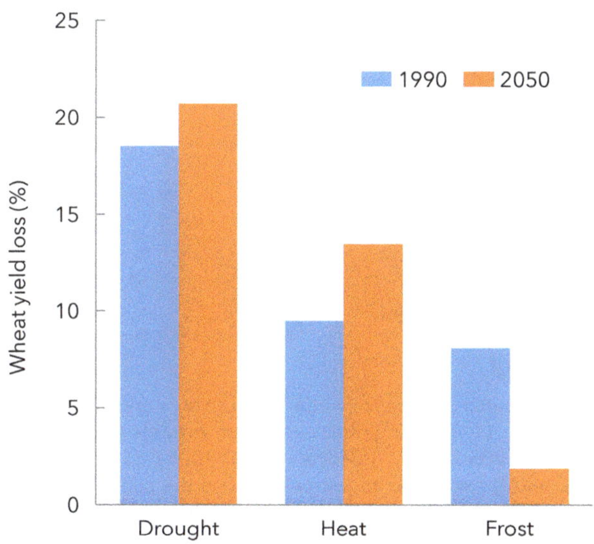

Figure 11 Estimated yield loss in Australian wheat due to drought, heat and frost in 1990 and 2050. Simulations for sowing dates and cultivars that were locally optimised for 1990. Source: Adapted from: Collins and Chenu (2021).

Drought-induced yield loss has increased by 3.6% per decade in Australia from 1981 to 2018, with an increased occurrence of severe water limitations (Collins and Chenu, 2022). By 2050, the frequency of severe drought environments is projected to further increase in western Australia but decrease in the eastern part of the wheatbelt due to (1) shorter crop life cycle associated with warmer temperature and (2) greater water-use efficiency associated with increased CO_2 (Lobell et al., 2015; Watson et al., 2017; Collins et al., 2021). Nevertheless, all regions are still expected to be subjected to frequent severe droughts in the coming decades. Without adaptation, wheat yield losses in Australia are projected to increase by 2% by 2050, thus causing 21% yield loss for crops managed with optimal historic sowing × cultivar combinations (Fig. 12; (Collins et al., 2021)).

5.1.5 Impact of projected climate change on future wheat yield

Future climate projections vary significantly depending on emission scenarios. The impact of future climate change on wheat yield is also highly influenced by adaptation strategies. Thus, the model set ups to reflect changes in crop

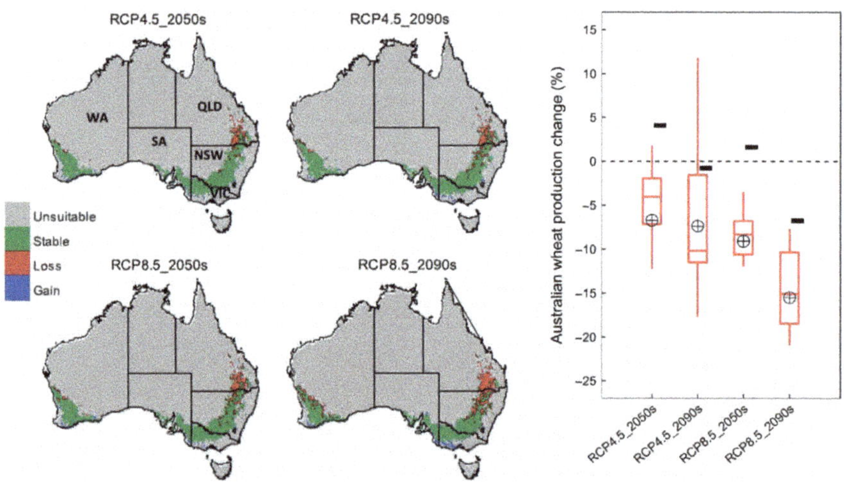

Figure 12 Projected spatial changes in wheat suitable area and production change from an ensemble mean of 11 CMIP5 global circulation models (GCMs). Changes are for RCP4.5 and RCP8.5 for 2050s (2041-2060) and 2090s (2081-2100) compared to baseline (1961-2000). Red and green areas together make up the suitable area for 1961-2000. Box boundaries indicate the 25th and 75th percentiles across GCMs, whiskers below and above the box indicate the 10th and 90th percentiles. The red lines and black crosshairs within each box indicate the multi-model median and mean, respectively. Black rectangles represent multi-model mean values from simulations with optimum adaptation. Source: Modified from: Wang et al. (2018).

varieties and management practices considered have a substantial impact on the estimates (Chenu et al., 2017). Without agronomic or breeding adaptation, yields of wheat crops are projected to stagnate or decrease by 2050 for most climate projections and most regions across Australia (Lobell et al., 2015; Wang et al., 2018; Collins et al., 2021).

At a regional scale, the modelled impact of the projected high-emissions scenario, assuming no adaptation, on wheat yields is likely to become more severe (−30% to −5%) in most of northeastern Australia as well as the drier rainfall zones of parts of southern Australia by 2050 (Potgieter et al., 2012). Under the RCP 8.5, which is the pathway most consistent with the current pace of global emissions, and without considering the impact of frost and heat stress on grain development, Wang et al. (2018) simulated negative impacts of climate change on wheat yield in eastern and western Australia for the 2050s, but positive impacts in southern and southeastern Australia for most of the 11 studied GCMs. Lobell et al. (2015) also simulated the negative impacts of 2050 climate scenarios (averaged across 33 GCMs) on wheat yield in eastern Australia, with simulations that did not account for frost stress. When considering both frost and heat stress on grain development, Collins and Chenu (2021) simulated a relatively unchanged yield for the 2050s in most of Australia, with a projected yield increase in southeastern Australia.

Future climate change could also change the land area available for wheat growth, impacting on total wheat production. There is an overall tendency for a decrease in the areas suitable for growing wheat and a decline in the yield of the northeast Australian wheat belt, resulting in reduced national wheat production (Fig. 12) although future climate change may benefit South Australia and Victoria (Wang et al., 2018).

5.1.6 Informed adaptation in agronomy and breeding

Effective adaptation to climate change could lead to increased crop yield and yield stability in the future. Agronomic measures and development of new cultivars can help by (1) improving the environment experienced by crops (e.g. stress avoidance for earlier sowing, or reduced water stress due to increased water supply) and (2) increasing crop tolerance to relevant stresses. Adaptation strategies related to management practices and trait genetics have been developed to enhance productivity both in current and projected climates.

The recent focus on Australian cropping systems has resulted in identifying sets of best-management practices to maximise long-term productivity and profitability (Kirkegaard et al., 2014). With recent changes in climate, farmers' practices have already evolved and wheat sowing now tends to occur earlier in all Australian regions (Flohr et al., 2018a). Sowing earlier and/or adapting cultivar maturity have been found as potential solutions to minimise the impact of abiotic stress on wheat crops in future climates (Ghahramani et al., 2015; Zheng et al., 2016; Hunt et al., 2019; Collins et al., 2021). Adaptation strategies based on cultivar maturity and sowing date have also been proposed to adapt to year-to-year climate variability, e.g. ENSO (Zheng et al., 2018). In sorghum, global and local options to adapt to future climate were explored to inform breeding and management options at the regional scale (Hammer et al., 2020).

With declining autumn rainfall and earlier sowing, traits such as long coleoptiles would reduce the risk of poor or late crop establishment (Kirkegaard and Hunt, 2010; Flohr et al., 2018a, 2021; Zhao et al., 2022). Other traits have the potential to increase wheat productivity in Australia (e.g. Casadebaig et al., 2016). Those include early vigour (Zhao et al., 2019; Bourgault et al., 2020), increased transpiration efficiency (Chenu et al., 2018; Christopher et al., 2018; Collins et al., 2021), greater stay-green (Christopher et al., 2014, 2016, 2018, 2021), adapted root architecture (Manschadi et al., 2006; Lilley and Kirkegaard, 2011, 2016), and beneficial morphological traits (Rebetzke et al., 2012; Hunt et al., 2018). The value of these traits is expected to change with global warming. For instance, benefits from limited transpiration at high evaporative demand and transpiration efficiency are expected to increase in the future, in major Australian-producing regions (Christopher et al., 2018; Collins et al., 2021). A crop-design study in sorghum (Hammer et al., 2014, 2020) also found transpiration efficiency as offering an avenue for advancing adaptation relevant to current and future climates. Other traits have the potential to enhance adaptation to climate change, such as traits involved in heat stress adaptation (Hunt et al., 2018; Hammer et al., 2020). While the impact of frost stress is expected to decrease in most regions by 2050 (Collins et al., 2021), breeding frost-tolerant wheat cultivars would allow earlier flowering and drought and heat avoidance without risking devastating impacts from frost events. Growing such frost-tolerant cultivars sown at the appropriate time would increase the yield potential of wheat by up to 20% in Australia (Zheng et al., 2015). The adaptation values were spatially varied in different regions with direct or direct plus indirect impacts (Fig. 13, Zheng et al., 2015).

Research on promising traits is progressing mainly to adapt to heat and drought stress. Methods have been developed to screen large populations for key adaptive traits and assist crop improvement in changing environments (e.g. Richard et al., 2015; Chenu et al., 2018; Fletcher et al., 2018).

Modelling studies have identified some practices which allow adaptation to recent climate change but may be unsuitable in future climates. Hunt et al. (2019) used field experiments and simulation to show that long-cycle wheat genotypes sown early could increase whole farm yields throughout most of southern and western Australia under recent climate change. However, Collins

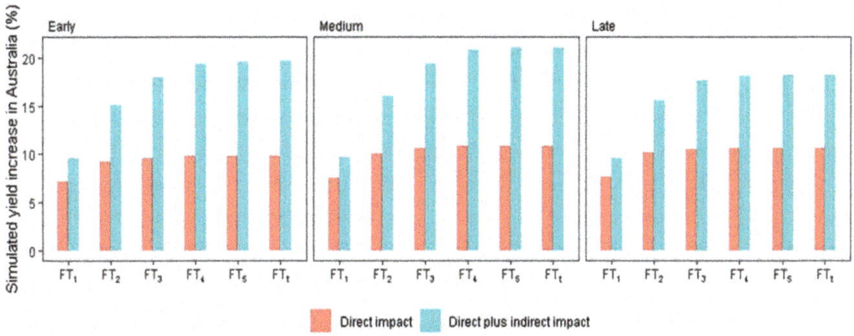

Figure 13 National yield advantage of increased frost tolerance to −1°C (FT1) to total frost tolerance (FTtot) for early-, mid-, and late maturing cultivars. Yield advantages were considered when changing either the frost-damage threshold temperature (direct impact; red) or both the frost-damage threshold temperature and the management (direct plus indirect effect; blue). Data correspond to the average over the 60 locations for simulations from 1957 to 2013. Source: Adapted from: Fig. 7 in Zheng et al. (2015)

and Chenu (2021) showed that this practice would be of little geographic benefit under a predicted climate of 2050.

5.2 Temperate grazing systems in Australia

Projected changes in climate factors, e.g. temperature, rainfall, and atmospheric CO_2 concentration are estimated to affect both plants and livestock directly across Australian temperate grasslands (Ghahramani et al., 2013). Moore and Ghahramani (2013) predicted that in 2050, the annual net primary productivity (ANPP) of southern Australian grasslands will change in the range between −18% and +6% across a range of plausible global circulation models (GCMs). This impact was predicted to vary across regions and livestock enterprises with the greatest negative impact in dry region zones of southern Australia, as a result of combined effects of changes in rainfall, temperature, and elevated CO_2 (eCO_2). A moderate increase in eCO_2 is expected to favour plant productivity (e.g. legumes). This effect is expected to offset the negative impact of changes in rainfall and increased temperature by 8% in high rainfall regions and +11% in dry regions but is not enough to offset in drier regions (Ghahramani and Moore, 2016). In another study by Cullen et al. (2009), a reduction in grassland productivity was predicted to decline in Australia mainland only if rainfall decreases more than 10%, but in Tasmania, current production would sustain even rainfall declines by 20% and this was related mostly to projections for vapour pressure deficit that could impact on evaporation and transpiration for a greater water-use efficiency.

Climate change was modelled to shorten pasture growing season, consequently reducing the stocking rate (Cullen et al., 2009). This impact was predicted to be compensated to some degree through a boost to production at the start of the growing season (Moore and Ghahramani, 2013). Although some increase in ANPP was predicted for Tasmania, modelling studies also suggested a potential decline in forage quality (Perring et al., 2010). Plant composition was also expected to change under the impact of climate change (Yang et al., 2011) as changes in eCO_2 and other factors may favour some plants, e.g. legume but there have been little modelling works about this impact. In southern Australia, it is likely that legume proportion of grasslands would increase (Moore and Ghahramani, 2013) and changes in plant species would affect the overall nutritional value of grassland. The observed progressive nitrogen limitation (PNL) in soil under eCO_2 was considered to be a potential issue, but the mechanism and degree of impact are not well known yet (Ghahramani et al., 2019). There is observed evidence of PNL (Newton et al., 2006, 2010) which can negatively affect the positive response to eCO_2 (Luo et al., 2004), in particular, if associated with a decrease in temperature and decline in rainfall (Ghahramani et al., 2020). However, this is likely to be offset by modelled increase in legume proportion in the composition of the plant communities, introduction of livestock (instead of hay-cut pasture), and application of N fertiliser, but the latter may not financially be feasible everywhere.

The impact of climate change on livestock will be through direct impacts on the animal, e.g. heat stress (IPCC, 2014) and indirect impact by disease (Özkan et al., 2016) and changes in grassland production and forage quality. Heat stress is expected to be the main direct impact that animals will experience for current genetic selections (Renaudeau et al., 2011). This is expected to be a major problem as a greater number of heat wave events are projected for coming years. Moore and Ghahramani (2013) modelled impacts on the animal production system through altered grassland productivity, reduced growing season, and changes in temperature, indicating an indirect impact of climate change on animals with decreased profitability of livestock enterprises. When considering all temperate grasslands of southern Australia, meat production was projected to decline by 20%, 26%, and 38% in 2030, 2050, and 2070, respectively, compared to historical production when no adaptation option was applied (Moore and Ghahramani, 2013).

5.3 Farm profitability in Australia

5.3.1 Climate projection scenario results

The *farmpredict* model was applied to assess the effects of both historical and projected changes in climate on farm profits (Hughes et al., 2022a). Projections for rainfall and temperature by 2050 were obtained from *Climate Change in*

Australia (Whetton et al., 2015), covering two RCPs: a 'low global emissions' scenario *RCP4.5* and a 'high global emissions' scenario *RCP8.5* across six different GCMs. Future climate scenarios were compared against both a long-term historical reference period of 1949–50 to 1999–2000 and the recent period 2000–1 to 2019–20, resulting in four climate scenarios in the study: *Historical (1950 to 2000)*, *Recent (2001 to 2020)*, *Future (RCP4.5 2050)*, and *Future (RCP8.5 2050)*. The prediction and analysis did not account for the offsetting positive effects of farm adaptation or technological improvement (or any changes in global commodity prices). As such, the results are not projections of likely outcomes in 2050, but rather estimates of 'adaptation pressure': identifying which regions, sectors, and farm types are likely to be under more pressure to adapt to climate change.

Climate projections for 2050 show a wide range of outcomes (largely due to variation in projected rainfall across the included GCMs) with simulated changes in average farm profits under the *Future (2050 RCP4.5)* scenario ranging from −31.9% to −2.0%, and for the *RCP8.5* scenario ranging from −49.9% to −10.7% (Table 1).

Cropping farms in southwestern Australia are more heavily impacted than other regions under most climate scenarios, due largely to the more substantial projected declines in winter rainfall and the resulting effects on crop yield (Fig. 14). For these farms, declines in profit under the future 2050 projection scenarios are (on average across the six GCMs) larger than those observed over the recent period (for both RCP4.5 and RCP 8.5). Modelled future impacts for cropping farms in eastern Australia (Southern and Northern cropping regions) are relatively modest, in comparison to the effects of the recent period. There remains uncertainty over rainfall projections in eastern Australia, with limited agreement among GCMs (Whetton et al., 2015), and a disconnect emerging in some regions between projections and observations, with recent trends tracking the extreme dry end of the projected range (The Bureau of Meteorology and CSIRO, 2020).

In the beef and sheep sectors projected impacts under the low emissions case (RCP 4.5) remain relatively modest, such that for most scenarios changes in profit are smaller than those observed under the recent climate period. However, impacts in the livestock sector become much more significant under the high-emissions scenario due to the larger projected temperature increases.

In general, more severe farm effects are observed in the 'marginal' (lower rainfall/in-land) sections of the Australian agricultural zone (Fig. 15). The largest impacts are estimated for the northern edge of the Western Australian cropping zone, parts of western New South Wales and central Queensland. Relatively mild impacts are observed in Tasmania and in many coastal high-rainfall locations on the mainland.

Table 1 Percentage change in farm profits relative to Historical (1950–2000) climate

Industry group	Historical (1950 to 2000) ($)	Recent (2001 to 2020)	Future (RCP4.5 2050)			Future (RCP8.5 2050)		
			Minimum	Mean	Maximum	Minimum	Mean	Maximum
Beef - Northern	152 815	−3.1	−22.1	−11.7	−3.0	−54.5	−27.6	−16.3
Beef - Southern	20 968	−22.5	−26.6	+0.5	+10.3	−63.8	−18.0	−2.7
Sheep - Lamb	108 234	−14.9	−16.6	−5.8	−0.1	−31.6	−12.9	−5.6
Sheep - Mixed	58 817	−26.7	−37.3	−13.2	−6.3	−66.3	−28.1	−15.9
Cropping - Northern	212 491	−36.2	−23.7	−9.8	−3.6	−43.1	−20.1	−4.8
Cropping - Southern	179 423	−21.7	−27.7	−3.3	+11.5	−30.8	−8.5	+4.1
Cropping - Western	437 227	−26.8	−55.9	−31.6	−5.1	−68.1	−50.5	−7.3
All farms	129 187	−22.6	−31.9	−13.1	−2.0	−49.9	−25.6	−10.7

Notes: Change in simulated weighted average farm profit for broadacre farms, assuming current farms and commodity prices (2015–6 to 2018–9), relative to historical climate conditions (1949–50 to 1999–2000). Farm profit is defined as 'farm business profit at full equity', including imputed unpaid family labour and depreciation expenses, changes in the value of stocks (including inventory and livestock) but excluding rent, interest, and other financing costs. Source: ABARES *farmpredict* model (Hughes et al., 2022a).

5.3.2 Climate change adaptation in the Australian broadacre farm sector

There is already evidence of strong farm adaptation responses to the recent climate shifts with improvements in technology and management practices helping to increase farm productivity (see Hughes et al., 2017; Hochman et al., 2017). Fig. 15 present estimates of farm Total Factor Productivity (TFP) which measure the performance of the industry over time, including gains achieved through the adoption of new technologies.

Climate-adjusted TFP numbers show an increase in farm productivity across the broadacre sector of around 28% since 1989, with much larger gains in the cropping sector (68% since 1989). These gains in productivity have offset the negative effects of climate over the last 30 years, such that actual (unadjusted) industry productivity levels have still increased or at least remained stable (see Hochman et al., 2017; Hughes et al., 2017).

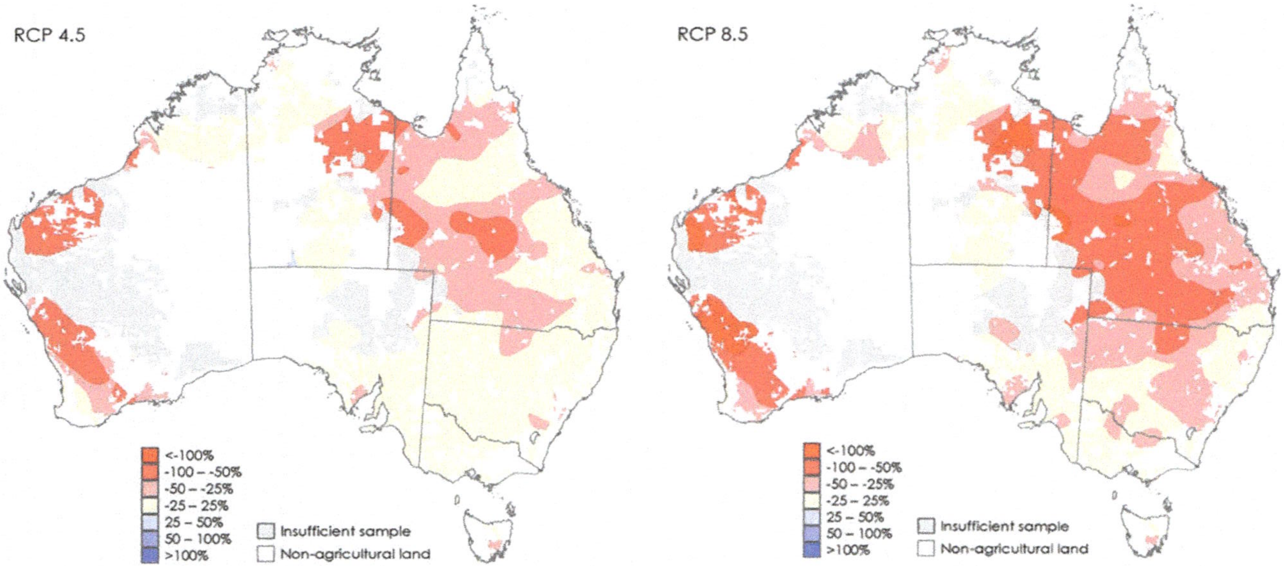

Figure 14 Percentage change in farm profits relative to the Historical (1950 to 2000) climate, average of six global circulation models. Simulated broadacre farm profit with current (2015–2016 to 2018–2019) farms and commodity prices and Future RCP4.5 2050 climate. The map presents interpolated farm-level percentage changes (relative to 1949–1950 to 1999–2000 climate), calculated using a Symmetric Mean Absolute Percentage Error (SMAPE) metric.

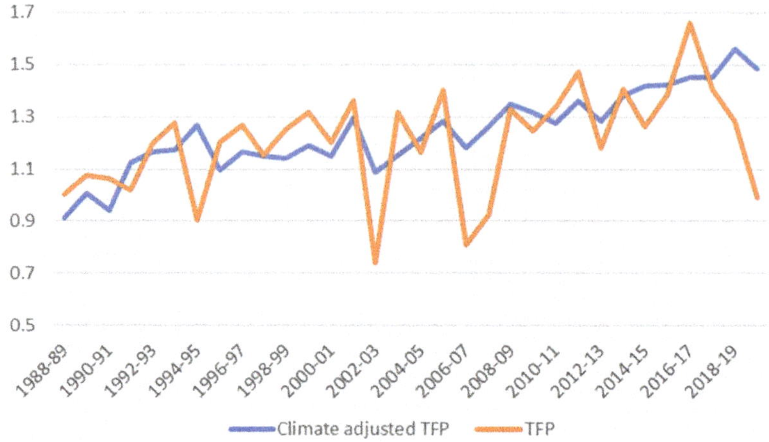

Figure 15 Total Factor Productivity Australian cropping farms 1988–1989 to 2019–2020. Climate-adjusted productivity under the average of 1988--1989 to 2019–2020 climate conditions.

Much of the recent adaptation effort on farms has been directed towards improving performance under dry conditions, particularly in maintaining crop yield crops under low in-season rainfall (Hughes et al., 2017). Modelling presented by Hughes et al. (2022a) suggests that since 2007–8 wheat yields under 'dry' (tenth percentile) conditions have increased at more than double the rate of wheat yields under 'normal' (median) conditions over the same period.

This trend is also partly due to changes in the location of cropping activity, including a long-term shift towards increased cropping in higher rainfall areas (see Hughes et al., 2017). Recent AAGIS data provide evidence of a shift away from cropping towards livestock farming in some more 'marginal' parts of the Australian cropping zone (Fig. 16), which likely reflects both the long-term shift in rainfall patterns and recent increases in livestock prices.

5.4 Broadacre cropping in New Zealand

Broadacre cropping systems are managed intensively in New Zealand. This implies that more realistic representations of management interventions are critical when designing modelling studies to assess responses to climate change. For example, by considering the importance of crop rotations, instead of single crops alone, when assessing climate change impacts on broadacre systems (Teixeira et al., 2018). This allows capturing both climate impacts on individual crops and carry-over effects across sequential crops, such as from residual soil conditions (e.g. water and N) and shifts in crop calendars (e.g. sowing and harvesting times). These responses at the cropping system level

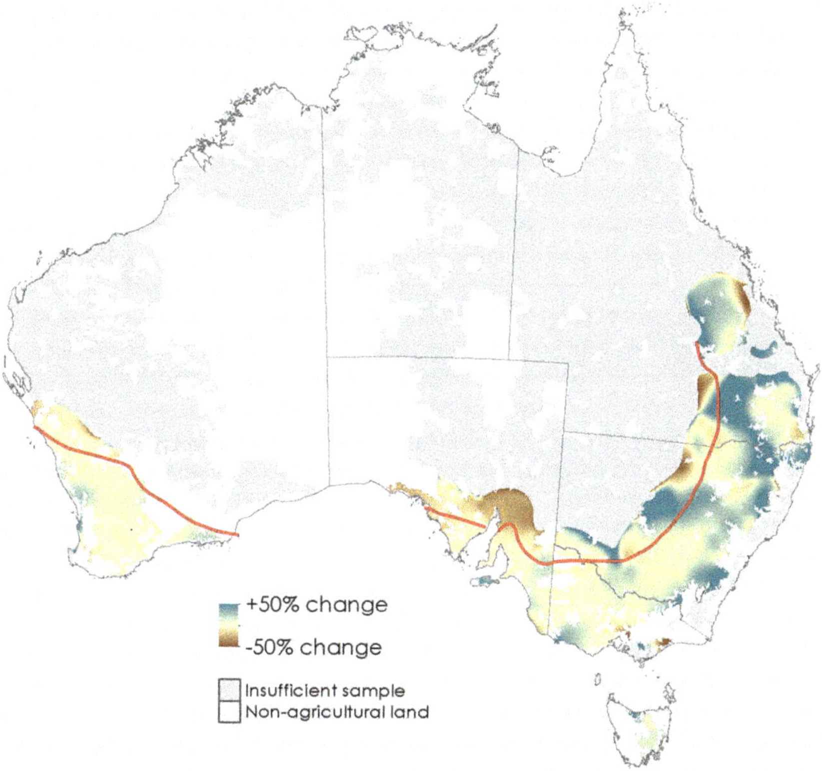

Figure 16 Percentage change in farmland set-up for cropping, 2016-2020 relative to 2006-2010, based on AAGIS data. Excludes beef farms and farms with less than 10% of land area set-up for cropping. Average of 2015-2016 to 2019-2020 compared to 2005-2006 to 2009-2010. Areas in-land of the 220 mm rainfall contour (red line) received less than 220 mm average rainfall per year during the period 1950 to 2020.

can only be accounted for when analysing effects across the full extent of crop rotation sequences. Such analysis also allows a better understanding of temporal and spatial variability in crop responses to climate change, particularly when species with contrasting sensitivities to temperature and CO_2 fertilisation are grown in a sequence.

For instance, a modelling study across the Kaituna catchment used the APSIM model to investigate the impact of future climate by the end of the century (2080-2100; RCP 8.5 in relation to 1985-2004 with six GCMs) on crop productivity for a rotation with a spring-sown C4 crop (maize for silage) followed by an autumn-sown C3 crop (wheat as cover crop) in New Zealand (Teixeira et al., 2018). The simulated magnitude and direction of changes in crop biomass differed depending on local climate and carry-over effects of one crop into the next (Fig. 17). In general, both crops grown in lowlands (elevations

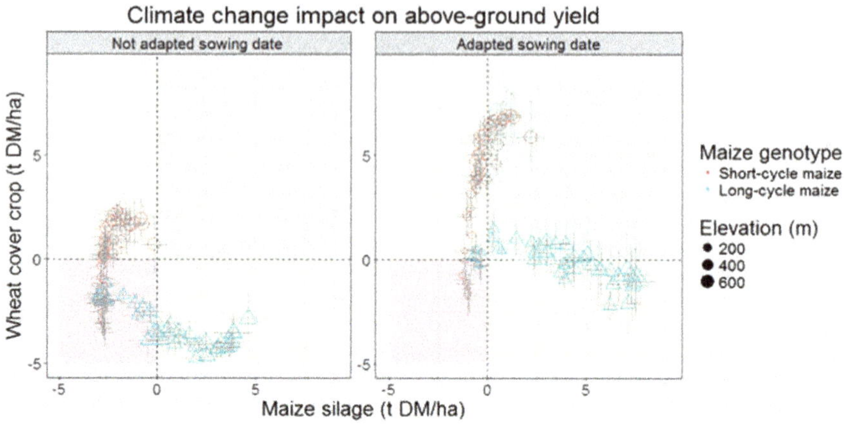

Figure 17 Simulated climate change impacts on aboveground biomass (t DM/ha) for individual 5-km resolution grid cells for short- (triangle) and long-cycle (circle) maize genotypes. Points represent ~20-year averages for a continuous crop rotation in the Kaituna catchment, New Zealand (silage maize followed by catch crop wheat). Grey bars represent uncertainty from the six global circulation models (GCMs) as one standard deviation. Graph quadrants show areas of yield loss for both crops (red), yield gain for both crops (green), and trade-off areas for climate change impact on yields (yellow). Symbol size represents altitude. Source: Adapted from: Teixeira et al. (2018).

< 200 m above see level) were more negatively impacted than under higher elevations, where low temperatures currently limit growth rates. By adapting sowing dates, these negative impacts were minimised and some locations in the catchment showed yield increases for one or both crops in the rotation. The selection for short- or long-cycle maize crops influenced how the following cover crop responded to climate change, with short-cycle maize favouring the following cover crop.

In addition to changes in average climate conditions, the potential increase in extreme events with climate change was shown to negatively impact a range of agricultural systems in New Zealand (Salinger et al., 2020). For instance, during three record heat wave summers (1934–5, 2017–8, and 2018–9) spring wheat grain simulations showed an increased risk of yield losses both in response to average warming (i.e. early flower leading to shortening of cycle length) and exposure to heat stress events during the reproductive stage.

Such negative climate change impacts can to some degree be reduced through a shift or expansion of cropping areas depending on the broadacre system (Teixeira et al., 2021). For silage maize crops, large-scale APSIM simulations showed there may be potential to expand towards more southern regions (Fig. 18), where C4 crops are currently constrained by low temperatures, assuming that favourable logistical and economic conditions allow (Rutledge et al., 2017; Teixeira et al., 2017). These results also highlighted that, when

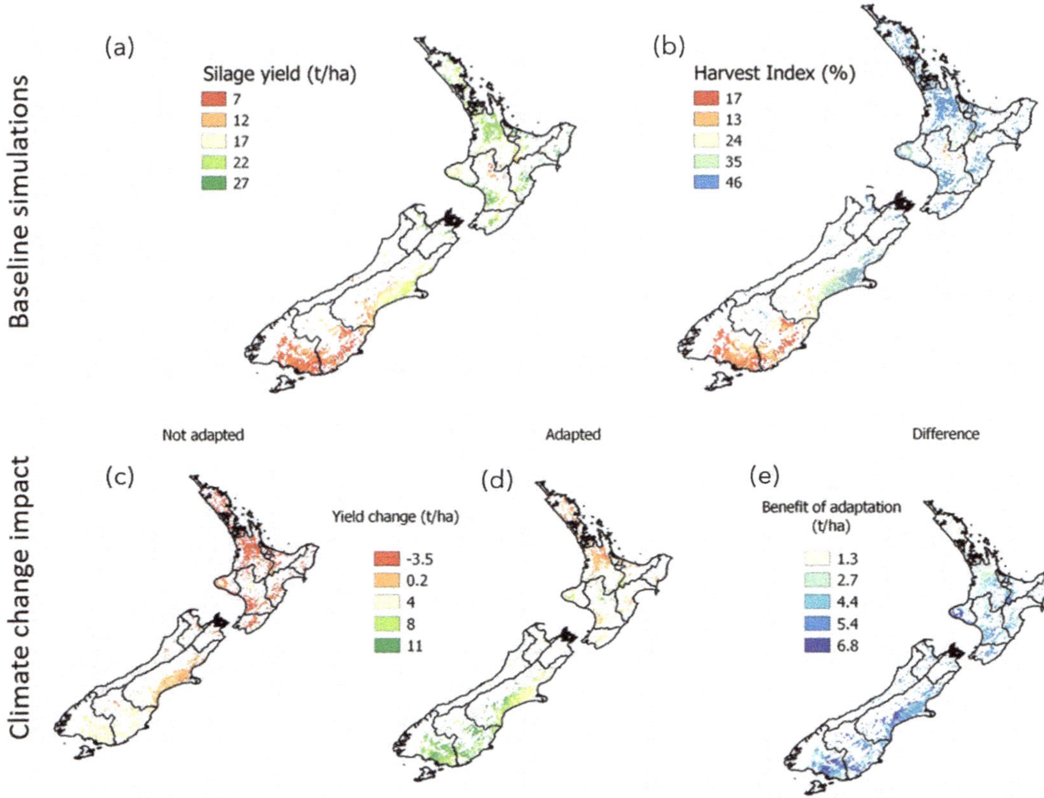

Figure 18 Historical climate (1971-2000; ERA-40 dataset, panels a and b) simulations of silage maize biomass yield (t dry matter DM/ha) and harvest index (HI; % yield) across New Zealand's arable lands. Climate change impact on maize silage yields for end-of-century climate (2080–2099, RCP 8.5 with HadGEM2-ES climate model; panels c-e) and the effect of adapting sowing dates and genotypes. Source: Adapted from: Rutledge et al. (2017).

considering current maize cropping areas in the country, tactic adaptations of agronomic practices (e.g. sowing dates and genotype selection) are likely to be required to maintain economic yields under future climate change. The effectiveness of such tactic adaptations might differ across regions due to the large spatial variability in climate change impacts and the non-linear physiological responses to future environmental conditions (Teixeira et al., 2021).

The influence of climate change on the effectiveness of management options in the future was also shown for cover crops sown to reduce risks of N leaching losses to groundwater. A simulation study with APSIM identified agricultural locations in New Zealand where climate change increased risks of N leaching pollution, despite a concomitant increase in cover crop capacity to take up N through higher yield potentials (Teixeira et al., 2021). These results illustrate the importance of considering multiple impact metrics to capture potential synergies and trade-offs among productivity, environmental and socio-economic goals in agricultural production systems (Ausseil et al., 2019).

5.5 Horticulture system in New Zealand

In New Zealand, statistical modelling of grapevine yields (Zhu et al., 2020) suggests that climate change might have contrasting effects depending on shifts in phenology, seasonal weather patterns and which yield components are mostly affected. For example, varieties in cool climate regions may initially respond positively to higher maximum temperatures, as long as optimal temperature thresholds are not crossed, through an increase in bunch number, berry number and bunch mass. Climate change impacts on seasonal weather patterns are particularly important because critical periods for defining bunch numbers occurred before 50% flowering of the 'previous season', while berry numbers and mass mainly responded to temperature and rainfall before 50% flowering of the 'current season'. Increasing rainfall near flowering was found detrimental to berry number and berry mass, but had a positive effect on berry mass post-flowering under water-limited conditions.

In all cases, these simulations highlight the importance of environmental conditions during the pre-flowering period, and the overall model including maximum and mean temperature, radiation, and rainfall could explain 60% to 85% of the seasonal variability of yield and its components (Zhu et al., 2020). These responses, originally developed as a statistical model, have been integrated into the recently developed APSIM grapevine model (Zhu et al., 2021), which can be further used to test future climate change scenarios. This is a critical step to advance the understanding of adaptive components of the cropping system (e.g. number of retained buds in winter) and the behaviour of different underlying plant processes that influence yield changes (e.g.

phenology, canopy expansion, light interception, and carbohydrate assimilation allocation). It is worth to note that although warming has a positive effect on yield in general, advancing phenology may push the flowering time to a cooler period because of the annual pattern of the temperature (Zhu et al., 2020). Thus, the final effects of warming on yield under New Zealand conditions depend on whether warming can compensate the cooling effects caused by advanced phenology and the environmental conditions at the new flowering time.

The diversification of grapevine varietals was studied as a means to increase the resilience of the wine industry to negative climate change impacts (Ausseil et al., 2021). Sauvignon blanc is currently the main variety grown in New Zealand, representing 61% of the total production area and 80% of the yield (New Zealand Winegrowers, 2020). About 90% of Sauvignon blanc in New Zealand is grown in Marlborough. Temperatures in this region have increased at 0.1°C per decade, with seven out of the ten warmest years occurring since 2010. Downscaled CMIP5 projections for RCP 8.5 by the New Zealand National Institute of Water and Atmospheric Research Institute at 5 km suggest that mean daily temperature in Marlborough may increase by 2.0–2.5°C by 2090 for the majority of wine growing areas.

One modelling study used the Grapevine Flowering Véraison model (Parker et al., 2011) and the Grapevine Sugar Ripeness model (Parker et al., 2020) to assess the effects of climate change on key phenological stages, flowering, Véraison, and the time to 200 g/L sugar concentrations of different varietals in New Zealand. Flowering time was shown to advance by about 7 days in mid-century (2046-2065) and by about 12 days in the end-of-century (2081–2100) under RCP 8.5 compared to the baseline period of 1971–2005 depending on the genotype (Fig. 19). Similarly, Véraison projections advanced

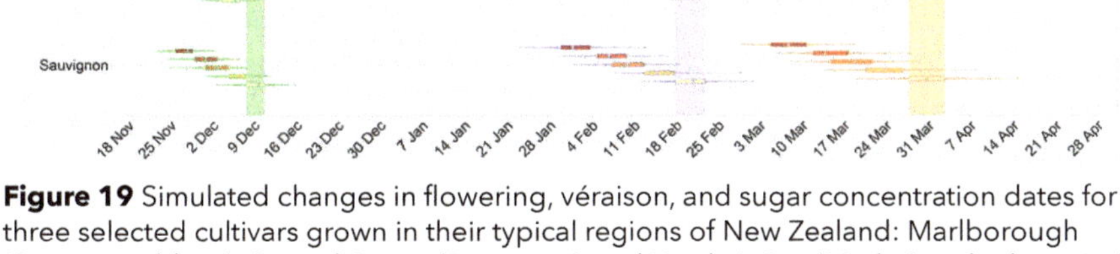

Figure 19 Simulated changes in flowering, véraison, and sugar concentration dates for three selected cultivars grown in their typical regions of New Zealand: Marlborough (Sauvignon blanc), Central Otago (Pinot noir), and Hawke's Bay (Merlot) under historical (1971-2005) and the mid-century (2046-2065) period considering RCP 2.6, 4.5, 6.0, and 8.5. Source: Adapted from: Ausseil et al. (2021).

by about 11 days in mid-century (2046-2065) and by about 20 days in the end-of-century (2081–2100). As a metric of quality affecting wine making processes, the time to 200 g/L sugar concentrations was projected to advance by about 14 days in mid-century (2046–2065) and by about 24 days in the end-of-century (2081–2100) in relation to historical climate.

For the New Zealand wine industry, modelling studies have mostly suggested the potential for long-term adaptation through varietal change, reallocation, or expansion of production areas. For example, suitable areas for viticulture in New Zealand were projected to increase by 170% depending on the climate scenario considered (Hannah et al., 2013). However, it is unclear which adaptive measures and socio-economic conditions would be required for establishing financially and environmentally sustainable production systems in these new areas. The availability of infrastructure to provide water supply to vineyards becomes critical in regions where severe droughts are becoming more intense and frequent (Sturman et al., 2017), as illustrated for the Marlborough in recent dry seasons (Ministry for Primary Industries, 2019). Similarly, the potential benefits of shifting to more adapted grapevine varietals under climate change (Ausseil et al., 2021) are subjected to complex market and socio-economic aspects, such as market preferences and traditional branding of local wines.

5.6 Pasture production system in New Zealand

Simulations of climate change impact on pastoral systems in New Zealand have been quite limited. One of the early works using relationships derived from historic data suggested considerable gains in biomass production in areas with cool climates which have either large rainfall or access to irrigation (Baars et al., 1990). For warmer areas in the North Island, the same work predicted no changes in annual pasture production, but a considerable change in the seasonal pattern. These trends have been generally corroborated by later studies using a decision-tree approach and geographic information system (Zhang et al., 2007). More recent work has shown that the impacts of climate variations on pastures can be more complex, affecting the species mix in the sward and feed quality. Especially for the North Island, C4 grasses are predicted to become more prevalent (Field and Forde, 1990; Dodd et al., 2009; Beukes et al., 2021) altering feed availability and quality over the seasons. Changes in carbon and N cycling due to elevated atmospheric CO_2 have also been suggested to alter (reduce) the protein content in pastures (Dodd et al., 2009; Li et al., 2014). Climate change studies focusing on animal welfare are still incipient, but likely negative impacts have been suggested for facial eczema (Dennis et al., 2014) and milk production due to heat stress (Schütz et al., 2014). Nearly all the studies suggest that the North Island will be much more impacted compared to the South Island, where the temperature is the most limiting factor. However, the impact of potential water shortages and reduction in production in non-irrigated areas, especially in hill country farming has not been much explored yet and could affect both islands.

Potential adaptation and actions to mitigate the impact of climate change in the pastoral sector have been studied in a few projects, but work is still incipient in New Zealand. Considerable effort has been spent in understanding and devising mitigation actions for the emission of greenhouse gases (e.g. Beukes et al., 2010; Giltrap et al., 2010; Veneman et al., 2015). A framework to aid in discussions about adaptations of farm management to climate change has been proposed (Kalaugher et al., 2013). Changes in pasture composition and their implications for future management have been investigated for dairy systems (Keller et al., 2014; Kalaugher et al., 2017; Beukes et al., 2021) and sheep and beef farming (Lieffering et al., 2016). These studies suggest that as the impact of climate change is likely to be highly localised; consequently, adaptation strategies will have to be tailored for each region and farming system. Most of the adaptations investigated so far are well-known actions and mostly encompass changes in farm management (e.g. feed conservation, improved rotations, and enhancing water-use efficiency). These studies suggest that the use of alternative forage species, to either increase drought and pest tolerance or to maintain production levels under warmer weather, needs to be further explored. The introduction of trees to pastoral systems, to supply shade to animals or as developed agroforestry systems (e.g. Tozer et al., 2021), have been suggested, but modelling analyses under climate scenarios are still lacking.

5.7 Oil palm production in Papua New Guinea

Published research using process-based simulation of oil palm systems in PNG is limited. Early model development and application of APSIM within PNG highlighted the difficulties in applying such models locally (Huth et al., 2014). However, recent applications on related topics can provide insights into the impacts of climate change. For example, the initial APSIM model development for oil palm in PNG (Huth et al., 2014) was able to capture differences in average productivity, and annual variation in productivity, in response to climate variability in space and time. In achieving this, the model demonstrated the importance of various yield development processes in such highly productive but drought-sensitive crops. The modelling highlighted the effect of clouds on potential growth rates, suggesting areas predicted to receive increased rainfall may be impacted via some reduction in photosynthesis. Flowering dynamics is important for yield formation but are highly sensitive to drought per subsequent modelling (Manoli et al., 2018) using an ecohydrological model. Fatichi et al. (2012) using data from the previous APSIM model development explored differences in surface water and energy balances as the rainforest is replaced by plantations. Young plantations were predicted to result in increases in surface air temperatures by 2°C to 3°C. The land transformation would thus accentuate increases due to climate change. Furthermore, mature plantations were predicted to maintain higher transpiration rates which may impact groundwater levels and stream flows, especially during drought periods. Again, this suggests that land use change for oil palm production may exacerbate the hydrological impacts of increased climate variability under climate change.

The issue of climate change mitigation has received some attention through the study of greenhouse gas emissions and its drivers (Manoli et al., 2018). In this study, sensitivity analysis highlighted the impact of climate variability on emissions. Climate change is expected to increase inter-annual climate variability and so it may be expected that variation in emissions will increase further, making field assessments or predictions more difficult. However, this work also found that the main parameters driving variation in emissions were often not fixed by the environment, but were related to plantation management, or interventions that could potentially be used to manage emissions. For example, fertiliser rates, cover crop growth and composition, and recycling of empty fruit bunches after processing at the mill were factors that often rates as more in driving emissions than inherent soil properties. These findings, though not originally targeted at climate change impacts, do suggest that management interventions may be available to not only respond to increased climate variability under climate change but to also provide more effective mitigation efforts to avoid the causes of climate change.

6 Future trends in research

For modelling crop response to climate change, while there have been many studies focusing on crop response to rising temperature, there seems still a lack of data to quantify crop response to extreme temperatures and acclimation, at both low and high temperature ranges. In addition, key processes related to pollination, sterility, senescence, or abortion of plant organs as affected by abnormal temperatures have not been well quantified and simulated but can become important under rising temperature, and thus needs more attention (Chenu et al., 2017; Wang et al., 2017). Improvements are also essential to account for the combined effects of different stresses, such as heat and drought stress, which often occur simultaneously.

Previous studies in Australia have shown that capturing synergies from combinations of pre- and in-crop management together with different rotations could increase crop yield at the farm scale by 11–47% (Kirkegaard et al., 2014). However, how to develop the best crop rotations in response to future climate change in various ecological zones of Australia has not been studied. It is important to consider crop rotations, instead of single crops alone, when assessing climate change impacts on broadacre systems, as demonstrated in the case studies in New Zealand (Teixeira et al., 2018). The use of alternative plant (crops/forages) species and planting them in combination (multicropping and agroforestry) may have potential, but we lack models for many of the alternative species as well as for properly describing them in competition.

An important indirect impact of climate change, not fully integrated into biophysical modelling assessments, is the risk of yield losses due to biotic stresses from insects, pathogens, and weeds (Sutherst et al., 2011). The increase in drought events projected for the eastern agricultural areas of New Zealand was highlighted as an important predisposing factor to increase the risk of biotic stress damage on many arable and horticultural crops of relevance in the country (Wakelin et al., 2018). This has also been raised as an important impact on current and future climates for oil palm (Abubakar et al., 2021). Pests and diseases are major concerns for many other crops as well under current and future climate conditions. It is not clear how these could be captured in process-based models. However, the importance of this issue has resulted in extensive basic research upon which to base initial model considerations.

The effects of climate variations on animal welfare and diseases are also incipient. For example, facial eczema is recognised as a significant issue for New Zealand (Dennis et al., 2014). Temperature control and shading (maybe water access too) are potential mitigation measures that are not well studied, especially in modelling scenarios.

Future climate impact assessments will benefit from further integrating modelling data pipelines across diverse research domains and institutions, given the multi-disciplinary nature of studies when considering productivity and environmental and socio-economic aspects of climate change impacts. In New Zealand, recent efforts to upscale APSIM from point-based to landscape scales were able to integrate daily climate projections at ~5 km resolution (Tait et al., 2016) with the S-map soil databases (Lilburne et al., 2004) within high-performance computing environments (Teixeira et al., 2020). This framework has been applied to inform spatial and temporal variability of climate change impacts and adaptation at both catchment and national scales (Ausseil et al., 2017; Rutledge et al., 2017).

Many of the important factors driving the impacts of climate change on oil palm production systems are yet to receive consideration within modelling efforts. The impact of CO_2 fertilisation on palm production is not verified in the models used in previous analyses and this has been highlighted as a deficiency and restriction in its use for previous work in other oil palm regions (Okoro et al., 2017). Internal carbon allocation dynamics are complex within the oil palm plant, and an extensive effort would be required in developing and testing model responses to increased CO_2 levels.

It is also essential to keep in mind that there is considerable uncertainty in climate projections which translates to high variability in the simulated crop results. Uncertainties related in part to future CO_2 level depends on a range of geological, economic, and technological factors and decisions that are difficult to anticipate (IPCC, 2021). There are also large uncertainties in projected daily wheat variations in the future, particularly in regards to extreme events such as heat waves, floods, and tornados. Such variations are likely to have major effects on crop growth and adaptation strategies.

These uncertainties over the future climate have the potential to constrain adaptation, particularly adaptation requiring long-term investment. Here, the development of climate information remains crucial in supporting adaptation. Advances in short- and long-term weather forecasting and climate change projection, better communication of such information, and attribution studies could all help farmers and policymakers to adapt to climate change more decisively.

Improvements in data could also support the development of drought insurance markets. Effective drought insurance could help farmers to manage their exposure to drought risk while encouraging long-term adaptation to climate change (Kennedy et al., 2018; Hughes et al., 2022a). Drought insurance markets have struggled to achieve commercial viability to date; however, improvements in temporal and spatial resolutions of data could support new approaches including parametric or 'index-based' insurance. Furthermore, harnessing the recent advances in digital technologies such as biophysical crop models, global climate models, remote sensing, and artificial intelligence technologies across G × E × M and integration with commodity markets is likely to generate more informed decisions and aid in enhancing global food production (Potgieter et al., 2021).

7 Conclusion

The Oceania region has diverse terrestrial and marine systems that are responsive to climate change. Agriculture is a major form of land use, particularly in countries with larger land areas such as Australia, New Zealand and Papua New Guinea. In the past few decades, the Oceania region has experienced a general trend of rising temperature, increased rainfall variability, declining rainfall in parts of Australia and New Zealand, and increased rainfall (particularly extreme events) in other parts of the Oceanian region, forcing agricultural management to adapt. Farming systems modelling has played a key role in understanding how climate variability and change impact agricultural systems. The Agricultural Production Systems model (APSIM) has become an important and effective tool to assess impact of climate change on crop development, growth and yield and to inform the development of adaptation strategies in agronomy and breeding. Other models are also used to assess impact grazing systems (GRAZPLAN, DairyMod etc), farming businesses (*farmpredict*) and horticulture (statistical model). While most of these studies are in Australia and New Zealand, some modelling studies were also carried out for local crops in other countries, e.g. oil palm in PNG. Future research will need to focus on improvement of modelling key crop physiological processes and yield formation in response to climate drivers, rotation systems with multiple crop species, impact of biotic stresses (pests and diseases), and implications for economy, environment and government policy.

8 Where to look for further information

To meet the demand for modelling farming systems to inform decision-making under variable and changing climate, the APSIM farming systems model is continuously updated with new components and functionalities. The newest development and applications can be found at:

- https://www.apsim.info/.

A comprehensive review on the impacts of climate change on the Australia economy was released in 2008 and updated in 2011 (The Garnaut Climate Change Review), details can be found at:

- www.garnautreview.org.au/index.html.

More recent updates on climate change and the agricultural sector in Australia including adaptation and mitigation can be found at:

- https://www.agriculture.gov.au/agriculture-land/farm-food-drought/climatechange.
- https://www.agriculture.gov.au/abares/products/insights/climate-change-impacts-and-adaptation.
- https://theconversation.com/changing-climate-has-stalled-australian-wheat-yields-study-71411.

Description of general climate and climate change in Australia can be found at: General climate:

- https://webarchive.nla.gov.au/awa/20090330051442/http://pandora.nla.gov.au/pan/96122/20090317-1643/www.bom.gov.au/lam/climate/levelthree/ausclim/zones.html.

Australia's climate - drivers, variability and extremes:

- https://coastadapt.com.au/australias-climate-drivers-variability-and-extremes.

Climate change in Australia:

- https://www.climatechangeinaustralia.gov.au/en/.

Australia climate change: trends and extremes:

- http://www.bom.gov.au/climate/change/#tabs=Tracker&tracker=timeseries&tQ=graph%3Dtmean%26area%3Daus%26season%3D0112%26ave_yr%3D2.
- http://www.bom.gov.au/state-of-the-climate/.

Information on climate changes, impacts and implications for New Zealand can be found at:

- https://niwa.co.nz/climate/research-projects/climate-changes-impacts-and-implications-for-new-zealand.
- http://www.nzagrc.org.nz/.
- https://www.royalsociety.org.nz/what-we-do/our-expert-advice/all-expert-advice-papers/climate-change-implications-for-new-zealand/.
- https://environment.govt.nz/publications/our-atmosphere-and-climate-2020/.
- https://www.mpi.govt.nz/funding-rural-support/environment-and-natural-resources/climate-change-primary-industries/.
- https://www.dairynz.co.nz/environment/environment-policy-and-leadership/climate-change/climate-change-101/.

Information on Oceania countries, land mass, and population (Commission on Ecosystem Management):

- https://www.iucn.org/commissions/commission-ecosystem-management/regions/oceania.

Pacific Islands climate change:

- https://www.climatelinks.org/resources/climate-risk-profile-pacific-islands.

9 References

Ababaei, B. and Chenu, K. (2020). Heat shocks increasingly impede grain filling but have little effect on grain setting across the Australian wheatbelt. *Agricultural and Forest Meteorology* 284, 107889.

ABS (2021). Agricultural Commodities: Small Area Data Canberra, Australia, 2020-2021: Australian Bureau of Statistics; 2021 [Time series data for the crops you requested

- by SD for the years 1983 to 2011, and 4 to 20.]. Available from: http://www.abs.gov.au.

Abubakar, A., Ishak, M. Y. and Makmom, A. A. (2021). Impacts of and adaptation to climate change on the oil palm in Malaysia: A systematic review. *Environmental Science and Pollution Research International* 28(39), 54339-54361.

Allan, R. J. (2000). El Niño and the southern oscillation: Multiscale variability and its impacts on natural ecosystems and society. In: Diaz, H. and Markgraf, V. (Ed.), *ENSO and Climatic Variability in the Last* (pp. 3-55). Cambridge University Press.

Anderson, W. K., Stephens, D. and Siddique, K. H. (2016). Dryland agriculture in Australia: Experiences and innovations. In: *Innovations in Dryland Agriculture* (pp. 299-319). Springer.

Angus, J. F., Grace, P. R., Angus, J. F. and Grace, P. R. (2017). Nitrogen balance in Australia and nitrogen use efficiency on Australian farms. *Soil Research* 55(6), 435-450.

Angus, J. F., Kirkegaard, J. A., Hunt, J. R., Ryan, M. H., Ohlander, L. and Peoples, M. B. (2015). Break crops and rotations for wheat. *Crop and Pasture Science* 66(6), 523-552.

Armstrong, R., Hochman, Z., Waldner, F., Bell, K. L., Perris, R., Dunsford, K., Hekmeijer, P. and Munn, M. (2019). Nitrogen supply, rotation and variety are critical predictors of the water use efficiency of wheat in grower's paddocks in Victoria. In: *Proceedings of the 2019 Agronomy Australia Conference* (pp. 25-29). Wagga Wagga.

Ausseil, A., Daigneault, A. J., Frame, B. and Teixeira, E. I. (2017). Climate change impacts and implications for New Zealand to 2100: Synthesis report RA2 lowland case study. In: *Climate Change Impacts and Implications for New Zealand to 2100 Synthesis Report LC2714*, Wellington.

Ausseil, A. G. E., Daigneault, A. J., Frame, B. and Teixeira, E. I. (2019). Towards an integrated assessment of climate and socio-economic change impacts and implications in New Zealand. *Environmental Modelling and Software* 119, 1-20.

Ausseil, A.-G. E., Law, R. M., Parker, A. K., Teixeira, E. I. and Sood, A. (2021). Projected wine grape cultivar shifts due to climate change in New Zealand. *Frontiers in Plant Science* 12, 618039.

Baars, J. A., Radcliffe, J. E. and Rollo, M. D. (1990). Climatic change effects on seasonal patterns of pasture production in New Zealand. In: *Proceedings of the New Zealand Grassland Association* (pp. 43-46).

Bell, L., Lilley, J., Hunt, J. and Kirkegaard, J. (2015). Optimising grain yield and grazing potential of crops across Australia's high rainfall zone: A simulation analysis. 1. Wheat. *Crop and Pasture Science* 66, 332-348.

Beukes, P., Babylon, A., Griffiths, W., Woodward, S., Kalaugher, E., Sood, A. and Chapman, D. (2021). Modelling perennial ryegrass (Lolium perenne) persistence and productivity for the Upper North Island under current and future climate. *NZGA: Research and Practice Series* 17, 297-306.

Beukes, P. C., Gregorini, P., Romera, A. J., Levy, G. and Waghorn, G. C. (2010). Improving production efficiency as a strategy to mitigate greenhouse gas emissions on pastoral dairy farms in New Zealand. *Agriculture, Ecosystems and Environment* 136(3-4), 358-365.

Beukes, P. C., Palliser, C. C., Macdonald, K. A., Lancaster, J. A. S., Levy, G., Thorrold, B. S. and Wastney, M. E. (2008). Evaluation of a whole-farm model for pasture-based dairy systems. *Journal of Dairy Science* 91(6), 2353-2360.

Bourgault, M., Webber, H. A., Chenu, K., O'Leary, G. J., Gaiser, T., Siebert, S., Dreccer, F., Huth, N., Fitzgerald, G. J., Tausz, M. and Ewert, F. (2020). Early vigour in wheat: Could it lead to more severe terminal drought stress under elevated atmospheric [CO_2] and semi-arid conditions? *Global Change Biology* 26(7), 4079–4093.

Bourke, R. (2018). Impact of climate change on agriculture in Papua New Guinea. In book: Climate Change: Our Environment, Livelihoods and Sustainability. Climate Change Conference 2018. *A. R. Quartermain (ed)* pp.(35–50). University of Goroka, Papua New Guinea.

Brown, T. B., Cheng, R., Sirault, X. R., Rungrat, T., Murray, K. D., Trtilek, M., Furbank, R. T., Badger, M., Pogson, B. J. and Borevitz, J. O. (2014). TraitCapture: Genomic and environment modelling of plant phenomic data. *Current Opinion in Plant Biology* 18, 73–79.

Cai, W. and Cowan, T. (2013). Southeast Australia autumn rainfall reduction: A climate-change-induced poleward shift of ocean-atmosphere circulation. *Journal of Climate* 26(1), 189–205.

Cai, W., Cowan, T. and Thatcher, M. (2012). Rainfall reductions over Southern hemisphere semi-arid regions: The role of subtropical dry zone expansion. *Scientific Reports* 2, 702.

Cann, D. J., Hunt, J. R. and Malcolm, B. (2020). Long fallows can maintain whole-farm profit and reduce risk in semi-arid south-eastern Australia. *Agricultural Systems* 178, 102721.

Casadebaig, P., Zheng, B., Chapman, S., Huth, N., Faivre, R. and Chenu, K. (2016). Assessment of the potential impacts of wheat plant traits across environments by combining crop modeling and global sensitivity analysis. *PLoS ONE* 11(1), e0146385.

Chapman, D. F., Edwards, G. R. and Nie, Z. N. (2011). Plant responses to climate and relationships with pasture persistence. *NZGA: Research and Practice Series* 15, 99–107.

Chauhan, Y., Allard, S., Williams, R., Williams, B., Mundree, S., Chenu, K. and Rachaputi, N. C. (2017). Characterisation of chickpea cropping systems in Australia for major abiotic production constraints. *Field Crops Research* 204, 120–134.

Chen, T. and Guestrin, C. (2016). Xgboost: A scalable tree boosting system. In: *Proceedings of the 22nd* ACM SIGKDD *International Conference on Knowledge Discovery and Data Mining* (pp. 785–794). Academic Medicine.

Chenu, K., Cooper, M., Hammer, G. L., Mathews, K. L., Dreccer, M. F. and Chapman, S. C. (2011). Environment characterization as an aid to wheat improvement: Interpreting genotype-environment interactions by modelling water-deficit patterns in North-Eastern Australia. *Journal of Experimental Botany* 62(6), 1743–1755.

Chenu, K., Deihimfard, R. and Chapman, S. C. (2013). Large-scale characterization of drought pattern: A continent-wide modelling approach applied to the Australian wheatbelt – Spatial and temporal trends. *New Phytologist* 198(3), 801–820.

Chenu, K., Porter, J. R., Martre, P., Basso, B., Chapman, S. C., Ewert, F., Bindi, M. and Asseng, S. (2017). Contribution of crop models to adaptation in wheat. *Trends in Plant Science* 22(6), 472–490.

Chenu, K., Van Oosterom, E. J., McLean, G., Deifel, K. S., Fletcher, A., Geetika, G., Tirfessa, A., Mace, E. S., Jordan, D. R., Sulman, R. and Hammer, G. L. (2018). Integrating modelling and phenotyping approaches to identify and screen complex traits – Illustration for transpiration efficiency in cereals. *Journal of Experimental Botany* 69, 3181–3194.

Christopher, J. T., Christopher, M. J., Borrell, A. K., Fletcher, S. and Chenu, K. (2016). Stay-green traits to improve wheat adaptation in well-watered and water-limited environments. *Journal of Experimental Botany* 67(17), 5159–5172.

Christopher, J. T., Veyradier, M., Borrell, A. K., Harvey, G., Fletcher, S. and Chenu, K. (2014). Phenotyping novel stay-green traits to capture genetic variation in senescence dynamics. *Functional Plant Biology: FPB* 41(11), 1035–1048.

Christopher, M., Chenu, K., Jennings, R., Fletcher, S., Butler, D., Borrell, A. and Christopher, J. (2018). QTL for stay-green traits in wheat in well-watered and water-limited environments. *Field Crops Research* 217, 32–44.

Christopher, M., Paccapelo, V., Kelly, A., Macdonald, B., Hickey, L., Richard, C., Verbyla, A., Chenu, K., Borrell, A., Amin, A. and Christopher, J. (2021). QTL identified for stay-green in a multi-reference nested association mapping population of wheat exhibit context dependent expression and parent-specific alleles. *Field Crops Research* 270, 108181.

Clothier, B., Hall, A. and Green, S. (2012). Adapting the horticultural and vegetable industries to climate change. In: Clark, A. J. (Ed.), *Impacts of Climate Change on Land-Based Sectors and Adaptation Option*. Ministry for Primary Industries, Wellington, New Zealand.

Collins, B., Chapman, S., Hammer, G. and Chenu, K. (2021). Limiting transpiration rate in high evaporative demand conditions to improve Australian wheat productivity. *In Silico Plants* 3(1), diab006.

Collins, B. and Chenu, K. (2021). Improving productivity of Australian wheat by adapting sowing date and genotype phenology to future climate. *Climate Risk Management* 32, 100300. /doi.org/10.1016/j.crm.2021.100300.

Collins, B. and Chenu, K. (2022). Change in abiotic stress and atmospheric CO_2 concentration significantly affected Australian wheat productivity over 1981–2018. In: *Australian Agronomy Conference Toowoomba*.

Collins, M. and Norton, R. (2019). Diversifying the cropping phase. In: Pratley, J. (Ed.), *Australian Agriculture in 2020: From Conservation to Automation* (pp. 307-322). Agronomy Australia and Charles Sturt University: Wagga Wagga, Agronomy Australia and Charles Sturt University Wagga Wagga, NSW.

Cradock-Henry, N. A. (2017). New Zealand kiwifruit growers' vulnerability to climate and other stressors. *Regional Environmental Change* 17(1), 245–259.

Cradock-Henry, N. A., Blackett, P., Hall, M., Johnstone, P., Teixeira, E. and Wreford, A. (2020). Climate adaptation pathways for agriculture: Insights from a participatory process. *Environmental Science and Policy* 107, 66–79.

Crimp, S. J., Flood, N. R., Carter, J. O., Conroy, J. P. and McKeon, G. M. (2002). *Evaluation of the Potential Impacts of Climate Change on Native Pasture Production: Implications for Livestock Carrying Capacity*. Australian Greenhouse Office.

Crimp, S. J., Gobbett, D., Kokic, P., Nidumolu, U., Howden, M. and Nicholls, N. (2016a). Recent seasonal and long-term changes in southern Australian frost occurrence. *Climatic Change* 139(1), 115–128.

Crimp, S. J., Zheng, B., Khimashia, N., Gobbett, D. L., Chapman, S., Howden, M. and Nicholls, N. (2016b). Recent changes in southern Australian frost occurrence: Implications for wheat production risk. *Crop and Pasture Science* 67(8), 801–811.

CSIRO, Australian Bureau of Meteorology and SPREP (2015). *Climate in the Pacific: A Regional Summary of New Science and Management Tools, Pacific-Australia Climate*

Change Science and Adaptation Planning Program Summary Report, Report 978-1-4863-0481-3. CSIRO, Melbourne, Australia.

Cullen, B. R., Johnson, I. R., Eckard, R. J., Lodge, G. M., Walker, R. G., Rawnsley, R. P. and McCaskill, M. R. (2009). Climate change effects on pasture systems in south-eastern Australia. *Crop and Pasture Science* 60(10), 933–942.

DairyNZ. (2018). *Plant Breeding for the Future*. DairyNZ, Hamilton, New Zealand.

Dennis, N., Amer, P. and Meier, S. (2014). Predicting the impact of climate change on the risk of facial eczema outbreaks throughout New Zealand. *Proceedings New Zealand Society of Animal Production, Napier, New Zealand* 74, 161–163.

Dodd, M., Lieffering, M., Newton, P. and Dongwen, L. (2009). *Tomorrow's Pastures: Subtropical Grass Growth under Climate Change*. Ministry of Agriculture and Forestry, Wellington.

Fatichi, S., Ivanov, V. Y. and Caporali, E. (2012). A mechanistic ecohydrological model to investigate complex interactions in cold and warm water-controlled environments: 1. Theoretical framework and plot-scale analysis. *Journal of Advances in Modeling Earth Systems* 4(2).

Field, T. R. O. and Forde, M. B. (1990). Effects of climate warming on the distribution of C4 grasses in New Zealand. In: *Proceedings of the New Zealand Grassland Association* (pp. 47–50).

Fletcher, A., Christopher, J., Hunter, M., Rebetzke, G. and Chenu, K. (2018). A low-cost method to rapidly and accurately screen for transpiration efficiency in wheat. *Plant Methods* 14, 77.

Fletcher, A. L., Robertson, M. J., Abrecht, D. G., Sharma, D. L. and Holzworth, D. P. (2015). Dry sowing increases farm level wheat yields but not production risks in a Mediterranean environment. *Agricultural Systems* 136, 114–124.

Flohr, B. M., Hunt, J. R., Kirkegaard, J. A., Evans, J. R. and Lilley, J. M. (2018a). Genotype × management strategies to stabilise the flowering time of wheat in the south-eastern Australian wheatbelt. *Crop and Pasture Science* 69(6), 547–560.

Flohr, B. M., Hunt, J. R., Kirkegaard, J. A., Evans, J. R., Swan, A. and Rheinheimer, B. (2018b). Genetic gains in NSW wheat cultivars from 1901 to 2014 as revealed from synchronous flowering during the optimum period. *European Journal of Agronomy* 98, 1–13.

Flohr, B. M., Ouzman, J., McBeath, T. M., Rebetzke, G. J., Kirkegaard, J. A. and Llewellyn, R. S. (2021). Redefining the link between rainfall and crop establishment in dryland cropping systems. *Agricultural Systems* 190, 103105.

Freer, M., Moore, A. D. and Donnelly, J. R. (1997). GRAZPLAN: Decision support systems for Australian grazing enterprises—II. The animal biology model for feed intake, production and reproduction and the GrazFeed DSS. *Agricultural Systems* 54(1), 77–126.

Garnaut, R. (2008). *The Garnaut Climate Change Review*. Cambridge University Press, Cambridge.

Ghahramani, A., Howden, S. M., del Prado, A., Thomas, D. T., Moore, A. D., Ji, B. and Ates, S. (2019). Climate change impact, adaptation, and mitigation in temperate grazing systems: A review. *Sustainability* 11(24), 7224.

Ghahramani, A., Kingwell, R. S. and Maraseni, T. N. (2020). Land use change in Australian mixed crop-livestock systems as a transformative climate change adaptation. *Agricultural Systems* 180, 102791.

Ghahramani, A., Kokic, P. N., Moore, A. D., Zheng, B., Chapman, S. C., Howden, M. S. and Crimp, S. J. (2015). The value of adapting to climate change in Australian wheat farm systems: Farm to cross-regional scale. *Agriculture, Ecosystems and Environment* 211, 112–125.

Ghahramani, A. and Moore, A. D. (2016). Impact of climate changes on existing crop-livestock farming systems. *Agricultural Systems* 146, 142–155.

Ghahramani, A., Moore, A. D., Ghahramani, A. and Moore, A. D. (2013). Climate change and broadacre livestock production across southern Australia. 2. Adaptation options via grassland management. *Crop and Pasture Science* 64(6), 615–630.

Giltrap, D. L., Singh, J., Saggar, S. and Zaman, M. (2010). A preliminary study to model the effects of a nitrification inhibitor on nitrous oxide emissions from urine-amended pasture. *Agriculture, Ecosystems and Environment* 136(3–4), 310–317.

Hammer, G., Cooper, M., Tardieu, F., Welch, S., Walsh, B., van Eeuwijk, F., Chapman, S. and Podlich, D. (2006). Models for navigating biological complexity in breeding improved crop plants. *Trends in Plant Science* 11(12), 587–593.

Hammer, G. L., McLean, G., Chapman, S., Zheng, B., Doherty, A., Harrison, M. T., van Oosterom, E. and Jordan, D. (2014). Crop design for specific adaptation in variable dryland production environments. *Crop and Pasture Science* 65(7), 614–626.

Hammer, G., McLean, G., Doherty, A., van Oosterom, E., Chapman, S., Ciampitti, I. and Prasad, V. (2016). *Sorghum Crop Modeling and Its Utility in Agronomy and Breeding*. American Society of Agronomy and Crop Science Society of America, Inc.

Hammer, G. L., McLean, G., van Oosterom, E., Chapman, S., Zheng, B., Wu, A., Doherty, A. and Jordan, D. (2020). Designing crops for adaptation to the drought and high temperature risks anticipated in future climates. *Crop Science* 60, 605–621.

Hannah, L., Roehrdanz, P. R., Ikegami, M., Shepard, A. V., Shaw, M. R., Tabor, G., Zhi, L., Marquet, P. A. and Hijmans, R. J. (2013). Climate change, wine, and conservation. *Proceedings of the National Academy of Sciences of the United States of America* 110(17), 6907–6912.

Hennessy, K., Power, S. and Cambers, G. (2011). *Climate Change in the Pacific: Scientific Assessment and New Research (Vol. 1),* Regional Overview.

Hochman, Z., Gobbett, D. L. and Horan, H. (2017). Climate trends account for stalled wheat yields in Australia since 1990. *Global Change Biology* 23(5), 2071–2081.

Hochman, Z. and Horan, H. (2018). Causes of wheat yield gaps and opportunities to advance the water-limited yield frontier in Australia. *Field Crops Research* 228, 20–30.

Holzworth, D. P., Huth, N. I., deVoil, P. G., Zurcher, E. J., Herrmann, N. I., McLean, G., Chenu, K., van Oosterom, E. J., Snow, V., Murphy, C., Moore, A. D., Brown, H., Whish, J. P. M., Verrall, S., Fainges, J., Bell, L. W., Peake, A. S., Poulton, P. L., Hochman, Z., Thorburn, P. J., Gaydon, D. S., Dalgliesh, N. P., Rodriguez, D., Cox, H., Chapman, S., Doherty, A., Teixeira, E., Sharp, J., Cichota, R., Vogeler, I., Li, F. Y., Wang, E., Hammer, G. L., Robertson, M. J., Dimes, J. P., Whitbread, A. M., Hunt, J., van Rees, H., McClelland, T., Carberry, P. S., Hargreaves, J. N. G., MacLeod, N., McDonald, C., Harsdorf, J., Wedgwood, S. and Keating, B. A. (2014). APSIM – Evolution towards a new generation of agricultural systems simulation. *Environmental Modelling and Software* 62, 327–350.

Holzworth, D., Huth, N. I., Fainges, J., Brown, H., Zurcher, E., Cichota, R., Verrall, S., Herrmann, N. I., Zheng, B. and Snow, V. (2018). APSIM next generation: Overcoming challenges in modernising a farming systems model. *Environmental Modelling and Software* 103, 43–51.

Horticulture and Food Research Institute of New Zealand, Plant & Food Research. (2020). *Fresh Facts: New Zealand Horticulture*. Plant & Food Research, Auckland, New Zealand.

Howden, S. M., Reyenga, P. J. and Meinke, H. (1999). Global change impacts on Australian wheat cropping. Report to the Australian Greenhouse Office. Working Paper 99/04. CSIRO Wildlife and Ecology, Canberra.

Hughes, N., Lawson, K. and Valle, H. (2017). *Farm Performance and Climate: Climate-Adjusted Productivity for Broadacre Cropping Farms*. ABARES Research Report.

Hughes, N., Lu, M., Soh, W. Y. et al. (2022a). Modelling the effects of climate change on the profitability of Australian farms. *Climatic Change* 172, 12. https://doi.org/10.1007/s10584-022-03356-5.

Hughes, N., Soh, W. Y., Lawson, K. and Lu, M. (2022b). Improving the performance of micro-simulation models with machine learning: The case of Australian farms. *Economic Modelling* 115: 105957.

Hunt, J. R., Hayman, P. T., Richards, R. A. and Passioura, J. B. (2018). Opportunities to reduce heat damage in rain-fed wheat crops based on plant breeding and agronomic management. *Field Crops Research* 224, 126–138.

Hunt, J. R. and Kirkegaard, J. A. (2011). Re-evaluating the contribution of summer fallow rain to wheat yield in southern Australia. *Crop and Pasture Science* 62(11), 915–929.

Hunt, J. R., Kirkegaard, J. A., Harris, F. A., Porker, K. D., Rattey, A. R., Collins, M. J., Celestina, C., Cann, D. J., Hochman, Z., Lilley, J. M. and Flohr, B. M. (2021). Exploiting genotype × management interactions to increase rainfed crop production: A case study from south-eastern Australia. *Journal of Experimental Botany* 72(14), 5189–5207.

Hunt, J. R., Lilley, J. M., Trevaskis, B., Flohr, B. M., Peake, A., Fletcher, A., Zwart, A. B., Gobbett, D. and Kirkegaard, J. A. (2019). Early sowing systems can boost Australian wheat yields despite recent climate change. *Nature Climate Change* 9(3), 244–247.

Huth, N. I., Banabas, M., Nelson, P. N. and Webb, M. (2014). Development of an oil palm cropping systems model: Lessons learned and future directions. *Environmental Modelling and Software* 62, 411–419.

Iese, V., Halavatau, S., N'Yeurt, A. D. R., Wairiu, M., Holland, E., Dean, A., Veisa, F., Patolo, S., Havea, R., Bosenaqali, S. and Navunicagi, O. (2020). Agriculture under a changing climate. In: Kumar, L. (Ed.), *Climate Change and Impacts in the Pacific* (pp. 323–357). Springer International Publishing, Cham.

IPCC. (2014). *Climate Change 2014 Impacts, Adaptation, and Vulnerability Part B: Regional Aspects: Working Group II Contribution to the Fifth Assessment Report of the Intergovernmental Panel on Climate Change*. Barros, V. R., Field, C. B., Dokken, D. J., Mastrandrea, M. D., Mach, K. J., Bilir, T. E., Chatterjee, M., Ebi, K. L., Estrada, Y. O. and Genova, R. C. (Eds) Cambridge University Press, Cambridge, UK; New York, NY.

IPCC. (2021). 2021: The physical science basis. contribution of working group I to the sixth assessment report of the intergovernmental Panel on Climate Change. *Climate Change*. Edited by Masson-Delmotte, V., Zhai, P., Pirani, A., Connors, S. L., Pean, C., Berger, S., Caud, N., Chen, Y., Goldfarb, L., Gomis, M. I. Cambridge University Press, Cambridge.

Johnson, I. R., Chapman, D. F., Snow, V. O., Eckard, R. J., Parsons, A. J., Lambert, M. G. and Cullen, B. R. (2008). DairyMod and EcoMod: Biophysical pasture-simulation models for Australia and New Zealand. *Australian Journal of Experimental Agriculture* 48(5), 621-631.

Jones, J. W., Hoogenboom, G., Porter, C. H., Boote, K. J., Batchelor, W. D., Hunt, L. A., Wilkens, P. W., Singh, U., Gijsman, A. J. and Ritchie, J. T. (2003). The DSSAT cropping system model. *European Journal of Agronomy* 18(3-4), 235-265.

Kalaugher, E., Beukes, P., Bornman, J. F., Clark, A. and Campbell, D. I. (2017). Modelling farm-level adaptation of temperate, pasture-based dairy farms to climate change. *Agricultural Systems* 153, 53-68.

Kalaugher, E., Bornman, J. F., Clark, A. and Beukes, P. (2013). An integrated biophysical and socio-economic framework for analysis of climate change adaptation strategies: The case of a New Zealand dairy farming system. *Environmental Modelling and Software* 39, 176-187.

Keating, B. A., Carberry, P. S., Hammer, G. L., Probert, M. E., Robertson, M. J., Holzworth, D., Huth, N. I., Hargreaves, J. N. G., Meinke, H., Hochman, Z., McLean, G., Verburg, K., Snow, V., Dimes, J. P., Silburn, M., Wang, E., Brown, S., Bristow, K. L., Asseng, S., Chapman, S., McCown, R. L., Freebairn, D. M. and Smith, C. J. (2003). An overview of APSIM, a model designed for farming systems simulation. *European Journal of Agronomy* 18(3-4), 267-288.

Keller, E. D., Baisden, W. T., Timar, L., Mullan, B. and Clark, A. (2014). Grassland production under global change scenarios for New Zealand pastoral agriculture. *Geoscientific Model Development* 7(5), 2359-2391.

Kennedy, R. E., Ohmann, J., Gregory, M., Roberts, H., Yang, Z., Bell, D. M., Kane, V., Hughes, M. J., Cohen, W. B., Powell, S., Neeti, N., Larrue, T., Hooper, S., Kane, J., Miller, D. L., Perkins, J., Braaten, J. and Seidl, R. (2018). An empirical, integrated forest biomass monitoring system. *Environmental Research Letters* 13(2), 025004.

Kirkegaard, J. A. and Hunt, J. R. (2010). Increasing productivity by matching farming system management and genotype in water-limited environments. *Journal of Experimental Botany* 61(15), 4129-4143.

Kirkegaard, J. A., Hunt, J. R., McBeath, T. M., Lilley, J. M., Moore, A., Verburg, K., Robertson, M., Oliver, Y., Ward, P. R., Milroy, S. and Whitbread, A. M. (2014). Improving water productivity in the Australian Grains industry–A nationally coordinated approach. *Crop and Pasture Science* 65(7), 583-601.

Lake, L., Chenu, K., Sadras, V. O., Lake, L., Chenu, K. and Sadras, V. O. (2016). Patterns of water stress and temperature for Australian chickpea production. *Crop and Pasture Science* 67(2), 204-215.

Li, F. Y., Newton, P. C. D. and Lieffering, M. (2014). Testing simulations of intra- and inter-annual variation in the plant production response to elevated CO_2 against measurements from an 11-year FACE experiment on grazed pasture. *Global Change Biology* 20(1), 228-239.

Lieffering, M., Newton, P. C. D., Vibart, R. and Li, F. Y. (2016). Exploring climate change impacts and adaptations of extensive pastoral agriculture systems by combining biophysical simulation and farm system models. *Agricultural Systems* 144, 77-86.

Lilburne, L., Hewitt, A., Webb, T. H. and Carrick, S. (2004). S-map: A new soil database for New Zealand. In: *SuperSoil 2004. Proceedings of the 3rd* Australian New Zealand Soils *Conference*, Sydney, Australia, pp. 5-9.

Lilley, J. M., Flohr, B. M., Whish, J. P. M., Farre, I. and Kirkegaard, J. A. (2019). Defining optimal sowing and flowering periods for canola in Australia. *Field Crops Research* 235, 118-128.

Lilley, J. M. and Kirkegaard, J. A. (2011). Benefits of increased soil exploration by wheat roots. *Field Crops Research* 122(2), 118–130.

Lilley, J. M. and Kirkegaard, J. A. (2016). Farming system context drives the value of deep wheat roots in semi-arid environments. *Journal of Experimental Botany* 67(12), 3665–3681.

Llewellyn, R. and Ouzman, J. (2019). Conservation agriculture in Australia: 30 years on. In: *Australian Agriculture in 2020: From Conservation to Automation* (pp. 21–33). Agronomy Australia and Charles Sturt University, Wagga Wagga.

Lobell, D. B., Hammer, G. L., Chenu, K., Zheng, B., McLean, G. and Chapman, S. C. (2015). The shifting influence of drought and heat stress for crops in Northeast Australia. *Global Change Biology* 21(11), 4115–4127.

Luo, Y., Su, B., Currie, W. S., Dukes, J. S., Finzi, A., Hartwig, U., Hungate, B., McMurtrie, R. E., Oren, R., Parton, W. J., Pataki, D. E., Shaw, M. R., Zak, D. R. and Field, C. B. (2004). Progressive nitrogen limitation of ecosystem responses to rising atmospheric carbon dioxide. *BioScience* 54(8), 731–739.

Manoli, G., Meijide, A., Huth, N., Knohl, A., Kosugi, Y., Burlando, P., Ghazoul, J. and Fatichi, S. (2018). Ecohydrological changes after tropical forest conversion to oil palm. *Environmental Research Letters* 13(6), 064035.

Manschadi, A. M., Christopher, J., deVoil, P. and Hammer, G. L. (2006). The role of root architectural traits in adaptation of wheat to water-limited environments. *Functional Plant Biology : FPB* 33(9), 823–837.

Ministry for Primary Industries. (2019). *Vineyard Benchmarking Report*. Marlborough, New Zealand, New Zealand Wine.

Ministry for the Environment. (2018). *Climate Change Projections for New Zealand: Atmosphere Projections Based on Simulations from the IPCC Fifth Assessment, 2nd Edition*. Ministry for the Environment, Wellington.

Ministry for the Environment and Stats, N. Z. (2017). *New Zealand's Environmental Reporting Series*: Our *Atmosphere and Climate*. Ministry for the Environment and Stats NZ.

Ministry for the Environment and Stats, N. Z. (2020). *New Zealand's Environmental Reporting Series. Our Atmosphere and Climate 2020*. Ministry for the Environment and Stats NZ.

Moore, A. D. (2014). The case for and against perennial forages in the Australian sheep-wheat zone: Modelling livestock production, business risk and environmental interactions. *Animal Production Science* 54(12), 2029–2041.

Moore, A. D., Donnelly, J. R. and Freer, M. (1997). GRAZPLAN: Decision support systems for Australian grazing enterprises. III. Pasture growth and soil moisture submodels, and the GrassGro DSS. *Agricultural Systems* 55(4), 535–582.

Moore, A. D. and Ghahramani, A. (2013). Climate change and broadacre livestock production across southern Australia. 1. Impacts of climate change on pasture and livestock productivity, and on sustainable levels of profitability. *Global Change Biology* 19(5), 1440–1455.

Mullan, B., Sood, A., Stuart, S. and Carey-Smith, T. (2016). *Climate Change Projections for New Zealand: Atmosphere Projections Based on Simulations From the IPCC Fifth Assessment*. Ministry for the Environment, Wellington.

Murphy, D. J., Goggin, K. and Paterson, R. R. M. (2021). Oil palm in the 2020s and beyond: Challenges and solutions. *CABI Agriculture and Bioscience* 2(1), 39.

New Zealand Winegrowers. (2020). *NZ Winegrowers Annual Report*. Auckland, New Zealand.

Newton, P. C. D., Allard, V., Carran, R. A. and Lieffering, M. (2006). Impacts of elevated CO_2 on a grassland grazed by sheep: The New Zealand FACE experiment. In: Nösberger, J., Long, S. P., Norby, R. J., Stitt, M., Hendrey, G. R., Blum, H. (eds), *Managed Ecosystems and* CO_2 (pp. 157–171). Springer, Berlin, Heidelberg.

Newton, P. C. D., Lieffering, M., Bowatte, W. M. S. D., Brock, S. C., Hunt, C. L., Theobald, P. W. and Ross, D. J. (2010). The rate of progression and stability of progressive nitrogen limitation at elevated atmospheric CO_2 in a grazed grassland over 11 years of free air CO_2 enrichment. *Plant and Soil* 336(1-2), 433–441.

NIWA. (2021). *Overview of New Zealand's Climate.*

Okoro, S. U., Schickhoff, U., Boehner, J., Schneider, U. A. and Huth, N. I. (2017). Climate impacts on palm oil yields in the Nigerian Niger Delta. *European Journal of Agronomy* 85, 38–50.

Özkan, Ş., Vitali, A., Lacetera, N., Amon, B., Bannink, A., Bartley, D. J., Blanco-Penedo, I., de Haas, Y., Dufrasne, I., Elliott, J., Eory, V., Fox, N. J., Garnsworthy, P. C., Gengler, N., Hammami, H., Kyriazakis, I., Leclère, D., Lessire, F., Macleod, M., Robinson, T. P., Ruete, A., Sandars, D. L., Shrestha, S., Stott, A. W., Twardy, S., Vanrobays, M. L., Ahmadi, B. V., Weindl, I., Wheelhouse, N., Williams, A. G., Williams, H. W., Wilson, A. J., Østergaard, S. and Kipling, R. P. (2016). Challenges and priorities for modelling livestock health and pathogens in the context of climate change. *Environmental Research* 151, 130–144.

Parker, A. K., De Cortázar-Atauri, I. G., Van Leeuwen, C. and Chuine, I. (2011). General phenological model to characterise the timing of flowering and veraison of Vitis vinifera L.: Grapevine flowering and veraison model. *Australian Journal of Grape and Wine Research* 17(2), 206–216.

Parker, A. K., García de Cortázar-Atauri, I., Gény, L., Spring, J., Destrac, A., Schultz, H., Molitor, D., Lacombe, T., Graça, A., Monamy, C., Stoll, M., Storchi, P., Trought, M. C. T., Hofmann, R. W. and van Leeuwen, C. (2020). Temperature-based grapevine sugar ripeness modelling for a wide range of Vitis vinifera L. cultivars. *Agricultural and Forest Meteorology* 285–286, 107902.

Peel, M. C., Finlayson, B. L. and McMahon, T. A. (2007). Updated world map of the Köppen-Geiger climate classification. *Hydrology and Earth System Sciences* 11(5), 1633–1644.

Perring, M. P., Cullen, B. R., Johnson, I. R. and Hovenden, M. J. (2010). Modelled effects of rising CO2 concentration and climate change on native perennial grass and sown grass-legume pastures. *Climate Research* 42(1), 65–78.

Polsky, L. and von Keyserlingk, M. A. G. (2017). Invited review: Effects of heat stress on dairy cattle welfare. *Journal of Dairy Science* 100(11), 8645–8657.

Pook, M., Lisson, S., Risbey, J., Ummenhofer, C. C., McIntosh, P. and Rebbeck, M. (2009). The autumn break for cropping in southeast Australia: Trends, synoptic influences and impacts on wheat yield. *International Journal of Climatology* 29(13), 2012–2026.

Potgieter, A., Meinke, H., Doherty, A., Sadras, V. O., Hammer, G., Crimp, S. and Rodriguez, D. (2012). Spatial impact of projected changes in rainfall and temperature on wheat yields in Australia. *Climatic Change*, 1–17.

Potgieter, A. B., Hammer, G. L. and deVoil, P. (2005). A simple regional-scale model for forecasting sorghum yield across North-Eastern Australia. *Agriculture and Forest Meteorology* 132(1), 143–153.

Potgieter, A. B., Hammer, G. L., Doherty, A. and deVoil, P. (2006). Oz-Wheat: A Regional-Scale Crop Yield Simulation Model for Australian Wheat. Brisbane, Australia. (ISSN 0727-6273): Queensland Department of Primary Industries & Fisheries; 2006 September Report No.: Information Series No, QI06033.

Potgieter, A. B., Lobell, D. B., Hammer, G. L., Jordan, D. R., Davis, P. and Brider, J. (2016). Yield trends under varying environmental conditions for sorghum and wheat across Australia. *Agricultural and Forest Meteorology* 228–229, 276–285.

Potgieter, A. B., Zhao, Y., Zarco-Tejada, P. J., Chenu, K., Zhang, Y., Porker, K., Biddulph, B., Dang, Y. P., Neale, T., Roosta, F. and Chapman, S. (2021). Evolution and application of digital technologies to predict crop type and crop phenology in agriculture. *In Silico Plants* 3(1), diab017.

Rebetzke, G. J., Chenu, K., Biddulph, B., Moeller, C., Deery, D. M., Rattey, A. R., Bennett, D., Barrett-Lennard, E. G. and Mayer, J. E. (2012). A multisite managed environment facility for targeted trait and germplasm phenotyping. *Functional Plant Biology: FPB* 40(1), 1–13.

Renaudeau, D., Gourdine, J. L. and St-Pierre, N. R. (2011). A meta-analysis of the effects of high ambient temperature on growth performance of growing-finishing pigs. *Journal of Animal Science* 89(7), 2220–2230.

Richard, C. A., Hickey, L. T., Fletcher, S., Jennings, R., Chenu, K. and Christopher, J. T. (2015). High-throughput phenotyping of seminal root traits in wheat. *Plant Methods* 11, 13.

Rickert, K. G., Stuth, J. W. and McKeon, G. M. (2000). Modelling pasture and animal production. In: *Field and Laboratory Methods for Grassland and Animal Production Research* (pp. 29–66).

Rutledge, D., Ausseil, A. G. E., Baisden, W. T., Bodeker, G., Booker, D., Cameron, M. P. and Zammit, C. (2017). Identifying feedbacks, understanding cumulative impacts and recognising limits: A national integrated assessment. Synthesis Report RA3. Climate Changes, Impacts and Implications for New Zealand to 2100.

Sadras, V. O., Lake, L., Chenu, K., McMurray, L. S. and Leonforte, A. (2012). Water and thermal regimes for field pea in Australia and their implications for breeding. *Crop and Pasture Science* 63(1), 33–44.

Sadras, V. O. and Lawson, C. (2011). Genetic gain in yield and associated changes in phenotype, trait plasticity and competitive ability of South Australian wheat varieties released between 1958 and 2007. *Crop and Pasture Science* 62(7), 533–549.

Salinger, M. J., Diamond, H. J., Behrens, E., Fernandez, D., Fitzharris, B. B., Herold, N., Johnstone, P., Kerckhoffs, H., Mullan, A. B., Parker, A. K., Renwick, J., Scofield, C., Siano, A., Smith, R. O., South, P. M., Sutton, P. J., Teixeira, E., Thomsen, M. S. and Trought, M. C. T. (2020). Unparalleled coupled ocean-atmosphere summer heatwaves in the New Zealand region: Drivers, mechanisms and impacts. *Climatic Change* 162(2), 485–506.

Schütz, K. E., Cox, N. R. and Tucker, C. B. (2014). A field study of the behavioral and physiological effects of varying amounts of shade for lactating cows at pasture. *Journal of Dairy Science* 97(6), 3599–3605.

Spyromitros-Xioufis, E., Tsoumakas, G., Groves, W. and Vlahavas, I. (2016). Multi-target regression via input space expansion: Treating targets as inputs. *Machine Learning* 104(1), 55–98.

Stats, N. Z. (2020). *Which Industries Contributed to New Zealand's GDP?*

Stone, R. C., Hammer, G. L. and Marcussen, T. (1996). Prediction of global rainfall probabilities using phases of the Southern Oscillation Index. *Nature* 384(6606), 252–255.

Sturman, A., Zawar-Reza, P., Soltanzadeh, I., Katurji, M., Bonnardot, V., Parker, A. K., Trought, M. C., Quénol, H., Le Roux, R., Gendig, E. and Schulmann, T. (2017). The application of high-resolution atmospheric modelling to weather and climate variability in vineyard regions. *OENO One* 51, 99.

Sturman, A. P. and Tapper, N. J. (2005). Regional and smaller scale climates. In: *Weather and Climate of Australia and New Zealand* (pp. 343–402). Oxford University Press.

Sutherst, R. W., Constable, F., Finlay, K. J., Harrington, R., Luck, J. and Zalucki, M. P. (2011). Adapting to crop pest and pathogen risks under a changing climate. *WIREs Climate Change* 2(2), 220–237.

Tait, A., Sood, A., Mullan, B., Stuart, S., Bodeker, G., Kremser, S. and Lewis, J. (2016). *Updated Climate Change Projections for New Zealand for Use in Impact Studies*. Synthesis Report RA1, Climate Changes, Impacts and Implications (CCII) for New Zealand to (Vol. 2100).

Teixeira, E., Ausseil, A.-G., Burgueño, E., Brown, H., Cichota, R., Davy, M., Ewert, F., Guo, J., Holmes, A., Holzworth, D., Hu, W., de Ruiter, J., Hume, E., Jesson, L., Johnstone, P., Powell, J., Kersebaum, K. C., Kong, H., Liu, J., Lilburne, L., Meiyalaghan, S., Storey, R., Richards, K., Tait, A. and van der Weerden, T. (2020). A spatial analysis framework to assess responses of agricultural landscapes to climates and soils at regional scale. Edited by L. Lothar Mueller (pp. 495–508). Springer, Berlin.

Teixeira, E. I., de Ruiter, J., Ausseil, A. G., Daigneault, A., Johnstone, P., Holmes, A., Tait, A. and Ewert, F. (2018). Adapting crop rotations to climate change in regional impact modelling assessments. *Science of the Total Environment* 616–617, 785–795.

Teixeira, E. I., Fischer, G., van Velthuizen, H., Walter, C. and Ewert, F. (2013). Global hot-spots of heat stress on agricultural crops due to climate change. *Agricultural and Forest Meteorology* 170, 206–215.

Teixeira, E., Kersebaum, K. C., Ausseil, A. G., Cichota, R., Guo, J., Johnstone, P., George, M., Liu, J., Malcolm, B., Khaembah, E., Meiyalaghan, S., Richards, K., Zyskowski, R., Michel, A., Sood, A., Tait, A. and Ewert, F. (2021). Understanding spatial and temporal variability of N leaching reduction by winter cover crops under climate change. *Science of the Total Environment* 771, 144770.

Teixeira, E. I., Zhao, G., Ruiter, Jd de, Brown, H., Ausseil, A.-G., Meenken, E. and Ewert, F. (2017). The interactions between genotype, management and environment in regional crop modelling. *European Journal of Agronomy* 88, 106–115.

The Bureau of Meteorology and CSIRO. (2020). *State of Climates 2020*. Bureau of Meteorology and CSIRO.

Thornton, P., Running, S. and Hunt, R. (2021). *Biome-BGC: Terrestrial Ecosystem Process Model Version 4.2 Final Release*. Numerical Terradynamic Simulation Group, School of Forestry, University of Montana 2005, Missoula.

Tozer, K., Douglas, G., Dodd, M. and Müller, K. (2021). Vegetation options for increasing resilience in pastoral hill country. *Frontiers in Sustainable Food Systems* 5, 8.

VanDerWal, J., Murphy, H. T., Kutt, A. S., Perkins, G. C., Bateman, B. L., Perry, J. J. and Reside, A. E. (2013). Focus on poleward shifts in species' distribution underestimates the fingerprint of climate change. *Nature Climate Change* 3(3), 239–243.

Vanuytrecht, E. and Thorburn, P. J. (2017). Responses to atmospheric CO_2 concentrations in crop simulation models: A review of current simple and semicomplex representations and options for model development. *Global Change Biology* 23(5), 1806–1820.

Veneman, J. B., Muetzel, S., Hart, K. J., Faulkner, C. L., Moorby, J. M., Perdok, H. B. and Newbold, C. J. (2015). Does dietary mitigation of enteric methane production affect rumen function and animal productivity in dairy cows? *PLoS ONE* 10(10), e0140282.

Wakelin, S. A., Gomez-Gallego, M., Jones, E., Smaill, S., Lear, G. and Lambie, S. (2018). Climate change induced drought impacts on plant diseases in New Zealand. *Australasian Plant Pathology* 47(1), 101–114.

Wang, E., Brown, H. E., Rebetzke, G. J., Zhao, Z., Zheng, B. and Chapman, S. C. (2019). Improving process-based crop models to better capture genotype×environment×management interactions. *Journal of Experimental Botany* 70(9), 2389–2401.

Wang, B., Liu, L., O'Leary, G. J., Asseng, S., Macadam, I., Lines-Kelly, R., Yang, X., Clark, A., Crean, J., Sides, T., Xing, H., Mi, C. and Yu, Q. (2018). Australian wheat production expected to decrease by the late 21st century. *Global Change Biology* 24(6), 2403–2415.

Wang, E., Martre, P., Zhao, Z., Ewert, F., Maiorano, A., Rötter, R. P., Kimball, B. A., Ottman, M. J., Wall, G. W., White, J. W., Reynolds, M. P., Alderman, P. D., Aggarwal, P. K., Anothai, J., Basso, B., Biernath, C., Cammarano, D., Challinor, A. J., De Sanctis, G., Doltra, J., Dumont, B., Fereres, E., Garcia-Vila, M., Gayler, S., Hoogenboom, G., Hunt, L. A., Izaurralde, R. C., Jabloun, M., Jones, C. D., Kersebaum, K. C., Koehler, A. K., Liu, L., Müller, C., Naresh Kumar, S., Nendel, C., O'Leary, G., Olesen, J. E., Palosuo, T., Priesack, E., Eyshi Rezaei, E., Ripoche, D., Ruane, A. C., Semenov, M. A., Shcherbak, I., Stöckle, C., Stratonovitch, P., Streck, T., Supit, I., Tao, F., Thorburn, P., Waha, K., Wallach, D., Wang, Z., Wolf, J., Zhu, Y. and Asseng, S. (2017). The uncertainty of crop yield projections is reduced by improved temperature response functions. *Nature Plants* 3, 17102.

Wang, E., Robertson, M. J., Hammer, G. L., Carberry, P. S., Holzworth, D., Meinke, H., Chapman, S. C., Hargreaves, J. N. G., Huth, N. I. and McLean, G. (2002). Development of a generic crop model template in the cropping system model APSIM. *European Journal of Agronomy* 18(1–2), 121–140.

Wang, J., Wang, E. and Liu, D. L. (2011). Modelling the impacts of climate change on wheat yield and field water balance over the Murray–Darling Basin in Australia. *Theoretical and Applied Climatology* 104(3–4), 285–300.

Wang, X., Hunt, C., Cruickshank, A., Mace, E., Hammer, G. and Jordan, D. (2020). The impacts of flowering time and tillering on grain yield of sorghum hybrids across diverse environments. *Agronomy* 10(1), 135.

Watson, J., Zheng, B., Chapman, S. and Chenu, K. (2017). Projected impact of future climate on water-stress patterns across the Australian wheatbelt. *Journal of Experimental Botany* 68(21–22), 5907–5921.

Weragoda, A. and Duver, A. (2021). *Snapshot of Australian Agriculture 2021*.

Whetton, P., Ekström, M. and Gerbing, C. (2015). *Climate Change in Australia Information for Australia's Natural Resource Management Regions: Technical Report*. CSIRO and Bureau of Meteorology.

Yang, H., Wu, M., Liu, W., Zhang, Z., Zhang, N. and Wan, S. (2011). Community structure and composition in response to climate change in a temperate steppe. *Global Change Biology* 17(1), 452–465.

Zhang, B., Valentine, I. and Kemp, P. D. (2007). Spatially explicit modelling of the impact of climate changes on pasture production in the North Island,New Zealand. *Climatic Change* 84(2), 203–216.

Zhao, Z., Rebetzke, G. J., Zheng, B., Chapman, S. C. and Wang, E. (2019). Modelling impact of early vigour on wheat yield in dryland regions. *Journal of Experimental Botany* 70(9), 2535–2548.

Zhao, Z., Wang, E., Kirkegaard, J. A. and Rebetzke, G. J. (2022). Novel wheat varieties facilitate deep sowing to beat the heat of changing climates. *Nature Climate Change* 12(3), 291–296.

Zheng, B., Chapman, S. and Chenu, K. (2018). The value of tactical adaptation to el Niño-southern oscillation for East Australian wheat. *Climate* 6(3), 77.

Zheng, B., Chapman, S. C., Christopher, J. T., Frederiks, T. M. and Chenu, K. (2015). Frost trends and their estimated impact on yield in the Australian wheatbelt. *Journal of Experimental Botany* 66(12), 3611–3623.

Zheng, B., Chenu, K. and Chapman, S. C. (2016). Velocity of temperature and flowering time in wheat - Assisting breeders to keep pace with climate change. *Global Change Biology* 22(2), 921–933.

Zheng, B. Y., Chenu, K., Dreccer, M. F. and Chapman, S. C. (2012). Breeding for the future: What are the potential impacts of future frost and heat events on sowing and flowering time requirements for Australian bread wheat (*Triticum aestivum*) varieties? *Global Change Biology* 18(9), 2899–2914.

Zhu, J., Fraysse, R., Trought, M. C. T., Raw, V., Yang, L., Greven, M., Martin, D. and Agnew, R. (2020). Quantifying the seasonal variations in grapevine yield components based on pre- and post-flowering weather conditions. *OENO One* 54(2).

Zhu, J., Parker, A., Gou, F., Agnew, R., Yang, L., Greven, M., Raw, V., Neal, S., Martin, D., Trought, M. C. T., Huth, N. and Brown, H. E. (2021). Developing perennial fruit crop models in APSIM next generation using grapevine as an example. *In Silico Plants* 3(2), diab021.

Chapter 16

Modelling the impact of climate change on agriculture in South Asia

S. Naresh Kumar, ICAR-Indian Agricultural Research Institute, India

1 Introduction
2 Climate-change scenarios for agriculture: the case of India
3 How climactic stresses affect field crops: key mechanisms
4 Methods for quantifying the impact of climate change on crop production: overview
5 Methods for quantifying the impact of climate change on crop production: the role of simulation modelling
6 Projected impacts of climate change on key crops: India
7 Projected impacts of climate change on key crops: Pakistan, Bangladesh, Nepal, Sri Lanka, Bhutan and the Maldives
8 Projected impacts of climate change on other horticultural crops
9 Assessing the impact of climate change on pests and diseases
10 Conclusion and future trends
11 Where to look for further information
12 Acknowledgements
13 References

1 Introduction

South Asia (SA) is comprised of India, Afghanistan, Pakistan, Bangladesh, Nepal, Sri Lanka, Bhutan and the Maldives. Geographically, it provides unique variation in terms of islands, plains and high mountains and is home to ~1.8 billion people, with around 70% depending on agriculture for their livelihood. South Asian agriculture covers ~57% of total SA land, which predominantly contains smallholder farms, with around 60% of the agricultural workforce comprising of women. Being a tropical region, SA is highly susceptible to climatic risks, with agriculture in India historically known as the 'gamble with

http://dx.doi.org/10.19103/AS.2022.0115.18

the summer monsoon'. Agriculture in SA is highly dependent upon climatic conditions, and, as it is one of the major pillars of the SA economic system, its performance is critical for a healthy economy. The SA population is projected to increase by 40% by 2050, and the population in India projected to stabilise at 1.6–1.7 billion by 2060. The rising population will require increasing the production of food grain in India to 335–350 Mt from the current 310 Mt, which needs to be achieved despite constraints exacerbated by stresses including climatic extremes.

Agriculture is a major sector in the economies of SA countries. In India e.g. the contribution of agriculture to GDP is ~17.5%, with the gross value added by agriculture, forestry and fishing amounting to about US$0.25 trillion. Although climatic stresses have been increasing, the volatility (in terms of the coefficient of variation) of agricultural output growth declined from 2.7 (1961–1988) to 1.6 (1989–2004) and subsequently to 0.8 (2005–2018). During the period 1961–2021, food grain production also increased considerably from 51 Mt to 310 Mt, hence transitioning from an importing country to an exporting country. India is now a major producer on a global scale with significant contributions to the global production of rice (22%), wheat (13%), pulses (25%) and cotton (25%). It is also the highest producer of milk and second on a global scale for fruit and vegetable production. The importance of agriculture exacerbates the potential impacts of climate change on countries such as India.

A decreasing trend in the summer monsoon precipitation over SA has been reported since the 1950s (Douville et al., 2021); however, this is also coupled with an increase in heavy precipitation events (Seneviratne, 2021). In addition, rising temperatures, increased frequency and intensity of heat and cold waves, cyclonic storms, etc. have also been reported (Inter-Governmental Panel on Climate Change 6th Assessment Report, Working Group II – IPCC AR 6 WGI, 2021). The SA region is prone to damaging weather extremes such as droughts, floods and cyclones, in addition to seasonal monsoon aberrations (Sivakumar and Stefanski, 2011; Senaratne and Rodrigo, 2014). The typical maximum temperature in this region reaches 53°C (Jacobabad, Pakistan). In India, maximum temperatures from March to April 2022 breached the long-term records, with the annual mean temperature increasing by 0.71°C throughout India during 1880–2010 (Indian Institute of Tropical Meteorology IITM, 2010). Similarly, the annual mean temperature during 1871–1990 increased by ~2°C over the central highlands of Sri Lanka (Chandrapala, 1996; Fernando and Chandrapala, 2002). Historical warming trends have also been reported in other locations in SA (Naveendrakumar and Methhika Vithanagec 2019). All these climatic stresses have significantly affected various sectors of the economy, and agriculture in particular, subsequently affecting the gross domestic product

(GDP). Climate change is projected to affect ~2% of the GDP by 2050 in SA, which may rise to 9% by 2100 (Ahmed and Suphachalasai, 2014).

Agriculture is vulnerable to climate change in many ways. As an example, ~52% (or 73.2 Mha out of 141.4 Mha net sown area) of farm fields are unirrigated and dependent on rainfall. The area operated by marginal- and smallholdings is ~47.4% (146 million holdings). Smallholders typically lack the resources to implement effective mitigation techniques in the face of climate change. These agriculture-related issues are similar in other SA countries. Additionally, the agricultural growth rates have not been very encouraging in India and the climatic stress-related losses to agriculture are increasing spatially, exacerbating stress on the economy. The cost of agricultural adaptation is estimated to be around US$275 billion between 2015 and 2030 for the SA region (Amjath-Babu et al., 2018), with the bulk of this cost (US$206 billion) being for India.

A detailed household-level analysis in a village indicated that the cost of adaptation and gains vary with the type of agriculture and farm size (Naresh Kumar et al., 2014, 2016). The analysis indicated that income from agriculture alone could not support the small and marginal farm families; however, adaptation to climatic risks can improve income sustainability.

Although farmers can incur adaptation costs, it is not always the case, particularly for middle- and large-farm holdings. Science- and technology-based crop management can rationalise the cost of adaptation to climatic risks (Naresh Kumar et al., 2014, 2016). The study further shows that although additional cost is not imperative for adaptation, the cost of adaptation is not directly proportional to gains across different groups of farmers. In the remaining SA nations, agricultural production has increased, although the growth rates have not been so spectacular across the spectrum. All these nations are also

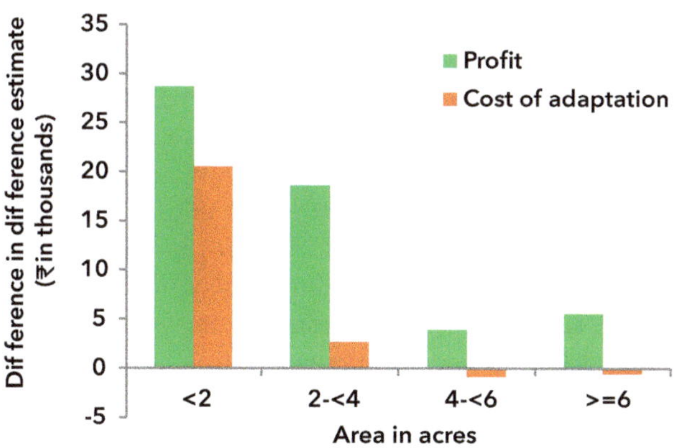

Figure 1 Adaptation costs and profits in different groups of farmers. Source: Naresh Kumar et al. (2016) in terms of Indian Rupees (₹). One United States Dollar = approximately ₹82.

facing climatic stress-related vulnerability in farming systems and farmers in SA are increasingly incurring adaption costs (Fig. 1).

Climate change projections for the SA region are alarming. The IPCC AR6 WG I (2021) and WG II (2022) reports project more intense and frequent heatwaves and humid heat stress over SA during the twenty-first century. Although annual and summer monsoon precipitation is projected to increase, the increased extreme events, inter-annual variability, and heat and cold waves will have a significant impact on agriculture. Climate change over SA has been projected to affect agriculture due to warming and extreme temperature events, extreme rainfall events, and a skewed monsoon, causing an increased risk of drought-related water and food shortage (AR 5 WG II, 2015). The risk level, however, can be moderately minimised with current adaptations and with high/intensive adaptation can be significantly minimised (AR 5 WG II, 2015). The AR 6 WG II (2022) report emphasises adaptation to climate change but warns about maladaptation, which may further worsen the situation. Thus, a detailed understanding of the impact is crucial to develop specific adaptation strategies. Before detailing impact studies, a summary of climate change scenarios for SA will contextualise the following sections.

2 Climate-change scenarios for agriculture: the case of India

Climate change projections are available for each country; however, this chapter focuses on the effect of climate change on agriculture, therefore, climate scenarios over India are discussed as a test case.

For SA as a whole, the IPCC AR 6 WG II report (2021) projected an increase in extreme events such as heat waves, cold snaps, floods, sea level rise and droughts, which challenge food production systems in SA and may undermine food security in the region.

Climate change is projected to cause a risk of food insecurity in SA due to warming and an increase in extreme temperature and rainfall events, droughts and a skewed monsoon (IPCC, 2021). The IPCC 2014 report mentions that the risk level can be moderately minimised with current adaptations, while the risk level can be significantly minimised with high and/or intensive adaptation. Over the past 120 years, the rate of increase in maximum temperatures has been 0.71°C/100 years, while that of minimum temperature has been 0.27°C/year (INCCA, 2010). This increase in temperature accelerated post-1970 at a rate of 0.17 and 0.29°C/10 years, with a higher rate of increase in the minimum temperature. In addition, an analysis of rainfall data from 1871 indicated a significant increase in drought year events post-1960. The CMIP5-Global Climate Models (GCMs)-based climate change projections for the Indian region indicated mean warming of 1.7–2°C by 2030s and 3.3–4.8°C by

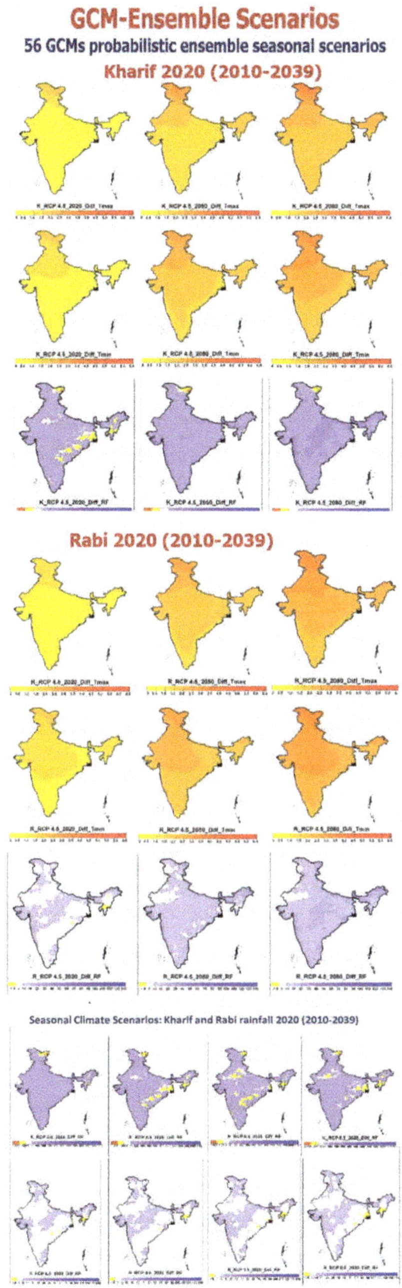

Figure 2 Seasonal projections for the spatiotemporal change in temperature and rainfall in India. Source: Naresh Kumar et al. (2019). GCM - Global Climate Model.

2080s, in business-as-usual scenarios (representative concentration pathway - RCP 6 and 8.5), compared with temperatures during the preindustrial period (Chaturvedi et al., 2012). The study further reported an increase in all-India precipitation by 4-5% in the 2030s and 6-14% in the 2080s compared with the 1961-1990 period. An increase in the frequency of extreme precipitation days is also projected.

From an agricultural perspective, an analysis of ensemble probabilistic scenarios for India (Fig. 2), derived from CMIP5-GCMs climate projections for 2020, 2050 and 2080 in RCP 2.6, 4.5, 6.0 and 8.5 scenarios, was performed by the Indian Agricultural Research Institute, New Delhi. The analysis projected:

1. a higher increase in minimum temperatures than maximum temperatures;
2. a higher increase in temperatures during Rabi (October-April) than during Kharif (June-September);
3. an increase in minimum temperatures of 0.946-4.067°C in the 2020-2080 period over baseline (1976-2005 period) in Kharif and by 1.096-4.652°C in Rabi;
4. an increase in maximum temperatures of 0.741-3.533°C (2020-2080) during Kharif and 0.882-4.01°C in Rabi;
5. a higher increase in temperatures in northern parts of India compared with southern parts;
6. an increase in rainfall of 2.3-3.3% (2020) and 4.9-10.1% (2050) during Kharif and an increase of 12% (2020) and 12-17% (2050) in Rabi rainfall; and
7. variability in minimum and maximum temperatures during Rabi projected to be significantly greater than during Kharif in future climates.
8. Variability in rainfall intensity and distribution are projected to increase future climates.

These climate change projections have significant spatiotemporal variations with a significant impact on agriculture in India (Naresh Kumar et al., 2019).

3 How climactic stresses affect field crops: key mechanisms

The growth, development and yield performance of a plant are dependent upon the environment in which it is grown. Among the major environmental factors, soil, water, nutrients and pests are manageable via agricultural practices; however, the climate is relatively difficult to manage. In the absence of other stresses, weather plays a pivotal role in determining the growth,

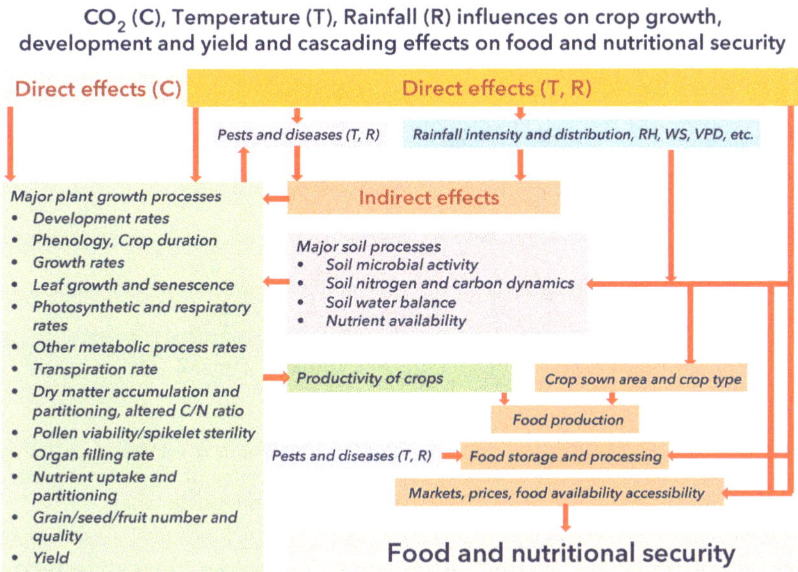

Figure 3 Summary of direct and indirect effects of CO_2 (C), temperature (T) and rainfall (R) on crop growth, development and yield, soil processes, pests and diseases and cascading effects on food and nutritional security. RH - relative humidity, WS - wind speed, VPD - vapour pressure deficit.

development and yield of crop plants. The major climate drivers of plant growth and development are temperature regimes, rainfall, CO_2 and solar radiation, with wind speed and relative humidity also playing an important role. Climate change, due to unabated greenhouse gases (GHGs) emission, causes a rise in atmospheric CO_2 levels leading to the warming of the atmosphere (the current warming potential of atmospheric gases is 2.3 W m^{-2}). The rise in CO_2, temperatures and rainfall variability, apart from other cascading weather events, is the major factor affecting plant growth and development (Fig. 3). The response of plant growth and development depends on species-specific and stage-specific thresholds (Fig 4).

The factor range is taken from lower lethal level to upper lethal level (ULL), where the system does not survive. The system performance depends on its normal baseline level (b1) or altered baseline (b2) as well as its normal maximum (max1) limit or altered maximum (max2) for a given factor. The increase in factor from b1 or b2 to O (optimum) will benefit system performance. Any increase in the factor beyond O will adversely affect the performance. A normal system may be working between the factor range b1 and max1. Adaptation can alter the system to a lower base value (b2) or higher maximum value (max2) making the system perform either in lower factor conditions (from b1 to b2) or in high

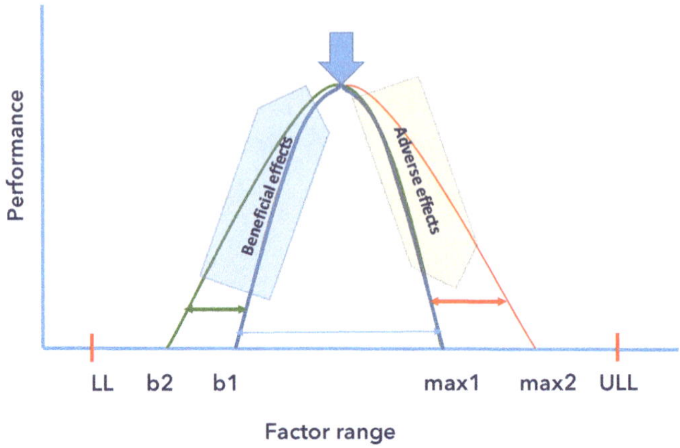

Figure 4 The response of the system to an influencing factor (e.g. temperature, nitrogen status or application, soil water and relative humidity) in terms of its performance (e.g. biomass production, yield, photosynthetic rates and other metabolic rates).

factor conditions (max1 to max2). As the crop final yield is the manifestation of the overall effects and responses of several factors and processes (inside and outside the plant), it is important to know the b2, b1, O, max1 and max2 values for major factors and processes (developmental, physiological and metabolic) so as to improve the quantification of effects under various environments. It is worth noting that this is also applicable to all indirect factors (pests, soil processes etc.). Crop productivity, production and other outputs are the final results.

4 Methods for quantifying the impact of climate change on crop production: overview

Approaches for quantifying the effects of climate change have been evolving. Some commonly followed approaches include:

1 past data analysis;
2 multi-location, multi-seasonal experiments;
3 surveys;
4 controlled environment experiments;
5 meta-analysis; and
6 simulation modelling (hydrological, crop modelling, crop-pest modelling, economic modelling).

All these are employed alone or in combination to assess the impact of climate change on agriculture.

4.1 Past data analysis

As mentioned earlier, by correlating the observed data on climatic parameters with previous crop yields, several studies have attempted to determine the impact of climatic parameters on crop yields. Mechanistic models used include:

1. linear model;
2. linear model on de-trended data;
3. linear model on aggregated data;
4. fixed-effects model;
5. linear model on differenced data;
6. generalized linear model;
7. Monte Carlo simulation;
8. ARIMA;
9. ARIMAX;
10. normalised deviation correlations;
11. fuzzy regression;
12. thumb rules;
13. principle component regression;
14. discriminant functional analysis,
15. artificial neural networks; and
16. machine learning.

Some of these methods have serious limitations in terms of:

1. not taking other factors into consideration that influence yield;
2. not explaining the processes behind yield response;
3. the relationships do not work in situations other than their development;
4. do not take daily stochastic weather variations into consideration; and
5. generally, they do not consider major plant processes and their dynamic response to climatic, management and soil conditions.

However, they do provide some important insights into the relationship between cause and effect as reported for yield (Chmielewski and Potts, 1995; Sarker et al., 2012; Leng and Huang, 2017; Kukal and Irmak, 2018). Weather parameters and pest and disease incidence are correlated to be effectively used as forewarning models for several pests and diseases. Some of the new approaches, such as artificial intelligence (AI), machine learning (ML) and big data analytics, have immense potential and are being increasingly employed.

The effects of climate change are projected based on statistical approaches including the Ricardian approach, wherein farmers maximise profit by allocating land to different crops (Kumar and Parikh, 2001). Another study indicated that during 1969–2005, an increase in maximum temperature adversely affected crop yields in India. Climate change is projected to affect the rice yield by 15% and wheat yield by 22%, whereas the yield of coarse cereals may be less affected and that of pulses may be affected to a greater extent (Birthal et al., 2014). Correlating past temperature and rainfall data with the yield of major crops for the 1980–2008 period, Lobell et al. (2011) indicated a net positive impact of around 1.5% on maize yield in India, while rice and wheat yields were negatively impacted by ~2% and 6%, respectively. An analysis of 60 years (1956–2015) of data on weather and sugarcane yield indicated that the Upper Gangetic Plain (UGP) of India has experienced enhanced drought frequency since 1986. Droughts coinciding with the grand growth phase of sugarcane are a major constraint to achieving high productivity (Verma et al., 2019).

In Pakistan, using the production function approach, where agricultural production is a function of climatic and non-climatic inputs, and the vectors of inputs include capital, climate and other controls, a district-level analysis was carried out using panel data for the 1980–2010 period (Sajid et al., 2014). Results indicated that warming has been adversely affecting agriculture, especially in warmer and rain-fed regions of the country, with a spatially varying impact of climate variables. Using the Ricardian model, it was estimated that an increase of 1% in temperature would result in a loss of 4180/ha to the net revenue per annum for a wheat farmer in arid regions of Pakistan (Shakoor et al., 2011). In another study, using feasible generalised least square (FGLS) and heteroscedasticity and autocorrelation (HAC) consistent standard error, the effects of climatic parameters (maximum temperature, minimum temperature, rainfall, relative humidity and sunshine) on the major crops of Pakistan (e.g. wheat, rice, maize and sugarcane) during the 1989–2015 period were estimated (Sajjad Ali et al., 2017). The results indicated the negative impact of maximum temperature on wheat production and the positive effect of minimum temperature on all crops. Rainfall negatively impacted the yield of the selected crops, except that of wheat.

4.2 Multi-location, multi-seasonal and long-term experiments

Multi-location experiments provide valuable information on the crop/variety response to climatic and soil conditions even under similar agronomic management regimes. India has several All India Coordinated Projects that play an important role in identifying the varieties suitable for different agroecological regions. In addition to these multi-location trials, experiments conducted by sowing the same crop/variety in different seasons also provide important information on the response of the crop/variety to growing season weather conditions (Table 1). Several studies have revealed that late sowing

Table 1 Major sources (indicative) of data sources for use in crop model validation

S. No	Experimental data sources	Brief details about data
1	All India Coordinated Research Projects for various crops/systems	Crop/cropping systems data, varietal information and varietal performance under different management and weather conditions
2	National Bureau of Soil Survey and Land Use Planning, Nagpur, India	Soil data for benchmark sites for crop models
3	India Meteorological Department	Long-term daily weather data (weather station-based and gridded data)
4	Agricultural Census Data Portal	Crop yield and agricultural data in India
5	Nepal Open Data portal	Weather, crop and fertiliser data
6	National Soil Science Research Centre, Nepal	Soil and crop maps
7	Department of Meteorology, Sri Lanka	Weather data
8	Department of the Census and Statistics, Sri Lanka	Crop-related data
9	Ministry of Agriculture, Sri Lanka	Soil maps
10	Bangladesh Bureau of Statistics Agriculture Wing	Crop-related data

affected tillering, dry matter accumulation, grain number, and weight and yield in wheat (Singh et al., 2010; Ghosh et al., 2000; Behera et al., 2000). In maize, the leaf area index, leaf area duration, dry matter accumulation, plant nitrogen content, number and weight of grains/cob, grain-filling period, harvest index and yield were higher in winter season grown crop while crop growth rate, relative growth rate, net assimilation rate and grain yield/ha/day were increased during the monsoon season (Naresh Kumar and Singh, 1999). Long-term experiments provide vital information on how climatic conditions influence crop yield and soil fertility under different management conditions. Although each country performs numerous experiments, the concerted effort to desegregate the effects of weather parameters on crop performance is very rare.

4.3 Surveys

Several survey studies have been conducted to assess yield losses due to climatic risks, including proximate analysis and remote-sensing analysis. In proximate analysis, the surveyor physically visits and estimates the damage due to climatic risks and various other factors (Savary et al., 2015; Johnson et al., 2018). Numerous reports on climatic risk-related crop loss exist in India and other nations of SA; further details are given in Section 4.4.

4.4 Use of satellite data, ground data and models

Satellite data are being used to assess crop conditions at the national level and to provide near real-time crop health indices (CRAMS website: http://creams.iari.res.in/). Highly specific studies are conducted using satellite data to assess the crop condition in farmers' fields in selected villages and link with ground truth data for two parameters viz. normalized difference vegetation index (NDVI) and leaf area index (LAI) (Naresh Kumar et al., 2022, personal communication). In addition, the leaf area index from satellite data is assimilated into the InfoCrop v2.1 model to simulate the wheat yield in selected villages (Dhakar et al., 2019a,b, 2021, 2022). Furthermore, crop-cutting experiments are used to assess the crop yield forecast and crop loss in combination with satellite data and yield modelling. In Nepal, district-level rice yield data, satellite data, soil and weather data were used in a 3D Convolutional Neural Network (CNN) to estimate yield in a specific district (Ruben et al., 2021). Similar specific examples exist for other SA countries.

The use of integrated approaches using satellite data in combination with crop simulation models is becoming increasingly important for various aspects of crop management, including crop insurance (Aggarwal et al., 2016). Despite the significant potential uses of remote-sensing linked crop simulation modelling/yield estimation modelling approaches; they can be used for past or current yield analysis and at best for an in-season yield forecast. However, simulation models remain indispensable tools in determining the impact of climate change, and adaptation and vulnerability assessments (Naresh Kumar, 2011). The crop simulation modellers and models benefit greatly from controlled environment experiments for fine-tuning the models for temperature and CO_2 effects alone or in combination with water and nitrogen stress on crops, in order to simulate climate change effects. These are in addition to the benefits accrued from conventional field experiments providing the effects of sowing time, organic matter, water and nitrogen effects, as well as pest and disease effects.

4.5 Controlled environment experiments

Controlled environment experiments are designed to delineate the effect of major factors of climate change, viz. CO_2, temperature and rainfall, on crop growth and yield. First-generation experiments used temperature gradient tunnels (TGTs) wherein the concept that the ambient air, while passing through a plastic-covered tunnel, heats up and forms a gradation of temperature regimes in the tunnel. When crops are grown in this setup, they are exposed to different levels of temperature; thus, the crop response to temperature is quantified. To quantify the effect of elevated CO_2, open-top chamber technology is used in which CO_2 is supplied through a controlled monitoring and release system to maintain the set level of CO_2. In both these systems, the crop microclimatic conditions are modified significantly, which leads to a magnified response of plants to either temperature or CO_2. To overcome this limitation, the response is generally compared with the control chamber, without an additional increase in CO_2 or temperature. To overcome such limitations, free atmospheric CO_2 enrichment (FACE) rings and free atmospheric temperature enhancement (FATE) and T-FACE (free atmospheric temperature and CO_2 enhancement) facilities are increasingly used to quantify the crop response to either elevated CO_2 or temperature or both. Free atmospheric ozone and CO_2 (O_3-FACE) rings have also been established in India. Similarly, rain simulators are being used to quantify the response of crops to differing amounts and intensities of rainfall. These are also used to study soil erosion, run-off and infiltration rates. In all these systems, the major advantage is that the CO_2, temperature and rainfall levels can be artificially altered without waiting for the natural occurrence of such events. Numerous global studies describe the response of crops to elevated CO_2 (Long et al., 2007; Ainsworth et al., 2008a,b; Leakey et al., 2009), temperature (Hatfield and Prueger, 2015) and a combination of both (Wang et al., 2012). A summary of results from Indian studies is presented as test cases.

FACE and TGT facilities have been used to grow rice, wheat, chickpea, green gram, groundnut, mustard and potato under elevated CO_2 (550 ppm) and elevated temperature (1–4°C above ambient) separately. Results indicated a decrease in yield due to an increase in temperature while elevated CO_2 increased the yield. Rice and chickpea exhibited greater thermal tolerance compared with other crops. Wheat and groundnut were more sensitive to temperature stress. The highest yield response to elevated CO_2 was in chickpea and the least was in rice and wheat (Singh et al., 2013). The FACE experiments have shown an increase in the yield of green gram (~12%), soybean (~17%), chickpea (~22%) and wheat (~17%) at an elevated CO_2 level of 550 ppm (Chakrabarti et al., 2012). Elevated CO_2 levels caused an increase in leaf area, photosynthetic rate, leaf chlorophyll, number of pods as well as the number of seeds per pod in green gram (Dey et al., 2017a). The application of phosphorus and cyanobacterial inoculation further increased the growth and yield of the green gram crop. Results from a study conducted with a cowpea crop indicated the synergistic effect of increased CO_2 concentration and cyanobacterial inoculation in enhancing the growth of cowpea, as well as increasing the availability of nitrogen in the soil mediated by biological nitrogen fixation under varying phosphorus levels (Dey et al., 2017b). An experiment carried out inside the FACE facility showed that soil hydrothermal regimes were significantly different under elevated CO_2 conditions compared with ambient conditions (Pramanik et al., 2018a).

Two oil seed crops viz. sunflower and groundnut grown under elevated CO_2 (600 ppm) in open-top chambers (OTCs) responded with a significant increase in dry matter of plant parts and total dry matter as well as leaf area (Venkteshwara Rao, 1999; Vanaja et al., 2006). In mustard, elevated CO_2 increased the net photosynthetic rates, foliage, root growth and number of siliqua, and seed oil content and reduced the saturated fatty acid pool under elevated CO_2 of 550 $\mu mol\ mol^{-1}$ (Uprety et al., 1995, 2007). Elevated CO_2 ameliorated the reduction in seed carbohydrate and oil content under moisture stress in mustard (Uprety et al., 1997). However, the protein content in seeds was adversely affected by CO_2 enrichment in many crops. The elevated CO_2 levels (550 and 700 ppm) increased growth, biomass and seed yield of castor (Vanaja et al., 2006). However, in other oil seeds (sunflower and groundnut) the effect of elevated CO_2 (600 ppm) on the root:shoot ratio at the vegetative stage was not significant (Vanaja et al., 2006).

A high temperature inside the TGT reduced the yield of wheat and chickpea crops (Chakrabarti et al., 2013). Another experiment on rice grown inside TGTs showed that with a rise in the temperature of 3.9°C, the rice yield reduced by more than 10% even with a recommended dose of nitrogen (N). The application of an additional (25%) N dose lowered the yield loss by 6% (Raj et al., 2016). A study inside TGT showed that high temperatures increased soil temperature, which further amplified the stress on the wheat crop and reduced the photosynthetic rate, stomatal conductance and root growth of the crop (Pramanik et al., 2018b). An experiment conducted in OTCs to study the effects of elevated atmospheric CO_2 and temperature on a maize crop indicated that elevated CO_2 decreased concentrations of nitrogen and phosphorus in maize grain (Abebe et al., 2016). Temperature stress during flowering and a combination of temperature and water stress significantly affected kidney bean seed yield (Kimani et al., 2022).

4.6 Meta-analysis

Meta-analysis is the statistical procedure for deriving information based on data from a large number of studies. This approach becomes particularly important in understanding a complex issue such as the impact of climate change on agriculture, as conclusions and decisions based on just one or two studies are difficult. A good and comprehensive meta-analysis looks for heterogeneity of factors and the sensitivity of a system to those. Several studies have followed this approach to quantify the impact of elevated CO_2 on crop yields (Kimball, 1983), climate change on crop yields and adaptation (Knox et al., 2012; Challinor et al., 2014), and yield and quality of pastures (Dellar et al., 2018). These analyses indicate that 2°C of local warming led to a significant decrease in the yield of wheat, rice and maize. Projected yield reduction is greater for tropical regions than for temperate regions. Crop-level adaptations improve yields by 7–15% and gains are more for wheat and rice; however, inter-annual yield variability may increase in the future.

5 Methods for quantifying the impact of climate change on crop production: the role of simulation modelling

Elevated CO_2 can minimise the negative impacts of rising temperatures, while in regions where a temperature increase does not cross the optimal level, the temperature rise will enhance growth. Thus, the net impact of both will depend on the temperature regimes of specific locations. In addition, daily temperatures, temperature extremes, wind, solar radiation, rainfall intensity and amount, soil moisture-holding capacity, soil characteristics and crop management interventions such as sowing time, method and amount of irrigation and nutrient supply, among other factors influence crop performance in changing climates.

To integrate all these with climate change scenarios, future water availability, pest incidence and other factors, simulation models are used to mimic the natural situation. In climate change studies, simulation models have an indispensable role in quantifying impacts, developing adaptation strategies and identifying vulnerable regions (Naresh Kumar, 2011). Climate change studies use more than a one-level model to quantify the impact, vulnerability and adaptation (IVA) gains. The outline of the sequence as given in Fig. 5 is an indicative approach (Naresh Kumar, 2010).

Climate projections from either GCMs or RCMs are typically used after bias correction. The outputs from GCMs/RCMs are input into the crop model to quantify IVAs for crops. A few studies have also attempted to derive irrigation

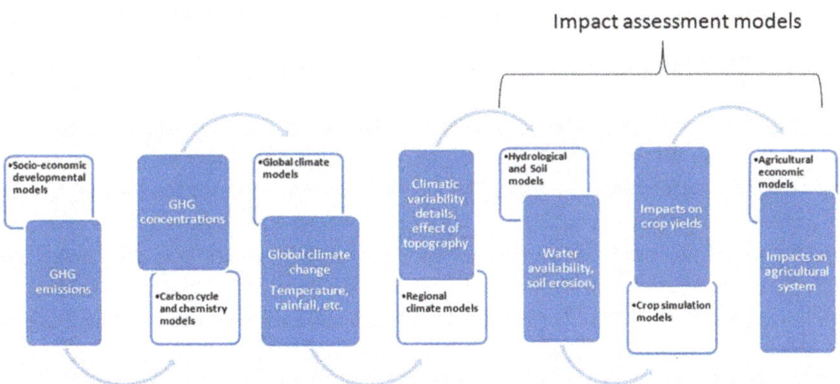

Figure 5 Cascading role of different simulation models in climate change studies. Source: Naresh Kumar (2011).

water availability scenarios (using models such as Soil and Water Assessment Tool - SWAT and Precipitation Runoff Modelling System - PRMS) for selected regions and then those are input into crop models for the IVA assessment. A few other studies have also used crop model (ex InfoCrop, Decision Support System for Agrotechnology Transfer - DSSAT, The Agricultural Production Systems Simulator - APSIM, AquaCrop etc.) inputs into economic models for the assessment of IVA in economic terms. Similarly, the crop-pest models (InfoCrop, Climex) are used in tandem with crop models and climate models. In addition, the ecological niche models, such as Ecopath, Bioclim and Domain, map climate suitability for specific species in current and future climates.

Models, being simplified representations of complex natural, biophysical and socioeconomic interactions, have limitations, which include:

1 lack of quality data sets;
2 model structure; and
3 lack of research on all the key processes in agroecosystems.

In addition, the climate change IVA assessments have uncertainties such as:

1 future GHG emissions and related global climate change;
2 technological developments in the future;
3 availability and quality of natural resources for agricultural production; and
4 the robustness of the model design.

Despite all these limitations and uncertainties, simulation models are by far the most robust option to study IVA in future climate scenarios, particularly at a regional level for agriculture. The regional-scale IVA assessments can be used to improve preparedness, prioritise research and plan development and investment.

These methods can also provide short-term weather-based forecasts, yield forecasts, crop monitoring and crop management information. IVA studies on agricultural crops follow an integrated path to provide the information required for different stakeholders (Fig. 6). The total integrated studies, however, are limited to a few locations/river basins. Most commonly and widely used are climate-crop models or climate-hydrological model-based analysis. The impact of climate change on soil factors including soil carbon is studied using models such as DeNitrification-DeComposition model - DNDC and Rothamsted Carbon Model - RothC (Table 2).

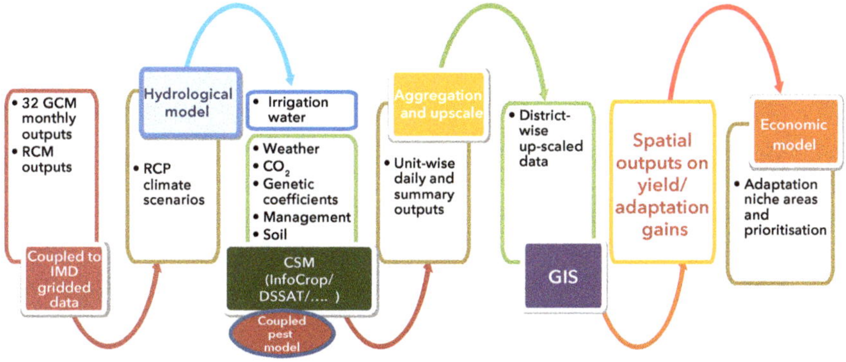

Figure 6 Modeling framework used to assess climate change impacts on agriculture and adaptation gains in India.

Table 2 Commonly used simulation models for climate change studies in South Asia

Type of model	Model names	Methods/types of analysis for climate change studies
Climate models	Global Climate Models Regional Climate Models	Directly used Bias corrected and used Bias corrected and ensembles Seasonal climate projections
Hydrological models	SWAT PRMS MODFLOW IHACRES	Surface water resources, flow rates, river or water basin analysis, flood zone mapping, groundwater source modelling, water scenarios at river/sub-river basin level, water resource management, etc.
Ecological niche models	Ecopath Bioclim Domain	Mapping the climate suitability for specific spices in current and future climates Shift in climate suitability for specific species
Crop models	DSSAT InfoCrop APSIM Aquacrop	Climate change impact, adaptation and vulnerability analysis for major crops (rice, wheat, maize, sorghum, pearl millet, mustard, groundnut, soybean, chickpea, pigeon pea, potato, cotton, sugarcane etc.) at national, state, agro-ecological zone, district, grid, village and farmer level. With static management and with alternate management for adaptation gains
Crop-pest models	InfoCrop v2.1 Climax Other models	Pest and disease scenarios
Soil dynamics models	DNDC RothC	Soil carbon changes in current future climates
Economic models	IMPACT IAT ToA Difference in Difference model	Economic impact at the national, state, household and farm level

The majority of climate change projection studies on agriculture provide quantified effects on major crops. Earlier studies mostly used data from one or a few GCMs from SRES-based climate scenarios. These were also of very coarse resolution (4 × 4° or 2.5 × 2.5°) and were down-scaled to 1 × 1°. However, the RCP-based scenarios have been used from 2013 onwards (IPCC, 2013) and are robust in terms of quantifiable effects of GHGs on climate warming. For instance, RCP2.6 indicates scenarios where the atmospheric concentration of GHGs has a radiate forcing of 2.6 W/m^2. In addition, earlier studies were mostly based on one crop model. However, recent global studies have been emphasising the use of an ensemble of models to reduce uncertainty (Asseng et al., 2013). Using this approach, global wheat production is projected to reduce by 6% with every 1°C increase in temperature (Asseng et al., 2013). A similar emphasis on the use of multi-models (at least 2–3) is reported for maize (Simona et al., 2014) and potato (Fleisher et al., 2017).

Thus, there is a requirement to assess the impact of climate change and adaptation gains for crops at finer resolutions using GCMs and RCMs. Moreover, the science on pest and disease scenarios in future climates is in the development phase and only recently have methodologies either using empirical models for leaf blast in rice (Viswanath et al., 2017) or simulation models for current (Donatelli et al., 2017) or future scenarios been used (Gebauer et al., 2015). In addition, using the model ensemble approach is required to reduce the uncertainty in assessments to make the studies more reliable in order to help develop the state and national action plans on climate change with respect to the agricultural sector.

Several global- and national-level studies have focused on simulating the impact of climate change on several crops in SA and also at national and subnational levels. A few have also given adaptation strategies and adaptation costs and identified vulnerable regions for specific crops or for agriculture. These details, though not an exhaustive list, are summarised in the following sections.

6 Projected impacts of climate change on key crops: India

Though several approaches are followed to project the impact of climate change on agricultural crops, the use of simulation models is the most accepted and robust method as they can integrate the effects of soil, weather and agronomic management on crop variety. As mentioned earlier, using simulation models in combination, it is possible to provide comprehensive assessments of the impact of climate change on crops and cropping systems, their adaptation and vulnerability and more. In India, a few models are mainly used for impact assessments, which include InfoCrop, DSSAT and APSIM. Indian studies have projected climate change effects on major crops such as rice, wheat, sorghum, maize, mustard, groundnut, soybean, potato, cotton and coconut (Sinha and Swaminathan, 1991; Sinha and Swaminathan, 1991; Aggarwal and Sinha, 1993; Mall et al., 2006; Aggarwal, 2006, 2009; Naresh Kumar, 2011, 2012, 2013, 2014, 2015a,b; Naresh Kumar and Aggarwal, 2013; National Network Project on Climate Change Project - NPCC reports, 2004–2012; National Initiative on Climate Resilient Agriculture Project - NICRA reports, 2012–2016; National Innovations in Climate Resilient Agriculture Project - NICRA reports, 2016 - onwards; Naresh Kumar et al., 2019; Rao et al., 2022). Several studies have shown the spatial variation in magnitude and direction of climate change effects on crops (Table 3).

6.1 Cereals

- ***Rice:*** the negative effects of climate change on rice yield have been projected, with ~4% loss in irrigated rice yield and 6% loss in rain-fed rice yield projected for the 2020 (2010–2039) scenario in India. However, the projected yield loss in rain-fed rice is <2.5 % in 2050 (2041–2070) and 2080 (2070–2099) scenarios. With adaptation, however, the irrigated rice yield increases by ~17%, while the increase in yield of rain-fed rice is ~20% (Naresh Kumar and Aggarwal, 2013). A recent study also indicated a reduction in rice yield in different agroclimatic regions in India (Singh et al., 2017). A 1°C increase in annual mean temperature can cause a 6% decrease in rice yield (Saseendran et al., 2000). Earlier studies reported a negative impact of climate change on rice yield in some parts of India (Lal et al., 1998; Saseendran et al., 2000; Aggarwal and Mall, 2002; Krishnan et al., 2007; Aggarwal and Swaroopa Rani, 2009; Kumar et al., 2016).
- ***Wheat:*** climate change is projected to affect wheat yield in India ranging from 6% to 23% by 2050, without any specific adaptation. The yield would reduce in areas with current seasonal maximum mean temperatures in excess of 27°C and 13°C for minimum, respectively. Sowing a suitable variety, adjusting sowing time and fertiliser and irrigation management may be a low-cost and practical adaptation strategy that can significantly increase wheat yield in changing climates (Naresh Kumar et al., 2014). Initial studies from India have projected a reduction in wheat yield in the range of 28-68% due to climate change, without considering how the beneficial effects of CO_2 and a 2°C increase in temperature would affect wheat yield in most parts of India (Aggarwal and Sinha, 1993; Rao and Shina, 1994 a,b). Adaptation was projected to improve wheat yield (Attri and Rathore, 2003). Earlier studies have indicated that a 1°C increase in growing

Table 3 Summary of studies with projections of the impact of climate change and adaptation gains in major field crops in India

Crop	Region	Impact without adaptation (percent yield deviation from baseline)	Impact with adaptation (percent yield deviation from baseline)	Scenario	Reference
Cereals and millets					
Rice	Irrigated rice in India	~ -3	~ +28	(RCP 2.6, 4.5, 6.0 and 8.5)	Naresh Kumar et al., 2019; Naresh
	Rain-fed rice in India	~ -2 to 2.5	~ +28	2020	Kumar and Aggarwal 2013;
	Irrigated rice in India	~2-5	~ +17	2050	Aggarwal and Mall, 2002; Naresh
	Rain-fed rice in India	7 to -28	~ +14	2080	Kumar et al., 2011; Birthal et al.,
	Irrigated rice in India	2 to -20	~ +8	2020	2014
	Irrigated rice in North	-10 to -47	~15	2050	Saravanakumar Sandee, 2015
	Eastern states	~ -4	~25	2080	Aggarwal et al., 2010; Krishnan
	India	~ -7	~25	2020	et al., 2007; Mishra et al., 2013
	Tamil Nadu	~ -10	–	2050	
	Irrigated rice in Upper	~ -6	11-15	2080	
	Ganga Basin, India	- <2.5	~7	2020	
	Upper Ganga Basin	- <2.5	6.6	2050	
	Indian Ganga Basin	-1 to -16.8	4.1	2080	
		-3.5 to -33.8	~ -9.8	Pessimistic scenario	
		+5% to -11%		Optimistic scenario	
		-6		A1b 2030	
		-12		2035	
		-16		2065	
		-10		2100	
		~ -5		2011-2100	
		~9		B2-2080	
		-7.63 to -9.02		A2-2080	
		-9.38 to -11.30		A1b-2030	
		-15.86 to -21.35			
		-43.2 to -24.8			

Wheat	India	-3.2 to 5.3	10–40%	(RCP 2.6, 4.5, 6.0 and 8.5)	Naresh Kumar et al., 2019; Naresh
	India	-8.4 to -19.3	9	2020	Kumar et al., 2014; Aggarwal et al.,
	Upper Ganga Basin	-18.9 to -41%	17	2050	2010; Mishra et al., 2013; Manju
	Indian Ganga Basin	-6 to -23	4-9	2080	Zacharias et al., 2014
	Indo Gangetic Plains	-15 to -25	6-12	2050	
		-20		2080	
		-6		A2-2080	
		-20.9 to -17.2		B2-2080	
		~ -30		A1b-2030	
		~ -22		A2-2030	
				B2-2030	
Maize	Kharif	No effect	+17	2020	Byjesh et al., 2010; Naresh Kumar
	upper Indo Gangetic	-5	+15	2050	et al., 2012
	Plains	-13	Offset	2020	Birthal et al., 2014
	Middle IGP	-17	Offset	2050	
	Rabi	+3	-9	2080	
	upper IGP	+6	+21	2020	
	All India	~ -18	+10	2080	
	irrigated Kharif	~ -18	+4	2020	
	India	~ -23		2050	
		0.2 to -1.2		2080	
		up to -3.7		2035	
		0.4 to -4.2		2065	
				2100	
Sorghum	Monsoon Sorghum	~ -7	1-2	2020	Aditi Srivastva et al., 2010
	All India	~ -11	3-8	2050	Naresh Kumar et al., 2012
	rain-fed Kharif	~ -32	4-9	2080	Birthal et al., 2014
	India	~ -2.5	8	2020	
		~ -8	2035	2050	
		-1.2 to -3.3	2065		
		-3.1 to -5.3	2100		
		-2.8 to -7.1			

(Continued)

Table 3 (*Continued*)

Crop	Region	Impact without adaptation (percent yield deviation from baseline)	Impact with adaptation (percent yield deviation from baseline)	Scenario	Reference
Pearl millet	Hisar	34	4 to 9.5	2030	Piara Singh et al., 2017; Yadav et al., 2013
	Aurangabad	8		2050	
	Gujarat (Kharif)	-1		2030	
		-14		2050	
		-12 to -15		A2	
Oil seeds					
Mustard	Irrigated	-21.4	~25	A1-2080	Boomiraj et al., 2010; Naresh Kumar et al., 2014
	Rain-fed	-15.7	~19	A2-2080	
		-11.2	~17	B2-2080	
		29.8–40.4 in north India		2080	
		31-42.5 in central India		2020	
		~ -2		2050	
		~ -7.9		2080	
		~ -15			
Groundnut	Middle Gujarat	-21 to -31	7-16	A2	Yadav et al., 2013; Naresh Kumar et al., 2012
	India, rain-fed	+4 to -7		A1b 2030	Birtal et al., 2014
	India	-2.2 to -5.6		2035	
		-5.5 to -8.6		2065	
		-5.1 to -11.8		2100	
Soybean	India	-13%		With current CO_2	Mall et al., 2004
		-19%		With doubled CO_2	
		-21%			
		-12%			
		-18%			
		-21%			

	India	~ +8 to 13		2030 and 2080	Naresh Kumar et al., 2012
	Madhya Pradesh	-10%		(+ or - over	Mohanty et al., 2017
		-10 to -15		1980–2010 values)	
		+10 to 15%		-10% rainfall +1°C temperature	
				+1°C temperature +10% in rainfall	
Pulses					
Chickpea	Cooler sites (Hisar,	+17 to +25		2050	Piara Singh et al., 2014
	Indore)	-7 to -16		2050	Birthal et al., 2014
	Warmer sites	-1.1 to -10		2035	
	(Nandyal, Ukiriguru)	-4.6 to -18.6		2065	
	India	-10.4 to -26.2		2100	
Pigeon pea	India	-3.2 to -10.1		2035	Birthal et al., 2014
		-8.6 to -17.7		2065	
		-5.1 to -11.8		2100	
Other commercial field crops					
Cotton	Rain-fed cotton	-10 A2,		A2	Hebbar et al., 2013
	at Nagpur	-18%		B2	
	Irrigated cotton	0		A1b	
	at Coimbatore	+9		A2	
		+10		B2	
		+12		A1b	
Potato	India	~ -2.5	~11	2020	Naresh Kumar, et al., 2015
		~ -6	~6	2050	
		~ -11	~3	2080	
Sugarcane	IGP	Negative impacts			Bhaskaran and Nair, 2014; Kumar and Sharma, 2014; Shrivastava et al., 2011

RCP, Representative Concentration Pathways; A1, B1, A2, B2 and A1B are emission scenarios.

season temperature can result in a loss of 4–5 Mt of wheat yield (Rao and Sinha, 1994 a,b; Agarwal and Sinha, 1993; Aggarwal and Mall, 2002; Mall and Aggarwal, 2002; Aggarwal and Swaroopa Rani, 2009). Other studies have indicated a reduction of 450 kg/ha wheat yield with a 0.5°C increase in temperature during winter (Sinha and Swaminathan, 1991; Lal et al., 1998; Kalra et al., 2008). Wheat yield is also projected to reduce by 5% with a warming of 0.5–1.5°C (Aggarwal, 2003).

- ***Maize:*** climate change is projected to affect the yield of maize in the Kharif season by about 18% in the 2020 and 2050 scenarios, but an approximate 21% increase in yield is possible in the 2020 scenario. In the 2050 scenario, up to a 10% increase is projected in the 2050 scenario (Byjesh et al., 2010; Naresh Kumar et al., 2012). Earlier studies suggest a 5% decrease in the potential yield of maize with a 0.5–1.5°C temperature rise in India (Aggarwal, 2003). The Rabi maize is projected to have an increased yield in the range of 8.4–18.2% in the 2020 scenario in Bihar (Haris et al., 2013).
- ***Sorghum:*** the projected effects of climate change on rain-fed sorghum show a reduction in yield of about 2.5% in the 2020 scenario (2010–2039) and 8% in the 2050 scenario. Simple adaptation options, however, can improve productivity by ~8% in 2020 (Srivastava et al., 2010; Naresh Kumar et al., 2012). Earlier studies projected a decrease in rain-fed sorghum yield in parts of Telangana and Maharashtra (Rao et al., 1995). This study also showed that sorghum grown in the post-rainy season on residual moisture may have a marginal improvement in yield under changing climates.
- ***Pearl millet:*** climate change is projected to negatively impact the pearl millet yield (–1 and –14%) in parts of Maharashtra, while it may improve the yield (34 and 8%) in Haryana in the scenarios for 2030 and 2050, respectively (Piara Singh et al., 2017).

6.2 Oil seeds

- ***Mustard:*** in India, climate change is projected to affect mustard seed yield by about 2% in the 2020 (2010–2039) scenario. The projected negative impact on seed yield is ~7.9 % in the 2050 scenario (2040–2069) and 15% in the 2080 (2070–2099) scenario. Areas with current mean seasonal temperatures (max/max) in excess of 25/10°C will lose mustard yield due to the temperature rise. A reduced climatically suitable window for the cultivation of mustard is also projected. Therefore, short-duration cultivars (<130 days) with a 63% period for pod filling can adapt better (Naresh Kumar et al., 2015). Earlier studies also indicated a reduction in mustard yield due to climate change with significant spatial variation. Mustard in eastern parts of India is projected to be affected to a greater degree (Boomiraj et al., 2010; Naresh Kumar et al., 2015). Studies have also indicated a decrease in mustard seed yield up to 201 kg/ha for every 1°C increase in temperature in Haryana (Kalra et al., 2008).

- ***Soybean:*** the seed yield of soybean is projected to increase by 8–13% in the 2030 and 2080 climate scenarios, respectively (Naresh Kumar et al., 2012).
- ***Groundnut:*** climate change is projected to reduce the seed yield of rain-fed groundnut by ~4% in the A1B 2080 scenario. However, in the climate scenario of A1B 2080 (–5%), and the remaining scenarios, it is projected that seed yield may increase in the range of 4–7% (Naresh Kumar et al., 2012).

6.3 Pulses

- ***Chickpea:*** the seed yield of chickpeas is projected to increase by 17-25% in Haryana and central Madhya Pradesh, but the yield is projected to decrease by 7-16% in southern Andhra Pradesh in the 2050 climate change scenario (Piara Singh et al., 2014).
- ***Pigeon pea:*** regression analysis indicated a plausible decrease in pigeon pea yield by ~ −3.2 to −10.1% in the 2035 scenario; the decrease in yield may be greater in the 2065 scenario (Birthal et al., 2014). However, these estimates did not consider the positive effects of CO_2 in future climates and other factors, which may give improved estimates.

6.4 Other commercial crops

- ***Cotton:*** climate change is projected to cause a marginal decline in cotton productivity in northern India but productivity may increase in central and southern India. However, at the national level, projections indicated no significant effect on cotton productivity in future climates (Hebbar et al., 2013).
- ***Sugarcane:*** projected impacts on sugarcane yield in India are mainly attributed to an increase in dry spells and water scarcity (Srivastava et al., 2011).

In addition to the impact on yield, climate change is projected to affect grain/seed quality due to a reduced concentration of protein in grain (particularly in low-input situations) and some minerals (such as zinc and iron) due to elevated CO_2 (Porter et al., 2014). Elevated CO_2 causes a reduction in protein and secondary metabolite concentration; conversely, high temperatures resulted in an increase in their concentration in several cereal and pulse crops. In addition, climate change is projected to affect other factors that directly or indirectly influence crop production. For instance, an increased frequency of extreme rainfall events increases soil erosion and affects water availability and quality (IPCC, 2014). However, a temperature rise may decrease the occurrence of cold waves and frost events resulting in less damage to some of the frost-sensitive field crops such as mustard, chickpea, potato and other vegetables. The enhanced use of technology, including simulation modelling, to manage agriculture in India is encouraged (Aggarwal et al., 2016).

7 Projected impacts of climate change on key crops: Pakistan, Bangladesh, Nepal, Sri Lanka, Bhutan and the Maldives

7.1 Pakistan

Pakistan basically has two major crop seasons, as in India (Kharif-Monsoon-JJASO and Rabi-Winter-NDJFM), with rice and wheat being the primary crops (Table 4).

7.2 Bangladesh

Rice is the major crop in Bangladesh. A recent report indicated that the domestic output deviation (%) in 2030 from 2006/2007 would be less than 1% for agricultural production (Yu et al., 2010; Banerjee et al., 2015). Some studies are summarized in Table 5.

7.3 Nepal

Nepal's agriculture is projected to be affected by climate change and the effects on a few major crops are detailed below (Table 6).

7.4 Sri Lanka

Most other crops such as maize, coarse grains, legumes, fruits, vegetables and tuber crops also seem to be adversely affected by the impact of climate change (Titumil and Basak, 2010; Eeswaran, 2018). Some studies are summarized in Table 7.

7.5 Bhutan

Studies on climate change impacts on major crops and adaptation gains are summarized in Table 8.

Table 4 Summary of studies with projections of the impact of climate change on major field crops in Pakistan

Crop	Region	Impact without adaptation (percent yield deviation from baseline)	Impact with adaptation (percent yield deviation from baseline)	Scenario	Reference
			Cereals and millets		
Rice	Pakistan	-20.5%		2035	Ahmad et al., 2015
	Punjab	-10%		2035	Khan and Tahir, 2018
		-1.7		2020	CIAT, 2017
		-3.2		(SSP2, RCP4.5)	Siddiqui et al., 2022Khan et al.,
		-4.6		2030	2020
		-5.7		2040	
		-3.08		2050	
		-20		2030	
				2050	
Wheat	Pakistan	-14.7%		**2035**	Ahmad et al., 2015
		-5.67		**2035**	Khan and Tahir, 2018
		-1.6		2020	CIAT 2017Khan et al.,
		-3.1		2030	2020Siddiqui et al., 2022
		-4.6		2040	
		-5.9		2050	
		−14.6		2050	
		2.49		2030	
Maize		-3.2		2020	CIAT 2017
		-6.1		2030	
		-8.9		2040	
		-11.7		2050	

(*Continued*)

Table 4 (*Continued*)

Crop	Region	Impact without adaptation (percent yield deviation from baseline)	Impact with adaptation (percent yield deviation from baseline)	Scenario	Reference
			Pulses		
Chickpea		-3.2		2020	CIAT 2017
		-6.3		2030	
		-9.5		2040	
		-12.6		2050	
			Other commercial field crops		
Cotton		-1.6		2020	CIAT 2017
		-3.1		2030	
		-4.6		2040	
		-6.1		2050	
		-28.0		2008–2030	
Sugarcane		-40.1		2008–2030	

Table 5 Summary of studies with projections of the impact of climate change on major field crops in Bangladesh

Crop	Region	Impact without adaptation (percent yield deviation from baseline)	Impact with adaptation (percent yield deviation from baseline)	Scenario	Reference
		Cereals and millets			
Rice	Bangladesh	-28 -8 -1.8 -12.67 -10.4 -22.9 -10 to -15		4°C increase 2050 2030 2050 2°C increase 4°C increase 2050	Climate Change Country Study Bangladesh Easterling et al., 2007 Yu et al., 2010 Ashrafun Nahar, 2016; Basak et al., 2010; Nelson et al., 2010
Wheat	Bangladesh	-68 -32 <1 -2.05		4°C increase 2050	Ibid Easterling et al., 2007 Banerjee et al., 2015 Yu et al., 2010

Table 6 Summary of studies with projections of the impact of climate change in major field crops in Nepal

Crop	Region	Impact without adaptation (percent yield deviation from baseline)	Impact with adaptation (percent yield deviation from baseline)	Scenario	Reference
		Cereals and millets			
Rice	Dudh Kosi basin	-21 to -51 -39 -3.4 in tarai 17.9 in hills 36.1 in mountains		2040-50 2100 4°C increase	Bocchiola et al. (2019); Malla (2008)
Wheat	Dudh Kosi basin	-13 to -19 -1.8 in tarai 5.3 in hills 33.3 in mountains		2040-50 4°C increase	Bocchiola et al. (2019); Malla (2008)
Maize	Dudh Kosi basin	-38 -46 -26.4 in tarai -9.3 in hills 26.8 in mountains		2040-50 2100	Bocchiola et al. (2019); Malla (2008)

Table 7 Summary of studies with projections of the impact of climate change on rice in Sri Lanka

Crop	Region	Impact without adaptation (percent yield deviation from baseline)	Impact with adaptation (percent yield deviation from baseline)	Scenario	Reference
		Cereals and millets			
Rice	Major rice-growing district-Kurunegala	-20		2050	Weerakoon (2013)

Table 8 Summary of studies with projections of the impact of climate change on major crops in Bhutan

Crop	Region	Impact without adaptation	Impact with adaptation	Scenario	Reference
		Cereals and millets			
Rice	Bhutan	+8% in a climatically suitable area		2050 RCP8.5	Parker et al. (2017)
Wheat					
Maize		+25% in a climatically suitable area		2050 RCP8.5	Parker et al. (2017)
		Other commercial field crops			
Cotton					
Potato	Bhutan	+5% in a climatically suitable area		2050 RCP8.5	Parker et al. (2017)

7.6 The Maldives

The Maldives is about 38 km^2 and the nation faces a challenge from rising sea levels as it is only ~1 m above mean sea level. It has more than 1000 coral islands and the agricultural area is over 23% of area contributing about 4% to GDP. About 21.33% of Maldives' total land area is under agriculture, primarily coconut and fruit trees, and with just about 4000 ha of agricultural land under sweet potato, maize, vegetables, fruit crops, etc. The country has experienced a devastating Asian Tsunami in 2004. It has put National Recovery and Reconstruction Plan to place the recovery.

8 Projected impacts of climate change on other horticultural crops

There are few simulation studies projecting the impact of climate change and adaptation gains. Simulation analysis using the InfoCrop Potato indicated that potato crop duration in the IGP may decrease due to climate change. Evapotranspiration (ET) is projected to increase, while the water use efficiency (WUE) for potato yield is projected to decline in future climates. Climate change is projected to reduce potato yields by ~2.5, ~6 and ~11% in the IGP region in 2020 (2010–2039), 2050 (2040–2069) and 2080 (2070–2099), respectively. Changing planting time is the single most important adaptation option which may lead to yield gains of ~6% in 2020 and its combination with an improved variety or additional nitrogen may be required to adapt to climate change leading to positive gains of ~8% in 2020 and ~5% in 2050. However, the adoption of all three adaptation strategies by 2080 may be needed for positive gains. Intra-regional differences in the impact of climate change and adaptation gains are projected, with a positive impact in northwestern IGP, gains in central IGP with adaptation and yield loss in eastern IGP even with adaptations (Naresh Kumar et al., 2015).

The assessment of climate change impact on coconuts, a perennial plantation crop, is challenging; however, the development of a simulation model (InfoCrop-COCONUT) (Naresh Kumar et al., 2008) has enabled a simulation study to be conducted. Climate change is projected to increase coconut productivity in the western coastal region, Kerala, parts of Tamil Nadu, Karnataka and Maharashtra (provided the current level of water and management is also made available in future climates) and also in northeastern states, islands of Andaman and Nicobar and Lakshadweep, while negative impacts are projected for Andhra Pradesh, Orissa, West Bengal, Gujarat and parts of Karnataka and Tamil Nadu (Fig. 4). Throughout India, even with current management, climate change is projected to increase coconut productivity by 4.3% in the A1B 2030, 1.9% in A1B 2080, 6.8% in A2 2080 and 5.7% in B2 2080 scenarios of Providing Regional Climate for Impact Studies (PRECIS) compared with the mean productivity of the 2000–2005 period. Agronomic adaptations, such as soil moisture conservation, summer irrigation, drip irrigation and fertiliser application, can minimise losses in the majority of coconut-growing regions, in addition to substantially improving productivity. Furthermore, genetic adaptation measures, such as growing improved local *Tall* cultivars and hybrids under improved crop management, are needed for the long-term adaptation of plantations to climate change, particularly in regions that are projected to be negatively impacted by climate change. Such strategies can increase productivity by about 33% in 2030 and 25–32% in 2080 climate scenarios. In fact, productivity can be improved by 20% to almost double if all plantations in India are provided with the abovementioned management, even in the current climate. In locations where positive impacts are projected, current poor management may become a limiting factor in reaping the benefits of CO_2 fertilisation, while in negatively affected regions adaptation strategies can reduce the impact. Thus, intensive genetic and agronomic adaptation to climate change can substantially benefit coconut production in India (Naresh Kumar and Aggarwal, 2013).

9 Assessing the impact of climate change on pests and diseases

The climate is an important determinant of the abundance and distribution of biological species, which has profound effects on populations of invertebrate pests, such as insects and mites, affecting their development, reproduction and dispersal. Climate change is expected to have a significant impact on the distribution, phenology and abundance of many species over the next few decades. Changes in the global climate may thus affect crop yields, the incidence of pests and the economic costs of agricultural production. The rising CO_2 concentrations may not only have a variety of direct effects on plants but may indirectly affect herbivores and their natural enemies. The impact of climate change on insects may include shifts in species distributions with a shift in geographic ranges to higher latitudes and elevations, changes in phenology with life cycles beginning earlier in spring and continuing later in autumn, an increase in population growth rates and the number of generations, changes in migratory behaviour, alterations to crop–pest synchrony and natural enemy-pest interactions and changes in inter-specific interactions. Changes in community structure and the extinction of some species are also expected. Climate change may alter the interactions between insect pests and their host plants. Plant–herbivore interactions are of particular significance because of their agricultural importance, as well as their potential to affect ecosystem nutrient and carbon fluxes. The relative abundance of different insect species may alter rapidly owing to climate change, and those unable to withstand the stress may be lost in the near future (Jump and Penuelas, 2005; Thomas et al., 2004).

Extreme weather events such as intense rainstorms, wind or high temperatures also affect the survival of insect populations. For species to survive changing climates, they must either adapt *in situ* to new conditions or shift their distributions in pursuit of more favourable ones. Many insects have large population sizes and short generation times, and their phenology, fecundity, survival, selection and habitat use can respond rapidly to climate change. These changes to insect life history may, in turn, produce rapid changes in abundance and distribution. Due to recent climate change, widespread generalist species at their cool range margins have expanded their distribution ranges, whereas ranges of localised, habitat-specialist species and those at their warm margins have narrowed. Changes in rainfall pattern also have implications for insect survival. It is predicted that rainfall frequency would decrease, but intensity would increase, which may lead to floods on the one hand and long dry spells on the other. More intense rainfall as projected under climate change may thus reduce the incidence of small pests on crops. An aphid population on barley was negatively related to the mean minimum temperature in January and total rainfall in February.

Pest management has the potential to provide an eco-friendly sustainable solution to pest problems. However, the relative efficacy of pest management components such as host-plant resistance, biopesticides, natural enemies and synthetic chemicals is liable to change as a result of global warming (Sharma, 2010). Therefore, it is important to assess the impact of climate change on insect populations and adopt suitable pest management adaptations for their effective management.

9.1 Climate change impact on crop pests

Climate change will have both direct and indirect effects on insect populations. Temperature is the major factor in global climate change that directly affects insect development, reproduction and survival. Although insect responses to global climate change will vary, the effect of global warming in general has been predicted to increase the intensity of herbivore pressure on plants. Climate change will also affect insects indirectly via their host plants.

Direct effects of climate change on insects include:

- expansion of habitat range;
- changes in migrating behaviour;
- changes in inter-specific interactions; and
- changes in population growth rates.

Indirect effects through host plants/predators include:

- CO_2-related increase in the C:N ratio in leaves leading to increased leaf feeding;
- increased temperature-related changes in the biochemical composition of leaves (increased phenolic, etc.);
- non-synchrony of the crop-sensitive stage and suitable weather for pest incidence;
- changes in host plant spatial and temporal distribution;
- changes in predator populations, synchrony of presence, etc.

The impact of global climate change on crop productivity and pest populations can be determined via experiments, and crop growth and insect population models.

As with qualifying the impact on crops, the effects on pests are also studied using multi-location data on population dynamics, pest damage scores, controlled facilities such as incubators, growth chambers, OTC, and FACE facilities. The modelling approaches include weather-related empirical models, thumb rules, niche models and dynamic pest population models.

Aphid incidence on the barley crop variety 'DL-70' during the Rabi season from 1985–1986 to 1999–2000 was related to weather parameters, with an appreciable inter-annual variation in aphid incidence probably due to climatic variability (Chander et al., 2003). The aphid population on barley exhibited a declining trend with time and it had a negative relationship with the mean minimum temperature in January and total rainfall in February. Under the changing climate situation, existing fungal pathogens, bacteria and viruses may cause increased damage and some minor pests may become major pests in the future. Added to these, the vector population may increase and new pathogens may emerge. An increase in temperature with relative humidity (RH) of less than 60% reduced leaf spot diseases in bananas in the subhumid region. The advancement of aphid appearance by 2 weeks with an increase of 1°C temperature reduced the growing period for seed crops in potatoes. Increasing the temperature beyond 20°C during winter affects the cultivation of seasonal button mushrooms and increases the incidence of insect pests and diseases. A temperature rise from 20°C to 22°C increases the incidence of pests and diseases in the case of the *Cymbadium* orchid. The tea mosquito bug populations and damage score in cashews were negatively correlated with minimum temperature (15–20°C), RH (85–90%) and positively bright sunshine hours (9–10 h). Less than 2 h per day of bright sunshine hours triggers the incidence of inflorescence blight in cashews. Changes in rainfall and temperature patterns caused an increased incidence of apple scabs, powdery mildew in apples and flee beetle in almost all temperate crops (Ahmad et al., 2015).

An insect population simulation model can be developed based on various bioecological factors of a species viz. fecundity, sex ratio, migration, abiotic and biotic mortality factors, development thresholds and thermal constants (Reji and Chander, 2008; Sujithra and Chander, 2013; Selvaraj and Chander, 2015). A population simulation model can be coupled to a crop growth model at relevant plant processes depending on pest damage mechanisms. A crop-pest model can then be used to analyse the impact of climate change on insect dynamics as well as crop-pest interactions (Kaukoranta, 1996; Luo et al., 1997; Sujithra and Chander, 2013; Selvaraj and Chander, 2015).

9.2 Pest risk assessment and adaptation planning

Pest risk analysis can be performed through agroecological pest zoning. This is a concept that is particularly applicable to large-area pest management, in which both tactics and strategies can be merged to achieve optimal management. A pest population dynamic model can be run with the requisite weather data and the probability of a pest outbreak for a site can be determined. Site predictions can then be extrapolated through geographical information system to determine the zones of equal epidemic potential for a pest. Yadav et al. (2013) undertook agroecological zoning of rice brown plant hopper (BPH), *N. lugens*, in Andhra Pradesh, which was divided into zones of differential pest epidemic potential. Knowledge of the pest epidemic potential in different zones allows strategic decisions with respect to the selection of crop cultivars and appropriate management options.

10 Conclusion and future trends

The studies so far have indicated a largely negative impact on crops and a positive impact in a few regions and some crops in SA.

A summary of the simulation studies indicating the impact on crops in climate change scenarios are as follows:

1. Negative in regions or a crop yield even with adaptation;
2. Negative in regions or a crop yield but with adaptation the productivity levels can be increased;
3. Positive in regions or a crop yield but with adaptation the productivity levels can be further increased by harnessing positive effects;
4. Positive in regions or a crop yield in the mid-century scenario but will maintain or increase productivity with adaptation towards the end of the century; and
5. Positive in regions or a crop yield in the mid-century scenario but will lose productivity even with adaptation towards the end of the century.

This indicates significant spatiotemporal variation based on region and crop species in climate change scenarios.

The adaptation strategies and gain scenarios indicated significant improvement in yield by implementing location-specific low-cost and easy-to-adopt adaptation technologies. The studies also highlight the vulnerable areas despite these interventions, indicating the need for the development of new and tolerant genotypes and improved crop management practices for implementation in these areas.

The simulation modelling efforts to date have enabled an understanding of the spatiotemporal variations in climate change impact and adaptation gains for major crops. Moving forward, the models, once well calibrated and validated, can also be used to develop and strategize the adaptation and mitigation measures and quantify the co-benefits. The integrated modelling efforts can lead to a significant increase in understanding and help implant the 'Ecosystem-based-adaptation' approach for sustainable intensification of agriculture to ensure **f**ood, env**ir**onment and income **s**ecuri**t**y (FIRST).

The studies also indicate the need for the following:

1 High-resolution impact studies particularly in hill and mountainous regions;
2 Studies on economically important but less studied crops, farming systems, cropping systems, etc.;
3 Deriving the water and pest scenarios;
4 Integrated impact studies such as climate-hydrology-crop-pest-economic analysis for more regions;
5 Integrated adaptation studies;
6 Adaptation gain evidence;
7 Adaptation and mitigation co-benefit analysis; and
8 Seasonal weather forecast-related impact studies.

This list is not exhaustive but indicative for studies to comprehensively understand climate change impact, adaptation and vulnerability assessments.

11 Where to look for further information

Each of the South Asian Countries provide information on soils, crops, land use and climate change impacts, adaptation and vulnerability of agriculture as well as the National Communications to UNFCCC. Some of the links, though not exhaustive, provide more information and data.

1. Bhattacharyya, T., Ray, S. K., SarkarDipak, Balbuddhe, D. V., Dasgupta, D., Chandran, P., Tiwari, P., Pal, D. K., Mandal, C., Nimje, A. M., Deshmukh, A. S., Deshmukh, R. R., Telpande, B. A., Lokhande, M. A., Wadhai, K. N., Dongare, V. T., Thakre, S. W., Likhar, C. K., Sheikh, S. M., Aggarwal, P. K., Pathak, H., Venkateswarlu, B., Rao, V. U. M., Naresh Kumar Soora and Singh, A. K. (2011). *Soil Resource Information of Different Agro-Eco Subregions of India for Crop and Soil Modelling. National Project on Climate Change (ICAR Network Project), National Bureau of Soil Survey and Land Use Planning*, Nagpur, Maharashtra, p. 302–188 (two volumes).
2. India-Second National Communication to the UNFCCC- 2012, MoEF, GOI New Delhi Publication, p. 310.
3. India. Biennial update report (BUR). BUR3. A Govt of India Pub.

4. Krishnan R. J. Sanjay Chellappan Gnanaseelan, Milind Mujumdar, Ashwini Kulkarni, Supriyo Chakraborty (Editors) (2021). *Assessment of Climate Change over the Indian Region: A Report of the Ministry of Earth Sciences (MoES), Government of India*. Springer Pub. p. 226.
5. Naresh Kumar, S., Bandyopadhyay, S. K., Padaria, R. N., Singh, A. K., Rashid, M., Wasim, M. and Prasad, S. (2014). Climatic risks and strategizing agricultural adaptation in climatically challenged regions. *IARI New Delhi* Publication. TB-ICN: 136/2014: 106.
6. Naresh Kumar, S, Islam, A., Swaroopa Rani, D. N., Panjwani, S., Sharma, K., Lodhi, N. K., Chander, S., Sinha, P., Khanna, M., Singh, D. K. and Bandyopadhyay, S. K. 2019. Seasonal Climate Change Scenarios for India: Impacts and Adaptation Strategies for Wheat and Rice. *ICAR- IARI Pub. TB-ICN*: 233/2019 p. 56.
7. Naresh Kumar, S., Singh, A. K., Aggarwal, P. K., Rao, V. U. M. and Venkateswarlu, B. (2012). Climate change and Indian Agriculture: Salient achievements from ICAR network project. *IARI Pub*., p. 32.
8. National Innovations in Climate Resilient Agriculture Annual Reports\
9. Indian Council of Agricultural Research Reports.
10. https://icar.org.in/.
11. https://mausam.imd.gov.in/.
12. https://ccafs.cgiar.org/regions/south-asia.
13. http://www.nicra-icar.in/nicrarevised/index.php/home1.
14. https://opendatanepal.com/.
15. http://www.bbs.gov.bd/site/page/3e838eb6-30a2-4709-be85-40484b0c16c6/Yearbook-of-Agricultural-Statistics.
16. https://www.pbs.gov.pk/.
17. http://www.statistics.gov.lk/Agriculture/StaticalInformation/new.
18. https://www.doa.gov.bt/.
19. https://www.agrimin.gov.lk/web/index.php/en/about-us-3.
20. http://kms.narc.gov.np/.
21. https://bangladeshbiosafety.org/website/bangladesh-agricultural-research-council-barc/.

12 Acknowledgements

The author thankfully acknowledges the funding support by the National Innovations in Climate Resilient Agriculture (NICRA) project under the National Mission on Sustainable Agriculture (NMSA), Indian Council of Agricultural Research, New Delhi; MRDP project entitled 'Agricultural productivity in climate change scenarios: impacts and adaptation pathways (A National Facility for Capacity Building on Simulation Modelling in Agriculture)' funded by Department of Science and Technology, under National Mission of Strategic

Knowledge on Climate Change (NMSKCC), GoI; and National Mission for Sustaining the Himalayan Eco system-Agriculture Taskforce project funded by Department of Science and Technology, under National Mission for Sustaining the Himalayan Ecosystem (NMSHE), GoI.

13 References

Abebe, A., Pathak, H., Singh, S. D., Bhatia, A., Harit, R. C. and Kumar, V. (2016). Growth, yield and quality of maize with elevated atmospheric carbon dioxide and temperature in north-west India. *Agric. Ecosyst. Environ.* 218: 66–72.

Aggarwal, P. K. (2003). Impact of climate change on Indian agriculture. *J. Plant Biol.* 30(2): 189–198.

Aggarwal, P. K., Chand, R., Bhutani, A., Kumar, V., Goel, S. K., Rao, K. N., Poddar, M. K., Sud, U. C., Krishna Murthy, Y. V. N., Ray, S. S., Murthy, C. S., Sikka, A., Shrisath, P. B., Mishra, J. P., Choudhary, M., Chaudhary, O. P., Joshi, A. K., Sen, G., Tarpara, C. N., Kumar, D., Sahoo, R. N., Kumar, N. S., Yadav, A. K., Singh, T. P., Khannan, H., Satish Raju, G., Sudhakar, M., Singh, K. K., Aggarwal, N., Reddy, P. K., Shukla, A., Shanbaug, A., Srivastava, A. K., Mehta, S., Gupta, P. K., Gopalakrishnan, A., Chandel, B. S., Duggal, R. K., Deshmukh, V., Kotwaliwale, N., Vora, P., Kumar, S. and Lochan, R. (2016). *Report of the Task Force on 'Enhancing Technology Use in Agriculture Insurance' NITI (National Institute for Transforming India) Aayog*. Govt of India, p. 24.

Aggarwal, P. K., Chand, R., Bhutani, A., Kumar, V., Goe, S. K., Rao, K. N., Poddar, M. K., Sud, U. C. and Lochan, R. (2016). *Report of the Task Force on 'Enhancing Technology Use in Agriculture Insurance' NITI (National Institute for Transforming India) Aayog*. Govt of India, p. 24.

Aggarwal, P. K. and Mall, R. K. (2002). Climate change and rice yields in diverse agro-environments of India. II. Effect of uncertainties in scenarios and crop models on impact assessment. *Clim. Change* 52(3): 331–343.

Aggarwal, P. K., Naresh Kumar, S. and Pathak, H. (2010). Impacts of Climate Change on Growth and Yield of Rice and Wheat in the Upper Ganga Basin. WWF Report, pp. 1–44.

Aggarwal, P. K. and Sinha, S. K. (1993). Effect of probable increase in carbon dioxide and temperature on wheat yields in India. *J. Agric. Meteorol.* 48(5): 811–814.

Aggarwal, P. K. and Swaroopa Rani, D. N. (2009). Assessment of climate change impacts on wheat production in India. *Glob. Clim. Change Indian Agric. Case Stud. ICAR Netw. Proj.*: 5–12.

Ahmad, A., Ashfaq, M., Rasul, G., Wajid, S. A., Khaliq, T., Rasul, F. and Ahmad Baig, I. (2015). *Impact of Climate Change on the Rice-Wheat Cropping System of Pakistan Handbook of Climate Change and Agroecosystems: The Agricultural Model Intercomparison and Improvement Project Integrated Crop and Economic Assessments, Part 2*. World Scientific Publishing, pp. 219–258.

Ahmed, M. and Suphachalasai, S. (2014). *Assessing the Costs of Climate Change and Adaptation in South Asia. Asian 23 Development Bank.*

Ainsworth, E. A., Beier, C., Calfapietra, C., Ceulemans, R., Durand-Tardif, M., Farquhar, G. D., Godbold, D. L., Hendrey, G. R., Hickler, T., Kaduk, J., Karnosky, D. F., Kimball, B. A., Körner, C., Koornneef, M., Lafarge, T., Leakey, A. D., Lewin, K. F., Long, S. P., Manderscheid, R., McNeil, D. L., Mies, T. A., Miglietta, F., Morgan, J. A., Nagy, J., Norby, R. J., Norton, R. M., Percy, K. E., Rogers, A., Soussana, J. F., Stitt, M., Weigel,

H. J. and White, J. W. (2008a). Next generation of elevated CO_2 experiments with crops: A critical investment for feeding the future world. *Plant Cell Environ.* 31(9): 1317–1324.

Ainsworth, E. A., Leakey, A. D. B., Ort, D. R. and Long, S. P. (2008b). FACE-ing the facts: Inconsistencies and interdependence among field, chamber and modeling studies of elevated [CO_2] impacts on crop yield and food supply. *New Phytol.* 179(1): 5–9.

Ali, S., Liu, Y., Ishaq, M., Shah, T., Ilyas, A., Ilyas, A. and Din, I. U. (2017). Climate change and its impact on the yield of major food crops: Evidence from Pakistan. *Foods* 6(6): 2–19. DOI: 10.3390/foods6060039.

Amjath-Babu, T., Aggarwal, P. K. and Vermeulen, S. (2018). Climate action for food security in South Asia? Analyzing the role of agriculture in nationally determined contributions to the Paris agreement. *Clim. Policy* 19: 1–16.

Asseng, S., Ewert, F., Rosenzweig, C., Jones, J. W., Hatfield, J. L., Ruane, A. C., Boote, K. J., Thorburn, P. J., Rötter, R. P., Cammarano, D., Brisson, N., Basso, B., Martre, P., Aggarwal, P. K., Angulo, C., Bertuzzi, P., Biernath, C., Challinor, A. J., Doltra, J., Gayler, S., Goldberg, R., Grant, R., Heng, L., Hooker, J., Hunt, L. A., Ingwersen, J., Izaurralde, R. C., Kersebaum, K. C., Müller, C., Naresh Kumar, S., Nendel, C., O'Leary, G., Olesen, J. E., Osborne, T. M., Palosuo, T., Priesack, E., Ripoche, D., Semenov, M. A., Shcherbak, I., Steduto, P., Stöckle, C., Stratonovitch, P., Streck, T., Supit, I., Tao, F., Travasso, M., Waha, K., Wallach, D., White, J. W., Williams, J. R. and Wolf, J. (2013). Uncertainty in simulating wheat yields under climate change. *Nat. Clim. Change* 3(9): 827–832.

Attri, S. D. and Rathore, L. S. (2003). Simulation of impact of projected climate change on wheat in India. *International Journal of Climatology* 23(6): 693–705.

Banerjee, O., Mahzab, M., Raihan, S. and Islam, N. (2015). An economy-wide analysis of climate change impacts on agriculture and food security in Bangladesh (2015). *Climate Change Economics* 6(1): 1–17.

Basak, J. K., Ali, M. A., Islam, M. N. and Rashid, M. A. (2010). Assessment of the effect of climate change on Boro rice production in Bangladesh using DSSAT model. *Civ. Eng.* 38: 95–108.

Behera, U. K., Chougule, B. A., Thakur, R. S., Ruwali, K. N., Bhawsar, R. C. and Pandey, H. N. (2000). Influence of planting dates and nitrogen levels on yield and quality of durum wheat (Triticum durum). *Indian J. Agric. Sci.* 70(7): 434–436.

Bhaskaran, A. and Nair, N. V. (2014). Challenges and opportunities in sugarcane cultivation under climate change scenario. *J. Sugarcane Res.* 4(1): 1–18.

Birthal, P. S., Khan, T., Negi, D. S. and Agarwal, S. (2014). Impact of climate change on yields of major food crops in India: implications for food security. *Agric. Econ. Res. Rev.* 27(2): 145–155.

Bocchiola, D., Brunettia, L., Soncinia, A., Polinellib, F. and Gianinetto, M. (2019). Impact of climate change on agricultural productivity and food security in the Himalayas: A case study in Nepal. *Agric. Syst.* 171: 113–125. DOI: 10.1016/j.agsy.2019.01.008.

Boomiraj, K., Chakrabarti, B., Aggarwal, P. K., Choudhary, R. and Chander, S. (2010). Assessing the vulnerability of Indian mustard to climate change. *Agril. Ecosys Environ.* 138(3–4): 265–273.

Byjesh, K., Naresh Kumar, S. and Aggarwal, P. K. (2010). Simulating impacts, potential adaptation and vulnerability of maize to climate change in India. *Mitigation Adapt. Strateg. Glob. Change* 15: 413–431.

Chakrabarti, B., Singh, S. D., Kumar, V., Harit, R. C. and Misra, S. (2013). Growth and yield response of wheat and chickpea crops under high temperature. *Indian J. Plant Physiol.* 18(1): 7–14.

Chakrabarti, B., Singh, S. D., Naresh Kumar, S., Aggarwal, P. K., Pathak, H. and Nagarajan, S. (2012). Low-cost facility for assessing impact of carbon dioxide on crops. *Curr. Sci.* 102(7): 1035–1040.

Challinor, A. J., Watson, J., Lobell, D. B., Howden, S. M., Smith, D. R. and Chhetri, N. (2014). A meta-analysis of crop yield under climate change and adaptation. *Nat. Clim. Chang.* 4(4): 287–291.

Chandrapala, L. 1996. Long term trends of rainfall and temperature in Sri Lanka. In: *Climate Variability and Agriculture*. Narosa Publishing House, New Delhi, India.

Chaturvedi, R. K., Joshi, J., Jayaraman, M., Bala, G. and Ravindranath, N. H. (2012a). Multi-model climate change projections for India under representative concentration pathways. *Curr. Sci.*: 791–802.

Chaturvedi, R. K., Joshi, J., Jayaraman, M., Bala, G. and Ravindranath, N. H. (2012b). Multi-model climate change projections for India under representative concentration pathways. *Curr. Sci.* 103(7). Available at: http://wcrp-cordex.ipsl.jussieu.fr/images/pdf/peer_rev_pub/2012_Chaturvedi.pdf.

Chmielewski, F.-M. and Potts, J. M. (1995). The relationship between crop yields from an experiment in southern England and long-term climate variations. *Agric. For. Meteorol.* 73(1–2): 43–66.

CIAT and World Bank. (2017). Climate-smart agriculture in Pakistan. In: *CSA Country Profiles for Asia Series*. International Center for Tropical Agriculture (CIAT). The World Bank, Washington, DC, p. 28.

Dellar, M., Topp, C. F. E., Banos, G. and Wall, E. (2018). A meta-analysis on the effects of climate change on the yield and quality of European pastures. *Agric. Ecosyst. Environ.* 265: 413–420.

Dey, S. K., Chakrabarti, B., Prasanna, R., Singh, S. D., Purakayastha, T. J., Datta, A. and Pathak, H. (2017a). Productivity of mungbean (Vigna radiata) with elevated carbon dioxide at various phosphorus levels and cyanobacteria inoculation. *Legume Research* 40(3): 497–505.

Dey, S. K., Chakrabarti, B., Prasanna, R., Pratap, D., Singh, S. D., Purakayastha, T. J. and Pathak, H. (2017b). Elevated carbon dioxide level along with phosphorus application and cyanobacterial inoculation enhances nitrogen fixation and uptake in cowpea crop. *Archives of Agronomy and Soil Science* 63(13): 1927–1937. http://doi.org/10.1080/03650340.2017.1315105.

Dhakar, R., Sehgal, V. K., Chakraborty, D., Mukherjee, J. and Kumar, S. N. (2019a). Evaluating infocrop model for growth, development and yield of spring wheat at farmers' field in semi-arid environment. *J. Agrometeorol.* 21: 254–261.

Dhakar, R., Sehgal, V. K., Chakraborty, D., Sahoo, R. N. and Mukherjee, J. (2019b). Field scale wheat LAI retrieval from multispectral sentinel 2A-MSI and LandSat 8-OLI imagery: Effect of atmospheric correction, image resolutions and inversion techniques. *Geocarto Int.* 1–21. DOI: 10.1080/10106049.2019.1687591.

Dhakar, R., Sehgal, V. K., Chakraborty, D., Mukherjee, J. and Kumar, S. N. (2021). Calibration and validation of InfoCrop model for phenology, LAI, dry matter and yield of wheat. *Indian J. Agric. Sci.* 91(5): 115–119.

Dhakar, R., Sehgal, V. K., Chakraborty, D., Sahoo, R. N., Mukherjee, J., Ines, A. V. M., Kumar, S. N., Shirsath, P. B. and Roy, S. B. (2022) Field scale spatial wheat yield forecasting system under limited field data availability by integrating crop simulation model with weather forecast and satellite remote sensing. *Agric. Syst.* 195. DOI: 10.1016/j.agsy.2021.103299.

Donatelli, M., Magarey, R. D., Bregaglio, S., Willocquet, L., Whish, J. P. M. and Savary, S. (2017). Modelling the impacts of pests and diseases on agricultural systems. *Agric. Syst.* 155: 213–224.

Douville, H. (2021). The physical science basis. contribution of 33 working group I to the sixth assessment report of the intergovernmental panel on climate change [Masson34 Delmotte, V., P. Zhai, A. Pirani, S. L. Connors, C. Péan, S. Berger, N. Caud, Y. Chen, L. Goldfarb, M. I. Gomis, 35 M. Huang, K. Leitzell, E. Lonnoy, J. B. R. Matthews, T. K. Maycock, T. Waterfield, O. Yelekçi, R. Yu and B. 36 Zhou (eds.)]. Water cycle changes. In: *Climate Change*. Cambridge University Press, Cambridge.

Easterling, W. E. (2007). Impacts, adaptation and vulnerability. Contribution of Working Group II to the Fourth Assessment Report of the Intergovernmental Panel on Climate Change. Food, fibre and forest products. In: Parry, M. L. (Ed.), *Climate Change*. Cambridge University Press, Cambridge, pp. 273–313.

Econ. Surv. (2018). Ministry of Finance, Government of India. Available at: http://mofapp.nic.in:8080/economicsurvey/pdf/082-101_Chapter_06_ENGLISH_Vol_01_2017-18.pdf: 2017–2018.

Eeswaran, R. (2018). Climate change impacts and adaptation in the agriculture sector of Sri Lanka: What we learnt and way forward. In: Leal Filho, W. (Ed.). *Handbook of Climate Change Communication* (Vol. 2, pp. 97–110). DOI: 10.1007/978-3-319-70066-3_8.

Fernando, T. K. and Chandrapala, L. (2002). *Global Warming and Rainfall Variability - Sri Lankan Situation*. Sri Lanka Association for the Advancement of Science (SLASS), p. 138.

Fleisher, D. H., Condori, B., Quiroz, R., Alva, A., Asseng, S., Barreda, C., Bindi, M., Boote, K. J., Ferrise, R., Franke, A. C., Govindakrishnan, P. M., Harahagazwe, D., Hoogenboom, G., Naresh Kumar, S., Merante, P., Nendel, C., Olesen, J. E., Parker, P. S., Raes, D., Raymundo, R., Ruane, A. C., Stockle, C., Supit, I., Vanuytrecht, E., Wolf, J. and Woli, P. (2017). A potato model intercomparison across varying climates and productivity levels. *Glob. Change Biol.* 23(3): 1258–1281.

Gebauer, K., Hemerik, L. and Meyhöfer, R. (2015). Effects of climate change on pest-parasitoid dynamics: Development of a simulation model and first results. *J. Plant Dis. Prot.* 122(1): 28–35.

Ghosh, D. C., Nandi, P. and De, B. (2000). Phenological development and productivity of wheat (*Triticum aestivum*) at different dates of sowing. *Indian J. Agric. Sci.* 70(6): 393–395.

Haris, A. V. A., Biswas, S., Chhabra, V., Elanchezhian, R. and Bhatt, B. P. (2013). Impact of climate change on wheat and winter maize over a sub-humid climatic environment. *Curr. Sci.* 104(2): 206–214.

Hatfield, J. L. and Prueger, J. H. (2015). Temperature extremes: Effect on plant growth and development. *Weather Clim. Extrem* 10: 4–10.

Hebbar, K. B., Venugopalan, M. V., Prakash, A. H. and Aggarwal, P. K. (2013). Simulating the impacts of climate change on cotton production in India. *Clim. Change* 118(3–4): 701–713.

INCCA Report. (2010). *Climate Change and India: A 4x4 Assessment- A Sartorial and Regional Analysis for 2030s: Chapter on Agriculture, Ministry of Environment and Forestry*. Under Indian Network for Climate Change Assessment, pp. 67–88.

IPCC (2013). The physical science basis working Group I contribution to the Fifth Assessment Report of the Intergovernmental Panel on Climate Change. *Climate Change*.

IPCC (2014). Impacts, adaptation, and vulnerability. Part A: Global and sectoral aspects. Contribution of working group II to the fifth assessment report of the intergovernmental panel on climate change. *Climate Change*.

IPCC (2021). The physical science basis working Group I contribution to the Sixth Assessment Report of the Intergovernmental Panel on Climate Change. *Climate Change*.

IPCC (2022). Impacts, adaptation, and vulnerability. Contribution of working group II to the sixth assessment report of the intergovernmental panel on climate change. *Climate Change*.

Javed, S. A., Ahmad, M. and Iqbal, M. (2014). Impact of climate change on agriculture in Pakistan: A district level analysis: Climate change Working Papers No. 3. *Pak. Inst. Dev. Econ. Islamabad*, p. 40.

Johnson, L. K., Dunning, R. D., Bloom, J. D., Gunter, C. C., Boyette, M. D. and Creamer, N. G. (2018). Estimating on-farm food loss at the field level: A methodology and applied case study on a North Carolina farm. *Resour. Conserv Recy.* 137: 243–250.

Kalra, N., Chakraborty, D., Sharma, A., Rai, H. K., Jolly, M., Chander, S., Kumar, P. R., Bhadraray, S., Barman, D., Mittal, R. B., Lal, M. and Sehgal, M. (2008). Effect of increasing temperature on yield of some winter crops in northwest India. *Curr. Sci.* 94(1): 82–88.

Khan, M. A., Tahir, A., Khurshid, N., ul Husnain, M. I., Ahmed, M. and Boughanmi, H. (2020). Economic effects of climate change-induced loss of agricultural production by 2050: A case study of Pakistan. *Sustainability. Clim. Change country* Study Bangladesh, under the United States Climate Change Study Programme, carried out by the Bangladesh Centre for Advanced Studies in 1996 12: 1216; 10.3390/su12031216, in Vancour, I. A. A. E. Pub 2 Aug, 2018. conjunction with BIDS and BUP, with support from the US government Khan, MA, A. Tahir 2018. Economic Effects of Climate Change on Agriculture Productivity by 2035: A case study of Pakistan. 30th International Conference on Agricultural Economists July 28.

Kimani, P. N., Kumar, Soora N., Harit, R. and Kumar, Y. (2022). Interactive effect of irrigation and temperature regimes on growth and development of kidney bean (*Phaseolus vulgaris* L.). *J. Agrometeorol.*

Kimball, B. A. (1983). Carbon dioxide and agricultural yield: An assemblage and analysis of 430 prior observations 1. *Agron. J.* 75(5): 779–788.

Knox, J., Hess, T., Daccache, A. and Wheeler, T. (2012). Climate change impacts on crop productivity in Africa and South Asia. *Environ. Res. Lett.* 7(3): 034032-034037.

Krishnan, P., Swain, D. K., Chandra Bhaskar, B., Nayak, S. K. and Dash, R. N. (2007). Impact of elevated CO_2 and temperature on rice yield and methods of adaptation as evaluated by crop simulation studies. *Agric. Ecosyst. Environ.* 122(2): 233–242.

Kukal, M. S. and Irmak, S. (2018). Climate-driven crop yield and yield variability and climate change impacts on the US Great Plains agricultural production. *Sci. Rep.* 8(1): 3450. DOI: 10.1038/s41598-018-21848-2.

Kumar, A. and Sharma, P. (2014). Climate change and sugarcane productivity in India: An econometric analysis. *J. Soc. Dev. Sci.* 5(2): 111–122.

Kumar, K. K. and Parikh, J. (2001). Indian agriculture and climate sensitivity. *Glob. Environ. Change* 11(2): 147–154.

Lal, M., Singh, K. K., Rathore, L. S., Srinivasan, G. and Saseendran, S. A. (1998). Vulnerability of rice and wheat yields in NW - India to future changes in climate. *Agric. For. Meteorol.* 89(2): 101–114.

Leakey, A. D., Ainsworth, E. A., Bernacchi, C. J., Rogers, A., Long, S. P. and Ort, D. R. (2009). Elevated CO_2 effects on plant carbon, nitrogen, and water relations: Six important lessons from FACE. *J. Exp. Bot.* 60(10): 2859–2876. DOI: 10.1093/jxb/erp096.

Leng, G.and Huang, M. (2017). Crop yield response to climate change varies with crop spatial distribution pattern. *Sci. Rep.* 7(1): 1463. DOI: 10.1038/s41598-017-01599-2.

Lobell, D. B., Schlenker, W. and Costa-Roberts, J. (2011). Climate trends and global crop production since 1980. *Science* 333(6042): 616–620. DOI: 10.1126/science.1204531.

Long, S. P., Ainsworth, E. A., Leakey, A. D., Ort, D. R., Nosberger, J. and Schimel, D. (2007). Crop models, CO_2, and climate change-response. *Science* 315: 459–460.

Mall, R. K. and Aggarwal, P. K. (2002). Climate change and rice yields in diverse agro-environments of India. I. Evaluation of impact assessment models. *Clim. Change* 52(3): 315–330.

Mall, R. K., Lal, M., Bhatia, V. S., Rathore, L. S. and Singh, R. (2004). Mitigating climate change impact on soybean productivity in India: A simulation study. *Agric. For. Meteorol.* 121(1–2): 113–125.

Malla, G. (2008). Climate change and its impact on Nepalese agriculture. *J. Agric. Environ.* 9: 62–71.

Mishra, A., Singh, R., Raghuwanshi, N. S., Chatterjee, C.and Froebrich, J. (2013). Spatial variability of climate change impacts on yield of rice and wheat in the Indian Ganga Basin. *Sci. Total Environ.* 468–469(Suppl.): S132–S138.

Mohanty, M., Sinha, N. K., McDermid, S. P., Chaudhary, R. S., Sammi Reddy, K., Hati, K. M., Somasundaram, J., Lenka, S., Patidar, R. K., Prabhakar, M., Cherukumalli, S. R. and Patra, A. K. (2017). Climate change impacts vis-a-vis productivity of soybean in vertisol of Madhya Pradesh. *J. Agrometeorol.* 19(1): 10–16.

Nahar, A. 2016. *The impact of climate change in Bangladesh on the rice market and farm households*. MSc Thesis submitted to University of Arkansas.

Naresh Kumar, S. (2011). Climate change and Indian agriculture: Current understanding on impacts, adaptation, vulnerability and mitigation. *J. Plant Biol.* 37(2): 1–16.

Naresh Kumar, S. and Aggarwal, P. K. (2013). Climate change and coconut plantations in India: Impacts and potential adaptation gains. *Agril. Syst.* 117: 45–54. DOI: 10.1016/j.agsy.2013.01.001.

Naresh Kumar, S., Aggarwal, P. K., Rani, S., Jain, S., Saxena, R. and Chauhan, N. (2011). Impact of climate change on crop productivity in Western Ghats, coastal and north-eastern regions of India. *Curr. Sci.*: 332–341.

Naresh Kumar, S., Aggarwal, P. K., Rani, D. S., Saxena, R., Chauhan, N. and Jain, S. (2014). Vulnerability of wheat production to climate change in India. *Clim. Res.* 59(3): 173–187.

Naresh Kumar, S., Aggarwal, P. K., Uttam, K., Surabhi, J., Rani, D. N. S., Chauhan, N. and Saxena, R. (2016). Vulnerability of Indian mustard (Brassica juncea (L.) Czernj. Cosson) to climate variability and future adaptation strategies. *Mitig. Adapt. Strateg. Glob. Change* 21(3): 403–420.

Naresh Kumar, S., Anuja, R. Md., Bandyopadhyay, S. K., Padaria, R. and Khanna, M. (2016). Adaptation of farming community to climatic risk: Does adaptation cost for sustaining agricultural profitability? *Curr. Sci.* 110(10): 1216–1224.

Naresh Kumar, S., Bandyopadhyay, S. K., Padaria, R. N., Singh, A. K., Rashid, M., Wasim, M. and Prasad, S. (2014). Climatic risks and strategizing agricultural adaptation in climatically challenged regions. *IARI New Delhi* Publication. TB-ICN: 136/2014: 106.

Naresh Kumar, S., Govindakrishnan, P. M., Swarooparani, D. N., Nitin, C., Surabhi, J. and Aggarwal, P. K. (2015). Assessment of impact of climate change on potato and potential adaptation gains in the Indo-Gangetic Plains of India. *Int. J. Plant Prod.* 9(1): 151–170.

Naresh Kumar, S., Islam, A., Swaroopa Rani, D. N., Panjwani, S., Sharma, K., Lodhi, N. K., Chander, S., Sinha, P., Khanna, M., Singh, D. K. and Bandyopadhyay, S. K. (2019). Seasonal climate change scenarios for India: Impacts and adaptation strategies for wheat and rice. *ICAR-IARI Pub. TB-ICN* 56: 233.

Naresh Kumar, S., Singh, A. K., Aggarwal, P. K., Rao, V. U. M. and Venkateswarlu, B. (2012). *Climate Change and Indian Agriculture: Salient Achievements From ICAR Network Project. Pub.*, p. 32.

Naresh Kumar, S. and Singh, C. P. (1999). Yield and yield components of maize (*Zea mays* L.): Physiological analysis on seasonal variations. *Indian J. Plant Physiol.* 4(2): 90–94.

Naresh Kumar, S., Singh, A. K., Aggarwal, P. K., Rao, V. U. M. and Venkateswarlu, B. (2012). *Climate Change and Indian Agriculture: Salient Achievements From ICAR Network Project. IARI Pub.*, p. 32.

Naresh Kumar, S. N., Bai, K. V., Rajagopal, V. and Aggarwal, P. K. (2008). Simulating coconut growth, development and yield with the InfoCrop-coconut model. *Tree Physiol.* 28(7): 1049–1058.

Naveendrakumar, G. and Meththika Vithanagec, S. (2019). South Asian perspective on temperature and rainfall extremes: A review. *Atmos. Res.* 225: 110–120.

Nelson, G. C., Rosegrant, M. W., Palazzo, A., Gray, I., Ingersoll, C., Robertson, R. and Tokgoz, S. (2010). *Food Security, Farming, and Climate Change to 2050: Scenarios, Results, Policy Options*. International Food Policy Research Institute, Washington, DC.

NPCC. (2012). *Indian Council of Agricultural Research Network Project on Impact of Climate Change and Vulnerability of Indian Agriculture to Climate Change* (Vol. 128p). CRIDA, Hyderabad, India, pp. 2004–2012.

Onil, B. and Mahzab, M. (2015). An economy-wide analysis of climate change impacts on agriculture and food security in Bangladesh. *Clim. Change Econ.* 6(1): 1550003. DOI: 10.1142/S2010007815500037.

Parker, L., Guerten, N., Thi Nguyen, T., Rinzin, C., Tashi, D., Wangchuk, D., Bajgai, Y., Subedi, K., Phuntsho, L., Thinley, N., Chhogyel, N., Gyalmo, T., Katwal, T. B., Zangpo, T., Acharya, S., Pradhan, S. and Penjor, S. (2017). *Climate Change Impacts in Bhutan: Challenges and Opportunities for the Agricultural Sector*. Working Paper No. 191. Available at: www.ccafs.cgiar.org. CGIAR Research Program on Climate Change, Agriculture and Food Security (CCAFS), Wageningen, The Netherlands.

Porter, J. R., Xie, L., Challinor, A. J., Cochrane, K., Howden, S. M., Iqbal, M. M., Lobell, D. B.and Travasso, M. I. (2014). Food security and food production systems. In: *Impacts, Adaptation, and Vulnerability. Part, A. Global and Sectoral Aspects. Contribution of Working Group II to the Fifth Assessment Report of the Intergovernmental Panel on Climate Change*. Cambridge University Press, Cambridge, UK and New York, pp. 485–533.

Pramanik, P., Chakrabarti, B., Bhatia, A., Singh, S. D., Mridha, N. and Krishnan, P. (2018a). Effect of elevated carbon dioxide on soil hydrothermal regimes and growth of maize crop (*Zea mays* L.) in semi-arid tropics of Indo-Gangetic Plains. *Environ. Monit. Assess.* 190(11): 661. DOI: 10.1007/s10661-018-6988-5.

Pramanik, P., Chakrabarti, B., Bhatia, A., Singh, S. D., Maity, A., Aggarwal, P. and Krishnan, P. (2018b). Effect of elevated temperature on soil hydrothermal regimes and growth of wheat crop. *Environ. Monit. Assess.* 190(4): 217.

Raj, A., Chakrabarti, B., Pathak, H., Singh, S. D., Mina, U. and Mittal, R. (2016). Growth, yield components and grain yield response of rice to temperature and nitrogen levels. *J. Agrometeorol.* 18(1): 1-6.

Rana, B. R. M. and Kalia, R. S. (2009). Weather changes related shift in apple belt in Himachal Pradesh. In: Aggarwal, P. K. (Ed.). *Glob. Clim. Change Indian Agric.-Case Stud. ICAR Netw. Proj.* ICAR, New Delhi Pub, pp. 48–53.

Rao, G. D. and Sinha, S. K. (1994a). Climate changes and agriculture. *Nature* 437: 102–109.

Rao, G. D. and Sinha, S. K. (1994b). Impact of climate change on simulated wheat production in India. Implications of climate change for international agriculture. *Crop Model. Stud.* 2(3): 4–10.

Sajid, A. J., Munir, A.and Iqbal, M. (2014). *Impact of Climate Change on Agriculture in Pakistan: A District Level Analysis*. Pakistan Institute of Development Economics Islamabad. Climate Change Working Paper Series No. 3. Pp 39.

Saravanakumar, V. (2015). *Impact of Climate Change on Yield of Major Food Crops in Tamil Nadu, India*. South Asian Network for Development and Environmental Economics (SANDEE) Working Papers, ISSN 1893-1891, WP 91–WP 15.

Sarker, M. A. R., Alam, K.and Gow, J. (2012). Exploring the relationship between climate change and rice yield in Bangladesh: An analysis of time series data. *Agric. Syst.* 112: 11–16.

Saseendran, S. A., Singh, K. K., Rathore, L. S., Singh, S. V. and Sinha, S. K. (2000). Effects of climate change on rice production in the tropical humid climate of Kerala, India. *Clim. Change* 44(4): 495–514.

Savary, S., Stetkiewicz, S., Brun, F. and Willocquet, L. (2015). Modelling and mapping potential epidemics of wheat diseases–Examples on leaf rust and Septoria tritici blotch using EPIWHEAT. *Eur. J. Plant Pathol.* 142(4): 771-790.

Senaratne, A., and Rodrigo, C., (2014). *Agriculture Adaptation Practices: Case Study of Sri Lanka*. South Asia Watch on Trade, Economics and Environment (South Asia Watch on Trade, Economics, and Environment), Nepal.

Seneviratne, S. I. (2021). Weather and climate extreme events in a changing climate. In: *Climate Change*. Cambridge University Press 2021: 24 The Physical Science Basis. Contribution of Working Group I to the Sixth Assessment Report of the 25 Intergovernmental Panel on Climate Change [Masson-Delmotte, V., P. Zhai, A. Pirani, S. L. Connors, C. Péan, S. 26 Berger, N. Caud, Y. Chen, L. Goldfarb, M. I. Gomis, M. Huang, K. Leitzell, E. Lonnoy, J. B. R. Matthews, T. K. 27 Maycock, T. Waterfield, O. Yelekçi, R. Yu and B. Zhou (eds.)].

Shakoor, U., Saboor, A., Ali, I. and Mohsin, A. Q. (2011). Impact of climate change on agriculture: Empirical evidence from arid region. *Pak. J. Agric. Sci.* 48(4): 327-333.

Shrivastava, A. K., Srivastava, A. K. and Solomon, S. (2011). Sustaining sugarcane productivity under depleting water resources. *Curr. Sci.* 101(6): 748-754.

Singh, P., Boote, K. J., Kadiyala, M. D. M., Nedumaran, S., Gupta, S. K., Srinivas, K. and Bantilan, M. C. S. (2017). An assessment of yield gains under climate change due to genetic modification of pearl millet. *Sci. Tot. Environ.* 601: 1226-1237.

Singh, P., Nedumaran, S., Boote, K. J., Gaur, P. M., Srinivas, K. and Bantilan, M. C. S. (2014). Climate change impacts and potential benefits of drought and heat tolerance in chickpea in South Asia and East Africa. *Eur. J. Agron.* 52: 123–137.

Singh, S., Gupta, A. K., Gupta, S. K.and Kaur, N. (2010). Effect of sowing time on protein quality and starch pasting characteristics in wheat (*Triticum aestivum* L.) genotypes grown under irrigated and rain-fed conditions. *Food Chem.* 122(3): 559–565.

Singh, S. D., Chakrabarti, B., Muralikrishna, K. S., Chaturvedi, A. K., Kumar, V., Mishra, S. and Harit, R. (2013). Yield response of important field crops to elevated air temperature and CO_2. *Indian J. Agric. Sci.* 83(10): 1009–1012.

Sinha, S. K. and Swaminathan, M. S. (1991). Deforestation, climate change and sustainable nutrition security. *Clim. Change* 16: 33–45.

Sivakumar, M. V. K. and Stefanski, R., (2011). Chapter 2: Climate change in south Asia. In: *Climate Change and Food Security in South Asia* (Vol. 2011). Springer Netherlands, pp. 13–30.

Soora, N. K., Aggarwal, P. K., Saxena, R., Rani, S., Jain, S. and Chauhan, N. (2013). An assessment of regional vulnerability of rice to climate change in India. *Clim. Change* 118(3–4): 683–699.

Srivastava, A., Naresh Kumar, S. and Aggarwal, P. K. (2010). Assessment on vulnerability of sorghum to climate change in India. *Agric. Ecosyst. Environ.* 138(3–4): 160–169.

Sumit, D., Chakrabarti, B., Prasanna, R., Singh, S. D., Purakayastha, T. J., Datta, A. and Pathak, H. (2017a). Productivity of mung bean (Vigna radiata) with elevated carbon dioxide at various phosphorus levels and cyanobacteria inoculation. *Legume Res.* 40(3): 497–505.

Sumit, D. K., Chakrabarti, B., Prasanna, R., Pratap, D., Singh, S. D., Purakayastha, T. J. and Pathak, H. (2017b). Elevated carbon dioxide level along with phosphorus application and cyanobacterial inoculation enhances nitrogen fixation and uptake in cowpea crop. *Arc. Agron. Soil. Sci.* 63(13): 1927–1937.

Titumil, R. A. M. and Basak, J. K. (2010). *Agriculture and Food Security in South Asia: A Historical Analysis and a Long Run Perspective*. The Innovators, Dhaka.

Uprety, D. C., Das, R., Lutheria, D., Barade, P. V. and Dutt, B. (2007). Effect of elevated CO_2 and water stress on the seed quality in Brassica species. *Physiol. Mol. Biol. Plants* 13: 253–258.

Uprety, D. C., Dwivedi, N. and Mohan, R. (1997). Interactive effect of rising atmospheric carbondioxide and drought on the nutrient constituents of Brassica juncea seeds. *Sci. Cult.* 11–12: 291–292.

Uprety, D. C., Mishra, R. S. and Abrol, Y. P. (1995). Effect of elevated CO_2 on the photosynthesis, growth and water relation of Brassica species. *J. Agron. Crop Sci.* 175(4): 231–237.

Vanaja, M., Vagheera, P., Ratnakumar, P., Jyothi Lakshmi, N., Raghuram Reddy, P., Yadav, S. K., Maheswari, M. and Venkateswarlu, B. (2006). Evaluation of certain rainfed food and oil seed crops for their response to elevated CO_2 at vegetative stage. *Plant Soil Environ.* 52(4): 164–170.

Venkteshwara Rao, K. (1999). The combined effect of elevated CO_2 levels and temperature on growth characteristics of groundnut (*Arachis hypogaea* L.). *Indian J. Plant Physiol.* 4: 297–301.

Verma, R. R., Srivastava, T. K. and Singh, P. (2019). Climate change impacts on rainfall and temperature in sugarcanegrowing Upper Gangetic Plains of India. *Theor. Appl. Climatol.* 135(1–2): 279–292. DOI: 10.1007/s00704-018-2378-8.

Viswanath, K., Sinha, P., Kumar, S. N., Sharma, T., Saxena, S., Panjwani, S., Pathak, H. and Shukla, S. M. (2017). Simulation of leaf blast infection in tropical rice agro-ecology under climate change scenario. *Clim. Change* 142(1-2): 155-167.

Wang, D., Heckathorn, S. A., Wang, X.and Philpott, S. M. (2012). A meta-analysis of plant physiological and growth responses to temperature and elevated CO_2. *Oecologia* 169(1): 1-13.

Weerakoon, W. M. W. (2013). Impacts of climate change on food security in Sri Lanka. In: *Proceedings of the International Conference on Climate Change Impacts and Adaptation*, Coconut Research Institute of Sri Lanka, Lunuwila, Sri Lanka.

Yadav, S. B., Patel, H. R., Lunagaria, M. M., Parmar, P. K., Chaudhari, N. J., Karande, B. I. and Pandey, V. (2013). Impact assessment of projected climate change on pearl millet in Gujarat. In: *National Seminar on Climate Change Impacts on Water Resources Systems*. International Crops Research Institute for the Semi-arid Tropics India, pp. 33-38.

Yu, W. H., Alam, M., Hassan, A., Khan, A. S., Ruane, A. C., Rosenzweig, C., Major, D. C. and Thurlow, J. (2010). *Climate Change Risks and Food Security in Bangladesh*. Earthscan Publications, London, New York.

Zacharias, M., Naresh Kumar, S., Singh, S. D. and Aggarwal, P. K. (2014). Assessment of impacts of climate change on rice and wheat in the Indo-Gangetic plains. *J. Agrometeorol.* 16(1): 9-17.

Chapter 17

Modelling the impact of climate change on agriculture in East Asia

Bing Liu, Zi Ye, Yuan Cao, Bo Liu and Yan Zhu, Nanjing Agricultural University, China

1 Introduction
2 Modelling the impact of climate change on rice
3 Modelling the impact of climate change on wheat
4 Modelling the impact of climate change on maize
5 Conclusion and future trends
6 Where to look for further information
7 References

1 Introduction

The changing global climate, including rising temperatures, elevated CO_2, changing precipitation patterns and global dimming, has substantially altered several important aspects of the Earth's ecosystems. The climate environment, including temperature, precipitation, radiation and CO_2, can play significant roles in determining crop yield and product quality, which directly affect food security. Agriculture production, especially crop production, is highly vulnerable to climate change.

East Asia includes China, Japan, South Korea and North Korea. Food security in East Asia, which has the most dense and rising population, could be heavily impacted by climate change. Temperatures in East Asia have increased significantly over the past five decades and to a greater degree in high-latitude regions, typically higher than other global regions. Over the last few decades, annual precipitation, with large decadal-scale variability, have shown divergent trends among different regions in East Asia. In addition, the frequency and intensity of agricultural droughts have increased over the last 50 years, mostly in the main agricultural production region of East Asia (Liu et al., 2022). A recent IPCC report indicates that climate-related risks to agriculture and food security in Asia will progressively escalate as global warming reaches 1.5°C, and higher, above pre-industrial levels (IPCC, 2022). Therefore, assessing the impact of climate change on agriculture has always been a priority in order to

http://dx.doi.org/10.19103/AS.2022.0115.19

inform local and regional crop production strategies to adapt to future climate change in East Asia.

Due to its vast area, East Asia has been and remains the major production region for the most important global food crops such as rice, wheat, maize and soybean. Climate conditions during crop seasons exhibit large variations, as the climate in East Asia covers several different climate zones, e.g. an arid and semi-arid climate in west China, a hot-humid climate in south China and a temperate climate in north China, Japan and Korea. Therefore, the same crop type can be planted in totally different climate areas within East Asia, resulting in diverse and complex cropping systems. In addition, with relatively small-sized farms, crop production in Asia is the global centre for intensified crop production, especially in China. Considerable effort has focused on understanding the impact of climate change on these diverse and complex cropping systems over the last few years.

Process-based crop models facilitate understanding of the physiological growth process, taking into account interactions with genotype, soil, management and weather conditions (Lobell et al., 2009; Sumberg, 2012; van Ittersum et al., 2013) and are widely used to simulate the dynamics of cropping systems from the local up to global scales in climate change impact assessments (Chenu et al., 2017). When investigating responses of cropping systems to different climate change and management scenarios, crop modelling approaches can be applied at different spatial scales, including location, regional, country and global. This approach can help fill knowledge gaps, e.g. there are currently very few model-based impact studies on soybean in East Asia. This chapter briefly describes recent research involving climate change impact assessments for the production of primary food crops (e.g. rice, wheat, maize) in different major production regions across East Asia through the application of process-based crop models.

2 Modelling the impact of climate change on rice

Rice is the most important staple food crop for over half the world's population. In East Asia, including China, Japan and Korea, rice provides up to 50% of the dietary caloric supply, especially for millions living in poverty and is therefore critical for food security (Muthayya et al., 2014). However, climate change poses considerable challenges in meeting future rising rice production demands (Peng et al., 2004; Zhao et al., 2016b). Rice production is subject to different regional climate conditions (temperate regions, subtropical regions) across East Asia, while cultivar types (Japonica rice, Indica rice) and cropping systems (single-season rice, double-season rice) are also diverse in different production regions. These spatial variations could result in highly divergent effects of climate change on rice productivity.

2.1 Historical impact of climate change on rice

Dissecting the historical impact of climate change on rice yield could help to understand the responses of rice-cropping systems to differing climate variables. Climate conditions during the rice-growing season across the main rice-producing region of China have exhibited consistent and considerable warming trends (0.41°C per decade) from 1980 to 2010, with stronger warming trends in northern and eastern China (Fig. 1). Substantial decreases in radiation during the rice-growing season have been demonstrated for northern and eastern China, while some areas of northeastern China have received more radiation since 1980 (Fig. 1) (Zhang et al., 2014). However, there were no consistent trends for different subregions when considering precipitation during the rice-growing season.

Significant increases in observed rice yield in China have been reported over the last few decades, as a combined result of climate change, cultivar shifting and improved management. With regard to the spatial pattern, actual local yields of single rice were higher in central China and southern parts of northeast China than south China and southwest China, ranging from 5.2 to 7.4 t ha^{-1}. While the observed yields of the double-season rice were higher in the south compared with the north across the double rice-producing regions. Rice crop models (e.g. ORYZA2000, RiceGrow, CERES-Rice) have been widely applied to explore the impact of historical climate change on rice yield potential. On a regional scale, the simulations from CERES-Rice captured most of the annual variability of rice production from census data (with a root mean square error (RMSE) of 13.9%), and the model performed well in capturing

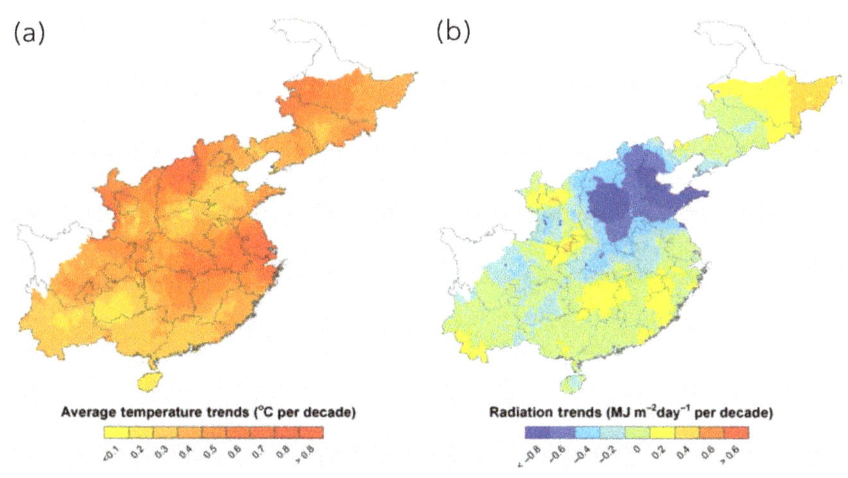

Figure 1 Trends of the average temperature (a) and radiation (b) during the rice-growing season from 1980–2010 in China. Source: Zhang et al. (2014).

the spatial patterns of reported yield variation (with RMSE less than 10%), particularly in the main rice cultivation areas, i.e. the eastern and southern parts of China; however, the model performed poorly in marginal and cold rice areas due to the availability and quality of input data and the inability of the model to simulate the impact of cold stresses, e.g. in northeast China (Yang et al., 2014).

Using a scenario analysis approach, several studies quantified the individual impact of climate change, cultivar shifting and improved management on historical rice yields in China. On a site scale, for single rice at four sites (Wuchang, Xinyang, Zhenjiang and Hanyuan) across the main producing region of China, the average temperature during the rice-growing season increased significantly from 1980 to 2009 (mostly due to increased minimum temperatures), while radiation and precipitation showed decreasing trends only in sections of the four sites. Climate change has resulted in reduced rice-growing duration and grain yield at all study sites, without cultivar changes (Liu et al., 2012). For double-season rice in China, significant warming from 1981 to 2009 mainly occurred before the jointing stage of early rice and after the jointing of later rice at three sites (Nanchang, Hengyang and Gaoyao), leading to reduced rice growth duration and grain yield potential both in early and late rice. The adoption of new rice cultivars could only partly mitigate the negative impact of warming on rice growth duration, biomass growth and grain yield (Liu et al., 2013a). On a regional scale, rice yield potential due to climate change has shown decreasing trends from 1980 to 2010 across all seven rice-producing subregions of China, except for the north-central section of the northeast subregion (Zhang et al., 2014). The northern and eastern subregions of China exhibited the greatest negative impact of climate change on rice yield, while the northeast subregion showed the least negative impact of climate change. The area-weighted national yield potential has decreased at a rate of 0.26 t ha^{-1} per decade over the last three decades.

Crop models have demonstrated the separate effects of climate change which usually have a negative impact on rice yield among different rice-cropping systems and differing subregions in China. When dissecting the impact of climate change into different climate variables using the CERES-Rice model, Yang found a significant decline of 11.5% for simulated national rice yield from 1961 to 2010 in China; the reasons for rice yield decline under climate change varied across different rice-producing subregions (Yang et al., 2014). In areas with a mean growing season temperature of 12–14°C and above 20°C, the decrease in rice yield was partly attributed to increased heat stresses and a shorter growth period under the warmer climate. In parts of southern China and the Yangtze River Basin with a rice-growing season temperature higher than 20°C, a decrease in the growing season radiation partly explained the rice yield decline, indicating the negative contribution of recent global dimming on rice yield in China's main rice-producing regions. However, in the northern rice-producing subregions with low growing season temperatures, decreases in the diurnal temperature range were identified as the main reason for the simulated rice yield decline (Yang et al., 2014).

When considering the CO_2 fertilization effects (from 340 ppm in 1981 to 388 ppm in 2010), a study using the EPCI and CERES models suggested that climate change has increased the national rice yield by 2% over the last three decades in China (Xiong et al., 2014). Similarly, from 1960–1990 to 2005–2009, the effects of increasing temperatures and CO_2 concentration estimated by the Agro-C model are positive (less than 1.0 t ha^{-1}) in most of the rice-producing subregions of China. Only in the Middle Yangtze Plain has the rice yield decreased slightly due to climate change. On average, climate change has increased the single-rice yield of China by 3.4%, whereas early- and late-rice yields have increased by 4.8% and 7.8%, respectively. The overall impact of climate change increased the national rice yield in China by 4.4% (Yu et al., 2012). However, even considering the CO_2 effects, climate change from 1961 to 2010 was estimated to reduce the national rice yield by 4.2% with the CERES model (Xiong et al., 2012).

In Japan, climate warming from 1964 to 2003 has advanced the rice heading stage by 2 weeks, resulting in a lower air temperature before heading and a higher air temperature and solar radiation after heading. The elevated CO_2 concentration (from 322 ppm to 367 ppm) was estimated to increase rice productivity by 2.5%, and climate change (including temperature and solar radiation) was estimated to affect rice yield by −6.0 to 3.1%, depending on the region (Kawatsu et al., 2007). This indicated that the positive effect of increased solar radiation on rice yield was usually offset by the negative effect of rising temperatures.

Frequent extreme heat events are a serious threat to rice production, especially during the reproductive stage. Based on the analysis of observed climate and phenology data from 228 stations in south China during 1981–2010, the spatial-temporal variation of post-heading heat stress was investigated by Shi et al. (2015). In general, post-heading heat stress increased but with large annual fluctuations from 1981 to 2010 in most study regions, and more significant increases were found in the east part of the early-rice subregion and west part of the southwest single-rice subregion. A phenological shift during 1981–2010 mitigated the increasing trends of heat stress in most areas but not in the west part of the southwest single-rice subregion. The rice grain yield was more sensitive to post-heading heat stress in the double-season early-rice region than that in the single-season rice region. The rice yield decreased by 1.5–9.7% due to post-heading heat stress, which was subregion dependent. Therefore, specific adaptation or mitigation strategies are necessary for different subregions to stabilize rice production under heat stress. In addition, without phenology shifting, global warming from 1961 to 2008 reduced cold stress during the seedling and heading-flowering stages of single rice in northeast China, the seedling stage of early rice and the heading-flowering stage of late rice in the double-rice regions. However, global warming did not increase heat stress during the heading-flowering stage, except in early rice in the mid-lower Yangtze River Valley (Sun and Huang, 2011). In addition, at four locations in northern Japan where the rice-growing season has relatively cool temperatures, the risk of potential rice yield loss due to cold stress at the sensitive booting stage has increased from 1961 to 2010, as a result of an earlier heading stage due to global warming (Shimono, 2011).

2.2 Projected impact of future climate change on rice

The impact of future climate change on rice production has been actively investigated in China and Japan under different warming scenarios. In China, climate warming tended to shorten the rice-growing season and mainly occurred during the vegetative stage rather than the reproductive stage (Lv et al., 2018). With the latest global warming scenarios from the Half a degree Additional warming, Projections, Prognosis and Impacts (HAPPI) project, Liu et al. (2020) assessed the impact of the two HAPPI scenarios (1.5°C and 2.0°C warming above pre-industrial levels) on rice production for the major rice-cropping systems of China, using the multi-model ensemble approach (three crop models and four climate models). Results show that increases in the mean rice-growing season temperature (GST) under 1.5°C and 2.0°C warming scenarios are highest in central China. Likewise, the increase in GST during the reproductive growth stage (GST-r) is greater than that during the vegetative growth stage (GST-v). The duration of the entire rice-growing season (GSD-w) is shortened by about 3–15 and 4–18 days under 1.5°C and 2.0°C warming scenarios, respectively. Most of the shortening in GSD-w occurs due to a decreased vegetative growth stage (GSD-v).

Projected changes in rice yield varied across the main producing region of China. Under 1.5°C and 2.0°C warming scenarios, projected yield changes of single-season rice and late-season rice were significantly negatively correlated with both GST-r and GSD-v, while yield changes of early-season rice were significantly negatively correlated with both GST-v and GSD-r. Without CO_2 fertilization effects, the impact of warming has a negative effect (11.6% and 14.5% under 1.5°C and 2.0°C warming scenarios, respectively) across the main rice-growing regions of China (Liu et al., 2020). The yield reduction is greater in the southern regions than in the northern regions, and the 2.0°C warming scenario would cause greater rice yield reductions than the 1.5°C warming scenario (Fig. 2). CO_2 fertilization will not offset the negative effects of climatic warming on rice production across the whole region, except for the north parts of the northeast region, where the rice GST is relatively low. Under

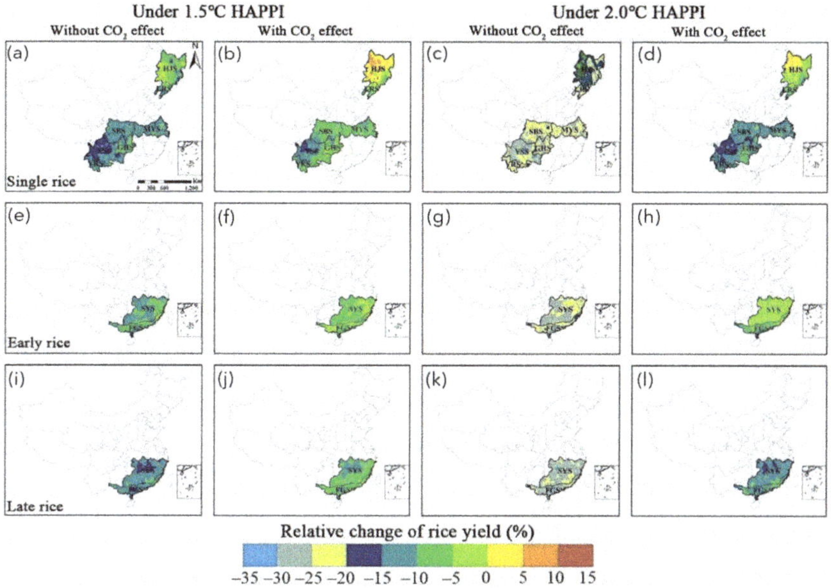

Figure 2 Projected rice yield changes for single rice (a, b, c, d), early rice (e, f, g, h) and late rice (i, j, k, l) in China under 1.5°C and 2.0°C HAPPI. The rice yield changes under 1.5°C and 2.0°C HAPPI were calculated as the means of the simulated rice yield changes of three crop models (CERES-Rice, ORYZA v3 and RiceGrow) and from four global climate models (GCMs), including CanAM4, CAM4, MIROC5 and NorESM1. The CO_2 concentration was 423 ppm and 487 ppm for 1.5°C and 2.0°C HAPPI, respectively. Source: Liu et al. (2020).

Representative Concentration Pathways (RCP) 4.5 and RCP8.5 scenarios from the Coupled Model Intercomparison Project Phase 5 (CMIP5), climate warming also tended to increase rice yield in northeast China, mostly due to elevated CO_2 and radiation effects (Zhang et al., 2019). The aggregated yield impact on the entire rice-producing region of China resulted in total rice production decreasing by approximately 3.6% and 6.2% with CO_2 fertilization effects and by 5.3% and 11.4% without CO_2 fertilization effects under 1.5°C and 2.0°C HAPPI scenarios, respectively.

In Japan, the projected increase in atmospheric CO_2 and global warming with the SIMRIW model under climate scenarios of Geophysical Fluid Dynamics Laboratory (GFDL), Goddard Institute for Space Studies (GISS) and Integrated Micrometeorology Model for Panicle and Canopy Temperature (UKMO) will substantially increase rice yields in northern Japan but reduce the yield in south-central and southwestern Japan (Horie, 2019). The positive effects of climate change in northern Japan were due to the relatively low base temperature, and the direct effect of elevated CO_2 exceeding the negative effects of shortened growth duration via the projected warming. The decrease in yield in south-central and southwestern Japan was due to a temperature rise not only shortening the duration of rice growth but also causing extensive spikelet sterility. Another recent assessment with H/H projected a similar spatial distribution of yield response to climate change using 6 global climate models (GCMs) and 3 RCPs in Japan, but with a slightly higher impact of CO_2 effects (Ishigooka et al., 2017). When aggregated to national level production, total rice production in Japan was estimated to increase slightly in most climate change scenarios.

In Korea, recent studies have assessed the potential impact of future climate change on rice yield. Choi et al. (2021) employed the CERES-Rice model to simulate the potential influences of climate change for the target years 2050, 2070 and 2100, under RCP4.5 and RCP8.5 scenarios, on rice yields in the Chonnam Province, southwest of the Republic of Korea. They found under the RCP4.5 scenario, rice yields were projected to decrease by 4.0% and 6.9% in 2050 and 2070, respectively, showing a 1.3% increase in 2100, and were projected to increase slightly under the RCP8.5 scenario, compared with baseline. Rice yields were projected to be higher in the coastal areas of Chonnam Province, but lower in the interior mountainous regions. Another simulation study using CERES-Rice suggested the rice yield at two locations in central Korea is expected to decrease by 13.0% (RCP4.5) and 16.1% (RCP8.5) in 2050 and 14.7% (RCP4.5) and 23.6% (RCP8.5) in 2080. The rice yield at two locations in south Korea is expected to decrease by 15.9% (RCP4.5) and 14.3% (RCP8.5) in 2050 and 17.5% (RCP4.5) and 17.1% (RCP8.5) in 2080. When adapting to the optimal growing season, the paddy rice yield was estimated to increase by approximately 8% in the RCP 4.5 scenario and 6–13% in RCP 8.5, compared with a fixed growing season. There is currently no study on modelling rice yields under climate change at the national level for Korea.

Climate change could affect the associated changes in extreme climate events during critical rice-growing periods. Using daily temperature and precipitation model outputs under three different scenarios of Shared Socioeconomic Pathways (SSP) (SSP1-2.6, SSP2-4.5 and SSP5-8.5) from the Coupled Model Intercomparison Project Phase 6 (CMIP6), Chou et al. (2021) characterized the changes of extreme climate events in the rice-growing regions of China. Under three warming scenarios, the risk of extreme high-temperature events (number of warm days, heat damage index), drought events (consecutive dry days) and extreme precipitation events (5-day maximum precipitation and the number of annual total extreme precipitation events) will increase in the future, while the risk of extreme low-temperature events (number of cold damage events) will decrease. The increase in extreme climate events under the high-emission scenario was more obvious, and there was quite a large spatial variability for the changes of extreme climate events. In particular, north and central China will experience a significant increase in the risk of extreme heat events. The risk of low-temperature events in northeast

China will be reduced. The risk of extreme precipitation events will increase, especially in southwest China and south China.

In Japan, compared with the 1971–2000 period, the projected future climate (2081–2100) could result in a 1.6–2.0°C increase in water temperature throughout the country. This means that the rice cultivation period will be prolonged by approximately 25–30 days, which could reduce the frequency of cold stress damage in the northern districts, but also increase heat stress in one-fifth of the current total cultivation area (Ohta and Kimura, 2007).

Global warming is predicted to have a significant impact not only on rice grain yield but also on rice grain quality. Focusing on the grain quality of rice at Kyushu in western Japan, projections of rice quality grade under A1B and A2 scenarios suggested that the percentage of first-grade rice would significantly decrease by 28% during the 2046–2065 period and 41% during the 2081–2100 period (Okada et al., 2011). These reductions were caused by the indirect effect induced by the shift of the grain-filling period rather than the direct effect of increased temperature. Recently, Masutomi et al. (2019) assessed the impact of global warming on the occurrence of chalky rice grain, using ten future climate scenarios and a statistical model for chalky rice grain. The results showed that global warming will double the occurrence of chalky rice grain by the 2040s under RCP8.5, mostly due to increasing temperature during the rice grain-filling period (Fig. 3). The decrease of rice grain grade under climate warming in Japan has also been confirmed by the projections from a process-based crop model H/H (Ishigooka et al., 2017). This indicated that one-third of paddy rice fields in Japan show a decreasing grain grade, indicating urgent need for the implementation of national and local adaptation strategies for rice production in Japan.

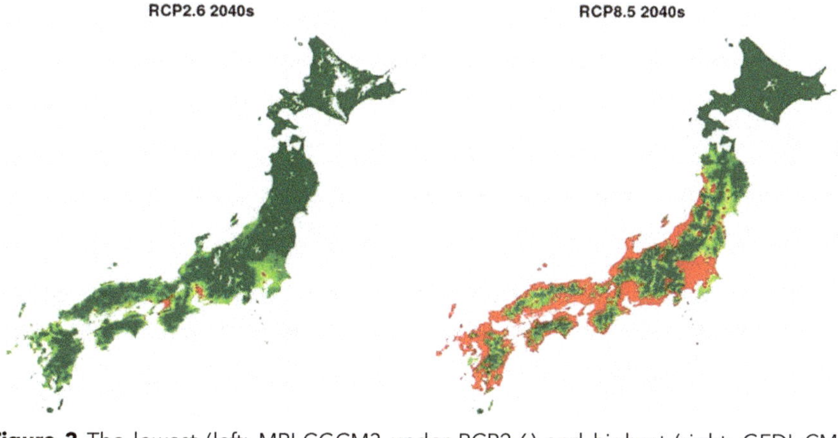

Figure 3 The lowest (left: MRI-CGCM3 under RCP2.6) and highest (right: GFDL-CM3 under RCP8.5) occurrence of chalky rice grain across Japan in the 2040s. Source: Masutomi et al. (2019).

3 Modelling the impact of climate change on wheat

China is the world's largest wheat producer, which accounts for 18% of the global wheat production (FAO, 2020). In China, more than 88% of the total wheat production is from winter wheat in the main wheat production region. There are four subregions, according to eco-climate conditions and geographical location, across the main wheat production region, including the north subregion (NS), the Huang-Huai subregion (HHS), the middle-lower reaches of Yangzi River subregion (MYS) and the southwest subregion (SWS). Modelling the impact of historical and future climate change on wheat production is essential to ensure a stable wheat supply and food security in China and East Asia. Wheat crop simulation models were among the first to show the importance of climate variability and extreme events on production (Semenov and Porter, 1995).

3.1 Historical impact of climate change on wheat

Over the last few decades, climate conditions during the wheat-growing season have changed significantly. An early study with 25 years of observed crop data from three locations (Nanyang, Zhengzhou and Luancheng) across the North China Plain (NCP, the most important wheat-producing area, accounting for more than 60% of the total wheat production in China) showed that climate warming mainly occurred during the vegetative growth stage of wheat, while there was a cooling trend or no significant change in temperature during the wheat reproductive stage (Liu et al., 2010). Similarly, with observations across the whole wheat-producing region of China during 1981–2009, significant climate change advanced the wheat anthesis date and maturity date in most of the stations, and as a result the growth period before anthesis and entire growth period were shortened, while the reproductive growth period was significantly prolonged despite the temperature increase (Tao et al., 2014).

Historical climate change has caused a significant impact on wheat yield across China. Xiao and Tao (2014) assessed the impact of changes in individual climate variables on wheat yield at four stations in the NCP using the APSIM-Wheat model, the results indicated that during the 1980–2009 period, rising temperatures increased winter wheat yield by 3.0–6.0% in the northern part of the NCP, whilst reducing rain-fed winter wheat yield by 9.0–12.0% in the southern part of the NCP. Decreased solar radiation reduced the wheat yield by 3.0–12.0% across the four stations. The impact of precipitation change on winter wheat yield was slight, as there were no significant changes in precipitation. When investigating the impact on wheat production on a region scale, the statistical modelling indicated that changes in temperature, precipitation and solar radiation over the past three decades jointly increased wheat yield in northern China by 0.9–12.9%; however, it reduced wheat yield in southern

China by 1.2–10.2% with large spatial variations (Tao et al., 2014). At the national production level, when using a statistical function and EPIC model to dissect the contributions of different factors to China's national wheat production growth from 1981 to 2015, climate change exerted a negative impact in both approaches (-2.2% and −1.2%), while fertiliser (57.4% and 79.8%), cultivar (37.9% and 16.1%) and irrigation (7.9% and 8.0%) contributed the most to yield growth (Liu et al., 2021c).

Different aspects of climate change, such as higher atmospheric CO_2 concentration, increasing temperature and altered rainfall (reduced or increased) have different effects on wheat. Among these changes, a temperature increase has the most likely negative impact on wheat yields (Porter et al., 1999; Ottman et al., 2012), and regional temperature changes can be projected from climate models with more certainty than precipitation. It is thus necessary to quantify the impact of increasing temperature on wheat yields, including its spatial variations. Liu et al. (2016) recently showed that grid-based and point-based simulations and statistical regressions (from historic yield records), without deliberate adaptation or CO_2 fertilization effects, produce similar estimates of temperature impact on wheat yields at global and national scales, with a 1°C global increase, the average wheat yield is projected to decline by 3.0% in China (Fig. 4a); similar observations were found by Zhao et al. (2017). There were also very large spatial variations in yield sensitivity to increasing temperature ($S_{Y,T}$, yield % change per °C increase in T) for the different wheat-growing regions. Namely, the simulations from local crop models indicated large negative $S_{Y,T}$ values in northwest China, north China and southeast China (−6.3, −4.6 and −9.8% per °C, respectively), but small negative values in northeast China (−0.4% per °C), and differed from results using statistical models and warming experiments (Fig. 4b) (Zhao et al., 2016a). In southeast China, in particular, local process-based crop models give obvious negative $S_{Y,T}$ values, while the warming experiments suggest a large positive effect of temperature on yield. This inconsistency was mostly due to the considerable differences in the three impact assessment approaches, e.g. different approaches have different management and presumably differing definitions of growing season, making a rigorous comparison difficult. In addition, simulations from crop models usually average results over several years or decades, while field warming experiments only reflect the response of yield to temperature over a few years. Thus, background climate probably has more effect on estimates of $S_{Y,T}$ from warming experiments than from models (Zhao et al., 2016a).

Historical climate change also affected the occurrence of extreme climate events during wheat-growing seasons in China. With several heat stress indices based on measured maximum temperature records from 1960 to 2009 at the 166 stations in the main wheat-growing region of China, the post-heading heat stress level was more severe in the cooler northern wheat-growing regions

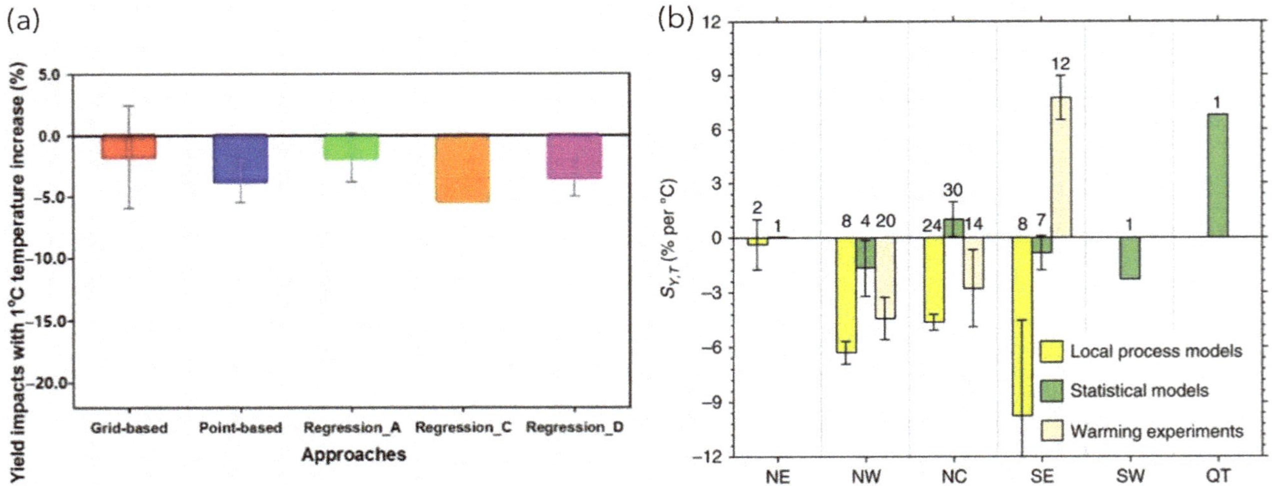

Figure 4 (a) Estimated impact of 1°C global increase on wheat yield in China with different modelling approaches, including grid-based simulations, point-based simulations and three different statistical regressions. Source: Liu et al. (2016). (b) Regional differences of temperature sensitivity of wheat yields ($S_{Y,T}$) from process-based crop models, statistical models and warming experiments in China. NE, NW, NC, SE, SW, QT NE, NW, NC, SE, SW, QT represent northeast China, northwest China, north China, southeast China, southwest China and Qinghai-Tibet, respectively. Source: Zhao et al. (2016a).

than the warmer southern regions of China. Heat stress between heading and maturity has increased over the last few decades in most of the main winter wheat production areas of China, but the increasing rate is higher in the south than in the north (Fig. 5). Heat stress or average temperature changes explained approximately 29% of the observed spatial and temporal yield variability (Liu et al., 2014). Furthermore, the estimated negative effects of heat stress on wheat yields from the WheatGrow model in the NSs were larger than in the southern subregions. The highest negative effects were observed in the border area of north Hebei, south Shaanxi and northwestern Henan, where the average and maximum potential yield reductions due to heat stress were 2.9% (236 $kg \bullet ha^{-1}$) and 29% (1899 $kg \bullet ha^{-1}$) during the baseline period (Liu et al., 2021a).

3.2 Projected impact of future climate change on wheat

Assessing the potential future warming impact on regional crop production can help to address agricultural adaptation more effectively. Projected changes in wheat phenology, grain yield and water use have been assessed under different climate change scenarios in China. The projected wheat anthesis date under future warming scenarios usually shows an obvious advance across the whole wheat-producing region, but with a more homogeneous spatial pattern, while the wheat grain-filling duration is relatively stable, compared with the baseline period (Lv et al., 2013; Ye et al., 2020). The advancement of the anthesis date under A2 (a high-greenhouse-gas-emission scenario), A1 (a low-emission scenario) and B1 (a medium-emission scenario) was about 3-6 days, 6–9 days and more than 9 days for 2030s, 2050s and 2070s, respectively (Lv et al., 2013). Recent projections with an ensemble of four wheat models for the HAPPI scenarios showed that the wheat-growing season duration was shortened by 6–15 days for the 1.5°C warming scenario and 8–18 days for the 2.0°C warming scenario across the entire primary wheat-producing region of China. Greater reductions in the duration of wheat growth, mostly due to the reduced vegetative period, were projected for southern subregions compared with NSs (Ye et al., 2020).

Most projections indicate that wheat yield in China under future climate scenarios would further increase when considering the warming impact and CO_2 fertilization effects, especially for regions with low GSTs. However, the magnitude of yield enhancement differed among warming scenarios and crop models. Compared with the 1961–1990 period, projected wheat yields under two greenhouse gas emission scenarios (A1FI and B1), with (and without) CO_2 fertilization effects, in the NCP would increase on average by up to 37.7% (18.6%), 67.8% (23.1%) and 87.2% (34.4%) during 2020s, 2050s and 2080s, respectively (Tao and Zhang, 2013). Most of the projected yield enhancement could be attributed to increased photosynthesis due to climate change and

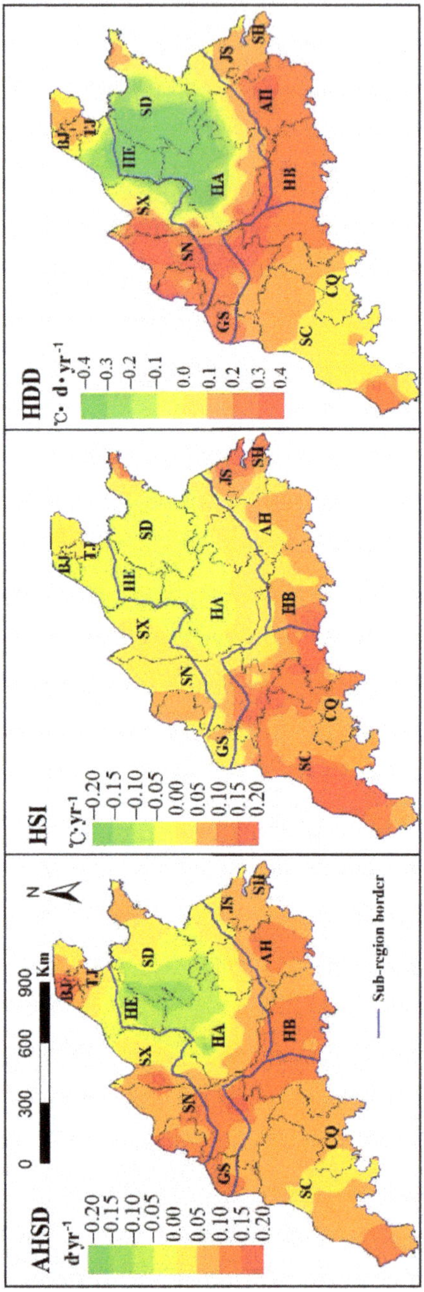

Figure 5 Spatial distribution for the trends of heat stress indices between heading and maturity from 1960 to 2009 in the winter wheat-growing region of China. AHSD, accumulated heat stress days with T_{max} >30°C between heading and maturity; HSI, heat stress intensity, defined as the average T_{max} for heat stress days between heading and maturity; HDD, heat degree-days between heading and maturity.

elevated CO_2, and avoidance of high-temperature stress with an accelerated maturity date. For the entire winter wheat-producing region, without the CO_2 fertilization effect, global warming of 1.5°C and 2.0°C scenarios are projected to increase grain yield by less than 5% in most of NS and HHS, but decrease grain yield by 0–28% in southern subregions (MYS and SWS, with a large spatial variability) (Fig. 6). The average changes of wheat potential yield with the CO_2 fertilization effect in the NS, HHS, MYS and SWS were 4.5% (2.2% to 6.8%), 3.8% (1.6% to 6.0%) , 0.3% (-1.5% to 1.6%) and −7.2% (-14.8% to 2.1%) under the 1.5°C scenario, and 10.7% (7.5% to 13.9%), 9.6% (5.1% to 12.6%), 4.8% (2.5% to 6.7%) and −3.6% (-11.7% to 8.2%) under the 2.0°C scenario, respectively. Generally, the negative impact of 1.5°C and 2.0°C global warming would be fully cancelled out by the positive effects of elevated CO_2 for regions with a GST greater than 11°C. For wheat production, 1.5°C and 2.0°C warming scenarios (with CO_2 fertilization effects) projected increased total winter wheat production in China by 2.8% and 8.3%, with the greatest increase in the NSs, which is the largest wheat-planting area (Ye et al., 2020).

The projected climate change could also alter evapotranspiration and water utilization efficiency in wheat. Under rain-fed conditions, the projected evapotranspiration using the WheatGrow model in the main wheat-producing region of China under A2, A1 and B1 scenarios (2030s, 2050s and 2070s, without CO_2 effects) decreased in the northern regions, but increased in the southern regions. The projected evapotranspiration under irrigated conditions would increase across the entire wheat-producing region, thus the utilization efficiency of irrigation water would increase in most of the region, except for western Shandong, southern Sichuan, northern Henan, Shanxi and Shaanxi (Lv et al., 2013). Similarly, projected changes in wheat evapotranspiration under insufficient irrigation in the NCP generally ranged from −6% to 6%, −10% to 8% and −17% to 4% with CO_2 effects, and from −0.6% to 10%, −1.0% to 17% and 7% to 12% without CO_2 effects, for 2020s, 2050s and 2080s, respectively (Tao and Zhang, 2013).

The alteration in temperature due to global climate change has a profound impact on crop yields, and this includes the impact from the rising mean GST and heat stress events. Climate warming is also projected to increase the frequency and extent of heat stress events in the main wheat-producing region of China, especially in NSs (Liu et al., 2021a). However, higher warming levels (2050s under RCP8.5) did not result in greater yield-impacting heat stress compared with low warming (1.5°C and 2.0°C warming scenarios), due to an advanced phenology with mean warming and avoiding heat stress events during grain filling in summer. The overall impact of temperature change (including changes of mean temperature and heat stress) on wheat yield was further quantified, from historical periods and warming scenarios (Liu et al., 2021a). When comparing the impact of increasing mean temperature and heat

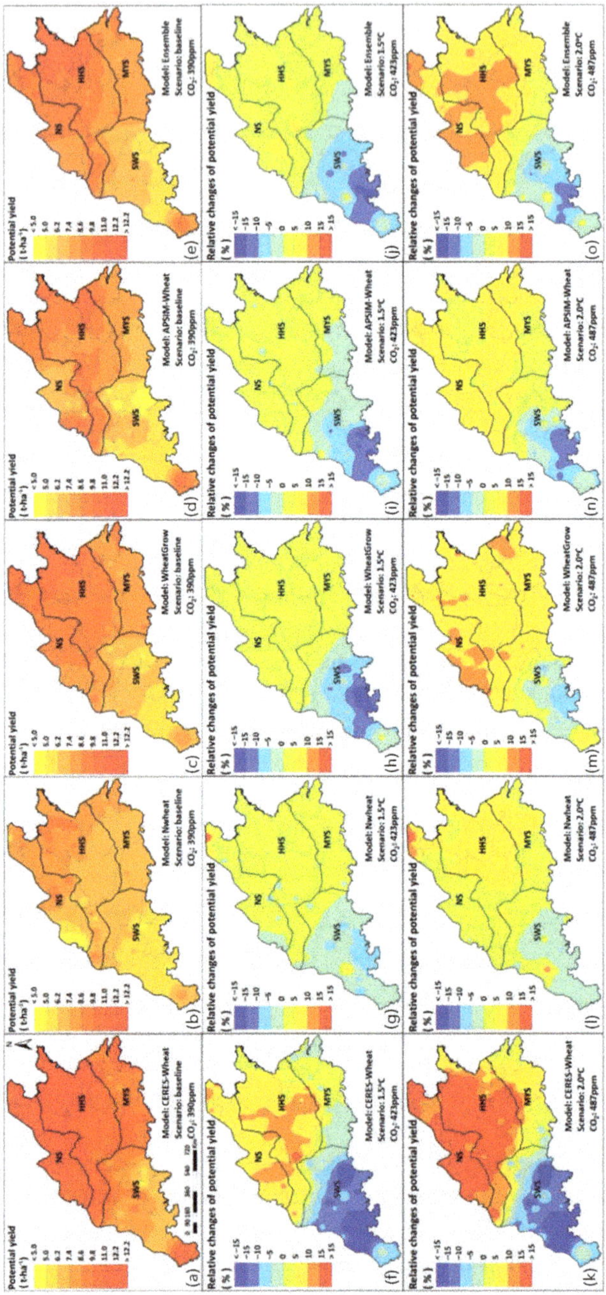

Figure 6 Spatial distribution of simulated potential yield under baseline (a–e) and relative changes of potential yield under 1.5°C (f–j) and 2.0°C (k–o) scenarios for CERES-Wheat (a, f, k), Nwheat (b, g, l), WheatGrow (c, h, m), APSIM-Wheat (d, i, n) and the ensemble (e, j, o) mean value of four models with CO_2 fertilization effects. The potential yield changes for 1.5°C and 2.0°C scenarios were the mean value of simulated potential yield changes from four global climate models (GCMs), including CanAM4, CAM4, MIROC5 and NorESM1. The CO_2 concentration was 390 ppm, 423 ppm and 487 ppm for baseline, 1.5°C and 2.0°C scenarios, respectively. Source: Ye et al. (2020).

stress during the growing season, most negative effects of climate-warming scenarios (3–5% yield reductions) are attributed to increasing mean GSTs, while changes in heat stress are projected to reduce wheat yields by an additional 0.7% to 1% in NSs (Fig. 7).

4 Modelling the impact of climate change on maize

China is the world's second-largest producer and consumer of maize, and the fluctuations in China's maize production have significantly affected food security and maize consumption on a global scale. The main maize-planting area in China, known as China's Maize Belt, is located in the northeast region (NER) for spring maize and the Huang-Huai-Hai region (HR) and the southwest region (SMR) for summer maize. Assessing the impact of climate change using crop modelling has provided important insight for adapting maize production in China.

4.1 Historical impact of climate change on maize

Due to climate change over the last few decades, the available thermal and radiation resources during the maize-growing season increased significantly, while decreasing the available water resources in the three major maize-producing regions of China (Zhao et al., 2015). Most maize modelling studies in China have focused on NER, considered to be one of the three 'Golden Maize Belts' across the globe. NER has experienced a warming trend in average temperature of 0.38°C per decade, decreased annual precipitation by 9 mm per decade and sunshine duration has decreased by 42 h per decade over the past 50 years (Liu et al., 2013b, 2009; Zhao et al., 2015). In addition, the chilling injury was the main extreme agrometeorological event during the growth stages of maize before the 2000s, while drought became the main extreme agrometeorological event after the 2000s (Zhang et al., 2020).

Over the past three decades, maize yields in NER have increased under changes in climate, cultivar selection and crop management. Using the APSIM-Maize model to separate the effects of changes in climate, sowing dates and cultivar types at six sites in NER, Liu et al. (2013b) indicated that climate change trends alone would have led to a negative impact on maize, with the maize yield decreasing by 4–22% for each 1°C increase in growing season mean temperature. In addition, the earlier sowing dates increased yields by up to 4%, and adoption of longer season cultivars caused a substantial increase in yield ranging from 13% to 38% over the past three decades (Liu et al., 2013b). In addition, a combination of new cultivars introduced with higher thermal time requirements and a longer season (Liu et al., 2013b; Su et al., 2021), higher fertiliser applications (Gao et al., 2010), more pest control and other management improvements in NER have overcome the negative effects of

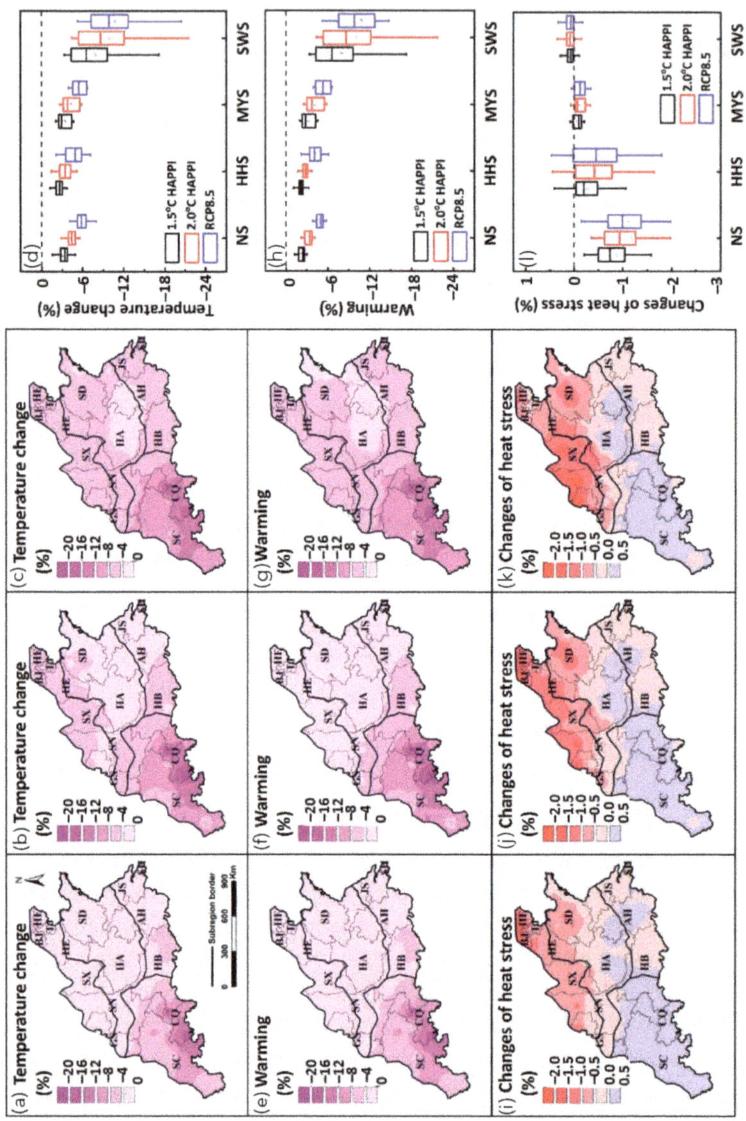

Figure 7 Projected relative yield impact of temperature change (including changes of mean temperature and heat stress) (a–d), warming (including an increasing mean temperature during the growing season) (e–h) and changes of heat stress on grain yield averaged over the 30-year period (i–l) under 1.5°C HAPPI (a, e, i), 2.0°C HAPPI (b, f, j) and RCP8.5 (c, g, k) scenarios across the main winter wheat-producing region of China. NS, northern winter wheat region; HHS, Huanghuai winter wheat region; MYS, middle- and low-Yangtze Valleys winter wheat region; SWS, southwest winter wheat region. Source: Liu et al. (2021a).

climate change, resulting in a significant increase in actual maize yield. Zhao et al. (2016c) analysed the variations in potential climatic suitability zones of spring maize and the possible effects of these variations on maize production potential during 1961-2010 in NER. The results show that historical climate change has increased growing degree-days during the growing season for spring maize universally and decreased accumulated sunshine duration during the growing season over the last five decades. Under the background of climate change, the potential climatic suitability for spring maize in this region was reduced during the most recent five decades, and both the yield potential and yield stability declined (Fig. 8). In particular, most of the highly suitable zones in Jilin and Liaoning province during the 1961-1980 period turned into a suitable or moderately suitable zone during the 1981-2010 period. Meanwhile, the total area of a marginally suitable zone and not suitable zone increased by 16% during the 1981-2010 period, compared with 1961-1980.

The HR has a typical monsoon climate, with large variations in annual precipitation. Droughts and floods occur frequently in this region, which has affected the stability of local maize production. In recent decades, the agroclimatic variables in the HR have changed significantly with global warming (Zhao and Yang, 2018a). During the growing season of summer maize, the regional accumulated temperature showed an overall increasing trend. This increase provided an opportunity to switch from early-to-middle-maturing varieties to late-maturing varieties, which could be translated into an increase in yield potential of summer maize (Ma et al., 2010). Nevertheless, heat and drought have also increased in the HR over the past few decades (Wang et al., 2015). Within the growing season, summer maize crops have different

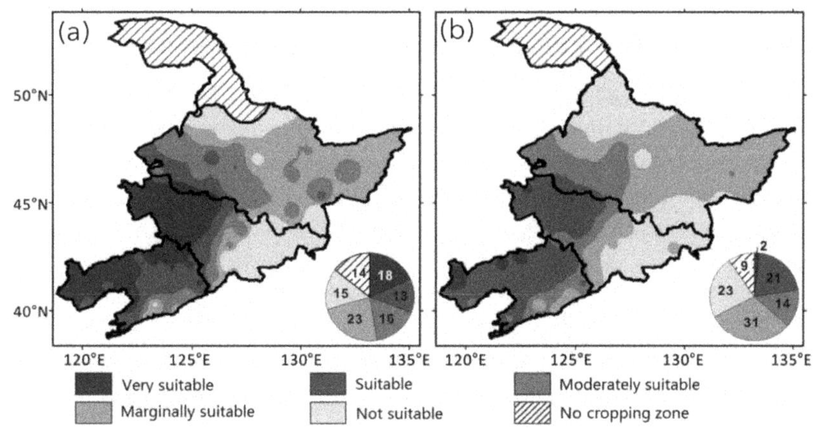

Figure 8 Spatial distribution of different potential climatic suitability zones for spring maize during the period of 1961-1980 (a) and 1981-2010 (b) in northeast China. The two pie charts show the percentages of different potential climatic suitability zones and no cropping zone in the entire NER. Source: Zhao et al. (2016c).

responses to changes in climatic elements. By combining crop modelling and historical yield data analysis, Liu et al. (2010) separated the impact of climate and crop varietal changes on wheat and maize yield in the HR and showed that cultivar selection was able to compensate for the negative impact of climate change, leading to an increased crop yield since the 1980s. The increasing temperatures before the over-wintering stage of winter wheat enabled a later harvesting of maize and a delayed sowing of winter wheat. Therefore, simulations in the NCP by Wang et al. (2012) explored an adaptation strategy in response to recent warming termed 'Double-Delay' technology, i.e. the delay of both the sowing of wheat and the harvesting of maize, which led to a 4–6% increase in total grain yield of wheat-maize double cropping systems.

For the SMR, Zhao and Yang (2018b) showed that yield stability was less negatively affected by changes in rain in the SMR (8.2%) than in NCR and HR (38.9% and 34.3%, respectively) during 1981–2010, but the yield level was reduced by rain in 19.1% of areas in the SMR. Liu et al. (2021b) recently reported that maize grain yield could be raised by 10.9–20.2% by increasing plant density by matching the solar radiation across the main maize-planting region of China. In addition, the yield gap could be reduced by 10.4–33.7% in these regions. Therefore, dense cultivation by matching maize growth and solar radiation is an effective approach to reduce yield gaps among the three regions of China.

At the national level, climate change including (temperature, precipitation and radiation) from 1961 to 2010 was estimated to reduce maize production by 1.2% but not significantly, while the CO_2 fertilization effects (from 318 ppm to 388 ppm) significantly increased maize production by 3.2% (Xiong et al., 2012).

4.2 Projected impact of future climate change on maize

Maize production will also be impacted by climate change in the future, including effects on maize phenology, yield and water use efficiency (WUE). The duration of maize growth will decrease under the warming scenarios across the main maize-planting region in China (Liu et al., 2021d; Zhang et al., 2021). Compared with the baseline period, the probability of advanced maturity for maize under three climate change scenarios (RCP2.6, RCP4.5, RCP8.5) was 62.96–64.50%, and the probability of reduced WUE was 35.86–37.93% using the CERES-Maize model (Liu et al., 2021d).

Mostly negative effects of climate change were projected for maize production in China. Lin et al. (2017) using CERES-Maize at several sites in northern China showed that the average maize yield would decrease by 2.1% in the 2020s, 12.9% in the 2050s and 22.7% in the 2080s under RCP4.5. Moreover, they projected a greater reduction of 6.3%, 18.4% and 47.5% in the respective periods under RCP8.5 without considering the effects of CO_2 fertilization. If planting days were advanced or delayed by 15 days in this region, the maize yields would decrease by 6.9% and increase by 15.9%, respectively (Lin et al., 2017). On a regional scale, using a novel model which coupled CERES-Maize with machine learning, Zhang et al. (2021) indicated that climate change would have a mostly negative impact on maize productivity (Fig. 9), and the magnitude of yield reduction would be highly dependent on the duration of cultivar growth. Without adaptation, national production would suffer a 7.6% decrease under RCP4.5 and RCP8.5. Similarly, when the CERES-Maize model was used for ensemble climate projections of 2011–2040 from five GCMs under three emission scenarios (RCP2.6, RCP4.5, RCP8.5), the maize yield reduction probability under three climate change scenarios was 40.44–41.41%. Spatially, the median value of yield reduction in HR was 10.77–11.68%, but an increasing yield was projected in the northwest inland maize region with a range of 40.84–48.49% (Liu et al., 2021d).

Other studies have also considered the impact of extreme climate events in future projections. Using the MCWLA model, the effects of 1.5°C and 2.0°C warming scenarios on maize growth and productivity in the maize-producing region of China included a decreased crop growth duration and increased impact of extreme events (e.g. heat stress and drought), which would have a major negative impact for maize across the entire cultivation area, except for some parts of NER and northwest China, even in the moderate climate-warming scenarios with CO_2 fertilization effects. On a country scale, the maize production projected via the ensemble median approach in China could increase by 0.2% and decrease by 1.7% under 1.5°C and 2.0°C scenarios, respectively (Chen et al., 2018). At a higher warming level, Huang et al. (2021) investigated the impact of climate change on maize yield across China's Maize Belt using the CERES-Maize model driven by the latest future climate scenario of SSP5-8.5 from CMIP6 during the 2050s and 2080s. The simulated results indicated that the shortened growth period resulting from rising temperatures, decreased photosynthesis caused by heat stress and a decreased grain-filling rate due to heat stress were the main reasons causing future maize yield decreases in all subregions of China's Maize Belt. Specifically, the potential to adapt to rising temperatures is generally higher than the potential to adapt to heat stress in all subregions, except the NCP where the potential to adapt to heat stress is higher than the potential to adapt to rising temperature.

5 Conclusion and future trends

The impact of historical and future climate changes on major food crop production in East Asia has been widely studied, from the effects on crop phenology, water use, biomass production to impacts on yield and regional production. The most obvious impact is that climate warming has accelerated

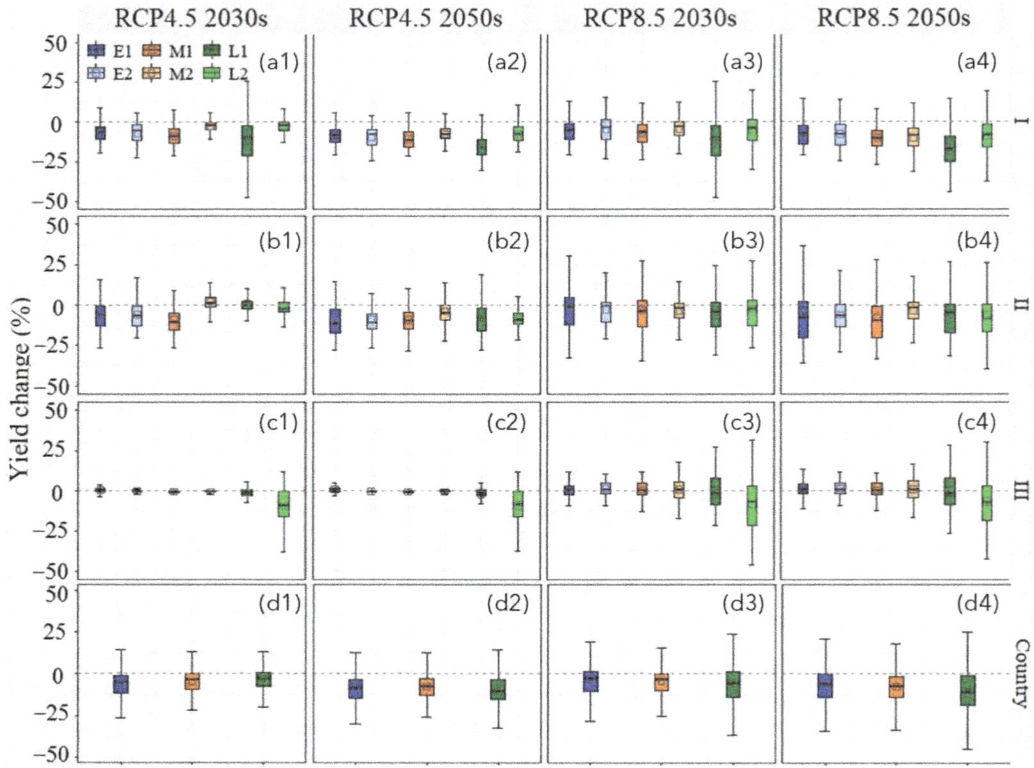

Figure 9 Projected hybrid-specific yield changes for the three agroecological maize-producing regions and the whole country in the 2030s (2021–2040) and 2050s (2041–2060) under RCPs 4.5 and 8.5 relative to the baseline period (1986–2005). The letters *a*, *b* and *c* indicate the north China spring maize region, the Huang-Huai-Hai summer maize region and southwest mountain maize region. E, M and L represent early-maturing hybrids, medium-maturing hybrids and late-maturing hybrids. Source: Zhang et al. (2021).

crop development and shortened crop growth duration. Generally, the negative impact of historical climate change since the 1980s on major crop yields has been quantified in most of the main crop-producing regions, including the NCP, south China, northwest China and Japan. However, in some temperate regions, e.g. northeast China, climate warming over the last few decades has benefited rice production (Table 1). As for the projected impacts under future climate change, a larger negative impact on yield are projected especially under climate change scenarios without the effects of CO_2 fertilization. In the cool regions of north China, wheat production could potentially show increased yields with CO_2 effects, and rice production in East Asia will be more negatively impacted than wheat and maize.

The uncertainty of climate change impact on crop production, especially for future projections, is of increasing concern, and recent efforts have tried to reduce uncertainties via future climate projections, crop model parameterizations and model structures. A multi-model ensemble, including several Agricultural Model Intercomparison and Improvement Project (AgMIP) studies in other regions, has been used as a robust approach to increase our confidence in quantifying the impact of climate change in East Asia. Meanwhile, due to the complexity of both cropping systems and climate change, considerable knowledge gaps remain in the assessment of climate change impact on crop production.

5.1 Modelling extreme climate impact on crop production

Climate change not only affects the mean state of climate variables but could also alter climate variability. As a result, increasing extreme climate events due to shifting climate variability (IPCC, 2012), such as heat stress, drought stress and extreme precipitation events, are projected to have an increasingly adverse impact on crop production (Deryng et al., 2014; Semenov and Shewry, 2011; Teixeira et al., 2013; Trnka et al., 2014), even under moderate climate warming (e.g. 1.5°C warming above the pre-industrial period) (Tebaldi and Lobell, 2018). Extreme climate events usually cause a spike in food prices and have a profound negative impact on local food security. For example, with reported records of extreme weather disasters, Lesk et al. (2016) showed that droughts and extreme heat significantly reduced national cereal production by 9–10% during the 1964–2007 period. Therefore, investigating the impact of climate warming on crop production, including changes in mean climate state and extreme climate events, is critical to understand climate effects and to adapt cropping systems to climate change, which is crucial to ensure regional and global food security.

Most of the current impact assessments have focused on the effects of changes in mean climate state, with less attention on the impact of extreme

Table 1 Summary of the estimated impact of historical and projected future climate change on regional crop yields in East Asia

		Historical impact (%)			Projected impact (%)		
Country	Region	Rice	Wheat	Maize	Rice	Wheat	Maize
China	Northeast China	−4.9 to −5.3	na	−5.0 to −18.0	−2.0 to −21.6	1.6 to 12.4	−6.4 to −10.0
	North China	na	0.9 to 12.9	−11.8 to 22.6	na	2.2 to 13.9	−2.9 to −11.7
	Central China	−6.6 to −27.0	na	na	−5.3 to −23.6	−1.4 to 12.6	na
	South China	−5.8 to −10.16	−1.2 to −10.2	na	−0.4 to −26.6	na	na
	Southwest China	−3.8 to −10.9	na	−25.8 to −13.9	−8.8 to −26.3	−18.1 to 8.2	< −2.0
	Whole country	−11.5 to 4.4	−1.2 to −3.0	−1.2 to 3.2	−3.6 to −5.3	−15 to 14	−4.7 to −8.5
Japan	Whole country	−6.0 to 3.1	na	na	10 to 24	na	na
South Korea	Whole country	na	na	na	−4.0 to −23.6	na	na

na, no regional yield impact assessments based on a crop modelling approach.

climate events on crop production; however, adaptation to these two effects could be substantially different. Although some recent studies have quantified the impact of extreme climate events, systematically isolating the individual effects of extreme climate events from the compounding effects of climate change on crop production is yet to be resolved, and the overall contribution of these two factors to the effects of climate warming on crop yield is unknown.

5.2 Modelling climate change impact on crop disease, weeds and pests

Crop production can be severely affected by diseases, weeds and pests. The effects quantified by recent studies usually only considered the direct impact of climate change. However, climate change will also affect the occurrence of crop diseases, weeds and pests. The biggest concern is that crop pathogens, weeds and pests may change their geographical distribution and severity due to changes in rainfall, increasing temperatures and CO_2 concentrations. For example, while crop yields may increase at high latitudes due to climate change, these benefits could be severely impacted by shifts in disease risk, due to the effects of climate change on pathogen infection (Chaloner et al., 2021). Increases in climate variability have already exacerbated maize yield losses due to weeds, and warmer and drier summers in the US maize-growing regions coupled with the growing prevalence of herbicide resistance, could increase the risk of future maize yield loss due to weeds without changes in weed management (Landau et al., 2021). Modelling the indirect effects of climate change due to diseases, weeds and pests on crop production could provide more information for future adaptation through crop management transformation.

5.3 Modelling climate change impact on crop quality

The impact of climate change on food security is not only limited to effects on crop yield but also the nutritional quality of the crop. Both climate warming and elevated CO_2 could affect crop quality. Increasing temperature and heat stress events could change the crop carbon and nitrogen dynamics, and could also interact with the dilution effects of elevated CO_2. The models used in the AgMIP-Wheat show that the ensemble of current crop wheat models could simulate wheat protein quality parameters under various climate change observations (Asseng et al., 2019). With simulations from 60 global wheat-producing locations, the potential benefits of elevated CO_2 by 2050 on global wheat grain and protein yield are likely to be counteracted by the impact of rising temperature and changes in rainfall, and the overall effects of adaptation strategies on protein concentration were negative (Asseng et al., 2019). The correlation analysis using observed data in China indicated that the effects of climate variables on wheat grain protein concentration were spatially heterogeneous (Zhou et al., 2021). However, in East Asia, the effects of climate change on crop nutritional quality, particularly grain protein concentration, on a large geographical scale have rarely been assessed.

On the other hand, the effects of climate change on crop micro-nutritional quality (e.g. mineral elements, amino acids), which have been observed in Free Air CO_2 Enrichment (FACE) or other environment-controlled experiments (Guo et al., 2021), are still missing in climate change impact assessments. The possible contrary effects of increased temperature and elevated CO_2 on nutrient concentration and bioavailability in rice and wheat grains reported by Guo et al. (2021) indicated an urgent demand for a comprehensive assessment under different climate change scenarios. Although some crop models have tried to simulate the micro-nutritional elements in the grain (Liu et al., 2019), the effects of climate change, especially under extreme climate events or elevated CO_2, on crop micro-nutritional elements on a regional scale remain to be determined. Understanding the potential impact of climate change on the nutritional quality of major food crops could have profound implications for developing regional or national nutritional policies.

5.4 Modelling climate change impact for diverse and intensified cropping systems

One of the unique characteristics of the cropping systems in East Asia is the intensification and diversification, due to smallholder farms and intrinsic pressure upon food supply. Intensified crop production, defined as producing more food from existing farmland, has been an important strategy in modern agricultural development (Garnett et al., 2013). In China, to meet domestic food demand for approximately 20% of the global population with only 8% of world's arable land (FAO, 2020), there is great interest in adopting intensified crop production approaches, e.g. cultivating crops on every possible area of farmland and in all potential growing seasons. For example, wheat–maize rotation systems in the NCP, wheat–rice rotation systems in south and central China raise considerable concern regarding the sustainability of water resources and soil characteristics. In the rotation systems, the impact of pre-crops on soil water and nutrient properties should not be ignored (Seidel et al., 2019). The intensified crop production system could also interact with climate variables, which may affect its climate resilience. For example, Mueller et al. (2016) suggest that cropland intensification could increase the evapotranspiration potential, leading to cooler crop-GSTs and thus alter the responses of maize-cropping systems to extreme climate events. Testing the existing crop models under different intensified cropping systems is important to model the impact of climate change on the productivity of intensified systems.

The diverse cropping system, including the combination of different crops through the calendar year and intercropping of different crops, was mostly in small-sized farms. The resilience and adaptability of cropping systems to climate change are of growing concern and are critical for designing future cropping systems. Crop diversification (e.g. intercropping or mixing different cultivars) may stabilize the productivity of cropping systems and reduce any negative environmental impact (Malézieux et al., 2009), but a shared understanding of the differences in climate change on intercropping and conventional monoculture systems across the regional level is lacking. To be more representative of actual production scenarios, quantifying the impact of climate change on these diverse systems will provide scientific guidance for local farmers and agricultural policy-makers.

6 Where to look for further information

The recently published reports from the Intergovernmental Panel on Climate Change (IPCC) have summarized potential climate change impacts on agriculture across the globe, and these reports can help us to have a general understanding of climate impact on regional agriculture. Two chapters from the Working Group II Contribution to the IPCC AR6 reports should be reviewed for further information (https://www.ipcc.ch/report/ar6/wg2/):

- Chapter 5: Food, Fibre and Other Ecosystem Products.
- Chapter 10: Asia.

The Agricultural Model Intercomparison and Improvement Project (AgMIP) have conducted several scientific activities for assessing impacts of climate variability and change and other driving forces on agriculture, food security, and poverty at local to global scales. Latest research from AgMIP can be found at: https://agmip.org/.

7 References

Asseng, S., Martre, P., Maiorano, A., Rötter, R. P., O'Leary, G. J., Fitzgerald, G. J., Girousse, C., Motzo, R., Giunta, F., Babar, M. A., Reynolds, M. P., Kheir, A. M. S., Thorburn, P. J., Waha, K., Ruane, A. C., Aggarwal, P. K., Ahmed, M., Balkovič, J., Basso, B., Biernath, C., Bindi, M., Cammarano, D., Challinor, A. J., De Sanctis, G., Dumont, B., Eyshi Rezaei, E., Fereres, E., Ferrise, R., Garcia-Vila, M., Gayler, S., Gao, Y., Horan, H., Hoogenboom, G., Izaurralde, R. C., Jabloun, M., Jones, C. D., Kassie, B. T., Kersebaum, K. C., Klein, C., Koehler, A. K., Liu, B., Minoli, S., Montesino San Martin, M., Müller, C., Naresh Kumar, S., Nendel, C., Olesen, J. E., Palosuo, T., Porter, J. R., Priesack, E., Ripoche, D., Semenov, M. A., Stöckle, C., Stratonovitch, P., Streck, T., Supit, I., Tao, F., Van der Velde, M., Wallach, D., Wang, E., Webber, H., Wolf, J., Xiao, L., Zhang, Z., Zhao, Z.,

Zhu, Y. and Ewert, F. 2019. Climate change impact and adaptation for wheat protein. *Global Change Biology* 25(1): 155–173.

Chaloner, T. M., Gurr, S. J. and Bebber, D. P. 2021. Plant pathogen infection risk tracks global crop yields under climate change. *Nature Climate Change* 11(8): 710–715.

Chen, Y., Zhang, Z. and Tao, F. 2018. Impacts of climate change and climate extremes on major crops productivity in China at a global warming of 1.5°C & 2.0°C. *Earth System Dynamics Discussions* 1–27.

Chenu, K., Porter, J. R., Martre, P., Basso, B., Chapman, S. C., Ewert, F., Bindi, M. and Asseng, S. 2017. Contribution of crop models to adaptation in wheat. *Trends in Plant Science* 22(6): 472–490.

Choi, J., Ko, J., An, K.-N., Qaisrani, S. A., Ban, J. and Kim, D. 2021. Simulation of staple crop yields for determination of regional impacts of climate change: A case study in Chonnam Province, Republic of Korea. *Agronomy* 11(12): 2544.

Chou, J., Zhao, W., Li, J., Xu, Y., Yang, F., Sun, M. and Li, Y. 2021. Changes in extreme climate events in rice-growing regions under different warming scenarios in China. *Frontiers in Earth Science* 9(200).

Deryng, D., Conway, D., Ramankutty, N., Price, J. and Warren, R. 2014. Global crop yield response to extreme heat stress under multiple climate change futures. *Environmental Research Letters* 9(3).

FAO 2020. *Food and Agriculture Organization of the United Nations*. http://fao.org.

Gao, Q., Feng, G. Z. and Wang, Z. G. 2010. Present situation of fertilizer application on spring maize in Northeast China. *Chinese Agricultural Science Bulletin* 8(4): 1524–1534.

Garnett, T., Appleby, M. C., Balmford, A., Bateman, I. J., Benton, T. G., Bloomer, P., Burlingame, B., Dawkins, M., Dolan, L., Fraser, D., Herrero, M., Hoffmann, I., Smith, P., Thornton, P. K., Toulmin, C., Vermeulen, S. J. and Godfray, H. C. 2013. Agriculture. Sustainable intensification in agriculture: Premises and policies. *Science* 341(6141): 33–34.

Guo, X., Huang, B., Zhang, H., Cai, C., Li, G., Li, H., Zhang, Y., Struik, P.C., Liu, Z., Dong, M. and Ni, R. 2021. T-FACE studies reveal that increased temperature exerts an effect opposite to that of elevated CO_2 on nutrient concentration and bioavailability in rice and wheat grains. *Food and Energy Security* e336.

Horie, T. 2019. Global warming and rice production in Asia: Modeling, impact prediction and adaptation. *Proceedings of the Japan Academy Series B, Physical and Biological Sciences* 95(6): 211–245.

Huang, M., Wang, J., Wang, B., Liu, D. L., Feng, P., Yu, Q., Pan, X. and Waters, C. 2021. Assessing maize potential to mitigate the adverse effects of future rising temperature and heat stress in China. *Agricultural and Forest Meteorology* 311: 108673.

IPCC 2012a. *Managing the Risks of Extreme Events and Disasters to Advance Climate Change Adaptation*. Cambridge University Press, Cambridge, UK.

IPCC 2022b. Global Warming of 1.5°C: IPCC Special Report on Impacts of Global Warming of 1.5°C above Pre-industrial Levels in Context of Strengthening Response to Climate Change, Sustainable Development, and Efforts to Eradicate Poverty. Cambridge University Press, Cambridge.

Ishigooka, Y., Fukui, S., Hasegawa, T., Kuwagata, T., Nishimori, M. and Kondo, M. 2017. Large-scale evaluation of the effects of adaptation to climate change by shifting transplanting date on rice production and quality in Japan. *Journal of Agricultural Meteorology* 73(4): 156–173.

Kawatsu, S., Homma, K., Horie, T. and Shiraiwa, T. 2007. Change of weather condition and its effect on rice production during the past 40 years in Japan. *Japanese Journal of Crop Science* 76(3): 423–432.

Landau, C. A., Hager, A. G. and Williams Ii, M. M. 2021. Diminishing weed control exacerbates maize yield loss to adverse weather. *Global Change Biology* 27(23): 6156–6165.

Lesk, C., Rowhani, P. and Ramankutty, N. 2016. Influence of extreme weather disasters on global crop production. *Nature* 529(7584): 84–87.

Lin, Y., Feng, Z., Wu, W., Yang, Y., Zhou, Y. and Xu, C. 2017. Potential impacts of climate change and adaptation on maize in Northeast China. *Agronomy Journal* 109(4): 1476–1490.

Liu, B., Asseng, S., Müller, C., Ewert, F., Elliott, J., Lobell, D., Martre, P., Ruane, A., Wallach, D., Jones, J., Rosenzweig, C., Aggarwal, P., Alderman, P., Anothai, J., Basso, B., Biernath, C., Cammarano, D., Challinor, A., Deryng, D., Sanctis, G., Doltra, J., Fereres, E., Folberth, C., Garcia-Vila, M., Gayler, S., Hoogenboom, G., Hunt, L., Izaurralde, R., Jabloun, M., Jones, C., Kersebaum, K., Kimball, B., Koehler, A., Kumar, S., Nendel, C., O'Leary, G., Olesen, J., Ottman, M., Palosuo, T., Prasad, P., Priesack, E., Pugh, T., Reynolds, M., Rezaei, E., Rötter, R., Schmid, E., Semenov, M., Shcherbak, I., Stehfest, E., Stöckle, C., Stratonovitch, P., Streck, T., Supit, I., Tao, F., Thorburn, P., Waha, K., Wall, G., Wang, E., White, J., Wolf, J., Zhao, Z. and Zhu, Y. 2016. Similar estimates of temperature impacts on global wheat yield by three independent methods. *Nature Climate Change* 6(12): 1130–1136.

Liu, B., Liu, L., Tian, L., Cao, W., Zhu, Y. and Asseng, S. 2014. Post-heading heat stress and yield impact in winter wheat of China. *Global Change Biology* 20(2): 372–381.

Liu, B., Zhang, D., Zhang, H., Asseng, S., Yin, T., Qiu, X., Ye, Z., Liu, L., Tang, L., Cao, W. and Zhu, Y. 2021a. Separating the impacts of heat stress events from rising mean temperatures on winter wheat yield of China. *Environmental Research Letters* 16(12): 124035.

Liu, G.-Z., Liu, W.-M., Hou, P., Ming, B., Yang, Y., Guo, X., Xie, R., Wang, K. and Li, S. 2021b. Reducing maize yield gap by matching plant density and solar radiation. *Journal of Integrative Agriculture* 20(2): 363–370.

Liu, H., Xiong, W., Mottaleb, K. A., Krupnik, T. J., Burgueño, J., Pequeno, D. N. L. and Wu, W. 2021c. Contrasting contributions of five factors to wheat yield growth in China by process-based and statistical models. *European Journal of Agronomy* 130: 126370.

Liu, J., Feng, H., He, J., Chen, H., Ding, D., Luo, X. and Dong, Q. 2019. Modeling wheat nutritional quality with a modified CERES-wheat model. *European Journal of Agronomy* 109: 125901.

Liu, L., Wang, E., Zhu, Y. and Tang, L. 2012. Contrasting effects of warming and autonomous breeding on single-rice productivity in China. *Agriculture, Ecosystems and Environment* 149: 20–29.

Liu, L., Wang, E., Zhu, Y., Tang, L. and Cao, W. 2013a. Effects of warming and autonomous breeding on the phenological development and grain yield of double-rice systems in China. *Agriculture, Ecosystems and Environment* 165: 28–38.

Liu, Z., Hubbard, K. G., Lin, X. and Yang, X. 2013b. Negative effects of climate warming on maize yield are reversed by the changing of sowing date and cultivar selection in Northeast China. *Global Change Biology* 19(11): 3481–3492.

Liu, M., Guo, Y., Wang, Y. and Hao, J. 2022. Changes of extreme agro-climatic droughts and their impacts on grain yields in rain-fed agricultural regions in China over the past 50 years. *Atmosphere* 13(1): 4.

Liu, Y., Tang, L., Qiu, X., Liu, B., Chang, X., Liu, L., Zhang, X., Cao, W. and Zhu, Y. 2020. Impacts of 1.5 and 2.0°C global warming on rice production across China. *Agricultural and Forest Meteorology* 284: 107900.

Liu, Y., Wang, E., Yang, X. and Wang, J. 2010. Contributions of climatic and crop varietal changes to crop production in the North China Plain, since 1980s. *Global Change Biology* 16(8): 2287–2299.

Liu, Y., Zhang, J., Pan, T., Chen, Q., Qin, Y. and Ge, Q. 2022. Climate-associated major food crops production change under multi-scenario in China. *The Science of the Total Environment* 811: 151393.

Liu, Z. J., Yang, X. G., Wang, W. F., Li, K. N. and Zhang, X. Y. 2009. Characteristics of agricultural climate resources in three provinces of northeast China under global climate change. *Chinese Journal of Applied Ecology* 20(9): 2199–2206.

Lobell, D. B., Cassman, K. G. and Field, C. B. 2009. Crop yield gaps: Their importance, magnitudes, and causes. *Annual Review of Environment and Resources* 34(1): 179–204.

Lv, Z., Zhu, Y., Liu, X., Ye, H., Tian, Y. and Li, F. 2018. Climate change impacts on regional rice production in China. *Climatic Change* 147(3–4): 523–537.

Lv, Z. F., Liu, X. J., Cao, W. X. and Zhu, Y. 2013. Climate change impacts on regional winter wheat production in main wheat production regions of China. *Agricultural and Forest Meteorology* 171-172: 234–248.

Ma, J., Liu, Y., Yang, X. G., Wang, W. F., Xue, C. Y. and Zhang, X. Y. 2010. Characteristics of climate resources under global climate change in the North China Plain. *Acta Ecologica Sinica* 30(14): 3818–3827.

Malézieux, E., Crozat, Y., Dupraz, C., Laurans, M., Makowski, D., Ozier-Lafontaine, H., Rapidel, B., Tourdonnet, S. and Valantin-Morison, M. 2009. Mixing plant species in cropping systems: concepts, tools and models: A review. *Agronomy for Sustainable Development* 29(1): 43–62.

Masutomi, Y., Takimoto, T., Shimamura, M., Manabe, T., Arakawa, M., Shibota, N., Ooto, A., Azuma, S., Imai, Y. and Tamura, M. 2019. Rice grain quality degradation and economic loss due to global warming in Japan. *Environmental Research Communications* 1(12): 121003.

Mueller, N. D., Butler, E. E., McKinnon, K. A., Rhines, A., Tingley, M., Holbrook, N. M. and Huybers, P. 2016. Cooling of US Midwest summer temperature extremes from cropland intensification. *Nature Climate Change* 6(3): 317–322.

Muthayya, S., Sugimoto, J. D., Montgomery, S. and Maberly, G. F. 2014. An overview of global rice production, supply, trade, and consumption. *Annals of the New York Academy of Sciences* 1324(1): 7–14.

Ohta, S. and Kimura, A. 2007. Impacts of climate changes on the temperature of paddy waters and suitable land for rice cultivation in Japan. *Agricultural and Forest Meteorology* 147(3–4): 186–198.

Okada, M., Iizumi, T., Hayashi, Y. and Yokozawa, M. 2011. Projecting climate change impacts both on rice quality and yield in Japan. *Journal of Agricultural Meteorology* 67(4): 285–295.

Ottman, M. J., Kimball, B. A., White, J. W. and Wall, G. W. 2012. Wheat Growth Response to Increased Temperature from Varied Planting Dates and Supplemental Infrared Heating. *Agronomy Journal* 104: 7–16

Peng, S., Huang, J., Sheehy, J. E., Laza, R. C., Visperas, R. M., Zhong, X., Centeno, G. S., Khush, G. S. and Cassman, K. G. 2004. Rice yields decline with higher night temperature from global warming. *Proceedings of the National Academy of Sciences of the United States of America* 101(27): 9971–9975.

Porter, J. R. and Gawith, M. 1999. Temperatures and the growth and development of wheat: a review. *European Journal of Agronomy* 10: 23–36

Seidel, S. J., Gaiser, T., Kautz, T., Bauke, S. L., Amelung, W., Barfus, K., Ewert, F. and Athmann, M. 2019. Estimation of the impact of precrops and climate variability on soil depth-differentiated spring wheat growth and water, nitrogen and phosphorus uptake. *Soil and Tillage Research* 195: 104427.

Semenov, M. A. and Porter, J. R. 1995. Climatic variability and the modelling of crop yields. *Agricultural and Forest Meteorology* 73(3–4): 265–283.

Semenov, M. A. and Shewry, P. R. 2011. Modelling predicts that heat stress, not drought, will increase vulnerability of wheat in Europe. *Scientific Reports* 1: 66.

Shi, P., Tang, L., Wang, L., Sun, T., Liu, L., Cao, W. and Zhu, Y. 2015. Post-heading heat stress in rice of South China during 1981–2010. *PLoS ONE* 10(6): e0130642.

Shimono, H. 2011. Earlier rice phenology as a result of climate change can increase the risk of cold damage during reproductive growth in northern Japan. *Agriculture, Ecosystems and Environment* 144(1): 201–207.

Su, Z.-R., Liu, Z.-K., Bai, F., Zhang, Z., Sun, S., Huang, Q., Liu, T., Liu, X. and Yang, X. 2021. Cultivar selection can increase yield potential and resource use efficiency of spring maize to adapt to climate change in Northeast China. *Journal of Integrative Agriculture* 20(2): 371–382.

Sumberg, J. 2012. Mind the (yield) gap(s). *Food Security* 4(4): 509–518.

Sun, W. and Huang, Y. 2011. Global warming over the period 1961–2008 did not increase high-temperature stress but did reduce low-temperature stress in irrigated rice across China. *Agricultural and Forest Meteorology* 151(9): 1193–1201.

Tao, F. and Zhang, Z. 2013. Climate change, wheat productivity and water use in the North China Plain: A new super-ensemble-based probabilistic projection. *Agricultural and Forest Meteorology* 170: 146–165.

Tao, F., Zhang, Z., Xiao, D., Zhang, S., Rötter, R. P., Shi, W., Liu, Y., Wang, M., Liu, F. and Zhang, H. 2014. Responses of wheat growth and yield to climate change in different climate zones of China, 1981–2009. *Agricultural and Forest Meteorology* 189–190: 91–104.

Tebaldi, C. and Lobell, D. 2018. Differences, or lack thereof, in wheat and maize yields under three low-warming scenarios. *Environmental Research Letters* 13(6): 065001.

Teixeira, E. I., Fischer, G., van Velthuizen, H., Walter, C. and Ewert, F. 2013. Global hot-spots of heat stress on agricultural crops due to climate change. *Agricultural and Forest Meteorology* 170: 206–215.

Trnka, M., Rötter, R. P., Ruiz-Ramos, M., Kersebaum, K. C., Olesen, J. E., Žalud, Z. and Semenov, M. A. 2014. Adverse weather conditions for European wheat production will become more frequent with climate change. *Nature Climate Change* 4(7): 637–643.

van Ittersum, M. K., Cassman, K. G., Grassini, P., Wolf, J., Tittonell, P. and Hochman, Z. 2013. Yield gap analysis with local to global relevance–A review. *Field Crops Research* 143(1): 4–17.

Wang, S., Wang, E., Wang, F. and Tang, L. 2012. Phenological development and grain yield of canola as affected by sowing date and climate variation in the Yangtze River Basin of China. *Crop and Pasture Science* 63(5): 478.

Wang, Z., Wang, M., Yin, X., Zhang, H., Chu, Q., Wen, X. and Chen, F. 2015. Spatiotemporal characteristics of heat and rainfall changes in summer maize season under climate change in the North China Plain. *Chinese Journal of Eco-Agriculture* 23(4): 473–481.

Xiao, D. and Tao, F. 2014. Contributions of cultivars, management and climate change to winter wheat yield in the North China Plain in the past three decades. *European Journal of Agronomy* 52: 112–122.

Xiong, W., Holman, I., Lin, E., Conway, D., Li, Y. and Wu, W. 2012. Untangling relative contributions of recent climate and CO_2 trends to national cereal production in China. *Environmental Research Letters* 7(4): 044014.

Xiong, W., van der Velde, M., Holman, I. P., Balkovic, J., Lin, E., Skalský, R., Porter, C., Jones, J., Khabarov, N. and Obersteiner, M. 2014. Can climate-smart agriculture reverse the recent slowing of rice yield growth in China? *Agriculture, Ecosystems and Environment* 196: 125–136.

Yang, J., Xiong, W., Yang, X.-g., Cao, Y. and Feng, L. 2014. Geographic variation of rice yield response to past climate change in China. *Journal of Integrative Agriculture* 13(7): 1586–1598.

Ye, Z., Qiu, X., Chen, J., Cammarano, D., Ge, Z., Ruane, A. C., Liu, L., Tang, L., Cao, W., Liu, B. and Zhu, Y. 2020. Impacts of 1.5 °C and 2.0 °C global warming above pre-industrial on potential winter wheat production of China. *European Journal of Agronomy* 120: 126149.

Yu, Y., Huang, Y. and Zhang, W. 2012. Changes in rice yields in China since 1980 associated with cultivar improvement, climate and crop management. *Field Crops Research* 136: 65–75.

Zhang, F., Yang, X., Sun, S., Gao, J., Liu, Z., Zhang, Z. and Liu, T. 2021. A spatiotemporal analysis of extreme agrometeorological events during selected growth stages of maize (Zea mays L.) from 1960 to 2017 in Northeast China. *Theoretical and Applied Climatology* 143(3–4): 943–955.

Zhang, H., Zhou, G., Liu, L., Wang, B., Xiao, D. and He, L. 2019. Climate-associated rice yield change in the Northeast China Plain: A simulation analysis based on CMIP5 multi-model ensemble projection. *Science of the Total Environment* 666: 126–138.

Zhang, L., Zhang, Z., Tao, F., Luo, Y., Cao, J., Li, Z., Xie, R. and Li, S. 2021. Planning maize hybrids adaptation to future climate change by integrating crop modelling with machine learning. *Environmental Research Letters* 16(12): 124043.

Zhang, T., Yang, X., Wang, H., Li, Y. and Ye, Q. 2014. Climatic and technological ceilings for Chinese rice stagnation based on yield gaps and yield trend pattern analysis. *Global Change Biology* 20(4): 1289–1298.

Zhao, C., Liu, B., Piao, S., Wang, X., Lobell, D. B., Huang, Y., Huang, M., Yao, Y., Bassu, S., Ciais, P., Durand, J. L., Elliott, J., Ewert, F., Janssens, I. A., Li, T., Lin, E., Liu, Q., Martre, P., Müller, C., Peng, S., Peñuelas, J., Ruane, A. C., Wallach, D., Wang, T., Wu, D., Liu, Z., Zhu, Y., Zhu, Z. and Asseng, S. 2017. Temperature increase reduces global yields of major crops in four independent estimates. *Proceedings of the National Academy of Sciences of the United States of America* 114(35): 9326-9331.

Zhao, C., Piao, S., Huang, Y., Wang, X., Ciais, P., Huang, M., Zeng, Z. and Peng, S. 2016a. Field warming experiments shed light on the wheat yield response to temperature in China. *Nature Communications* 7(1): 13530.

Zhao, C., Piao, S., Wang, X., Huang, Y., Ciais, P., Elliott, J., Huang, M., Janssens, I. A., Li, T., Lian, X., Liu, Y., Müller, C., Peng, S., Wang, T., Zeng, Z. and Peñuelas, J. 2016b. Plausible rice yield losses under future climate warming. *Nature Plants* 3(1): 16202.

Zhao, J., Yang, X., Liu, Z., Lv, S., Wang, J. and Dai, S. 2016c. Variations in the potential climatic suitability distribution patterns and grain yields for spring maize in Northeast China under climate change. *Climatic Change* 137(1–2): 29–42.

Zhao, J. and Yang, X. 2018a. Average amount and stability of available agro-climate resources in the main maize cropping regions in China during 1981–2010. *Journal of Meteorological Research* 32(1): 146–156.

Zhao, J. and Yang, X. 2018b. Distribution of high-yield and high-yield-stability zones for maize yield potential in the main growing regions in China. *Agricultural and Forest Meteorology* 248: 511–517.

Zhao, J., Yang, X., Lin, X., Sassenrath, G. F., Dai, S., Lv, S., Chen, X., Chen, F. and Mi, G. 2015. Radiation interception and use efficiency contributes to higher yields of newer maize hybrids in Northeast China. *Agronomy Journal* 107(4): 1473–1480.

Zhou, W., Liu, Y., Ata-Ul-Karim, S. and Ge, Q. 2021. Spatial difference of climate change effects on wheat protein concentration in China. *Environmental Research Letters* 16(12).

Chapter 18

Modelling the impact of climate change on agriculture in North Africa and Southwest Asia

Ehsan Eyshi Rezaei , Leibniz Centre for Agricultural Landscape Research (ZALF), Germany; and Sevim Seda Yamaç, Konya Food and Agriculture University, Turkey

1 Introduction
2 North Africa: geography, economy, climate and agriculture
3 Southwest Asia: geography, economy, climate and agriculture
4 Future climate projections and their effects on regional crop production
5 Modelling tools to quantify crop response to climate change
6 Current climate change impact assessments in North Africa and Southwest Asia
7 Accounting for the potential of adaptation strategies
8 Limitations of climate change impact assessments
9 Conclusion and future trends
10 Where to look for further information
11 References

1 Introduction

Since the 1970s, there has been increasing warming in North Africa and Southwest Asia and a rise in intra-annual and inter-annual precipitation variability and extreme events such as heatwaves and flooding. Those regions face food security challenges as a consequence of the combination of population growth and climate change. Consequently, adaptation strategies are essential.

Process-based crop models coupled with climate projections are the most suitable tools for evaluating interactions between climate, management, and genotypes on crop yield. Simulations can also be undertaken across spatial and temporal scales. Climate change adaptation strategies for different environments have been developed and tested with the help of such tools. However, there are considerable knowledge gaps in projecting the effects of climate change on cropping systems in North Africa and Southwest Asia

http://dx.doi.org/10.19103/AS.2022.0115.20

compared to other regions of the world due to the lack of investment and data availability.

This chapter systematically reviews published studies on climate change projections and suggested adaptation strategies for North Africa and Southwest Asia cropping systems using modelling tools. We then examine the limitations of impact assessment studies in these regions and potential improvements needed in planning field and modelling experiments. Finally, we discuss future trends and research gaps that need to be addressed.

2 North Africa: geography, economy, climate and agriculture

Located in the northernmost section of the African continent, North Africa embraces Western Sahara, Tunisia, Morocco, Algeria, Libya, and Egypt, as defined by the United Nations classification of Subregions (Fig. 1). It encompasses both the Arab World and Africa from a socio-economic perspective. Geographically, North Africa is characterized by the Sahara Desert in the south, the Nile River and its delta in the east, and the Atlas Mountains in the west. Since the Atlas Mountains dominate their northern landscapes, Morocco, Algeria, and Tunisia are also called the Atlas Lands. Every country includes significant portions of the Sahara, particularly Algeria. Nearly 7 million km^2 of North Africa is inhabited by more than 234 million people. Algeria is the largest country geographically covering 2.4 million km^2 area; however, with 95 million people, Egypt is the region's most populated country. The total gross domestic product (GDP) of individual North African countries is between US$33 billion (Libya) and US$332 billion (Egypt) (World Bank, 2017).

Climatic conditions in North Africa are generally characterized by hot-dry to hot-extremely dry conditions (Fig. 2). The annual mean temperature across North Africa ranges between 10°C and 32°C, but it is generally above 19°C (Fig. 2). There are temperate areas in the northern part of North Africa and desert areas in central and southern parts. Precipitation regimes are quite diverse because of the wide range of latitudes between 19°N and 37° N and altitudes up to 4165 m. Precipitation in most of the Sahara is below 50 mm, but it can reach up to 1500 mm in northern Tunisia as well as up to 2000 mm in the mountains of Morocco (Harris et al., 2014). Only 10% of the region receives precipitation exceeding 300 mm per year. Water is the most important limiting factor for sustainable development in North Africa with major implications for agriculture, industry, and regional politics (Droogers et al., 2012).

Analysis of trends in precipitation using various gridded data sets from 1901 to 2016 suggests a decline of up to −3.4 mm per decade across North Africa (Zittis, 2018). Other studies also show an increase in the region’s aridity

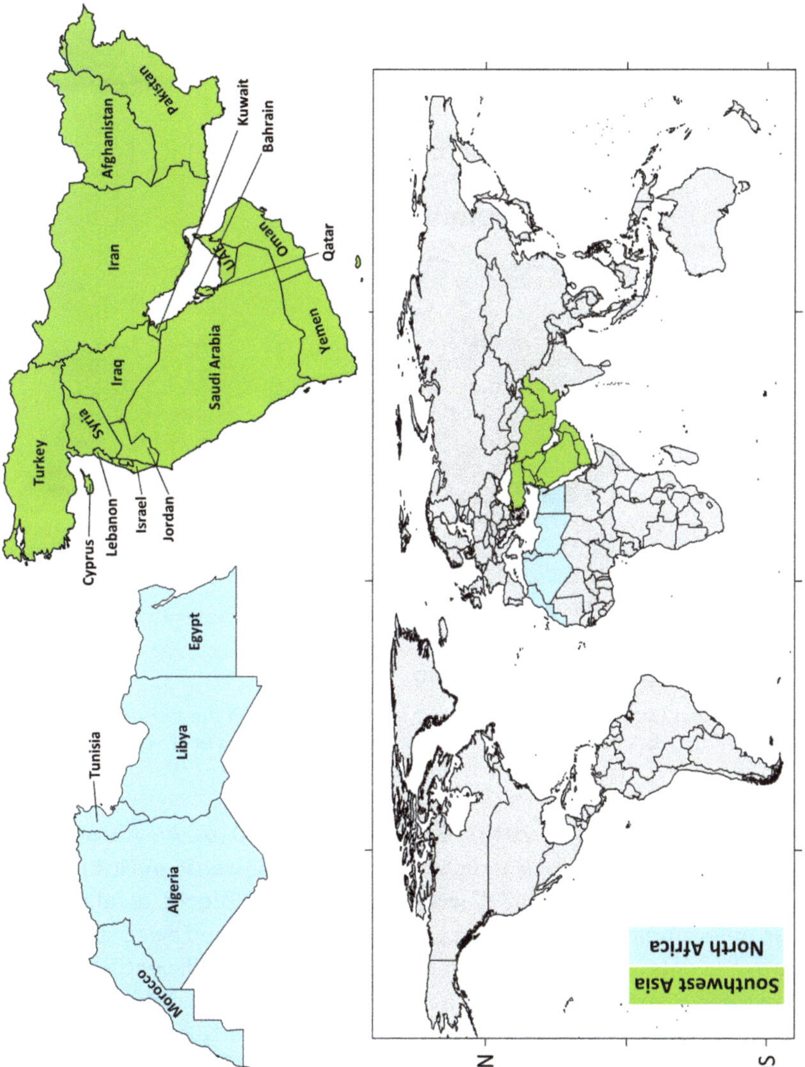

Figure 1 The countries are located in North Africa and Southwest Asia.

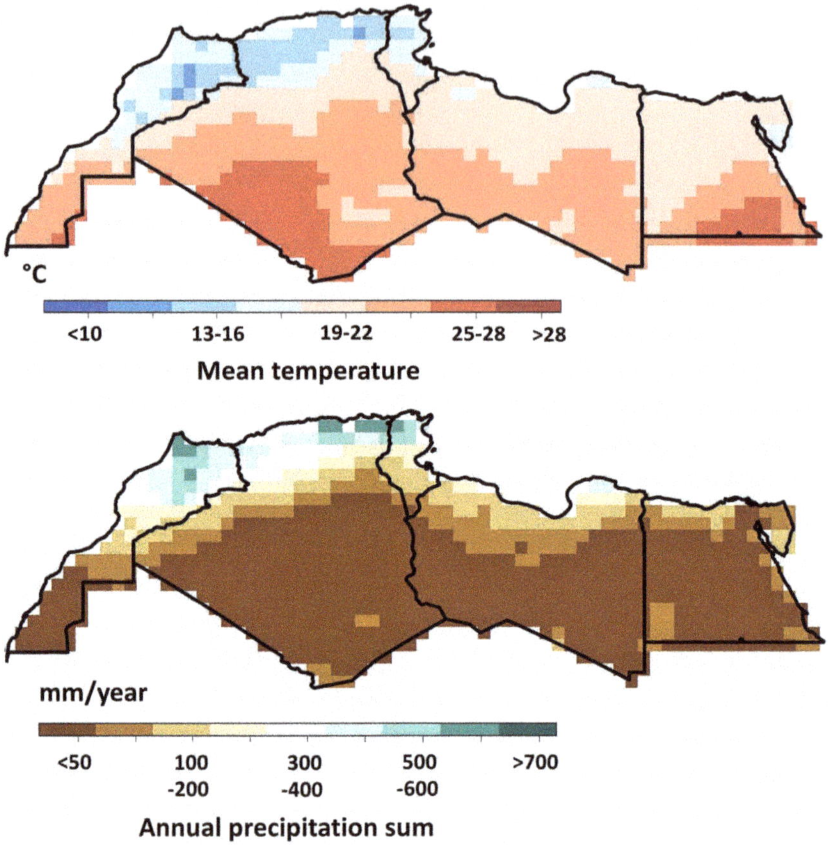

Figure 2 The mean temperature and annual precipitation sum in the period 1985-2016 across North Africa obtained from the CRU climate data set (Harris et al., 2014).

over a 54-year period (1960-2013) (El Kenawy et al., 2016). An African-wide long-term temperature analysis indicated the highest increase in North Africa (from 1.6°C per century to 3.2°C per century) (Engelbrecht et al., 2015). However, other studies using station-based climate data found a negative trend in heatwave events, although the number of stations is insufficient to prove a declining trend in extreme temperatures in the region (Qiu and Yan, 2020).

Agriculture plays a vital role in North Africa, particularly in the economies of Morocco and Tunisia, though the economies of other countries such as Algeria are heavily dependent on the oil industry (IFPRI, 2018). A key issue for the region is that agricultural production is struggling to keep pace with increasing population (Omidvar et al., 2019). The growing gap between food production and population growth is a particular issue in Egypt, Morocco, and Algeria (Abdelhedi and Zouari, 2020). Estimates of self-sufficiency in wheat range between 62% (Morocco) and 9% (Libya) (Araujo-Enciso and Fellmann, 2020).

Less than 3.5% of the entire area is cropland, primarily in Tunisia, Algeria, and Egypt (Abdelhedi and Zouari, 2020). More than 80% of croplands in North Africa is farmed in holdings smaller than 10 ha, with 70% of all farms smaller than 2 ha (Lowder et al., 2016). Based on the MIRCA2000 data set (Portmann et al., 2010), rainfed farming (including grasslands and pastures) predominates in Morocco (38 M–ha^{-1}), Algeria (24 M–ha^{-1}), Libya (3 M–ha^{-1}), and Tunisia (16 M–ha^{-1}) (Fig. 3). However, irrigated farming is the main cropping system in Egypt (34 M–ha^{-1}) (Fig. 3). The number of rainy days is generally less than 120 days per year, resulting in high frequency and magnitude of drought stress which, together with extreme temperatures, limits crop production (El-Gharras et al., 2017). The majority of agriculture in the rainfed regions of North Africa is based on integrated crop-livestock systems, especially cereals and sheep (El-Gharras et al., 2017).

Cereals, perennials, and grassland/pasture are the most important cropping systems across rainfed and irrigated farming in North Africa (Fig. 4). Wheat and barley predominate in the rainfed system in all North African countries except Egypt. Irrigated systems are used mainly for perennials (Fig. 4). The harvested area of cereals grew during 1960–1985, then declined dramatically up to 2000 before growing again to 2010 (Fig. 5). Based on FAO statistics,

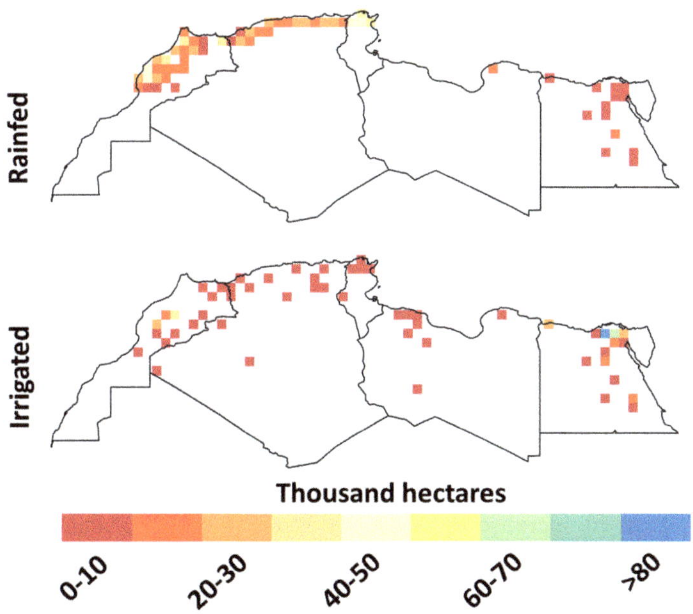

Figure 3 The size of the rainfed and irrigated areas per 0.5° × 0.5° grid across North Africa obtained from the MIRCA2000 data set (Portmann et al., 2010).

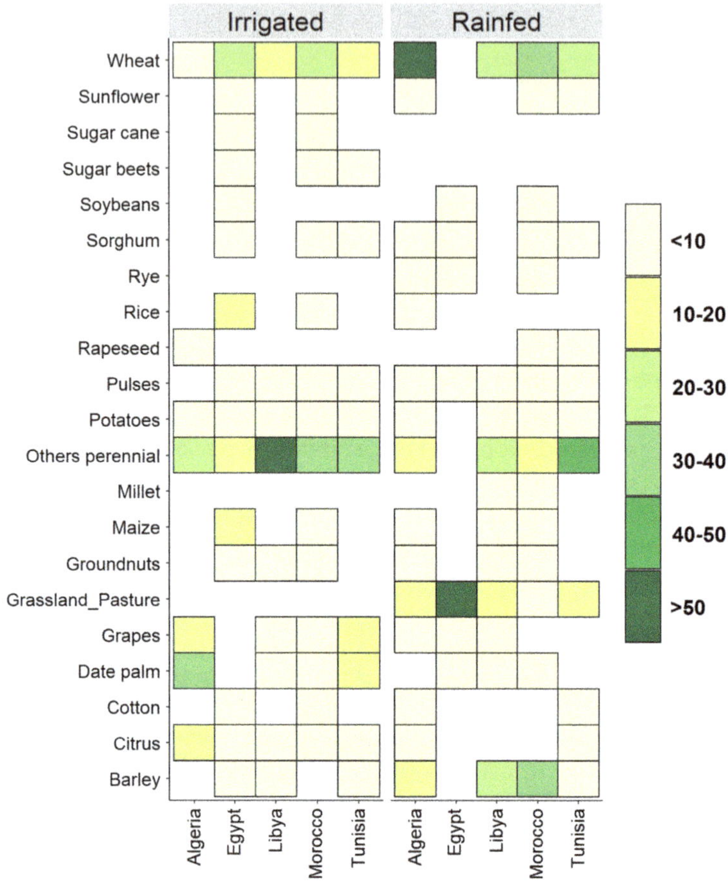

Figure 4 The share of the specific crops under rainfed and irrigated conditions in North African countries (Portmann et al., 2008).

the improvement in cereal yields in North Africa during 1960–2020 (179 kg per decade) is much lower than the global average (461 kg per decade) (Fig. 5) (FAO, 2021). Lack of investment in agricultural infrastructure (Lezzaik et al., 2018), the high price of inputs (e.g. fertilizer and pesticides) (Shimeles et al., 2015), and severe droughts (Fragaszy et al., 2020) are the main causes of low agricultural production in North Africa. Agricultural production is also threatened by the increase in frequency and magnitude of extreme climatic events (Waha et al., 2017). The region needs effective adaptation strategies to mitigate the negative impacts of climate change on crop production (Schilling et al., 2020).

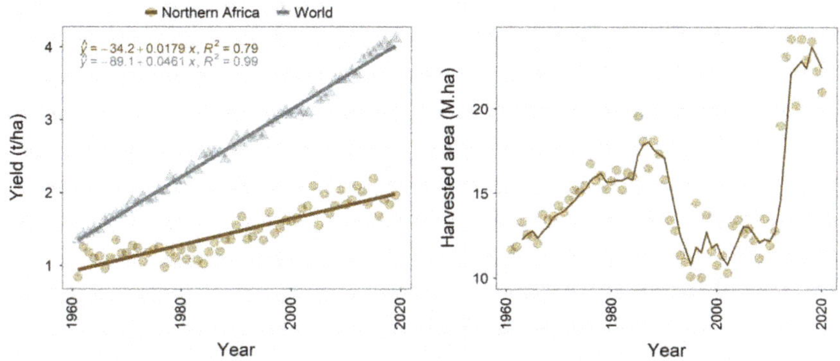

Figure 5 The trend and variability of yield and harvested area of cereals as the most important crops cultivated in North Africa and the world in 1960–2020 (FAO, 2021). The classification of the regions is based on the United Nations definition. The yield plot indicated the linear trend, and the harvest area plot showed 3 years moving average.

3 Southwest Asia: geography, economy, climate and agriculture

A wide range of classifications have been developed to categorize countries located in the southwestern part of Asia. We include all countries covered in these different classifications: Iraq, Turkey, Iran, Syria, Cyprus, Lebanon, Israel, Jordan, Saudi Arabia, UAE, Bahrain, Kuwait, Oman, Yemen, Qatar, Afghanistan, and Pakistan (Fig. 1). Southwest Asia borders the Mediterranean Sea to the west, the Sinai Peninsula and the Red Sea to the south, and the Indian Ocean and Persian Gulf to the southeast (Britannica, 2021). In the north, the region reaches to the Caspian and Black Seas. In the north, Turkey is located on the Anatolian Peninsula connecting the European and Asian continents. However, the vast majority of croplands of Turkey are located in Southwest Asia.

Southwest Asia is situated at the crossroads of Asia, Africa, and Europe. This region covers 7.7 million km^2, with countries ranging in size from Saudi Arabia (2.1 million km^2) to Bahrain (780 km^2). The population of Southwest Asia is around 0.6 billion, with Pakistan being the most populated (227 million) country in the region (World Bank, 2021). Saudi Arabia ($792B) and Turkey ($761B) have the highest total GDP (World Bank, 2021). Qatar ($62 276) has the highest per capita GDP (World Bank, 2021). With the exception of Turkey, the primary income of the richest countries is from oil. Historically, Southwest Asia is seen as the region where agriculture first developed (Shennan, 2018). Agriculture remains the primary source of income for 10–35% of the population (depending on the country) (Sivakumar et al., 2013).

The climate of Southwest Asia is mainly characterized as semi-arid and arid (Sivakumar et al., 2013), with drought a significant issue in most agricultural areas (Meza et al., 2020). The annual mean temperature gradually increases from <10°C in higher latitudes (e.g. Turkey, Iran, etc.) to >28°C in lower latitudes (e.g. Saudi Arabia, Yemen, etc.) (Fig. 6). Most areas of Southwest Asia receive 100–300 mm of precipitation per year. Eastern and western parts of Turkey, as well as northern Pakistan, receive the highest amount of precipitation (>700 mm per year) (Fig. 6). A strong correlation exists between the distribution of precipitation and topography in the region (Rana et al., 2017). The rainy season typically begins in autumn (October–November), peaks in January–February, and then declines in April–May. Inter-annual variability in precipitation is relatively high (Mariotti, 2007). The length of the plant-growing period in Southwest Asia is limited to 149 days (FAO, 2007).

Rainfed systems cover almost 54% of agriculture, with the remaining 46% irrigated (Fig. 7). Turkey and Pakistan have the largest rainfed and irrigated production areas, respectively (Fig. 7). Cropping systems across rainfed and irrigated systems in Southwest Asia mainly include cereals, perennials, pulses, and grassland (Fig. 8). Wheat, maize, and barley are the most cultivated cereals in the region in both rainfed and irrigated systems (Fig. 8). Cereal yield improvement in Western Asia between 1960 and 2020 (306 kg per decade) is lower than the global average (461 kg per decade) (Fig. 9) (FAO, 2021). The

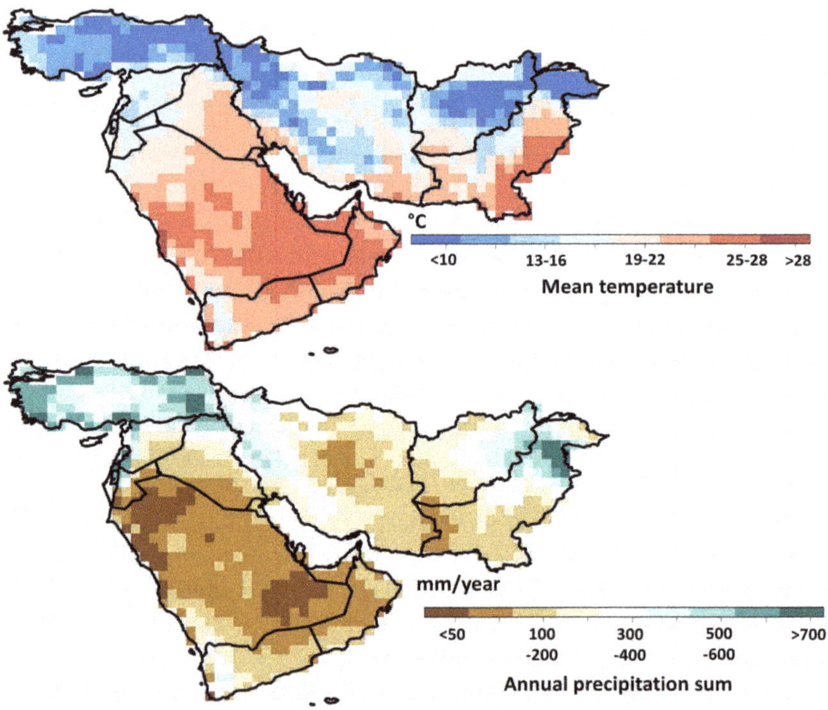

Figure 6 The mean temperature and annual precipitation sum in the period 1985–2016 across Southwest Asia obtained from the CRU climate data set (Harris et al., 2014).

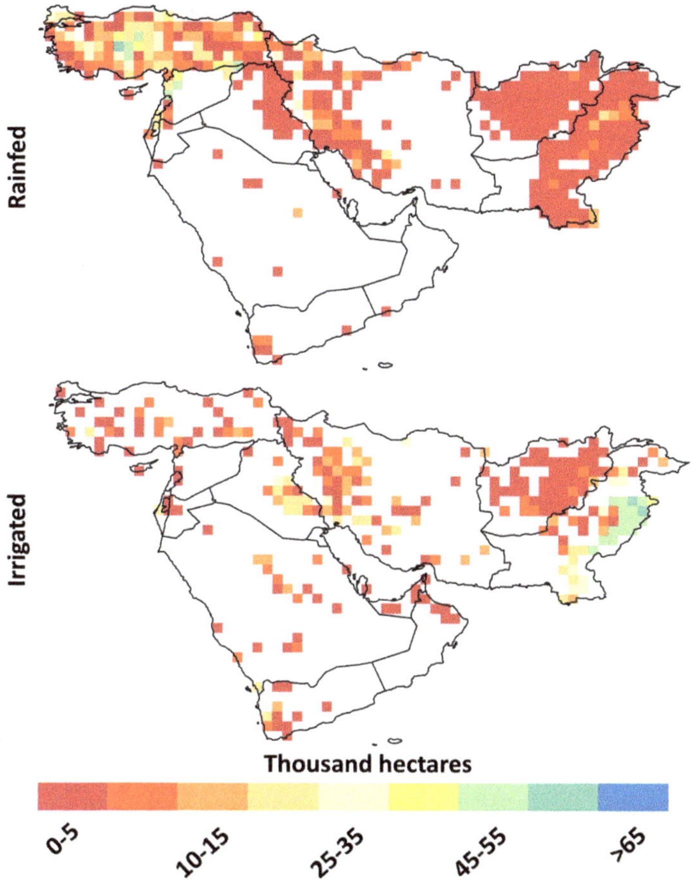

Figure 7 The size of the rainfed and irrigated areas per 0.5° × 0.5° grid across Southwest Asia obtained from the MIRCA2000 data set (Portmann et al., 2010).

harvested area for cereals increased from 19 million ha in 1960 to 25 million ha in the late 1990s but declined to 18 million ha by 2020 (Fig. 9).

Multiple stresses (e.g. flooding, heat, frost) affect agricultural production in Southwest Asia, particularly drought (Barlow et al., 2016). Evidence suggests a substantial contribution of drought to the change in harvested area (Rezaei et al., 2021). Climate models project that the share of the area covered by torrid-arid climate conditions will increase from 13.8% to 29.3% (Representative Concentration Pathway (RCP)4.5) and 36.4% (RCP8.5) (Rahimi et al., 2019). Large-scale climate variability appears to be a major factor contributing to recent droughts which are among the worst (2001 and 2008) in the last 50 years (Barlow et al., 2016). Eleven extreme drought events have been reported nationally for Western Asia in the period 1990–2021 (EM-DAT, 2022). Some studies suggest that the effect of extreme drought (2007–2010) on food

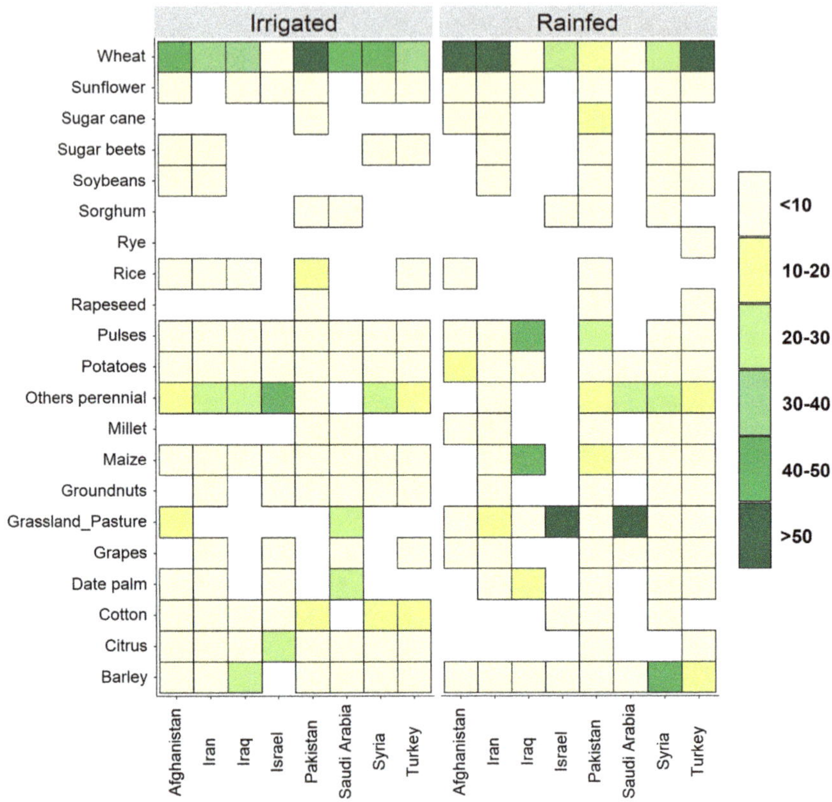

Figure 8 The share of the specific crops under rainfed and irrigated conditions in Southwest Asian countries (Portmann et al., 2008).

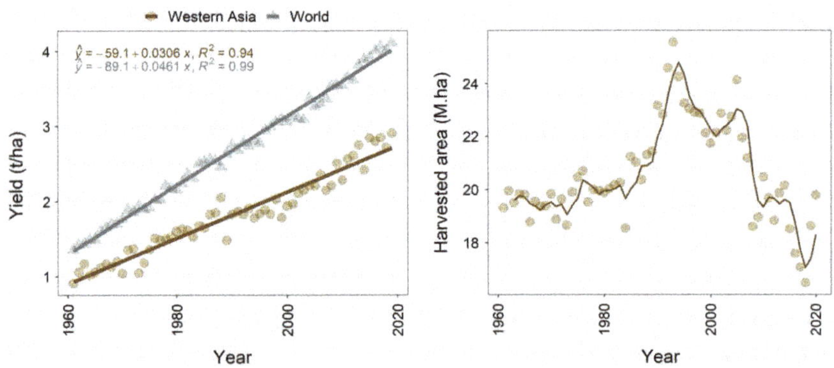

Figure 9 The trend and variability of yield and harvested area of cereals as the most important crops cultivated in West Asia and the world in 1960–2020 (FAO, 2021). The classification of the regions is based on the United Nations definition. The yield plot indicated the linear trend, and the harvest area plot showed 3 years moving average.

insecurity was a potential cause of the Syrian conflict (Kelley et al., 2015; Ide, 2018). Drought conditions are now experienced by more than 30% of the land surface area compared to 10–15% in the early 1970s (Sivakumar et al., 2013)

4 Future climate projections and their effects on regional crop production

Global warming triggered by greenhouse gas (GHG) emissions has led to recent climate change affecting North Africa and Southwest Asia, with further changes predicted by climate models in the future (Figs. 10–12). The magnitude of global warming trends depends on emission scenarios (IPCC, 2021). Compared to baseline, the mean temperature is projected to increase between 1.5°C and 4.5°C compared to baseline until 2100 under sustainability (SSP1) to fossil fuel development (SSP5) scenarios, respectively (Meinshausen et al., 2020). Climate change is expected to cause an increase in the frequency and magnitude of extreme events such as heat and drought, flooding, and fire events, particularly in semi-dry and dry regions such as North Africa and Southwest Asia.

Continuing current emission trends will lead to super- and ultra-extreme heatwave conditions in the MENA (North Africa + Middle East) region. Extreme temperatures (up to 56°C) over an extended period of time may pose a threat to life potentially affecting 600 million inhabitants in the region by the end of the century (Zittis, 2018). A multi-model ensemble showed that a 3°C increase in temperature would result in a 32% increase in drought-prone areas across North Africa (Elkouk et al., 2021). Similar increases in length of the dry period (up to 90%) have been projected across 80% of the Middle East area by the end of the century (Tabari and Willems, 2018).

Co-occurrence of temperature rise and extreme events in those regions may have a severe impact on food security in the coming decades (Al-Delaimy, 2020). However, food security is also determined by other factors such as the dynamics and openness of international trade and fluctuations in international food prices that can directly or indirectly affect food security irrespective of regional food production or oil prices (Mahmah and Amar, 2021).

Crop production in North Africa and Southwest Asia has been severely affected by the increase in temperatures and the lengthening of dry spells, as well as periods of heat stress, defined as short, extremely high-temperature episodes. Cropland suitability for commonly grown crops across semi-arid and arid regions declines significantly as a result of climate change (with shifting temperature and precipitation patterns) (Shi et al., 2014; Kumar et al., 2020). Climate change effects have influenced crop yield in two ways:

- First, the rise in mean temperature accelerates crop development and reduces the length of critical phases such as grain filling, resulting in significant yield loss (Mamrutha et al., 2020).

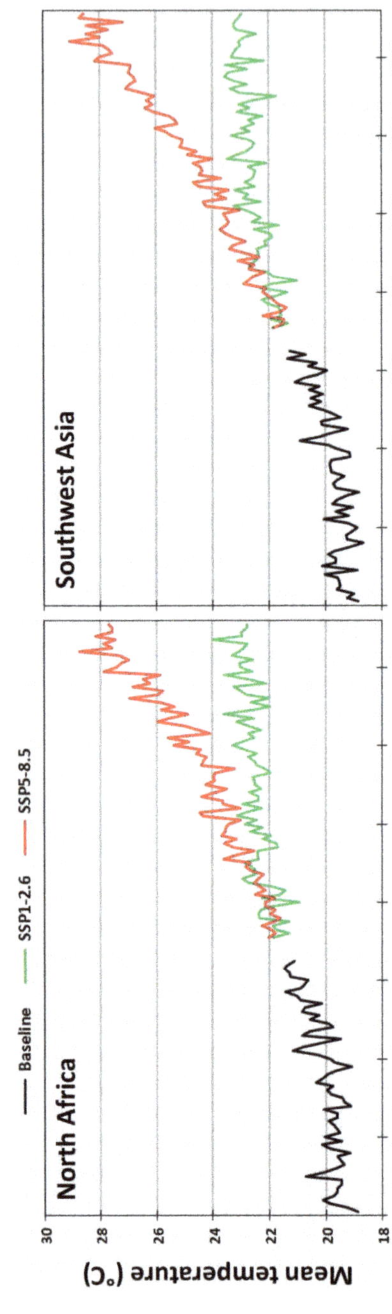

Figure 10 The yearly near-surface temperature in the baseline period (1950–2014) and future projections (2020–2100) under sustainability (SSP1-2.6) and fossil fuel development (SSP5-8.5) scenarios across North Africa and Southwest Asia. Data obtained from Coupled Model Intercomparison Project Phase 6 portal (O'Neill et al., 2016).

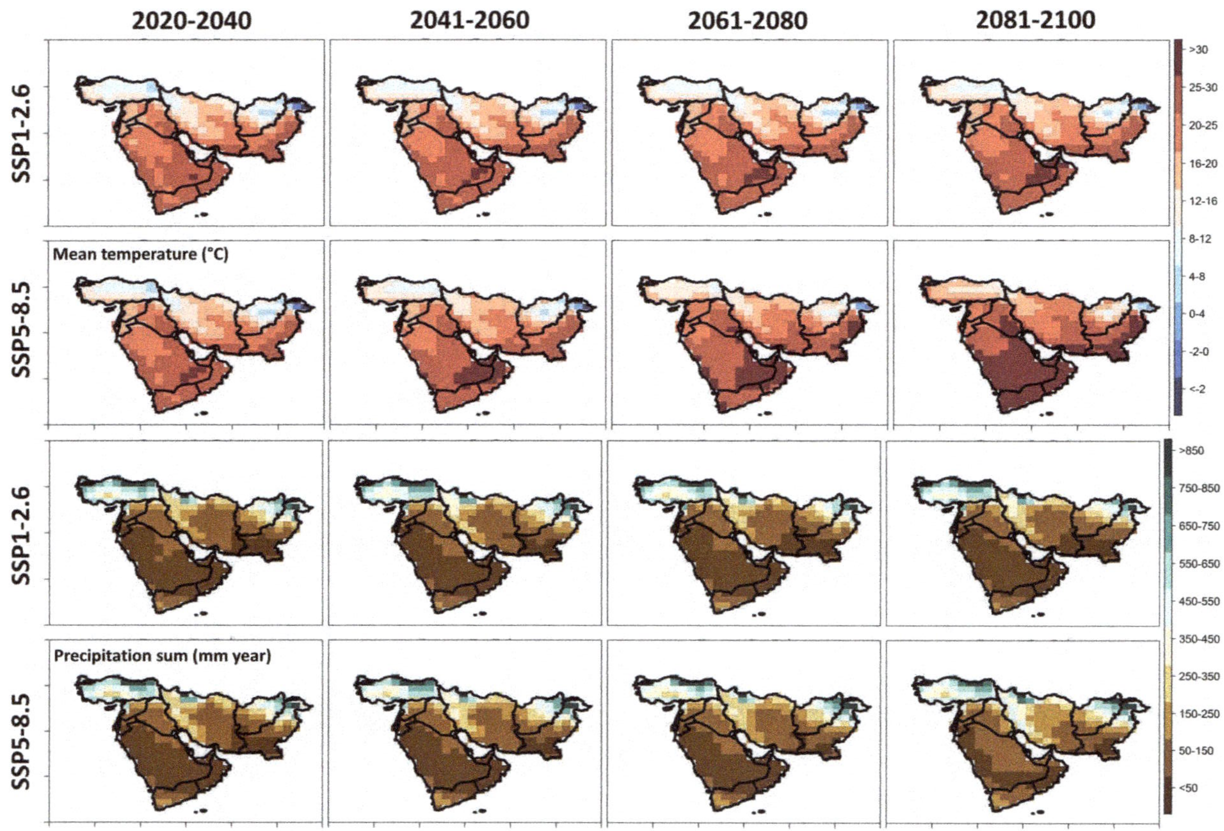

Figure 11 The spatial pattern of mean temperature and precipitation projections in four time steps (2020–2040, 2041–2060, 2061–2080, and 2081–2100) under sustainability (SSP1-2.6) and fossil fuel development (SSP5-8.5) scenarios in Southwest Asia. Data obtained from Coupled Model Intercomparison Project Phase 6 portal (O'Neill et al., 2016).

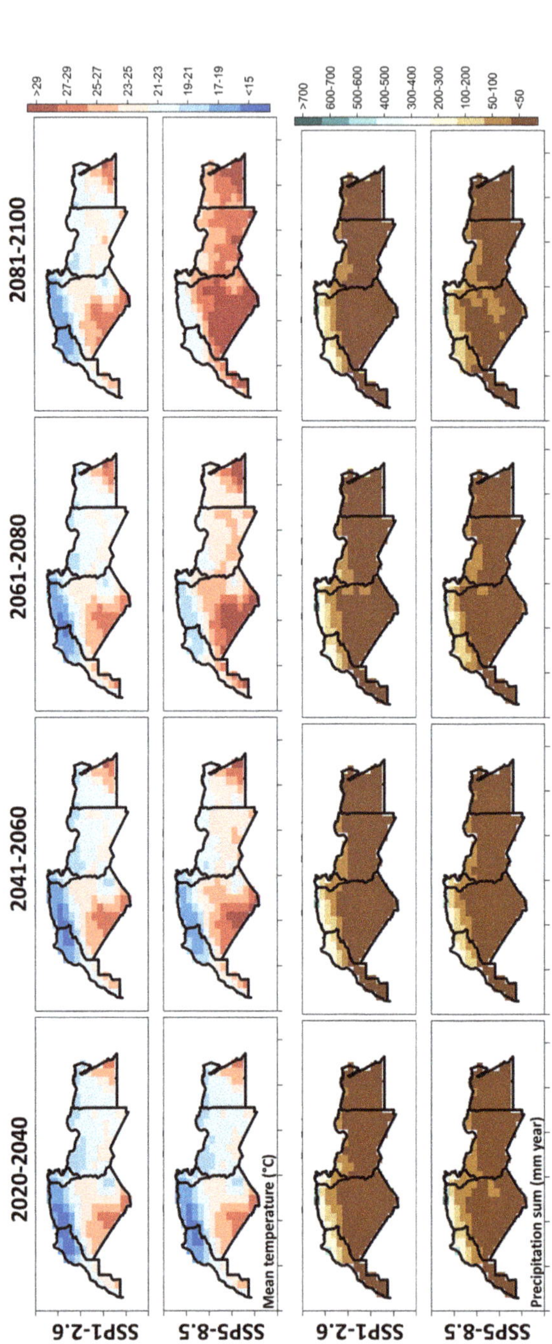

Figure 12 The spatial pattern of mean temperature and precipitation projections in four time steps (2020–2040, 2041–2060, 2061–2080, and 2081–2100) under sustainability (SSP1-2.6) and fossil fuel development (SSP5-8.5) scenarios in North Africa. Data obtained from Coupled Model Intercomparison Project Phase 6 portal (O'Neill et al., 2016).

- Second, more frequent heat stress generally occurs around anthesis and grain filling (Eyshi Rezaei et al., 2015b). Such episodes decrease grain number and single grain weight through grain abortion (Akter and Rafiqul Islam, 2017) and accelerated leaf senescence (Rossi et al., 2017), respectively.

When temperatures and heat intensities increase, protein concentrations in grains increase, affecting protein-starch balance and starch accumulation (Spiertz et al., 2006; Li et al., 2013; Hernández-Espinosa et al., 2018). Total protein yield is thus projected to decline under climate change (Asseng et al., 2019).

Drought and heat stress generally co-occur (Tricker et al., 2018). Drought events result in stomatal closure to save water, restricting transpiration cooling (Zhu et al., 2020) and heating the canopy surface (Elsayed et al., 2015). An increase in drought intensity affects (Ilyas et al., 2021):

1. leaf area expansion;
2. photosynthesis; and
3. altering assimilate partitioning between above- and below-ground organs.

All these processes limit crop yield.

Elevated concentrations of atmospheric CO_2 resulting from climate change can benefit crop growth by promoting photosynthesis (Amthor, 2001) and reducing water loss through transpiration (Wall et al., 2006). Doubling concentration of CO_2 increases photosynthesis of C_3 crops such as wheat and rice by approximately 19% (Kimball, 2016). Under elevated CO_2, stomatal conductance declines significantly, resulting in less water loss through the stomata and higher water use efficiency by up to 36% (O'Leary et al., 2015). Carbon and nitrogen metabolism in plants are altered due to the increase in photosynthesis caused by elevated CO_2 (Adavi and Sathee, 2021). This can significantly reduce protein accumulation in grains, reducing crop quality (Kimball et al., 2001). Overall, however, the positive impacts of CO_2 on crop yield in semi-arid regions are counterbalanced by the combination of high temperature and drought (Asseng et al., 2019).

Climate change warming trends and changes in precipitation patterns can also increase the risks from pests and diseases (in terms of population size, development rate, survival rate, and geographical distribution) (Newton et al., 2011). However, extremely high temperatures and intense drought in semi-arid environments can also slow down disease and pest development, making risk potential environment-specific (Chakraborty and Newton, 2011). Temperature increases since 1960 have caused crop pests and diseases to move at an average of 2.7 ± 8 km per year both north and south (Bebber et al., 2013). Since the mid-1980s, climate change has produced increasingly conducive conditions for the development of wheat stem rust, a devastating disease of wheat and barley (Lewis et al., 2018). It is also possible that rising levels of CO_2 will affect pests indirectly by altering the architecture of host plants and reducing their defences (Robinson et al., 2012).

5 Modelling tools to quantify crop response to climate change

Crop models are mathematical tools which use climate, soil, genotypic characteristics, and other factors to simulate crop growth and development (Asseng et al., 2014) . Some crop models include detailed, mechanistic representations of growth processes (White and Hoogenboom, 2010; Yin and Struik, 2010). Most mechanistic models rely on empirical data (Yin and Struik, 2010). Strictly speaking, empirical models involve constant regression-based relationships that link crop yield changes to climatic factors or other variables (Lobell and Burke, 2010). The main advantage of empirical models is that they can provide a more accurate prediction of variables such as pest and disease pressure than process-based models (Roberts et al., 2017). Mechanistic crop models can represent driving mechanisms of growth processes from simple to complex traits such as grain yield (Webber et al., 2019).

Process-based crop models are the most widely used tools to quantify the impacts of climate change (temperature, precipitation, and CO_2 shift) and the potential of adaptation strategies (change in genotypes, sowing date, and irrigation) to improve resilience of cropping systems (Rötter et al., 2011). Growth and development processes in any crop model can differ greatly in terms of detail and complexity (Hammer et al., 2006). Most models simulate crop growth and development, corresponding soil water and nutrient status (primarily nitrogen) on at least a daily basis, linking to radiation interception, water, and nutrient stress (Stöckle and Kemanian, 2020). Crop models are predominantly driven by temperature, precipitation, and incoming radiation (Chenu et al., 2017). Additional input information, including location characteristics, agronomic management, genotype traits, and chemical/physical characteristics of different soil layers, is also necessary (Silva and Giller, 2020). Crop models that take into account cumulative state variables require an initial state value specification (Basso et al., 2015).

Most crop models simulate phenological phases. Simulation of phenology is vital since it determines the period when crops can effectively intercept radiation and the points at which they are most vulnerable to biotic and abiotic stressors (Wallach et al., 2021). Generally, crop models use thermal time as the driver for crop development from sowing through maturity, which can then be adjusted by vernalization and photoperiod. The effects of temperature rise on acceleration of phenological development is adequately captured in crop models owing to the sensitivity of the thermal time concept to temperature change (Parent et al., 2018; Rezaei et al., 2018). However, a large number of temperature and photoperiod related parameters make parameterization of phenology modules challenging (White et al., 2011).

Two mechanisms mainly drive simulation of biomass accumulation:

- Photosynthesis/respiration; and
- Light use efficiency (LUE).

Simulating photosynthesis/respiration uses semi-empirical functions to simulate net accumulation rate driven by light intensity, temperature, CO_2 level, water, and nitrogen availability (Wu et al., 2016). A significant number of crop models use LUE to simulate crop growth and biomass. In addition to crop species/genotypes and development stage, LUE varies with the concentration of CO_2 in the atmosphere and with the temperature (van Ittersum et al., 2003). Biomass accumulation responds strongly to temperature and water status. The temperature sensitivity response targets the functioning of Rubisco in restricting photosynthesis (Bernacchi et al., 2001). When models simulate LUE effects, they commonly incorporate temperature responses that increase LUE between a base temperature and an optimum temperature before decreasing again at more elevated temperatures (Wang et al., 2017). Crop models not only address the increase in mean temperature on growth and development but also capture the effects of extreme heat episodes on grain number and single grain weight (Eyshi Rezaei et al., 2015b).

The impacts of drought are simulated in most crop models but waterlogging stress is rarely considered in modelling frameworks (Webber et al., 2019). Several crop models account for water shortage when demand for water exceeds supply (Boote et al., 2013). Others utilize the remaining water fraction in soil between a threshold of water uptake without stress and wilting pointing (Sinclair et al., 2005).

The ratio between actual and potential evapotranspiration is frequently considered as an indicator of drought intensity in crop models (van Ittersum et al., 2003). Drought stress primarily affects leaf area expansion, LUE or photosynthesis, carbon partitioning, and leaf senescence in crop models (Webber et al., 2019). A limited number of models account for waterlogging effects on transpiration rate, root, and canopy development (Webber et al., 2019).

In process-based crop models, higher CO_2 is related to an increase in LUE, while mechanistic models simulate increased assimilation as a function of increasing CO_2 (Ewert, 2004). Reduction in transpiration is also simulated using empirical approaches to represent CO_2 effects on stomatal behaviour (Bassu et al., 2014).

A climate change impact assessment is defined as a change in yield (or biomass) because of a climate scenario projection compared to a baseline climate (Webber et al., 2019). An adaptation strategy refers to a management practice to compensate for the adverse impact of climate change. Crop models can assess yield benefits or penalties from implementing specific management changes. Models are increasingly taking into account adaptation strategies (Silva and Giller, 2020).

6 Current climate change impact assessments in North Africa and Southwest Asia

A global meta-analysis found that 2°C of warming would reduce wheat, rice, and maize production (Challinor et al., 2014). Most studies of climate change impacts on agriculture projected significant negative impacts on hot-dry environments, including North Africa and Southwest Asia (Selvaraju, 2013). Such impacts potentially influence food security, quality, and prices as well as biodiversity and land use (Lobell et al., 2008; Thomas, 2008; Ibrahim and Lal, 2013; Ianchovichina et al., 2014; Abdelhedi and Zouari, 2020). These assessments suggest that the existing gap between food demand and production in these regions will get wider due to climate change (Tian and Yu, 2019).

To get a better understanding of available climate change impact assessments related to North Africa and Southwest Asia, we performed a meta-analysis of studies exploring the magnitude of projected yield change for different crops with and without considering CO_2 fertilization effects as well as studies relating to key crops grown in the region. The screening process involved three main steps:

1. Web of Science was used as a database with the search limited to predefined terms covering the period 2000–2022. The search term was TS= ((“Climate change*”) AND (“crop*”) AND (“adaptation*” OR ”yield*”) AND (“model*”)) AND (“North Africa*” OR ”West Asia*” OR ”Southwest Asia*” OR ”Middle East*”)) (number of records = 93).
2. All records were filtered peer-reviewed studies used climate change projections with at least one general circulation model (GCM) and two climate change scenarios included in the analyses (number of records = 41).
3. The last level of filtering was to identify process-based crop models providing yield data whether they considered CO_2 effects or not (number of records = 27).

The results of the meta-analysis showed that wheat and maize are the most popular crops for which there are climate change impact assessments relating to North Africa and Southwest Asia (Fig. 13). The percentage of climate change projections (number of studies × climate scenarios) employed process-based crop models in Southwest Asia (155) was noticeably higher than in North Africa (63) (Fig. 13). Most of those projections did not consider the effects of CO_2 on crop growth, mainly because of a lack of field data for model calibration in those regions (Sultan and Gaetani, 2016).

The potential climate change impact without considering CO_2 effects on mean yield was relatively similar (−18%) for both regions (Fig. 14). However, the projected range was greater for Southwest Asia (+5% to −60% compared to baseline) than North Africa (0% to −55% compared to baseline) (Fig. 14). The increase in relative yield without considering CO_2 fertilization is related to higher precipitation in GCMs (Ruane and McDermid, 2017). Models accounting for higher CO_2 elevation showed a significant increase in yield for Southwest Asia (from 0% to +20%) but not in North Africa (from −26% to +10%) (Fig. 14). The relatively low impacts of CO_2 in North Africa are related to extremely high projected temperatures for the region which already suffers temperature stress at current conditions (Sivakumar et al., 2013; Ahmadalipour et al., 2019). DSSAT (Jones et al., 2003) was the most used modelling platform (70% of the studies) for projecting yield change in both regions, followed by AquaCrop (Steduto et al., 2009) and WOFOST (Wit et al., 2019).

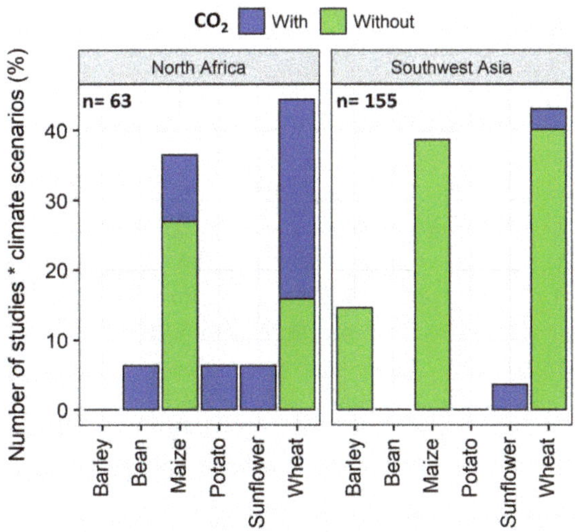

Figure 13 The percentage of projected climate change scenarios (number of studies × climate scenarios) using process-based crop models for different crops with and without considering CO_2 across North Africa and Southwest Asia was obtained from the meta-analysis (Section 2.2).

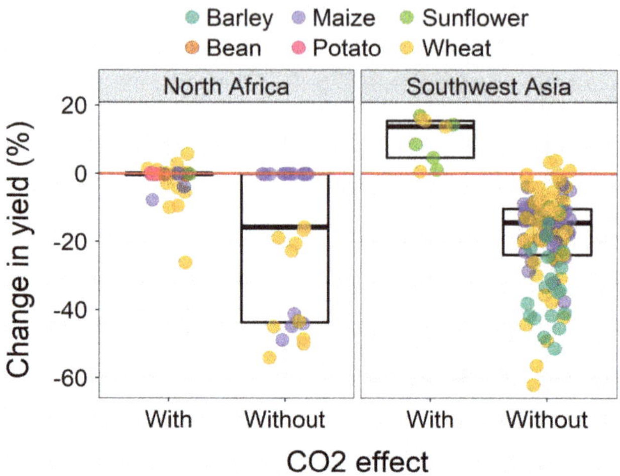

Figure 14 The relative yield change (compared to baseline) projected for different crops using process-based crop models with and without considering CO_2 across North Africa and Southwest Asia was obtained from the meta-analysis (Section 2.2).

7 Accounting for the potential of adaptation strategies

Adaptation strategies seek to adjust agricultural systems to cope with unfavourable climatic conditions and extreme weather events such as heat stress related to climate change (Mechler et al., 2010; Aguilera et al., 2013; Waha et al., 2017; Lamichhane et al., 2015; Atlin et al., 2017). Implementing effective adaptation measures is essential to increasing crop production and quality in the face of a changing climate (Challinor et al., 2014; Sloat et al., 2020). The concept of adaptation is sometimes confused with mitigation (Jagers and Duus-Otterström, 2008). Mitigation refers to preventing or reducing GHGs to lessen the impacts of climate change (Zhao et al., 2018).

Crop models can act as robust tools for assessing ways agriculture could adapt to climate change (Chenu et al., 2017). Crop models have been employed to test a wide range of agronomic practices compensating for the adverse effects of higher temperatures and more frequent extreme weather events on cropping systems. These practices include

1 adapting irrigation schemes to maximize water use efficiency whilst limiting impact on yield in drought-prone regions (Kamali et al., 2022);
2 modifying nitrogen management to maintain grain protein content under elevated CO_2 (Nuttall et al., 2017);
3 adjusting sowing time (shifting to earlier in the growing season) to avoid crop exposure to hotter months (Bourgault et al., 2020);
4 testing the potential of new crops (particularly forage crops) to improve biomass production (Eyshi Rezaei et al., 2015a); and

5 using late maturity genotypes to offset the shortening of the growing season length under higher temperatures (Martín et al., 2014).

The use of crop models can complement field experiments and be a vital component in exploring the adaptive potential of different combinations of crop genetic traits to changing climatic conditions (Hernandez-Ochoa et al., 2019). Models can assist breeding programmes developing climate-resilient genotypes, disentangling the complex interactions between environment, management, and genotype (Paleari et al., 2017).

Studies suggest that adaptation strategies for individual crops could increase simulated yields by 7–15% on average, with wheat and rice adaptation more effective than maize (Challinor et al., 2014). The following sections look at studies focusing specifically on North Africa and Southwest Asia.

7.1 North Africa

Climate change adaptation studies on cropping systems in North Africa have been focused on Algeria, Egypt, Morocco, and Tunisia.

Algeria's rainfed area is planted primarily with durum wheat, representing half of cereal cultivation (Bensemane et al., 2011). A modification in sowing dates would improve yields by 8–135% (depending on scenarios) when CO_2 fertilization effect is considered using AquaCrop (Kourat et al., 2021). Using a generic crop model without including CO_2 effects showed that an earlier sowing date for wheat could compensate for climate-induced yield reduction by −23% to −10% (Chourghal et al., 2016).

In Egypt, crop production is heavily dependent on irrigation due to low precipitation. Adaptation studies in Egypt have focused on improving irrigation efficiency. Using drip irrigation and adjusting the timing of watering would largely counteract the negative impacts of climate change on maize productivity (Ouda et al., 2015). However, deficit irrigation would not improve crop yield under climate change conditions (Attaher et al., 2010). Introducing earlier sowing dates could also compensate for projected wheat yield reductions by −22% to −5% by 2100 (Mostafa et al., 2021). Modelling suggests that new genotypes of barley with less time required to reach heading and a more extended grain filling period would be a possible avenue for developing climate-resilient genotypes (Mansour et al., 2018).

Though effective climate adaptation measures for the cropping systems of Morocco are a priority, there have been few studies modelling their potential benefits (Hirich et al., 2016). Modelling sowing 10 days earlier and no-tillage management found these techniques could maintain water use efficiency in the face of climate change (Brouziyne et al., 2018).

Climate change is expected to adversely impact crop production in Tunisia, where increased temperatures and aridity are expected to increase (Attiaoui and Boufateh, 2019). An increase in irrigation combined with mulching would be a possible adaptation strategy for improving wheat and tomato yields (Saadi et al., 2015; Bird et al., 2016). Mixing zero-tillage and crop residue management could result in a 15% wheat yield improvement under climate change conditions (Bahri et al., 2019).

7.2 Southwest Asia

The number of studies to assess the potential of adaptation studies on cropping systems in Southwest Asia is significantly higher than for North Africa. Most studies have focused on Iran, Pakistan, and Turkey with much less coverage of other countries. Papadaskalopoulou et al. (2020) tested effects of change in sowing period for four crops (barley, wheat, potato, and tomato) in Cyprus. They found that early sowing can fully compensate for the negative impacts of climate change under two different RCP scenarios for all the crops studied. However, early sowing did not compensate for yield losses in wheat and cotton yield in Israel, though improving irrigation had more potential (Haim et al., 2008).

Damaging effects of climate change on crop productivity (−1% to −77%) without implementation of adaptation measures have been reported in several studies in Iran (Eyshi Rezaie and Bannayan, 2012; Mansouri et al., 2015; Sayari et al., 2015; Mosammam et al., 2016; Nouri et al., 2017; Shahvari et al., 2019; Raoufi and Soufizadeh, 2020). Studies of climate change impacts on cropping systems in Iran suggest that changes in precipitation patterns would have a more significant impact than temperature rises, suggesting targeting improvements in irrigation as a key adaptation strategy (Karimi et al., 2018; Rezaei et al., 2021). One study found that increasing the amount of irrigation level was the only effective adaptation measure (compared to late maturity and increased nitrogen application) in maintaining current maize yields in the face of climate change (Rezaei and Lashkari, 2019). Deihimfard et al. (2018) found that improving water use efficiency could increase wheat yield significantly under future climate change conditions.

However, Moradi et al. (2013) found that the combination of earlier sowing dates and increasing the amount of irrigation water could improve maize yields by up to 48%. There have been several studies testing the effects of modified sowing dates and early maturity types of wheat and maize as climate change adaptation strategies across different regions of Iran (Moradi et al., 2014; Hesam Arefi et al., 2017; Rahimi-Moghaddam et al., 2018; Khordadi et al., 2019). Increasing of planting density of wheat together with irrigation could improve yields by 26%, offsetting the adverse effects of climate change in northeast Iran (Paymard et al., 2018). The potential of crop substitution was also tested for eastern parts of Iran (Eyshi Rezaei et al., 2015a). Substituting common genotypes of maize with hybrid pearl millet showed strong potential for improving forage production under climate change conditions. However, this strategy was less effective under more extreme scenarios (Eyshi Rezaei et al., 2015a).

A wide range of modelling studies have been conducted to test different adaptation strategies in Pakistan. Most studies focused on maize, wheat, cotton, sunflowers, and rice as primary crops across Pakistan (Ahmad et al., 2016; Abbas et al., 2017; Ahmed et al., 2018; Awais et al., 2018; Ahmad et al., 2020b; Gul et al., 2020; Alvar-Beltrán et al., 2021; Hussain et al., 2021). Ahmad et al. (2020a) assessed a range of adaptation strategies on maize production in Pakistan. They projected that the development of tolerant genotypes (6.1% yield improvement), increased plant density (5.9%), and improving fertilizer application (4.8%) are the most effective adaptation strategies.

A combination of an increase in planting density, nitrogen application, and early transplanting would increase the rice yield in Pakistan by 8.7% under RCP 8.5 through to mid-century compared to a −7.3% decline in yield projected without adaptation measures (Shabbir et al., 2020). Another study found that using modern genotypes with higher nutrient use efficiencies and higher irrigation levels resulted in a 9–15% increase in simulated rice yield under climate change conditions in the Faisalabad region (Nasir et al., 2020). Similar results have been obtained for winter wheat in the Islamabad region by implementing early sowing and modified irrigation under climate change conditions (Azmat et al., 2021). Another study showed that increasing the nitrogen application rate would not mitigate the negative impacts of climate change, but modifying the sowing date showed a significant wheat yield improvement (Shahid et al., 2021). A combination of adaptation measures, including sowing 5 weeks earlier, 30% increase in nitrogen application, 18% higher planting density, and 17% enhancement in genotype potential, would collectively compensate for adverse effects of climate change on cotton production in Pakistan (Rahman et al., 2018). Increment in phosphorus application rate would also be considered as an adaptation strategy for cotton production (Amin et al., 2018).

A significant yield decline has been projected for wheat, maize, cotton, and sunflowers in Turkey without implementing adaptation measures (Baydar, 2010; Koc, 2011; Sen et al., 2012; Durdu, 2013; Caylak, 2015; Vanli et al., 2019). Gurkan et al. (2020) showed that the implementation of supplementary irrigation is the most effective adaptation strategy to compensate for the negative effects of future climate on sunflowers (39% increase in yield) in the semi-arid region of Turkey. Climate-resilient genotypes have the potential to offset projected yield decline (−35%) in northwestern Turkey (Özdoğan, 2011).

8 Limitations of climate change impact assessments

North Africa and Southwest Asia are among regions with least research conducted on climate change impact assessments on cropping systems (Klingelhöfer et al., 2020). According to our meta-analysis, Pakistan, Iran, and Turkey have the most significant number of published studies employing process-based crop models to test climate change impacts on crop growth. However, most of the studies conducted in those regions suffer from different levels of limitations and uncertainties (Mulla et al., 2020; Silva and Giller, 2020).

Limitations in impact assessment studies in those regions can be classified into two categories (Fig. 15):

- Lack of data; and
- Incompatibility of crop models.

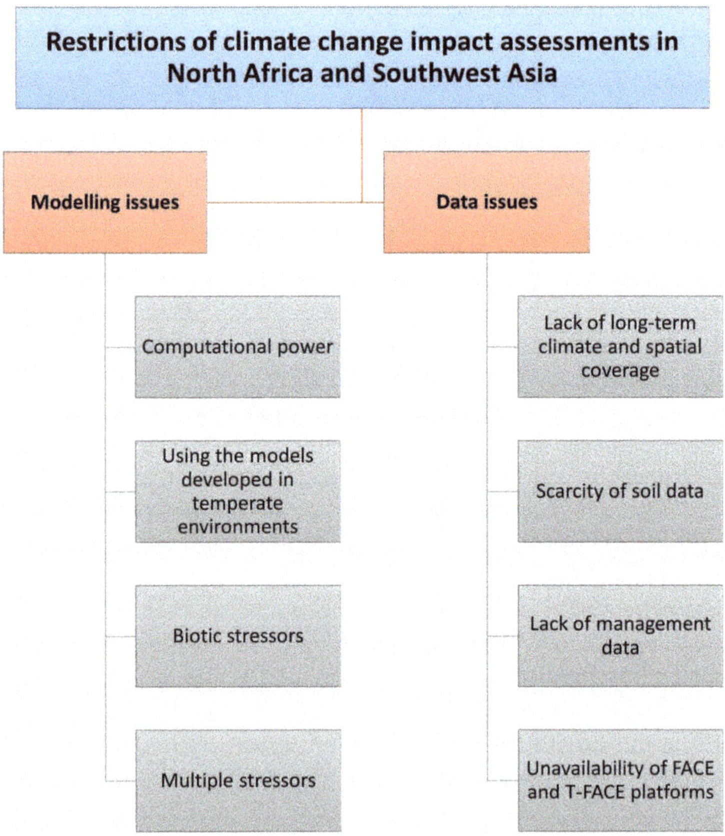

Figure 15 The schematic representation of limitations of climate change impact assessments on cropping systems in North Africa and Southwest Asia.

Limited available data sets for model inputs, including climate, soil, management, and experimental data for model calibration/testing, have a significant influence on crop yield projections. The lack of long-term climate data and heterogeneity in the spatial distribution of stations collecting data mean a considerable level of uncertainty in climate data for crop modelling, particularly in North Africa (Xie et al., 2015; Nordling, 2019). The scarcity of high-resolution soil (chemical and physical properties) and management data (sowing, harvest dates, fertilization and irrigation scheme, etc.) in those regions (Greiner et al., 2017) also limit the accuracy of crop model results.

There is insufficient investment in experimental platforms to assess effects of extreme temperatures, increases in CO_2, and their combination with drought stress in field conditions on genotypes grown in North Africa and Southwest Asia. Free-air CO_2 enrichment (FACE) experiments have allowed exposure of crops to elevated CO_2 concentrations under field conditions since the early 1990s (Ainsworth and Long, 2005). The absence of such experimental data for North Africa and Southwest Asia has meant relying on calibration of crop models based on the response of genotypes developed in other regions (mainly temperate environments) to changes in CO_2 concentrations and CO_2-temperature combinations. Genetic variation in crop response to CO_2 concentration has been explored for crops such as soybean (Bishop et al., 2015), rice (Hasegawa et al., 2013), and wheat (Houshmandfar et al., 2016), based on changes in photosynthesis transpiration and carbon partitioning. It is questionable to base models on the response of genotypes in other environments to climate change.

The costs and advanced engineering required to establish and maintain FACE (Kimball, 2016) and T-FACE (temperature free-air controlled enhancement) (Kimball et al., 2015) systems have limited their use in developing countries. The operational cost of a robust FACE experiment could be US$81–398 m^{-2} $year^{-1}$, depending on crop and environmental conditions (Chakrabarti et al., 2012). FACE system construction and maintenance costs can be up to US$2.8 million (BERAC, 2002). Another limitation is the lack of trained personnel to handle complex engineering design and maintain FACE experiments. One possible option for overcoming those challenges is to extend collaborations between North African and Southwest Asian–based researchers and institutions with access to FACE facilities to evaluate the response of locally grown crop genotypes to climate change. Another option is the use of probabilistic modelling approaches (Awoye et al., 2017) in regional impact assessment studies to cover the possible range of crop responses to combined changes in CO_2 and temperature range instead of using single crop parameter sets in the crop models.

The problem of incompatibility in crop models is mainly due to the fact that most models used in semi-arid and arid regions were developed in temperate environments (Rötter et al., 2018). The significant computational capacity, resources, and cost of large-scale simulations have restricted what indigenous researchers are able to achieve (Guan et al., 2017). Crop models often cannot capture the impacts of biotic stressors on crop growth (Donatelli et al., 2017). Ignoring biotic stress may not significantly change model results under optimum management practices in developed cropping systems (Senapati and Semenov, 2019). However, cropping systems in North Africa and Southwest Asia are generally not optimally managed (Marrou et al., 2021; Wajid et al., 2021). Consequently, the adverse effects of pests and disease on yield are significant, resulting in inaccuracy in crop model outcomes.

There is also a lack of crop model-based ideotyping studies (Paleari et al., 2019) in those regions to guide breeding efforts for developing genotypes adapted for specific environmental conditions. An ideotype as 'a combination of morphological and/or physiological traits, or their genetic bases, optimizing crop performance to a particular biophysical environment, crop management, and end-use' (Martre et al., 2015). Crop models are widely used for testing the impact of different crop traits related to phenology, root, resource use efficiency, canopy structure, and photosynthesis under different environmental conditions (Hammer et al., 2002; Raza et al., 2013; Drewry et al., 2014).

There have been few regional studies testing the effects of phosphorus and potassium management on yield, particularly in Africa (van der Velde et al., 2014; Rötter et al., 2018). More detailed crop models need to be developed to capture the effects of modifications in phosphorus application under changing climatic conditions (Dzotsi et al., 2010).

9 Conclusion and future trends

A series of changes are needed for more reliable projections of the response of cropping systems to climate change in North Africa and Southwest Asia. The first step is to invest in establishing targeted experimental platforms to capture the response of local crops, genotypes, and adaptation management practices to combined changes in CO_2, temperature, and agronomic management. A limited number of studies have tested the effects of elevated CO_2 on crop growth under controlled conditions (growth chambers) in these regions (Ahmadi Lahijani et al., 2018; Kazemi et al., 2018). However, the effect of CO_2 on crop yield is more pronounced in chamber conditions than in field conditions (Kimball, 2016). Calibrating the crop models for climate change impact assessments based on chamber experiments can overestimate CO_2 effects (Hasegawa et al., 2017). FACE experimental results are key for calibration and validation for crop models (Wang et al., 2021). It is also critical to test CO_2 effects interacting with other climatic variables such as temperature and management practices (Burkart et al., 2009).

Modelling should also expand from cereals into perennial crops and grasslands which cover significant areas in North Africa and Southwest Asia. However, the modelling solutions for those crops are not as advanced as available modelling routines for cereal crops (van Oijen et al., 2018). New crop models need to be developed simulating perennial crops and grasslands under semi-arid conditions (Oomen et al., 2016). Capturing the impacts of multiple stressors which generally co-occur (e.g. heat and drought) should be one of the priorities for developing new modelling routines in impact assessments. Experimental evidence has shown that crop response to multiple stressors is significantly altered compared to response to individual stresses (Pandey et al., 2017). Due to complexities in dealing with pests and diseases, only a few numerical solutions are available to simulate relevant crop responses (Fernandes et al., 2019). Crop responses to biotic stressors should be integrated into crop modelling, particularly for the regions where crop protection measures are not strong (Donatelli et al., 2017).

There is also a need for better allowance for differences in soil conditions (Hawkins et al., 2013; Maraun et al., 2017) and better soil sampling (Hengl et al., 2015), particularly in North Africa. The initiatives such as Africa Soil Information Service (AfSIS) are trying to address the scarcity of soil data in Africa (Lawrimore, 2012), but data quality and coverage still need improvement (Leenaars et al., 2018). It is fundamental to determine the primary source of uncertainty in impact assessments (Laux et al., 2021). The main source of uncertainty in projecting future crop yield varies across regions. For instance, GCMs are the dominant source of uncertainty in Australian wheat yield projections, whereas crop model structures are the primary source of uncertainty in China (Wang et al., 2020).

The lack of high-resolution spatial information regarding canopy state variables (e.g. leaf area index and chlorophyll content), biotic or abiotic stress intensity, soil characteristics, and land-cover data can be covered by remote sensing technologies. Sensing information from multiple platforms can play a critical role in developing multi-scale solutions for developing crop models developed initially at the field scale of specific environments, making them more suitable for large-scale multi-environments assessments. Integration of remotely sensed information and crop growth models also need to be implemented in adaptation planning for climate change programmes (Ginaldi et al., 2019) without proper experimental and ground-based measurements such as North Africa and Southwest Asia (Shiff et al., 2021).

Satellite-based sensing can provide high spatiotemporal resolution data used for estimation of heat and drought intensity (e.g. land surface temperature), agronomic management practices (e.g. nitrogen fertilization management), and improve the parametrization of crop model in combination with data-driven approaches (machine and deep learning) (Boote et al., 2011; Kasampalis et al., 2018; Silva and Giller, 2020). It can significantly decrease crop model simulation uncertainty due to a lack of high-resolution input data and yield reduction factors from other sources (Kasampalis et al., 2018). Several approaches have been suggested to link remote sensing data and crop models (Benami et al., 2021), including data assimilation techniques (Jin et al., 2018). Algorithms for assimilating data link crop model simulations with sensing data to develop an estimate that is an optimum combination of both inputs (Huang et al., 2019). Such approaches are particularly important for data-scarce regions such as North Africa and Southwest Asia.

10 Where to look for further information

Contribution of agriculture to the economy of North Africa and Southwest Asia:

North Africa: https://longreads.tni.org/towards-a-just-agricultural-transition-in-north-africa.

Southwest Asia: https://www.mei.edu/publications/agricultural-technology-middle-east-sowing-seeds-future.

Climate change in North Africa and Southwest Asia:

https://www.worldbank.org/en/programs/mena-climate-change#1.

https://unfccc.int/news/climate-change-is-an-increasing-threat-to-africa.

https://www.unep.org/regions/west-asia/regional-initiatives/responding-climate-change.

Impacts of climate change on agricultural systems of North Africa and Southwest Asia:

https://www.fao.org/publications/card/en/c/CA1439EN/.

https://www.iemed.org/publication/climate-change-in-the-mena-region-environmental-risks-socioeconomic-effects-and-policy-challenges-for-the-future/.

https://link.springer.com/book/10.1007/978-94-007-6751-5.

11 References

Abbas, G., Ahmad, S., Ahmad, A., Nasim, W., Fatima, Z., Hussain, S., Rehman, M. H. u., Khan, M. A., Hasanuzzaman, M., Fahad, S., Boote, K. J. and Hoogenboom, G. 2017. Quantification the impacts of climate change and crop management on phenology of maize-based cropping system in Punjab, Pakistan. *Agricultural and Forest Meteorology* 247, 42–55.

Abdelhedi, I. T. and Zouari, S. Z. 2020. Agriculture and food security in North Africa: A theoretical and empirical approach. *Journal of the Knowledge Economy* 11(1), 193–210.

Adavi, S. B. and Sathee, L. 2021. Elevated CO_2 alters tissue balance of nitrogen metabolism and downregulates nitrogen assimilation and signalling gene expression in wheat seedlings receiving high nitrate supply. *Protoplasma* 258(1), 219–233.

Aguilera, E., Lassaletta, L., Gattinger, A. and Gimeno, B. S. 2013. Managing soil carbon for climate change mitigation and adaptation in Mediterranean cropping systems: A meta-analysis. *Agriculture, Ecosystems and Environment* 168, 25–36.

Ahmad, I., Ahmad, B., Boote, K. and Hoogenboom, G. 2020a. Adaptation strategies for maize production under climate change for semi-arid environments. *European Journal of Agronomy* 115, 126040.

Ahmad, M. J., Iqbal, M. A. and Choi, K. S. 2020b. Climate-driven constraints in sustaining future wheat yield and water productivity. *Agricultural Water Management* 231, 105991.

Ahmad, S., Nadeem, M., Abbas, G., Fatima, Z., Zeb Khan, R. J., Ahmed, M., Ahmad, A., Rasul, G. and Azam Khan, M. 2016. Quantification of the effects of climate warming and crop management on sugarcane phenology. *Climate Research* 71(1), 47–61.

Ahmadalipour, A., Moradkhani, H. and Kumar, M. 2019. Mortality risk from heat stress expected to hit poorest nations the hardest. *Climatic Change* 152(3–4), 569–579.

Ahmadi Lahijani, M. J., Kafi, M., Nezami, A., Nabati, J., Mehrjerdi, M. Z., Shahkoomahally, S. and Erwin, J. 2018. Variations in assimilation rate, photoassimilate translocation, and cellular fine structure of potato cultivars (*Solanum tuberosum* L.) exposed to elevated CO_2. *Plant Physiology and Biochemistry: PPB* 130, 303–313.

Ahmed, I., Ur Rehman, M. H., Ahmed, S., Hussain, J., Ullah, A. and Judge, J. 2018. Assessing the impact of climate variability on maize using simulation modeling under semi-arid environment of Punjab, Pakistan. *Environmental Science and Pollution Research International* 25(28), 28413–28430.

Ainsworth, E. A. and Long, S. P. 2005. What have we learned from 15 years of free-air CO_2 enrichment (FACE)? A meta-analytic review of the responses of photosynthesis, canopy properties and plant production to rising CO_2. *The New Phytologist* 165(2), 351–371.

Akter, N. and Rafiqul Islam, M. 2017. Heat stress effects and management in wheat: A review. *Agronomy for Sustainable Development* 37(5), 1–17.

Al-Delaimy, W. K. 2020. Vulnerable populations and regions: Middle East as a case study. In: Al-Delaimy, W. K., Ramanathan, V. and Sánchez Sorondo, M. (Eds.), *Health of People, Health of Planet and Our Responsibility*. Cham: Springer International Publishing, pp. 121–133.

Alvar-Beltrán, J., Heureux, A., Soldan, R., Manzanas, R., Khan, B. and Dalla Marta, A. 2021. Assessing the impact of climate change on wheat and sugarcane with the AquaCrop model along the Indus River Basin, Pakistan. *Agricultural Water Management* 253, 106909.

Amin, A., Nasim, W., Mubeen, M., Ahmad, A., Nadeem, M., Urich, P., Fahad, S., Ahmad, S., Wajid, A., Tabassum, F., Hammad, H. M., Sultana, S. R., Anwar, S., Baloch, S. K., Wahid, A., Wilkerson, C. J. and Hoogenboom, G. 2018. Simulated CSM-CROPGRO-cotton yield under projected future climate by SimCLIM for southern Punjab, Pakistan. *Agricultural Systems* 167, 213–222.

Amthor, J. S. 2001. Effects of atmospheric CO_2 concentration on wheat yield: Review of results from experiments using various approaches to control CO_2 concentration. *Field Crops Research* 73(1), 1–34.

Araujo-Enciso, S. R. and Fellmann, T. 2020. Yield variability and harvest failures in Russia, Ukraine and Kazakhstan and their possible impact on food security in the Middle East and North Africa. *Journal of Agricultural Economics* 71(2), 493–516.

Asseng, S., Martre, P., Maiorano, A., Rötter, R. P., O'Leary, G. J., Fitzgerald, G. J., Girousse, C., Motzo, R., Giunta, F., Babar, M. A., Reynolds, M. P., Kheir, A. M. S., Thorburn, P. J., Waha, K., Ruane, A. C., Aggarwal, P. K., Ahmed, M., Balkovič, J., Basso, B., Biernath, C., Bindi, M., Cammarano, D., Challinor, A. J., Sanctis, G. de, Dumont, B., Eyshi Rezaei, E., Fereres, E., Ferrise, R., Garcia-Vila, M., Gayler, S., Gao, Y., Horan, H., Hoogenboom, G., Izaurralde, R. C., Jabloun, M., Jones, C. D., Kassie, B. T., Kersebaum, K.-C., Klein, C., Koehler, A.-K., Liu, B. and Minoli, S. 2019. Climate change impact and adaptation for wheat protein. *Global Change Biology* 25, 155–173.

Asseng, S., Zhu, Y., Basso, B., Wilson, T. and Cammarano, D. 2014. Simulation modeling: Applications in cropping systems. In: *Encyclopedia of Agriculture and Food Systems*. Elsevier, pp. 102–112.

Atlin, G. N., Cairns, J. E. and Das, B. 2017. Rapid breeding and varietal replacement are critical to adaptation of cropping systems in the developing world to climate change. *Global Food Security* 12, 31–37.

Attaher, S. M., Medany, M. A. and El-Gindy, A. 2010. Feasibility of some adaptation measures of on-farm irrigation in Egypt under water scarcity conditions. *Options: Mediterraneennes* 95, 12.

Attiaoui, I. and Boufateh, T. 2019. Impacts of climate change on cereal farming in Tunisia: A panel ARDL-PMG approach. *Environmental Science and Pollution Research International* 26(13), 13334–13345.

Awais, M., Wajid, A., Saleem, M. F., Nasim, W., Ahmad, A., Raza, M. A. S., Bashir, M. U., Mubeen, M., Hammad, H. M., Habib Ur Rahman, M., Saeed, U., Arshad, M. N. and Hussain, J. 2018. Potential impacts of climate change and adaptation strategies for sunflower in Pakistan. *Environmental Science and Pollution Research International* 25(14), 13719–13730.

Awoye, O. H. R., Pollinger, F., Agbossou, E. K. and Paeth, H. 2017. Dynamical-statistical projections of the climate change impact on agricultural production in Benin by means of a cross-validated linear model combined with Bayesian statistics. *Agricultural and Forest Meteorology* 234–235, 80–94.

Azmat, M., Ilyas, F., Sarwar, A., Huggel, C., Vaghefi, S. A., Hui, T., Qamar, M. U., Bilal, M. and Ahmed, Z. 2021. Impacts of climate change on wheat phenology and yield in Indus Basin, Pakistan. *The Science of the Total Environment* 790, 148221.

Bahri, H., Annabi, M., Cheikh M'Hamed, H. and Frija, A. 2019. Assessing the long-term impact of conservation agriculture on wheat-based systems in Tunisia using APSIM simulations under a climate change context. *The Science of the Total Environment* 692, 1223–1233.

Barlow, M., Zaitchik, B., Paz, S., Black, E., Evans, J. and Hoell, A. 2016. A review of drought in the Middle East and southwest Asia. *Journal of Climate* 29(23), 8547–8574.

Basso, B., Hyndman, D. W., Kendall, A. D., Grace, P. R. and Robertson, G. P. 2015. Can impacts of climate change and agricultural adaptation strategies be accurately quantified if crop models are annually re-initialized? *PLoS ONE* 10(6), e0127333.

Bassu, S., Brisson, N., Durand, J. L., Boote, K., Lizaso, J., Jones, J. W., Rosenzweig, C., Ruane, A. C., Adam, M., Baron, C., Basso, B., Biernath, C., Boogaard, H., Conijn, S., Corbeels, M., Deryng, D., Sanctis, G. de, Gayler, S., Grassini, P., Hatfield, J., Hoek, S., Izaurralde, C., Jongschaap, R., Kemanian, A. R., Kersebaum, K. C., Kim, S. H.,

Kumar, N. S., Makowski, D., Müller, C., Nendel, C., Priesack, E., Pravia, M. V., Sau, F., Shcherbak, I., Tao, F., Teixeira, E., Timlin, D. and Waha, K. 2014. How do various maize crop models vary in their responses to climate change factors? *Global Change Biology* 20(7), 2301–2320.

Baydar, A. 2010. *Climate Change Effects on Cotton Production Under the Condition of Seyhan Plain*. MSc, Adana, Turkey.

Bebber, D. P., Ramotowski, M. A. T. and Gurr, S. J. 2013. Crop pests and pathogens move polewards in a warming world. *Nature Climate Change* 3(11), 985–988.

Benami, E., Jin, Z., Carter, M. R., Ghosh, A., Hijmans, R. J., Hobbs, A., Kenduiywo, B. and Lobell, D. B. 2021. Uniting remote sensing, crop modelling and economics for agricultural risk management. *Nature Reviews Earth and Environment* 2(2), 140–159.

Bensemane, L., Bouzerzour, H., Benmahammed, A. and Mimouni, H. 2011. Assessment of the phenotypic variation within two- and six-rowed barley (*Hordeum vulgare* L.) breeding lines grown under semi-arid conditions. *Advances in Environmental Biology* 5, 1454–1460.

BERAC. 2002. An evaluation of the Department of Energy's Free-Air Carbon Dioxide Enrichment (FACE) experiments as scientific user facilities. Available at: https://science.osti.gov/-/media/ber/berac/pdf/Face_report.pdf (accessed 6 April 2022).

Bernacchi, C. J., Singsaas, E. L., Pimentel, C., Portis, A. R., Jr. and Long, S. P. 2001. Improved temperature response functions for models of RuBisCO-limited photosynthesis. *Plant, Cell and Environment* 24(2), 253–259.

Bird, D. N., Benabdallah, S., Gouda, N., Hummel, F., Koeberl, J., La Jeunesse, I., Meyer, S., Prettenthaler, F., Soddu, A. and Woess-Gallasch, S. 2016. Modelling climate change impacts on and adaptation strategies for agriculture in Sardinia and Tunisia using AquaCrop and value-at-risk. *The Science of the Total Environment* 543(B), 1019–1027.

Bishop, K. A., Betzelberger, A. M., Long, S. P. and Ainsworth, E. A. 2015. Is there potential to adapt soybean (*Glycine max* Merr.) to future CO_2? An analysis of the yield response of 18 genotypes in free-air CO_2 enrichment. *Plant, Cell and Environment* 38(9), 1765–1774.

Boote, K. J., Ibrahim, A. M. H., Lafitte, R., McCulley, R., Messina, C., Murray, S. C., Specht, J. E., Taylor, S., Westgate, M. E., Glasener, K., Bijl, C. G. and Giese, J. H. 2011. Position statement on crop adaptation to climate change. *Crop Science* 51(6), 2337–2343.

Boote, K. J., Jones, J. W., White, J. W., Asseng, S. and Lizaso, J. I. 2013. Putting mechanisms into crop production models. *Plant, Cell and Environment* 36(9), 1658–1672.

Bourgault, M., Webber, H. A., Chenu, K., O'Leary, G. J., Gaiser, T., Siebert, S., Dreccer, F., Huth, N., Fitzgerald, G. J., Tausz, M. and Ewert, F. 2020. Early vigour in wheat: Could it lead to more severe terminal drought stress under elevated atmospheric CO_2 and semi-arid conditions? *Global Change Biology* 26(7), 4079–4093.

Britannica. 2021. Southwest Asia. Available at: https://www.britannica.com/place/Southwest-Asia (accessed 18 January 2021).

Brouziyne, Y., Abouabdillah, A., Hirich, A., Bouabid, R., Zaaboul, R. and Benaabidate, L. 2018. Modeling sustainable adaptation strategies toward a climate-smart agriculture in a Mediterranean watershed under projected climate change scenarios. *Agricultural Systems* 162, 154–163.

Burkart, S., Manderscheid, R. and Weigel, H.-J. 2009. Canopy CO_2 exchange of sugar beet under different CO_2 concentrations and nitrogen supply: Results from a free-air CO_2 enrichment study. *Plant Biology (Stuttgart, Germany)* 11(Suppl. 1), 109–123.

Caylak. 2015. Examination of possible effects of climate change on wheat growth and yield by a crop-climate simulation model. *Medicine Science, Istanbul, Turkey*.

Chakrabarti, B., Singh, S. D., Kumar, S. N., Aggarwal, P. K., Pathak, H. and Nagarajan, S. 2012. Low-cost facility for assessing impact of carbon dioxide on crops. *Current Science* 102, 1035–1040.

Chakraborty, S. and Newton, A. C. 2011. Climate change, plant diseases and food security: An overview. *Plant Pathology* 60(1), 2–14.

Challinor, A. J., Watson, J., Lobell, D. B., Howden, S. M., Smith, D. R. and Chhetri, N. 2014. A meta-analysis of crop yield under climate change and adaptation. *Nature Climate Change* 4(4), 287–291.

Chenu, K., Porter, J. R., Martre, P., Basso, B., Chapman, S. C., Ewert, F., Bindi, M. and Asseng, S. 2017. Contribution of crop models to adaptation in wheat. *Trends in Plant Science* 22(6), 472–490.

Chourghal, N., Lhomme, J. P., Huard, F. and Aidaoui, A. 2016. Climate change in Algeria and its impact on durum wheat. *Regional Environmental Change* 16(6), 1623–1634.

Deihimfard, R., Eyni-Nargeseh, H. and Mokhtassi-Bidgoli, A. 2018. Effect of future climate change on wheat yield and water use efficiency under semi-arid conditions as predicted by APSIM-wheat model. *International Journal of Plant Production* 12(2), 115–125.

Donatelli, M., Magarey, R. D., Bregaglio, S., Willocquet, L., Whish, J. P. M. and Savary, S. 2017. Modelling the impacts of pests and diseases on agricultural systems. *Agricultural Systems* 155, 213–224.

Drewry, D. T., Kumar, P. and Long, S. P. 2014. Simultaneous improvement in productivity, water use, and albedo through crop structural modification. *Global Change Biology* 20(6), 1955–1967.

Droogers, P., Immerzeel, W. W., Terink, W., Hoogeveen, J., Bierkens, M. F. P., van Beek, L. P. H. and Debele, B. 2012. Water resources trends in Middle East and North Africa towards 2050. *Hydrology and Earth System Sciences* 16(9), 3101–3114.

Durdu, Ö. F. 2013. Evaluation of climate change effects on future corn (*Zea mays* L.) yield in western Turkey. *International Journal of Climatology* 33(2), 444–456.

Dzotsi, K. A., Jones, J. W., Adiku, S. G. K., Naab, J. B., Singh, U., Porter, C. H. and Gijsman, A. J. 2010. Modeling soil and plant phosphorus within DSSAT. *Ecological Modelling* 221(23), 2839–2849.

El-Gharras, O., El-Mourid, M. and Boulal, H. 2017. Conservation agriculture in North Africa: Experiences, achievements and challenges. In: Kassam, A. H., Mkomwa, S. and Friedrich, T. (Eds.), *Conservation Agriculture for Africa: Building Resilient Farming Systems in a Changing Climate*. Wallingford: CABI Publishing, pp. 127–138.

El Kenawy, A., McCabe, M. F., Vicente-Serrano, S. M., Robaa, S. M. and Lopez-Moreno, J. I. 2016. Recent changes in continentality and aridity conditions over the Middle East and North Africa region, and their association with circulation patterns. *Climate Research* 69(1), 25–43.

Elkouk, A., El Morjani, Z. E. A., Pokhrel, Y., Chehbouni, A., Sifeddine, A., Thober, S. and Bouchaou, L. 2021. Multi-model ensemble projections of soil moisture drought over North Africa and the Sahel region under 1.5, 2, and 3°C global warming. *Climatic Change* 167(3–4).

Elsayed, S., Rischbeck, P. and Schmidhalter, U. 2015. Comparing the performance of active and passive reflectance sensors to assess the normalized relative canopy temperature and grain yield of drought-stressed barley cultivars. *Field Crops Research* 177, 148–160.

EM-DAT. 2022. *The International Disaster Darabase*. Available at: https://www.emdat.be/ (accessed 27 January 2022).

Engelbrecht, F., Adegoke, J., Bopape, M.-J., Naidoo, M., Garland, R., Thatcher, M., McGregor, J., Katzfey, J., Werner, M., Ichoku, C. and Gatebe, C. 2015. Projections of rapidly rising surface temperatures over Africa under low mitigation. *Environmental Research Letters* 10(8), 85004.

Ewert, F. 2004. Modelling plant responses to elevated CO_2: How important is leaf area index? *Annals of Botany* 93(6), 619–627.

Eyshi Rezaie, E. and Bannayan, M. 2012. Rainfed wheat yields under climate change in northeastern Iran. *MET Apps* 19(3), 346–354.

Eyshi Rezaei, E., Gaiser, T., Siebert, S. and Ewert, F. 2015a. Adaptation of crop production to climate change by crop substitution. *Mitigation and Adaptation Strategies for Global Change* 20(7), 1155–1174.

Eyshi Rezaei, E., Webber, H., Gaiser, T., Naab, J. and Ewert, F. 2015b. Heat stress in cereals: Mechanisms and modelling. *European Journal of Agronomy* 64, 98–113.

FAO. 2007. Length of growing period (LGP) zones of the world (FGGD). Available at: https://data.apps.fao.org/map/catalog/srv/eng/catalog.search?id=14057#/metadata/e6f969e0-74a6-11db-b9b2-000d939bc5d8 (accessed 27 January 2022).

FAO. 2021. FAOSTAT statistical database. *FAOSTAT Statistical Database* (accessed 10 January 2022).

Fernandes, J. M., Pavan, W., Pequeno, D., Wiest, R., Holbig, C. A., Oliveira, F. and Hoogenboom, G. 2019. Improving crop pest/disease modeling. In: Boote, K. (Ed.), *Advances in Crop Modelling for a Sustainable Agriculture. Burleigh Dodds Series in Agricultural Science*. Cambridge: Burleigh Dodds Science Publishing, pp. 127–148.

Fragaszy, S. R., Jedd, T., Wall, N., Knutson, C., Fraj, M. B., Bergaoui, K., Svoboda, M., Hayes, M. and McDonnell, R. 2020. Drought monitoring in the Middle East and North Africa (MENA) region: Participatory engagement to inform early warning systems. *Bulletin of the American Meteorological Society* 101(7), E1148–E1173.

Ginaldi, F., Bajocco, S., Bregaglio, S. and Cappelli, G. 2019. Spatializing crop models for sustainable agriculture. In: Farooq, M. and Pisante, M. (Eds.), *Innovations in Sustainable Agriculture*. Cham: Springer International Publishing, pp. 599–619.

Greiner, L., Keller, A., Grêt-Regamey, A. and Papritz, A. 2017. Soil function assessment: Review of methods for quantifying the contributions of soils to ecosystem services. *Land Use Policy* 69, 224–237.

Guan, K., Sultan, B., Biasutti, M., Baron, C. and Lobell, D. B. 2017. Assessing climate adaptation options and uncertainties for cereal systems in West Africa. *Agricultural and Forest Meteorology* 232, 291–305.

Gul, F., Ahmed, I., Ashfaq, M., Jan, D., Fahad, S., Li, X., Wang, D., Fahad, M., Fayyaz, M. and Shah, S. A. 2020. Use of crop growth model to simulate the impact of climate change on yield of various wheat cultivars under different agro-environmental conditions in Khyber Pakhtunkhwa, Pakistan. *Arabian Journal of Geosciences* 13(3), 1–14.

Gurkan, H., Shelia, V., Bayraktar, N., Ersoy Yildirim, Y., Yesilekin, N., Gunduz, A., Boote, K., Porter, C. and Hoogenboom, G. 2020. Estimating the potential impact of climate change on sunflower yield in the Konya Province of Turkey. *Journal of Agricultural Science* 158(10), 806–818.

Haim, D., Shechter, M. and Berliner, P. 2008. Assessing the impact of climate change on representative field crops in Israeli agriculture: A case study of wheat and cotton. *Climatic Change* 86(3–4), 425–440.

Hammer, G., Cooper, M., Tardieu, F., Welch, S., Walsh, B., van Eeuwijk, F., Chapman, S. and Podlich, D. 2006. Models for navigating biological complexity in breeding improved crop plants. *Trends in Plant Science* 11(12), 587–593.

Hammer, G. L., Kropff, M. J., Sinclair, T. R. and Porter, J. R. 2002. Future contributions of crop modelling–From heuristics and supporting decision making to understanding genetic regulation and aiding crop improvement. *European Journal of Agronomy* 18(1–2), 15–31.

Harris, I., Jones, P. D., Osborn, T. J. and Lister, D. H. 2014. Updated high-resolution grids of monthly climatic observations - The CRU TS3.10 dataset. *International Journal of Climatology* 34(3), 623–642.

Hasegawa, T., Li, T., Yin, X., Zhu, Y., Boote, K., Baker, J., Bregaglio, S., Buis, S., Confalonieri, R., Fugice, J., Fumoto, T., Gaydon, D., Kumar, S. N., Lafarge, T., Marcaida Iii, M., Masutomi, Y., Nakagawa, H., Oriol, P., Ruget, F., Singh, U., Tang, L., Tao, F., Wakatsuki, H., Wallach, D., Wang, Y., Wilson, L. T., Yang, L., Yang, Y., Yoshida, H., Zhang, Z. and Zhu, J. 2017. Causes of variation among rice models in yield response to CO_2 examined with free-air CO_2 enrichment and growth chamber experiments. *Scientific Reports* 7(1), 14858.

Hasegawa, T., Sakai, H., Tokida, T., Nakamura, H., Zhu, C., Usui, Y., Yoshimoto, M., Fukuoka, M., Wakatsuki, H., Katayanagi, N., Matsunami, T., Kaneta, Y., Sato, T., Takakai, F., Sameshima, R., Okada, M., Mae, T. and Makino, A. 2013. Rice cultivar responses to elevated CO_2 at two free-air CO_2 enrichment (FACE) sites in Japan. *Functional Plant Biology: FPB* 40(2), 148–159.

Hawkins, E., Osborne, T. M., Ho, C. K. and Challinor, A. J. 2013. Calibration and bias correction of climate projections for crop modelling: An idealised case study over Europe. *Agricultural and Forest Meteorology* 170, 19–31.

Hengl, T., Heuvelink, G. B. M., Kempen, B., Leenaars, J. G. B., Walsh, M. G., Shepherd, K. D., Sila, A., MacMillan, R. A., Mendes de Jesus, J., Tamene, L. and Tondoh, J. E. 2015. Mapping soil properties of Africa at 250 m resolution: Random forests significantly improve current predictions. *PLoS ONE* 10(6), e0125814.

Hernández-Espinosa, N., Mondal, S., Autrique, E., Gonzalez-Santoyo, H., Crossa, J., Huerta-Espino, J., Singh, R. P. and Guzmán, C. 2018. Milling, processing and end-use quality traits of CIMMYT spring bread wheat germplasm under drought and heat stress. *Field Crops Research* 215, 104–112.

Hernandez-Ochoa, I. M., Luz Pequeno, D. N., Reynolds, M., Babar, M. A., Sonder, K., Milan, A. M., Hoogenboom, G., Robertson, R., Gerber, S., Rowland, D. L., Fraisse, C. W. and Asseng, S. 2019. Adapting irrigated and rainfed wheat to climate change in semi-arid environments: Management, breeding options and land use change. *European Journal of Agronomy* 109, 125915.

Hesam Arefi, I., Saffari, M. and Moradi, R. 2017. Evaluating planting date and variety management strategies for adapting winter wheat to climate change impacts in arid regions. *IJCCSM* 9(6), 846–863.

Hirich, A., Fatnassi, H., Ragab, R. and Choukr-Allah, R. 2016. Prediction of climate change impact on corn grown in the south of morocco using the Saltmed model. *Irrigation and Drainage* 65(1), 9–18.

Houshmandfar, A., Fitzgerald, G. J., Macabuhay, A. A., Armstrong, R., Tausz-Posch, S., Löw, M. and Tausz, M. 2016. Trade-offs between water-use related traits, yield components and mineral nutrition of wheat under free-air CO_2 enrichment (FACE). *European Journal of Agronomy* 76, 66–74.

Huang, J., Gómez-Dans, J. L., Huang, H., Ma, H., Wu, Q., Lewis, P. E., Liang, S., Chen, Z., Xue, J.-H., Wu, Y., Zhao, F., Wang, J. and Xie, X. 2019. Assimilation of remote sensing into crop growth models: Current status and perspectives. *Agricultural and Forest Meteorology* 276-277, 107609.

Hussain, J., Khaliq, T., Rahman, M. H. u., Ullah, A., Ahmed, I., Srivastava, A. K., Gaiser, T. and Ahmad, A. 2021. Effect of temperature on sowing dates of wheat under arid and semi-arid climatic regions and impact quantification of climate change through mechanistic modeling with evidence from field. *Atmosphere* 12(7), 927.

Ianchovichina, E. I., Loening, J. L. and Wood, C. A. 2014. How vulnerable are Arab countries to global food price shocks? *The Journal of Development Studies* 50(9), 1302–1319.

Ibrahim, M. and Lal, R. 2013. Climate change and land use in the WANA region with a specific reference to morocco. In: Sivakumar, M. V., Lal, R., Selvaraju, R. and Hamdan, I. (Eds.), *Climate Change and Food Security in West Asia and North Africa*. Dordrecht: Springer Netherlands, pp. 89–113.

Ide, T. 2018. Climate war in the middle east? Drought, the Syrian civil war and the state of climate-conflict research. *Current Climate Change Reports* 4(4), 347–354.

IFPRI. 2018. *Agriculture and Economic Transformation in the Middle East and North Africa: A Review of the Past With Lessons for the Future*. Washington, DC: International Food Policy Research Institute.

Ilyas, M., Nisar, M., Khan, N., Hazrat, A., Khan, A. H., Hayat, K., Fahad, S., Khan, A. and Ullah, A. 2021. Drought tolerance strategies in plants: A mechanistic approach. *Journal of Plant Growth Regulation* 40(3), 926–944.

IPCC. 2021. *IPCC, 2021: Summary for Policymakers*. Cambridge: Cambridge University Press.

Jagers, S. C. and Duus-Otterström, G. 2008. Dual climate change responsibility: on moral divergences between mitigation and adaptation. *Environmental Politics* 17(4), 576–591.

Jin, X., Kumar, L., Li, Z., Feng, H., Xu, X., Yang, G. and Wang, J. 2018. A review of data assimilation of remote sensing and crop models. *European Journal of Agronomy* 92, 141–152.

Jones, J. W., Hoogenboom, G., Porter, C. H., Boote, K. J., Batchelor, W. D., Hunt, L. A., Wilkens, P. W., Singh, U., Gijsman, A. J. and Ritchie, J. T. 2003. The DSSAT cropping system model. *European Journal of Agronomy* 18(3–4), 235–265.

Kamali, B., Lorite, I. J., Webber, H. A., Rezaei, E. E., Gabaldon-Leal, C., Nendel, C., Siebert, S., Ramirez-Cuesta, J. M., Ewert, F. and Ojeda, J. J. 2022. Uncertainty in climate change impact studies for irrigated maize cropping systems in southern Spain. *Scientific Reports* 12(1), 4049.

Karimi, V., Karami, E. and Keshavarz, M. 2018. Climate change and agriculture: Impacts and adaptive responses in Iran. *Journal of Integrative Agriculture* 17(1), 1–15.

Kasampalis, D., Alexandridis, T., Deva, C., Challinor, A., Moshou, D. and Zalidis, G. 2018. Contribution of remote sensing on crop models: A review. *Journal of Imaging* 4(4), 52.

Kazemi, S., Eshghizadeh, H. R. and Zahedi, M. 2018. Responses of four rice varieties to elevated CO_2 and different salinity levels. *Rice Science* 25(3), 142–151.

Kelley, C. P., Mohtadi, S., Cane, M. A., Seager, R. and Kushnir, Y. 2015. Climate change in the fertile crescent and implications of the recent Syrian drought. *Proceedings of the National Academy of Sciences of the United States of America* 112(11), 3241–3246.

Khordadi, M. J., Olesen, J. E., Alizadeh, A., Nassiri Mahallati, M., Ansari, H. and Sanaeinejad, H. 2019. Climate change impacts and adaptation for crop management of winter wheat and maize in the semi-arid region of Iran. *Irrigation and Drainage* 68(5), 841-856.

Kimball, B. A. 2016. Crop responses to elevated CO_2 and interactions with H_2O, N, and temperature. *Current Opinion in Plant Biology* 31, 36–43.

Kimball, B. A., Morris, C. F., Pinter, P. J., Wall, G. W., Hunsaker, D. J., Adamsen, F. J., LaMorte, R. L., Leavitt, S. W., Thompson, T. L., Matthias, A. D. and Brooks, T. J. 2001. Elevated CO_2, drought and soil nitrogen effects on wheat grain quality. *New Phytologist* 150(2), 295–303.

Kimball, B. A., White, J. W., Ottman, M. J., Wall, G. W., Bernacchi, C. J., Morgan, J. and Smith, D. P. 2015. Predicting canopy temperatures and infrared heater energy requirements for warming field plots. *Agronomy Journal* 107, 129–141.

Klingelhöfer, D., Müller, R., Braun, M., Brüggmann, D. and Groneberg, D. A. 2020. Climate change: Does international research fulfill global demands and necessities? *Environmental Sciences Europe* 32(1), 137.

Koc, E. 2011. Investigation of possible impacts of climate change on agriculture by WOFOST crop-climate model. *Medicine Science, Istanbul, Turkey*.

Kourat, T., Smadhi, D., Mouhouche, B., Gourari, N., Mostofa Amin, M. G. and Bryant, C. R. 2021. Assessment of future climate change impact on rainfed wheat yield in the semi-arid Eastern High Plain of Algeria using a crop model. *Natural Hazards* 107(3), 2175–2203.

Kumar, S., Roshni, T., Kahya, E. and Ghorbani, M. A. 2020. Climate change projections of rainfall and its impact on the cropland suitability for rice and wheat crops in the Sone river command, Bihar. *Theoretical and Applied Climatology* 142(1–2), 433–451.

Lamichhane, J. R., Barzman, M., Booij, K., Boonekamp, P., Desneux, N., Huber, L., Kudsk, P., Langrell, S. R. H., Ratnadass, A., Ricci, P., Sarah, J.-L. and Messéan, A. 2015. Robust cropping systems to tackle pests under climate change: A review. *Agronomy for Sustainable Development* 35(2), 443–459.

Laux, P., Rötter, R. P., Webber, H., Dieng, D., Rahimi, J., Wei, J., Faye, B., Srivastava, A. K., Bliefernicht, J., Adeyeri, O., Arnault, J. and Kunstmann, H. 2021. To bias correct or not to bias correct? An agricultural impact modelers' perspective on regional climate model data. *Agricultural and Forest Meteorology* 304–305, 108406.

Lawrimore, J. 2012. *Africa Soil Profiles Database Version 1.0: A Compilation of Georeferenced and Standardised Legacy Soil Profile Data for Sub-Saharan Africa (With Dataset)*. Technical Report. ISRIC – World Soil Information.

Leenaars, J. G. B., Claessens, L., Heuvelink, G. B. M., Hengl, T., Ruiperez González, M., van Bussel, L. G. J., Guilpart, N., Yang, H. and Cassman, K. G. 2018. Mapping rootable depth and root zone plant-available water holding capacity of the soil of sub-Saharan Africa. *Geoderma* 324, 18–36.

Lewis, C. M., Persoons, A., Bebber, D. P., Kigathi, R. N., Maintz, J., Findlay, K., Bueno-Sancho, V., Corredor-Moreno, P., Harrington, S. A., Kangara, N., Berlin, A., García, R., Germán, S. E., Hanzalová, A., Hodson, D. P., Hovmøller, M. S., Huerta-Espino, J., Imtiaz, M., Mirza, J. I., Justesen, A. F., Niks, R. E., Omrani, A., Patpour, M., Pretorius, Z. A., Roohparvar, R., Sela, H., Singh, R. P., Steffenson, B., Visser, B., Fenwick, P. M., Thomas, J., Wulff, B. B. H. and Saunders, D. G. O. 2018. Potential for re-emergence of wheat stem rust in the United Kingdom. *Communications Biology* 1, 13.

Lezzaik, K., Milewski, A. and Mullen, J. 2018. The groundwater risk index: Development and application in the Middle East and North Africa region. *Science of the Total Environment* 628–629, 1149–1164.

Li, Y., Wu, Y., Hernandez-Espinosa, N. and Peña, R. J. 2013. The influence of drought and heat stress on the expression of end-use quality parameters of common wheat. *Journal of Cereal Science* 57(1), 73–78.

Lobell, D. B. and Burke, M. B. 2010. On the use of statistical models to predict crop yield responses to climate change. *Agricultural and Forest Meteorology* 150(11), 1443–1452.

Lobell, D. B., Burke, M. B., Tebaldi, C., Mastrandrea, M. D., Falcon, W. P. and Naylor, R. L. 2008. Prioritizing climate change adaptation needs for food security in 2030. *Science (New York, N.Y.)* 319(5863), 607–610.

Lowder, S. K., Skoet, J. and Raney, T. 2016. The number, size, and distribution of farms, smallholder farms, and family farms worldwide. *World Development* 87, 16–29.

Mahmah, A. E. and Amar, A. 2021. Food security in the MENA region: Does agriculture performance matter? In: Behnassi, M., Barjees Baig, M., El Haiba, M. and Reed, M. R. (Eds.), *Emerging Challenges to Food Production and Security in Asia, Middle East, and Africa*. Cham: Springer International Publishing, pp. 101–125.

Mamrutha, H. M., Rinki, K., Venkatesh, K., Gopalareddy, K., Khan, H., Mishra, C. N., Kumar, S., Kumar, Y., Singh, G. and Singh, G. P. 2020. Impact of high night temperature stress on different growth stages of wheat. *Plant Physiology Reports* 25(4), 707–715.

Mansour, E., Moustafa, E. S. A., El-Naggar, N. Z. A., Abdelsalam, A. and Igartua, E. 2018. Grain yield stability of high-yielding barley genotypes under Egyptian conditions for enhancing resilience to climate change. *Crop and Pasture Science* 69(7), 681.

Mansouri, H., Raei, Y. and Nokhbe Zaeim, A. 2015. Simulating future wheat yield under climate change, carbon dioxide enrichment and technology improvement in Iran. Case study: Azarbaijan region. *Spanish Journal of Agricultural Research* 13(4), e0306.

Maraun, D., Shepherd, T. G., Widmann, M., Zappa, G., Walton, D., Gutiérrez, J. M., Hagemann, S., Richter, I., Soares, P. M. M., Hall, A. and Mearns, L. O. 2017. Towards process-informed bias correction of climate change simulations. *Nature Climate Change* 7(11), 764–773.

Mariotti, A. 2007. How ENSO impacts precipitation in southwest central Asia. *Geophysical Research Letters* 34(16).

Marrou, H., Ghanem, M. E., Amri, M., Maalouf, F., Ben Sadoun, S., Kibbou, F. and Sinclair, T. R. 2021. Restrictive irrigation improves yield and reduces risk for faba bean across the Middle East and North Africa: A modeling study. *Agricultural Systems* 189, 103068.

Martín, M. M.-S., Olesen, J. E. and Porter, J. R. 2014. A genotype, environment and management (GxExM) analysis of adaptation in winter wheat to climate change in Denmark. *Agricultural and Forest Meteorology* 187, 1–13.

Martre, P., Quilot-Turion, B., Luquet, D., Memmah, M.-M. O.-S., Chenu, K. and Debaeke, P. 2015. Model-assisted phenotyping and ideotype design. In: *Crop Physiology*. Elsevier, pp. 349–373.

Mechler, R., Hochrainer, S., Aaheim, A., Salen, H. and Wreford, A. 2010. Modelling economic impacts and adaptation to extreme events: Insights from European case studies. *Mitigation and Adaptation Strategies for Global Change* 15(7), 737–762.

Meinshausen, M., Nicholls, Z. R. J., Lewis, J., Gidden, M. J., Vogel, E., Freund, M., Beyerle, U., Gessner, C., Nauels, A., Bauer, N., Canadell, J. G., Daniel, J. S., John, A., Krummel,

P. B., Luderer, G., Meinshausen, N., Montzka, S. A., Rayner, P. J., Reimann, S., Smith, S. J., van den Berg, M., Velders, G. J. M., Vollmer, M. K. and Wang, R. H. J. 2020. The shared socio-economic pathway (SSP) greenhouse gas concentrations and their extensions to 2500. *Geoscientific Model Development* 13(8), 3571-3605.

Meza, I., Siebert, S., Döll, P., Kusche, J., Herbert, C., Eyshi Rezaei, E., Nouri, H., Gerdener, H., Popat, E., Frischen, J., Naumann, G., Vogt, J. V., Walz, Y., Sebesvari, Z. and Hagenlocher, M. 2020. Global-scale drought risk assessment for agricultural systems. *Natural Hazards and Earth System Sciences* 20(2), 695-712.

Moradi, R., Koocheki, A. and Nassiri Mahallati, M. 2014. Adaptation of maize to climate change impacts in Iran. *Mitigation and Adaptation Strategies for Global Change* 19(8), 1223-1238.

Moradi, R., Koocheki, A., Nassiri Mahallati, M. and Mansoori, H. 2013. Adaptation strategies for maize cultivation under climate change in Iran: Irrigation and planting date management. *Mitigation and Adaptation Strategies for Global Change* 18(2), 265-284.

Mosammam, H. M., Mosammam, A. M., Sarrafi, M., Nia, J. T. and Esmaeilzadeh, H. 2016. Analyzing the potential impacts of climate change on rainfed wheat production in Hamedan Province, Iran, via generalized additive models. *Journal of Water and Climate Change* 7(1), 212-223.

Mostafa, S. M., Wahed, O., El-Nashar, W. Y., El-Marsafawy, S. M. and Abd-Elhamid, H. F. 2021. Impact of climate change on water resources and crop yield in the Middle Egypt region. *Journal of Water Supply: Research and Technology-Aqua* 70(7), 1066-1084.

Mulla, S., Singh, S. K., Singh, K. K. and Praveen, B. 2020. Climate change and agriculture: A review of crop models. In: Venkatramanan, V., Shah, S. and Prasad, R. (Eds.), *Global Climate Change and Environmental Policy: Agriculture Perspectives*. Singapore: Springer, pp. 423-435.

Nasir, I. R., Rasul, F., Ahmad, A., Asghar, H. N. and Hoogenboom, G. 2020. Climate change impacts and adaptations for fine, coarse, and hybrid rice using CERES-rice. *Environmental Science and Pollution Research International* 27(9), 9454-9464.

Newton, A. C., Johnson, S. N. and Gregory, P. J. 2011. Implications of climate change for diseases, crop yields and food security. *Euphytica* 179(1), 3-18.

Nordling, L. 2019. Scientists struggle to access Africa's historical climate data. *Nature* 574(7780), 605-606.

Nouri, M., Homaee, M., Bannayan, M. and Hoogenboom, G. 2017. Towards shifting planting date as an adaptation practice for rainfed wheat response to climate change. *Agricultural Water Management* 186, 108-119.

Nuttall, J. G., O'Leary, G. J., Panozzo, J. F., Walker, C. K., Barlow, K. M. and Fitzgerald, G. J. 2017. Models of grain quality in wheat–A review. *Field Crops Research* 202, 136-145.

O'Leary, G. J., Christy, B., Nuttall, J., Huth, N., Cammarano, D., Stöckle, C., Basso, B., Shcherbak, I., Fitzgerald, G., Luo, Q., Farre-Codina, I., Palta, J. and Asseng, S. 2015. Response of wheat growth, grain yield and water use to elevated CO_2 under a free-air CO_2 enrichment (FACE) experiment and modelling in a semi-arid environment. *Global Change Biology* 21(7), 2670-2686.

Omidvar, N., Ahmadi, D., Sinclair, K. and Melgar-Quiñonez, H. 2019. Food security in selected Middle East and North Africa (MENA) countries: An inter-country comparison. *Food Security* 11(3), 531-540.

O'Neill, B. C., Tebaldi, C., van Vuuren, D. P., Eyring, V., Friedlingstein, P., Hurtt, G., Knutti, R., Kriegler, E., Lamarque, J.-F., Lowe, J., Meehl, G. A., Moss, R., Riahi, K. and Sanderson, B. M. 2016. The scenario model intercomparison project (ScenarioMIP) for CMIP6. *Geoscientific Model Development* 9(9), 3461–3482.

Oomen, R. J., Ewert, F. and Snyman, H. A. 2016. Modelling rangeland productivity in response to degradation in a semi-arid climate. *Ecological Modelling* 322, 54–70.

Ouda, S. A., Taha, A. M. and Ibrahim, M. M. 2015. Water management for maize grown in sandy soil under climate change conditions. *Archives of Agronomy and Soil Science* 61(3), 299–311.

Özdoğan, M. 2011. Modeling the impacts of climate change on wheat yields in Northwestern Turkey. *Agriculture, Ecosystems and Environment* 141(1–2), 1–12.

Paleari, L., Movedi, E. and Confalonieri, R. 2017. Trait-based model development to support breeding programs. A case study for salt tolerance and rice. *Scientific Reports* 7(1), 4352.

Paleari, L., Movedi, E., Vesely, F. M. and Confalonieri, R. 2019. Tailoring parameter distributions to specific germplasm: Impact on crop model-based ideotyping. *Scientific Reports* 9(1), 18309.

Pandey, P., Irulappan, V., Bagavathiannan, M. V. and Senthil-Kumar, M. 2017. Impact of combined abiotic and biotic stresses on plant growth and avenues for crop improvement by exploiting physio-morphological traits. *Frontiers in Plant Science* 8, 537.

Papadaskalopoulou, C., Moriondo, M., Lemesios, I., Karali, A., Konsta, A., Dibari, C., Brilli, L., Varotsos, K. V., Stylianou, A., Loizidou, M., Markou, M. and Giannakopoulos, C. 2020. Assessment of total climate change impacts on the agricultural sector of Cyprus. *Atmosphere* 11(6), 608.

Parent, B., Leclere, M., Lacube, S., Semenov, M. A., Welcker, C., Martre, P. and Tardieu, F. 2018. Maize yields over Europe may increase in spite of climate change, with an appropriate use of the genetic variability of flowering time. *Proceedings of the National Academy of Sciences of the United States of America* 115(42), 10642–10647.

Paymard, P., Bannayan, M. and Haghighi, R. S. 2018. Analysis of the climate change effect on wheat production systems and investigate the potential of management strategies. *Natural Hazards* 91(3), 1237–1255.

Portmann, F., Siebert, S., Bauer, C. and Döll, P. 2008. Global data set of monthly growing areas of 26 irrigated crops, Frankfurt Hydrology, Paper 06. Institute of Physical Geography, University of Frankfurt, Frankfurt am Main, Germany.

Portmann, F. T., Siebert, S. and Döll, P. 2010. MIRCA2000-Global monthly irrigated and rainfed crop areas around the year 2000: A new high-resolution data set for agricultural and hydrological modeling. *Global Biogeochemical Cycles* 24(1).

Qiu, W. and Yan, X. 2020. The trend of heatwave events in the Northern Hemisphere. *Physics and Chemistry of the Earth, Parts A/B/C* 116, 102855.

Rahimi, J., Khalili, A. and Butterbach-Bahl, K. 2019. Projected changes in modified Thornthwaite climate zones over southwest Asia using a CMIP5 multi-model ensemble. *International Journal of Climatology* 39(12), 4575–4594.

Rahimi-Moghaddam, S., Kambouzia, J. and Deihimfard, R. 2018. Adaptation strategies to lessen negative impact of climate change on grain maize under hot climatic conditions: A model-based assessment. *Agricultural and Forest Meteorology* 253-254, 1–14.

Rahman, M. H. u., Ahmad, A., Wang, X., Wajid, A., Nasim, W., Hussain, M., Ahmad, B., Ahmad, I., Ali, Z., Ishaque, W., Awais, M., Shelia, V., Ahmad, S., Fahd, S., Alam, M., Ullah, H. and Hoogenboom, G. 2018. Multi-model projections of future climate and climate change impacts uncertainty assessment for cotton production in Pakistan. *Agricultural and Forest Meteorology* 253–254, 94–113.

Rana, S., McGregor, J. and Renwick, J. 2017. Wintertime precipitation climatology and ENSO sensitivity over central southwest Asia. *International Journal of Climatology* 37(3), 1494–1509.

Raoufi, R. S. and Soufizadeh, S. 2020. Simulation of the impacts of climate change on phenology, growth, and yield of various rice genotypes in humid sub-tropical environments using AquaCrop-Rice. *International Journal of Biometeorology* 64(10), 1657–1673.

Raza, A., Friedel, J. K., Moghaddam, A., Ardakani, M. R., Loiskandl, W., Himmelbauer, M. and Bodner, G. 2013. Modeling growth of different lucerne cultivars and their effect on soil water dynamics. *Agricultural Water Management* 119, 100–110.

Rezaei, E. E., Ghazaryan, G., Moradi, R., Dubovyk, O. and Siebert, S. 2021. Crop harvested area, not yield, drives variability in crop production in Iran. *Environmental Research Letters* 16(6), 64058.

Rezaei, E. E. and Lashkari, A. 2019. The consequences of change in management practices on maize yield under climate warming in Iran. *Theoretical and Applied Climatology* 137(1–2), 1001–1013.

Rezaei, E. E., Siebert, S., Hüging, H. and Ewert, F. 2018. Climate change effect on wheat phenology depends on cultivar change. *Scientific Reports* 8(1), 4891.

Roberts, M. J., Braun, N. O., Sinclair, T. R., Lobell, D. B. and Schlenker, W. 2017. Comparing and combining process-based crop models and statistical models with some implications for climate change. *Environmental Research Letters* 12(9), 95010.

Robinson, E. A., Ryan, G. D. and Newman, J. A. 2012. A meta-analytical review of the effects of elevated CO_2 on plant-arthropod interactions highlights the importance of interacting environmental and biological variables. *The New Phytologist* 194(2), 321–336.

Rossi, S., Burgess, P., Jespersen, D. and Huang, B. 2017. Heat-induced leaf senescence associated with chlorophyll metabolism in bentgrass lines differing in heat tolerance. *Crop Science* 57(S1).

Rötter, R. P., Appiah, M., Fichtler, E., Kersebaum, K. C., Trnka, M. and Hoffmann, M. P. 2018. Linking modelling and experimentation to better capture crop impacts of agroclimatic extremes–A review. *Field Crops Research* 221, 142–156.

Rötter, R. P., Carter, T. R., Olesen, J. E. and Porter, J. R. 2011. Crop-climate models need an overhaul. *Nature Climate Change* 1(4), 175–177.

Ruane, A. C. and McDermid, S. P. 2017. Selection of a representative subset of global climate models that captures the profile of regional changes for integrated climate impacts assessment. *Earth Perspectives* 4(1), 1–20.

Saadi, S., Todorovic, M., Tanasijevic, L., Pereira, L. S., Pizzigalli, C. and Lionello, P. 2015. Climate change and Mediterranean agriculture: Impacts on winter wheat and tomato crop evapotranspiration, irrigation requirements and yield. *Agricultural Water Management* 147, 103–115.

Sayari, N., Bannayan, M., Alizadeh, A., Farid, A., Hessami Kermani, M. R. and Eyshi Rezaei, E. 2015. Climate change impact on legumes' water production function in the northeast of Iran. *Journal of Water and Climate Change* 6(2), 374–385.

Schilling, J., Hertig, E., Tramblay, Y. and Scheffran, J. 2020. Climate change vulnerability, water resources and social implications in North Africa. *Regional Environmental Change* 20(1), 1-12.

Selvaraju, R. 2013. Implications of climate change for agriculture and food security in the Western Asia and northern Africa region. In: Sivakumar, M. V., Lal, R., Selvaraju, R. and Hamdan, I. (Eds.), *Climate Change and Food Security in West Asia and North Africa*. Dordrecht: Springer Netherlands, pp. 27-51.

Sen, B., Topcu, S., Türkeş, M., Sen, B. and Warner, J. 2012. Projecting climate change, drought conditions and crop productivity in Turkey. *Climate Research* 52, 175-191.

Senapati, N. and Semenov, M. A. 2019. Assessing yield gap in high productive countries by designing wheat ideotypes. *Scientific Reports* 9(1), 5516.

Shabbir, G., Khaliq, T., Ahmad, A. and Saqib, M. 2020. Assessing the climate change impacts and adaptation strategies for rice production in Punjab, Pakistan. *Environmental Science and Pollution Research International* 27(18), 22568-22578.

Shahid, M. R., Wakeel, A., Ishaque, W., Ali, S., Soomro, K. B. and Awais, M. 2021. Optimizing different adaptive strategies by using crop growth modeling under IPCC climate change scenarios for sustainable wheat production. *Environment, Development and Sustainability* 23(8), 11310-11334.

Shahvari, N., Khalilian, S., Mosavi, S. H. and Mortazavi, S. A. 2019. Assessing climate change impacts on water resources and crop yield: A case study of Varamin plain basin, Iran. *Environmental Monitoring and Assessment* 191(3), 134.

Shennan, S. (Ed.). 2018. *The First Farmers of Europe: An Evolutionary Perspective*. Cambridge: Cambridge University Press.

Shi, W., Tao, F., Liu, J., Xu, X., Kuang, W., Dong, J. and Shi, X. 2014. Has climate change driven spatio-temporal changes of cropland in northern China since the 1970s? *Climatic Change* 124(1-2), 163-177.

Shiff, S., Lensky, I. M. and Bonfil, D. J. 2021. Using satellite data to optimize wheat yield and quality under climate change. *Remote Sensing* 13(11), 2049.

Shimeles, A., Gurara, D. Z. and Birhanu Tessema, D. 2015. Market distortions and political rent: The case of fertilizer price divergence in Africa journal. *SSRN Electronic Journal*.

Silva, J. V. and Giller, K. E. 2020. Grand challenges for the 21st century: What crop models can and can't (yet) do. *Journal of Agricultural Science* 158(10), 794-805.

Sinclair, T. R., Hammer, G. L. and van Oosterom, E. J. 2005. Potential yield and water-use efficiency benefits in sorghum from limited maximum transpiration rate. *Functional Plant Biology: FPB* 32(10), 945-952.

Sivakumar, M. V., Lal, R., Selvaraju, R. and Hamdan, I. (Eds.). 2013. *Climate Change and Food Security in West Asia and North Africa*. Dordrecht: Springer Netherlands.

Sloat, L. L., Davis, S. J., Gerber, J. S., Moore, F. C., Ray, D. K., West, P. C. and Mueller, N. D. 2020. Climate adaptation by crop migration. *Nature Communications* 11(1), 1243.

Spiertz, J. H. J., Hamer, R. J., Xu, H., Primo-Martin, C., Don, C. and van der Putten, P. E. L. 2006. Heat stress in wheat (*Triticum aestivum* L.): Effects on grain growth and quality traits. *European Journal of Agronomy* 25(2), 89-95.

Steduto, P., Hsiao, T. C., Raes, D. and Fereres, E. 2009. AquaCrop–The FAO crop model to simulate yield response to water: I. Concepts and underlying principles. *Agronomy Journal* 101(3), 426-437.

Stöckle, C. O. and Kemanian, A. R. 2020. Can crop models identify critical gaps in genetics, environment, and management interactions? *Frontiers in Plant Science* 11, 737.

Sultan, B. and Gaetani, M. 2016. Agriculture in West Africa in the twenty-first century: Climate change and impacts scenarios, and potential for adaptation. *Frontiers in Plant Science* 7, 1262.
Tabari, H. and Willems, P. 2018. More prolonged droughts by the end of the century in the Middle East. *Environmental Research Letters* 13(10), 104005.
Thomas, R. J. 2008. Opportunities to reduce the vulnerability of dryland farmers in Central and West Asia and North Africa to climate change. *Agriculture, Ecosystems and Environment* 126(1–2), 36–45.
Tian, X. and Yu, X. 2019. Crop yield gap and yield convergence in African countries. *Food Security* 11(6), 1305–1319.
Tricker, P. J., ElHabti, A., Schmidt, J. and Fleury, D. 2018. The physiological and genetic basis of combined drought and heat tolerance in wheat. *Journal of Experimental Botany* 69(13), 3195–3210.
van der Velde, M., Folberth, C., Balkovič, J., Ciais, P., Fritz, S., Janssens, I. A., Obersteiner, M., See, L., Skalský, R., Xiong, W. and Peñuelas, J. 2014. African crop yield reductions due to increasingly unbalanced nitrogen and phosphorus consumption. *Global Change Biology* 20(4), 1278–1288.
van Ittersum, M. K., Leffelaar, P. A., van Keulen, H., Kropff, M. J., Bastiaans, L. and Goudriaan, J. 2003. On approaches and applications of the Wageningen crop models. *European Journal of Agronomy* 18(3–4), 201–234.
van Oijen, M., Bellocchi, G. and Höglind, M. 2018. Effects of climate change on grassland biodiversity and productivity: The need for a diversity of models. *Agronomy* 8(2), 14.
Vanli, Ö., Ustundag, B. B., Ahmad, I., Hernandez-Ochoa, I. M. and Hoogenboom, G. 2019. Using crop modeling to evaluate the impacts of climate change on wheat in southeastern turkey. *Environmental Science and Pollution Research International* 26(28), 29397–29408.
Waha, K., Krummenauer, L., Adams, S., Aich, V., Baarsch, F., Coumou, D., Fader, M., Hoff, H., Jobbins, G., Marcus, R., Mengel, M., Otto, I. M., Perrette, M., Rocha, M., Robinson, A. and Schleussner, C.-F. 2017. Climate change impacts in the Middle East and northern Africa (MENA) region and their implications for vulnerable population groups. *Regional Environmental Change* 17(6), 1623–1638.
Wajid, A., Hussain, K., Ilyas, A., Habib-ur-Rahman, M., Shakil, Q. and Hoogenboom, G. 2021. Crop models: Important tools in decision support system to manage wheat production under vulnerable environments. *Agriculture* 11(11), 1166.
Wall, G. W., Garcia, R. L., Kimball, B. A., Hunsaker, D. J., Pinter, P. J., Long, S. P., Osborne, C. P., Hendrix, D. L., Wechsung, F., Wechsung, G., Leavitt, S. W., LaMorte, R. L. and Idso, S. B. 2006. Interactive effects of elevated carbon dioxide and drought on wheat. *Agronomy Journal* 98, 354–381.
Wallach, D., Palosuo, T., Thorburn, P., Gourdain, E., Asseng, S., Basso, B., Buis, S., Crout, N., Dibari, C., Dumont, B., Ferrise, R., Gaiser, T., Garcia, C., Gayler, S., Ghahramani, A., Hochman, Z., Hoek, S., Hoogenboom, G., Horan, H., Huang, M., Jabloun, M., Jing, Q., Justes, E., Kersebaum, K. C., Klosterhalfen, A., Launay, M., Luo, Q., Maestrini, B., Mielenz, H., Moriondo, M., Nariman Zadeh, H., Olesen, J. E., Poyda, A., Priesack, E., Pullens, J. W. M., Qian, B., Schütze, N., Shelia, V., Souissi, A., Specka, X., Srivastava, A. K., Stella, T., Streck, T., Trombi, G., Wallor, E., Wang, J., Weber, T. K. D., Weihermüller, L., Wit, A. de, Wöhling, T., Xiao, L., Zhao, C., Zhu, Y. and Seidel, S. J. 2021. How well do crop modeling groups predict wheat phenology, given

calibration data from the target population? *European Journal of Agronomy* 124, 126195.

Wang, B., Feng, P., Liu, D. L., O'Leary, G. J., Macadam, I., Waters, C., Asseng, S., Cowie, A., Jiang, T., Xiao, D., Ruan, H., He, J. and Yu, Q. 2020. Sources of uncertainty for wheat yield projections under future climate are site-specific. *Nature Food* 1(11), 720-728.

Wang, E., Martre, P., Zhao, Z., Ewert, F., Maiorano, A., Rötter, R. P., Kimball, B. A., Ottman, M. J., Wall, G. W., White, J. W., Reynolds, M. P., Alderman, P. D., Aggarwal, P. K., Anothai, J., Basso, B., Biernath, C., Cammarano, D., Challinor, A. J., Sanctis, G. de, Doltra, J., Dumont, B., Fereres, E., Garcia-Vila, M., Gayler, S., Hoogenboom, G., Hunt, L. A., Izaurralde, R. C., Jabloun, M., Jones, C. D., Kersebaum, K. C., Koehler, A. K., Liu, L., Müller, C., Naresh Kumar, S., Nendel, C., O'Leary, G., Olesen, J. E., Palosuo, T., Priesack, E., Eyshi Rezaei, E., Ripoche, D., Ruane, A. C., Semenov, M. A., Shcherbak, I., Stöckle, C., Stratonovitch, P., Streck, T., Supit, I., Tao, F., Thorburn, P., Waha, K., Wallach, D., Wang, Z., Wolf, J., Zhu, Y. and Asseng, S. 2017. The uncertainty of crop yield projections is reduced by improved temperature response functions. *Nature Plants* 3, 17102.

Wang, W., Yuan, S., Wu, C., Yang, S., Zhang, W., Xu, Y., Gu, J., Zhang, H., Wang, Z., Yang, J. and Zhu, J. 2021. Field experiments and model simulation based evaluation of rice yield response to projected climate change in Southeastern China. *The Science of the Total Environment* 761, 143206.

Webber, H., Hoffmann, M. and Rezaei, E. E. 2019. Crop models as tools for Agroclimatology. In: Hatfield, J. L., Siva Kumar, M. V. K. and Prueger, J. H. (Eds.), *Agroclimatology. Linking Agriculture to Climate*. Madison, WI: American Society of Agronomy, pp. 519-546.

White, J. W. and Hoogenboom, G. 2010. Crop response to climate: Ecophysiological models. In: Lobell, D. and Burke, M. (Eds.), *Climate Change and Food Security. Adapting Agriculture to a Warmer World*. Dordrecht, London: Springer, pp. 59-83.

White, J. W., Kimball, B. A., Wall, G. W., Ottman, M. J. and Hunt, L. A. 2011. Responses of time of anthesis and maturity to sowing dates and infrared warming in spring wheat. *Field Crops Research* 124(2), 213-222.

Wit, A. de, Boogaard, H., Fumagalli, D., Janssen, S. and Knapen, R. 2019. 25 Years of the WOFOST cropping systems model. *Agricultural Systems* 168, 154-167.

World, B. 2017. *World Bank Open Data*. Available at: https://data.worldbank.org/ (accessed 17 May 2019).

World, B. 2021. *The World Bank Dataset*. Available at: https://data.worldbank.org (accessed 20 February 2022).

Wu, A., Song, Y., van Oosterom, E. J. and Hammer, G. L. 2016. Connecting biochemical photosynthesis models with crop models to support crop improvement. *Frontiers in Plant Science* 7, 1518.

Xie, S.-P., Deser, C., Vecchi, G. A., Collins, M., Delworth, T. L., Hall, A., Hawkins, E., Johnson, N. C., Cassou, C., Giannini, A. and Watanabe, M. 2015. Towards predictive understanding of regional climate change. *Nature Climate Change* 5(10), 921-930.

Yin, X. and Struik, P. C. 2010. Modelling the crop: From system dynamics to systems biology. *Journal of Experimental Botany* 61(8), 2171-2183.

Zhao, C., Yan, Y., Wang, C., Tang, M., Wu, G., Ding, D. and Song, Y. 2018. Adaptation and mitigation for combating climate change - from single to joint. *Ecosystem Health and Sustainability* 4(4), 85-94.

Zhu, L., Cernusak, L. A. and Song, X. 2020. Dynamic responses of gas exchange and photochemistry to heat interference during drought in wheat and sorghum. *Functional Plant Biology: FPB* 47(7), 611–627.

Zittis, G. 2018. Observed rainfall trends and precipitation uncertainty in the vicinity of the Mediterranean, Middle East and North Africa. *Theoretical and Applied Climatology* 134(3–4), 1207–1230.

Chapter 19

Modelling the impact of climate change on agriculture in West Africa

D. S. MacCarthy, University of Ghana, Ghana; P. B. I. Akponikpe, Université de Parakou (UP), Benin; F. M. Akinseye, International Crop Research Institute for the Semi-Arid Tropics (ICRISAT), Nigeria and Centre d'Etude Régional pour l'Amélioration de l'Adaptation à la Sécheresse (CERAAS), Sénégal; M. Ly, Cheikh, Anta Diop University, Senegal; E. C. Timpong-Jones, University of Ghana, Ghana; I. Hathie, Initiative Prospective Agricole et Rurale (IPAR), Senegal; and S. G. K. Adiku, University of Ghana, Ghana

1 Introduction
2 The West African region
3 The West African climate: measurement, characteristics and trends
4 Modelling climate change in West Africa
5 Crop modelling in West African farming systems
6 Modelling cereal and legume production in West Africa
7 Accounting for the evolution of agricultural systems in climate change impact assessments
8 Modelling climate change adaptation strategies
9 Key issues in modelling climate change impacts on livestock production
10 Modelling livestock systems
11 Integrated climate change impact assessments
12 Conclusion and future trends
13 Where to look for further information
14 References

1 Introduction

The overall objective of this chapter is to highlight the manner in which climate change continues to be a major challenge to agricultural development and productivity in West Africa. In particular, the current state of knowledge and understanding of the West African climate, and the changes in patterns and trends that have occurred over the years, is discussed. The chapter also presents

http://dx.doi.org/10.19103/AS.2022.0115.21

the tools (especially models) that are available as well as those still required to assess the impact of climate change on agriculture in the region.

This chapter begins with an overview of West Africa as a region, followed by an overview of West African climatic conditions. The chapter then provides a meta-analysis of studies investigating the effects of climate change on the productivity of cereals and legumes in West Africa. It also highlights challenges in modelling the varied and distinctive characteristics of West African farming systems as well as potential adaptation options. The next section reviews the impact of climate change on livestock production, followed by a brief section on integrated climate change impact assessment. The chapter ends with future developments and links to further reading.

2 The West African region

West Africa comprises a landmass of approximately 8 million km^2 and is made up of 17 countries encompassing a wide range of landscapes. In 2022, the population was estimated at a little over 420 million, representing 5.43% of the world's population (World Population Review, 2022), and with a population growth rate of 3% per annum which is projected to almost double by 2050.

The landscape is mainly flat with coastal plains in the west and south with hilly low plateaus of less than 500 m in the interior. There are alluvial valleys in Senegal and Ghana, sandy plains and low plateaus across the Sahel, rolling hills of Togo to rugged mountains in the densely forested Guinea Highlands in Sierra Leone (1945 m), with Mount Idoukal-n-Taghès (2022 m) located in Niger. It is bounded in the west by the Atlantic Ocean, by the Gulf of Guinea in the south and the Sahara Desert and the Sahel in the north. The Niger River is one of its longest rivers stretching over 2600 miles from the Guinean Highlands through Mali, Niger and Nigeria prior to flowing into the Atlantic Ocean in the Gulf of Guinea.

The distribution of soils in the region is influenced by differences in age, parent material, vegetation and climatic conditions (Bationo et al., 2006); i.e. the soils vary based on bioclimatic conditions. The humid zone consists of predominantly Ferralsols and Acrisols, while the sub-humid zones have Ferralsols and Lixisols and to a lesser extent, Acrisols and Arenosols. The semi-arid zone has Regosols followed by Arenosols to a lesser extent, while the arid zone is dominated by Lithosols (Bationo et al., 2006).

West Africa contains a broad range of ecosystems, bioclimatic regions and habitats from rainforest in the southern section to desert in the northern section. The most dramatic transitions in natural features and land use occur as one moves north or south across these belts (bioclimatic regions). Precipitation increases from the north to the south while temperatures generally decrease moving from the north to the south.

Agriculture continues to be the backbone of the economy of West African countries. In Ghana, about 50% of the population engages directly in agricultural production (Ghana Statistical Service, 2016). However, the proportion of the population engaged in agriculture-related activities (i.e. traders, marketers, middlemen, agro-processors, policy-makers) may be as high as 78% (Tanko et al., 2019) and approximately 75% in Cote d'Ivoire (Kouassi et al., 2021), and the proportions are similar in other West African countries. Agriculture provides inputs for industry, e.g. with animal husbandry providing hide and skin to leather factories.

Agriculture contributes substantially to the gross domestic product (GDP) of most West African countries, ranging from 17% (Senegal) to 60% (Sierra Leone) (World Bank data, 2020). Agriculture in West Africa has long been weather dependent (De Longueville et al., 2020). According to Akpoti et al. (2022), only 4–6% of the total arable lands in sub-Saharan Africa are equipped with irrigation infrastructure. Crop processing such as drying is largely dependent on sunshine and winnowing the chaff from cereal grains is wind dependent.

Scientific analysis of climate change has shown increasing trends in variables such as temperature, while other variables (such as rainfall) have become unpredictable. In general, the frequency of weather extremes (drought, flooding) has increased (Seneviratne et al., 2012), beyond prediction through indigenous knowledge. Even though farmers may have indigenous knowledge of historic weather patterns to use in farm operations and planning, this knowledge is less useful when climate becomes more unpredictable. A number of studies (Barffour-Ata et al., 2021; Roncoli et al., 2009) have investigated the value of indigenous knowledge in the management of West African agriculture. Though the knowledge system appears to have worked historically, the content has been eroded over time and the combination of this knowledge with mythology hinders its formal study for wider application.

3 The West African climate: measurement, characteristics and trends

Climate is defined as the average weather conditions over a climatological period, usually 30 years. Climate comprises several variables including temperature, rainfall, wind (gust, velocity and direction), solar radiation and humidity. In West Africa, climate variables are measured on varying timescales. Until recently, when automated recording became common in West Africa, climate variables were largely recorded manually, using instruments such as rain gauges, minimum and maximum thermometers, hygrometers, sunshine recorders, anemometers, etc.

Most of the documentation detailing climate variables is derived from Meteorological Agencies or bureaus of the various West African countries. The spatial density of weather stations also varies from country to country. Not every weather station is equipped with all the measurement instruments and weather stations are classified according to the frequency and scope of weather variables measured. The most common type of station in the West African region is the climatological station where variables such as rainfall, wind velocity, temperature and sunshine hours are recorded. Rainfall is measured on a daily basis, while the temperature is measured at mid-afternoon and early morning.

A survey published by Adiku et al. (2011) on the state of weather stations in Ghana was sponsored by the German International Cooperation, also known as the GIZ. For the northern locations, which constituted about 45% of the total land area of Ghana, the findings indicated that for the period between 2010 and 2011, there were 47 total active stations of which 53% was rated between good and satisfactory. In the case of the southern Ghana locations, Volta Region had 19 out of a total of 38 stations active, with all the active stations rated between good and satisfactory. The Eastern Region had 30 active stations, with about 70% rated between good and satisfactory.

The situation in other West African countries is similar although Nigeria, despite its vast land area, boasts of a relatively high density and network of weather stations. In the Francophone West African countries, Cote d'Ivoire has the highest density and network. In general, the rainfall observation network in the region has a poor spatial distribution and an average density often less than 2 gauges for each 1° × 1° square (less than 6000 km² per gauge) (Lebel et al., 2000). Furthermore, the interconnectivity of weather stations in the West African region, which is important for information exchange, is handicapped by language differences. Climate data and information exchange between anglophone and francophone countries are extremely limited.

The climate system of West Africa is unique and primarily tropical though the far northern portion of West Africa is arid and extends into the Sahara Desert (Sultan and Gaetani, 2016; Emetere, 2017). The West African climate zones are divided into four broad regions from north to south (Fig. 1):

- Sahel;
- Sudan;
- Guinean; and
- Guineo-Congolian.

These regions follow a latitudinal rainfall gradient, with the lowest rainfall in the Sahel (as little as 100 mm annually) and increasing rainfall to the south, with the highest rainfall along the southern coast (up to 5000 mm annually).

The atmospheric circulation in the four zones is controlled by features including the African Easterly Jet, Inter-tropical Convergence Zone, Inter-tropical Discontinuity, associated heat low, Subtropical Jet, as well as troughs and cyclonic centres associated with African Easterly Waves and the Tropical

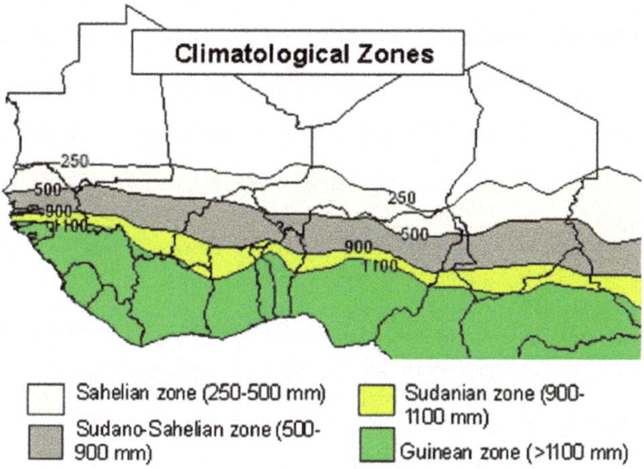

Figure 1 Climatic zones in West Africa. Source: Adapted from Emetere (2017).

Easterly Jet. This chapter also takes into account bioclimatic zones but groups them into three sub-regions (United States Agency for International, 2018):

- Sahel;
- Inland Forests, which encompass most of the Sudan zone and the inland portions of the Guinean and Guineo-Congolian zones; and
- Coastal sub-region, which comprises the coastal areas across bioclimatic zones.

Table 1 details the temperature and rainfall of each climate zone (Committee for Drought Control in the Sahel, 2016; United States Agency for International, 2018).

West Africa is a region potentially vulnerable to climate variability and changes due to its high exposure to climate stress and low adaptive capacity (poor infrastructure, limited access to markets, high illiteracy rates, accelerated coastal erosion with habitat loss on the coast mainly induced by sea level rise, an upsurge in the magnitude of floods, an increase in surface temperatures and heatwaves, a decline in agricultural productivity often accompanied by a strong movement of populations, etc.). This structural vulnerability to climate was particularly amplified at the end of the twentieth century (Intergovernmental Panel on Climate Change (IPCC), UNFCC, 2015), following a reduction in rainfall coinciding with an increase in population.

Climate change, coupled with the typical high interannual and spatial variability of precipitation in the region, poses serious challenges to the development of most countries in the region. West Africa has already experienced a warming of 1°C (IPCC, 2013) since 1950 (Guichard et al., 2015).

Table 1 Bioclimatic zones: vegetation, rainfall and temperature

Bioclimatic zones	Rainfall and seasonality	Temperature
Sahel: grassland, shrubs and small trees	Average annual rainfall of 100-200 mm (north) to 500-600 mm (south), with a rainy season ranging from 1 to 5 months during June-October; maximum rainfall in August	Average annual temperatures of 22-36°C
Sudan: savannah and open woodlands	One rainy season (May-October), with an annual rainfall of 600-1200 mm	Average annual temperatures above 22-28°C
Guinean: dense, seasonally wet and dry forests	Two rainy seasons (April-July and September-October), with an annual rainfall of 1200-2200 mm. Most rain, 75-80%, falls during April-July	
Guineo-Congolian: dense rainforests	Two rainy seasons (April-July and September-October) or year-round rainfall; annual averages of 2200-5000 mm	

Source: Adapted from USAID (2018).

In the case of the Sahel, the average warming since 1950 has been even higher, especially in spring, at more than 2°C (Ly et al., 2013). The last few decades have been characterised by the emergence of a warming trend independent of precipitation behaviour, by a partial resumption of rainfall since the persistent drought of the late twentieth century (Ly et al., 2013; Lodoun et al., 2013) and by increased variability at all timescales, with alternating years of drought and abundant rain (Lodoun et al., 2013; Sarr et al., 2015).

Climate change is likely to increase the frequency of extreme events such as floods or droughts, with various negative impacts: limited access to drinking water, famines, epidemics, large migrations and regional conflicts. The analysis of climate change projections, according to two climate change scenarios, shows the warming of temperatures for West Africa is estimated to be 3-6°C in the representative concentration pathway (RCP) scenario 8.5, which represents a warming 60% higher than the average global warming of the planet (Deme et al., 2015). With regard to precipitation, considerable uncertainty remains for the region; however, the western Sahel could expect a decrease in total annual rainfall, while the eastern Sahel could see an increase. Thus, limiting end-of-century warming in Africa to 1.5°C will greatly reduce the frequency of extreme heat compared with a warming of 2°C (Deme et al., 2015; Nagombe et al., 2018).

4 Modelling climate change in West Africa

Climate models provide a key way to understand the processes that underpin variability and long-term changes in key meteorological parameters such as precipitation and temperature. Crop models also enhance our ability to analyse and understand the way in which several factors (edaphic, soil, climate and management) interact to affect agricultural productivity. Crop models have the advantage that once validated, their use is non-location or time specific and can therefore be applied wherever the required input data are available. Crop modelling began from studies by de Wit (1961) and evolved through the work of Loomis et al. (1979), Nix and Fitzpatrick (1969) and McCown (1973), among others. In West Africa, agricultural research using crop modelling can be traced to works including Adiku et al. (1998); Naab et al. (2004); MacCarthy et al. (2009); Akponikpè et al. (2010) and Fosu et al. (2012).

However, the use of models to project adaptation strategies that could be adopted to minimize climate change impact at different locations is often handicapped by the general lack and the often low quality of the required data at the locations of interest. Improving the quality of climate information is important for better adaptation to climate change (Ly et al., 2019). Another challenge is the choice of the appropriate climate scenarios that match the West African conditions, though the Paris Agreement has identified potential climate impacts from temperature increases of 1.5°C and 2°C compared with pre-industrial levels as assessed in the Special Report 1.5°C of the IPCC.

The last few decades have seen more effort focused on understanding climate variability and change in West Africa. The African Monsoon Multidisciplinary Analysis Programme (http://amma-international.org/), launched in 2002, involving a number of research institutions in the international scientific community, was the first large-scale coordinated programme aimed at improving the understanding of the West African rainfall system and its influence on the physical, chemical and biological environment, regionally and globally. More recently, the Agricultural Model Intercomparison and Improvement Project (AgMIP) was launched with a focus to significantly improve the scientific and technological capabilities of regional agricultural models, to assess the impact of climate variability and change and other driving forces on agriculture, food security and poverty at local to global scales.

There are currently a number of other initiatives underway to improve the performance of models to meet user expectations. The IPCC Sixth Assessment Report 2021 (AR6) has showcased new state-of-the-art CMIP6 models (higher spatial resolutions, greatly improved physical processes and biogeochemical cycles). There are also regional model intercomparison projects or regional climate models RCMs (CORDEX) and global model intercomparison projects or general circulation models (GCMs) (CMIP5, CMIP6), which aim to improve the capacity to analyse climate models in different regions of the world (United Nations Framework Convention on Climate Change, 2015; Ly et al., 2019). The CMIP6 project currently includes a database of 100 separate climate models produced by different modelling groups. These new versions of the climate models appear to have significantly higher sensitivity values than those of the CMIP5 simulations; they have a wider range of emission scenarios based on different socio-economic assumptions and at differing time horizons.

The expected average change in temperature and precipitation during the main months of the rainy season (from June to September) relative to the baseline period 1981–2010 has been evaluated from 29 GCMs. Overall, in the region, according to the RCP8.5 scenario, temperatures are expected to increase in the future by 2°C. The projected changes in temperature and mean precipitation in June, July, August and September in West Africa predicted for the mid-twenty-first century (2046–65) relative to the late twentieth century (1981–2010) have been based on two alternative greenhouse gas emission scenarios: RCP4.5 and RCP8.5. Contrasting model projections based on these two pathways are shown here, with RCP4.5 assumed to be a medium stabilisation scenario of the radiative forcing at 4.5 Wm^{-2} in the year 2100 without exceeding that value, and RCP8.5 assuming that emissions continue to increase throughout the twenty-first century. The projections, which are multi-model averages, tend to level out the considerable variability and disagreement between the individual models.

As the maps (Fig. 2) show, both RCP scenarios predict a warming trend and predominantly positive changes in annual rainfall for most of West Africa. While most changes are small and insignificant, a wetter future is predicted for Niger and Chad, whereas the scenarios indicate a possible drying trend for the western part of West Africa. The availability of reliable climate simulations at high-spatial temporal resolution is crucial for a robust assessment of climate impact at the regional scale, and the CORDEX-Africa exercise shows encouraging results for West Africa (Table 2).

Looking ahead, temperatures are predicted to continue to rise across the region, along with an increased frequency and duration of heatwaves. Rainfall will continue to be variable, with increased rainfall in some areas and decreased rainfall in others. The onset of the rainy season will occur later in the spring, and heavy rainfall events will increase in frequency and intensity. Sea levels and surface temperatures will also continue to rise. The dynamic downscaling of GCMs operated at higher resolution by RCMs leads to improvements in the simulation of atmospheric circulation, temperature and precipitation patterns, as well as the occurrence of wet and dry spells, the frequency of heavy rain events and geographical distribution of droughts (Laprise et al., 2013; Bucchignani et al., 2016; Diasso and Abiodun, 2015; Dosio et al., 2015).

The biases in the lateral boundary conditions provided by the driving GCMs may, however, significantly affect the output of the RCMs (Laprise et al., 2013; Dosio et al., 2015). With the GCM biases more pronounced over

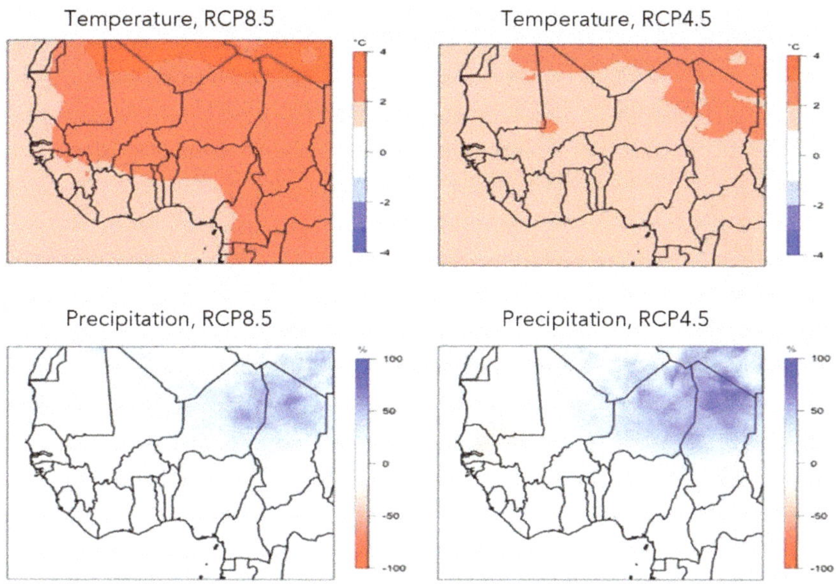

Figure 2 Expected average change (from 29 GCMs) in JJAS temperature and precipitation in West Africa for two scenarios, RCP8.5 and RCP4.5, mid-century time slice (2040-69). Source: Adapted from MacCarthy et al. (2021a), courtesy of the AgMIP project. JJAS, June, July, August, September.

the tropical Atlantic, RCM performances are generally better over the Sahel than in the Guinean coast, which is influenced to a greater extent by local sea surface temperature (SST) variability (Paxian et al., 2016). Uncertainties in the simulation of daily precipitation are also observed, mainly related to the diverse convection schemes utilised in the CORDEX-Africa models (Klutse et al., 2016). However, the spread in the performance of the individual model is substantially improved when the ensemble mean is computed (Klutse et al., 2016).

In conclusion, the climate in West Africa is expected to show variations with time and space. While it is clear that temperature will continue to rise in all areas, rainfall variability remains uncertain. Even though there is a general expectation of wetting in the northern Sahel, there will also be some drying along the southwestern parts of the region.

For farmers, it is the short-term seasonal forecast that will be of immediate benefit, rather than the forecast for the near future (2050s) or far future (2100). Little research effort has focused on the seasonal forecast except for work by Opoku-Ankomah and Cordery (1994), Adiku and Stone (1995), Adiku et al. (1997) and, more recently, Owusu and Waylen (2009, 2013) and MacCarthy et al. (2017). Adiku and Stone (1995) showed a significant correlation between

Table 2 Future climate projections across West Africa

Parameters	General future projections by 2050s	Examples (for 2050s)
Temperature 	• Increased temperatures region-wide, with the greatest warming in the Sahel	• +1.6°C to 2.9°C (Niger, Burkina Faso, Mali, Niger, Mauritania) • +1.4°C to 2.5°C (Senegal, The Gambia) • +1.3°C to 2.3°C (Côte d'Ivoire, Ghana, Togo, Benin)
Hot days/nights; heatwaves 	• Increased duration of long-lasting heatwaves (+6–28 days), with generally a greater increase in the east	• +8–28 days (Burkina Faso, Mali, Niger) • +8–26 days (Ghana, Benin, Togo, Mauritania) • +6–23 days (Nigeria) • +6–18 days (Senegal, The Gambia)
Rainfall 	• Uncertain rainfall trends with models suggesting: • decreased rainfall in western Mali, Senegal and The Gambia • no change or increased rainfall in the rest of the region • Likely delay in the spring rainy season onset (about 5 days by 2100)	• Western Sahel: most models show decreased rainfall with a range of −16% to +6% (Senegal, The Gambia) • Remainder of the region: most models show increased rainfall with a range of 0% to +38% (Niger), −1% to +12% (Nigeria), −2% to +7% (Côte d'Ivoire, Ghana, Togo, Benin) and −3% to +11% (Burkina Faso, Mali)
Heavy rainfall 	• Increased frequency and intensity of heavy rainfall events	• Increased frequency (+1% to 43%) and intensity (+1% to 12%) of heavy rainfall events in much of the region (i.e. Nigeria, Ghana, Benin, Togo, Côte d'Ivoire, Burkina Faso, Mali) • Increased frequency (+16% to 75%) and uncertain trends for intensity (−4% to +21%) in Niger • Uncertain trends in frequency of (−10% to +31%) and intensity (−2% to +14%) in Senegal and The Gambia

Table 2 *(Continued)*

Parameters	General future projections by 2050s	Examples (for 2050s)
Dry spell 	• Generally increased dry spells in western Sahel (Senegal, The Gambia) • Generally reduced dry spells in the rest of the region	• Change in dry spells, with a range of −4 to +22 days (Senegal, The Gambia, Mauritania) • Change in length of dry spells, with a range of −11 to 0 days (Nigeria) and −8 to +1 days (Côte d'Ivoire, Ghana, Benin, Togo)
Sea level rise 	Sea level rise of +17 to 45 cm	+17 cm to 43 cm (Senegal, Nigeria) +18 cm to 43 cm (Ghana, Benin, Togo) +18 cm to 45 cm (Côte d'Ivoire)

Source: Adapted from USAID (2018).

El Nino-Southern Oscillation (ENSO) and rainfall in Ghana, especially with the south Atlantic SST or the warm and cold phases of the ENSO of the tropical Pacific.

In particular, they highlighted the October to December ENSO phase, providing a 3- to 6-month pre-season planning period for farming operations. Using simulation modelling, MacCarthy et al. (2017) showed that cereal yields from planting dates based on seasonal forecasts in combination with various soil management practices (mulching, etc.) were significantly higher than yields from traditional planting dates. However, actual field or on-farm validation of the seasonal forecast impact is scarce. To date, the only work has been in Ghana by Adiku et al. (2007), where groundnut farmers in the Akatsi District, Volta Region, who chose planting dates based on seasonal forecasts, had significantly higher yields than those who based planting dates on traditional indigenous knowledge.

Climate change forecasts for the near and long term (2050s to 2100s) would be useful for governments to position themselves for future agriculture development planning. For medium-term policy formulation and planning with a time horizon of 5–10 years, considerably more research needs to be focused on weather forecasting, as weather patterns may change rapidly from decade to decade.

5 Crop modelling in West African farming systems

Over the past three decades, crop modelling has been used by the agricultural sector of several West African countries as research and decision support. The initial focus of attention was to adapt the models to tropical crops. Therefore, most modelling studies have focused on:

1 crop physiology research leading to the derivation of genetic coefficients to enable calibration of models; and
2 soil processes research leading to the simulation of soil organic matter (SOM), water and nutrient dynamics to enable the simulation of nutrient responses by crops.

The majority of the calibration and validation work has been at research stations, where biophysical model input data are readily available or can be fairly easily derived. With respect to decision support, model applications are relatively recent, and this has been made largely possible through projects such as AgMIP.

The Climate Change Impact on West African Agriculture, a Regional Assessment project (Adiku et al., 2015) which encompassed five West African countries, perhaps represented the first regional attempt to apply integrated methods for the assessment of climate change on West African agriculture. The extent to which recommendations from the project have fed into the agricultural policies of participating countries is unknown. In general, crop modelling has remained a research-level activity with little effort to improve the biophysical content of the models once the models have been calibrated and validated.

The application of crop models for on-farm agricultural planning and decision-making has been met with many challenges (Akponikpè et al., 2014; MacCarthy et al., 2018) as on-farm data are often obtained via farmer surveys with emphasis on socio-economic rather than biophysical information. Thus, the effective use of crop modelling as one of the tools to develop the management of enhanced tropical farming is often handicapped by the paucity of input data as reviewed by Adiku et al. (2021a). To date, we are not aware of any such model development for management within the West African region and capacity for crop model use in this region is still generally lacking (MacCarthy et al., 2018). A synthesis by Adiku (unpublished) on crop modelling-related publications from South Saharan Africa showed that by 2009, about 25, 15 and 18 papers were published using Decision Support System for Agrotechnology Transfer (DSSAT), Agricultural Production Systems sIMulator (APSIM) and NUTrient MONitoring, respectively. Since then, on average, only around two modelling papers have been published annually from the region, with respect to these four models (MacCarthy et al., 2018).

Even for limited cases where models have been applied to on-farm situations, the main recommendations seem to lie within the scope of fertiliser management, because farming systems in West Africa are characterised by no or low inputs of fertiliser (0–50 kg/ha). However, fertiliser application also has cost dimensions, which are often beyond the reach of most farmers. Furthermore, crop models may show the superiority of residue retention for SOM improvement but fail to recognise that minimal crop residues are left on farms after grain harvest and that there is often competition for the residue for household uses such as fencing, fuel, animal feeding, among others (Bationo et al., 1995); hence, farmers may fail to adopt model-based recommendations.

It should also be noted that the mere application of models developed for temperate systems to tropical conditions has not adequately addressed the farming challenges of the West African region, due to a number of reasons, which for simplicity are grouped into the following classes:

1 Agricultural landscape structures;
2 Diversity of tropical cropping systems;
3 Soil variability and rapid soil degradation;
4 Diversity of tropical tillage systems;
5 Complexity of animal-crop interactions;
6 Climate variability and change; and
7 Low-level modelling capacity and data requirements.

The West African agricultural landscape, unlike the uniform contiguous forms that are commonly found in temperate countries, is highly varied and complex. The shifting cultivation system, which still dominates the landscape (Morton et al., 2020), results in a patchy mixture of croplands and other vegetation such as trees and wild bushes. There are significant interactions among the various vegetation patches and the croplands, especially on hill slopes, where the runoff from a given patch located at a higher position may influence the hydrology of the lower-positioned cropland (Adiku et al., 2021b). The effects of such landscape geometry on the overall crop productivity are not captured in most comprehensive models, except for a few such as the Agricultural Policy Environmental eXtender (APEX; Williams et al., 1995). In addition, the role of trees in the overall productivity of the agricultural landscape is not represented in the majority of models.

A better understanding of the tropical farming systems found in the West African region is also crucial. Farms are typically small and often interspersed among forest (tree-crop systems) and other vegetation, and not contiguous as in the case of many temperate region farmlands. Farms in the West African region are characterised by crops associated with scattered agroforestry trees, which are usually selected according to their socio-economic value (food, fruits, timber, fodder, house building materials, medicinal, etc.) and local cultural conditions (belief and rituals) (Diallo et al., 2019). Tree/shrub species are also used in farmers' fields to conserve and restore degraded lands, improve soil fertility and increase crop yield.

Whether the effect is positive (facilitation) or not (competition), trees contribute to yield variation within farmers' fields (Diallo et al., 2021). Crop association and agroforestry are yet to be accounted for in most crop modelling tools available in the region. Cropping systems also vary in the West African region, with examples including sole and mixed or intercropping. More complex tillage systems include the Zai, which is commonly found in the Sahel zones. Except for a few models, such as APSIM (McCown et al., 1996), most crop models do not simulate intercropping systems.

Cropping systems vary widely in the tropics. Not only is multiple cropping the normal practice, but there are also other varied cropping forms including ratooning, relay cropping, strip and mixed intercropping. To date, only a few models have included some intercropping aspects (e.g. APSIM: Adiku et al., 1993; Carberry et al., 1996; Adiku et al., 1998; and *Simulateur mulTIdisciplinaire pour les Cultures Standard*: Brisson et al., 2003). There has been little effort to improve the modelling concepts over the years. Most West African farming systems integrate small ruminants (sheep, goats) and poultry with cropping systems. Heve et al. (2015) observed that droppings from these animals constitute a major nutrient source when applied as manure to crop fields, which are close to homesteads. Yet, the system is ill-defined as rigorous quantification that would enable model development is lacking. With regard to large animals, models such as LivSim (see Section 10.1) are emerging to provide useful modelling output such as manure and milk production; however, these models are yet to be validated widely.

Tropical tillage practices vary, e.g. within the Sahel zones of West Africa, Zai tillage, which is designed to harvest rainfall (Zougmoré et al., 2014), and stone-bunding, designed to reduce runoff and erosion, are common. These types of practices cannot be implemented within the currently available comprehensive models. Research is required to understand and describe how the soil surface configuration affects the hydrological and water balance to assess crop productivity under such systems.

Soil challenges are paramount to the sustainable productivity of tropical agricultural systems. Apart from the generally low organic matter content of croplands, there is a rapid loss of SOM when new land is cleared and cultivated. More than 50% of SOM is lost within the first 4 years of cultivating new land (Adiku et al., 2009). The dynamics are often more complex than has been previously described by SOM models in current crop models. In the moist forested zones, which have high rainfall, high leaching of the basic cations results in soil acidity.

Acid soil infertility is a major constraint to phosphorus availability and this effect is not implemented in the commonly applied models. Apart from these, there is a high variability of soil properties even over short distances (Akponikpè et al., 2011; Diallo et al., 2019), which must be accounted for in the models because soil variability also results in yield variability (Akponikpè et al., 2014). In general, land degradation leading to loss of the nutrient-rich topsoils as well as the reduction of the overall soil depth are major factors that must be addressed in applying crop modelling in the West African region.

The adverse impact of climate change on crop productivity is a major issue of concern in the West African tropics. Although studies have shown that temperature increases can be ascertained for all locations in the region (see Section 4), much of the research effort in crop improvement has not been geared towards improving heat-tolerant varieties. Since temperature accelerates the life cycle of crops, long-duration varieties may be more appropriate under climate change. Yet, much more attention has been placed on the effect of drought, even though there is also evidence of an increased frequency of flooding. Indeed, flash flooding resulting in field inundation has caused considerable devastation to croplands with crops at different stages of growth. The impact of inundation can far outweigh that of drought. Currently, crop models do not represent the effect of flash flooding on crop yields.

Perhaps the most critical factor that handicaps the progress of crop modelling in the West African region is the lack of capacity (MacCarthy et al., 2018). Expertise is generally lacking, even though there have been several training workshops in the West African region. Few institutions and universities teach and offer formal tuition in crop modelling and even in such situations, participation by students is scant. Thus, issues of model development and improvement, scale and data availability, application of models for decision-making and many desired aspects of crop modelling are not implemented. There is no dedicated research funding and support from any of the West African governments for modelling work; it is, therefore, dependent on external donor support, or at best, through individual initiative and interest.

6 Modelling cereal and legume production in West Africa

Despite the challenges and limitations in the use of models for yield simulation in West African agriculture, they are useful for projecting future yields, especially within the context of climate change. In this part of the chapter, we review studies that have modelled the impact of climate change on the productivity of major cereals (maize, millet, sorghum and rice) and legumes in the region. The outputs of these studies are, however, often difficult to compare given that there are variations in terms of time horizons, GCMs used, type and number of crop models used, as well as management practices used in the simulations. Thus, there is considerable uncertainty in the magnitude and direction of projected yield changes. The analysis of 30 published articles, with approximately 650 data sets (different GCMs, crop models, varieties, etc.) focusing on climate impact on crop yield across the sub-region, indicates DSSAT, APSIM and SARA-H as the most commonly used crop simulation models.

The available studies on the projected impact of climate change on maize productivity indicate a wide range of yield change between −72% and +48% with a median of −17% (Fig. 3). The cases of positive yield outcomes were due, in most cases, to projected increases in rainfall, particularly in the Sahelian region (Akumaga et al., 2018) and the extension of the rainfall into drier months in certain seasons (Tachie-Obeng et al., 2013). Reduction in

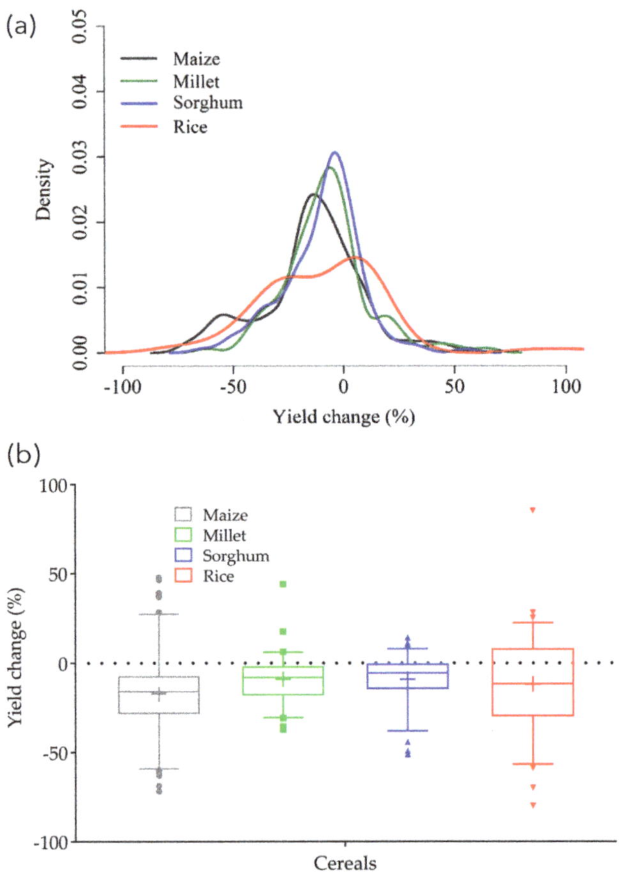

Figure 3 Distribution and variation of projected climate change impact on the productivity of cereals (maize, millet, rice and sorghum) in West Africa.

grain yield is attributable to increased temperatures which shorten the growth cycle of the crop and hence reduce total resource uptake. The reduction in rainfall also contributes to yield reductions. Our analysis of the literature also indicates that maize will be more adversely impacted than millet and sorghum. This is corroborated by the results of a study by Lobell et al. (2008), which indicated that maize is more vulnerable to climate change than millet in West Africa.

In the case of rice, the projected climate change impact shows a wide variation of yield across 10 countries (Oort and Zwart, 2017); projected yields varied from between +85% and −80% with a median of −12%. Rice yield reductions were more severe (−19 to −33%) in the Sahelian region (Niger, Mali, Burkina Faso), followed by the Guinean zone (Côte d'Ivoire, Ghana, Benin, Nigeria), ranging between −12% and 18%. For Senegal and The Gambia, the average impacts were positive (9% and 13%, respectively). The spread of yield change was much higher than in maize. Lobell et al. (2008), in prioritising climate change adaptation for food security, also indicated a decline in rice productivity but to a lower magnitude.

With regard to sorghum, a smaller variation in projected relative yields was simulated compared with maize and rice; projected yield changes ranged from +15% to −57% with a median yield of −9%. As with maize and rice, most studies projected a decline in sorghum yields with climate change. The projections for millet are similar to that for sorghum with a median of −9% and relative yields in the range of +44% to −37%. In the main, median yield changes projected for the four cereals suggest that sorghum and millet are less susceptible to climate change impact than maize and rice, which is supported by other studies (Adiku et al., 2015; MacCarthy et al., 2021b; Faye et al., 2018b).

In the case of legumes, the projected impact of climate change on the yield of groundnut was largely positive across most studies. The yield outcomes were widespread, ranging from +47% to −50% with a median of 8% (Fig. 4). Unlike cereals, the positive impact could be attributed to CO_2 fertilisation; hence, studies that excluded the effect of CO_2 often simulated yield loss (MacCarthy et al., 2022; Faye et al., 2018a). Groundnut yield simulations for reduced rainfall scenarios, or under water-limited conditions, would not reflect the CO_2 fertilisation effect which is often masked by water stress conditions. For soybean, relatively few studies also indicated productivity gains under climate change and the range of impact was less varied (Fig. 3); projected yield increases ranged between +10% and +40% with a median of +23%. As with most legumes, soybean also benefits from CO_2 fertilisation. Without CO_2 fertilisation, soybean will decline under warmer temperatures especially under a drier climate as shown by Ogunkanmi et al. (2022) in a screen house experiment.

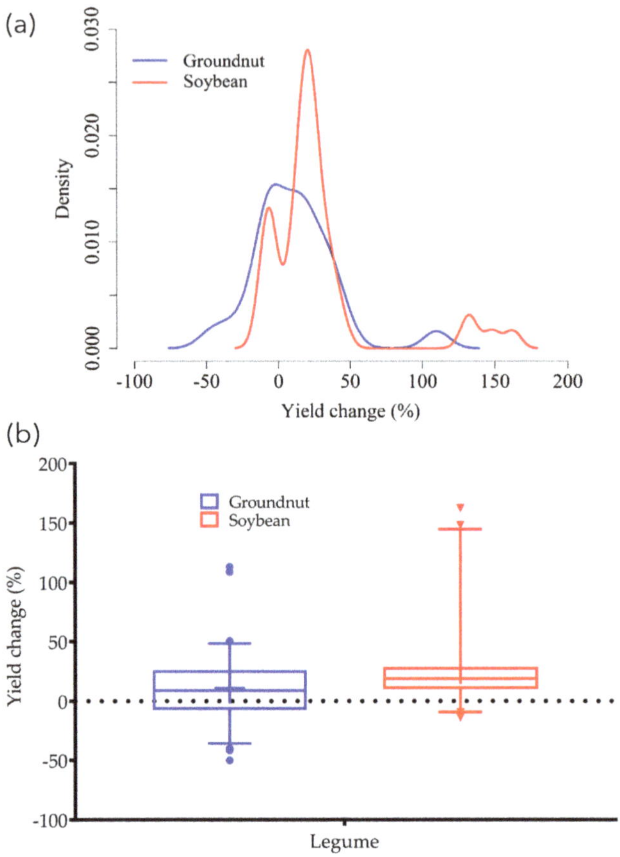

Figure 4 Distribution and variation of projected climate change impact on the productivity of groundnut and soybean in West Africa.

7 Accounting for the evolution of agricultural systems in climate change impact assessments

Agricultural development is not static and, over time, new technologies evolve, e.g., the development of new seeds. Whether there is climate change or not, countries have developmental goals that also extend to the agriculture sector. There are efforts to improve the yield potentials and resilience of several crops to different stress factors, which might lessen the impact of climate change. Yet, most climate change impact assessment studies (Akinseye et al., 2020; Freduah et al., 2019; Akumaga et al., 2018; Faye et al., 2018b; Sultan et al., 2013; Tachie-Obeng et al., 2013) assume that current production systems will also prevail in the future.

This leads to an overestimation of the impact of climate change on crop productivity, as the dynamic trajectory of development in the agricultural sector is often not captured in these studies. Such impact assessments can be best described as sensitivity analysis of the current production system under future climate conditions (MacCarthy et al., 2021b; Adiku et al., 2015). To address this defect, scenario development is increasingly used to forecast potential trajectories in agricultural systems as a way to capture the dynamic nature of the agriculture sector.

A number of climate change impact assessments have reported the influence of crop management practices on the magnitude of yield changes under climate change. Faye et al. (2018b) observed differences in the magnitude of yield changes when using the current levels of fertilisation compared with applying more fertilisers. An AgMIP project titled 'Climate Change Impact on West African Agriculture: A Regional Assessment: CIWARA' showed that the impact of climate change on crops growing on poor soils is less than the soil fertility impact since the latter is the predominant limiting factor (Adam et al., 2020).

The agricultural sector in West Africa is dominated by smallholder farms and characterised by diverse management practices that are governed by availability and access to the appropriate inputs and socio-economic factors. Although farming systems, management practices and soil types are heterogeneous; most climate change impact assessments assume uniformity in the management of each farm in the analyses of farm productivity (Tachie-Obeng et al., 2013; Amouzou et al., 2019; Sultan et al., 2013). Such approaches do not adequately capture the heterogeneity and the differential vulnerability of the farms to climate change. As part of the Department for International Development (DiFID)-funded AgMIP project, attempts were made to capture the heterogeneity in farming systems.

When farmer heterogeneity was considered, Freduah et al. (2019) showed substantial variation in climate change impact among smallholders. It was observed that the relative standard deviation for climate change impact was as high as 8–117% for maize productivity at Nioro, Senegal, whereas for Navrongo and Tamale in Ghana, the relative standard deviation was 13–64% and 9–37%, respectively. Adiku et al. (2015) earlier reported wide variability in the impact of climate change among other crops (millet, peanut) at Nioro and Navrongo. The wide variation in climate change impact has been attributed to factors including variation in planting date, soil type and amount of fertiliser used (Fig. 5). Therefore, a blanket or wholesale management strategy to minimise the impact of climate change on different farms would be erroneous.

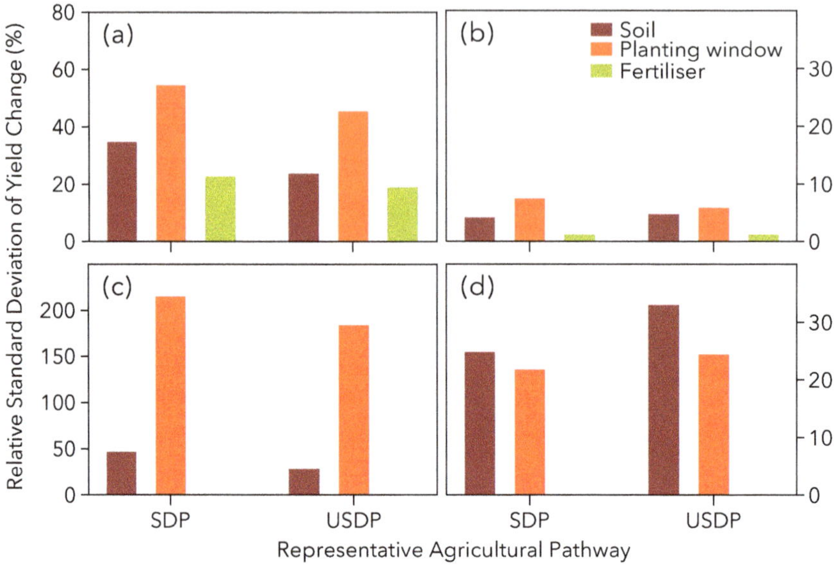

Figure 5 The relative contribution of crop and soil management to the variation in climate change impact on maize, sorghum and pearl millet among smallholder farmers in Navrongo, Ghana, and Nioro du Rip, Senegal. (a–d) are maize in Nioro du Rip, maize in Navrongo, pearl millet in Nioro du Rip and sorghum in Navrongo, respectively. Source: Adapted from MacCarthy et al. (2021b). SDP and USDP are sustainable development pathways and unsustainable development pathways, respectively.

8 Modelling climate change adaptation strategies

Adaptation is recognised worldwide as one of the major ways to minimise the adverse impact of climate change. For West African agriculture, a number of climate change adaptation strategies have been discussed in the literature; Table 3 presents some of them. These are, however, largely based on surveys conducted in the different farming systems and countries within the sub-region. Depending on the agro-ecology and farming system, farmers adapt using different strategies that suit their conditions to minimise the negative impacts of changing climate on the productivity of farming activities. Most of these strategies are currently in use to some extent to address current variability in weather that limits crop productivity.

In spite of the increasing number of adaptation strategies, only a few have been assessed using crop models. The strategies simulated include optimising planting dates (Tachie-Obeng et al., 2013), use of heat- and drought-tolerant varieties (Singh et al., 2014; Adiku et al., 2015), optimising fertiliser application (Akumaga et al., 2018) and optimising cultivar cycles (Sultan and Gaetani, 2016). Tachie-Obeng et al. (2013) in their study of the Guinea savanna and the

Table 3 Common adaptation strategies used by farmers in the West African sub-region

Category	Adaptation strategies	References
Improved varieties	Heat- and drought-tolerant varieties, shorter cycle	Singh et al. (2014); Adiku et al. (2015); Guan et al. (2017); Tofa et al. (2021)
Adjusting planting calendar and cropping systems	Optimising planting dates, crop rotation, mix cropping, land rotation	Fagariba et al. (2018); Adzawla and Alhassan, 2021)
Soil and water management	Mulching, crop residue retention, manure application, Zai, ridging, irrigation	Zougomoré et al. (2016); Akponikpè et al. (2010)
Weather and climate forecasting	Early warning systems, forecast of season onset	Zougomoré et al. (2016)
Fertiliser efficiency	Optimising fertiliser application, integrated soil fertility management	Fagariba et al. (2018)
Crop diversification	Use of different crops and varieties	Yegbemey et al. (2013)
Livelihood diversification	Bee keeping, trading	Yegbemey et al. (2013)

savanna-forest transition zones assessed the potential of shifting planting dates and use of heat-tolerant varieties to mitigate the impact of climate change.

They showed that delaying planting by 8 weeks in Wa increased yields by 13.9% across 3 GCMs. In Wenchi (a forest-savannah transition zone), delaying planting by 4 weeks in the major season while maintaining the planting dates in the minor season produced the highest yield gain of 3.9%. While shifting planting dates yielded a positive outcome, the use of a heat-tolerant variety produced better results and could be a more effective adaptation strategy in the study area. Using the heat-tolerant variety restored the crop phenology that had been lost due to climate change, which allowed the plants to mobilise more resources and hence increased yield.

The average gain in yield in Wa was 30% across the 3 GCMs used. In Wenchi, the yield gain across GCMs was 59% in the minor rainy season compared with a marginal gain (0.2%) in the main growing season. The effectiveness of the strategies varied across the GCMs, sites and seasons. In another study to assess the planting date as an adaptation strategy under rainfed and climate change conditions in Cameroon, Laux et al. (2010) observed yield gains (6–31% for maize and 7–20% for groundnut) in Garoua, but yield loses for Bamenda (1–2% for maize and 2–11% for groundnut).

However, for another site (Ngaoundere), no benefits were projected for both crops; i.e., the main yield-limiting factor differed among the sites. While water availability limited the yields at Garoua, temperature appeared to be the greater yield-limiting factor at Bamenda. Waongo et al. (2015) also combined climate, soil and crop data to derive planting dates for several locations in

Burkina Faso to show that significantly higher yields could be obtained when the planting dates were optimised.

According to Lobell (2014), adaptation should be described strictly as actions that reduce the impact of climate change. Thus, the benefit of the action provided under climate change should outweigh the benefits under the historical climate. For instance, in a study by Singh et al. (2014), the potential of virtual heat- and drought-tolerant varieties used as climate change adaptation strategies satisfy this definition, as the gain under climate change using the virtual varieties (especially the heat-tolerant sorghum) was higher under climate change than under baseline conditions.

The same magnitude of yield was produced in Cinzana under both climates when a drought-tolerant sorghum variety was used; however, it did not produce any benefit under the climate change scenario in Samanko (both in Mali). Using a heat-tolerant sorghum variety yielded greater benefits (5% more yields) under climate change than under the historical climate at both locations (Fig. 6). Combining tolerance of the two stress factors also produced greater benefits under climate change, translating into yield benefits of 4% and 5% according to Lobell (2014) compared with 6% and 13% per the conventional approach. This suggests heat stress, and not water stress, was the main yield-limiting factor.

An assessment of the benefits of five adaptation strategies: (i) delayed sowing; (ii) intensification (i.e. increased plant population and fertiliser use); (iii) longer cycle cultivar; (iv) water harvesting and (v) heat-tolerant varieties of cereals across West Africa was undertaken by Guan et al. (2017). This study revealed that most adaptation strategies were less effective under future climate conditions than under historical climate conditions and hence, do not effectively reduce the impact of climate change. The use of a heat-tolerant variety produced the highest benefit of 4.5%, implying that temperature stress was a greater limitation to crop yield than water stress in the study area under climate change conditions. Even though optimising plant population and fertiliser use increased yields under climate change conditions, the yield gain was not significantly different from that obtained under the historical climate and hence, this strategy cannot reduce the impact of climate change on sorghum yields.

It could be concluded that most countries in the West African sub-region have climate change policies that utilise adaptation strategies to either minimise the adverse effects or maximise positive impacts. There are, however, implementation challenges in most countries (Sorgho et al., 2020). The extent of adoption of these adaptation strategies is limited by several factors (Yegbemey et al., 2013). In some studies in northern Ghana (e.g. Fagariba et al., 2018; Antwi-Agyei et al., 2013), a number of constraints were listed as impeding the level of adoption of climate change adaptation strategies including: (i)

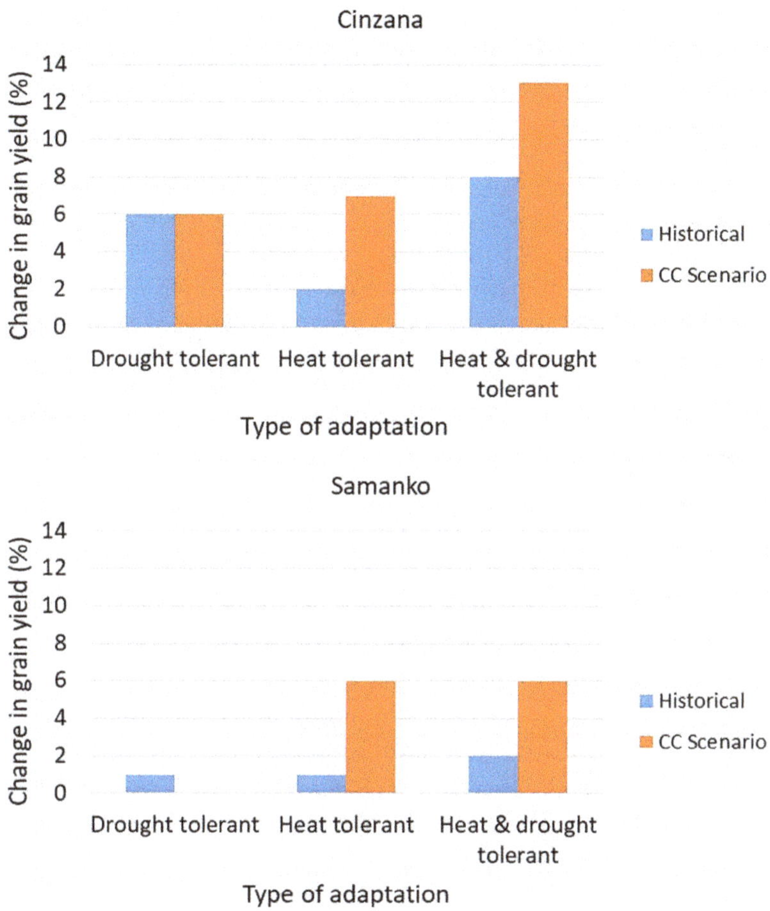

Figure 6 Performance of adaptation strategies under historical and climate change scenarios at two sites in Mali, West Africa. Source: Data extracted from Singh et al. (2014). CC is climate change.

inadequate policy support from the government; (ii) land tenure issues; (iii) inadequate extension officers; (iv) poor weather forecast information; and (v) poor soil fertility and poverty among farmers.

9 Key issues in modelling climate change impacts on livestock production

As with crops, climate change has adverse impacts on livestock, directly (e.g., via heat stress) and indirectly via resource (forage, nutrients and water) availability. It is generally known that feed intake by cattle reduces with increasing temperature (Timpong-Jones et al., 2019). The effect is particularly severe for exotic breeds often introduced from temperate regions into the West African region to raise the productivity of the local and more adaptive livestock breeds. Generally, these exotic breeds do not perform to their potential if ambient temperatures become excessive (Timpong-Jones et al., 2019).

Many studies have also shown the impact of climate change on forage growth. As forage is often a mixture of grasses, forbs and legumes, some components of the plant mix benefit, while others are adversely affected (Taub, 2010). The work by Taub (2010) and the United States Global Change Research Program (USGCRP) (2009) have shown that although the legume components of herbaceous cover benefit from a high CO_2 concentration in the atmosphere, by way of increased dry matter production, quality often declined due to a reduction in nitrogen concentration and subsequent crude protein content.

The drivers of climate change that directly or indirectly affect livestock production include rise in atmospheric temperature, variation in rainfall amount, rise in sea level, rise in atmospheric CO_2 concentration and extreme weather events (Timpong-Jones et al., 2014). The combination of two or more of these factors can affect livestock production by altering primary production, herd and plant species composition and also the quality of herbaceous species. Other effects of these climatic drivers on livestock production include challenges in livestock genetics and breeding, challenges in health and diseases as well as herder–farmer conflicts (Timpong-Jones et al., 2014).

Natural pastures left at the mercy of the weather play a very important role in livestock feeding in West Africa. A rise in temperature, extreme weather events like floods and prolonged droughts coupled with a rise in atmospheric CO_2 concentration affect the natural ecosystem and limit primary productivity. A rise in temperature and prolonged droughts also influence soil moisture content. A reduction in soil moisture (which is essential to dissolve soil nutrients for plant uptake) leads to a reduction in forage biomass yield and crop production. According to Thornton et al. (2009), crop yields will fall by 10–20% by 2050 in the tropics and subtropics due to a rise in temperature and recurrent droughts, with some locations expected to have an even higher reduction in yields than others. Lower forage biomass and crop yields translate into a reduction in grains, fodder, stover and metabolisable energy, thus causing challenges in livestock feeding, particularly in the dry season.

Climate change, driven by high temperatures and recurrent droughts has led to an increase in the ratio of actual to potential evapotranspiration in the semi-arid rangelands of Sahelian West Africa, hampering plant growth and reducing the length of the growing season (Timpong-Jones et al., 2023). This situation has led to a reduction in distribution and productivity and subsequently a reduction in carrying capacity and loss of livestock. Under this situation, pastoralists face difficult choices regarding where to move their livestock for feed and water in the dry season (McCartney, 2010).

The effect of climate change on primary plant production sometimes compels livestock farmers to change the animal species in their herds. The Tuareg (in Mali) who have been historically identified with camel culture switched from camels to cattle due to relatively high rainfall in the 1960s. However, the onset of prolonged droughts driven by climate change in the 1970s and 1980s onward demonstrated that this was not a good strategy. Changes in herd composition also occurred within the same breed of livestock. One good example involves the Fulbe herders in Nigeria. Faced with the declining quality and quantity of herbaceous plants in the semi-arid zone, they switched from the Bunaji cattle breed, which depends heavily on grass, to Sokoto Gudali, as this breed utilises browse species more efficiently (Blench, 1999).

Changes in animal species can also alter the woody species-grass mix over an extended period. Small ruminants are poor dispersal agents of many browse species due to their complete mastication of seeds (Brown and Havstad, 2004). Reduction of the cattle population due to the difficulty of managing them, because of feed scarcity resulting from climate change, may further aggravate the plight of the rural livestock farmer by causing a decline in browsing capacity. However, Polley et al. (2002) reported that the rise in CO_2 levels (from the current 350 ppm to a projected 700 ppm by 2100) might favour C3 shrubs and legumes over C4 grasses.

Similarly, rising temperatures and CO_2 levels are expected to alter the optimum growth ranges of different grass species leading to changes in the productivity and mix of grass species in rangelands (Calvosa et al., 2009). The Coastal Savannah Plains of Ghana were formerly dominated by grass species such as *Panicum maximum*, *Brachiaria falcifera*, *Andropogon gayanus* var. *bisquamulatus*, *Hyparrhenia dissoluta* and *Schizachyrium schweinfurthii* (Alhassan et al., 1999), which was an indication of good range condition.

However, a study by Timpong-Jones (2010) revealed that apart from *Panicum maximum*, the dominance of the other four species was replaced by species such as *Sporobolus pyramidalis* and *Vetiveria fulvibarbis* indicating a change in the mix of grass species towards reduced quality. This corroborates the findings of Oppong-Anane (2006) that forage species with mediocre nutritive quality have invaded the Coastal Savannah Plains of Ghana at the expense of more nutritive and palatable species. It is expected that there will be a change in the grass species spectrum towards species tolerant to drought, high temperatures and high CO_2 levels.

The rise in temperature, CO_2 concentration and recurrent droughts experienced in various parts of West Africa due to climate change adversely affects the quality of forage species. The lack of adequate water in the soil coupled with high atmospheric temperatures cause a reduction in nutrient uptake and increased evapotranspiration, respectively. This leads to increased lignification of forage species reducing the intake and digestibility by ruminant livestock that feed on them (Minson, 1990).

According to Taub (2010), an increase in the concentration of atmospheric CO_2 leads to higher rates of photosynthesis, increased growth, decreased water use and lowered tissue concentrations of nitrogen and protein in most plant species (species with the C_4 photosynthetic pathway show less response to elevated CO_2 concentration). This reduction in the concentration of plant tissue nitrogen and protein led to a decline in forage quality in pastures and rangelands (USGCRP, 2009). The resource-poor rural farmer with livestock depending on such forage without supplementation have recorded a reduction in meat and milk production impacting negatively their income and food security.

Pastoralists and small-scale livestock farmers in West Africa mainly keep local breeds adapted to their environment to deliver a wide range of products and services. With extreme weather events such as floods and prolonged droughts driven by climate change, these local breeds risk being completely wiped out (Carson et al., 2009). The cross-breeding of temperate high-yielding breeds with local breeds for economic purposes in recent times is a threat to animal genetic diversity. It may also lead to genetic erosion - indirectly increasing the risk of extinction of the less productive but adaptive local breeds (Hoffmann, 2010). According to van de Steeg and Tibbo (2012), animal genetic biodiversity is essential for food security and rural development.

A rise in atmospheric temperature, variable precipitation and extreme events promote the growth and survival of disease agents and their intermediate hosts of economic importance to the livestock industry (van den Bossche and Coetzer, 2008). A rise in atmospheric temperature and wetter conditions promote the growth of snails, the intermediate host of the liver fluke disease, biting midges and tick-borne encephalitis viruses. The incidence of bluetongue disease in cattle and sheep is expected to increase in areas of higher altitude as the temperature rise allows the virus to thrive in such areas.

Floods will cause faecal material to spill across fields and water bodies spreading disease pathogens over a wider geographic area. This will cause a rise in the incidence of dermatophilosis, anthrax, foot rot and blackleg diseases (van den Bossche and Coetzer, 2008). Recurrent droughts will cause overgrazing and mass movement of livestock to areas with feed and water. This concentration of livestock around feed and water resources will increase the incidence of foot and mouth disease, contagious bovine pleuropneumonia and rinderpest (van den Bossche and Coetzer, 2008). The rise in the prevalence of livestock mortality from these diseases caused by recurrent extreme climatic events will overburden the resource-poor livestock farmer in West Africa, making it difficult for them to care for and replenish their stock.

Climate change does not inherently cause herder–farmer conflicts. However, climate change is an important factor that influences resource availability for agricultural and pastoral production. Extreme climatic events such as a rise in temperature, prolonged droughts and floods heighten the vulnerability of the natural resource-dependent pastoral farmers. Depending on their level of resilience and coping strategies, the response can lead to insecurity through conflicts over natural resources like water and forage for their livestock. Climate change is a driver for pastoralist migration, pushing pastoralists in West Africa towards the south, which has resulted in the shrinkage of environmental space and increased natural resource scarcity (Dary et al., 2017).

Fulani herders in West Africa traditionally migrate with their cattle southwards towards the coast during the post-harvest period to feed on crop residues. Increased competition for natural resources drives the herders and their cattle to areas where the cattle destroy crops. This has led to a number of herder-farmer conflicts in West Africa. Climate change has been identified as a 'threat multiplier' rather than a direct trigger (Baptista et al., 2013). Contrary to many opinions, however, Hsiang and Burke (2014) reported a causal association between climatological changes and conflicts. The migration of the Fulbes to the south due to drought in countries like Mali and Nigeria resulted in herder-farmer conflicts (Bukari and Schareika, 2015).

During such conflicts, farmers, villagers and cattle have been killed and large markets attacked, leading to a loss of produce (Pugliese, 2014). In 2016, over 2000 people were killed and tens of thousands were displaced in Nigeria's Benue and Kaduna states due to herder-farmer conflicts (International Crisis Group, 2017). In Ghana, Konkomba farmers carried out a night raid in Damdaboli, Zamashigu, Batiga and Naboya farming communities in the Gushiegu district in 2011. Thirteen Fulani pastoralists were killed, their houses burned and their cattle rustled as a result of the raid (Olaniyan et al., 2015). These conflicts are a major risk to ruminant livestock production and a threat to food security in the West African sub-region.

10 Modelling livestock systems

Modelling the impact of climate change on livestock production is challenging because both animal and plant growth must be simulated. The feedback between the animals and crop systems can be complex. For temperate regions, a number of livestock models have been developed including GRAZPLAN (a suite comprising GrazFeed and GrassGro) (Donnelly et al., 2002), Grass Production (GRASP) (McKeon et al., 2000) and Great Plains Framework for Agricultural Resource Management (GPFARM)-Range that was applied by Adiku et al. (2010) to assess rangeland productivity in the Great Plains of the USA. A number of crop models such as DSSAT and APSIM now have forage growth components. Some of the few forage productivity and livestock models used in the West African sub-region and their specific applications and limitations are described in the following subsections.

10.1 The LivSim model

Within the West African agricultural context, one of the livestock models being evaluated is the LIVestock SIMulator (LivSim), (Rufino et al., 2009). The LivSim is designed to simulate the impact of different management strategies on the long-term productivity of cattle systems. The model quantifies the production of milk, manure and offspring of individual animals of small herds common in smallholder farming systems. The purpose of analysing this type of mixed system is geared towards optimising the use of resources instead of the maximisation of one single production trait.

The cattle model is linked to the farm model to share nutrient, labour and cash resources and allow farm-scale exploration of different livelihood strategies. The West African AgMIP team parameterised LivSim in order to simulate the growth of two local cattle breeds in Mali and Ghana, namely Mere and Ghana Shorthorn (GhSh) cattle, respectively. The Ghana study was conducted in Tamale in the northern region of Ghana. It simulated the performance of these indigenous West African cattle breeds using various climate scenarios. The study also quantified milk production to determine the trend in herd size dynamics under various climate scenarios. Aspects of the study on the GhSh breed are detailed later.

Agriculture in the Ghanaian town of Tamale is dominated by a large number of smallholder farmers who mainly cultivate a range of cereals and legumes; in addition to crop production, many households also raise livestock. Farmers rely on rainwater and the soil's inherent fertility for crop production. Based on this, the system of farming in Tamale can be described as a low-input cereal/legume-livestock integrated system. In a study, a total of 261 households were interviewed and data were collected on their livestock production system with emphasis on the GhSh breed (Table 4). This information was required to establish the herd composition as an input to LivSim. Pasture production (biomass yield) was simulated using APSIM and DSSAT and daily availability ranged from 1.5 kg to 8.4 kg dry matter (DM)/day.

Table 4 Summary of livestock survey data

Type of data	No. or %
No. of households interviewed	261
No. of households with cattle	108
Cattle herd size between 1 and 10	53%
Cattle herd size between 11 and 20	24.8%
Cattle herd size between 21 and 30	16.7%
Cattle herd size more than 30	5.4%

The LivSim model consists of three main parts (Fig. 7): (i) a user-defined input component; (ii) modules containing algorithms describing different biological processes; and (iii) an output component indicating the results of the simulations.

LivSim was parameterised for the GhSh breed, and Table 5 details some of the crucial parameters used.

With breed parameters and feed inputs (both from crop residues and biomass from rangeland simulations) provided, the performance of the GhSh breed was simulated using RCP8.5 and RCP4.5 under five different scenarios (cool dry, cool wet, hot dry, hot wet and middle) using two different models (DSSAT and APSIM). Milk yield using DSSAT ranged from 55.6 to 2528.3 L per annum, which was slightly higher than the simulation from APSIM with a range of 55.6–1925.7 L per annum (Fig. 8).

While milk yield differences were not pronounced among scenarios for a given model, the comparison of milk yields between models showed some numeral differences with DSSAT simulations recording higher yields than APSIM simulations (Fig. 9).

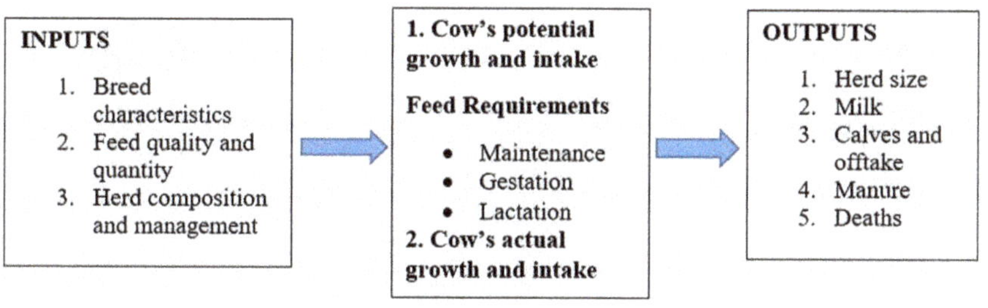

Figure 7 An overview of the LivSim structure and data used.

Table 5 Parameters of GhSh used for LivSim simulations

Parameter	Set limits
Maximum lactation	4
Gestation length	282 days
Mature body weight for male	300 kg
Mature body weight for female	275 kg
Calf birth weight	15 kg
Maximum lactation length	10 months
Milk ME content	23.3 MJ/kg
Fat content of milk	46 g/kg
Crude protein content of milk	3.7%
Energy used for walking	1 KJ/km per kg body weight

ME: metabolisable energy; MJ: megajoules; KJ: kilojoules.

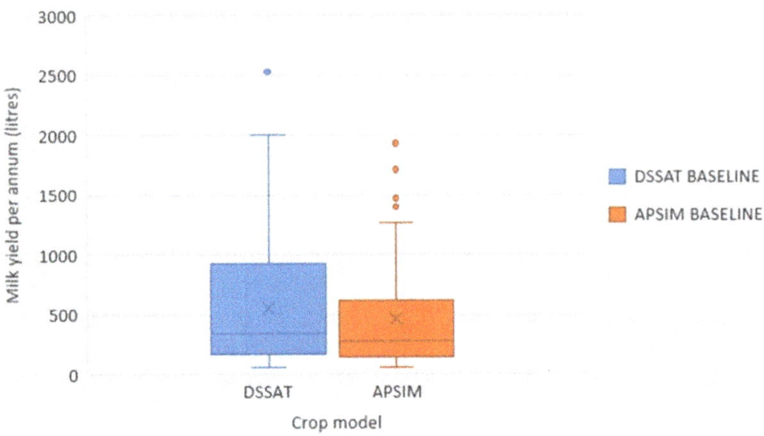

Figure 8 Simulated annual milk yield using forage input from DSSAT and APSIM.

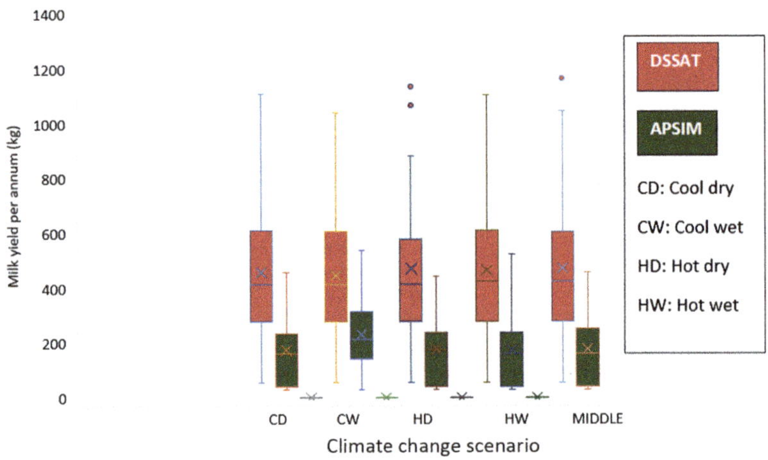

Figure 9 Milk yield per annum using DSSAT and 4 GCMs under RCP4.5 and RCP8.5. GCM and RCP are a general circulation model and representative concentration pathway, respectively.

Milk yields for the five scenarios for RCP4.5 were significantly higher than that of RCP8.5 for DSSAT (Fig. 10).

Manure production was not sensitive to differences in the various scenarios with an annual production of 12 439.3 kg and a range of 12 429.1-12 463.8 kg for the other scenarios for DSSAT. There was no significant change in manure yield between RCP4.5 and RCP8.5 for both DSSAT and APSIM.

Some of the limitations of the LivSim model include (i) the model depends on the input of biomass production simulated using other models such as DSSAT and APSIM. Outputs may differ greatly depending on the source of biomass data; (ii) many local African and West African cattle breeds have

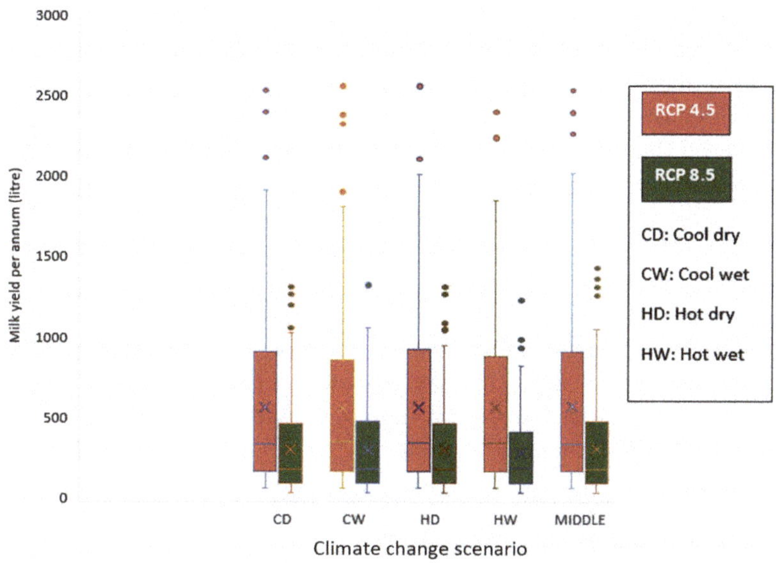

Figure 10 Annual milk yield using DSSAT and 4 GCMs under RCP4.5 and RCP8.5. GCM and RCP are a general circulation model and representative concentration pathway, respectively.

not been parameterised using LivSim, which will have to be done prior to simulations; and (iii) if a breed is not parameterised and the model used, voluntary dry matter intake and daily milk yields will be poorly predicted (Bateki and Dickhoefer, 2020).

10.2 The Landscape DNDC model

In another study by Rahimi et al. (2021), the process-based biogeochemical model, Landscape DNDC, was used on a regional scale to determine the livestock carrying capacity (LCC) in the Sahelian and Sudanian agro-ecological zones of West Africa. Two frameworks were used: one to convert the main types of feed biomass, namely, natural pasture, browse and crop residues, into energetic values (energy supply, S) and the second to determine the energy required to support a given population of livestock (energy demand, D). This was done for the period 1981–2020.

Comparison of the energy supply (S) with energy demand (D) of the livestock population was to assess regional surpluses (S > D) or deficits (S < D) in feed availability. The data sets used in the simulation were climate, soil, livestock production, livestock production system, land use/cover, harvest area, woody cover, crop calendar, fertiliser use and *Action Contre la Faim* biomass production data sets (Rahimi et al., 2021). The results of the study showed that, over 40 years, there had been a large-scale transition of feed availability from surplus to deficit.

The LCC was exceeded by 27% of the area in the first decade (1981–90), but this increased to 72% in the last decade (2011–20). This trend was attributed to an 8% reduction in the total feed supply and a 37% increase in feed demand per decade, driven by climate change and increased livestock population, respectively. The simulation helped to identify the 'hotspots' with livestock populations exceeding their self-sufficient sustainable feed supply, which is useful in locating areas needing interventions to supplement the feed supply.

The main limitations of the Landscape DNDC model include (i) simulated dry matter production was about half the amount obtained using remotely sensed data; (ii) the use of an average instead of a year-specific land use map, resulted in an underestimation of feed production in the wet season for some years and overestimation in the dry season and (iii) since several data sets are used, inherent errors in the data sets will result in an underestimation or overestimation of feed supply or demand

10.3 The Structural Ricardian model

Seo and Mendelsohn (2007) modelled the impact of climate change on livestock management using the Structural Ricardian model with the aim of understanding how farmers change their behaviour in response to climate. Three climate-sensitive elements of livestock production, namely, animal species selected, the number of animals per farm and the net revenue per animal were estimated. A three-equation model was used to determine the sensitivity of African livestock production decisions to climate based on the named elements. This study was conducted with data from 10 countries on the African continent including Burkina Faso, Cameroon, Ghana, Niger and Senegal.

Assuming a rise in temperature of 2.5°C or 5°C and a change in precipitation of +15% or −15%, the net income across all animals will fall but, in particular, for beef cattle. A 2.5°C warming resulted in a 32% loss in expected net income and a 5°C warming led to a 70% loss in expected net income. An increase in precipitation by 15% resulted in a 1% loss in expected net income. When models were used separately for small and large livestock farms, results also varied. With warming, small farms are expected to shift away from dairy cattle and chickens to goats and sheep.

Net incomes are expected to fall for all animals except sheep, with animal numbers also expected to reduce. Income is expected to decline by 13% with a warming of 2.5°C but to recover with more warming to current levels of income. A 15% decrease in precipitation is expected to increase small livestock farm incomes by 6%. For large farms, warming will cause a shift from goats, chickens and particularly beef cattle to dairy cattle and sheep. The income per animal is expected to decline for all species as temperatures rise. With higher temperatures, large farms will prefer to have fewer beef cattle, chickens and sheep but more goats and dairy cattle.

Large farmers' incomes are expected to fall by an average of 26% with a 2.5°C warming and by 67% with a 5°C warming, but a 15% decrease in precipitation is expected to increase these farmers' incomes by 2%. Some limitations of the Structural Ricardian model include (i) although five types of animals (beef cattle, dairy cattle, chicken, sheep and goats) were identified as being profitable, farmers could only choose one and not a combination of animals and (ii) it is difficult to calculate the variance of climate change impact parametrically using the model.

11 Integrated climate change impact assessments

The assessment of climate change impact on the West African region cannot be limited to crop yields and productivity alone. It is increasingly realised that a more holistic approach must be taken. Various factors that affect the livelihoods of farmers must also be considered in such assessments because the impact of climate change on crop productivity will affect the revenue and socio-economic status of farmers. Other factors such as institutional (access to market, access to credit and input subsidies), socio-economic (household size, farm size, off-farm income), biophysical (soil degradation, livestock productivity, herd sizes, biodiversity) and technology (access to energy, mechanisation, use of improved varieties) would influence the magnitude and direction of the impact of climate change on the livelihood of farmers (Valdivia et al., 2015).

A simulation study in Nioro, Senegal, by MacCarthy et al. (2021a) using the AgMIP Regional Integrated Assessment (RIA) approach (Fig. 11) showed for RCP4.5 and RCP8.5 (when temperatures are expected to increase between 1°C and 3°C and rainfall change would be between −32% and 7% by 2040-69), cereal, especially maize yields, would be severely impacted, but groundnut would benefit from CO_2 fertilisation. However, if future socio-economic, institutional and technological conditions improve, the impact of climate change on farmers' livelihoods would be positive, depending on the price and proportion of groundnut in the crop mix.

In spite of the overall positive outcome, a significant proportion of farms would remain vulnerable to climate change, especially under the low-price scenarios (Fig. 12). In terms of farmers' vulnerability to poverty, regression studies by Bawayelaazaa Nyuor et al. (2016) showed that the combination of declining rainfall and increasing temperature impacted negatively on maize and sorghum yields in northern Ghana, resulting in increased poverty of farmers and creating new vulnerabilities for farmers (Zougmoré et al., 2016). The change in climate negatively impacted net revenue and varied across seasons and among crops.

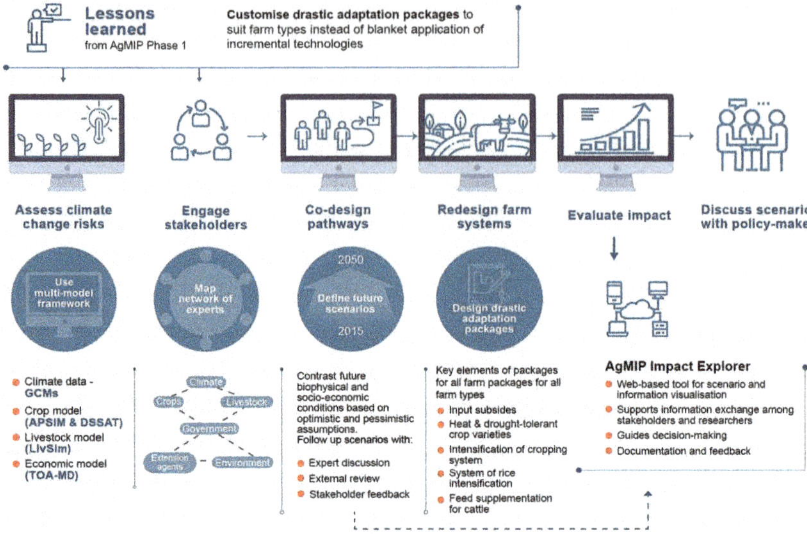

Figure 11 The AgMIP Regional Integrated Assessment of Climate Change Impact, Vulnerability and Adaptation of Agricultural Systems and the process of co-designing RAPs and adaptation packages. Source: Adapted from Valdivia et al. (2021) and ICRISAT (2016).

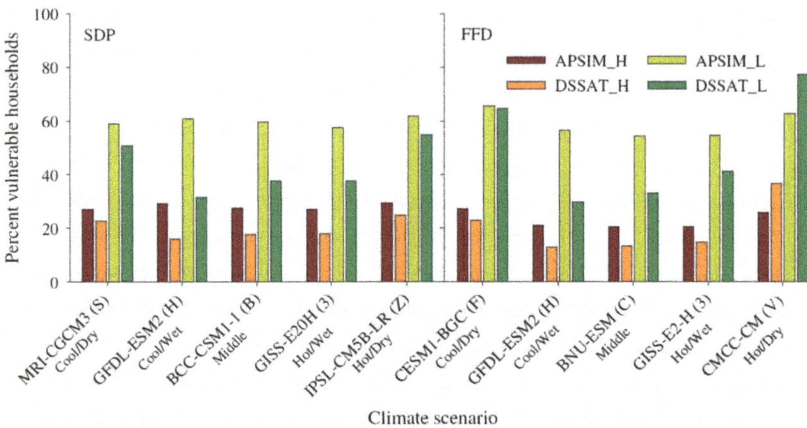

Figure 12 Percent of vulnerable households in Nioro, Senegal, under high (H) and low (L) prices, five climate scenarios, two crop models (DSSAT and APSIM) and two Representative Agricultural Pathways (Sustainable Development Pathway (SDP) and Fossil Fuel Driven Development Pathway (FFD). Source: Adapted from MacCarthy et al. (2021a).

In another study on the impact of climate change on agricultural net revenue, a case study in the Fouta Djallon, Wood and Mendelsohn (2015) suggested that climate change is associated with agricultural net revenue. They indicated that increased temperature and rainfall led to reduced revenue in the main cropping season while revenue increased in the dry season (secondary importance) (Wood and Mendelsohn, 2015). Another study by Dasmani and Dadzie (2021), using the Ricardian approach, reported that increases in temperature negatively impacted food crop production and hence, net revenue and whole-farm net revenue.

MacCarthy et al. (2021c) showed that for the current production system in Navrongo, Ghana, climate change would negatively affect net farm returns and increase poverty and food insecurity. A similar finding was earlier reported in Nioro, Senegal, by Adiku et al. (2015) using AgMIP's RIA framework, whereby the mean net farm returns decreased by between 10% and 38%, resulting in decreased per capita income and increased poverty of smallholder farmers. The decline in the economic indicators was higher under maize-based farming systems due to the fact that maize yield suffered the highest reduction in yield compared with the other crops being farmed. Thus, the type of crops in the farming system will influence the magnitude and direction of climate change on the livelihood of farmers.

Nhemachena et al. (2010) also performed an analysis of the economic impact of climate change on agriculture in West, East and southern Africa by controlling for the effects of key socio-economic, technology, soil and hydrological factors influencing agricultural production. Their results show that net farm revenues were generally negatively impacted under warmer and drier climates. The farming systems characterised by small-scale mixed crops and livestock, which is predominantly typical in Africa, were more tolerant to warming and lower rainfall impact compared with specialised crop production which was highly vulnerable. They also indicated that mixed crop and livestock farming and irrigation offered higher economic returns compared with the mono-cultural production system and, hence, better adaptation options for farmers against further warming and drying predicted under various future climate scenarios used in their study.

12 Conclusion and future trends

The West African economy is largely dependent on agriculture, which engages approximately 60% of the population and significantly contributes to the GDP of most of the countries in the sub-region. Agriculture in the West African region is weather dependent and it is likely to remain so for the foreseeable future. Though several governments continue to expand irrigation, the proportion of arable lands that are currently irrigated is negligible. Furthermore, as most of the irrigation projects depend on surface water, which is harvested from rainfall, the dependence on climate remains. Climate projections indicate a significant increase in temperatures across the region, especially by 2100. Projections for rainfall are somewhat uncertain, but there is a tendency for wetting especially in the north-eastern locations.

It has been shown that the impact of climate change on West African agriculture can be assessed using modelling approaches. The challenges in the direct application of currently published models have been highlighted, especially as most of them may not address some unique features of tropical cropping and farming systems. Further challenges, including the paucity of input data, have been indicated. Furthermore, the time horizons of most climate change studies do not fit into the planning horizons of farmers, who require seasonal forecasts rather than near-future (2050s) or far-future (2100s) forecasts. There is a need for governments in the sub-region to fund efforts aimed at incorporating the unique features that are currently unavailable in the models. Additionally, work needs to be done to ensure that a meta-database is created for all scientific studies in the sub-region to enhance accessibility to data already in the system.

In a similar vein, a medium-term (5–10 years) forecast will be required for policy planning. These aspects have not received sufficient research attention. Despite modelling challenges, it has been shown to be the appropriate tool to examine a range of climate effects on crop and animal growth and performance. Simulated results showed that the combination of increasing temperature and declining rainfall would adversely affect cereal productivity, especially maize. Simulated results also show that the legumes, by virtue of CO_2 fertilisation can benefit from climate change, resulting in increased growth and yield. The overall impact on farmers' income will depend, in part, on the proportion of land allocated to legumes in the farming enterprise and the price of legumes. The assessment of several adaptation options suggested that improving heat tolerance by crop varieties was most effective in offsetting the adverse effects of climate change. Climate change impact assessment approaches that integrate institutional, socio-economic, technological and biophysical factors should be encouraged to ensure a holistic understanding of climate change impact and the benefits of adaptation options.

Climate change can affect livestock production either directly via heat stress or indirectly through forage growth and water availability. Heat stress distorts the physiological functions of the animals resulting in reduced feed intake. Declining rainfall and high temperature stress would reduce forage growth. Even if the legume component of the forage benefits from CO_2 fertilisation, the quality of the legume biomass declines; hence, the overall forage quality is lowered. It was shown via simulation that animal (cattle) performance in terms of milk production declined under climate change. Drying conditions imply that animals have to trek long distances for water, resulting in loss of energy and reduction in live weight. Expertise in this area in the sub-region needs to be developed to ensure that the livestock sector is adequately captured in climate change impact and adaptation discussions.

The major limitation of the assessment of climate change impact on West African agriculture is related to the limited modelling capacity and the lack of resources for the development of tropical cropping system models. A number of previous studies have highlighted the challenge but support from governments is still lacking. Thus, a concerted effort is required to train crop modellers from the region that will focus on addressing challenges unique to the sub-region.

Though most countries in the sub-region have climate change policies and adaptation strategies to either minimise the adverse effects or maximise positive impacts, a number of challenges handicap the implementation of these strategies including issues relating to inadequate policy support from the government, land tenure, inadequate extension officers, poor weather forecast information and ineffective approaches for sustained soil fertility enhancement. These policy and infrastructural challenges need to be effectively addressed to support climate change impact assessment and adaptation efforts.

13 Where to look for further information

The Agricultural Model Improvement and Inter-comparison Project (AgMIP) has published some detailed studies on regional integrated assessment of climate change impact on the livelihoods of smallholder farmers. Two chapters that provide more insight into climate change impact in West Africa are:

- Adiku, S. G. K., MacCarthy, D. S., Haithie, I., Diancoumba, M., Freduah, B. S., Amikuzuno, J., Traore, P. C. S., Traore, S., Koomson, E., Agali, A., Lizaso, J. I., Fatondji, D., Adams, M., Tigana, L., Diarra, D. Z., N'diaye, O. and Valdiva, R. O. (2015). Climate change impacts on West African agriculture: An integrated regional assessment. In Rosenzweig, C. and Hillel D. (Eds.), *Handbook of Climate Change and Agroecosystems: The Agricultural Model Intercomparison and Improvement Project (AgMIP) Integrated Crop and Economic Assessments - Joint Publication with the American Society of Agronomy, Crop Science Society of America, and Soil Science Society of America*. London: Imperial College Press, pp. 25–73. https://doi.org/10.1142/9781783265640_0014.
- MacCarthy, D. S., Hathie, I., Freduah, B. S., Ly, M., Adam, M, Ly A., Nenkam, A., Traore, P. S. and Valdivia, R. O. (2021). Potential impacts of agricultural intensification and climate change on the livelihoods of farmers in Nioro, Senegal, West Africa. In *Handbook of Climate Change and Agro-Ecosystem: Climate Change and Farming System Planning in Africa and South Asia: AgMIP Stakeholder-driven Research, Part 2*, pp. 3-31. https://doi.org/10.1142/9781786348814_0001.

The International Food Policy Research Institute also commissioned a study on the impact of climate change on agriculture in a number of West African countries that was published as West African agriculture and climate change:

A comprehensive analysis. This book also provide additional information on climate change impact assessment at the country level and hence serves as an additional source for more information on the subject.

- Jalloh, A., Nelson, G. C., Thomas, T. S., Zougmore, R. and Roy-Macauley, H. (Eds.). (2013). *West African Agriculture and Climate Change: A Comprehensive Analysis*. IFPRI Research Monograph. 408. International Food Policy Research Institute (IFPRI). doi: http://dx.doi.org/10.2499/9780896292048.

14 References

Adam, M., MacCarthy, D. S., Traoré, P. C. S., Nenkam, A., Freduah, B. S., Ly, M. and Adiku, S. G. K. (2020). Which is more important to sorghum production systems in the Sudano-Sahelian zone of West Africa: Climate change or improved management practices? *Agricultural Systems* 185, 102920. https://doi.org/10.1016/j.agsy.2020.102920.

Adiku, S. G. K., MacCarthy, D. S. and Kumahor, S. K. (2021a). A conceptual modelling framework for simulating the impact of soil degradation on maize yield in data-sparse regions of the tropics. *Ecological Modelling* 448(C).

Adiku, S. G. K., Adanu, S., MacCarthy, D. S. and Kumahor, S. K. (2021b). A preliminary evaluation of tree-crop systems on some hilly landscapes of Ghana. Landscape 2021 - Diversification for sustainable and resilient agriculture: 20 September (Vols. 20-23), Germany, 2021. (ZALF), Müncheberg.

Adiku, S. G. K., Atika, E., Jagtap, S. S., Nkansah, A., Atidoh, L., Jones, J. W., Duadze, S. and Naab, J. B. (2007). Climate variability and farmers' response in the groundnut producing areas of the Akatsi District of Ghana. In: *Conference Paper, African Soil Science Society*, Accra.

Adiku, S. K., Carberry, P. S., Rose, C. W., McCown, R. L. and Braddock, R. (1993). Assessing the performance of maize (*Zea mays*) - cowpea (*Vigna unguiculata*) intercrop under variable soil and climatic factors in the tropics. In: *Proceedings of the 7th Australian Agronomy Conference*, 19-26 September, Adelaide, Australia, p. 382.

Adiku, S. G. K., Dayananda, P. W. A., Rose, C. W. and Dowuona, G. N. N. (Eds.). (1997). An analysis of the within-season rainfall characteristics and simulation of the daily rainfall in two savanna zones in Ghana. *Agricultural and Forest Meteorology* 86(1-2), 51-62.

Adiku, S. G. K., Duenow, L. and Naab, J. B. (2011). *State of Weather Stations in the Three Northern Regions of Ghana*. GIZ Consultancy Report, Ghana, p. 92.

Adiku, S. G. K., Dunn, G. H., Ahuja, L. R., Gunter, S., Bradford, J., Garcia, L. and Andales, A. A. (2010). Simulation of sandsage-bluestem forage growth under varying stocking rates. *Rangeland Ecology and Management* 63(5), 546-552.

Adiku, S. G. K., Jones, J. W., Kumaga, F. K. and Tonyigah, A. (2009). Effects of crop rotation and fallow residue management on maize growth, yield and soil carbon in a savannah-forest transition zone of Ghana. *The Journal of Agricultural Science* 147(3), 313-322. https://doi.org/10.1017/S002185960900851X.

Adiku, S. G. K., MacCarthy, D. S., Haithie, I., Diancoumba, M., Freduah, B. S., Amikuzuno, J. and N'diaye, O. (2015). Climate change impacts on West African agriculture: An integrated regional assessment. In: *Handbook of Climate Change and Agroecosystems, the Agricultural Model Intercomparison and Improvement Project*

(AgMIP) Integrated Crop and Economic Assessments-Joint Publication With the American Society of Agronomy, Crop Science Society of America, and Soil Science Society of America, pp. 25–73.

Adiku, S. G. K., Mawunya, F. D., Jones, J. W. and Yangyouru, M. (2007). Can ENSO help in agricultural decision-making in Ghana? In: *Climate Prediction and Agriculture*. Berlin, Heidelberg: Springer, pp. 205–212.

Adiku, S. G. K. and Stone, R. C. (1995). Using the Southern Oscillation Index for improving rainfall prediction and agricultural water management in Ghana. *Agricultural Water Management* 29(1), 85–100.

Adzawla, W. and Alhassan, H. (2021). Effects of climate adaptation on technical efficiency of maize production in Northern Ghana. *Agricultural and Food Economics* 9(1), 1–18. https://doi.org/10.1186/s40100-021-00183-7.

Akinseye, F. M., Ajeigbe, H. A., Traore, P. C. S., Agele, S. O., Zemadim, B. and Whitbread, A. (2020). Improving sorghum productivity under changing climatic conditions: A modelling approach. *Field Crop Research* 246(1), 1–11. https://doi.org/10.1016/j.fcr.2019.107685.

Akponikpè, P. B. I., Ganderard, B. and Bielders, C. L. (2014). Soil water crop modeling for decision support in millet-based systems in the Sahel: A challenge. *African Journal of Agricultural Research* 9(22), 1700–1713.

Akponikpè, P. B. I., Gérard, B., Michels, K. and Bielders, C. (2010). Use of the APSIM model in long term simulation to support decision making regarding nitrogen management for pearl millet in the Sahel. *European Journal of Agronomy* 32(2), 144–154.

Akponikpè, P. B. I., Minet, J., Gérard, B., Defourny, P. and Bielders, C. L. (2011). Spatial fields' dispersion as a farmer strategy to reduce agro-climatic risk at the household level in pearl millet-based systems in the Sahel: A modeling perspective. *Agricultural and Forest Meteorology* 151(2), 215–227.

Akpoti, K., Higginbottom, T. P., Foster, T., Adhikari, R. and Zwart, S. J. (2022). Mapping land suitability for informal, small-scale irrigation development using spatial modelling and machine learning in the Upper East Region, Ghana. *Science of the Total Environment* 803, 149959. https://doi.org/10.1016/j.scitotenv.2021.149959.

Akumaga, U., Tarhule, A., Piani, C., Traore, B. and Yusuf, A. A. (2018). Utilizing process-based modeling to assess the impact of climate change on crop yields and adaptation options in the Niger River Basin, West Africa. *Agronomy* 8(2), 11.

Alhassan, W. S., Karbo, N., Aboe, P. A. T. and Oppong-Anane, K. (1999). *Ghana's Savanna Rangeland: Agro-Ecology, Current Improvement and Usage Practices, Research Needs and Sustainable Management Criteria*. Accra, Ghana: National Agricultural Research Project. Council of Scientific and Industrial Research.

Amouzou, K. A., Lamers, J. P. A., Naab, J. B., Borgemeister, C., Vlek, P. L. G. and Becker, M. (2019). Climate change impact on water-and nitrogen-use efficiencies and yields of maize and sorghum in the northern Benin dry savanna, West Africa. *Field Crops Research* 235, 104–117.

Antwi-Agyei, P., Dougill, A. J. and Stringer, L. C. (2013). *Barriers to Climate Change Adaptation in Sub-Saharan Africa: Evidence From Northeast Ghana and Systematic Literature Review*.

Baffour-Ata, F., Antwi-Agyei, P., Apawu, G. O., Nkiaka, E., Amoah, E. A., Akorli, R. and Antwi, K. (2021). Using traditional agroecological knowledge to adapt to climate change and variability in the Upper East Region of Ghana. *Environmental Challenges* 4, 100205. https://doi.org/10.1016/j.envc.2021.100205.

Baptista, S., Brottem, L., de Sherbinin, A., Edquist, M., Fischer, A., Levy, M., Schnarr, E., Simon, C., Sundareshwar, P. V. and Trzaska, S. (2013). Background paper for the ARCC West Africa regional climate change vulnerability assessment. ARD Publication. USAID, Tetra Tech. Available at: https://pdf.usaid.gov/pdf_docs/PA00JNGV.pdf.

Bateki, C. A. and Dickhoefer, U. (2020). Evaluation of the modified livestock simulator for stall-fed dairy cattle in the tropics. *Animals: An Open Access Journal From MDPI* 10(5), 816.

Bationo, A., Buerkert, A., Sedogo, M. P., Christianson, B. C. and Mokwunye, A. U. (1995). A critical review of crop-residue use as soil amendment in the West African semi-arid tropics. In: *International Conference on Livestock and Sustainable Nutrient Cycling in Mixed Farming Systems of Sub-Saharan Africa*, Addis Ababa, Ethiopia, 22–26 November 1993.

Bationo, A., Hartemink, A., Lungu, O., Naimi, M., Okoth, P., Smaling, E. and Thiombiano, L. (2006). *African soils: Their productivity and profitability of fertilizer use: Background paper for the African Fertilizer Summit 9–13th June 2006*. Abuja, Nigeria: International Fertilizer Development Center. Available at: https://edepot.wur.nl/26759 (Accessed 20 March 2022).

Bawayelaazaa Nyuor, A., Donkor, E., Aidoo, R., Saaka Buah, S., Naab, J. B., Nutsugah, S. K., Bayala, J. and Zougmoré, R. (2016). Economic impacts of climate change on cereal production: Implications for sustainable agriculture in Northern Ghana. *Sustainability* 8(8), 724. https://doi.org/10.3390/su8080724.

Blench, R. (1999). The Nigerian national livestock resource survey: A personal account. In: *Colloques et Séminaires-Institut Français de Recherche Scientifique pour le Développement en Coopération*, 627–648.

Brisson, N., Gary, C., Justes, E., Roche, R., Mary,B., Ripoche, D., Zimmer, D., Sierra, J., Bertuzzi, P., Burger, P., Bussière, F., Cabidoche, Y. M., Cellier, P., Debaeke, P., Gaudillère, J. P., Hénault, C., Maraux, F., Seguin, B. and Sinoquet, H. (2003). An overview of the crop model STICS. *European Journal of Agronomy* 18(3–4), 309–332. https://doi.org /10.1016/S1161-0301(02)00110-7.

Brown, J. R. and Havstad, K. M. (2004). Monitoring to detect change on rangelands: Physical, social and economic/policy drivers. *African Journal of Range and Forage Science* 21(2), 115–121.

Bucchignani, E., Mercogliano, P., Rianna, G. and Panitz, H. -J. (2016). Analysis of ERA-Interim-driven COSMO-CLM simulations over Middle East–North Africa domain at different spatial resolutions. *International Journal of Climatology* 36(9), 3346–3369. https://doi.org/10.1002/joc.4559.

Calvosa, C., Chuluunbaatar, D. and Fara, K. (2009). Livestock and climate change. In: *Livestock and Climate Change*, p. 20.

Carson, A., Elliott, M., Groom, J., Winter, A. and Bowles, D. (2009). Geographical isolation of native sheep breeds in the UK—Evidence of endemism as a risk factor to genetic resources. *Livestock Science* 123(2–3), 288–299.

Dary, S. K., James, H. S. and Mohammed, A. S. (2017). Triggers of farmer-herder conflicts in Ghana: A non-parametric analysis of stakeholders' perspectives. *Sustainable Agriculture Research* 6(2), 141–151.

Dasmani, I. and Dadzie, S. K. N. (2021). Weather differentials and discount rate impact on farmers' net revenue: Modeling results from Ghanaian data focusing on different agro-climatic zones. *Journal of Extreme Events* 8(1), 1. https://doi.org/ 10.1142/ S2345737621500123.

De Longueville, F., Ozer, P., Gemenne, F., Henry, S., Mertz, O. and Nielsen, J. Ø. (2020). Comparing climate change perceptions and meteorological data in rural West Africa to improve the understanding of household decisions to migrate. *Climatic Change* 160(1), 123–141. https://doi.org/10.1007/s10584-020-02704-7.

Deme, A., Gaye, A. T. and Hourdin, F. (2015). Climate projections in West Africa, evidences and uncertainties. In: *Rural Societies in the Face of Climate and Environmental Change in West Africa*. Marseille Ed: Institut de Recherche Pour le Développement.

Diallo, M., Akponikpè, P. B. I., Abasse, T., Fatondji, D. and Agbossou, E. K. (2019). Why is the spatial variability of millet yield high at farm level in the Sahel? Implications for research and development. *Arid Land Research and Management* 33(4), 351–374.

Diallo, M. B., Akponikpè, P. B. I., Abasse, T., Fatondji, D. and Agbossou, E. K. (2021). Combined effect of trees and soil fertility management practices on millet yields in the Sahel. *Agroforestry Systems* 95(4), 717–730.

Diasso, U. and Abiodun, B. J. (2015). Drought modes in West Africa and how well CORDEX RCMs simulate them. *Theoretical and Applied Climatology* 1(1–2), 223–240.

Donnelly, J. R., Freer, M., Salmon, L., Moore, A. D., Simpson, R. J., Dove, H. and Bolger, T. P. (2002). Evolution of the GRAZPLAN decision support tools and adoption by the grazing industry in temperate Australia. *Agricultural Systems* 74(1), 115–139.

Dosio, A., Panitz, H. J., Schubert-Frisius, M. and Lüthi, D. (2015). Dynamical downscaling of CMIP5 global circulation models over CORDEX-Africa with COSMO-CLM: Evaluation over the present climate and analysis of the added value. *Climate Dynamics* 44(9–10), 2637–2661. https://doi.org/10.1007/s00382-014-2262-x.

Emetere, M. E. (2017). Investigations on aerosols transport over micro- and macro-scale settings of West Africa. *Environmental Engineering Research* 22(1), 75–86. https://doi.org/10.4491/eer.2016.080.

Fagariba, C. J., Song, S. and Soule Baoro, S. K. G. (2018). Climate change adaptation strategies and constraints in Northern Ghana: Evidence of farmers in Sissala West District. *Sustainability* 10(5), 1484. https://doi.org/10.3390/su10051484.

Faye, B., Webber, H., Diop, M., Mbaye, M. L., Owusu-Sekyere, J. D., Naab, J. B. and Gaiser, T. (2018a). Potential impact of climate change on peanut yield in Senegal, West Africa. *Field Crops Research* 219, 148–159.

Faye, B., Webber, H., Naab, J. B., MacCarthy, D. S., Adam, M., Ewert, F., Lamers, J. P. A., Schleussner, C., Ruane, A., Gessner, U., Hoogenboom, G., Boote, K., Shelia, V., Saeed, F., Wisser, D., Hadir, S., Laux, P. and Gaiser, T. (2018b). Impacts of 1.5 versus 2.0 C on cereal yields in the West African Sudan Savanna. *Environmental Research Letters* 13(3), 034014. http://doi.org/10.1088/1748-9326/aaab40.

Freduah, B. S., MacCarthy, D. S., Adam, M., Ly, M., Ruane, A. C., Timpong-Jones, E. C., Traore, P., Boote, K., Porter, C. and Adiku, S. G. (2019). Sensitivity of maize yield in smallholder systems to climate scenarios in semi-arid regions of West Africa: Accounting for variability in farm management practices. *Agronomy* 9(10), 639. http://doi.org/10.3390/agronomy9100639.

Ghana Statistical Service. (2016). Labour force report survey (vols. 112–113). Available at: https://www.statsghana.gov.gh/gssmain/fileUpload/Demography/LFS%20REPORT_fianl_21-3-17.pdf.

Guan, K., Sultan, B., Biasutti, M., Baron, C. and Lobell, D. B. (2017). Assessing climate adaptation options and uncertainties for cereal systems in West Africa. *Agricultural and Forest Meteorology* 232, 291–305. https://doi.org/10.1016/j.agrformet.2016.07.021.

Guichard, F., Hourdin, C., Léauthaud, J. and Barbier, E. (2015). Le réchauffement climatique observé depuis 1950. In: Sultan, B. (Ed.), *Evolutions Récentes et Futures du Climat en Afrique de l'ouest: Evidences, Incertitudes et Perceptions Les Sociétés Rurales Face aux Changements Climatiques et Environnementaux en Afrique de l'Ouest*. Institut de Recherche Pour le Développement, pp. 23–42.

Heve, W. K., Oleson, J. E., Ngonidzashe, C. and Adiku, S. G. K. (2015). Targeted management of organic resources for sustainably increasing soil organic carbon: Observations and perspectives for resource use and climate adaptations in northern Ghana. *Acta Agriculturae Scandinavica. Soil & Plant Science*. http://doi.org/10.1080/09064710.2015.1081396.

Hoffmann, I. (2010). Climate change and the characterization, breeding and conservation of animal genetic resources. *Animal Genetics* 41, 32–46.

Hsiang, S. M. and Burke, M. (2014). Climate, conflict, and social stability: What does the evidence say? *Climatic Change* 123(1), 39–55. https://doi.org/10.1007/s10584-013-0868-3.

IPCC. (2013). The physical science basis. Contribution of working group I to the fifth assessment report of the Intergovernmental Panel on Climate Change (IPCC). In: Stocker, T. F., Qin, D., Plattner, G.-K., Tignor, M., Allen, S. K., Boschung, J., Nauels, A., Xia, Y., Bex, V. and Midgley, P. M. (Eds.), *Climate Change*. Cambridge, UK and New York, NY: Cambridge University Press, p. 1535. https://doi.org/10.1017/CBO9781107415324.

Klutse, N. A. B., Sylla, M. B., Diallo, I., Sarr, A., Dosio, A., Diedhiou, A., Kamga, A., Lamptey, B., Ali, A., Gbobaniyi, E. O., Owusu, K., Lennard, C., Hewitson, B., Nikulin, G., Panitz, H. and Büchner, M. (2016). Daily characteristics of West African summer monsoon precipitation in CORDEX simulations. *Theoretical and Applied Climatology* 123(1–2), 369–386. https://doi.org/10.1007/s00704-014-1352-3.

Kouassi, J. L., Gyau, A., Diby, L., Bene, Y. and Kouamé, C. (2021). Assessing land use and land cover change and farmers' perceptions of deforestation and land degradation in South-West Côte d'ivoire, West Africa. *Land* 10(4), 429.

Laprise, R., Hernández-Díaz, L., Tete, K., Sushama, L., Šeparović, L., Martynov, A., Winger, K. and Valin, M. (2013). Climate projections over CORDEX Africa domain using the fifth-generation Canadian Regional Climate Model (CRCM5). *Climate Dynamics* 41(11–12), 3219–3246. https://doi.org/10.1007/s00382-012-1651-2.

Laux, P., Jäckel, G., Tingem, R. M. and Kunstmann, H. (2010). Impact of climate change on agricultural productivity under rainfed conditions in Cameroon—A method to improve attainable crop yields by planting date adaptations. *Agricultural and Forest Meteorology* 150(9), 1258–1271.

Lebel, T., Delclaux, F., Le Barbé, L. and Polcher, J. (2000). From GCM scales to hydrological scales: Rainfall variability in West Africa. *Stochastic Environmental Research and Risk Assessment* 14(4), 275–295.

Lobell, D. B. (2014). Climate change adaptation in crop production: Beware of illusions. *Global Food Security* 3(2), 72–76. http://doi.org/10.1016/j.gfs.2014.05.002.

Lobell, D. B., Burke, M. B., Tebaldi, C., Mastrandrea, M. D., Falcon, W. P. and Naylor, R. L. (2008). Prioritizing climate change adaptation needs for food security in 2030. *Science* 319(5863), 607–610.

Lodoun, T., Giannini, A., Traore, P. S., Some, L., Sanon, M., Vaksmann, M. and Rasolodimby, J. M. (2013). Changes in seasonal descriptors of precipitation in Burkina Faso associated with late 20th century drought and recovery in West Africa. *Environmental Development* 5, 96–108.

Loomis, R. S., Rabbinge, R. and Ng, E. (1979). Explanatory models in crop physiology. *Annual Review of Plant Physiology* 30(1), 339-367.

Ly, M., Segnon, A. C., D'haen, S., Totin, E., Noblet, M., Camara, I. and Pfleiderer, P. (2019). Understanding and interpreting the outputs of climate models for the conduct of vulnerability studies: A guide for practitioners. *Climate Analytics gGmbH, Berlin*. http://doi.org/10.13140/RG.2.2.27520.79365.

Ly, M., Traore, S. B., Agali, A. and Sarr, B. (2013). Evolution of Some Observed Climate Extremes in the West African Sahel. *Weather And Climate Extreme* 1, 19-25. http://dx.doi.org/10.1016/j.wace.2013.07.005.

MacCarthy, D. S., Hathie, I., Freduah, B. S., Ly, M., Adam, M., Ly, A., Nenkam, A., Traore, P. S. and Valdivia, R. O. (2021a). Potential impacts of agricultural intensification and climate change on the livelihoods of farmers in Nioro. In: *Handbook of Climate Change and Agroecosystems: Climate Change and Farming System Planning in Africa and South Asia: AgMIP Stakeholder-Driven Research (Part 2)*, West Senegal, Africa, pp. 3-31. https://doi.org/10.1142/9781786348814_0001.

MacCarthy, D. S., Adam, M., Freduah, B. S., Fosu-Mensah, B. Y., Ampim, P. A. Y., Ly, M., Traore, P. S. and Adiku, S. G. K. (2021b). Climate change impact and variability on cereal productivity among smallholder farmers under future production systems in West Africa. *Sustainability* 13(9), 5191. https://doi.org/10.3390/su13095191.

MacCarthy, D. S., Adiku, S. G., Freduah, B. S., Gbefo, F. and Kamara, A. Y. (2017). Using CERES-Maize and ENSO as decision support tools to evaluate climate-sensitive farm management practices for maize production in the northern regions of Ghana. *Frontiers in Plant Science* 8, 31. http://doi.org/10.3389/fpls.2017.00031.

MacCarthy, D. S., Kihara, J., Masikati, P. and Adiku, S. G. K. (2018). Decision support tools for site-specific fertilizer recommendations and agricultural planning in selected countries in sub-Sahara Africa. *Nutrient Cycling in Agroecosystems* 110(3), 343-359. Cham: Springer.

MacCarthy, D. S., Traore, P. S., Freduah, B. S., Adiku, S. G. K., Dodor, D. E. and Kumahor, S. K. (2022). Productivity of soybean under projected climate change in a semi-arid region of West Africa: Sensitivity of current production system. *Agronomy* 12(11), 2614. https://doi.org/10.3390/agronomy12112614.

MacCarthy, D. S., Valdivia, R., Clottey, J., Adiku, S., Anaglo, J., Madajewicz, M., Amoah, A. B., Amoako, K. K., Quist, E., Botchway, V. A., Klutse, N. A. B., Mutter, C. and Rosenzweig, C. (2021c). National agricultural development pathways influence climate change impacts on smallholders' livelihoods in Navrongo, Ghana. *IDRC-AgMIP-CLARE Policy Brief*, November 2021.

McCartney, M. (2010). Planning agricultural water shortage for climate change in Sub-Saharan Africa. *Nature and Fauna* 25(8), 35-38.

McCown, R. L. (1973). An evaluation of the influence of available soil water storage capacity on growing season length and yield of tropical pastures using simple water balance models. *Agricultural Meteorology* 11, 53-63.

McCown, R. L., Hammer, G. L., Hargreaves, J. N. G., Holzworth, D. P. and Freebairn, D. M. (1996). APSIM: A novel software system for model development, testing and simulation in agricultural systems research. *Agricultural Systems* 50(3), 255-271.

McKeon, G., Ash, A., Hall, W. and Smith, M. S. (2000). Simulation of grazing strategies for beef production in north-east Queensland. In: Hammer, G., Nicholls, N., Mitchell, C. (Eds.), Applications of Seasonal Climate Forecasting in Agricultural and Natural Ecosystems, pp. 227-252.

Minson, D. J. (1990). *Forage in Ruminant Nutrition*. San Diego, CA: Academic Press Inc.

Morton, O., Borah, J. R. and Edwards, D. P. (2020). Economically viable forest restoration in shifting cultivation landscapes. *Environmental Research Letters* 15(6), 064017.

Nhemachena, C., Hassan, R. and Kurukulasurya, P. (2010). Measuring the economic impact of climate change on African Agricultural production systems. *Climate Change Economics* 1(1), 33-55. https://doi.org/10.1142/S2010007810000066.

Nix, H. A. and Fitzpatrick, E. A. (1969). An index of crop water stress related to wheat and grain sorghum yields. *Agricultural Meteorology* 6(5), 321-337. https://doi.org/10.1016/0002-1571(69)90024-7.

Ogunkanmi, L., MacCarthy, D. S. and Adiku, S. G. K. (2022). Impact of extreme temperature and soil water stress on the growth and yield of soybean (*Glycine max* (L.) Merrill). *Agriculture* 12(1), 43. https://doi.org/10.3390/agriculture12010043.

Olaniyan, A., Francis, M. and Okeke-Uzodike, U. (2015). The cattle are "Ghanaians" but the Herders are strangers: Farmer-Herder conflicts, expulsion policy and pastoralist question in Agogo, Ghana. *African Studies Quarterly* 15(2), 39-51.

Opoku-Ankomah, Y. and Cordery, I. (1994). Atlantic sea surface temperatures and rainfall variability in Ghana. *Journal of Climate* 7(4), 551-558.

Oppong-Anane, K. (2006). Country pasture/forage resource profiles-Ghana. Grassland and pasture crops. *Food and Agricultural Organization of the United Nations*. Available at: https://www.scribd.com/document/346404450/FAO-Forage-Profile-Ghana.

Owusu, K. and Waylen, P. (2009). Trends in spatio-temporal variability in annual rainfall in Ghana (1951-2000). *Weather* 64(5), 115-120. https://doi.org/10.1002/WEA.255.

Owusu, K. and Waylen, P. R. (2013). The changing rainy season climatology of mid-Ghana. *Theoretical and Applied Climatology* 112(3-4), 419-430. https://doi.org/10.1007/s00704-012-0736-5.

Paxian, A., Sein, D., Panitz, H. -J., Warscher, M., Breil, M., Engel, T., Tödter, J., Krause, A., Cabos Narvaez, W. D., Fink, A. H., Ahrens, B., Kunstmann, H., Jacob, D. and Paeth, H. (2016). Bias reduction in decadal predictions of West African monsoon rainfall using regional climate models. *Journal of Geophysical Research: Atmospheres* 121(4), 1715-1735. https://doi.org/10.1002/2015JD024143.

Rahimi, J., Haas, E., Grote, R., Kraus, D., Smerald, A., Laux, P., Goopy, J. and Butterbach-Bahl, K. (2021). Beyond livestock carrying capacity in the Sahelian and Sudanian zones of West Africa. *Scientific Reports* 11(1), 22094.

Roncoli, C., Jost, C., Kirshen, P., Sanon, M., Ingram, K. T., Woodin, M., Somé, L., Ouattara, F., Sanfo, B. J., Sia, C., Yaka, P. and Hoogenboom, G. (2009). From accessing to assessing forecasts: An end-to-end study of participatory climate forecast dissemination in Burkina Faso (West Africa). *Climatic Change* 92(3-4), 433-460. https://doi.org/10.1007/s10584-008-9445-6.

Rufino, M. C., Herrero, M., Van Wijk, M. T., Hemerik, L., De Ridder, N. and Giller, K. E. (2009). Lifetime productivity of dairy cows in smallholder farming systems of the Central highlands of Kenya. *Animal* 3(7), 1044-1056.

Sarr, B., Atta, S., Ly, M., Salack, S., Ourbak, T., Subsol, S. and George, D. A. (2015). Adapting to climate variability and change in smallholder farming communities: A case study from Burkina Faso, Chad and Niger. *Journal of Agricultural Extension and Rural Development* 7(1), 16-27. DOI: 10.5897/JAERD2014.0595.

Seneviratne, S. I., Nicholls, N., Easterling, D., Goodess, C. M., Kanae, S., Kossin, J., Luo, Y., Marengo, J., McInnes, K., Rahimi, M., Reichstein, M., Sorteberg, A., Vera, C. and

Zhang, X. (2012). Changes in climate extremes and their impacts on the natural physical environment. In *Managing the Risks of Extreme Events and Disasters to Advance Climate Change Adaptation* Field, C. B., Barros, V., Stocker, T. F., Qin, D., Dokken, D. J., Ebi, K. L., Mastrandrea, M. D., Mach, K. J., Plattner, G.-K., Allen, S. K., Tignor, M. and Midgley, P. M. (eds), Cambridge and New York: Cambridge University Press, pp. 109–230.

Singh, P., Nedumaran, S., Ntare, B. R., Boote, K. J., Singh, N. P., Srinivas, K. and Bantilan, M. C. S. (2014). Potential benefits of drought and heat tolerance in groundnut for adaptation to climate change in India and West Africa. *Mitigation and Adaptation Strategies for Global Change* 19(5), 509–529.

Sorgho, R., Quiñonez, C. A. M., Louis, V. R., Winkler, V., Dambach, P., Sauerborn, R. and Horstick, O. (2020). Climate change policies in 16 West African countries: A systematic review of adaptation with a focus on agriculture, food security, and nutrition. *International Journal of Environmental Research and Public Health* 17(23), 8897.

Sultan, B. and Gaetani, M. (2016). Agriculture in West Africa in the twenty-first century: Climate change and impacts scenarios, and potential for adaptation. *Frontiers in Plant Science* 7, 1262. https://doi.org/10.3389/fpls.2016.01262.

Sultan, B., Roudier, P., Quirion, P., Alhassane, A., Muller, B., Dingkuhn, M., Ciais, P., Guimberteau, M., Traore, S. and Baron, C. (2013). Assessing climate change impacts on sorghum and millet yields in the Sudanian and Sahelian savannas of West Africa. *Environmental Research Letters* 8(1), 014040. http://doi.org/10.1088/1748-9326/8/1/014040.

Tachie-Obeng, E., Akponikpe, P. B. I. and Adiku, S. (2013). Considering effective adaptation options to impacts of climate change for maize production in Ghana. *Environmental Development* 5, 131–145.

Taub, D. (2010). Effects of rising atmospheric concentrations of carbon dioxide on plants. *Nature Education Knowledge* 3(10), 21.

Thornton, P. K., van de Steeg, J., Notenbaert, A. and Herrero, M. (2009). The impacts of climate change on livestock and livestock systems in developing countries: A review of what we know and what we need to know. *Agricultural Systems* 101(3), 113–127. https://doi.org/10.1016/j.agsy.2009.05.002.

Timpong-Jones, E. C., Adiku, S. G. K. and MacCarthy, D. S. (2019). Climate change and dairy cattle production in Ghana. In: Adiku, S. G. K. (Ed.), *Climate Change Variability and Biophysical Land Use in Ghana*. Ghana: Afram Publications Ltd, pp. 188–203.

Timpong-Jones, E. C., Adogla-Bessa, T., Naazie, A. and Adjorlolo, L. K. (2014). Chapter 14: Climate change: Implications for the resource poor livestock farmers in Africa. In: Yangyuoru, M., Naazie, A., MacCarthy, D. S., Nkansah, G. O. and Torkpo, S. K. (Eds.), *Agricultural Research for Food Security and Sustainable Livelihood: University of Ghana Readers Series*. Tema, Ghana: Digibooks Ghana Ltd, pp. 231–248.

Timpong-Jones, E. C., Owusu-Bremang, R., Mopipi, K. and Sarkwa, F. O. (2023). Climate change and variability affect rangeland quality and productivity- How? *African Journal of Food, Agriculture, Nutrition and Development* 23(3), 22711–22729. https://doi.org/10.18697/ajfand.118.21975.

Tofa, A. I., Kamara, A. Y., Babaji, B. A., Akinseye, F. M. and Bebeley, J. F. (2021). Assessing the use of a drought-tolerant variety as an adaptation strategy for maize production under climate change in the savannas of Nigeria. *Scientific Reports* 11, 8983. https://doi.org/10.1038/s41598-021-88277-6.

United Nations Framework Convention on Climate Change (UNFCCC). (2015). L.9/Rev.1. Adoption of the Paris agreement. *FCCC/CP*. Paris: UNFCCC. Available at: https://unfccc.int/resource/docs/2015/cop21/eng/l09r01.pdf.

USAID. (2018). Climate risks in West Africa: Regional risk profile. *Climate Fact Sheet*, pp. 1–10.

USGCRP. (2009). Global climate change impacts in the United States. In: Karl, T. R., Melillo, J. M. and Peterson, T. C. (Eds.), *Global Change Research Program*. New York: Cambridge University Press.

Valdivia, R. O., Antle, J. M., Rosenzweig, C., Ruane, A. C., Vervoort, J., Ashfaq, M., Hathie, I., Tui, S. H., Mulwa, R., Nhemachena, C., Ponnusamy, P., Rasnayaka, H. and Singh, H. (2015). Representative agricultural pathways and scenarios for regional integrated assessment of climate change impacts, vulnerability, and adaptation. *Handbook of Climate Change and Agroecosystems* 3, 101–145. https://doi.org/10.1142/9781783265640_0005.

Valdivia, R. O., Homann-Kee, T. S., Antle, J. M., Subash, N., Singh, H., Nedumaran, S. and Schönhart, M. (2021). *Handbook of Climate Change and Agro-Ecosystem: Climate Change and Farming System Planning in Africa and South Asia: AgMIP Stakeholder-Driven Research, Part 2*, pp. 47–102. https://doi.org/10.1142/9781786348791_0003.

Van den Bossche, P. and Coetzer, J. A. (2008). Climate change and animal health in Africa. *Revue Scientifique et Technique Office Internationale des Epizooties* 27(2), 551–562.

Waongo, M., Laux, P. and Kunstmann, H. (2015). Adaptation to climate change: The impacts of optimized planting dates on attainable maize yields under rainfed conditions in Burkina Faso. *Agricultural and Forest Meteorology* 205, 23–39.

Williams, J. R., Jones, C. A., Gassman, P. W. and Hauck, L. M. (1995). Simulation of animal waste management with APEX. In: *Innovations and New Horizons in Livestock and Poultry Manure Management*, 6–7 September, Austin, TX, pp. 22–26.

Wood, S. A. and Mendelsohn, R. O. (2015). The impact of climate change on agricultural net revenue: A case study in the Fouta Djallon, West Africa. *Environment and Development Economics* 20(1), 20–36. Available at: https://www.jstor.org/stable/26391869.

World Bank Data. (2020). Agriculture, forestry, and fishing, value added (% of GDP). Available at: https://data.worldbank.org/indicator/NV.AGR.TOTL.ZS?end=2020andlocations=GH-NG-BF-TG-BJ-CI-GN-NE-SL-SNandstart=2020andview=bar (Accessed 2 February 2022).

World Population Review. (2022). Western Africa population 2022. Available at: https://worldpopulationreview.com/continents/western-africa-population.

Yegbemey, R. N., Yabi, J. A., Tovignan, S. D., Gantoli, G. and Kokoye, S. E. H. (2013). Farmers' decisions to adapt to climate change under various property rights: A case study of maize farming in northern Benin (West Africa). *Land Use Policy* 34, 168–175. https://doi.org/10.1016/j.landusepol.2013.03.001.

Zougmoré, R., Jalloh, A. and Tioro, A. (2014). Climate-smart soil water and nutrient management options in semiarid West Africa: A review of evidence and analysis of stone bunds and zaï techniques. *Agriculture and Food Security* 3(1), 1–8. https://doi.org/10.1186/2048-7010-3-16.

Zougmoré, R., Partey, S., Ouédraogo, M., Omitoyin, B., Thomas, T., Ayantunde, A., Ericksen, P., Said, M. and Jalloh, A. (2016). Toward climate-smart agriculture in West Africa: A review of climate change impacts, adaptation strategies and policy developments for the livestock, fishery and crop production sectors. *Agriculture and Food Security* 5(1), 1–16. https://doi.org/10.1186/s40066-016-0075-3.

Index

AAGIS. *see* Australian Agricultural and Grazing Industry Survey (AAGIS)
Accounting for uncertainties
  climate risk and global demand 131-132
  crop model 133
  future trends 142
  managing uncertainties 134-140
  overview 131-134
  reducing model uncertainties 141
  rice productivity and water 132-133
  simulations GCMs *vs.* RCMs 132
  sources of 132
  sources of variability 133-134
  triple-ensemble probabilistic approach 134
Acid soil infertility 681
Acrisols 668
Adaptation strategy 640, 642-645
Adaptive responses types 331
Advances in integrating different models
  adaptation strategies 5
  changing climatic conditions 4
  climate-smart agriculture (CSA) 5
  different crop models, test on soya relative yield 16
  ensemble modelling 21-22
  five ways to couple models and modules 14
  future trends 22-26
  global temperature rises 3-4
  ideal conceptual model to deal with climate changes 9
  'ideal model' to assess the impact on agriculture 8-11
  from ideal to pragmatic modelling, coupling sub-models 13-15
  implementing the 'ideal model' current approaches 11-13
  incorporating inter- and intra-species diversity into models 19-20
  integrating adaptation management practices into models 17-19
  integrating new knowledge and facilitating interoperability 15-17
  livestock management strategies 5
  mitigation encompasses crop 5
  socio-economic impacts 4-5
  understanding models and application to problems
    knowledge and test assumptions 6
    purpose of model, knowledge/innovation 6
    representations 6
  from white to black box models, hybrid approaches and gridded simulations 20-21
The African Monsoon Multidisciplinary Analysis Programme 673
Agent-based models (ABMs) 385
Aggregated soil data 136
AgMIP. *see* Agricultural Model Intercomparison and Improvement Project (AgMIP)
Agricultural adaptation responses 329-330
Agricultural commodity markets
  economic adaptation potential 315-317
    Common Agricultural Policy Regional Impact Analysis (CAPRI) model 315-316
    crop model 315
    farmers production decisions and trade 317
    negative and positive yield adjustments 317
    trade feedback and market adjustments 317
    WOFOST *vs.* ISI-MIP approaches 315-316
    World Food Studies (WOFOST) 315
  future trends 321-323
  information flow, biophysical *vs.* agro-economic models 313
  measuring direct impacts of climate change on agriculture 313-315
    analysis of 313, 315
    ISI-MIP initiative 315
  overview 311-313
  role of technologies for mitigation and adaptation 317-320
    analysis 319-320

combined measures approach 319
EcAMPA 3 study 318
food system emissions 317
framework 320
marginal abatement cost curves (MACCs) 319
mitigated approaches 318–319
standalone measures approach 319
synopsis 320–321
total global crop productivity 314
Agricultural Model Intercomparison and Improvement Project (AgMIP) 15–16, 611, 615, 673, 685
Agricultural production systems simulator (APSIM) 496–497, 521, 524, 693–696
Agriculture in Europe
agricultural systems
cereal production 373–374
Common Agricultural Policy (CAP) 373
crop rotation systems 374
food 372–373
grasslands 372
land use 372
oilseeds and legumes 374
wine, olive oil and tomatoes production 373
air temperature, average values of differences 369
climate 362–364
effect of elevated CO_2 concentration 378
Europe 361–362
focus of 365
future trends and conclusion 393–395
importance of agriculture 365
lamellar open top chambers 377
modelling and experimental work
animal production systems 387–389
changes in soil organic carbon in Europe 381–383
crop yields 376–381
farm economics 384–387
grapevine and olives 389–391
greenhouse gas emissions 383–384
vegetables 391–393
models and platforms 374–376
observed climate zones, 1975–95 and 1996–2016 363
past climate and future projections
annual average temperatures 368
annual precipitation 368, 371
averaged values 368
December to February (DJF) 368–371
expected temperature and precipitation shift 366, 368
June, July and August (JJA) 366–371
March to May (MAM) 368–371
multi-proxy reconstructions 366
regional climate change 368
September to November (SON) 368–371
precipitation, average values of ratios 370
soybean maturity group 380
type of adverse events for croplands and grasslands and expected frequency 364
Agri-photovoltaic (agri-PV) 349
Agro-C model 593
Agro-ecology 345
Agro-ecosystem models 375
Agro-ecosystem services
categorization of countries, caloric sufficiency levels, trends, and trade situation 287–288
ecosystem services from agricultural landscapes 267–268
climate regulation 271–272
cultural values 272–273
food provision 270–271
regulation of water quality and timing of flow 271
three hypotheses 267–268
ecosystem services to agriculture
maintenance of soil fertility and structure 268
pollination and pest regulation 269
water provisioning 269–270
food-provisioning ecosystem services 275
crop production, increase globally despite climate change 279–282
global statistical model and limits 276–277
integrated approach to modeling nutrition, climate change and management conditions 278–279
future trends 293–294
maps of current status and projected changes, caloric yield for nutrition 282
modeling multiple ecosystem services in landscapes 273–275

models to decision-relevant metrics, integrating demand 283
approximating diets and demand, estimate caloric sufficiency 283-284
caloric sufficiency decreases consistently 286-291
large decline in global caloric sufficiency, production gains are outweighed by consumption 284-286
overview 265-267
quantitative, spatially-explicit projections 285
synopsis 291-293
trends in caloric sufficiency 290
Agroforestry systems 245-246
Agronomic weed 423
Algeria 624, 626
AMBAWA model 245
Analysis and Experimentation on Ecosystems (AnaEE) 377
APSIM. *see* Agricultural production systems simulator (APSIM)
APSIM model 81
APSIM-Sugar model 456, 457
APSIM-Tropical Pasture model 459, 460
AquaCrop Corn model 441
AquaCrop sugarcane 442
Arenosols 668
Asia cropping systems 624
Australia and Oceania, climate change impact on agriculture 481-483
agricultural systems modelling 495
agricultural production systems simulator (APSIM) 496-497
farmpredict micro-simulation model 497-499
agriculture challenges
Australia 492-493
New Zealand 493-495
Papua New Guinea 495
climate 483-485
farm profitability
climate change adaptation in Australian broadacre farm sector 512-513
climate projection scenario 510-512
future trends 522-524
importance of agriculture 485-487
New Zealand
broadacre cropping in 514-518
horticulture system in 518-520
pasture production system in 520-521
Oceania region 481-483
Papua New Guinea, oil palm production in 521-522
past climate change and future projections 487-488
Australia 488
New Zealand 488-491
Pacific Islands 491-492
temperate grazing systems 509-510
wheat and sorghum production
adaptation in agronomy and breeding 507-509
climate change impact on crop yield and production 499-502
growing season and crop phenology impact 503-504
impact on water, frost, heat stresses, and yield 504-506
projected climate change impact 506-507
technology advances and adaptation impact 502-503
technology advances and adaptation on crop yield 502-503
Australian Agricultural and Grazing Industry Survey (AAGIS) 497

Bayesian probability inversion 137
Bio-economic farm models (BEFMs) 385
Biomass accumulation 639
Brazil 442-446
Brazilian Meteorological Institute 448
Broadacre cropping 486
BRS Primavera 444

Caloric sufficiency 283-290
Caloric yield 278-279
Calories 283-284
Carbon accounting
Global Livestock Environmental Assessment Model (GLEAM) 212
IPCC guidelines, national greenhouse gas inventories 206-211
life-cycle assessment 211-212
Carbon sequestration capacity 389-390
CASUPRO 441
CENTURY model 381
CERES-Rice model 592, 593, 596, 608, 609
Chickpeas 563, 565
Chile 437-441
Climate-analogues 277

Climate bins 277
Climate change 3-4
Climate change impact, adaptation and vulnerability (CCIAV) 115
Climate change models 444
Climate-smart agriculture (CSA) 5, 327
CMIP5-Global Climate Models (GCMs) 544, 546
CMIP6 project 673
Coastal Savannah Plains of Ghana 691
Coefficient of variation (CV) 445
Colombian agriculture 441
Cool Farm Tool 384
Cotton 563, 565
CREFT 441
Crop diversification 615
CROPGRO-Perennial Forage model 458-460
Crop growth and yield formation
 assessing yield with model ensembles 169
  AgMIP barley 174
  AgMIP intercomparisons 171
  AgMIP maize 173-174
  AgMIP potato 175-176
  AgMIP rice 174-175
  AgMIP soybean 176
  AgMIP wheat 171-173
  early comparisons 169-171
  other AgMIP comparisons 176
  pest and disease model 177
 CM intercomparison activities 156
 concept of
  anthesis 159
  concentration and composition 161
  Farquhar model 159
  grain mass supply/demand 160
  grain yield 159
  Harvest Index (HI) approach 160
  photosynthesis 157, 159
  RUE approach 159
  simulation of quantity *vs.* quality 160-161
 crop modeling response 161-162
  data 165-166
  soil-crop-climate interactions 167-169
  structures of crop growth models 162-165
 future trends 177-178
 global food supply 155
 process-based model 156
 statistical-based model 155
Crop growth models (CGM) 80-81, 156
Cropland suitability 633
Crop lodging 457
Crop management practices 344-345
Crop modeling response
 data
  calibration and evaluation of grain yield 165-166
  limitation of 166
  phenology and growth 165
 multi-model approach 161-162
 single crop growth model 161
 soil-crop-climate interactions
  adaptation and mitigation strategies 167
  cumulative evapotranspiration 167-168
  limitations, CM applications 167
  potential evapotranspiration equations 167
  simulated yield 167, 169
  soil N mineralization 169
 structures of crop growth models
  actual crop transpiration (*Ta*) 163-164
  light utilization 162
  potential crop transpiration (*Tp*) 163-164
  potential evapotranspiration (*ETp*) 163
  soil evaporation (*Ep*) 163-164
  sub-routines implementation 162-163
  Water Stress Index 163-164
Crop modelling 375-376
Crop multi-model ensembles 141
Cropping systems 630
Crop production, increase globally despite climate change
 ecosystem services approach 279
 GDP per capita 280-281
 global yields 281-282
 inputs collected and pre-processed 280
 modeling crop yields 281
 representative concentration pathways (RCPs) 280
 shared socioeconomic pathways (SSPs) 279-280
Crop suitability approach 338-339
Crop yields 376
 crop rotations, simulations in Central Europe 379-380
 future of soybean 380-381

heat and drought stress, for wheat and maize 379
on-the-ground experiments, inform agro-ecosystem models 377-379
CSM-CROPGRO-DRYBEAN model 448, 449
CV. *see* Coefficient of variation (CV)

Dairy cattle 388
Decision Support System for Agrotechnology Transfer (DSSAT) 441, 695, 696
Decreased vegetative growth stage (GSD-v) 594
Deficit irrigation 643
Developing integrated approaches
addressing key challenges
enabling socio-economic scenarios to condition adaptation 114-115
incorporating socio-economic drivers of impacts 111-114
model structural uncertainty 108-111
representing adaptation, objective purposive response 115-116
European food system 107
food supply chain 105
future trends
climate and socio-economic scenarios, regional integrated modelling 120-121
importance of shocks and extreme events 121
multiscale model coupling 121
PC-based to cloud-based modelling infrastructure 121-122
IMPRESSIONS Integrated Assessment Platform (IAP) 117-120
integrated/holistic modelling approaches 107-108
modelling tools 106
non-climatic pressures 106
overview 105-108
resources *vs.* allocation *vs.* demand of water interactions 106-107
socio-economic changes 105-106
synopsis 120
Diverse cropping system 615
Double-delay technology 608
Drip irrigation 643
Dry growing season in high favorable environments (HFEd) 450
DSSAT. *see* Decision Support System for Agrotechnology Transfer (DSSAT)
DSSAT-CANEGRO model 456, 457
DSSAT/CROPGRO model 452-454

East Asia, climate change impact on agriculture 589-590
future trends, climate change impact 609-611
crop disease, weeds and pests 613
crop production 611-613
crop quality 613-614
diverse and intensified cropping systems 614-615
maize, impact modelling on
historical impact 605-608
projected impact 608-609
rice, impact modelling on
historical impact 591-594
projected impact 594-597
wheat, impact modelling on
historical impact 598-601
projected impact 601-605
'Ecosystem-based-adaptation' approach 243
Ecosystem services 265
Elevated CO_2 concentrations
biomass-dependent approach 139-140
LAI development 139-140
leaf area 139
modeling processes 138-139
performance of 16 rice models 139
performance of 21 maize models 139
productivity of C3 crops 138
El Nino-Southern Oscillation (ENSO) 485, 677
Emissions trading systems (ETS) 429
ENSO. *see* El Nino-Southern Oscillation (ENSO)
EPIC model 599
ET. *see* Evapotranspiration (ET)
ETS. *see* Emissions trading systems (ETS)
EU agricultural markets 312
EU's Emissions Trading Scheme (ETS) 214-215
Evaluating future ecosystem 266-267
Evapotranspiration (ET) 571

FACE. *see* Free air CO_2 Enrichment (FACE)
FarmDesign 386
FARMDYN 386-387
Farm economics 384-386
impacts and adaptation strategies, Southern France 386
impacts of Common Agricultural Practice diversification measures 387
model-aided life cycle sustainability assessment, European beef production 386-387

Farm-level models 389
Farmpredict model 510
Feasible generalised least square (FGLS) 550
Ferralsols 668
FGLS. *see* Feasible generalised least square (FGLS)
Field vegetable production 391
Food distribution and storage 217
Formalization of system concepts 77-79
Free-air CO_2 enrichment (FACE) 614, 647, 648
Free atmospheric CO_2 enrichment (FACE) rings 553
Free atmospheric ozone and CO_2 (O_3-FACE) rings 553
Fulani herders 693
Future risks 217-219

GCMs. *see* General circulation models (GCMs); Global circulation models (GCMs)
GD Arroz software 447
General circulation models (GCMs) 249, 558, 673, 674, 681, 687
Ghana Shorthorn (GhSh) breed 694, 695
Global agriculture principal target 187-188
Global circulation models (GCMs) 436, 449
Global Food Security Index (GFSI) 286
Global temperatures 215-217
Global warming potential (GWP) 429
Goiás (GO) 444
Grapevine Flowering Véraison model 519
Grapevine Sugar Ripeness model 519
GRAZPLAN 499
Greenhouse gases (GHG) emissions 3
Greenhouse Gas Protocol 430
Groundnut 562, 565
GSD-v. *see* Decreased vegetative growth stage (GSD-v)
GST-r. *see* Reproductive growth stage (GST-r)
GWP. *see* Global warming potential (GWP)

HAC. *see* Heteroscedasticity and autocorrelation (HAC)
Half a degree Additional warming, Projections, Prognosis and Impacts (HAPPI) project 594
Heat stress 388
Heavy precipitation 418
Heteroscedasticity and autocorrelation (HAC) 550
HFEd. *see* Dry growing season in high favorable environments (HFEd)
HFEw. *see* Wet growing season in high favorable environments (HFEw)
High-pressure systems 488
Hydrogenotrophic methanogens 194

Impact, vulnerability and adaptation (IVA) 555, 556
IMPRESSIONS Integrated Assessment Platform (IAP)
  adaptation strategies 117, 119-120
  land use, water and biodiversity simulations 118
  schematic illustration 117
Improving data flow and integration models on agriculture
  assimilation of data in spatiotemporal simulations 48-51
  concepts for distributed modelling 58-61
  future trends 61
  granularity and modular design for improvement, reuse, exchange and interoperability 56-58
  informing spatiotemporal simulations 44-45
    distributed data 47-48
    proximal sensing 46-47
    remote sensing (RS) 45-46
  model-data integration 40-41
    multi-criteria calibration of mechanistic agro-ecosystem models 41-44
  model-model integration 54
    model philosophies 56
    obstacles for model-model integration 54-56
  overview 39-40
  workflows for massive parallel computing 51-54
Incorporating genetics crop models, identify new phenotypes
  crop growth models, breeding to adapt and combat climate change 93-95
  future trends 95-96
  G × E × M interactions 76
  integrating environmental and genomic predictions 82-84
    Bayesian approaches 82-83
    CGM-WGP 84
    Genomic Best Linear Predictors (GBLUPs) 83-84

posterior distribution 83
RUE 83
static crop models 82
linking genetics and physiology, phenotypic prediction 80-81
managing genotype and environment interactions to optimize, genetic gain in current and future climates 86-89
models that predict emergent phenotypes 81
negative and inequitable impact 75
physiology and genetics 77
predictive breeding enables iterative impact assessment and adaptation 89-93
progressive intensification 77
systems analyses and models of crop growth 77-79
target population of environments (TPE) 76-77
traits and their underpinning genetics on crop performance 84-86
trait underpinning drought adaptation 76
Incorporating socio-economic drivers of impacts
five Eur-Agri- SSPs and directions 113-114
five scenarios, SSP1 to SSP5 113
global SSPs 113
multiple environmental and human system interactions 111
Incremental short-term response 329-330
Individual Farm Model for Common Agricultural Policy Analysis (IFM-CAP) 387
Innovative solutions, mitigate ruminant methane emissions 194-197
breeding and dietary innovations 194-195
breeding-based strategies 195
feeding systems 195-196
hydrogen 196
microbiome-targeted tools 196
in vivo study 196-197
Integrative perspective 248
accounting for farmer decision-making in models
AgMIP framework 251
optimization models 251
risks 253
utility-maximizers 251, 253
co-design adaptations with decision-makers
heterogeneity 253
investments and input 254
meteorological data and farmers' perception 254-255
Representative Agricultural Pathways (RAPs) 253-254
sustainable farming system 253
coupling models to assess climate change effect at farm scale
adaptation and mitigation 248
GCMs and RCPs 249-251
non-poor farms 249
West Africa 248-249
Interdecadal Pacific Oscillation 485
Intergovernmental Panel on Climate Change (IPCC) 436, 447, 615
Intergovernmental Panel on Climate Change (IPCC) assessment 487
Invertebrates and micro-organisms 268
IPCC. *see* Intergovernmental Panel on Climate Change (IPCC)
IPCC 2014 report 544
IPCC AR 6 WG II report 544
IPCC guidelines, national greenhouse gas inventories 206-211
default values 208
GE calculated for cattle equation 208-209
gross energy (GE) 206-207
LU/LUC management 210-211
N_2O emissions 209-210
net energy requirements 208
N mineralised 210
SOC stocks 211
tier 1-3 method 206-211
UK Greenhouse Gas Inventory 211
IPCC's Fifth Assessment Report (AR5) 135
Irrigated rice (IR) 443
Irrigated systems 627
Irrigation 392
IVA. *see* Impact, vulnerability and adaptation (IVA)

LAI. *see* Leaf area index (LAI)
Landscape DNDC model 698
Land-use change models 345
Latin America, climate change impact on agriculture 435-437
in Brazil

common bean production and climate change 448-451
grass forage production 458-460
irrigated rice production and climate change 446-448
pland rice production and climate change 442-446
soybean production and climate change 451-454
sugarcane production and climate change 454-458
in Chile 437-441
in Colombia 441-442
LCC. *see* Livestock carrying capacity (LCC)
Leaf area index (LAI) 552
Light use efficiency (LUE) 639
Livestock 387-388
Livestock carrying capacity (LCC) 697, 698
Livestock production
area (ha) of global grassland 206
economic value 188-189
emissions and carbon removals associated with farming 204
enteric methane mitigation
innovative solutions to mitigate ruminant methane emissions 194-197
methods to quantify methane emissions from ruminants 198-203
ruminant contribution to nitrous oxide emissions 197
ruminant enteric fermentation 191-194
ruminant food security and climate change 190-191
examples of bacterial species 193
food security role and challenges 189-190
future trends 219
global methane emissions 192
grassland agriculture, net zero livestock system
carbon accounting 206-212
carbon farming 204-206
carbon finance 214-215
measure, report and verification 212-214
milk and meat production 215-217
models and methods for estimating climate risks, vulnerability and impacts 217-219
net zero 203-204
measure, report and verification (MRV) 213
net emissions/removal, CO_2 207
overview 187-188
regional production of protein from different types 190
schematics of different systems 199
Livestock production systems (LPS) 188-189
Lixisols 668
Low-input smallholder farming systems
agent-based model exploration 244
agroforestry 245
modeling keys processes of particular relevance 246-248
trees and agroforestry productivity and adaptation strategies 245-246
cropping systems
crop productivity and adaptation strategies 237-238
modeling key processes of particular relevance 239-241
crops, livestock, grasslands and trees 237
livestock and grazing land 242
livestock productivity and adaptation strategies 242-243
modeling key processes of particular relevance 243-245
overview 235-237
percentage of farm households vulnerable 250
remote sensing-based model 247
representation of effects 236
representation of household decision-making 252
simulated N uptake and grain yield for 24 AgMIP models 240
synopsis 255-256
toward an integrative perspective 248
accounting for farmer decision-making in models 251-253
co-design adaptations with decision-makers 253-255
coupling models to assess climate change effect at farm scale 248-251
LUE. *see* Light use efficiency (LUE)

Machine learning 278, 345-346
Maize 561, 564

Maladaptation 331
Malnutrition 286
Managing uncertainties
adaptations in agricultural practices 140-141
crop model input data 136
greenhouse gas emission and climate change scenarios 134-136
model parameters and human error 136-137
model structures 137
simulate impacts of extreme temperature 138
simulation of crop responses to elevated CO_2 138-140
Markov Chain Monte Carlo (MCMC) technique 133, 137
Mathematical programming (MP) 385
Mathematical programming-based multiagent systems (MP-MAS) 385
Mato Grosso (MT) 444, 445
MCWLA model 609
Measure, report and verification (MRV) 212-214
Meat consumption 217
Mechanistic model 20
Methanobrevibacter (*Mbb.*) 194
Methanogenic archaea 191
Methods to quantify methane emissions from ruminants
Automated Head-Chamber System (AHCS) 200-201
inverse dispersion approach 201
long-chain fatty acids (LCFA) 202-203
mechanistic models 201-202
mitigation methods 198
open-circuit respiration chamber 198, 200
sulphur hexafluoride (SF6) tracer technique 200
in vivo measure 198, 202
MIRCA2000 data set 627
Mitigations challenge 187
MODAM 386
Model for Nitrogen and Carbon in Agro-ecosystems (MONICA) 454
Modeling multiple ecosystem services in agricultural landscapes
crop-water linkages 274-275
decision-makers 275
interconnections and feedback services 273-274
pest regulation 274
SmartScape 273
Modelling transition of agricultural systems
capturing transition in modelling approaches and climate change impact study 331
agro-economic models 335-336
empirical crop models 332
integrated assessment models 336-338
mechanistic models 332-333
modelling future pathways of agriculture 339-341
models of elements of the agricultural system 333-335
suitability models 338-339
future trends in research 344
improvements in data 346-347
model development 344-346
new concepts 347-348
scaling 347
global climate-driven agricultural frontiers, representative concentration pathways 340
overarching limitations of current modelling approaches 341
data limitations 343-344
interactions and feedbacks 342-343
uncertainty and uncertainty propagation 341-342
overview 327-328
prominent agriculture integrated assessment models simulation 337
role of models in informing adaptation responses 330-331
transition of agricultural systems in response to climate change
man-made responses 329-330
natural responses 328-329
Models of elements of the agricultural system
crop rotation models 333-334
nitrate leaching 335
ruminant digestion 334
soil organic matter turnover 334-335
Model structural uncertainty
European land-use simulations 109
example of 108-109
field-scale simulation 109
importance of 109, 111
individual biophysical/socio-economic models 111

subsystems 108
MONICA. *see* Model for Nitrogen and Carbon in Agro-ecosystems (MONICA)
Monogastric animals 388
MOSAICA model 12
MPPC models 441
Multi-criteria calibration of mechanistic agro-ecosystem models 41-44
Multi-model approach 161-162
Multi-model ensemble (MME) technique 22
Multiple ecosystem services 267
Multi-proxy reconstructions 366
Mustard 562, 564-565

National Climate Assessments 410
Natural Capital Project 276-277
NCP. *see* North China Plain (NCP)
New technology 348
New Zealand 484
Niger River 668
Nioro agriculture 248
Nitrogen fluxes 384
Nitrous oxide emissions 413
North Africa, climate change impact on agriculture
adaptation strategies 642-643
climate change impact assessment limitations 646-648
climate projections and effects on regional crop production 633-638
current climate change impact assessments 640-642
future trends 648-650
geography, economy, climate and agriculture 624-629
modelling tools, quantify crop response to climate change 638-640
overview 623-624
North China Plain (NCP) 598, 601, 603, 608

OECD. *see* Organisation for Economic Co-operation and Development (OECD)
OECD-FAO 436
Open-top chamber (OTC) technology 552, 554
Organic farming 349
Organisation for Economic Co-operation and Development (OECD) 486
ORYZA2000 443, 446
OTC. *see* Open-top chamber (OTC) technology

Pacific Islands 485, 486, 491-492
Papua New Guinea (PNG) 483, 485, 492, 495
PBCM. *see* Process-based crop models (PBCM)
PDS. *see* Plant developmental stage (PDS)
Pearl millet 562, 564
Photosynthesis 157-159
Photosynthesis/respiration 639
Pigeon pea 563, 565
PlanejArroz software 447
Plant developmental stage (PDS) 447
Plant growing degree days 413
Plant Modelling Framework 496-497
PNL. *see* Progressive nitrogen limitation (PNL)
Positive mathematical programming (PMP) 385
Precipitation 4
Process-based crop models (PBCM) 436-437, 455, 638
Progressive nitrogen limitation (PNL) 510

Quantitative trait loci (QTL) and genes 20

Rainfed systems 630
Random forest approach 279
RCM. *see* Regional climate models (RCM)
RCPs. *see* Representative concentration pathways (RCPs)
Reducing model uncertainties 141
Regional climate models (RCM) 673-675
Regional Greenhouse Gas Initiative (RGGI) 430
Representative concentration pathways (RCPs) 135, 430, 444, 445, 449, 490, 595, 672, 674
Representing adaptation, objective purposive response
conventional approach 115
pathways framing 116
transformative climate science aims 115
vision targets 115-116
Reproductive growth stage (GST-r) 594
Resilience of system 266
RGGI. *see* Regional Greenhouse Gas Initiative (RGGI)
Ricardian model 549, 550
Rice 559, 560
Rondônia (RO) 444
Ruminant animals 388-389
Ruminant production systems (RPS) 187

SALUS model 424
SAMUCA 456, 457
Satellite-based sensing 649
Sauvignon blanc 519
SEAMLESS-IF modelling framework 12
Second resilience strategy 278
Shared socioeconomic pathways (SSPs) 113, 218, 341, 596
SimulArroz model 446, 447
Sink-source model 160
Smallholder farming systems 235
SOC. *see* Soil organic carbon (SOC)
Soil erosion 419
Soil organic carbon (SOC) 381–383, 425, 426, 428
Soil productivity 268
Sorghum 561, 564
South Asia, climate change impact on agriculture
  climactic stresses affect field crops 546–548
  future trends 575–576
  horticultural crops 570–572
  in India 544–546
  on key crops 558–559
    Bangladesh 566, 569
    Bhutan 566, 570
    cereals 559–564
    Maldives 570
    Nepal 566, 569
    oil seeds 564–565
    Pakistan 566, 567–568
    pulses 565
    Sri Lanka 566, 570
  methods, climate change impact on crop production
    controlled environment experiments 552–554
    meta-analysis 554–555
    multi-location, multi-seasonal and long-term experiments 550–551
    past data analysis 549–550
    satellite data, ground data and models 552
    simulation modelling role 555–558
    survey 551
  overview 541–544
  on pests and diseases 572–573
    on crop pests 573–574
    risk assessment and adaptation planning 575
Southwest Asia, climate change impact on agriculture
  adaptation strategies 644–645
  climate change impact assessment limitations 646–648
  climate projections and effects on regional crop production 633–638
  current climate change impact assessments 640–642
  future trends 648–650
  geography, economy, climate and agriculture 629–633
  modelling tools, quantify crop response to climate change 638–640
  overview 623–624
Soybean 562–563, 565
Soybean and AusFarm Pasture 460
SOYGRO model 452
SPEI. *see* Standardized Precipitation Evapotranspiration Index (SPEI)
SSP. *see* Shared socioeconomic pathways (SSP)
Standardized Precipitation Evapotranspiration Index (SPEI) 415
Statistical models 21, 277
Sub-Saharan Africa (SSA) 235
Sugarcane 563, 565
Sustainable Development Indicators (SDIs) 120

Temperature gradient tunnels (TGTs) 552–554
TFP. *see* Total factor productivity (TFP)
TGTs. *see* Temperature gradient tunnels (TGTs)
Tocantins (TO) 444
Total factor productivity (TFP) 512, 514
Trade independency 286, 289
Transformative long-term response 329–330
Tropical tillage 680
Two bias correction (BC) methods 444

UGP. *see* Upper Gangetic Plain (UGP)
United Framework Convention on Climate Change 437
United Nations Framework Convention on Climate Change (UNFCCC) 212
United States, climate change impact on agriculture
  on crop-based agriculture 410
    climate change effect on soil and water 419–420

crop response to climate change 420-422
indirect climate change effects 423
spatial scales and their disparity 423
U.S. national and regional climate 410-419
incentives and payment mechanisms, carbon markets 428-429
sequester carbon and reduce greenhouse gas emissions 429-430
U.S. carbon markets for agriculture 430-431
new genetics mitigate climate change impacts
implications 428
methods 424-425
results 425-428
overview 409
United States Global Change Research Program (USGCRP) 690
Upland (rainfed) rice (UR) 443
Upper Gangetic Plain (UGP) 550
UR. *see* Upland (rainfed) rice (UR)
U.S. Department of Agriculture 410
USGCRP. *see* United States Global Change Research Program (USGCRP)
U.S. Midwest 428
U.S. Midwest. Model simulations 424
U.S. national and regional climate 410-411
extreme precipitation events 418-419
extreme temperature events and droughts 414-415
greenhouse gas emissions 411, 412
precipitation 415-418
temperature 411-414

Voluntary Carbon Markets (VCMs) 214-215

Water security 286
Web of Science 640
West Africa, climate change impact on agriculture
agricultural systems evolution 684-686
cereal and legume production 681-684
climate change adaptation strategies 686-689
climate change impacts on livestock production 689-693
crop modelling in farming systems 678-681
future climate projections 676-677
future trends 701-703
integrated climate change impact assessments 699-701
measurement, characteristics and trends 669-672
modelling climate change 672-677
modelling livestock systems 693
landscape DNDC model 697-698
LivSim model 694-697
Structural Ricardian model 698-699
overview 667-668
West African region 668-669
Wet growing season in high favorable environments (HFEw) 450
Wheat 559, 561, 564
World Bank's classification 289
Worldwide meat production 189

www.ingramcontent.com/pod-product-compliance
Ingram Content Group UK Ltd.
Pitfield, Milton Keynes, MK11 3LW, UK
UKHW021008120926
8556IPUK00028B/133

9 781801 461740